T0350241

CryptoSchool

Dedicated to Dorothea, Rafaela, Désirée.
For endless patience.

Frontispiece of Gaspar Schott's *Schola Steganographica*, 1665.

Joachim von zur Gathen

CryptoSchool

 Springer

Joachim von zur Gathen
b-it, Universität Bonn
Bonn, Germany

ISBN 978-3-662-48423-4   ISBN 978-3-662-48425-8   (eBook)
DOI 10.1007/978-3-662-48425-8

Library of Congress Control Number: 2015957084

Mathematics Subject Classification (2010): 68P25, 94A60, 11T71, 14G50, 68Q25; 11A51, 11H06, 11K45, 12Y05, 14H52 , 68Q12, 94A62

Springer Heidelberg New York Dordrecht London
© Springer-Verlag Berlin Heidelberg 2015
This work is subject to copyright. All rights are reserved by the Publisher, whether the whole or part of the material is concerned, specifically the rights of translation, reprinting, reuse of illustrations, recitation, broadcasting, reproduction on microfilms or in any other physical way, and transmission or information storage and retrieval, electronic adaptation, computer software, or by similar or dissimilar methodology now known or hereafter developed.
The use of general descriptive names, registered names, trademarks, service marks, etc. in this publication does not imply, even in the absence of a specific statement, that such names are exempt from the relevant protective laws and regulations and therefore free for general use.
The publisher, the authors and the editors are safe to assume that the advice and information in this book are believed to be true and accurate at the date of publication. Neither the publisher nor the authors or the editors give a warranty, express or implied, with respect to the material contained herein or for any errors or omissions that may have been made.

Printed on acid-free paper

Springer-Verlag GmbH Berlin Heidelberg is part of Springer Science+Business Media (www.springer.com)

Contents

By this contrivance, the most ignorant person, at a reasonable
charge, and with a little bodily labor, may write books in
philosophy, poetry, politics, law, mathematics, and theology,
without the least assistance from genius or study.
JONATHAN SWIFT (1726)

'Tis pleasant sure to see one's name in print;
A Book's a Book, altho' there's nothing in't.
LORD BYRON (1809)

In these cursory observations we have by no means
attempted to exhaust the subject of Cryptography.
With such object in view, a folio might be required.
EDGAR ALLAN POE (1840)

Die Kunst stille zu schweigen, geneigter leser, ist gar ein edles
Ding; noch edler aber wird meines Erachtens seyn die Kunst
redend zu schweigen und solche lehret dich meine alhier
in unserer Muttersprache an das Licht gegebene *Cryptographia*,
oder Kunst verborgene Brieffe und Schrifften zu machen.[1]
JOHANNES BALTHASAR FRIDERICI (1685)

I went into the Army at Fort Devens in Massachusetts and was
sorted into the Signal Corps, and so got in the Signal Corps
training camp at Fort Monmouth in New Jersey. They took
everybody who had no visible talent or aptitude whatsoever for
electrical work, or communications in the technological sense,
and if they had a certain level of testing score
under the Army General Classification Test,
bing! Crypt School before you could sneeze.
WILLIAM P. BUNDY (1987)

Will the study of cryptology become an epidemic that
even all the government's resources will be unable to stem?
DAVID KAHN (1979)

---

[1]The art of being quietly silent, estimated reader, is a truly noble thing; even nobler
is in my view the art of being silent while talking, and this art is taught to you by my
Cryptographia, or the art of making hidden letters and writings, brought here to the
light of day in our mother tongue.

Chapter 1

Introduction

This text is an introduction to *cryptology*, whose objective is to provide various aspects of security in electronic transactions. It consists of *cryptography*—the art of making secure systems—and *cryptanalysis*—the art of breaking them.

The text consists of two parts, entwined with each other. The modern part, with chapters numbered numerically, explains some of the basic systems used today and some attacks on them. The historical part, with chapters numbered alphabetically, recounts some inventions and episodes from the rich history of this subject, sometimes amusing, sometimes dramatic.

Chapters 2 through 5 explain some of the basic tools of modern cryptology, namely the AES and RSA cryptosystems and cryptography in groups and elliptic curves. While algebraic and arithmetic methods of cryptanalysis (factoring integers and computing discrete logarithms) are interwoven with these chapters, Chapter 6 deals purely with (Boolean) cryptanalysis. The next two chapters describe hash functions and digital signatures. Chapter 9 presents a core topic of modern cryptology: security reductions. Such a reduction usually shows that if some computational problem is hard, then the cryptographic system at hand is secure, since a successful adversary would perform a computationally infeasible feat. This allows the design of cryptographic systems that withstand any possible attack, not just the known ones. In fact, reductions of various types are a central tool in this text.

After discussing identification and authentication (Chapter 10), we come to more advanced topics: pseudorandom generation (Chapter 11), zero knowledge proof systems (Chapter 12), and lattice-based cryptography (Chapter 13). All these rely heavily on the concepts introduced in Chapter 9. Chapter 14 explains the power of quantum computers: if

© Springer-Verlag Berlin Heidelberg 2015

J. von zur Gathen, *CryptoSchool*, DOI 10.1007/978-3-662-48425-8_1

they could ever be built to scalable size, they would presumably devastate most cryptosystems discussed here, with the possible exception of AES, some hash functions, and the lattice cryptosystem in Section 13.13.

Chapter 15 at the end presents a toolbox, mainly of methods from computer algebra, that will be known to many but can be looked up at the reader's convenience.

A holistic view of cryptology includes its rich history. The historical parts provide a glimpse at this, sometimes amusing, sometimes scary. Chapter A begins with a general introduction and goes on to discuss frequency analysis. Chapter B deals with classical group cryptography, but in additive groups that are useless for modern methods. Chapter C explains a remarkable piece of cryptanalysis, namely, that of Kasiski on the "unbreakable" Vigenère system. Will anyone designing unbreakable systems heed this message? The codebooks of Chapter D were a staple of cryptography for over five centuries. Steganography—which survives today in attempts at digital watermarks and copyright protection—and transposition ciphers follow in Chapters E and F. Chapters G and H introduce some designers and some users of cryptography, from the Middle Ages to the nineteenth century. The last two historical Chapters I and J explain successes of British cryptanalysis over German cryptography, in the First and Second (hopefully last) World Wars. These played a prominent role in history, presumably saving countless lives—of Britons, Germans, and others.

Cryptology has a distinguished history of brilliant ideas and tools, some of which are described in this text. But for over two millennia, it was essentially confined to the *black chambers* of secret services, diplomacy, and the military.

The Diffie-Hellman revolution in 1976 brought public-key cryptography into power, with their key exchange, RSA, and most of the systems discussed here. But even more important than the technical advance was the sociological one: all of a sudden, leading researchers in mathematics and computer science realized the wealth of fascinating and difficult scientific questions whose answers could contribute to cryptology. The field remains deeply rooted in computer science, mathematics, and electrical engineering, but it has matured into a discipline of its own.

How to use this text.   This book is suitable for one- or two-semester courses for graduate or advanced undergraduate students in computer science and mathematics, and also for students in computer engineering or security with a mathematical inclination. The prerequisite mathematical facts are minimal and largely explained in the toolbox of Chapter 15. What is required is the mental acceptance that rigorous proofs are more desirable than hand-waving plausibilities, and the ensuing drudgery (better: joy) of going through those proofs. The basic ideas, protocols, and facts of modern cryptography that are presented try to challenge those who enjoy the thrill of convincing themselves of a stated truth rather than those who are willing to take someone's (like this author's) word for it. This is underlined by the more than 240 exercises, some of which deeply challenge the student's understanding of the material. The fun generated in class is visible in the various contributions of students, duly acknowledged, to this text.

A ubiquitous predicament of a conscientious instructor is that on the one hand, one would like to teach all details including proofs, but on the other hand, the material is too rich to do this in the allotted time. The usual way out is to point to the relevant literature. This is ok for some bright students, but many are confused by different notations and an embedding into other material that is not relevant to the purpose at hand.

This text, based on lecture notes from the author's many courses on *the art of cryptography*, tries to alleviate this predicament by presenting as many details as possible. It is then easy to look them up if the lecture time is insufficient. The necessary exceptions to this quest for completeness are duly noted in the text. The author has taught one-semester introductory courses covering parts of the first nine chapters, namely, Sections 2.1–2.8, 3.1–3.5, 3.10–3.13, 4.1–4.8, 5.1, 5.3–5.6, 7.1–7.3, 7.6, 8.1–8.2, and 9.1–9.6. For a second course, one or two of the following topics are appropriate: Chapters 6 and 10, Sections 9.7–9.10 with the background of Chapter 11, and Chapters 12, 13, or 14. This can be supplemented by suitable sections that were previously left out. Furthermore, the historical Chapters A–J make up a nice course on the history of cryptology, with fewer mathematical challenges (if any) and appealing to a wider audience.

The quest for completeness in treating the protocols presented here entails, given any reasonable page limitation, incompleteness in the list of tools described. Some references for further reading are given, but even those do not aim at encyclopedic coverage. The text is largely structured in a linear fashion, whereby all prerequisites for certain material have been discussed previously. A natural exception is the toolbox of Chapter 15, which is used throughout the book.

Like most natural sciences, cryptology combines theoretical and practical aspects. In its theory, we have well-defined models, and claims are proven with mathematical rigor, under clearly stated assumptions.

An interesting new result usually either proposes a new direction or improves in a clear sense on previous results. Much of this text works within this framework. Some protocols such as AES and practical hash functions seem—in principle—not amenable to such an analysis, but may still be deemed secure.

The vibrancy and attraction of this theory are largely due to its multitude of practical applications, of central importance in today's digital world. The secure cryptosystems described here provide excellent building blocks for such applications. But beware! It is easy (and such has happened time and again) to build insecure systems based on secure subroutines. There are many dangers outside of the models considered here: bad secret keys—from passwords stuck on a screen or handed to a fake website to RSA moduli with a common factor, virus-installed key loggers, fault attacks, attacks on the operating system ("cold boot"), denial of service, and many others.

However, this seems the best we can do at this time, and it is a comfort to have secure cryptographic building blocks which will thwart any attack in the standard computational model. Then one "only" has to worry about the mortar holding the blocks together, and unwanted entry into the system through a back door.

The origins.   In this text, we use the word *cryptography* for the art and science of making cryptosystems, *cryptanalysis* for breaking them, and *cryptology* for both together. The words come from the Greek κρυπτός (kryptos), meaning *hidden* or *secret*, together with γράφειν (graphein) *to write*, λύσις (lysis) *solution*, and λόγος (lógos), which is *word, science*, and also has other meanings.

Webster's dictionary defines cryptography as *the act or art of writing in or deciphering secret characters*, and also *secret characters or cipher, or a system of writing with them. Cryptology* is a *secret or enigmatic language*, or simply *cryptography. Cryptanalysis* is the *art or science of deciphering a code or coded message without a prior knowledge of the key.* The word *cipher* comes from the Arabic صِفْر (sifr) *zero*, from صَفَر (safara), to be empty. One of its meanings is *a secret or disguised manner of writing meant to be understood only by the persons who have the key to it; a code; also, the key to such a code.*

Thanks.   Martina Kuhnert, Daniel Loebenberger, and Konstantin Ziegler have contributed substantially to the production of this book, always in a cheerful and supportive manner. Many thanks for help in many ways go to Michael Nüsken, and other present and former members of my research group: Raoul Blankertz, Jérémie Detrey, Nihal Dip, Laila El Aimani, Michael Heußen, Claudia Jakob, Cláudia Oliveira Coelho, Dejan Pejić, Alexander Pfister, Yona Raekow, Deniz Sarier, Ayush Sharma, Jamshid Shokrollahi, Damien Vergnaud, and Marianne Wehry.

The work was supported by the B-IT Foundation and the Land Nordrhein-Westfalen.

I am indebted to David Kahn, Pascal Paillier, and Igor Shparlinski for substantial insights into various aspects of cryptography, and also to the other instructors at our annual crypt@bit summer school: Joan Daemen, Max Gebhard, Gary McGuire, Dennis Hofheinz, Marc Joye, Alexander May, Phong Nguyễn, Valtteri Niemi, Kenny Paterson, Chris Peikert, Bart Preneel, Charles Rackoff, Vincent Rijmen, Gadiel Seroussi, and Hoeteck Wee. I am also grateful for help on various matters to Frank Bergmann, Klaus Lagally, Sihem Mesnager, and Werner Schindler.

I owe my introduction to the subject to some "early" cryptographers, including Shafi Goldwasser, Russell Impagliazzo, Mike Luby, Silvio Micali, and Charlie Rackoff, whose presentations and courses in the 1980s opened a new field for me.

The *Notation* section towards the end of this book explains some of the symbols used. The web page `https://cosec.bit.uni-bonn.de/cryptoschool/` of the book contains additional material and corrections if necessary (sigh). Readers are encouraged to send their comments to the address on that web page.

Bonn, July 2015
Joachim von zur Gathen

The *Characters* used to express *Numbers* by are either . . .
The Ten Numeral *Figures* of the *Arabians*:
. . . 0 *Nothing or a Cypher.*
WILLIAM JONES (1706)

Real mathematics has no effects on war.
No one has yet discovered any warlike purpose
to be served by the theory of numbers as relativity,
and it seems very unlikely that anyone will do so for many years.
GODFREY HAROLD HARDY (1940)

It may well be doubted whether human ingenuity can construct
an enigma of the kind which human ingenuity may not,
by proper application, resolve.
EDGAR ALLAN POE (1843)

There is, the cryptographic experts assure us,
no unbreakable cipher.
RICHARD WILMER ROWAN (1934)

It is extremely probable that an insoluble cipher
could be produced by mathematical means today.
FLETCHER PRATT (1939)

[The NSA's stranglehold on cryptography] ended abruptly
in 1975 when a 31-year-old computer wizard
named Whitfield Diffie came up with a new system,
called 'public-key' cryptography, that hit the world of cyphers
with the force of an unshielded nuke.
STEVEN LEVY (1993)

Solange ein Wissenszweig Überfluß an Problemen bietet,
ist er lebenskräftig; Mangel an Problemen bedeutet
Absterben oder Aufhören der selbstständigen Entwicklung.[1]
DAVID HILBERT (1900)

---

[1] As long as a branch of science offers an abundance of problems, so long it is alive; a lack of problems foreshadows extinction or the cessation of independent development.

Chapter 2

Basic cryptosystems

This chapter starts with a look at some of the most popular cryptosystems. The description in this chapter focusses on the fundamental properties and leaves out some details, in particular proofs why certain things work the way they do. The complete underpinnings for these methods are provided in later chapters.

We learn to ask the fundamental questions: does it work correctly? How easy is the system to use for its legitimate players? How hard is it to break for others? In other words: what can we say about its security?

The first system is the *Advanced Encryption Standard* (AES), chosen from 15 candidates in a competition launched in 1997 by the National Institute of Standards and Technology (NIST), a US government institution. This system is an example of a symmetric cryptosystem in which the two protagonists (sender and receiver) share the same key. AES is characterized by its simplicity, good structure, and efficiency.

We briefly discuss two fundamentally different types of cryptosystems that we will encounter: symmetric vs. asymmetric systems. In the first type, sender and receiver share the same secret key, while in the latter type, only the receiver needs a secret key to decrypt an encrypted message and all other information is publicly available. If you have not yet seen such systems, stop here for a moment! Does this not sound contradictory? How could it possibly work?

We describe the RSA system named after its inventors Rivest, Shamir & Adleman. The security of this *asymmetric* or *public key cryptosystem* is somewhat related to the difficulty of *factoring large integers* into their prime factors.

The third example is the Diffie & Hellman key exchange protocol. Here the goal is not to send a secret message, but slightly more modest: the two players just want to agree on a common secret key (which they may

© Springer-Verlag Berlin Heidelberg 2015
J. von zur Gathen, *CryptoSchool*, DOI 10.1007/978-3-662-48425-8_2

then use in some other cryptographic setting). This example introduces the idea of doing cryptography in groups. The security of such systems relies on the difficulty of computing *discrete logarithms* in these groups.

We then discuss Shamir's scheme for sharing a secret among many players so that together they know the secret but any coalition of fewer than all players has no knowledge about it. This is based on *polynomial interpolation*.

The final example is Naor & Shamir's *visual cryptography*. We include it here because of its striking effect: you have two random pictures, one on paper and one on a transparency, and when you overlay them, you can see a secret message.

2.1. Goals of cryptography

Electronic transactions and activities play an ever increasing role in our lives. Many of these need to be protected against all kinds of interference, accidental or malicious. This is the general task of *information technology security*. *Cryptography* provides some basic building blocks for secure electronic systems. Its most fundamental task is secure information transmission: someone wants to send a message over an insecure channel such as the internet to a recipient, and a third party listening in should not be able to understand the message. Following a long-standing tradition, the computers involved are personalized as BOB sending a message to ALICE, and EVE eavesdropping on the line. This is achieved by BOB *encrypting* his message x with a key K and sending the result $y = \text{enc}_K(x)$ to ALICE, and then ALICE *decrypting* y with her own key S to recover $x = \text{dec}_S(y)$. In some systems they share the same key: $K = S$. Both BOB and ALICE should be able to do their work efficiently, but EVE, knowing only y (and also K in some systems; then $K \neq S$), should not be able to decipher the message, that is, to recover x with reasonable effort.

In addition to this fundamental task, there are many other objectives in cryptography, such as securely signing a message or establishing one's identity. Can you imagine how one would do that over the internet, without meeting in person? These questions are discussed in later chapters.

Coming back to the fundamental task, we have to clarify what the "efficiency" of BOB's and ALICE's actions might mean, and the inability of EVE to recover the message. Some systems, such as AES in Section 2.2, are completely fixed. Then efficiency means that it has to be implementable for the purpose at hand, maybe for secure transmission of a pay-TV video signal at a rate of megabits per second. EVE's inability to decipher the message without the secret key might mean that it is, as far as we know, beyond the power of any adversary for as long as its secrecy is important.

In other systems, such as RSA or Diffie-Hellman (Sections 2.5 and 2.6), we have a security parameter n which may, in principle, be chosen arbitrarily. Then the standard notion from Theoretical Computer Science is that "efficient" should mean that BOB and ALICE work in time polynomial in n, and that EVE is not able to discover x in polynomial time.

The concepts of complexity theory provide a precise framework in which to state the latter task. But its basic questions, such as whether $P \neq NP$, are unresolved at this time, and the design of a system secure in this sense is an open and extremely difficult question. However, a reasonable modification asks not for an absolute proof that EVE is unable to recover the original message from the transmission, but to relate it to other problems: if she could find the message, then she would also be able to solve a well-studied open problem.

The question can be formulated as the quest for a *one-way function* f: given x, it should be easy to compute $y = f(x)$, but given some y, which occurs as an image under f, it should be hard to find an x with $f(x) = y$. Furthermore, ALICE has some (small) secret S with whose help it is actually easy to find x from y; then f is called a *trapdoor function*. An example of a one-way function is multiplication: it is easy to multiply two integers, say two large prime numbers p and q, and find $N = p \cdot q$. But computationally it is quite difficult to recover p and q from N, although they are completely determined by N. Thus $x = (p, q)$ with $p < q$ and $f(x) = p \cdot q$ is a one-way function. No trapdoor is known for this f, but a closely related trapdoor function is used in the RSA cryptosystem (Section 2.5).

There are many variations of what it means for EVE to break such a system. Clearly it should be infeasible to efficiently recover x or ALICE's secret key S from y. But also much weaker achievements might be considered fatal, for example if she can find out some information about the message x: is it an English text message? Does some specific word occur, such as "MasterCard" or "bomb"? It would even be dangerous if she could not do this all the time, but only for some messages, just slightly better than guessing.

EVE might have some knowledge about the possible values of x. For example, she might know (or guess) that x is a string of 1024 bits representing the Extended ASCII encoding of a 128-letter English text. Only a tiny fraction of the 2^{1024} possible x's are of that form: we have a sparse *message space*.

In Chapter 9, we study *security notions* which consist of *resources* and *attack goals*. One possible resource is that EVE may be able to see $\text{enc}_K(x)$ for many x's of her choice: *chosen plaintext attack (CPA)*. A strong and hence desirable concept is the *chosen ciphertext attack*, where she can see

x for several y's of her choice with $\mathrm{enc}_K(x) = y$. Among the various *attack goals*, recovering the secret key or the plaintext come to mind at first. A weak and hence desirable notion here is *indistinguishability*: EVE submits two plaintexts and receives an encryption y_0 of one of them, chosen with equal probability. As above, she may ask for as many pairs (x, y) with $y = \mathrm{enc}_K(x)$ as she wants, where she may specify either x or y; of course, with $y \neq y_0$. Then she has to distinguish which of the two plaintexts was chosen, with probability better than just guessing. Indistinguishability means resistance against such an attack.

A combination of allowed resources and desired attack goal defines a *security notion*. A cryptographic system is *secure* in this notion if the goal cannot be accomplished with those resources, for example, if encryptions cannot be distinguished as above even using chosen ciphertexts. These things are further explained in Chapter 9 and throughout the text. For a brief impression, the reader might take a peek at Figure 9.5. An important aspect of these notions is that no specific method of attack is assumed, rather just the tools allowed and the goal to be achieved.

2.2. Advanced Encryption Standard (AES)

In the early 1970's, a team at International Business Machines developed a cryptosystem which became known as the *Data Encryption Standard* (DES). The US *National Bureau of Standards* (NBS) published it in FIPS PUB 46 on 15 January 1977 as a standard for US government cryptography, for documents that are sensitive but not classified. (The *National Security Agency* (NSA) is responsible for higher levels of security.) As a consequence, any software or hardware system with cryptographic capabilities tendered to the US government had to be based on DES. Sales to government agencies can be highly lucrative, and any company interested in them had to use DES. Thus it quickly found widespread use.

Over the years, many attacks on DES were developed, most notably differential cryptanalysis and linear cryptanalysis (Chapter 6). In response to this and concerns about its small key space, DES was strengthened by tripling its number of "rounds": triple-DES or 3-DES. In DES, the so-called S-boxes provide the only nonlinear functions. They are optimized with respect to resistance to differential cryptanalysis, but their structure is rather opaque.

From the start, experts harbored suspicions—never substantiated—that the NSA might have built a "trapdoor" into DES that enabled it to decipher encrypted messages. Already in 1981, Deavours warned that

> *The agency [NSA] is currently capable of breaking DES*
> *using probable plaintext. The major cryptanalytic hardware*

> *involved is rumored to consist of 4 CRAY-1 computers. Analysis takes less than a day, on the average.*

Finally, on 17 July 1998 the *Electronic Frontier Foundation* (EFF) presented its US\$ 250,000 DES breaker. DES was dead, for most practical purposes. But it was still the standard and thus in heavy use ... The standard was finally withdrawn in 2005. The US NIST, successor agency of the NBS, opened on 12 September 1997 a competition for AES, to replace DES. The requirements were for a block cipher with blocks of 128 bits, and possible key lengths of 128, 192, and 256 bits. Not surprisingly, the specifications were rather more precise than in their 1973 competition which led to the adoption of DES. 15 candidates were submitted to NIST, and pared down to a short list of five systems by August 1999. These included *MARS* from IBM's Don Coppersmith, one of the chief designers of DES, *RC6*, developed by Ron Rivest and three collaborators from RSA Laboratories, *Serpent* by Anderson, Biham, and Knudsen, and *Twofish* by Bruce Schneier's Counterpane Company. On 2 October 2000, the NIST announced the winner: AES, a system developed by the Belgian cryptographers Joan Daemen and Vincent Rijmen and originally called *Rijndael*. NIST expects this system to be secure for at least thirty years.

NIST was generally lauded for an open and well-documented procedure. One of its requirements was to make plausible that there are no hidden trapdoors, thus alleviating some of the concerns that had surrounded the DES standardization in 1977.

The features that secured Rijndael's first place in a tough competition are security—resistance against all currently known attacks—and efficiency—on a wide variety of platforms, from 8-bit smartcards to 32- or 64-bit processors. Furthermore, it has a simple algebraic description with few unexplained choices (see the end of this section), and it is implausible that they could hide a trapdoor. No effective attack is known in 2015.

AES encrypts a message of 128 bits using a key of 128, 192, or 256 bits, distinguished by designations like AES-128. It is an *iterated cipher*, in which a sequence of four operations is applied a certain number of times. Namely, it consists of 10 *rounds* at key length 128 (12 rounds at 192 and 14 rounds at 256 bits) and each round performs these four operations, except that the last one leaves out MixColumns. Furthermore, there is an additional initial round, executing only AddRoundKey. Each operation turns a 128-bit word into another 128-bit word. To describe the operations, each 128-bit word (or *state* in AES) is treated as a 4×4 matrix (or array, or block) of 8-bit bytes:

$$(2.1) \qquad \begin{matrix} a_{00} & a_{01} & a_{02} & a_{03} \\ a_{10} & a_{11} & a_{12} & a_{13} \\ a_{20} & a_{21} & a_{22} & a_{23} \\ a_{30} & a_{31} & a_{32} & a_{33} \end{matrix}.$$

The four operations have the following features:

- SUBBYTES substitutes each single byte by another value,

- SHIFTROWS permutes the bytes in each row,

- MIXCOLUMNS performs a linear transformation on each column of the matrix,

- ADDROUNDKEY adds the round key to the whole matrix.

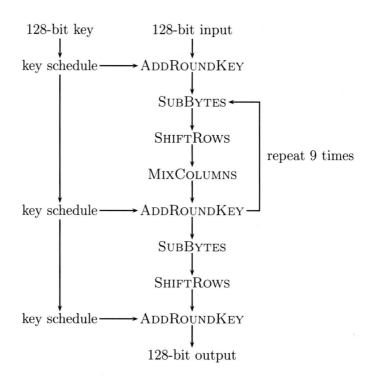

Figure 2.1: AES with a 128-bit key.

Figure 2.1 illustrates the global view. The four operations in the middle constitute one round. For the initial round, the round key is explicitly

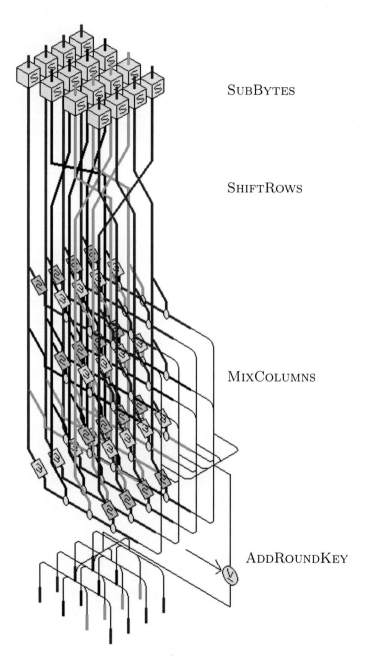

SUBBYTES

SHIFTROWS

MIXCOLUMNS

ADDROUNDKEY

Figure 2.2: One round of AES.

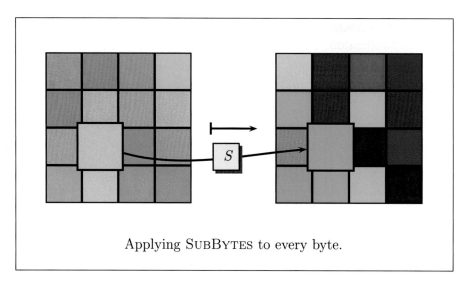

Applying SUBBYTES to every byte.

Figure 2.3: The SUBBYTES operation.

provided as the secret key to the procedure. From this, the round keys
for the later rounds are calculated by the *key schedule*; see Section 2.3.

We now describe in more detail the four operations, assuming that
the reader is familiar with the material in Sections 15.1 through 15.4. We
see many cryptosystems in this book, including RSA and group-based
cryptography, say with elliptic curves, which by their nature require some
algebra. But AES is the winner in a competition for bit-oriented (or
Boolean) cryptography. The elegant algebraic description that follows is
witness to the *unreasonable effectiveness of algebra* in cryptography.

SUBBYTES. The basic processing unit is an 8-bit byte

$$(2.2) \qquad a = (a_7, a_6, a_5, a_4, a_3, a_2, a_1, a_0) \in \{0, 1\}^8.$$

The fundamental operations on these bytes are addition and multiplication. The sum

$$c = a + b$$

of two bytes simply is the bitwise sum modulo 2 (exclusive-or, XOR):

$$c_i = a_i + b_i$$

for $0 \le i \le 7$. For example, if we take

$$(2.3) \qquad a = (10011011), b = (11001101),$$

then

$$(2.4) \qquad c = a + b = (01010110).$$

In hexadecimal notation, we have $a = \mathtt{9B}$, $b = \mathtt{CD}$, and $c = \mathtt{56}$.

For multiplication, we first consider the byte a to represent the polynomial

$$a_7 t^7 + a_6 t^6 + \cdots + a_1 t + a_0,$$

so that a as in (2.3) now represents

$$t^7 + t^4 + t^3 + t + 1 \in \mathbb{F}_2[t].$$

The product $a \cdot b$ of two bytes a and b is calculated by multiplying the two polynomials, giving a polynomial of degree not more than 14. For the two polynomials from (2.3), this is

$$p = t^{14} + t^{13} + t^{11} + t^{10} + t^8 + t^6 + t^5 + t^3 + t^2 + t + 1 \text{ in } \mathbb{Z}_2[t].$$

Since we work over $\mathbb{F}_2$, all coefficients are reduced modulo 2. More details are given in Section 15.1.

We have an obvious problem: the result has up to 15 bits, but we should come up with just one byte. Algebra provides an elegant solution: reduce modulo a polynomial of degree 8. Indeed, in AES we work in the finite field $\mathbb{F}_{256}$ defined by the irreducible polynomial

$$(2.5) \qquad m = t^8 + t^4 + t^3 + t + 1 \in \mathbb{F}_2[t],$$

so that $a \bmod m \in \mathbb{F}_2[t]/(m) = \mathbb{F}_{2^8} = \mathbb{F}_{256}$. Now we divide p by m with remainder, obtaining

$$(2.6) \qquad p = (t^6 + t^5 + t^3) \cdot m + (t^4 + t^2 + t + 1) \quad \text{in } \mathbb{F}_2[t],$$
$$\mathtt{9B} \cdot \mathtt{CD} = a \cdot b = (00010111) = 17 \quad \text{in } \mathbb{F}_{256}.$$

Thus we are back to degree at most 7, or 8 bits. Multiplication in $\mathbb{F}_{256}$ maps two bytes to one byte. But in SUBBYTES, we have only one byte as input. How can we use the arithmetic in $\mathbb{F}_{256}$? The answer is: inversion.

Since $\mathbb{F}_{256}$ is a field, every nonzero element $a \in \mathbb{F}_{256}^{\times}$ has an inverse $a^{-1} \in \mathbb{F}_{256}^{\times}$. This can be calculated by the Extended Euclidean Algorithm (Section 15.4). We extend this mapping to all of $\mathbb{F}_{256}$ by simply sending zero to itself:

$$(2.7) \qquad \operatorname{inv}(a) = \begin{cases} a^{-1} & \text{if } a \neq \mathtt{00}, \\ \mathtt{00} & \text{if } a = \mathtt{00}, \end{cases}$$

where $\mathtt{00} = (00000000)$. This is called the *patched inverse*. In our example (2.3), the Extended Euclidean Algorithm produces

$$(2.8) \qquad (t^7 + t^3) \cdot a + (t^6 + t^3 + t^2 + t + 1) \cdot m = 1 \text{ in } \mathbb{F}_2[t],$$

$\mathbb{F}_{256} = \mathbb{F}_{2^8} \ni a = a_7 t^7 + a_6 t^6 + a_5 t^5 + a_4 t^4 + a_3 t^3 + a_2 t^2 + a_1 t + a_0,$
with all $a_i \in \mathbb{F}_2 = \{0, 1\}$.

Representation: 8 bits for an element = 1 byte.

Addition: XOR, $(a + b)_i = a_i + b_i$.

Multiplication: as for polynomials modulo $t^8 + t^4 + t^3 + t + 1$.

Example $57 \cdot 83 = \text{C1}$:

$$
\begin{aligned}
(t^6 + t^4 + t^2 &+ t + 1) \cdot (t^7 + t + 1) \\
&= t^{13} + t^{11} + t^9 + t^8 + t^7 \\
&\quad + t^7 + t^5 + t^3 + t^2 + t \\
&\quad + t^6 + t^4 + t^2 + t + 1 \\
&= t^{13} + t^{11} + t^9 + t^8 + t^6 + t^5 + t^4 + t^3 + 1 \text{ in } \mathbb{Z}_2[t] \\
&= t^7 + t^6 + 1 \text{ in } \mathbb{Z}_2[t]/(t^8 + t^4 + t^3 + t + 1).
\end{aligned}
$$

Figure 2.4: The byte field $\mathbb{F}_{256}$.

as calculated in Example 15.20 (ii), so that indeed $\gcd(a, m) = 1$ in $\mathbb{F}_2[t]$, and

$$\text{inv}(a) = (10001000) = 88 \quad \text{in } \mathbb{F}_{256}.$$

In a surprising connecting with elliptic curves, we show in Section 6.4 that the patched inverse is nearly optimal in its resistance against a particular attack, namely, linear cryptanalysis.

AES also uses a similar, yet different, algebraic structure on bytes, namely the ring $R = \mathbb{F}_2[t]/(t^8 + 1)$. This is not a field, since $t^8 + 1 = (t+1)^8$ is not irreducible in $\mathbb{F}_2[t]$; see (15.4). Thus a byte a as in (2.2) now represents the element

$$a_7 t^7 + a_6 t^6 + a_5 t^5 + a_4 t^4 + a_3 t^3 + a_2 t^2 + a_1 t + a_0 \text{ in } R.$$

Addition is, again, just the bitwise addition (or XOR). Thus (2.4) is also valid in R. Multiplication of two such polynomials gives a polynomial of degree at most 14, whose remainder modulo $t^8 + 1$ has again degree at most 7. Reduction modulo $t^8 + 1$ is particularly easy, since it corresponds to just adding the lower and the upper halves of the polynomial, in the following sense. We split

$$c = c_1 t^8 + c_0$$

into its upper and lower parts $c_1, c_0 \in \mathbb{F}_2[t]$ of degree at most 7. Then

$$c = c_1(t^8 + 1) + (c_1 + c_0) = c_1 + c_0 \text{ in } R.$$

To multiply the two bytes a and b of (2.3) in this new representation, we write their product as

$$(2.9) \qquad p = (01101101) \cdot x^8 + (01101111) = \mathtt{6D} \cdot x^8 + \mathtt{6F},$$

and then their product in the ring R is the sum of these two bytes:

$$\mathtt{9B} \cdot \mathtt{CD} = (10011011) \cdot (11001101) = (00000010) = \mathtt{02} \text{ in } R.$$

In AES, actually only multiplication in R by the fixed polynomial

$$t_1 = (00011111) = \mathtt{1F} = t^4 + t^3 + t^2 + t + 1$$

is used, and only the polynomial

$$t_0 = (01100011) = \mathtt{63} = t^6 + t^5 + t + 1$$

is added to others. Since t_1 is invertible modulo $t^8 + 1$, multiplication of bytes by t_1 corresponds to an invertible linear transformation over $\mathbb{F}_2$. For a byte b, the bits in

$$c = t_1 \cdot b + t_0$$

can also be described by the affine linear transformation

$$
\begin{pmatrix} c_7 \\ c_6 \\ c_5 \\ c_4 \\ c_3 \\ c_2 \\ c_1 \\ c_0 \end{pmatrix}
=
\begin{pmatrix}
1 & 1 & 1 & 1 & 1 & 0 & 0 & 0 \\
0 & 1 & 1 & 1 & 1 & 1 & 0 & 0 \\
0 & 0 & 1 & 1 & 1 & 1 & 1 & 0 \\
0 & 0 & 0 & 1 & 1 & 1 & 1 & 1 \\
1 & 0 & 0 & 0 & 1 & 1 & 1 & 1 \\
1 & 1 & 0 & 0 & 0 & 1 & 1 & 1 \\
1 & 1 & 1 & 0 & 0 & 0 & 1 & 1 \\
1 & 1 & 1 & 1 & 0 & 0 & 0 & 1
\end{pmatrix}
\cdot
\begin{pmatrix} b_7 \\ b_6 \\ b_5 \\ b_4 \\ b_3 \\ b_2 \\ b_1 \\ b_0 \end{pmatrix}
+
\begin{pmatrix} 0 \\ 1 \\ 1 \\ 0 \\ 0 \\ 0 \\ 1 \\ 1 \end{pmatrix}.
$$

To sum up, SUBBYTES consists of applying to each byte a in the block individually the following steps:

$$(2.10) \qquad\qquad a \leftarrow \mathrm{inv}(a) \quad \text{in } \mathbb{F}_{256},$$
$$a \leftarrow t_1 \cdot a \quad \text{in } R,$$
$$a \leftarrow a + t_0.$$

Its description involves some algebra, but SUBBYTES is most efficiently implemented by a 256-byte look-up table. It is the only nonlinear operation in AES and is sometimes called its S-box, in analogy with DES.

$$\mathbb{F}_{256} \quad \longrightarrow \quad \mathbb{F}_{256} \quad \longrightarrow \quad \mathbb{F}_{256},$$

$$\boxed{S} : \quad a \quad \longmapsto \mathrm{inv}(a) = \begin{bmatrix} b_7 \\ b_6 \\ b_5 \\ b_4 \\ b_3 \\ b_2 \\ b_1 \\ b_0 \end{bmatrix} \longmapsto \begin{bmatrix} 1&1&1&1&1&0&0&0 \\ 0&1&1&1&1&1&0&0 \\ 0&0&1&1&1&1&1&0 \\ 0&0&0&1&1&1&1&1 \\ 1&0&0&0&1&1&1&1 \\ 1&1&0&0&0&1&1&1 \\ 1&1&1&0&0&0&1&1 \\ 1&1&1&1&0&0&0&1 \end{bmatrix} \cdot \begin{bmatrix} b_7 \\ b_6 \\ b_5 \\ b_4 \\ b_3 \\ b_2 \\ b_1 \\ b_0 \end{bmatrix} + \begin{bmatrix} 0 \\ 1 \\ 1 \\ 0 \\ 0 \\ 0 \\ 1 \\ 1 \end{bmatrix}$$

$$a \mapsto 05 \cdot a^{254} + 09 \cdot a^{253} + \mathsf{F9} \cdot a^{251} + 25 \cdot a^{247} + \mathsf{F4} \cdot a^{239}$$
$$+ \, 01 \cdot a^{223} + \mathsf{B5} \cdot a^{191} + \mathsf{8F} \cdot a^{127} + 63$$

Figure 2.5: The SUBBYTES S-box.

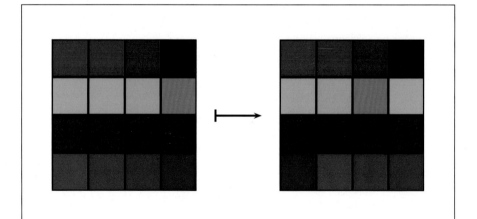

The four rows are shifted cyclically to the left by zero, one, two, and three bytes, respectively.

Figure 2.6: The SHIFTROWS operation.

SHIFTROWS. The operation SHIFTROWS shifts each of the four rows cyclically to the left by 0, 1, 2, and 3 places, respectively. Thus SHIFTROWS applied to the block (2.1) yields the array

(2.11)

$$\begin{array}{cccc} a_{00} & a_{01} & a_{02} & a_{03} \\ a_{11} & a_{12} & a_{13} & a_{10} \\ a_{22} & a_{23} & a_{20} & a_{21} \\ a_{33} & a_{30} & a_{31} & a_{32} \end{array} .$$

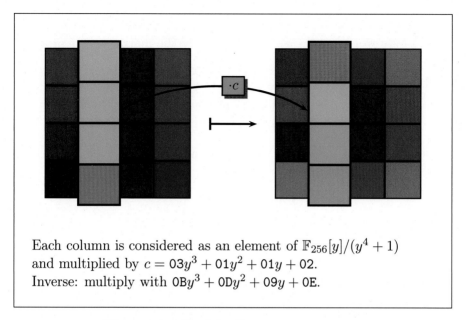

Each column is considered as an element of $\mathbb{F}_{256}[y]/(y^4+1)$ and multiplied by $c = 03y^3 + 01y^2 + 01y + 02$.

Inverse: multiply with $0By^3 + 0Dy^2 + 09y + 0E$.

Figure 2.7: The MIXCOLUMNS operation.

This is illustrated in Figure 2.6.

MIXCOLUMNS. Here we consider an array $a = (a_3, a_2, a_1, a_0)$ of four bytes a_3, a_2, a_1, and a_0 as a polynomial

$$a_3 s^3 + a_2 s^2 + a_1 s + a_0 \in \mathbb{F}_{256}[s]$$

of degree at most 3. Addition of such polynomials again corresponds to a bit-wise XOR. Multiplication gives a polynomial of degree at most 6 which is then decreased to degree at most 3 by reducing the result modulo $s^4 + 1 \in \mathbb{F}_{256}[s]$. Thus in effect we are working in the ring

$$S = \mathbb{F}_{256}[s]/(s^4+1)$$

with 256^4 elements. As $t^8 + 1$ above, $s^4 + 1 = (s+1)^4$ is not irreducible in $\mathbb{F}_{256}[s]$, hence S is not a field. Reduction modulo $s^4 + 1$ is again particularly easy. If $b_0, b_1 \in \mathbb{F}_{256}[s]$ have degree at most 3, then

$$b_1 s^4 + b_0 = b_1 + b_0 \text{ in } S.$$

In fact, this multiplication is only applied when one factor is the fixed polynomial
(2.12)
$$c = (00000011) \cdot s^3 + (00000001) \cdot s^2 + (00000001) \cdot s + (00000010)$$
$$= 03 \cdot s^3 + 01 \cdot s^2 + 01 \cdot s + 02$$

in $\mathbb{F}_{256}[s]$. The product of c with $a = (a_3, a_2, a_1, a_0)$ can also be described as the 4-byte word $b = (b_3, b_2, b_1, b_0)$ given by the matrix-vector product

$$\begin{pmatrix} b_3 \\ b_2 \\ b_1 \\ b_0 \end{pmatrix} = \begin{pmatrix} 02 & 01 & 01 & 03 \\ 03 & 02 & 01 & 01 \\ 01 & 03 & 02 & 01 \\ 01 & 01 & 03 & 01 \end{pmatrix} \cdot \begin{pmatrix} a_3 \\ a_2 \\ a_1 \\ a_0 \end{pmatrix}.$$

The operations on individual bytes are those in $\mathbb{F}_{256} = \mathbb{F}_2[t]/(m)$, as above. We take the example

$$\begin{pmatrix} a_3 \\ a_2 \\ a_1 \\ a_0 \end{pmatrix} = \begin{pmatrix} \texttt{A0} \\ \texttt{80} \\ \texttt{01} \\ \texttt{02} \end{pmatrix}.$$

Then

$$\begin{aligned} b_3 &= \texttt{02} \cdot \texttt{A0} + \texttt{01} \cdot \texttt{80} + \texttt{01} \cdot \texttt{01} + \texttt{03} \cdot \texttt{02} \\ &= t \cdot (t^7 + t^5) + 1 \cdot t^7 + 1 \cdot 1 + (t+1) \cdot t \\ &= t^8 + t^7 + t^6 + t^2 + t + 1. \end{aligned}$$

Since $t^8 = t^4 + t^3 + t + 1$ in $\mathbb{F}_{256}$, we have

$$b_3 = t^7 + t^6 + t^4 + t^3 + t^2 = (11011100) = \texttt{FC} \text{ in } \mathbb{F}_{256}.$$

It is interesting to note the three roles that the byte 11011100 plays here: first as an element of $\mathbb{F}_{256}$, represented by a polynomial in $\mathbb{F}_2[t]$ of degree 7, then as an 8-bit string, and finally a 2-letter hexadecimal word. It is also the binary representation of the decimal integer 220. Even more interesting is the fact that we consider the byte as elements of different domains, such as in the inversion in $\mathbb{F}_{256}$ or in the second step in SUBBYTES, and then a multiplication on the same data may yield completely different results depending on the underlying domain. This versatility is another aspect of the unreasonable effectiveness of algebra in cryptography.

ADDROUNDKEY. The 128-bit block and a round key of the same size are added bitwise.

This concludes our general description of the four AES operations. In a software implementation, it is ususally advantageous to replace calculations by table look-up as far as possible. Using a table of 4 kB, a round of AES can be executed with 16 table look-ups and 16 XORs of 32 bits.

AES evolved from earlier ciphers like *SHARK* (Rijmen *et al.* 1996) and *Square* (Daemen *et al.* 1997). Its design philosophy aimed at resistance against linear and differential cryptanalysis (Chapter 6) and high

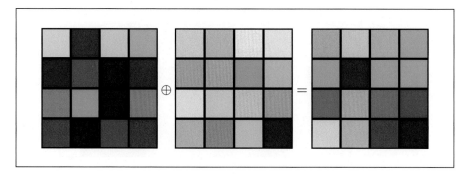

Figure 2.8: The ADDROUNDKEY operation.

throughput. The choices that this entailed are explained in Daemen & Rijmen (2002b). As examples, SUBBYTES using inversion was suggested in Nyberg (1994), and the modulus m is the first of 30 irreducible polynomials in Table C of Lidl & Niederreiter (1983). MIXCOLUMNS is based on matrices in which every square submatrix is nonsingular, a notion from the theory of error-correcting codes (MacWilliams & Sloane 1977, Chapter 11, Theorem 8). These have good diffusion properties. Namely, if F is a field, $M \in F^{n \times n}$ is MDS and $x, x^* \in F^n$ distinct, then the two vectors (x, Mx) and (x^*, Mx^*) in F^{2n} have Hamming distance at least $n + 1$, that is, the two vectors differ in at least $n + 1$ positions. The authors say convincingly: *"We believe that the cipher structure does not offer enough degrees of freedom to hide a trapdoor."*.

The omission of MIXCOLUMNS in the last round — or generally a final permutation step — is quite common, because it does not decrease security (ciphertext bits are just permuted in a publicly known way), but enables decryption with a similar structure; see Exercise 6.1.

As required in the AES competition, the algorithm is fast on a large variety of platforms. Software implementations can reach over 12 GB/sec.

Experts and the relevant standardization institutions consider AES secure. The strongest attack publicly known in 2015 (Bogdanov *et al.* 2011) has a cost of $2^{126.1}$ operations, compared to 2^{128} for exhaustive key search for AES-128. It does not constitute a serious threat to the security of AES. The most effective attacks are not on the system itself, but on specific implementations. Even the NSA seems stymied: *"Electronic codebooks, such as the Advanced Encryption Standard, are widely used and difficult to attack cryptanalytically. NSA has only a handful of in-house techniques."*

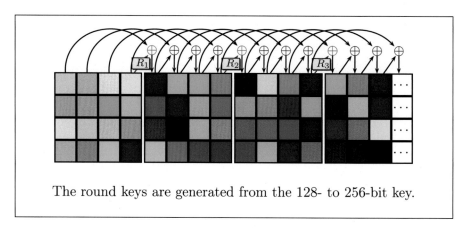

The round keys are generated from the 128- to 256-bit key.

Figure 2.9: The AES key schedule.

2.3. The AES key schedule

AES allows keys of 128, 192, or 256 bits, which corresponds to ℓ_k many 32-bit words for $\ell_k = 4$, 6, or 8. The number ℓ_r of rounds after the initial one is given in Table 2.10.

key length		
in bits	in ℓ_k words	ℓ_r rounds
128	4	10
192	6	12
256	8	14

Table 2.10: Key lengths and number of rounds in AES.

Each word has the format of a single column in an array like (2.11), but with 6 or 8 columns at the larger key lengths. In each round we need a *round key* array of ℓ_k words, and one more for the initial round. Thus we require a total of $\ell_k(\ell_r + 1)$ round key words.

We first explain this for 128-bit keys K, so that $\ell_k = 4$. The secret key K makes up the first four 4-byte words E_0, E_1, E_2, E_3 of the *extended key* $E_0, \ldots, E_{4(\ell_r+1)-1}$, consisting of $4(\ell_r+1)$ such words. We produce the others one by one, using the previous ones. Then our round keys consist of one block of four words after the other from the extended key.

For most indices $i \geq 4$, E_i is simply the sum in $\mathbb{Z}_2^{32}$ (bitwise XOR) of E_{i-1} and E_{i-4}:

$$E_i = E_{i-1} + E_{i-4}.$$

If i is a multiple of 4, then first a transformation is applied to E_{i-1}. Namely, the four bytes (a_3, a_2, a_1, a_0) of E_{i-1} are right-shifted cyclically

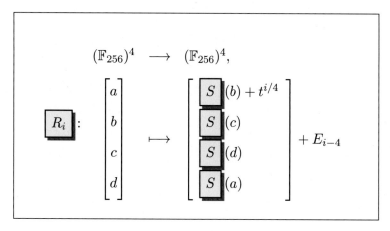

Figure 2.11: The nonlinear part of the key schedule for 128-bit keys and a multiple i of 4.

to give (a_0, a_3, a_2, a_1). If we think of the word as an element a of $S = \mathbb{F}_{256}[s]/(s^4+1)$, this is simply multiplication by s^3 in S. Then SUBBYTES is applied to each byte individually, and a constant round word $c_{i/4}$ is added. These c_j are defined as

$$(2.13) \qquad c_j = (0,0,0,t^{j-1}) = t^{j-1} \text{ in } S.$$

Thus c_j is constant in that it does not depend on the plaintext or the key. It is also a "constant" in S in that a general element of S is of the form $a_3 s^3 + a_2 s^2 + a_1 s + a_0$, but for c_j we have $a_3 = a_2 = a_1 = 0$.

We recall that ℓ_k can take the values $4, 6$, and 8. In the last case, a further transformation is applied. Namely, if i is $4 \bmod 8$, then E_{i-1} is replaced by SUBBYTES (E_{i-1}). Putting this together, the key expansion runs as follows for all three key lengths.

ALGORITHM 2.14. AES key expansion.

Input: A key $K_0, \ldots, K_{\ell_k-1}$ consisting of ℓ_k many 4-byte words K_i.

Output: An extended key $E_0, \ldots, E_{\ell_k(\ell_r+1)-1}$ consisting of $\ell_k(\ell_r+1)$ many 4-byte words E_i.

1.   For i from 0 to $\ell_k - 1$ do
2. $E_i \leftarrow K_i$,
3.   For i from ℓ_k to $\ell_k(\ell_r+1)$ do steps 4 to 10
4. $L \leftarrow E_{i-1}$,
5.   If ℓ_k divides i then
6. $c \leftarrow (0,0,0,t^{i/\ell_k-1})$
7. $L \leftarrow$ SUBBYTES $(s^3 \cdot L) + c$,

$$E_0 = K_0 = \begin{bmatrix} 00 & 00 & 00 & 00 \end{bmatrix}, \qquad E_1 = K_1 = \begin{bmatrix} 00 & 00 & 00 & 00 \end{bmatrix},$$
$$E_2 = K_2 = \begin{bmatrix} 00 & 00 & 00 & 00 \end{bmatrix}, \qquad E_3 = K_3 = \begin{bmatrix} 00 & 00 & 00 & 00 \end{bmatrix},$$
$$E_4 = K_4 = \begin{bmatrix} 00 & 00 & 00 & 00 \end{bmatrix}, \qquad E_5 = K_5 = \begin{bmatrix} 00 & 00 & 00 & 00 \end{bmatrix},$$
$$E_6 = \begin{bmatrix} 62 & 63 & 63 & 63 \end{bmatrix}, \qquad E_7 = \begin{bmatrix} 62 & 63 & 63 & 63 \end{bmatrix},$$
$$E_8 = \begin{bmatrix} 62 & 63 & 63 & 63 \end{bmatrix}, \qquad E_9 = \begin{bmatrix} 62 & 63 & 63 & 63 \end{bmatrix},$$
$$E_{10} = \begin{bmatrix} 62 & 63 & 63 & 63 \end{bmatrix}, \qquad E_{11} = \begin{bmatrix} 62 & 63 & 63 & 63 \end{bmatrix},$$
$$E_{12} = \begin{bmatrix} 9B & 98 & 98 & C9 \end{bmatrix}, \qquad E_{13} = \begin{bmatrix} F9 & FB & FB & AA \end{bmatrix},$$
$$E_{14} = \begin{bmatrix} 9B & 98 & 98 & C9 \end{bmatrix}, \qquad E_{15} = \begin{bmatrix} F9 & FB & FB & AA \end{bmatrix},$$
$$E_{16} = \begin{bmatrix} 9B & 98 & 98 & C9 \end{bmatrix}, \qquad E_{17} = \begin{bmatrix} F9 & FB & FB & AA \end{bmatrix},$$
$$E_{18} = \begin{bmatrix} 90 & 97 & 34 & 50 \end{bmatrix}, \qquad E_{19} = \begin{bmatrix} 69 & 6C & CF & FA \end{bmatrix},$$
$$E_{20} = \begin{bmatrix} F2 & F4 & 57 & 33 \end{bmatrix}, \qquad E_{21} = \begin{bmatrix} 0B & 0F & AC & 99 \end{bmatrix},$$
$$E_{22} = \begin{bmatrix} A6 & A7 & 44 & 62 \end{bmatrix}, \qquad E_{23} = \begin{bmatrix} 9F & A2 & 37 & 0C \end{bmatrix}, \quad \ldots$$

Figure 2.12: Key expansion in AES-192 for the key consisting of all zeroes.

8.   If $\ell_k = 8$ and i is 4 modulo 8 then
9. $L \leftarrow \text{SUBBYTES}(L)$,
10. $E_i \leftarrow E_{i-\ell_k} + L$.

EXAMPLE 2.15. For a 192-bit key, messages consist of four words of four bytes each, and the cipher key of $\ell_k = 6$ such words $K_0, \ldots, K_5$. From Table 2.10, we have $\ell_r = 12$ rounds, and thus need an extended key of $6 \cdot (12 + 1) = 78$ words. The first 24 are illustrated in Figure 2.12 for the key consisting of all zeroes. ◊

We have now described the operations for one round and the key schedule. It remains to specify the I/O convention and how the whole system is put together.

Input and output are arrays of sixteen 8-bit bytes as in (2.1), 128 bits in total. The conversion to and from a 4×4 block is columnwise, so that the array $a_{00}, a_{10}, a_{20}, a_{30}, a_{01}, \ldots$ corresponds to the block (2.1). In the same manner, the cipher key of $4\ell_k$ bytes is fed into the first words $E_0, \ldots, E_{\ell_k-1}$ of the extended key.

Finally, the whole system operates as follows, as in Figure 2.1:

1. inital round: ADDROUNDKEY

2. for i from 1 to $\ell_r - 1$ do

   o SUBBYTES

   o SHIFTROWS

   o MIXCOLUMNS

   o ADDROUNDKEY

3. final round:

   o SUBBYTES

   o SHIFTROWS

   o ADDROUNDKEY

The round keys are taken consecutively from the extended key.

2.4. Asymmetric vs. symmetric cryptography

In a *symmetric* (or *secret key*) *cryptosystem*, the same key is used for encryption and for decryption. All cryptographic methods were of this type until the 1970s, and so is AES. But then Diffie & Hellman made their revolutionary proposal for *asymmetric* (or *public key*) *cryptosystems*: each player uses a *public key* and a *secret key*. The public key of, say, BOB is used by everybody to encrypt messages to BOB. With his secret key, BOB can easily decrypt these messages, but without it, nobody should be able to do this.

	Cryptosystems	
	symmetric	asymmetric
examples	one-time pad, Caesar, DES, AES	RSA, Diffie-Hellman, ElGamal
speed	+	−
authentication	+	−
key exchange	−	+

Table 2.13: Symmetric vs. asymmetric cryptosystems.

At the current state of the art, both types of systems have their pros and cons. Table 2.13 lists some systems, to be explained later. The basic advantage of some symmetric systems is their speed, for example, over 12 GB/sec for AES. Their disadvantage is that a previous exchange of the common key is necessary; see Chapter I for drastic illustrations of the ensuing problems, sometimes costing the lives of thousands. A further advantage of symmetric cryptosystems is that the recipient can feel secure about the sender's identity, while in the asymmetric situation additional measures have to be taken against impostors; identification schemes and authentication are discussed in Chapter 10. But for modern cryptography, "symmetric vs. asymmetric" is not a competition, rather both sides win! We use asymmetric systems to share the common keys in a symmetric system, and then the latter for high-throughput communication.

Wonderful! Three examples illustrate this: we have already seen the symmetric system AES, next comes the asymmetric Rivest, Shamir and Adleman (RSA) system, and finally the asymmetric Diffie-Hellman key exchange.

2.5. The RSA cryptosystem

As the first realization of the abstract public key cryptosystem model suggested by Diffie & Hellman (1976), Rivest, Shamir & Adleman (1977) invented the *Rivest-Shamir-Adleman cryptosystem* (RSA).

We follow the long-standing tradition of calling the two players ALICE and BOB. Our scenario is that ALICE wants to send a message to BOB that he should be able to read, but nobody else. To this end, BOB generates a *secret key* sk and a *public key* pk. Anybody can read pk; imagine it is posted on the internet or in some large database. But BOB guards sk carefully as his secret. ALICE uses pk to encrypt her message for BOB. BOB uses sk to decrypt it. In a symmetric cryptosystem like AES, the encryption and decryption keys are (essentially) the same, but here pk and sk are different, and in fact sk cannot be computed easily from pk (hopefully).

The messages to be sent may be text, digitized pictures, movies, or corrrect, data or program files, etc. But we assume here and always in the future that the messages have been converted into some standard form, say into a (possibly very long) string of bits 0 and 1. How to perform this conversion best depends on the type of data. For text, a common way is to use ASCII or extended ASCII encoding of letters into 7-bit or 8-bit strings, respectively. In practice, RSA is mainly used to transmit a secret key, and the (short) message is derived from the key. Digital signatures can also be produced via RSA (Section 8.1).

Suppose that ALICE wants to send a (long) string of bits. There is a *security parameter* n to be explained in a minute. ALICE splits her string into blocks of $n-1$ bits each, and transmits each block separately. So we now explain how to transmit a single block $(x_0, \ldots, x_{n-2})$ of $n-1$ bits in the RSA system. We interpret the block as the binary representation of the natural number $x = \sum_{0 \le i \le n-2} x_i 2^i$. This number shall be transmitted.

The idea is the following. BOB chooses two prime numbers p and q at random with $n/2$ bits each, and so that their product $N = p \cdot q$ has n bits. He also chooses some random integer e with $1 \le e < N$ and $\gcd(e, (p-1)(q-1)) = 1$. BOB's public key is $\mathsf{pk} = (N, e)$. ALICE looks it up and sends the encryption $y = x^e$ in $\mathbb{Z}_N$ to BOB, that is, the remainder of x^e on division by N. The magic now is that BOB can recover x from ALICE's message with the help of his private information derived from (p, q). Here is the system described in full. The required algebraic terminology is explained in the *computer algebra toolbox* of Chapter 15. Throughout this text, the notation $a \xleftarrow{\boxtimes} A$ denotes a uniformly random choice of a from the finite set A, so that for any $b \in A$ the random variable a assumes the value b with probability $1/\#A$.

CRYPTOSYSTEM 2.16. RSA.

Key Generation keygen.

Input: Security parameter n.
Output: secret key sk and public key pk.

1. Choose two distinct primes p and q at random with $2^{(n-1)/2} < p$, $q < 2^{n/2}$.
2. $N \leftarrow p \cdot q$, $L \leftarrow (p-1)(q-1)$. [N is an n-bit number, and $L = \varphi(N)$ is the value of Euler's φ function.]
3. Choose $e \xleftarrow{\boxtimes} \{2, \dots, L-2\}$ at random, coprime to L.
4. Calculate the inverse d of e in $\mathbb{Z}_L$.
5. Publish the public key $\mathsf{pk} = (N, e)$ and keep $\mathsf{sk} = (N, d)$ as the secret key.

Encryption enc.

Input: $x \in \mathbb{Z}_N$, $\mathsf{pk} = (N, e)$.
Output: $\mathrm{enc}_{\mathsf{pk}}(x) \in \mathbb{Z}_N$.

6. $y \leftarrow x^e$ in $\mathbb{Z}_N$.
7. Return $\mathrm{enc}_{\mathsf{pk}}(x) = y$.

Decryption dec.

Input: $y \in \mathbb{Z}_N$, $\mathsf{sk} = (N, d)$.
Output: $\mathrm{dec}_{\mathsf{sk}}(y) \in \mathbb{Z}_N$.

8. $z \leftarrow y^d$ in $\mathbb{Z}_N$.
9. Return $\mathrm{dec}_{\mathsf{sk}}(y) = z$.

After the key generation, ALICE may forget p, q, and $\varphi(N)$, and erase them on her computer. N is included in the secret key only for symmetry. Here is a simple example.

EXAMPLE 2.17. We take $n = 6$. Literally, we would be looking for primes between 6 and 7, but at such small values we have to be a bit more liberal and choose $p = 5$ and $q = 11$. Thus $N = 55$ is a 6-bit number, and $L = 40$. We choose $e = 13$. Using the Extended Euclidean Algorithm 15.4, we find in a single step that $-3 \cdot 13 + 40 = 1$, so that $d = e^{-1} = -3 = 37$ in $\mathbb{Z}_{40}$. Thus BOB publishes his public key $\mathsf{pk} = (55, 13)$ and keeps his secret key $\mathsf{sk} = (55, 37)$. This finishes the key generation.

Now ALICE wants to send a message to BOB, say $x = 7$. Thus she has to calculate $y = x^e = 7^{13}$ in $\mathbb{Z}_{55}$. The obvious way to do this is to compute the integer 7^{13} and take its remainder modulo 55. This would be quite cumbersome here, and utterly infeasible at practical values of the security parameter n, where x^e would have more bits than there are elementary particles in the universe. But there is an easy way out: the repeated squaring Algorithm 15.48 uses fewer than $2m$ operations in $\mathbb{Z}_N$ for an m-bit exponent. This is illustrated in Figure 2.14. Its first column

instruction	value	exp	bit	in $\mathbb{Z}_{55}$
$y_0 \longleftarrow x$	x	1	1	7
$y_1 \longleftarrow y_0^2$	x^2	10		49
$y_2 \longleftarrow y_1 \cdot y_0$	x^3	11	1	13
$y_3 \longleftarrow y_2^2$	x^6	110	0	4
$y_4 \longleftarrow y_3^2$	x^{12}	1100		16
$y_5 \longleftarrow y_4 \cdot y_0$	x^{13}	1101	1	2

Figure 2.14: Computing x^{13}.

shows the instruction, the second one the value as a power of x, the third column the binary representation of the exponent in column two, the fourth column the corresponding bit in the binary representation 1101 of 13, and the last column the value for $x = 7$ in $\mathbb{Z}_{55}$. By a squaring, a 0 is appended to the right of the exponent's representation, and a subsequent multiplication by $y_0 = x$ turns this into a 1. This multiplication is done if and only if the corresponding bit is 1. Then the representation of the exponent is an initial segment of the representation of 13; this is the case just above the horizontal lines. All intermediate results are taken modulo

$N = 55$ and never get larger than N.

Now ALICE has done her share of calculation and sends $y = \text{enc}_{55,13}(7)$ $= 2$ to BOB. He decrypts in the same way, using the binary representation 100101 of 37, and computes the following sequence of results in $\mathbb{Z}_{55}$:

$$2, 4, 16, 36, 17, 14, 31, 7.$$

Thus $\text{dec}_{55,37}(2) = z = 2^{37} = 7$ in $\mathbb{Z}_{55}$ and indeed, this is the message that ALICE wanted to send to BOB. $\Diamond$

Because of its importance, we assemble the notation of Figure 2.15 for RSA, which will be used repeatedly in this text. The length of the public key (N, e) is $2n$, and also for the secret key.

---

security parameter n,
distinct random primes p and q of at least $n/2$ bits,
and so that $N = pq$ has n bits,
$L = \varphi(N) = (p - 1)(q - 1)$,
$e, d \in \mathbb{Z}_L$ with $ed = 1$ in $\mathbb{Z}_L$,
plaintext x, ciphertext y, decryption z, all in $\mathbb{Z}_N$,
$y = x^e,\ z = y^d$.

Figure 2.15: The RSA notation.

We have to address several questions.

Correctness: Is $z = x$?

Efficiency: How to calculate fast ...

  o ... large primes at random?

  o ... d from e?

  o ... powers modulo N? This has to be done for each message, and speed is even more a concern than for the previous two points.

Security: Suppose that an eavesdropper—traditionally called EVE—listens in to the communications between ALICE and BOB. Thus EVE knows y and, of course, (N, e), and she would like to compute x. In fact, x is uniquely determined! But how long does it take to calculate this value? Is this difficult enough to provide security?

Some of these questions are addressed in Section 3.1. The tools for an efficient implementation of RSA will only be discussed there, but we state a precise version of *efficient* now. It is the fundamental notion of polynomial time, which will be familiar to any student of computer science. In the key generation, we have to make random choices and thus require a *probabilistic algorithm*. This is the standard type of algorithm in this text. Its *expected runtime* for a fixed input is obtained by averaging over the algorithm's internal random choices. Thus a bound on the runtime has to hold for all inputs; there is no averaging over the inputs. For RSA encryption and decryption, no random choices are required. This special type of probabilistic algorithms is called *deterministic*. We recall the notions of "easy" and "hard" from Definition 15.29.

For the cryptographic protocols discussed in this text, efficiency always means that the operations performed by the legitimate players are easy. On the other hand, we want an adversary's problem to be hard. This leads to the security notions of Chapter 9. It turns out that RSA is not secure in the most demanding sense; see Example 9.22. For practical purposes, the current recommendation is to use a security parameter $n \geq 3000$; see Table 5.12.

Cryptographers at the British government agency CESG (Communications-Electronics Security Group) had invented, starting in 1970, many ingredients of public-key cryptography and the RSA and ElGamal cryptosystems before they were published in the open literature.

2.6. The Diffie-Hellman key exchange

For symmetric cryptosystems like AES, the shared secret key has to be chosen and communicated between all involved (and authorized) parties. In practice this agreement is a crucial difficulty in secret-key cryptography. A historical example is in Chapter I on the Zimmermann telegram. During the First World War, the Germans were cut off from their overseas embassies. They suspected that their old codes were broken, and sent a new code to Washington by U-boat. But it could not be forwarded to Mexico. An important secret message was sent safely to Washington in the new code, but then had to be transmitted in the old code to Mexico— and was duly broken. A safe public way of exchanging secret keys might have prevented this.

Can two players agree upon a shared secret key (for a symmetric cryptosystem) when they are forced to use an insecure channel? Stop here for a moment and convince yourself that this is impossible. The—somewhat surprising—correct answer is: yes, they can!

We cannot simply transmit the shared secret key, because the channel is insecure. Instead, each player sends a disguised version of the key from which the other player can assemble the shared key, but an eavesdropper cannot. To achieve this, Diffie and Hellman proposed to use a prime p and work with the integers that are not divisible by p, and their multiplicative properties. These numbers form the group $\mathbb{Z}_p^\times$ of *units* modulo p. There are $p - 1$ of them, the product of two of them is again not divisible by p, and any of them has a multiplicative inverse. This inverse can be computed via Euclid's algorithm. Furthermore, this group $G = \mathbb{Z}_p^\times$ is *cyclic*, meaning that there is an element $g \in G$, called a *generator*, whose powers comprise all of $G = \{1 = g^0, g, g^2, g^3, \ldots, g^{p-2}\}$. The details are discussed in the *computer algebra toolbox* of Chapter 15.

It turns out that the latter properties are also sufficient. Chapter 4 is dedicated to cryptography in groups, and we now describe the Diffie-Hellman key exchange in this general setting. So we have a finite cyclic group $G = \langle g \rangle$ and a generator $g \in G$. The group G, the generator g and all transmitted messages are public, so they are known not only to Alice and Bob but also to any unauthorized eavesdropper Eve. The *order* of G, that is, the number of elements of G, is denoted as $d = \#G$. There is a global security parameter n, and d is an n-bit number. Then the elements of G can be represented by n-bit strings. A basic requirement is that the description of G can be provided with polynomially in n many bits, and the group operations can be executed on the representations of the elements of G in polynomially in n many bit operations. The latter two requirements are usually satisfied in an obvious way.

Some examples for G are:

1. the multiplicative group $G = \mathbb{Z}_p^\times$ of units modulo a prime p,

2. the multiplicative group $G = \mathbb{F}_q^\times$ of a finite field $\mathbb{F}_q$,

3. cyclic subgroups of elliptic curves (Chapter 5) over finite fields.

We use the following example for illustration.

EXAMPLE 2.18. Let $G = \mathbb{Z}_{2579}^\times$ be the multiplicative group of units modulo the prime number 2579. Because $d = \varphi(2579) = 2578 = 2 \cdot 1289$ with 2 and 1289 prime, and $2^2 = 4$ and $2^{1289} = -1$ in $\mathbb{Z}_{2579}^\times$ are both different from 1, we conclude from Corollary 15.59 that $g = 2 \in G$ is a generator of G. The description of G consists of a few bits saying that "G is of the form $\mathbb{Z}_p^\times$", for example the ASCII encoding of the phrase in quotes plus the binary representation of p, 101000010011 in this case. G is identified with a subset of $\{0, \ldots, p - 1\}$, and the binary representation of these integers provides the representation of the elements of G. The tools for efficient arithmetic in Chapter 15 show that the group operations can be performed at cost quadratic (hence polynomial) in n, with $n = 12$ in this example. ◇

We now describe the *Diffie-Hellman key exchange*. First ALICE and
BOB individually choose a secret key a and b, respectively, randomly in
$\mathbb{Z}_d$. Then they publish their public keys $A = g^a$ and $B = g^b$, respectively,
maybe by posting them in a large internet database. After this asymmetric set-up, they can now both compute, maybe for use in a symmetric
cryptosystem, the shared secret key

$$g^{ab} = B^a = A^b,$$

since

(2.19) $$B^a = (g^b)^a = g^{ab} = (g^a)^b = A^b.$$

Figure 2.16 gives a graphical representation of the system.

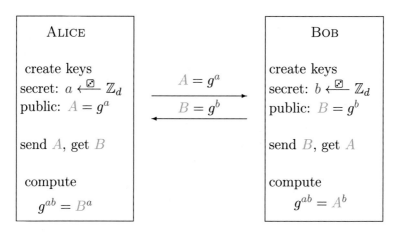

Figure 2.16: Diffie-Hellman key exchange.

Here is a formal description of the protocol.

PROTOCOL 2.20. Diffie-Hellman key exchange.

Set-up.

Input: security parameter n.
Output: G, g and d as below.

1. Determine a description of a finite cyclic group $G = \langle g \rangle$ with $d = \#G$
    elements and a generator g of G, where d is an n-bit integer.

Key exchange.

2. ALICE chooses her secret key $a \xleftarrow{\boxtimes} \mathbb{Z}_d$. She computes her public key $A \leftarrow g^a \in G$.
3. BOB chooses his secret key $b \xleftarrow{\boxtimes} \mathbb{Z}_d$. He computes his public key $B \leftarrow g^b \in G$.
4. ALICE and BOB exchange their public keys A and B.
5. ALICE computes the shared secret key $k_A = B^a$.
6. BOB computes the shared secret key $k_B = A^b$.

We continue our little example.

EXAMPLE 2.18 CONTINUED. Recall that $G = \mathbb{Z}_{2579}^{\times}$ and $g = 2 \in G$.

2. ALICE chooses her secret key $a = 765$ and computes her public key $A = 2^{765} = 949$ in G.

3. BOB chooses his secret key $b = 853$ and computes his public key $B = 2^{853} = 435$ in G.

4. ALICE and BOB exchange their public keys $A = 949$ and $B = 435$.

5. ALICE computes the shared secret key $k_A = B^{765} = 2424$ in G.

6. BOB computes the shared secret key $k_B = A^{853} = 2424$ in G.

And lo and behold, the system works not only in general, but also in this particular case: ALICE and BOB share the key $k_A = k_B$. ◇

In order to analyze this protocol, we should answer the three usual questions:

Correctness: Do ALICE and BOB get the same shared key?

Efficiency: Can ALICE and BOB do their computations efficiently?

Security : Is it hard for EVE to get information on the shared key?

The correctness is shown in (2.19), and ALICE and BOB indeed possess the same shared key $k_A = k_B$.

Concerning efficiency, the most costly operation is exponentiation in G. Figure 2.14 presents an example for the efficient algorithm of *repeated squaring* for computing $g^e \in G$, given $g \in G$ and $e \in \mathbb{N}$; see also Section 15.13. We may assume that $0 \le e < d$ by Corollary 15.57 and then it uses $O(n)$ multiplications in G.

In order to discuss security, we slip into the rôle of EVE who works on the following task.

DIFFIE-HELLMAN PROBLEM 2.21 (DH$_G$). *Given a group $G = \langle g \rangle$ of order d and A and B in G, compute $C \in G$ so that*

$$A = g^a, B = g^b, C = g^{ab},$$

for some a and b in $\mathbb{Z}_d$.

DEFINITION 2.22. *Let $G = \langle g \rangle$ be a group of order d, let $a, b, c \in \mathbb{Z}_d$ and $A = g^a$, $B = g^b$, and $C = g^c$ in G. If $ab = c$ in $\mathbb{Z}_d$, then (A, B, C) is a Diffie-Hellman triple.*

Thus DH$_G$ asks, for given A, $B \in G$, to find some $C \in G$ so that (A, B, C) is a Diffie-Hellman triple. There is also the following, possibly easier, version of DH$_G$.

DECISIONAL DIFFIE-HELLMAN PROBLEM 2.23 (DDH$_G$). *Given a group $G = \langle g \rangle$ of order d and A, B, C in G, decide whether (A, B, C) is a Diffie-Hellman triple.*

If DH$_G$ can be solved efficiently, then so can DDH$_G$. But the reverse is conjectured to be false in general.

Instead of sending the public versions $A = g^a$ and $B = g^b$ of a and b, ALICE and BOB might of course send the secret values a and b. But then they have no secret hidden from EVE. More generally, if EVE can compute a from A, g, and G, then she only has to compute $B^a \in G$ to get the common key, just like ALICE does. We have already noted that raising to a power can be done efficiently. The "inverse" task amounts to the following.

DISCRETE LOGARITHM PROBLEM 2.24 (DL$_G$). *Let G be a group, and g, $x \in G$ two group elements. Decide whether x is in the subgroup $\langle g \rangle \subseteq G$ generated by g, and if so, compute some $a \in \mathbb{Z}$ with $g^a = x$.*

A refined version of this problem is described in Definition 4.2. The Discrete Logarithm Problem 2.24 is of great importance in cryptography and occupies a prominent place in this book. In Chapter 4, we discuss various ways to solve it, and also the "generic model" of computation which says that unless you have a special insight into your group G, the discrete log problem is hard. Chapter 5 is devoted to a type of groups—elliptic curves—for which no-one seems to have the required "special insight" so far, and which are currently popular in cryptography. As noted above, if EVE can solve the discrete logarithm problem efficiently, then she also can break the Diffie-Hellman Problem fast. The reverse $DL_G \leq_p DH_G$ is also true for "most" groups G under some plausible assumptions.

We have discussed security under the assumption that EVE attacks our protocol in a passive way so far. The bad guy only listens to the transmission and tries to compute results without disturbing the communication between ALICE and BOB. But can we assume EVE to be passive? This leads to a more general question on security: Can EVE get secret information without solving the Diffie-Hellman Problem 2.21?

One attack that works in this case is the *woman-in-the-middle attack*: EVE pretends to be ALICE when talking to BOB and to be BOB when exchanging data with ALICE. Both legitimate players think they have the right partner. And EVE can act in the expectedly evil way by generating her own part of the used shared key(s). This is illustrated in Figure 2.17.

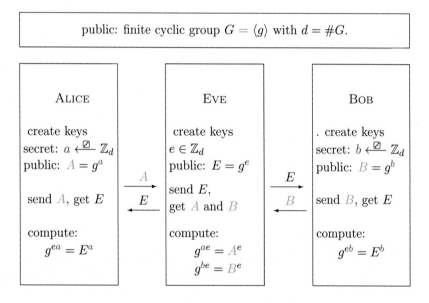

Figure 2.17: Woman-in-the-middle attack on the Diffie-Hellman key exchange.

BOB now believes that his shared key with ALICE is g^{be}, but the unsuspecting ALICE knows nothing about this; only EVE does. And similarly the other way around. In effect, EVE can read all information encrypted with the supposedly "shared" keys g^{eb} or g^{ea}, respectively. She can decrypt any message and re-encrypt it with the other key to hide the attack. BOB and ALICE may not recognize this active attack until EVE stops "translating" the encrypted messages.

This is clearly a killer for the system, and ALICE and BOB have to thwart this attack. Of course, they can simply compare whether the "shared" keys are identical. If they are not, they can be sure that someone has been messing around. But how can they be sure to compare to the right key without identifying their partner and without disclosing their common key? How does BOB tell ALICE from EVE? They have to find a way to identify each other. In the "non-cryptographic world", we know each other's voices or faces, and we can use identity cards or passports when introducing ourselves to a stranger. ALICE and BOB have to use similar tools within the protocol. Such electronic tools do indeed exist, and we discuss them in Chapter 10.

2.7. Block ciphers

We now give a more formal description of encryption. Some theoretical framework is necessary to put examples into perspective. It remains bloodless until infused with specific instances; we have already seen AES and RSA, and more examples will follow.

DEFINITION 2.25. *A block encryption scheme (block cipher) S consists of four probabilistic algorithms $S = (\text{set-up}, \text{keygen}, \text{enc}, \text{dec})$. We have a security parameter $n \in \mathbb{N}$ and for each n a key space K_n, a plaintext space P_n, and a ciphertext space C_n, all of which are finite and publicly known. Often they consist of all (or most) bit strings of a certain length. The value n may be restricted to some set of integers. We let $K = \bigcup_{n \in \mathbb{N}} K_n$.*

> set-up: *on input n, it outputs descriptions of K_n, P_n, C_n,*
>   *and of the encryption scheme used.*
>
> keygen: $\mathbb{N} \longrightarrow K^2$,
> $\text{keygen}(1^n) = (\text{sk}, \text{pk}) \in K_n^2$,

and for each n

$$\text{enc}: K_n \times P_n \longrightarrow C_n,$$
$$\text{enc}_{\text{pk}}(x) = y,$$
$$\text{dec}: K_n \times C_n \longrightarrow P_n,$$
$$\text{dec}_{\text{sk}}(y) = z.$$

The options for the descriptions are determined by the larger system that uses S. For example, set-up$(3000) = (3000, \text{"RSA"})$ might mean that 3000-bit RSA is used, and $P_n = \{0,1\}^n$ and $C_n = \{0,1\}^{n+1}$ consist of the binary representations of numbers modulo the RSA modulus. The input to keygen is the integer n, represented in unary as a string of n ones. The values sk, pk, x, and y are arbitrary elements of K_n, K_n, P_n, and C_n, respectively, and those in dec need not be related to those in enc. The "probabilistic encryption" enc is not a function in the usual sense, but rather a "function ensemble", that is, a probabilistic distribution on the functions $K_n \times P_n \longrightarrow C_n$. The distribution is given by enc's internal randomization. We require three properties of S.

Correctness: The decryption of an encrypted plaintext equals the plaintext. That is, for all $n \in \mathbb{N}$ and $x \in P_n$, if keygen(1^n) outputs $(\mathsf{sk}, \mathsf{pk})$, then

$$\mathrm{dec}_{\mathsf{sk}}(\mathrm{enc}_{\mathsf{pk}}(x)) = x.$$

Efficiency: The three "function ensembles" keygen, enc, and dec can be computed by probabilistic polynomial-time algorithms, where the algorithm for dec is further restricted to be deterministic.

Security: This is the most nontrivial requirement, and various aspects of it will be discussed later; see Chapter 9.

This text sometimes uses appropriate modifications of these notions. For the security, the resources at an adversary's disposal may be of different types. If neither pk nor sk is allowed to be known, we might as well replace both of them by the pair $(\mathsf{sk}, \mathsf{pk})$, so that $\mathsf{sk} = \mathsf{pk}$ afterwards.

DEFINITION 2.26. *In a* symmetric *(or secret key) encryption scheme, we have* $\mathsf{sk} = \mathsf{pk}$. *In an* asymmetric *(or public key) encryption scheme,* pk *is known to an adversary.*

An early block cipher is the *Caesar cipher* (Section A.3) in which the encryption replaces each letter by another one, three positions later in the alphabet. This is a secret key encryption scheme of block size 1, with $P_1 = C_1 = K_1$ being the alphabet (which here replaces $\{0, 1\}$). If we allow a shift by an arbitrary number of positions, then this number is the secret key sk. In Caesar's system, we have $\mathsf{sk} = 3$ and $\mathrm{enc}_3(\mathtt{crypto}) = \mathtt{fubswr}$ (using the English alphabet). Decryption consists of shifting back each letter by three positions, and $\mathrm{dec}_3(\mathtt{fubswr}) = \mathtt{crypto}$.

In a fixed system like AES, there is no security parameter, and we only have the three key lengths of 128, 192, and 256 bits. Then $K_n = \{0, 1\}^n$ and $P_n = C_n = \{0, 1\}^{128}$ for $n \in \{128, 192, 256\}$, since all plaintexts and ciphertexts consist of 128 bits. All other K_n, P_n and C_n are empty. Say, we consider $n = 256$. Then Definition 2.25 simplifies as follows:

$$
\begin{aligned}
\text{keygen} &\colon \{256\} \longrightarrow K_{256}, \\
\text{keygen}(256) &= \mathsf{sk} \in K_{256}, \\
\mathrm{enc}_{\mathsf{sk}} &\colon K_{256} \times P_{256} \longrightarrow C_{256}, \\
\mathrm{dec}_{\mathsf{sk}} &\colon K_{256} \times C_{256} \longrightarrow P_{256}.
\end{aligned}
$$

The last two requirements stated above have to be modified. Efficiency now means fast enough on current computers, and security means to resist the known attacks; see Chapter 6.

2.8. Stream ciphers and modes of operation

In a *stream cipher*, one has an input stream whose length is not known in advance, and encodes the message as it flows by. As an example, we can single out the *one-time pad*, in which the ciphertext bits are obtained by binary addition (XOR) of the plaintext and the key bits:

$$y_i = x_i \oplus k_i,$$

where y_i, x_i, and k_i denote the ith bit of the ciphertext, plaintext, and key, respectively. This system is *perfectly (information-theoretically) secure* provided the key is "truly random" (Section 9.4). Such a scheme is difficult to implement for long plaintexts, since the key is as long as the plaintext. We mention a related system, the synchronous additive stream cipher, in Section B.1. These ciphers played an important role in the history of cryptography, associated with the names of Trithemius, Vigenère, and Vernam, and include the German *Siemens Geheimschreiber* and the British *Typex* from the Second World War. Such a stream cipher usually produces a stream of pseudorandom bits (Chapter 11) and uses it like the one-time pad above.

In this text we discuss mostly *block ciphers*, encrypting blocks of some length n. Examples so far are AES (with $n = 128$) and RSA (with variable n). However, we have to consider how to encrypt long messages with one of our block ciphers (keygen, enc, dec) of fixed message length n. There are several ways of doing this. First, we might simply chop the message into blocks of length n and encrypt each block separately. This is called the *Electronic Codebook* (ECB) and actually not a good idea. A passive adversary, intercepting many encryptions, would then know which data are identical to those in other messages.

In order to mitigate this problem, we can chain the encryptions together, so that the encryption of each block depends on the previous blocks. We split our input, as it streams by, into blocks $x_0, x_1, x_2, \ldots$, each of n bits. For $i = 0, 1, 2, \ldots$, we calculate some n-bit z_i which is transmitted. In the *cipher block chaining* mode (CBC), z_i is obtained by encrypting the sum (bitwise XOR) of x_i and z_{i-1}. (With an initial value v for z_{-1}; often one takes $v = 0$, but this is less secure than a random v.) In the *cipher feedback* mode, z_i is the sum of x_{i-1} and $\mathrm{enc}_K(x_i)$, again with an initial value for x_{-1}. The same key is used for all encryptions. One might also consider transmitting $z_i = \mathrm{enc}_K(x_i) \oplus \mathrm{enc}_K(z_{i-1})$, or use the

$$\cdots x_{i-1}x_ix_{i+1} \cdots \qquad \cdots x_{i-1}x_ix_{i+1} \qquad \cdots x_{i-1}x_ix_{i+1} \cdots$$

$$\downarrow \qquad\qquad \downarrow \qquad\qquad \downarrow$$

$$\qquad\qquad y_i = x_i \oplus z_{i-1} \qquad y_i = \mathrm{enc}_K(z_{i-1})$$

$$\downarrow \qquad\qquad \downarrow \qquad\qquad \downarrow$$

$$z_i = \mathrm{enc}_K(x_i) \qquad z_i = \mathrm{enc}_K(y_i) \qquad z_i = y_i \oplus x_i$$

$$\text{ECB} \qquad\qquad \text{CBC} \qquad\qquad \text{CFB}$$

$$\cdots x_{i-1}x_ix_{i+1} \cdots \qquad \cdots x_{i-1}x_ix_{i+1} \cdots \qquad \cdots x_{i-1}x_ix_{i+1} \cdots$$

$$\downarrow \qquad\qquad \downarrow \qquad\qquad \downarrow$$

$$y_i = \mathrm{enc}_K(z_{i-1}) \qquad\qquad\qquad y_i = \mathrm{enc}_K(v \oplus i)$$

$$\downarrow \qquad\qquad \downarrow \qquad\qquad \downarrow$$

$$z_i = \mathrm{enc}_K(x_i) \oplus y_i \qquad z_i = \mathrm{enc}_{x_{i-1}}(x_i) \qquad z_i = y_i \oplus x_i$$

$$\text{OFB} \qquad\qquad \text{autokey} \qquad\qquad \text{CTR}$$

Figure 2.18: Modes of operation for block ciphers.

plaintext itself as key, assuming a key of length n: $z_i = \mathrm{enc}_{x_{i-1}}(x_i)$. The former is the output feedback mode (OFB) and the latter is called *autokey* and was invented by Vigenère; see Section C.4. The *counter mode* (CTR) uses an *initial vector* (or *nonce*) v and a counter which, in the simplest case, is increased by one for each new block. The ith ciphertext then is $z_i = \mathrm{enc}_K(v \oplus c) \oplus x_i$, where c is the binary representation of the counter, padded with leading zeroes to the appropriate length. CBC and OFB suffer from the problem of error propagation: an error in one encryption may contaminate subsequent blocks; see Exercise 2.10. Figure 2.18 summarizes these six modes of operation.

2.9. Secret sharing

The basic task for this section is the following: we want to distribute shares of a secret among n players so that together they can reconstruct the secret, but no proper subset of the players can. You may think of sharing access to your bowling club account with your friends, or of other people, possibly not your friends, sharing access to a nuclear first strike capability.

Shamir (1979) presents an elegant solution on just two pages. We identify possible secrets with elements of the finite field $\mathbb{F}_p = \mathbb{Z}_p$ for an appropriate prime p. Some bank cards for *Automatic Teller Machine* (ATM) access have four-digit decimal numbers as their secret *Personal Identifi-*

cation Number (PIN) codes. To distribute such a secret PIN to n players, we may take the smallest prime $p = 10\,007$ with five decimal digits. Then we choose $2n - 1$ random elements $f_1, \ldots, f_{n-1}, u_1, \ldots, u_n \xleftarrow{\boxtimes} \mathbb{F}_p$ uniformly and independently with all u_i nonzero, call our secret $f_0 \in \mathbb{F}_p$, set $f = f_{n-1}x^{n-1} + \cdots + f_1 x + f_0 \in \mathbb{F}_p[x]$, and give to player number i the value $f(u_i) \in \mathbb{F}_p$. If $u_i = u_j$ for some $i \neq j$, we have to make a new random choice; this is unlikely to happen if n is much smaller than $\sqrt{p}$, according to the birthday Theorem 3.41 and Exercise 3.18 (iv). Then all players together can determine the (unique) *interpolation polynomial* f of degree less than n, and thus f_0. But if any smaller number of them, say $n - 1$, get together, then the possible interpolation polynomials consistent with this partial knowledge are such that each value in $\mathbb{F}_p$ of f_0 is equally likely: they have no information on f_0. The tools for interpolation are described in Section 15.10.

We verify this security claim for a coalition of the $n - 1$ players $1, \ldots, n - 1$. There is a unique interpolation polynomial $g \in \mathbb{F}_p[x]$ of degree at most $n - 2$ for their values, so that $g(u_i) = f(u_i)$ for all $i < n \leq p$. All possible interpolation polynomials for n players are of the form $s = g + h \prod_{1 \leq i < n}(x - u_i)$, with $h \in \mathbb{F}_p[x]$. The degree of the product is $n - 1$, and the degree of the sum is $\deg s = n - 1 + \deg h$, which is less than n—as required for f—if and only if $\deg h \leq 0$, that is, if and only if h is a constant. Now the constant term of the product is $u = (-1)^{n-1} \prod_{1 \leq i < n} u_i \neq 0$, by the choice of the u_i's. Thus the constant term

$$s(0) = g(0) + h \cdot u$$

of s may be any value $v \in \mathbb{F}_p$ for the appropriate unique choice

$$h = (v - g(0))/u$$

of h. In other words, the $n - 1$ players have no information about $f_0 = f(0)$, since any value for it is consistent with their knowledge $f(u_1), \ldots, f(u_{n-1})$.

EXAMPLE 2.27. Suppose that $n = 3$, $p = 7$, and the secret is $f_0 = 6$. We choose randomly $u_1 = 2$, $u_2 = 3$, $u_3 = 5$, $f_1 = 4$, $f_2 = 2$. Then the three players receive:

player i	1	2	3
point u_i	2	3	5
value $f(u_i)$	1	1	6

Together they can compute the unique interpolation polynomial $f = f_2 x^2 + f_1 x + f_0 = 2x^2 + 4x + 6$ and, in particular, the secret $f_0 = 6$.

But if only, say, players 2 and 3 collaborate, their unique interpolation polynomial is $g = 6x + 4$ of degree 1. All quadratic polynomials $s_h = g + h \cdot (x - 3)(x - 5)$ with $h \in \mathbb{F}_7$ interpolate these two values, and the constant term of s_h is $s_h(0) = h + 4$. When h runs through $\mathbb{F}_7$, every value in $\mathbb{F}_7$ appears exactly once as $h + 4$; for example, $s_6 = 6 + 4 = 3$ in $\mathbb{Z}_7$. The "correct" value $h = 2$ is as likely as any other.   ◇

This secret sharing scheme is *information-theoretically secure*: no matter how large the computational resources of players 2 and 3 are, they cannot find out anything about the secret f_0.

We can extend this scheme to the situation where $k \leq n$ and each subset of k players is able to recover the secret, but no set of fewer than k players can do this. This is achieved by randomly and independently choosing $n + k - 1$ elements $u_0, \ldots, u_{n-1}, f_1, \ldots, f_{k-1} \in \mathbb{F}_p$ and giving $f(u_i)$ to player i, where $f = f_{k-1}x^{k-1} + \cdots + f_1 x + f_0 \in \mathbb{F}_p[x]$ and $f_0 \in \mathbb{F}_p$ is the secret as above. Again, it is required that $u_i \neq u_j \neq 0$ if $i \neq j$. Since f is uniquely determined by its values at k points, each subset of k out of the n players can calculate f and thus the secret f_0, but fewer than k players together have no information on f_0.

2.10. Visual cryptography

The system discussed here provides a visual representation of a secure symmetric cryptosystem such as the one-time pad; see Sections 2.8 and 9.4. In its simplest variant, this scheme of Naor & Shamir (1995) transmits an image by first creating a random image as secret key and then a second image depending on it and the message. By itself, this second image is again randomly generated, but the two images are highly correlated.

For illustration, suppose a company manager stays at a hotel for negotiations with another company. If she requires information from home, maybe a blueprint or picture, her company sends her the second image by fax. Anyone seeing this fax alone obtains no information. But she can superimpose her secret key slide, which she took with her, on the fax and see the message.

How is this achieved? The plaintext image is split into square pixels, each of which is either black or white. Each pixel is further divided equally into four square subpixels. Both in the random key and in the encrypted message, exactly two of the four subpixels are black, and two are white. There are six possible arrangements of two blacks in a 2×2 square as in Figure 2.19. For the random key, one of the six is chosen uniformly at random, and independently for each of the many pixels. For the encryption, we choose the same arrangement as on the key if the plaintext pixel is white, and the complementary one if the plaintext pixel is black. If we

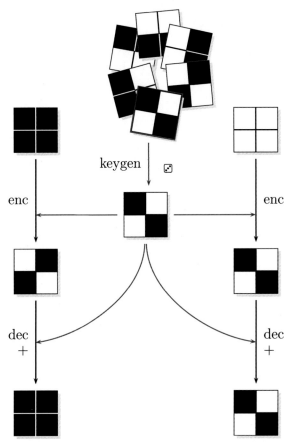

Figure 2.19: Visual cryptography—single pixel operation.

then superimpose the key and the encryption, we have exactly two or four
subpixels black if the plaintext pixel is white or black, respectively.

More information about the *VisKy* system of visual encryption is available on the book's website, including examples and javascript software
that allows you to encrypt your own images.

Notes 2.1. AES is not completely fixed, but rather there are three choices for the key
length.

2.2. EFF's website is `http://www.eff.org/`.

For the AES selection process, see `http://csrc.nist.gov/archive/aes/`. AES was
announced by NIST as US FIPS PUB 197 on 26 November 2001 which became effective
on 26 May 2002. It may be used in the United States up to the top secret level. *Rijndael*
is related to the Flemish word for *Rhine valley* and a play on the inventors' names.
A visual presentation of AES is available at the book's website. Matrix entries are
usually read row-before-column, as $a_{00}, a_{01}, \ldots$, but AES uses a different convention
for its 4×4 blocks.

Hamming (1980) coined *"the unreasonable effectiveness of mathematics"*.

Osvik *et al.* (2010) report a throughput of 12.9 GB/sec for AES on an NVIDIA Graphics Processing Unit 8800 GTX running at 1.35 GHz. See Bingmann (2008) for an extensive but slightly more dated comparison of various platforms and standard software packages.

The expected cost of exhaustive search for a 128-bit key is 2^{127} operations. The quote at the end of the section is from an undated NSA document provided by Edward Snowden and published in the article Appelbaum *et al.* (2014). Attacks on specific AES implementations are known that use differential power analysis, fault attacks, cache misses, or cold boot.

2.5. The three inventors of RSA were postdocs at MIT at that time. They later rose to prominent positions in science, at MIT, the Weizmann Institute, and the University of Southern California, respectively. They founded in 1982 a company now called RSA Labs, sold it in 1996, and its present owner acquired it in 2006 for 2.1 billion US\$.

ALICE and BOB were both born on 4 April 1977, or at least we found the first usage of these names as players in a cryptosystem in the technical memo MIT/LCS/TM-82 of that date, in which the inventors of RSA present their system and write about users "*A and B (also known as Alice and Bob)*".

There are several ways of prescribing the generation of the RSA modulus N precisely; see Loebenberger & Nüsken (2014) for a discussion.

According to the memos posted on the web page `http://www.cesg.gov.uk/about/nsecret.htm`, James Henry Ellis put forward in 1970 the idea of public key cryptography, which he called *non-secret encryption*. Clifford Christopher Cocks proposed in 1973 a practical implementation which was essentially equivalent to RSA (using $e = N$ in the RSA notation), and Malcolm Williamson suggested the Diffie-Hellman key exchange in $\mathbb{Z}_p^{\times}$ in 1976 (with a precursor in 1974). The CESG, based in Cheltenham, Gloucestershire, and today called the GCHQ is a descendant of the *Government Code and Cypher School* (GCCS) of ENIGMA-breaking fame; see Chapter J.

2.6. The key exchange idea was first published by Diffie & Hellman (1976). Their proposal used the only type of group then under consideration in cryptography, namely the group of units modulo a prime. The generalizaton to arbitrary cyclic groups is straightforward. In Figure 2.17, EVE might also use two different values of e, one for ALICE and one for BOB.

We do not discuss efficient pairings on an elliptic curve E in this text. In such groups E, DDH_E is easy, but before the advances in discrete logarithms mentioned in Section 4.8, DH_E looked hard.

It is not known whether $DL_G \leq_p DH_G$, but this was shown to be true under some plausible assumptions by Maurer & Wolf (1999). They take a cyclic group G with order $d = \#G$ and bit size $n \approx \log_2 d$, and assume that the set P of prime factors of d is given. According to Figure 4.6, even the case $P = \{d\}$ is of great interest. For $p \in P$ with p^2 dividing d, they assume that $p \in \text{poly}(n)$. For any prime p, they consider the Hasse interval $I_p = \{i: |i - p - 1| \leq 2\sqrt{p}\}$ of sizes of elliptic curves over $\mathbb{Z}_p$ (Theorem 5.10). Furthermore, $\nu(p)$ denotes the smallest integer s for which there exists an s-smooth integer in I_p, and $m = \max\{\nu(p): p \in P\}$. Maurer & Wolf exhibit a reduction from DL_G to DH_G using time $\sqrt{m} \cdot \text{poly}(n)$. Under the plausible assumption that $\nu(p) = (\log p)^{O(1)}$, the running time is polynomial in n. Proving this assumption seems out of reach for current methods in number theory. Even showing that $\nu(p) \leq p^{\alpha}$ for all $\alpha > 0$ is an open question; see Granville (2008), Section 4.4, for a description of the state of the art in smoothness bounds throughout his survey.

2.7. For an n-bit RSA modulus N, it would be more natural to take $\mathbb{Z}_N$ as plaintext and ciphertext spaces. The bit lengths n and $n+1$ only depend on n and are chosen because, identifying numbers and their binary representations, we have $\{n\text{-bit integers}\} \subseteq \mathbb{Z}_N = \{0, \ldots, N-1\} \subseteq \{(n+1)\text{-bit integers}\}$.

2.8. Here is a potential attack on ECB. The adversary sets up an internet shop and sells laptops at slightly lower prices than everybody else. He intercepts the encrypted versions of bank transfers from his clients to him, and might actually deliver the goods. Suppose that the beneficiary's account in a transfer is in the fourth of twenty blocks. He now has the encryptions of this fourth block and can insert it into later transfers that he intercepts. These monies would then end up in his account. (An attack might not work this way, but it illustrates the idea.)

A variant of the OFB mode is used in the ISO 10116 standard. A security reduction for the counter mode is in Rackoff (2012). Goldreich (2004), Chapter 5, discusses security aspects of various modes.

2.10. Visual cryptography was invented by Naor & Shamir (1995). Several variations and generalizations have been studied in the literature; see Cimato (2011).

Exercises.

EXERCISE 2.1 (Birth number).   *Let d be your date of birth in the format Year-MonthDay, so that 24 May 1990 gives $d = 19900524$. Let $N = 2^{16} + 1$, $a = d$ in $\mathbb{Z}_N$ with $0 \leq a < N$, $b = (d-a)/N = \lfloor d/N \rfloor$, and $c = a + b$ in $\mathbb{Z}_N$. Thus $c = 43116$ in the example. See Section 15.3 for division with remainder. We call c your birth number and will use it in several exercises. Now interpret the low-order eight bits a^*, b^*, c^* of the binary representations of a, b, c, respectively, as elements of $\mathbb{F}_{256}$, just as in AES. Compute*

   (i) *$a^* + b^*$, and compare to c^*,*

   (ii) *$a^* \cdot b^*$,*

  (iii) *$\operatorname{inv}(a^*)$. State your year of birth and give a date in the same year for which $a^* = 0$.*

EXERCISE 2.2 (One round of AES). *In this exercise we compute the first round of AES by hand. We start with an input matrix*

$$\begin{pmatrix} 01 & 11 & 21 & 31 \\ 02 & 12 & 22 & 32 \\ 03 & 13 & 23 & 33 \\ 04 & 14 & 24 & 34 \end{pmatrix}$$

and a key

$$\begin{pmatrix} AA & BB & CC & DD \\ AA & BB & CC & DD \\ AA & BB & CC & DD \\ AA & BB & CC & DD \end{pmatrix},$$

where all entries are in hexadecimal representation.

   (i) *Compute ADDROUNDKEY for the first two bytes.*

  (ii) *Compute SUBBYTES for the two bytes that result in (i).*

 (iii) *After SUBBYTES, the matrix looks like*

$$\begin{pmatrix} * & * & 55 & CE \\ C2 & D3 & 28 & DF \\ D3 & C2 & DF & 28 \\ E4 & 79 & 9B & 1E \end{pmatrix}.$$

Compute SHIFTROWS *of this matrix.*

(iv) *Compute* MIXCOLUMNS *for the last column of the matrix that results in (iii).*

EXERCISE 2.3 (Encryption and decryption with AES). *Consider* t_1, $t_0 \in R$ *as in* SUBBYTES.

(i) *Check that* $\gcd(t_1, x^8 + 1) = 1$ *in* $\mathbb{F}_2[x]$.

(ii) *Compute the inverse of* t_1 *in* $\mathbb{F}_2[x]/(x^8 + 1)$.

(iii) *Prove that* $t_1 \cdot a + t_0 = t_1 \cdot (a + (00000101))$ *for all bytes* $a \in \mathbb{F}_{256}$.

(iv) *Compute the inverse* P^{-1} *of the following matrix* P *modulo 2:*

$$P = \begin{bmatrix} 1 & 0 & 0 & 0 & 1 & 1 & 1 & 1 \\ 1 & 1 & 0 & 0 & 0 & 1 & 1 & 1 \\ 1 & 1 & 1 & 0 & 0 & 0 & 1 & 1 \\ 1 & 1 & 1 & 1 & 0 & 0 & 0 & 1 \\ 1 & 1 & 1 & 1 & 1 & 0 & 0 & 0 \\ 0 & 1 & 1 & 1 & 1 & 1 & 0 & 0 \\ 0 & 0 & 1 & 1 & 1 & 1 & 1 & 0 \\ 0 & 0 & 0 & 1 & 1 & 1 & 1 & 1 \end{bmatrix}$$

Is there any relation to your result from (ii)?

(v) *Show that for all bytes* $a \in \mathbb{F}_{256}$*, we have* $\mathrm{inv}(a) = a^{254}$*. Use Fermat's Little Theorem 15.56.*

(vi) *Given an output of the function* SUBBYTES*, how can you compute the corresponding 1-byte input?*

(vii) *Verify that* $(0By^3 + 0Dy^2 + 09y + 0E) \cdot (03y^3 + 01y^2 + 01y + 02) = 1$ *in* $\mathbb{F}_{256}[y]/(y^4 + 1)$.

(viii) SUBBYTES *is given by the polynomial expression* $a \mapsto 05 \cdot a^{254} + 09 \cdot a^{253} + F9 \cdot a^{251} + 25 \cdot a^{247} + F4 \cdot a^{239} + 01 \cdot a^{223} + B5 \cdot a^{191} + 8F \cdot a^{127} + 63$*. What does this mean? Where do the various operations take place? Illustrate each by an example which is not in this text.*

(ix) *Specify a decryption algorithm* AES^{-1} *that recovers the plaintext from an AES encryption and the secret key.*

EXERCISE 2.4 (The two byte rings). *The elements of the AES-rings* $\mathbb{F}_{256}$ *and* $R = \mathbb{F}_2[x]/(x^8 + 1)$ *can each be encoded using 8 bits per element. Show that the two rings are not isomorphic; see Section 15.2.*

EXERCISE 2.5 (Commuting steps?). *Do the following operations commute? That is, is the sequence in which you apply them irrelevant?*

(i) SUBBYTES *and* SHIFTROWS.

(ii) SHIFTROWS *and* MIXCOLUMNS.

(iii) MIXCOLUMNS *and* ADDROUNDKEY.

(iv) ADDROUNDKEY *and* SUBBYTES.

EXERCISE 2.6 (Skipping a step). *Consider the four operations (SUBBYTES, SHIFTROWS, MIXCOLUMNS and ADDROUNDKEY) performed in each round of AES. An obvious proposal to speed up the encryption is to omit one of the steps. Comment on the security of a modified 128-bit AES, where each round leaves out the following step.*

    (i) SUBBYTES,

    (ii) SHIFTROWS,

    (iii) MIXCOLUMNS,

    (iv) ADDROUNDKEY.

EXERCISE 2.7 (AES operations). *Take Baby-AES from Section 6.1 and do the following for all four AES operations $X \in \{$SUBBYTES, SHIFTROWS, MIXCOLUMNS, ADDROUNDKEY$\}$.*

    (i) For which blocks a is $X(a) = a$?

    (ii) Given a block b, describe how to compute a block a with $X(a) = b$.

    (iii) Do (i) and (ii) for the real AES of Section 2.2.

You may make reasonable assumptions about the key.

EXERCISE 2.8 (RSA keys). *We want to build an RSA system using the prime numbers $p = 251$ and $q = 337$. (In practice, these numbers are much too small.) Furthermore we choose $e = 54323$ and $N = p \cdot q$ for our public key. Use the Extended Euclidean Algorithm (Section 15.4) to compute the matching secret key d with $e \cdot d = 1$ in $\mathbb{Z}_{\varphi(N)}$.*

EXERCISE 2.9 (3RSA). *The RSA system can be generalized to allow for products of more than two distinct prime numbers. Consider the RSA system for products $N = p_1 \cdot p_2 \cdot p_3$ of three distinct primes:*

ALGORITHM. 3RSA key generation with security parameter n.

1. Choose three distinct prime numbers $p_1 < p_2 < p_3 < 2p_1$, so that

$$2^{n-1} < N = p_1 \cdot p_2 \cdot p_3 < 2^n.$$

2. Compute $\varphi(N) = (p_1 - 1) \cdot (p_2 - 1) \cdot (p_3 - 1)$.
3. Choose the public exponent $e \xleftarrow{\boxtimes} \{2, 3, \ldots \varphi(N) - 2\}$ so that $\gcd(e, \varphi(N)) = 1$.
4. Compute the secret exponent d so that $d \cdot e = 1$ in $\mathbb{Z}_{\varphi(N)}$.
5. Now (N, e) is the public key and (N, d) is the secret key.

With these keys, the remainder of the system is identical to the one using products of two distinct primes.

    (i) Prove that the system works correctly.

    (ii) Which advantages and disadvantages of this system over standard RSA do you see? Efficiency? Security?

  (iii*) Generalize the RSA system to products $N = p_1 \cdot p_2 \cdots p_r$ of r primes. Are there noteworthy differences?

EXERCISE 2.10 (Modes of operation).   *Answer the following questions concerning error propagation for each of the six modes of operation described in Section 2.8. Let t be the time to encrypt one block of n bits.*

  (i) *How many text blocks are incorrect if one of the transmitted blocks is corrupted?*

 (ii) *How many text blocks are incorrect if one of the transmitted blocks is dropped and this is not noticed?*

(iii) *How many text blocks are incorrect if one of the block encryptions outputs a wrong result?*

 (iv) *How long does it take to encrypt a message of $k \cdot n$ bits?*

  (v) *Draw conclusions from your observations.*

 (vi) *We define a further mode Plain Block Chaining that adds the plaintext x_{i-1} to the encryption of x_i as in the following picture.*

$$\cdots x_{i-1} x_i x_{i+1} \cdots$$
$$\downarrow$$
$$y_i = enc_K(x_i)$$
$$\downarrow$$
$$z_i = y_i \oplus x_{i-1}$$

PBC

Answer the questions (i)–(v) also for this mode.

EXERCISE 2.11 (Caesar).   *We consider the 1-letter block cipher Caesar$_3$ and the plaintext CryptoSchool. Calculate the six encryptions according to the six chaining methods in Section 2.8. Use the English alphabet and $IV = x$ where needed.*

EXERCISE 2.12 (Secret sharing).   *As an example for the secret sharing method of Section 2.9, we take $p = 10000019$, $n = 5$, and $u_1 = 1484998$, $u_2 = 8055552$, $u_3 = 412501$, $u_4 = 8994679$, $u_5 = 236054$.*

  (i) *Compute the secret from $f(u_1) = 2016419$, $f(u_2) = 951970$, $f(u_3) = 9707737$, $f(u_4) = 6395629$, $f(u_5) = 8552973$.*

 (ii) *Can you find a value for $f(u_5)$, so that the corresponding secret $f(0)$ is your birth number from Exercise 2.1, without changing the other values?*

We investigate which data yields sensitive information and which data does not. We set $p = 1009$ so that iterations over all of $\mathbb{F}_p$ are reasonable, and $n = 7$. Let f_0 be your birth number (Exercise 2.1) in $\mathbb{F}_{1009}$, choose $u_0, \ldots, u_6, f_1, \ldots, f_6 \in \mathbb{F}_{1007}$ at random with the u_i nonzero and pairwise distinct, and $u_1, \ldots, u_6$ not 1008.

(iii) *Suppose that $u_0 = 1008$ and that a coalition of the secret bearers 1 through 6 learns this by an indiscretion. Compute the distribution of possible secrets. That is, try all values for $f(u_0)$ and count how many times each possible secret occurs as the value $f(0)$.*

 (iv) *Now suppose that $f(u_0) = 1008$ and a coalition of secret bearers 1 through 6 learns this fact. Compute the distribution of possible secrets by trying all values for u_0 and counting the number of times that each possible secret occurs as the value $f(0)$.*

  (v) *Compare the results: is one of the indiscretions a problem for secret bearer number 0? Which one? Why? Can you describe "how much" information was disclosed?*

Hanc Graecis conscriptam litteris mittit,
ne intercepta epistola nostra ab hostibus consilia cognoscantur.[1]
CAESAR (c. 50 BC)

It must have been one of those ingenious secret codes which
mean one thing while they seem to mean another.
I must see this letter. If there were a hidden meaning in it,
I was confident that I could pluck it forth.
SIR ARTHUR CONAN DOYLE (1893)

La cifera semplice per ciò à lungo andare può più facilmente
essere intesa per coniettura, senza contracifra,
però si è trovato di formare le cifere più varie
et con caratteri doppii e tripplicati et con molte altre cose,
per iugannare li decifratori.[2]
MATTEO ARGENTI (c. 1605)

The inclusion of references in other languages may help
to break down the linguistic provincialism which,
ostrichlike, takes refuge in the mistaken impression
that everything worthwhile appeared in,
or has been translated into, the English language.
CARL BENJAMIN BOYER (1968)

'The letter?—Oh!—The letter! Keep looking at me between the
eyes, please. It was a string-talk letter, that we'd learned the way
of it from a blind beggar in the Punjab.' I remembered that there
had once come to the office a blind man with a knotted twig and
a piece of string which he wound round the twig according to
some cipher of his own. He could, after the lapse of days or
weeks, repeat the sentence which he had reeled up. He had
reduced the alphabet to eleven primitive sounds, and tried to
teach me his method, but I could not understand.
RUDYARD KIPLING (1888)

---

[1]He sent this letter written in Greek characters, lest by intercepting it the enemy
might get to know of our designs.

[2]The simple cipher can, because of this, in the long run rather easily be broken by
guessing, without the key, but one has found how to design more varied ciphers with
double and triple characters and many other things, to fool the decipherers.

Chapter A

Classical cryptology

or most of this text we are concerned with "modern" cryptography, whose current development started in the 1970s. But cryptology deals with such universal subjects—language and communication—that it has accumulated a rich history over the centuries of proud inventors and secretive cryptanalysts, famous people and amusing tales, redolent with fascinating characters and episodes, towering victories and abysmal failures. In this and nine other chapters, numbered alphabetically, we present an eclectic selection of such stories. On these few pages, the goal is not a complete or balanced account. Rather we concentrate on a few systems, individuals, and happenings. If you find these glimpses to your liking, you might turn to the real thing: David Kahn's monumental work *The Codebreakers* from 1967, still unsurpassed half a century later.

A.1. Cryptographic primitives

Over the millennia, people have invented and used a bewildering array of cryptosystems for the secret transmission of messages. In this section, we establish a general framework into which these systems fit. This is a scientific approach and rather ahistorical. To assess the contributions of individuals over the centuries in a fair way, one has to look at them in the context of contemporary knowledge, not with modern 20/20 hindsight. However, our hindsight helps us to sort ideas and see when new things have emerged.

© Springer-Verlag Berlin Heidelberg 2015

J. von zur Gathen, *CryptoSchool*, DOI 10.1007/978-3-662-48425-8_3

There are two fundamental cryptographic primitives:

○ substitution,

○ transposition.

In Claude Shannon's terminology, these generate *confusion* and *diffusion*, respectively, and the goal is to create enough of one of them, or preferably of both, to provide secrecy in communication. There is also a modern notion of cryptographic primitives which includes one-way and trapdoor functions; however, in this chapter we are only concerned with historical cryptography.

The modern notions of security (Chapter 9) started to be developed in the 1970s and are frequently based on concepts from computational complexity. A century earlier, Auguste Kerckhoffs (1883) had already stipulated a fundamental condition:

(A.1)
> *Il faut qu'il [le système] n'exige pas le secret, et qu'il puisse sans inconvénient tomber entre les mains de l'ennemi.*
>
> *It must not be required to keep the system secret, and it must be able to fall into the hands of the enemy without harm.*

Still today, some companies believe in *security by secrecy* and develop systems in secrecy, which are in due course reverse-engineered and broken.

In a *substitution*, we have some *alphabet* X. This might be the 26-letter English alphabet $A = \{a, b, c, \ldots, x, y, z\}$, or pairs of letters (*digrams*), so that $X = A^2$, or even longer *polygrams*, or bits $\mathbb{B} = \{0, 1\}$, or 128-bit words $X = \mathbb{B}^{128}$ for AES. In general, X is an arbitrary finite set. Furthermore, we have another alphabet Y, which might equal X or not.

We first explain some *simple substitutions*. In the examples that follow, we try to be brief and make liberal use of forward references. The neophyte reader should first get familiar with the forward material, and then go back and look at it from this general point of view.

We sometimes use a color-coding of `plaintext`, `key`, and `ciphertext`.

EXAMPLE A.2.

(i) A *(simple) substitution* (Section A.3) is a bijection $\sigma\colon A \longrightarrow Y$ from letters to some alphabet Y. See Section 15.2 for permutations.

(ii) AES-128 (Section 2.2) uses two substitutions. The first is the fixed substitution SUBBYTES which consists of the patched inversion σ on $\mathbb{F}_{256}$ with $\sigma(x) = x^{-1}$ if $x \neq 0$ and $\sigma(0) = 0$, followed by a linear map, and is applied individually to each of the 16 blocks. The second one is the key addition $\sigma = $ ADDROUNDKEY $: \{0,1\}^{128} \longrightarrow \{0,1\}^{128}$, where the 128-bit key (which we here consider as fixed) and the state are added bitwise.

(iii) RSA (Section 2.5) with public key (N, e) is the substitution $\sigma \colon \mathbb{Z}_N \longrightarrow \mathbb{Z}_N$ with $\sigma(x) = x^e$.

The conversion Table A.1 below may be useful for checking the letter addition in some of the following systems.

A	B	C	D	E	F	G	H	I	J	K	L	M
0	1	2	3	4	5	6	7	8	9	10	11	12
N	O	P	Q	R	S	T	U	V	W	X	Y	Z
13	14	15	16	17	18	19	20	21	22	23	24	25

Table A.1: Letter-to-number conversion in a 26-letter alphabet.

(iv) The Caesar cipher (Section A.3) identifies A with $\mathbb{Z}_{26}$ as in Table A.1 and uses the substitution $\sigma \colon \mathbb{Z}_{26} \longrightarrow \mathbb{Z}_{26}$ with $\sigma(x) = x + 3$. More generally, we might have any key $k \in \mathbb{Z}_{26}$ and use $\sigma_k(x) = x + k$.

Caesar's Latin alphabet had 24 letters, identifying I = J and U = V. It is not clear from the historical accounts how the cyclic shift is handled. Is $\sigma($Z$) = $ C? This is what we would use in a modern sense. The later *Augustus cipher* σ_1, the shift by one letter, uses $\sigma($Z$) = $ AA. Today's modular arithmetic is younger by 1800 years, basically due to Gauß (1801); see Chapter 15.

(v) The Vigenère cipher (Section B.1) with an ℓ-letter keyword k uses ℓ Caesar substitutions $\sigma_0, \dots, \sigma_{\ell-1}$. Alternatively, it can be viewed as a simple substitution $\sigma \colon A^\ell \longrightarrow A^\ell$ with $\sigma(x) = x + k$, using letterwise addition in A^ℓ with truncation at the end. For an example, we take the rather unimaginative keyword $k = $ KEY of length $\ell = 3$, and encrypt the plaintext $x = $ confuse the enemies as follows:

$$
\begin{array}{rcl}
x & = & \text{CONFUSETHEENEMIES} \\
k' & = & \text{KEYKEYKEYKEYKEYKE} \\
y & = & \text{MSLPYQOXFOILOQGOW}
\end{array}
$$

As is common in classical cryptography, blanks, spaces, and punctuation marks are ignored. Thus $\sigma(x) = x + k'$, where k' is the 17-letter key obtained from the Vigenère key $k = key$ by sufficiently many repetitions with the *Procrustes rule* of cutting off unneeded key letters at the end.

The attentive reader has noticed that in this short example, we have no fewer than four single-letter additions e + k = o. This is a general phenomenon, although usually not this frequent, and will be used in Chapter C to break this cryptosystem.

Generally a *polyalphabetic substitution* applies a fixed sequence of simple substitutions one after the other. When the sequence is exhausted, one starts again with the first one.

(vi) As a generalization of simple substitutions, a *homophonic substitution* works in the same way, only for each letter we do not have just a single possibility, but several ones. We see an example in Tranchedino's codebook from 1463 in Figure D.2. Its first line (after the heading) gives the 21 letters A, b, ..., z of the alphabet, plus the frequent words and, with, and of. Five of the letters get three possible encryptions, the others two. In general, the goal of the multiple possibilities is to even out the disparate frequencies of the various letters. The corresponding σ is now only a relation, not necessarily a function.

(vii) *Nomenclators* or *codebooks* (Chapter D) have large alphabets X and Y, with several hundreds (in the 17th century) or thousands (19th century) of elements each. Y has at least as many elements as X does, and the codebook is a simple substitution $\sigma \colon X \longrightarrow Y$, or a relation if homophones are used, as is often the case. The alphabet X usually comprises letters plus certain frequently occurring items, such as syllables, words, or names that were likely to appear in the correspondence. Their use is recorded from 1377 to the Second World War, where in one German submarine cipher each square of a grid covering the North Atlantic was given its code. More examples are in Chapter D.

(viii) The basic ingredient of the *one-time pad* (Section 9.4) is a substitution σ on single bits, where a (random) key bit k is chosen in $\mathbb{B}$, and a one-bit message $x \in \mathbb{B}$ is encrypted as $\sigma(x) = x + k$. Longer messages are encrypted by repeating this procedure, with keys chosen anew (independently at random) for each message bit.

(ix) The *Playfair cipher* (Section C.3) is a simple substitution $\sigma \colon A_0^2 \longrightarrow A_0^2$ on digrams, where $A_0 = A \setminus \{j\}$ is the standard alphabet with the letter j removed.

(ix) In the ENIGMA (Chapter J), the secret key determines (in a complicated fashion) a sequence of simple substitutions $\sigma_0, \sigma_1, \dots$, with $\sigma_i \colon A \longrightarrow A$ for all i.   ◊

A further classical security measure was the introduction of *dummies* (or *null values*, or *nulls*), These are encrypting symbols that will be discarded by the legitimate decryptor, but whose presence is intended to confuse the cryptanalyst. Tranchedino's system in Figure D.2 shows twelve dummies in the fifth line of the text. The Spanish cipher from around 1590 in Figure D.4 contains the line: *Las nullas tendran una raya enzima, exemplo* $\overline{19}$.[1] This provides a systematic way of introducing a large number of dummies.

For the second cryptographic primitive, we have a length parameter ℓ. A *transposition* is simply a bijection (or permutation) on the first ℓ numbers:

$$\tau \colon \{0, \dots, \ell - 1\} \longrightarrow \{0, \dots, \ell - 1\}.$$

EXAMPLE A.3.

(i) AES uses the transposition SHIFTROWS which performs certain cyclic shifts on the rows of the state matrix; see Section 2.2.

(ii) In a single columnar transposition (Section F.2) we write the plaintext in r rows of length c and read it off in columns as the ciphertext. Thus $x = $ COLUMN becomes $y = $ CLMOUN $= x_0 x_2 x_4 x_1 x_3 x_5$ in an $r \times c = 3 \times 2$ array:

$$x \;=\; \begin{matrix} C & O \\ L & U \\ M & N \end{matrix}$$

The transposition τ and its inverse α are given by

i	0	1	2	3	4	5
$\tau(i)$	0	3	1	4	2	5
$\alpha(i)$	0	2	4	1	3	5

---

[1] The nulls [dummies] will have a bar above them, for example $\overline{19}$.

and one checks that $\tau(i) = 3i - 5\lfloor i/2 \rfloor$.

(iii) The grille (Section F.4) is a transposition on a square array.

(iv) The skytale (Section F.1) can be viewed as a columnar transposition. $\Diamond$

From a transposition τ on ℓ numbers we obtain, for any alphabet A, a substitution $\tau_A \colon A^\ell \longrightarrow A^\ell$ by setting $\tau_A(x) = x_{\alpha(0)} x_{\alpha(1)} \cdots x_{\alpha(\ell-1)}$ for $x \in A^\ell$, where α is the inverse of τ. This is illustrated in Example A.3 (ii). Thus a transposition of length ℓ yields a simple substitution on ℓ-grams. However, it is profitable to keep the two primitives apart. For one, τ as above is much less "powerful" than a general substitution on A^ℓ, and furthermore, τ works for any A and might be called a "scheme" for such substitutions. From a higher point of view, substitutions are semantic objects and transpositions have a syntactical (or combinatorial) nature.

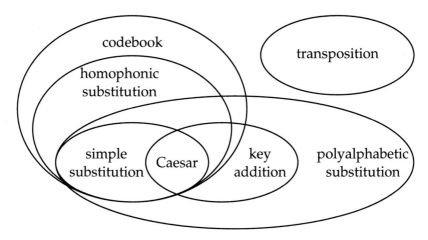

Figure A.2: A taxonomy of basic cryptosystems.

Once we have the primitives, we need two operations to work with them:

o *chaining,*

o *composition* (classically called *superencipherment*).

The primitives work on messages of a fixed length, maybe single letters, or digrams, or 128-bit strings. In order to transmit messages of arbitrary length, one has to "chain" such fixed-length primitives together.

The most common mode is to just repeat the primitive as often as necessary. Starting with simple substitutions, this yields polyalphabetic substitutions; it is the ECB mode of Section 2.8. See Example A.2 (v) and Section C.4 for examples. When the primitive is key-driven, there are other ways of chaining them together. For example, in the cipher-block chaining (or autokey) mode one uses the output of the previous application as key for the next one; see Section 2.8. A characteristic of modern cryptosystems is that they operate on uniform data formats for input, output, and key, so that the basic operations can be composed and iterated in many rounds.

In particular, when we have a substitution $\sigma \colon A \longrightarrow B$ and some number ℓ, we can apply σ independently to each of ℓ elements from A and thus obtain $\sigma^{\ell} \colon A^{\ell} \longrightarrow B^{\ell}$ with

$$\sigma^{\ell}(x_0, \ldots, x_{\ell-1}) = (\sigma(x_0), \ldots, \sigma(x_{\ell-1}))$$

for any $x_0, \ldots, x_{\ell-1} \in A$. Then σ^{ℓ} is called the *ℓ-fold product substitution* derived from σ. This is the electronic code book (ECB) from Section 2.8.

The second operation is the composition of two substitutions ρ and σ. For this to work, we have $\rho \colon A \longrightarrow B$ and $\sigma \colon B \longrightarrow C$, and then $\sigma \circ \rho \colon A \longrightarrow C$ with

$$(\sigma \circ \rho)(x) = \sigma(\rho(x))$$

for any $x \in A$ is their composition. When $\sigma \colon A \longrightarrow A$ is the Caesar cipher, shifting cyclically by three positions, then $\sigma^2 = \sigma \circ \sigma \colon A \longrightarrow A$ is the shift by six positions. More generally, ρ might encrypt strings over A by strings over B. We can then apply σ to each resulting letter in B. Classically, this is called *superencipherment*.

The most profitable application is when we start with a substitution $\sigma \colon A \longrightarrow A$ and a transposition τ on ℓ numbers. Then we have the product substitution $\sigma^{\ell} \colon A^{\ell} \longrightarrow A^{\ell}$, and can compose it with the transposition $\tau_A \colon A^{\ell} \longrightarrow A^{\ell}$ to obtain

$$(A.4) \qquad\qquad \tau_A \circ \sigma^{\ell} \colon A^{\ell} \longrightarrow A^{\ell}.$$

This leads to the more general modern concept of substitution-permutation networks.

EXAMPLE A.5.

(i) AES uses the four primitives SUBBYTES, MIXCOLUMNS, SHIFT-ROWS, and ADDROUNDKEY. The basic substitution $\sigma \colon \mathbb{B}^8 = \mathbb{F}_{256} \longrightarrow \mathbb{B}^8 = \mathbb{F}_{256}$ has been discussed in Example A.2 (ii), and

SUBBYTES $= \sigma^{16} \colon \mathbb{B}^{128} \longrightarrow \mathbb{B}^{128}$ is the 16-fold product of σ. SHIFT-ROWS consists of four different transpositions of the four blocks in each row of the state.

(ii) A German code from the First World War (Section I.1) involved a codebook $\sigma \colon X \longrightarrow Y \subseteq A^3$, with X and Y consisting of several thousand words, the latter being encoded by trigrams in a 29-letter älphabet A, which includes Ä, Ö and Ü. Furthermore, there was a simple substitution $\tau \colon A \longrightarrow A$, and the complete cipher was $\tau^3 \circ \sigma$, that is, the codebook superenciphered by the 3-fold product of a simple substitution.

(iii) We want to superencipher the Vigenère cipher σ with key length 3 from Example A.2 (v) by a transposition. We take the columnar transposition τ from Example A.3 (ii). Following the general recipe, we consider

$$\rho = \tau_{A^3}^3 \circ \sigma^6 \colon A^{18} \longrightarrow A^{18},$$

which first performs the Vigenère on six blocks of three letters each (appending a dummy X to the plaintext), and then interchanges the six blocks according to τ. Thus τ is applied to the matrix

$$\begin{array}{cc} \text{MSL} & \text{PYQ} \\ \text{OXF} & \text{OIL} \\ \text{OQG} & \text{OWV} \end{array}$$

to yield the ciphertext

$$z_1 = \text{MSLOXFOQGPYQOILOWV}.$$

However, we may also perform first the Vigenère and then the transposition separately on each sixpack of consecutive letters:

$$(\tau_A)^3 \circ \sigma^6 \colon A^{18} \longrightarrow A^{18}.$$

That is, τ is applied to each of the following three 3×2 blocks individually

$$\begin{array}{ccc} \text{MS} & \text{OX} & \text{OQ} \\ \text{LP} & \text{FO} & \text{GO} \\ \text{YQ} & \text{IL} & \text{WV} \end{array}$$

to yield the ciphertext

$$z_2 = \texttt{MLYSPQOFIXOLOGWQOV}.$$

See Exercise C.1 for another example. ◊

We have appended a dummy letter X to the plaintext in order to make its length divisible by 6. It encrypts as $\texttt{X} + \texttt{Y} = \texttt{V}$. The two options are not cryptographically equivalent: under the Kasiski attack of Section C.1, z_1 reveals the Vigenère key length 3, but z_2 does not.

There is one further general set of ingredients: the many tools for efficient encryption and decryption, and for remembering keys. The Vigenère table is already in Trithemius (1518) (Figure G.5) and reduces the three steps

$$\text{letter} \longrightarrow \text{number} \overset{\text{key add}}{\longrightarrow} \text{number} \longrightarrow \text{letter}$$

to a simple table look-up. The Alberti system helps to memorize a simple substitution and an example is shown in Section A.4; the Playfair cipher (Section C.3) has a mnemonic component for its digram substitution. And today, couldn't we do with a little help to remember all our passphrases?

A generally useful mnemonic aid is the alphabet Table A.1 for key addition systems like the Caesar and Vigenère ciphers.

The RSA cryptosystem (Section 2.5) is an amazing example. It is just a simple substitution but, with a current key size of 3000 bits, the alphabet of 2^{3000} letters is so huge that frequency analysis is hopeless. The winning point here is to encode a substitution on such an enormous alphabet in an extremely concise fashion, namely by its modulus and two exponents. We might call this a *key exponentiation system*: the plaintext has to be multiplied with itself as many times as the key indicates.

A.2. Brief history of cryptography

Over the centuries, several cryptographic systems have played major roles in professional use, mainly by the relevant government institutions: diplomatic, military, and secret services. The timeline in Figure A.3 tries to give an overview of the dominating systems throughout history. Of course, this has to leave out many of the finer points. In particular, it was not uncommon to mix two types of systems. A fundamental distinction is between the transposition systems, where individual letters are moved to other positions without being changed, and various types of substitution, where the units (letters, words, ...) are altered individually, but the flow of the message is not changed.

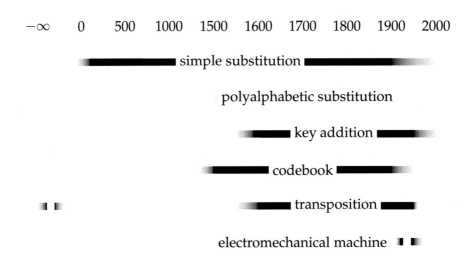

Figure A.3: Timeline of cryptography before computers.

Historical completeness cannot be achieved in such a concise pre-sentation, and some injustice to systems, attacks, and their inventors is inherent.

In the history of cryptography, we can distinguish several periods and indicate, very roughly, the corresponding time frames in Section A.2.

From antiquity, a few cryptographic tidbits have survived. In the beginning, the knowledge of writing was so exclusive that it did not require further protection. However, some texts were considered cosmic top secrets that should be accessible only to the highest level of royalty. We present one example from Egypt.

antiquity	1500 BC – 100 AD
Arab civilization	800 – 1400
European Middle Ages	1000 – 1500
Renaissance	1450 – 1600
Baroque, salon cryptography	1600 – 1850
mechanical devices	1580 – 1950
electromechanical devices	1920 – 1950
computers	1943 – present
public key systems	1976 – present

Table A.4: Cryptographic time periods.

Figure A.5: The second shrine of Tutankhamun.

The usual hieroglyphic writing system of Ancient Egypt employed symbols at three levels: meaning, word, and sound. As an example, the red crown of lower Egypt ⤳ can mean this crown, or be placed after other hieroglyphs spelling this crown letter by letter to categorize them. It is pronounced *dšrt* or *nt*, and can also (rarely) stand for the letter *n*. It is a remarkable achievement—akin to the cryptanalysis of an unknown system in an unknown language—that starting with Jean-François Champollion (1790–1832), egyptologists have learned how to understand and pronounce hieroglyphic texts.

Our example is from the famous tomb of Pharaoh Tutankhamun (ruled c. 1332–1322 BC) in the *Valley of the Kings* near Luxor in Southern Egypt. It was discovered by Howard Carter in 1922 and its treasures continue to amaze visitors at the Egyptian Museum in Cairo, and their replicas on worldwide exhibitions. The burial chamber of the 18-year old pharaoh contains four gilded wooden shrines nested matryoshka-like inside each other. Inside the fourth and smallest shrine rests a granite sarcophagus, in which are three mummiform coffins, the innermost one made of 110 kg of pure gold. This at last contains the mummy with its famous golden mask.

The second shrine is 3.75 m deep, 2.35 m wide, and 2.25 m high, with a sloping roof (Figure A.5). Its walls are decorated with images and texts from the Egyptian *Book of the Dead*, often used in tombs, and

Figure A.6: Two goddesses around a pole representing Re.

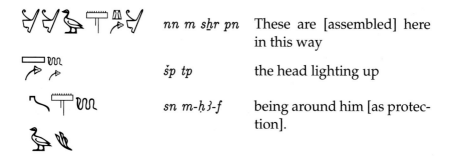

	nn m ẖr pn	These are [assembled] here in this way
	šp tp	the head lighting up
	sn m-ḥ3-f	being around him [as protection].

Figure A.7: The text at the top of Figure A.6.

a unique cryptographic composition now called the *Enigmatic Netherworld Book of the Solar-Osirian Unity*. Unlike the Book of the Dead, its text was kept as a royal secret and is only known from the tombs of Tutankhamun, Ramesses VI, and Ramesses IX. It uses secret spellings that only the cognoscenti would be able to read.

We explain the first field, at top right in Figure A.5 and enlarged in Figure A.6. It shows the sun god Re and two goddesses ready to protect the deceased. They are depicted by a post with ram head in the center and *Nephthys* (left) and *Isis* (right), two sisters of Osiris. The pole stands for Re and is imagined to be shining from the sun's rays. Left of the pole are the two names *head of Re* (head plus sun disk) and Nephthys, and to its right *is.t* = Isis. The two symbols representing the name of Isis depict a pool with sloping sides, usually pronounced *š* = *sh*. Here the first one is read as *s* and the second one as *t*; an English equivalent would be to write *she shall* for *sea tell*.

The inscription at the top is to be read in four lines (Figure A.7). The center column gives the pronounciation. The translation at right may be mysterious enough for us, but the designers have added another layer of secrecy, substituting standard letters by others, which are also standard but usually carry a different meaning. However, the reader is helped by a similar text in normal writing, just below this.

The first word *nn* = *these* is usually written as ‡‡. The two rushes with shoots ‡, pronounced *nn*, have been replaced by two which may be pronounced in the same manner. The next symbol is a pintail duck, but stands here for the owl , pronounced *m* and meaning *in*. Replacing one bird by another one only occurs in cryptography during this period. Within these enigmatic texts, can stand for eleven different sounds.

The snake stands for the usual symbol of a cobra. It encrypts the letter *d̲* in this cryptogram.

k	y	t	ḥ	z	w	h	d	g	b	ʾ (a)
כ	י	ט	ח	ז	ו	ה	ד	ג	ב	א
ל	מ	נ	ס	ע	פ	צ	ק	ר	ש	ת
l	m	n	s	ʿ	p	ṣ	q	r	š	t

Figure A.8: The Hebrew *atbash* substitution.

"Endless speculation would be possible for reading" the next five signs. One decipherment is to read them counterclockwise as [hieroglyphs] 𝔚𝔪. This transposition yields the translation given above.

In summary, this kind of Egyptian cryptography served to communicate secret religious information to the pharaoh in the netherworld. In addition to being buried in a sealed tomb, substitutions of signs by other ones and transpositions may have served to make them undecipherable by anyone outside the royal circle. This is quite successful, since egyptologists still debate details of how to read these texts. On the other hand, they may have been puzzles testing the reader's knowledge of the intricacies of the Netherworld.

In the Hebrew bible, a simple substitution occurs in a few places. The first letter is interchanged with the last one, the second with the last but one, and so on. Figure A.8 shows it with the modern Hebrew alphabet. In our alphabet this would be: $a \leftrightarrow z, b \leftrightarrow y, c \leftrightarrow x, \ldots, m \leftrightarrow n$. It is called the *atbash* system, after the replacements in the two rightmost columns, and an example is בָּבֶל = babel encrypted as שֵׁשַׁךְ = sheshakh. If we identify the 22 letters of the Hebrew alphabet with $\mathbb{Z}_{22} = \{0, \ldots, 21\}$, then this keyless encryption is $\text{enc}(x) = -x - 1$ in $\mathbb{Z}_{22}$ for any letter $x \in \mathbb{Z}_{22}$.

Our knowledge of Greek cryptography consists of a few isolated incidents such as the Polybius square and the *skytale*, of Spartan origin. The latter is a transposition cipher with a hardware implementation and described in Section F.1.

The historian Polybius (c. 200–c. 118 BC) described in his *Histories* the conquest of the Mediterranean world by the Romans, covering the period from 264 to 167 BC. King Philip V. of Macedonia (238–179 BC) had defended his territories in the First Macedonian War from 215 to 210, but lost everything except his home state in the Second Macedonian War, 200 to 197. Polybius describes his war preparations against Attalos I. Soter, King of Pergamon (269–197 BC), who had become an ally of the Romans in 211 BC.

They used a signalling system with lighted torches on hilltops; see Figure A.9 for a 19th century illustration. One example is the commu-

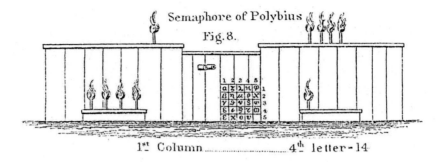

Figure A.9: Polybius' signalling system as interpreted by Myer (1879).

	1	2	3	4	5
1	A	B	C	D	E
2	F	G	H	I	K
3	L	M	N	O	P
4	Q	R	S	T	U
5	V	W	X	Y	Z

Figure A.10: The Polybius Square.

nication from the top of Mount Tisaion, 644 m high, across the Strait of Trikeri to Demetrios, a distance of about 7 km. On the tortuous mountain roads around the Bay of Pegasis, the land distance is about 140 km. Polybius writes that before his invention, only the few terms of a prearranged tiny "codebook" could be transmitted by fire signals. Of his own method, he says that *"Kleoxenus is the inventor, but that others think that Democrit proposed it; in any case, Polybius perfected it."* He uses an alphabet of 25 letters, and writes them in a 5×5 square. With our letters, leaving out J, this would look as in Figure A.10.

The person signalling a message has an arrangement of ten torches hidden behind a screen, five to the left and five to the right. For each letter, he raises as many left torches as are required to indicate the row, and then right torches for the column. Thus to transmit D, in row 1 and column 4, he raises one left torch and then four right torches. Formally, this is a simple substitution with elements from $\{1, 2, 3, 4, 5\}^2$. Both the system and the key are public. Polybius does not mention the possibility of arranging the letters in his square in a different sequence. The security depends on the enemy being unable to observe the light signal—an unexpected similarity to photon-based quantum cryptography.

Brigadier General Albert James Myer, United States signals officer, adapted Polybius' system and replaced torches by flags. This was used on both sides in the US Civil War, and the energetic up-and-down waving of flags earned the procedure the name *wigwag system*. Many other variations have been used, for example a prisoners' system where the torches are replaced by knocks on the jail walls.

The Romans perfected military technology in many respects, but apparently not in the area of cryptography. Caesar invented his famous cipher, and Augustus simplified it; see Example A.2 (iv) and Section A.3. This seems to have been used in private correspondence only.

The Arabs mastered already around 800 AD the major aspects of simple substitutions, including cryptanalysis based on frequency counts, and had a basic knowledge of transpositions; see Section G.1. The Western medieval world took some 700 years to catch up.

Medieval cryptography was usually not employed for secret communication, but for a different purpose. Secret writing often occurs in signatures and a scribe's request to pray for him. It does not make much sense for the latter, but may be related to an atavistic aversion against naming names. In signatures on legal documents, one may have thought that they add some security. And finally, bashfulness may be responsible for the scribes' use of cryptography in superstitious and pornographic writing.

In the Middle Ages, two systems of simple substitutions for the vowels only were popular: by one to five dots, and by the consonant following the vowel. Here are one plaintext and its two encryptions:

(A.6)
```
MEDIEVALCRIPTOHADNOKEI
M:D∴:V·LCR∴PT::H·DN::K:∴
MFDJFVBLCRJPTPHBDNPKFJ
```

There is no secret key between correspondents involved. Anyone who knows the system can decrypt any message.

Figure A.11 shows a page of the beautifully illustrated *Codex Aureus* from the monastery of St. Emmeran, written in 870. It was restored under the direction of abbot Ramwold (975–1001). The names Aripo and Adalpertus of the renovators are given in a cryptogram in the center of the right-hand column and enlarged in Figure A.12:

```
D::mn∴ ·bb·t∴s r·m::∴ld∴ ∴::ss∴::n : h::nc l·br∴m ·r∴p::
: t ·d·lp : rt∴:s r : n::v·v : r∴:nt.  S∴s m : m::r  ::::r

Dom[i]ni abbatis Ramuoldi iussione hunc librum Aripo
```

Figure A.11: Leaf 126^r of the *Codex Aureus* from the Bayrische Staatsbibliothek, München.

Figure A.12: The cryptogram below the center of the right column in Figure A.11.

```
et Adalpertus renovaverunt.   Sis memor eor[um].²
```

Figure A.13 shows a unique example of a different type. It comes from a biography of the English missionary Saint Willibald von Eichstätt (ca. 700–787), who tried hard—with his brother, Saint Wynnebald (701–761)—and ultimately successfully to prosyletize the heathens of Southern Germany. Willibald had spent about a decade of adventure travel in Italy, Asia Minor and the Holy Land. His *Vita* is the first travel book written by an English person, long before *lonely planet*. It was penned around 800 by an Anglo-Saxon *nun of Heidenheim*, whose name remained cryptographically hidden for a long time. Namely, after the last words *Amen. Finit* of the biography, the scribe inserted the text in Figure A.13 which reads literally as follows:

```
Secdgquar. quin. npri. sprix quar. nter.
cpri. nquar. mter. nsecun. hquin. gsecd
bquinrc. qarr. dinando hsecdc. scrter
bsecd. bprim.
```

The consonants are written in plaintext, and the five vowels are encrypted in order:

---

²At the order of abbot Ramwold, Aripo and Adalpertus repaired this book. Remember them.

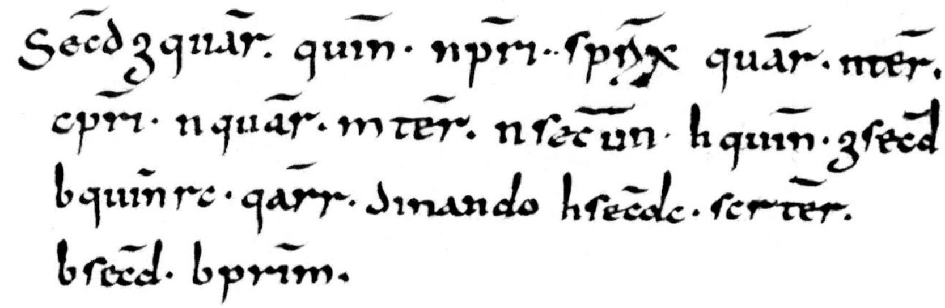

Figure A.13: Hugeburc's encrypted subscription to her *Life of Willibald*.

```
a  =  primum    =  first,
e  =  secundum  =  second,
i  =  tertium   =  third,
o  =  quartum   =  fourth,
u  =  quintum   =  fifth.
```

The ciphertext itself is abbreviated in a standard medieval fashion, indicated by a tilde. For example, the first word Secdgquar̄ means secundum g quartum = second g fourth, which decrypts as ego. Thus the Latin decryption is

Ego una Saxonica nomine Hugeburc ordinando hec scribebam.[3]

The last half-sentence presumably means that Hugeburc did not just copy the text, but restructured the material available to her.

Systematic, professional, and well-documented use of cryptography in Western Europe starts only in the early Renaissance, in Italy. The city states established permanent diplomatic missions in other states. These had to communicate regularly with their governments at home. Travel was insecure, and messengers were often attacked and had their letters (and sometimes lives) taken. To protect the secrets, cipher bureaus were established which produced codebooks for the various embassies. Tranchedino's nomenclator in Section D.1 is a sample output of such a code factory. Frequency cryptanalysis of a simple substitution was well understood, and protective measures such as dummies, several encryptions of a single letter (homophones), and long codebooks were commonly used.

The principles developed by these early Italian code builders formed the backbone of professional cryptography until the First World War, almost half a millennium later.

---

[3]*I am a Saxon lady by name of Hugeburc. I arranged and wrote this.*

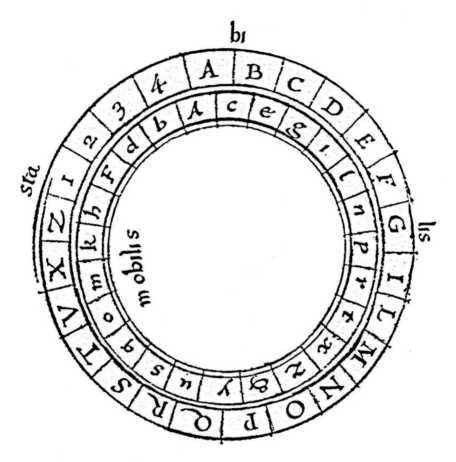

Figure A.14: Alberti's cipher disk.

Leon Battista Alberti (1404–1472), a brillant Renaissance Man, wrote about 1467 a remarkable treatise *Trattati in cifra* on cryptology. It is *"the West's oldest extant text on cryptanalysis"* and introduced two new tools: using several simple substitutions (*polyalphabetic substitution*) and superencipherment. His cipher disk in Figure A.14 allowed to turn the movable (*mobilis*) inner disk with 23 lower case letters and & against the fixed (*stabilis*) outer disk with 20 upper-case lettes and 1, 2, 3, 4, leaving out H, J, K, U, W, and Y. The secret key is one of the inner-disk letters, say m. The capital letters on the outer disk serve two purposes: to spell the plaintext and, when occurring in a ciphertext, to indicate the position of the inner disk. Say the initial position is m against X, as in Figure A.14 after slightly rotating the inner disk clockwise. Then the encryption of some capital plaintext letters is read off as the juxtaposed lower-case letters, until the next instruction to rotate the inner disk. Thus enc_m (GIOVANNI) could be XnpzCtqePqy. And what is dec_m (Xar4NmdEnfqTtpnq)?

This novel idea of *polyalphabetic encryption* presents a huge jump ahead in cryptology. Unlike a simple substitution, one and the same letter may be encrypted by different ciphertext letters.

Alberti's cipher disk is only printed, but meant to be made of two concentric brass plates. If τ is the permutation of the inner disk and σ the cyclic shift by 1, this gives a hardware implementation of the 24 substitutions $\{\tau\sigma^i : 0 \le i \le 23\}$. The inner disk in Figure A.14 is constructed in a fairly regular way, most adjacent letters being 2 apart in their alphabetical order. But in his text Alberti suggests to use an irregular sequence like a, g, q, ... We may take this as intending a random order, and also the capital-letter instructions to rotate the inner disk as random. Such a perspective was alien to contemporary minds, but we may take it as a randomized encryption system, the first of its kind in history.

Alberti then suggests to build a codebook with the pairs, triples, and quadruples from 1, 2, 3, 4 as ciphertexts. This allows for $4^2 + 4^3 + 4^4 = 336$ plaintext terms. His text includes a list of these ciphertexts, with plenty of space to fill in your own plaintext. Then the four numbers on the outer disk allow to incorporate a (lengthy) term from the codebook with up to four letters of ciphertext. This *superenciphered codebook* constitutes another first in cryptology. Weaker systems of this type were in use until World War I (Section I.1).

As a third major contribution, Alberti's book also contains the first description of frequency analysis outside of the Arab civilization.

Encryption by several simple substitutions was also proposed by the abbot Trithemius and published as the first printed book on cryptography in 1518; see Section G.3. He included a table which reduced encryption to a table look-up and which was later named after Vigenère, who used it in 1586. Its arithmetic nature—the encryption is the modular sum of plaintext and key—was recognized around 1690 and the system was completely broken by Kasiski in 1863, but continued as the "chiffre indéchiffrable" well into the 20th century. We discuss it at length in Chapters B and C. There is not much evidence of its use under practical conditions; one such application was on the Confederate side in the US Civil War. On the other hand, we typically only learn about cryptography gone awry and much less about successful uses.

Codebooks—small and large—continued to be the method of choice, but while the Renaissance had freed spirits from dogmatic confines, the flowering imagination of baroque and later mindsets brought about an exuberant multitude of cryptographic proposals, often beautifully illustrated and explained, and in general quite useless. This *salon crypto-*

graphy includes the elaborate image in Figure E.2, musical ciphers in Section G.4, knots on threads (but much simpler than Inca quipus), trumpets sounding (Notes B.2), flower arrangements, or arithmetic puzzles (Buck 1772; see von zur Gathen 2004b). These systems were often esthetically or intellectually pleasing, difficult to execute and easy to break (being simple substitutions), and showed off the imagination of their authors, who rarely failed to assert their absolute security.

Giovanni Battista della Porta (c. 1535–1615) also designed a cipher disk in 1563. Its beautiful reproduction in Paderborn (Figures A.15 and A.16) has a fixed outer disk with a 20-letter alphabet and Roman numerals on the outside and a movable inner disk with 20 phantasy symbols on its inside. A pointing hand coming out of a cloud indicates one of the fields, containing an ⊙. The twenty positions

Figure A.15: The allow twenty simple substitutions. Grabbing this
center of the Porta helping hand, one can move the inner disk accord-
disk.   ing to the rules for changing the substitutions and
read off encryptions or decryptions.

One of the movable Vigenère substitutions in Collange's 1561 edition of Trithemius' *Polygraphia* is shown in Figure G.7. These are called *volvelles*, printed on stiff but flexible paper and may be called the earliest cryptographic hardware. Later inventions had several metal wheels that rotated about a common axis; they were in military use as handy and robust field ciphers until the Second World War. As an example, the *US Signal Corps Converter M-209-B* based on the *Hagelin Cryptos C-38*, is shown in Figure A.17. Around 140 000 machines of this type were used by about 60 nations during World War Two.

In the early 1920s, the time was ripe for electromechanical devices, and four people independently invented rotor-based machines; see Chapter J. The most famous of these became the ENIGMA. Also in the 1920s, Vernam proposed his *one-time pad*. Here the message is represented as a string of bits, each either 0 or 1, a random bit string of equal length is generated, and the two are added, bit by bit. This provides perfect security; see Section 9.4. Alas, it is not easy to generate and distribute the required huge keys, but the system was employed extensively by Soviet and Eastern Bloc secret services during the Cold War. Variants of it were put to practical use, where the key is not really random but pseudorandom; see Chapter 11 on pseudorandom generation. Actually, Vernam's invention was of this type, and there were later electromagnetical implementations such as the German *Siemens Geheimschreiber* and the *Lorenz SZ* in World War Two. The hope presumably was that the

Figure A.16: The Porta disk.

Figure A.17: Looking inside the M-209.

seemingly minor change from random to pseudorandom would leave
the security intact—but the British cryptanalysts broke the system, in-
cidentally building the first computer, called *Colossus*, for this purpose;
see Section J.2.

From the 1950s on, computers took over much of the cryptographic
work. Shannon had developed a theory and identified confusion and
diffusion as fundamental goals. The *Data Encryption Standard* (DES), established in 1977, is a typical product of that era: a fairly complicated set
of bit operations performed in 16 rounds on the 64 bits of a message, and
which can be run by standard digital computer hardware at great speed.

In 1976, a 12-page paper by Diffie and Hellman brought about a revolution in cryptography. They proposed to consider systems where one
part of the key is kept secret and another part is made public. This
sounds rather strange, but it soon sparked the interest of a large community. Much of the present text is about various aspects of this new
public-key cryptography. On the technical side, it solves the problem of
sharing a secret key. More importantly, the new methods use a wide
variety of tools from computer science and mathematics, and attracted
leading scientists, in particular from computational complexity and from
number theory, The latter's influence is pervasive throughout this book,
from RSA and discrete logarithms to elliptic curves. Typical questions

in complexity theory are: What does it mean for a problem to be hard to solve? Can we prove problems to be hard? The ultimate answer to the latter question is still lacking, but the methodologies developed are essential for the modern theory of cryptography; we can look at formal notions of security, pseudorandom generation, and zero knowledge protocols as examples (Chapters 9, 11, 12).

A.3. Simple substitutions

In the traditional terminology of historical cryptography, a simple substitution is called an *alphabet*. Unfortunately, this clashes with the use of *alphabet* as a finite set of letters (or symbols) in which everything is written; this is its standard meaning in computer science, mathematics and natural languages. Thus we cannot use the traditional term or its derivates, but arrive at the following dictionary:

simple substitution = alphabet or monoalphabetic substitution,
homophonic substitution = monoalphabetic substitution with
homophones.

A *simple substitution* is a permutation enc: $A \longrightarrow A$ of an alphabet A, that is, to each letter $x \in A$ is associated a unique encrypting letter $\text{enc}(x) \in A$, and different letters have different encryptions. We now expand on the cipher of Gaius Iulius Caesar (100–44 BC) from Example A.2 (iv). There is no contemporary evidence as to whether Caesar actually used it, but the historian Gaius Suetonius Tranquillus (c. 70–c. 140) writes about Caesar's cryptography:

> *Extant & ad Ciceronem, item ad familiares domesticis de rebus: in quibus si qua occultiùs perferenda erãt, per notas scripsit, id est, sic structo litterarum ordine, vt nullum verbum effici posset: quæ si quis inuestigare & persequi vellet, quartam elementorum litteram, id est, d pro a, & perinde reliquas commutet.*[6]

Sueton also relates how Caesar's successor Augustus (63 BC–14 AD) used an even simpler version: shift by one, and no wrap around:

---

[6]There exist also [letters of Caesar] to Cicero, and to his family about domestic matters, in which he wrote in cipher if something was to be hidden. That is, in an arrangement of letters where no word was recognizable even to someone who wants to find out and read it. Namely, he turned a letter into the fourth element [following it], that is, *a* into *d*, and the others in the same way.

plaintext	G	A	L	L	I	A	O	M	N	I	S
numerical	6	0	11	11	8	0	14	12	13	8	18
Caesar num.	9	3	14	14	11	3	17	15	16	11	21
ciphertext	J	D	O	O	L	D	R	P	Q	L	V
Augustus num.	7	1	12	12	9	1	15	13	14	9	19
ciphertext	H	B	M	M	J	B	P	N	O	J	T

Table A.18: Caesar and Augustus encryption.

> *quotiens autem per notas scribit, B pro A, C pro B ac deinceps eadem ratione sequentis litteras ponit; pro X autem duplex A.*[7]

As an example, Caesar would send the plaintext $x = $ GALLIA OMNIS as $\text{enc}_{\text{Caesar}}(x) = $ JDOOLDRPQLV, and Augustus as $\text{enc}_{\text{Aug}}(x) = $ HBMMJB PNOJT. For simplicity, we are using a 26-letter alphabet (which the Romans did not), and arrive at the encryptions in Table A.18. In this and the following examples, the conversion Table A.1 may be helpful.

In his collection *Noctes Atticae*, Aulus Gellius (c. 125–180) has preserved excerpts from Greek and Roman writings, several of which are known to us only through this work. He mentions that

> *Libri sunt epistolarum C. Cæsaris ad C. Oppium, & Baltum Cornelium, qui res eius absentis curabant. In his epistolis quibusdam in locis inueniuntur literæ singulariæ, sine coagmentis syllabarum, quas tu putes positas incondité. Nam uerba ex his literis confici nulla possunt. Erat autē conuenium inter eos clandestinum, de commutando situ literarum, ut inscriptio quidem alia aliæ locū & nomen teneret: sed in legēdo locus cuiq, suus & potestas restitueretur.* [Marginal note:] *Id est, hāc latentē & occultā significationem literarum.*[8]

The meaning is not quite clear, but it may well be that Caesar also used either a codebook ("strange letters") or a transposition cipher. Roman cryptography may have been more imaginative than what we learn

[7]Often he writes in cipher and puts B for A, C for B, and the following letters in the same way; for X, he writes a double A. [X is the last letter of the Latin alphabet.]

[8]There are also collections of letters from Gaius Caesar to Gaius Oppius and Baltus Cornelius, who took care of his affairs in his absence. In these letters you find in some places strange letters, not connected into syllables, which you would think were placed at random. For no words can be formed from these letters. They also had arranged a secret key among them of changing the position of letters. Then although in the writing one letter has the position and meaning of another one, by reading it in its proper position, the real meaning is restored. [Marginal note:] That is the hidden and secret meaning of the letters.

from Sueton, although no original sources have survived. There are only 26 "Caesar ciphers" as above (in our alphabet), but if we consider arbitrary permutations on 26 letters, then there are $26! \approx 4.03 \cdot 10^{26}$ such permutations. If a cryptographer chooses a permutation at random among those 26! possibilities and a cryptanalyst wants to decrypt a message, it seems that he has to try out all of them—a hopeless task, at least by hand. Even if it were feasible on a computer, one would still have to choose one of the 26! outputs, most of which are nonsense. In the unlikely event that there are two or more that make sense, you would not even know which one is right, as for *bale* or *able*. A precise analysis of this problem is in Section A.5.

While the cryptanalyst has to find out the permutation, the legitimate users only have to agree on it, and then remember it. One of the most popular ways of facilitating this was invented by an Italian family of cryptographers, the Argentis. Giovanni Batista Argenti was cipher secretary to the popes Sixtus V and Gregorius XIV, at the end of the 16th century. His two nephews Matteo and Marcello Batista succeeded him in this post. After being sacked in 1605, Matteo Argenti wrote an important *manuale* on cryptanalysis.

The Argentis proposed the following way of memorizing a substitution enc: $A \longrightarrow A$, where A is an alphabet. You choose a key word k, map its letters in sequence to the first letters of A (removing duplicates in k), and then the rest of the alphabet in sequence. With the English alphabet for A and $k = $ GIOVANNI, this gives the following permutation:

(A.7)   plaintext G I O V A N B C D E F H J K L M P Q R S T U W X Y Z
          ciphertext A B C D E F G H I J K L M N O P Q R S T U V W X Y Z

and $\text{enc}_k(\text{BATISTA}) = $ GEUBTUE. As is visible in the example, most keywords provide only little change in the later part of the alphabet.

A.4. Frequency analysis

Any simple substitution is easy prey to *frequency analysis*, if only the message is long enough. This cryptanalysis requires as its main tools frequency tables for individual letters, but also for *digrams* (pairs of letters), *trigrams* (triples), and short words. Table A.19 gives eight lists of letter frequencies in percent, four for English in the first columns, and one each for German (D), French (F), Spanish (E), and Italian (I).

The first English column "COCA" refers to the corpus COCA, containing about 2 GB of English text, as analyzed in von zur Gathen &

letter	COCA	HP	Ch. 15	MM	D	F	E	I
a	8.28	7.94	7.83	8.04	5.26	7.75	12.25	10.71
b	1.54	1.59	1.53	1.54	1.85	0.99	1.48	0.74
c	3.18	1.99	2.91	3.06	3.62	2.67	3.63	5.14
d	3.90	4.96	3.47	3.99	5.05	3.35	5.33	3.73
e	12.17	11.91	12.27	12.51	17.41	16.61	14.01	12.04
f	2.14	2.05	2.70	2.30	1.50	1.08	0.46	1.29
g	2.12	2.57	1.68	1.96	2.94	1.29	1.05	1.82
h	5.03	6.73	4.43	5.49	5.90	0.93	1.22	1.81
i	7.29	6.21	8.80	7.26	8.85	7.33	5.50	10.26
j	0.19	0.11	0.10	0.16	0.15	0.71	0.64	
k	0.82	1.20	0.23	0.67	1.13			0.01
l	4.22	4.36	4.14	4.14	3.75	4.90	5.45	5.78
m	2.53	2.21	2.88	2.53	3.19	3.28	2.73	2.98
n	7.08	6.51	7.49	7.09	10.71	7.61	6.63	6.60
o	7.55	7.80	7.47	7.60	1.93	6.92	9.93	9.55
p	2.08	1.66	2.42	2.00	0.37	2.53	2.17	2.79
q	0.10	0.13	0.29	0.11	0.02	1.47	1.99	0.82
r	6.24	6.46	6.34	6.12	6.65	6.57	6.17	6.44
s	6.79	5.88	6.47	6.54	6.14	7.56	7.68	5.61
t	9.01	8.67	9.17	9.25	5.79	6.54	3.77	5.74
u	2.76	2.91	2.66	2.71	3.86	6.62	4.86	3.60
v	1.03	0.87	1.14	0.99	0.76	2.22	1.09	2.01
w	1.84	2.51	1.54	1.92	2.01			0.02
x	0.20	0.11	0.51	0.19	0.01	0.37	0.02	0.04
y	1.80	2.57	1.40	1.73	0.01	0.22	1.53	0.02
z	0.12	0.08	0.12	0.09	1.15	0.48	0.40	0.45
$\sum f_i^2$	6.50	6.36	6.65	6.58	7.77	7.55	7.51	7.22
$H(f)$	4.19	4.20	4.16	4.17	4.03	4.04	4.03	4.02

Table A.19: Four frequency tables for English, and one each for German, French, Spanish, and Italian, all in percent. Blank entries represent very infrequent letters.

Loebenberger (2015), the second one "HP" is from Joanne Rowling's (1998) *Harry Potter and the Philosopher's Stone*, the third one from Chapter 15 of this book, and the fourth one from Meyer & Matyas (1982). The last but one row is 100 times the sum of the squares of the frequencies; thus e contributes $100 \cdot 0.1217^2 \approx 1.48$ to the first entry 6.50. The last row presents the entropy (Definition A.9) of its column. More details are in the notes. We can observe material differences between the various

%	12	9	7-8	6	3–5	1–2	0
letters	e	t	aoins	r	hldcum	fgpwybvk	xjzq

Table A.21: The English alphabet grouped by frequencies.

distributions for English, for example, the frequency of c varies between 1.99% and 3.18%. The message of the four English columns is that there

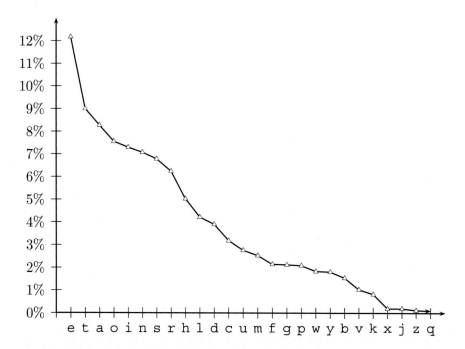

Figure A.20: Letter frequencies in COCA, ordered by frequency.

is some consistency across various types of texts, but variations do occur. Table A.21 sorts the letters into seven categories, according to a rough approximation of their frequencies. Thus *etaoinsr* is a useful mnemonic for English frequencies if you ever have to break a simple substitution.

If we want to analyze some encrypted message that we suspect to be a simple substitution, we set up the frequency table of the ciphertext, preferably in the frequency ordered way as in Figure A.20. If it matches roughly the English table, then this is a strong indication that we deal indeed with a simple substitution, and that the plaintext could be in English. The context will usually tell us the language, or leave a choice between two or three. Then we assume the language that matches best. If no language matches at all, as in the "flat" distribution of Figure C.2, then we conclude that this is not a simple substitution.

Then we substitute matching entries in the two frequency tables, starting with the ones that occur most often. Of course, we cannot expect the two tables to match exactly. For example, in Figure A.20, the rates for i, n, s, r, and h (near the 6% line) are so close to each other that we can, at best, expect them to match as a group. Thus one makes conjectures about individual letters. Some sections of the ciphertext will then have a substantial portion of plaintext guesses, and one tries to find actual words that fit. The search for individual words, such as one-letter words like a or I (if the word divisions are visible), or frequent words like *the*, helps along. The whole is a process of trial and error. Some experts have observed that as important as the technical tools is a certain degree of ingenuity and perseverance—virtues that are generally useful in life.

The great American poet Edgar Allan Poe (1809–1849) became interested in cryptography in late 1839, and had a forum as the editor of the weekly *Graham's Magazine*, where readers would send him ciphertexts and he would publish his solutions. The only systems he solved were simple substitutions. He soon achieved a reputation as a master cryptographer, but modern-day experts judge differently; see below.

Rather than quibble about his boastful self-aggrandization as master cryptographer, we follow the master story-teller's frequency analysis in his story *The Gold-Bug*, written in 1843. It deals with the hunt for the treasure of the pirate Captain Kidd, hidden on Sullivan's Island, near Charleston SC. The hero, William Legrand, has discovered a parchment with hidden characters on it. This is an example of superencipherment: the secret message was first encrypted (by a simple substitution, as it turns out), and this then superenciphered by steganographic use of sympathetic ink; see Section E.1. The superencipherment was stripped by accident: on a cold autumn evening, the narrator held the parchment close to the fire-place in order to examine it closely, and revealed the secret writing. His servant Juniper helps along.

```
53‡‡†305))6*;4826)4‡.)4‡);806*;  48†8¶60))
85;]8*:‡*8†83(88)5*†;46(;88*96*?;8)*‡(;485);
5*†2:*‡(;4956*2(5*--4)8¶8*;4069285);)6†8)
4‡‡;1(‡9;48081;8:8‡1;48†85;4)485†528806*81(
‡9;48;(88;4(‡?34;48)4‡;161;:188;‡?;
```

Legrand shows off:

> the solution is by no means so difficult as you might be
> led to imagine from the first hasty inspection of the charac-
> ters. These characters, as any one might readily guess, form

"*This is a strange scarabæus, I must confess*"

Figure A.22: Legrand, the narrator (seated), and Jupiter examine the parchment with the Gold-Bug cryptogram.

a cipher—that is to say, they convey a meaning; but then, from what is known of Kidd, I could not suppose him capable of constructing any of the more abstruse cryptographs. [...] Circumstances, and a certain bias of mind, have led me to take interest in such riddles, and it may well be doubted whether human ingenuity can construct an enigma of the kind which human ingenuity may not, by proper application, resolve.

Circumstantial evidence points to English as the language. He sets up the ciphertext's frequency table:

Of the character 8	there are	33.
;	"	26.
4	"	19.
‡)	"	16.
*	"	13.
5	"	12.
6	"	11.
†1	"	8.
0	"	6.
9 2	"	5.
: 3	"	4.
?	"	3.
¶	"	2.
]—	"	1.

In Figure A.23, we have overlain the graphs of this and the English frequency tables; the match is quite reasonable. Legrand goes on:

Now, in English, the letter which most frequently occurs is e. Afterwards, the succession runs thus: a o i d h n r s t u y c f g l m w b k p q x z. E however predominates so remarkably that an individual sentence of any length is rarely seen, in which it is not the prevailing character. Let us assume 8, then, as e. Now, of all words in the language, 'the' is most usual; let us see, therefore, whether there are not repetitions of any three characters, in the same order of collocation, the last of them being 8. If we discover repetitions of such letters, so arranged, they will most probably represent the word 'the.' On inspection, we find no less than seven such arrangements, the characters being ;48. We may, therefore,

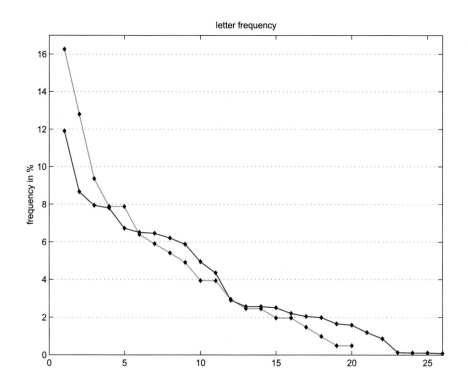

Figure A.23: English frequencies (red) and Gold Bug (green).

assume that the semicolon represents t, that 4 represents h, and that 8 represents e—the last being now well confirmed. Thus a great step has been taken.

But, having established a single word, we are enabled to establish a vastly important point; that is to say, several com- mencements and terminations of other words. Let us refer, for example, to the last instance but one, in which the com- bination ;48 occurs—not far from the end of the cipher. We know that the semicolon immediately ensuing is the com- mencement of a word, and, of the six characters succeeding this 'the,' we are cognizant of no less than five. Let us set these characters down, thus, by the letters we know them to represent, leaving a space for the unknown—

t eeth.

Here we are enabled, at once, to discard the 'th,' as forming no portion of the word commencing with the first t; since, by experiment of the entire alphabet for a letter adapted to the vacancy we perceive that no word can be formed of which this th can be a part. We are thus narrowed into

t ee,

and, going through the alphabet, if necessary, as before, we arrive at the word 'tree', as the sole possible reading.

His frequency positions of *t* and *n* are somewhat different from those in Table A.21. He notes the texts *"the tree thr‡?3h the."* and *"the word 'through' makes itself evident at once."* He then finds †83 88 or †egree, which gives † = *d*, and ;46;88∗ or th6rtee∗, and explains

an arrangement immediately suggestive of the word 'thir-teen', and again furnishing us with two new characters, i and n, represented by 6 and ∗. The first characters 5good yield 5 = a, and to avoid confusion, it is now time that we arrange our key, as far as discovered, in a tabular form. It will stand thus:

5	represents	a
†	"	d
8	"	e
3	"	g
4	"	h
6	"	i
∗	"	n
‡	"	o
(	"	r
;	"	t

We have, therefore, no less than ten of the most important let-ters represented, and it will be unnecessary to proceed with the details of the solution. I have said enough to convince you that ciphers of this nature are readily soluble, and to give you some insight into the rationale of their develop-ment. But be assured that the specimen before us appertains to the very simplest species of cryptograph. It now only re-mains to give you the full translation of the characters upon the parchment, as unriddled. Here it is: A good glass in the Bishop's hostel in the Devil's seat—twenty-one degrees and thirteen minutes—northeast and by north—main branch sev-enth limb east side—shoot from the left eye of the death's-head—a bee-line from the tree through the shot fifty feet out.

The secret message is deciphered, but Legrand still has a lot of figuring to do. Will he find Captain Kidd's treasure?

William F. Friedman, the leading US cryptographer of his days, says about Poe: *"The serious student of cryptography can, if he takes the trouble, see in Poe's essay and in his other writing on this subject many things which are not apparent to the layman. Against his will he is driven to the conclusion that Poe was only a dabbler in cryptography. At the same time it is only fair to say that as compared with the vast majority of other persons of his time in this or in foreign countries, his knowledge of the subject, as an amateur, was sufficient to warrant notice. Had he had opportunity to make cryptography a vocation, there is no doubt that he would have gone far in the profession."*

Wimsatt (1943) writes: *"Legrand's explanation of how he solved the cipher is a fine feat of exposition—as anybody will realize who undertakes to write a few paragraphs about ciphers. As we follow the steps of the argument, we have the impression of intricacy and precision, of Legrand's shrewdness and patience—each detail receives attention—and yet we are never lost, the main outlines remain clear, the reasoning turns where it should, the momentum, or rhythm, of the whole is sustained. The writing of this kind of prose was, as I see it, one of Poe's most impressive gifts."*

Many writers have commented on the intellectual capabilities that are useful for cryptanalysis. S'Gravesande (1748) goes one step further: he considers cryptanalysis (of a simple substitution) as a part of logic, which in turn is a branch of philosophy. Indeed, he develops on twelve pages the decipherment of a 109-letter text, first using letter frequencies and repetitions, then the word structure of Latin. Particularly instructive are his wrong turns, and explanations on how to backtrack from them.

Section C.1 explains some tools for breaking a simple substitution, even in the difficult situation where letter frequencies are known, but no digrams or longer polygrams appear. Professionals (then) and amateurs (still today) have burnt a lot of midnight oil figuring out messages encrypted in this kind of system, which must be considered perfectly insecure.

A.5. Information theory

Claude Elwood Shannon worked at the Bell Laboratories and published in 1948 and 1949 two treatises on a *mathematical theory of communication* and on a *communication theory of secrecy systems*. The first one became the foundation for the theory of error-correcting codes. The second one "transformed cryptography from an art to a science". He identified the two principal actions that provide security: *confusion* and *diffusion*. The first action is to scramble the alphabet thoroughly, as in the AES SUB-BYTES operation, and the second one is to diffuse information through-out the message, as MIXCOLUMNS and SHIFTROWS do in AES. Fur-thermore, Shannon quantified the notion of "information content" and derived a result saying that an encrypted message has to have at least a certain length for a cryptanalytic attack to be successful. An example is the one-time pad of Section 9.4, which is proven to be perfectly secure.

We now explain some of Shannon's theory. It only gives a lower bound on the required length of the ciphertext, but no method for ac-tually decrypting. However, for the simplest systems, like simple sub-stitutions or Vigenère ciphers, the cryptanalytic methods described in Sections C.1 and C.2 almost attain that bound in practice. This section requires more mathematical agility than the other ones in the present chapter.

We have an alphabet A of s letters. A letter $x \in A$ occurs with prob-ability p_x, so that $p_x \geq 0$ and $\sum_{x \in A} p_x = 1$. This can be abbreviated by the probability distribution $p = (p_x)_{x \in A}$. Using the alphabetic order, the COCA example of Table A.19 has $s = 26$, $A = (\mathsf{a}, \mathsf{b}, \mathsf{c}, \ldots, \mathsf{z})$, and $p = (8.28, 1.54, \ldots, 1.80, 0.12)$.

How much "information" do we provide by writing down one let-ter, or a long message? One of Shannon's contributions is to make this notion precise, in a useful way. Intuitively, his idea is to insist on writing everything in binary—using only 0 and 1—and to say that the shortest general way of specifying a message in binary is the *information* provided by the message. That is, we count the number of *bits*—a word coined by Shannon.

But to have a meaningful notion, we cannot allow any old way of presenting letters in binary, but must look at the cleverest one.

EXAMPLE A.8.

(i) In *Extended ASCII*, each letter is coded by 8 bits, and an n-letter message requires $8n$ bits.

(ii) The following is the International Morse Code:

letter	a	b	c	d	e	f	g	h	i	j	k	l	m
Morse code	·—	—···	—·—·	—··	·	··—·	——·	····	··	·———	—·—	·—··	——
length	2	4	4	3	1	4	3	4	2	4	3	4	2

letter	n	o	p	q	r	s	t	u	v	w	x	y	z
Morse code	—·	———	·——·	——·—	·—·	···	—	··—	···—	·——	—··—	—·——	——··
length	2	3	4	4	3	3	1	3	4	3	4	4	4

Besides — and ·, there is actually a third invisible symbol present: the space between adjacent letters. Without it, one could not distinguish between the encodings of ee and of i. Thus Morse coding is not a binary encoding. Dropping the spaces would violate a property called *prefix-freeness*, meaning that no letter code may be a prefix (an initial segment) of another code. But for the sake of illustration, suppose that we had a binary prefix-free encoding with the same lengths as above. Then the expected length of the code for a message of n letters would be

$$2np_a + 4np_b + 4np_c + \cdots = n \cdot \sum_{x \in A} \text{length}(\text{code}(x))p_x,$$

where A is the English alphabet, and p_x the frequency of some letter x. Thus we expect about $np_a \approx 0.0828n$ many a's in the message, which take $2 \cdot np_a$ bits, the first term in the sum.

Thus the expected length is a constant times n, with the constant depending on the alphabet and its letter frequencies.

(iii) Suppose that we have an (ordered) 3-letter alphabet $A = (a, b, c)$ with frequency distribution $p = (5/12, 1/3, 1/4)$. If we use a two-bit code like $(00, 01, 10)$, then this is prefix-free, and the expected length of an n-letter encoding is

$$n \cdot \left(2 \cdot \frac{5}{12} + 2 \cdot \frac{1}{3} + 2 \cdot \frac{1}{4}\right) = 2n.$$

In the prefix-free code $(1, 01, 00)$, the expected length is

$$n \cdot \left(1 \cdot \frac{5}{12} + 2 \cdot \frac{1}{3} + 2 \cdot \frac{1}{4}\right) = \frac{19}{12}n \approx 1.583n < 2n.$$

This is quite a bit better. To go even further, one of Shannon's ideas is that we might also encode digrams, that is $A^2 = (\text{aa}, \text{ab}, \text{ac}, \text{ba}, \text{bb}, \text{bc}, \text{ca}, \text{cb}, \text{cc})$, which occur with probabilities

$$p_2 = \left(\frac{25}{144}, \frac{20}{144}, \frac{15}{144}, \frac{20}{144}, \frac{16}{144}, \frac{12}{144}, \frac{15}{144}, \frac{12}{144}, \frac{9}{144}\right),$$

where we have not simplified the fractions. When we use the prefix-free code

$$(000, 001, 010, 011, 100, 101, 110, 1110, 1111),$$

then a message of n letters will consist of $n/2$ codewords (for even n) and have expected length

$$\frac{n}{2} \cdot \left(\frac{3}{144} \cdot (25 + 20 + 15 + 20 + 16 + 12 + 15) + \frac{4}{144} \cdot (12 + 9) \right)$$
$$= \frac{151}{96} n \approx 1.573n < \frac{19}{12}n.$$

When we code trigrams and longer polygrams, it turns out we can get smaller and smaller constant factors of n, but there is a limit, 1.555 in this case.   ◊

We now define the limit alluded to at the end of the example.

DEFINITION A.9. *Let $p = (p_1, p_2, \ldots, p_s)$ be a probability distribution. Then its* entropy $H(p)$ *is*

$$H(p) = \sum_{1 \le i \le s} p_i \log_2(p_i^{-1}).$$

We write p_i^{-1} to make the logarithm nonnegative, and interpret the summand as 0 when $p_i = 0$. The entropy has the following property:

(A.10) $$0 \le H(p) \le H\left(\frac{1}{s}, \frac{1}{s}, \ldots, \frac{1}{s}\right) = \log_2 s.$$

For $s = 2$, we have $p_2 = 1 - p_1$; $H(p_1, 1 - p_1)$ is shown in Figure A.24. Intuitively, the entropy corresponds to the uncertainty in finding the text. It is maximal for uniformly random messages.

A prefix-free code $c \colon A \longrightarrow \{0, 1\}^*$, or $c \colon A^2 \longrightarrow \{0, 1\}^*$, or $c \colon A^k \longrightarrow \{0, 1\}^*$ for some $k \ge 1$, gives an encoding $c \colon A^* \longrightarrow \{0, 1\}^*$ of messages over A of arbitrary length (padding the messages if necessary). Here,

$$A^* = \{x_1 \cdots x_n \colon n \ge 0, x_1, \ldots, x_n \in A\}$$

consists of the finite strings over A, and similarly for $\{0, 1\}^*$. We denote by $\lambda_c(x)$ the *length* of $c(x) \in \{0, 1\}^*$ for any message $x \in A^*$. The expected length $\lambda_c(n)$ for n-letter messages is

$$\lambda_c(n) = \sum_{x \in A^n} \lambda_c(x) \operatorname{prob}(x),$$

where $\operatorname{prob}(x) = p_{x_1} \cdot p_{x_2} \cdots p_{x_n}$ is the probability of $x = x_1 x_2 \cdots x_n$. Shannon proved the following fundamental theorem.

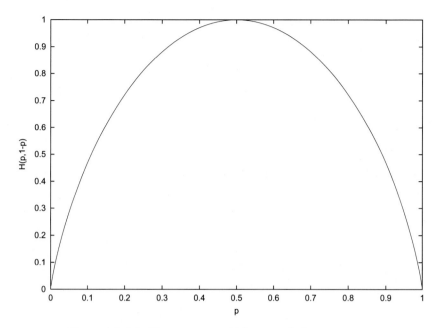

Figure A.24: The entropy $H(p, 1 - p)$ for $0 \leq p \leq 1$.

THEOREM A.11. *Let A be an alphabet with probability distribution p.*

(i) *For any $\epsilon > 0$, there exists a code c so that $\lambda_c(n) \leq n \cdot (H(p) + \epsilon)$ for all sufficiently large n.*

(ii) *For any code c, we have $\lambda_c(n) \geq n \cdot H(p)$.*

We interpret Shannon's Theorem as saying that an n-letter message contains $nH(p)$ bits of information, and thus one letter conveys $H(p)$ bits on average. Huffman trees provide a reasonable approximation to the upper bound in (i), namely with $\epsilon = 1$. The more general question is data compression, where one tries to get close to Shannon's bound under practical constraints; in the form of JPEG, MPEG, or MP3 coded files this theory is now part of daily life.

EXAMPLE A.8 CONTINUED.

(ii) The first column of Table A.19 yields the value

(A.12) $$h = H(p_{\text{Eng}}) \approx 4.19$$

for the letter entropy of English.

(iii) The entropy of this 3-letter alphabet is

$$H(p) = \frac{5}{12} \log \frac{12}{5} + \frac{1}{3} \log 3 + \frac{1}{4} \log 4 \approx 1.555$$
$$< 1.585 \approx \log 3 = H\left(\frac{1}{3}, \frac{1}{3}, \frac{1}{3}\right).$$

One letter of this alphabet contains about 1.555 bits of information, and the redundancy is about 0.03 bits per letter.   ◊

Shannon applied his theory to cryptanalysis. We have, as usual, an alphabet A of size s and with probability distribution $p = (p_x)_{x \in A}$, and a block encryption system (set-up, keygen, enc, dec) with keys k in the total key space K (Definition 2.25). We assume that $\text{enc}_k \colon A^n \longrightarrow A^n$ maps n-letter plaintexts to n-letter ciphertexts, for any $k \in K$. The ciphertext is supposed to look random and to have

$$H\left(\frac{1}{s}, \ldots, \frac{1}{s}\right) = \log_2 s$$

bits of information per letter. An n-letter message contains $nH(p)$ bits of information. We denote by $I(K)$ the average information in one randomly chosen key. If keys are m-bit strings chosen at random (as they should be), then $K = \{0, 1\}^m$ and $I(K) = m$. If K is an m-letter English text, then $I(K) = m \cdot H(p_{\text{Eng}}) \approx 4.19\,m$.

DEFINITION A.13. *The* unicity distance *of the above encryption system is*

$$d = \left\lceil \frac{I(K)}{\log_2 s - H(p)} \right\rceil.$$

THEOREM A.14. *Consider a ciphertext of length n.*

(i) *If $n \geq d$, then an exhaustive key search is likely to reveal the plaintext.*

(ii) *If $n < d$, then the plaintext cannot be derived.*

The idea of the proof is simple. The ciphertext contains $n \log_2 s$ bits of information, the plaintext $nH(p)$, and the key $I(K)$. In order to derive the plaintext and key from the ciphertext, we need

$$n \log_2 s \geq nH(p) + I(K),$$

which is the claim. Exhaustive key search is usually not feasible, but we happily ignore this fact.

Shannon also showed how to apply his theory to linguistic questions such as the entropy h_{Eng} of English. Given a large corpus of the language, one can calculate the letter entropy as in Table A.19 with reasonable precision. But the resulting letter entropy in (A.12) is rather different from the actual entropy of the English language. The digrams *th* and *ht* are assumed to be equally likely, and *qqqq* occurs with positive probability. Taking better account of the properties of the language, Shannon arrives at an estimate of h_{Eng} between 0.6 and 1.3. Brown *et al.* (1992) deduce an upper bound $h_{Eng} \leq 1.75$ from trigram statistics. However, computational linguists have yet to agree on a good approximation. As long as this is not in sight, we use the approximation

(A.15) $h_{Eng} \approx 1.7$

as our operational value. This says that one letter of an English text contains about 1.7 bits of information on average. However, for some cryptanalytic purposes like Kasiski's analysis of the Vigenère cipher (Section C.1), the letter entropy from (A.12) is relevant. Moreover, in a suitable sense it is computationally infeasible to even approximate a source entropy from samples.

If we ignore spaces, punctuation, foreign words with funny letters, numerals, the distinction between upper and lower case, etc., the maximal entropy of any distribution on 26 letters is $\log_2 26 \approx 4.7$, according to (A.10). The *redundancy* of English is the difference $\log_2 26 - 1.7 \approx 3$. This can be interpreted as saying that we "lose" about three bits of information per letter when we write English rather than some uniformly random nonsense. On the other hand, redundancy gains through better readability of texts.

EXAMPLE A.16.

(i) We consider arbitrary simple substitutions $\sigma \colon A \longrightarrow A$, where A is the English alphabet with probability distribution p_{Eng}. The key space $K = \mathrm{Sym}_{26}$ is the set of all $26! \approx 2^{88.382}$ permutations. Thus $I(K) \approx 88.382$. The unicity distance is

$$d = \left\lceil \frac{88.382}{\log_2 26 - H(p_{Eng})} \right\rceil = 30.$$

Thus messages of length at least 30 can usually be deciphered, but not when the length is less. When we have such a short length, then exhaustive key search will sometimes discover two or more pairs (plaintext x, key k) so that $\mathrm{enc}_k(x)$ equals the given plaintext. Thus the decipherment is not unique below the unicity distance.

(ii) We take our toy example with $A = \{a, b, c\}$ from Example A.8 (iii), and the same encryption method as in (i), namely by a random permutation on three letters. Then $I(K) = \log_2(3!) = \log_2 6 \approx 2.585$, and the unicity distance is

$$d = \left\lceil \frac{\log_2 6}{\log_2 3 - H(p)} \right\rceil = 86.$$

Thus ciphertexts of length at least 86 will usually have a unique solution (plaintext, key), and shorter ones may have several solutions.

(iii) The size of the ENIGMA key space is estimated at around 2^{67}; see (J.2). We are not aware of a serious estimate of the entropy of German military jargon, and simply take 1.7 as an estimate. Then we find the unicity distance as $d = 23$.

(iv) In the one-time pad (Section 9.4), we have an n-bit message (in English, coded in 8-bit Extended ASCII, say) of length n, so that $H(p) = H(p_{\text{Eng}})/8 \approx 0.2$, an n-bit random key, with $I(K) = n$, and the ciphertext of n bits. The alphabet is $\{0, 1\}^n$, so that $s = 2^n$, and the unicity distance is

$$d = \left\lceil \frac{n}{n - 0.2} \right\rceil = 2$$

for $n \geq 3$. Thus the "two-time pad" would be unsafe, and that is why the rules say that you may encrypt only a single message with the same key; see Section 9.4.

(v) What about a modern system like AES? Suppose we encrypt English plaintext in 8-bit Extended ASCII by 128-bit AES with a 128-bit random key. Each letter contains 1.7 bits of information, so that the ASCII message has $1.7 \cdot 128/8 \approx 27.2$ bits of information per 128-bit word of the alphabet $\{0, 1\}^{128}$. The key has $I(K) = 128$ bits, so that the unicity distance is

$$d = \left\lceil \frac{128}{\log_2 2^{128} - 27.2} \right\rceil = 2.$$

Thus two transmitted messages would be enough to determine the key... if only we could perform an exhaustive key search.   ◇

In summary, Shannon's theory tells us that for ciphertexts with a certain minimum length, namely his unicity distance, we can expect a single "solution" (plaintext, key), and below this minimum, we will usually have several solution. It does not tell us how to find these solutions.

Nor does it say much about modern cryptosystems with huge alphabets, say, of size 2^{128} in AES. It does say that very short messages are not uniquely deciphrable even in easy-to-break systems like short-keyword-driven alternations between random permutations.

Finding an appropriate version of the unicity distance for codebooks requires a bit of care; see Section D.6.

Notes. Kahn wrote his famous book before the story of breaking ENIGMA (Chapter J) became public. He closed the gap in his account with his *Seizing the Enigma* in 1991.

A.1. A famous example of failed *security by secrecy* is the widely deployed *KeeLoq* system for protecting radio access to cars and garages. After being reverse-engineered, it was broken by Indesteege *et al.* (2008), and the secret key can be recovered in seconds (Kasper *et al.* 2009). The system is still in use.

Procrustes (Προχρούστης) was a mythological Greek innkeeper, a blacksmith and son of Poseidon, whose establishment had only two beds, a short one and a long one. Any tall guest arriving would be assigned to the short bed and have his legs cut off in the night, and a short guest would be stretched to fit the long bed. Since then, Greek hospitality has vastly improved. The Procrustes rule is applied here by cutting a long key text to fit a short message. More usually in cryptography, a short text is stretched to a prescribed length by appending some dummy text; such padding must be done carefully.

Codebooks often have homophones, so that σ in Example A.2 (vii) is a relation.

A.2. Much of the Egyptian material is from Darnell (2004), including the *"endless speculation"* and the *"intricacies"*. Cosmic top secret is a high security classification for NATO documents.

The atbash encryption of babel occurs in Jeremiah's Book, 25:26 and 51:41. The form ך stands at the end of a word for the usual כ. Modern Hebrew does not normally use the diacritical marks (niqqud) under the letters in the example; for some reason, the first two marks are interchanged in this case.

Polybius' signalling system is in Book 10, Sections 42–47, of his *Histories*. The observer is supposed to watch through two narrow tubes in fixed positions, one for the left and the other for the right torches. The protocol also includes a *start of transmission* signal, where both parties raise two torches to indicate readiness. The signal station on Mount Tisaeus is also mentioned by Titus Livius in his book *Ab urbe condita*, namely in Book 28, 5:17; 7:1 and Book 29, 6:10. Myer (1879) describes Polybius' and his own system in great detail and with many variations. In one of them, for use at night, lanterns are swung—not much change and certainly no increase in the communication speed over the intervening 2000 years.

Religious prohibitions to pronounce the name of God are early instances of shyness to pronounce names, and a modern reflection may be the popular scientific citation style that produces unreadables like "[DH76]" or even "[13]". And what was the name that Hermione and Harry must not say?

A medieval scribe would be puzzled by our example (A.6)—his alphabet has no κ, and identifies I = J and U = V. See Swarzenski (1969), page 30, for the *cryptographia aurea*. Wynnebald was the founding abbot of the monastery of Heidenheim in Southern

Germany, about 40 km north-west of the older monastery at Eichstätt. The ciphertext in Figure A.13 was explained by Bischoff (1931). See also Levison (1946), pages 81 and 294.

Giovanni Battista Belaso proposed in 1553 an irregular rotation through the various alphabets, driven by keyphrases—he suggested *optare meliora* or *virtuti omnia parent*[4]—which can be repeatedly used if necessary.

According to Lange & Soudart (1935), four manuscript copies of the *Trattati in cifra* have survived. The first printed version of Alberti's book is *La cifera* in Alberti (1568), an Italian translation of the Latin original. Only the outer disk is printed there, on page 213. A sumptuous reprint of Alberti's *De componendis cyfris*[5] titled *Dello scrivere in cifra* was produced as Alberti (1994). This also shows the cipher disk as Tavola II *Formula*. The *"West's oldest"* is from Kahn (1967), Chapter 4, page 127.

Confederate use of the Vigenère cipher is explained in Gallagher (1989); see Section B.3.

The M-209-B was produced around 1944 by the LC Corona Smith Typewriter Inc. under a license from the Swedish A.B. Ingeniörsfirma Cryptoteknik. About 140 000 machines were manufactured. Apparently the system was broken by German cryptanalysts; see Notes J.2. Other mechanical wheel ciphers include the Jefferson cylinder and the Bazeries cylinder.

Porta (1602), Book 4, exhibits two disks similar to the one in Figure A.16 in print, but only with the outer disk. The inner disk was presumably meant to be attached as a volvelle, but this was not executed.

A.3. Quotations are reproduced with the original spelling if possible. In neo-Latin, $\bar{e}$ and $\bar{u}$ stand for en and un, respectively, and in the *Extant &* quote the *erāt* is erant = *(they) were*.

Isidor Hispalensis (ca. 560–636), Book I, Chapter 24, mentions Caesar's cryptoscheme. Aulus Gellius writes in Book 17, Chapter 9, about Caesar; the wording is from a 1549 edition. Two quotes are from Sueton's *Lives of Twelve Caesars*, in *Divus Iulius Caesar* 56 and *Divus Augustus* 88:1, as given in a 1615 edition.

A.4. The frequency table A.19 is relative to the 26-letter alphabet, ignoring spaces, punctuation marks, accents, etc. Von zur Gathen & Loebenberger (2015) analyze the COCA corpus of English text, produced by Brigham Young University (Davies 2008-2012). The online text of *Harry Potter and the Philosopher's Stone* contains 335007 letters, the LATEX file of Chapter 15 was cleared of LATEX commands and then contains 35920 letters, and the one by Meyer & Matyas (1982), on their page 616, was based on a sample of 8000 excerpts of 500 letters, a total of four million letters, taken from the Brown University Corpus of Present-Day American English. The tables of the other languages are based on online versions of some highlights of classical literature: Johann Wolfgang von Goethe described *Die Leiden des jungen Werther*, Parts I & II, with 187 765 letters, Jean-Baptiste Molière's *Le malade imaginaire* rolled in with 93 593 letters, the *Don Quijote* of Miguel de Cervantes takes 1 636 322 letters, and Dante Alighieri's *La Divina Commèdia* has 411 518 letters. The Molière text is on
http://www.toutmoliere.net/le-malade-imaginaire,51.html
and the other three are from the Project Gutenberg at http://promo.net/pg/.

According to Poe (1902), he wrote the *Gold-Bug* at Philadelphia in 1842, and sold it the following year to George Rex Graham for $52, for publication in *Graham's Magazine*. Mr. Graham, however, did not like it, and by exchanging some critical articles for it, Poe received back the manuscript and submitted it to judges of a $100 prize story competition offered by *The Dollar Newspaper*, which was edited in Philadelphia by N. P. Willis and

---

[4]*Choose the better choice* or: *everything is apparent to virtue.*
[5]How to design ciphers

Joseph Sailer. *The Gold-Bug* received the award, and was published in two parts; the first part on 22 June 1843, and the second part (together with the first) on 29 June 1843. On 7 July 1843, it was published again with other prize tales as a supplement to this newspaper. The first appearance in book form was in *Tales, by Edgar A. Poe*, New York, Wiley and Putnam, 1845; and later in the first collective edition, *Poe's Works*, with a memoir by R. W. Griswold, and notices by N. P. Willis and J. R. Lowell, New York, 1850. There are actually several versions of the ciphertext. Ours is from Poe's work edited by Harrison, which is based on the 1845 edition plus manuscript corrections from the period. Actually, the beginning of the second line decrypts as [s]eat weny one, with two t's missing, but Bauer (1995) presents a different ciphertext where our plaintext solution "twenty-one degrees" is replaced by "forty-one". This shows how error-prone it is to typeset text without visible redundancy, and we still wonder which way to look from the devil's seat. For Legrand's solution of *t eeth*, a sports or fashion moralist might also have considered *"tie ethics"*—even the *i* would match.

Willem Jacob 's Gravesande (1688–1742) studied law at Den Haag, became professor of mathematics and astronomy at Leiden in 1717, and of philosophy in 1734. He was the first proponent of Newton's ideas outside of England, and also a capable cryptanalyst. His *Introduction* appeared first in 1737 at Leiden, in Latin.

A.5. Shannon devised his theory (Definition A.9 and Theorem A.11) first for single letters in finite alphabets, and then adapted it to the English language in order to capture lexical and grammatical aspects of languages. This reduces the unicity distance for deciphering simple substitutions, as in Example A.16 (i), to about 23 in practice. The trees of Huffman (1952) are explained in Cormen *et al.* (1990), Section 17.3; Knuth (1997), §2.3.4.5; and in von zur Gathen & Gerhard (2013), Exercise 10.6.

There is a big gap between the letter entropy 4.19 in Table A.19 and the actual entropy of English, as the 1.7 in (A.15). This is due to the lexical and grammatical structure and gives a higher value of redundancy, which in turn makes texts easier to read and write.

The word "unicity distance" comes from the world of error-correcting codes: below this distance, different inputs (plaintext, key) may have encryptions with Hamming distance 0 (that is, equal encryptions), so that unique decoding is not possible.

Von zur Gathen & Loebenberger (2015) report on the difficulties of estimating the entropy of English by counting polygrams in a large corpus. More generally, Goldreich *et al.* (1999) show that estimating the entropy of a source from samples is complete for the class NISZK of non-interactive statistical zero knowledge and thus infeasible under standard complexity assumptions. The source is modelled by the outputs of a polynomial-size Boolean circuit on uniformly random inputs.

Exercises.

EXERCISE A.1. *We have a substitution* $\sigma \colon A \longrightarrow A$, *with* $\#A = s$, *and a transposition* τ *on* s *numbers, as described by (A.4). Compare* $\tau_A \circ \sigma^s$ *and* $\sigma^s \circ \tau_A$ *first in two examples of your choice and then in general.*

EXERCISE A.2 (Affine cryptosystem). *Hill (1931) proposed to encrypt* $x \in A^n$ *as* $Mx + b \in A^n$, *with secret matrix* $M \in A^{n \times n}$ *and* $b \in A^n$, *where the alphabet* A *of size* s *is identified with* $\mathbb{Z}_s$. *We consider a special case, namely, the usual 26-letter alphabet* $A = \{a, b, \ldots, z\}$ *and identify it with* $\mathbb{Z}_{26} = \{0, 1, \ldots, 25\}$ *as in Table A.1. An affine cryptosystem is given by the following encryption function, where* a, b *are chosen from*

$\mathbb{Z}_{26}$:

$$\text{enc}_{a,b} \colon \mathbb{Z}_{26} \to \mathbb{Z}_{26},$$
$$x \mapsto ax + b \text{ in } \mathbb{Z}_{26}.$$

(i) Encrypt the plaintext cryptography using the affine code $\text{enc}_{3,5}$. What is the decryption function corresponding to $\text{enc}_{3,5}$? Decrypt the ciphertext XRHLAFUUK.

(ii) A central requirement of cryptography is that the plaintext must be computable from the key and the ciphertext. Explain why $\text{enc}_{2,3}$ violates this rule. Show that the function $\text{enc}_{a,b}$ satisfies the rule if and only if $\gcd(a, 26) = 1$, that is, if a and 26 have no common divisor except 1.

In the following we consider only functions $\text{enc}_{a,b}$ with $\gcd(a, 26) = 1$.

(iii) Show that all affine codes with $b = 0$ map the letter a to a and the letter n to n.

(iv) How many matching pairs of plaintext/ciphertext $(x, \text{enc}_{a,b}(x))$ are needed so that an affine code is uniquely determined? [Hint: Consider the situation modulo 2 and modulo 13 and use the Chinese Remainder Theorem 15.44.]

EXERCISE A.3 (Random text and key). *We take an alphabet A of 26 letters, a uniformly random substitution $\sigma \xleftarrow{\boxdot} \text{Sym}_{26}$ from all 26! possibilities, some (large) n, a uniformly random plaintext $x \xleftarrow{\boxdot} A^n$, and its encryption $y = \text{enc}_\sigma(x)$, where each letter is encrypted with σ. What are the values of the parameters that determine the unicity distance? What can you say about an adversary with unlimited computing power who tries to decrypt y? Can there be several pairs (σ, x) that yield the same encryption? If so, how many? Explain and discuss your findings.*

[...] die Anzahl der Schritte gegenüber <u>dem blossen Probieren</u>
von N auf log N (oder (log $N)^n$) verringert werden kann.
Es wäre interessant zu wissen, wie es damit
z.B. bei der Feststellung, ob eine Zahl Primzahl ist,
steht u. wie stark <u>im allgemeinen</u> bei
finiten kombinatorischen Problemen die Anzahl der Schritte
gegenüber dem blossen Probieren verringert werden kann.[9]
KURT GÖDEL (1956)

Given any two numbers, we may by a simple and infallible
process obtain their product; but when a large number is given
it is quite another matter to determine its factors.
Can the reader say what two numbers multiplied together
will produce the number $8,616,460,799$?
I think it unlikely that anyone but myself
will ever know; for they are two large prime numbers,
and can ony be rediscovered
by trying in succession a long series of prime divisors
until the right one be fallen upon. The work would probably
occupy a good computer for many weeks, but it did not
occupy me many minutes to multiply the two factors together.
WILLIAM STANLEY JEVONS (1877)

Quel est le plus petit *non premier*
qui rend la formule $2^n - 2$ divisible par n?[10]
JOSEPH DIAZ GERGONNE (1820)

In many research problems in the theory of numbers
it is desirable to have a rapid, universal and positive method
for the identification of prime numbers.
DERRICK HENRY LEHMER (1936)

---

[9][...] the number of steps can be reduced, compared to <u>exhaustive trial</u>, from N to
log N (or (log $N)^n$). It would be interesting to know the situation, e.g., for determining whether a number is prime and by how much <u>in general</u> for finite combinatorial
problems the number of steps can be reduced compared to exhaustive trial.
[10]Which is the smallest *nonprime* number for which $2^n - 2$ is divisible by n?

Chapter 3

The RSA cryptosystem

Section 2.5 introduces one of the most famous and widely deployed public-key cryptosystems, namely RSA. The present chapter is devoted to the background required for a better understanding of RSA.

The first section proves that the system works correctly. The next two provide the algorithmic tools required for executing the system: primality testing and finding primes, using the Prime Number Theorem.

The remainder of this chapter deals with security aspects. We first relate various attack goals to each other via reductions. An attack by Wiener completely breaks the system when the secret exponent is small.

An obvious attack is by factoring the modulus. Integer factorization is a fascinating problem in its own right, and we present some algorithms. They incorporate brilliant ideas and perform much better than naive methods. But the question of whether a polynomial-time algorithm exists is unlikely to be resolved any time soon. Who knows? In the meantime, we may consider RSA to be secure at appropriate key lengths.

3.1. Analysis of RSA

We recall the RSA notation 2.15 and the three basic questions posed in Section 2.5: correctness, efficiency, and security of the RSA cryptosystem. Throughout this chapter, we use notions from the computer algebra toolbox in Chapter 15. We start with two proofs of correctness.

THEOREM 3.1. *RSA works correctly, that is, for every message x the decrypted encrypted message z equals x.*

FIRST PROOF.   By definition, $z = x^{ed}$ in $\mathbb{Z}_N$, and we know that

$$(3.2) \qquad\qquad ed = 1 + k \cdot \varphi(N)$$

© Springer-Verlag Berlin Heidelberg 2015
J. von zur Gathen, *CryptoSchool*, DOI 10.1007/978-3-662-48425-8_4

for some $k \in \mathbb{Z}$. We have to show that for any $x \in \mathbb{Z}$, the identity $z = x$ holds in $\mathbb{Z}_N$.

First, we consider the case where x is a unit modulo N, in other words, that $\gcd(x, N) = 1$ (Corollary 15.26 (i)). In this case, Euler's Theorem says $x^{\varphi(N)} = 1$, hence $x^{ed} = x^{1+k\varphi(N)} = x \cdot (x^{\varphi(N)})^k = x \cdot 1^k = x$ in $\mathbb{Z}_N^\times$. Furthermore, $\varphi(N) = (p-1)(q-1)$, by Corollary 15.65.

Next, we consider the case where $\gcd(x, N) \neq 1$. This gcd is then p, q, or N. We first take $\gcd(x, N) = p$. Then $x = u \cdot p$ for some unit $u \in \mathbb{Z}_N^\times$. Since $\varphi(N)$ is a multiple of $\varphi(q) = q - 1$, we find from (3.2) that $ed = 1 + \ell \cdot \varphi(q)$ for $\ell = k\varphi(p) \in \mathbb{Z}$. In $\mathbb{Z}_q$, we have $p^{ed} = p^{1+\ell\varphi(q)} = p \cdot (p^{\varphi(q)})^\ell = p \cdot 1^\ell = p$. Since $p^{ed} = 0$ in $\mathbb{Z}_p$, we infer from the Chinese Remainder Theorem 15.44 that $p^{ed} = p$ in $\mathbb{Z}_N$. Using the first case, we also have $u^{ed} = u$. Altogether, $x^{ed} = (up)^{ed} = u^{ed}p^{ed} = up = x$.

The case $\gcd(x, N) = q$ is analogous. The remaining case $\gcd(x, N) = N$ is trivial, since then $x = 0$. □

The cases where $\gcd(x, N) \neq 1$ are somewhat unrealistic, since then ALICE can find BOB's secret key easily and may henceforth read any encrypted message addressed to BOB. But fortunately, this happens for a random x only with the negligible probability

$$(3.3) \quad 1 - \frac{\varphi(N)}{N} = 1 - \frac{(p-1)(q-1)}{N} = \frac{p+q-1}{N} \approx \frac{2 \cdot 2^{n/2}}{2^n} = 2^{-n/2+1}.$$

The proof becomes simpler if we use Fermat's Little Theorem 15.56 and the Chinese Remainder Theorem 15.44 in their full beauty.

SECOND PROOF.   By Corollary 15.57 (iv), for any integer multiple m of $p - 1$ and $a \in \mathbb{Z}_p$ we have $a^{m+1} = a$ in $\mathbb{Z}_p$. Furthermore, $\varphi(N) = (p-1)(q-1)$ and hence $ed = 1$ in $\mathbb{Z}_{p-1}$. Thus, $a^{ed} = a$ for any $a \in \mathbb{Z}_p$, and similarly $b^{ed} = b$ for any $b \in \mathbb{Z}_q$.

We denote by $\chi : \mathbb{Z}_N \to \mathbb{Z}_p \times \mathbb{Z}_q$ the isomorphism of the Chinese Remainder Theorem 15.44, and let $\chi(x) = (a, b)$. Then $\chi(x^{ed}) = (a^{ed}, b^{ed}) = (a, b) = \chi(x)$. Since χ is an isomorphism, this means that $x^{ed} = x$. □

We note as a special property of RSA its *multiplicativity* (or *homomorphic property*), for any $x, w \in \mathbb{Z}_N$:

$$(3.4) \qquad\qquad (x \cdot w)^d = x^d \cdot w^d.$$

3.2. Primality testing

We face the following task from the efficiency question in Section 2.5: given a large B, say $B \geq 2^{1500}$, find a prime p with $B \leq p \leq \sqrt{2}B$.

We start with a famous theorem from the Στοιχεῖα (*Elements*) of Euclid (365–300 BC) about the number of primes. Its proof is "*as fresh and significant as when it was discovered — two thousand years have not written a wrinkle on [it],*" according to Hardy (1940).

THEOREM 3.5. *There are infinitely many primes.*

PROOF. We assume that the set of primes is finite and denote it as $P = \{p_1, ..., p_k\}$. The number $p = p_1 \cdot p_2 \cdots p_k + 1$ is coprime to each of the primes in P. On the other hand, according to the fundamental theorem of arithmetic on unique prime factorization, p has some prime factors, which must be in P. This contradiction proves the theorem. □

A famous type of prime numbers consists of some *Mersenne numbers*, of the form $M_n = 2^n - 1$, where $n \in \mathbb{N}$. Such a number is not necessarily prime, as shown by the example $M_4 = 15$. In fact, when n is composite, then so is M_n. When searching for Mersenne primes, one may therefore assume n to be prime. 48 of them are known (in 2015), and the largest one $M_{57\,885\,161}$ has over 17 million decimal digits.

In order to find (random) primes in a given range, we choose integers N randomly and test them for primality. By definition, N is prime if and only if it has no divisor between 2 and $N-1$. If it has any such divisor, then it has one up to $\sqrt{N}$, and it is sufficient to try all prime numbers up to $\sqrt{N}$ as divisors. This leads to about $\sqrt{N}$ operations, which is utterly infeasible in the cryptographic range. A brilliant idea is to replace the definitional property by a property that is true for prime numbers but (hopefully) sufficiently false for composite numbers. This property is Fermat's Little Theorem 15.56: $a^{N-1} = 1$ for all nonzero $a \in \mathbb{Z}_N$. We have to relinquish deterministic certainty and rather use a probabilistic algorithm. A first attempt—which fails—is the following.

ALGORITHM 3.6. Fermat test.

Input: A number $N \in \mathbb{Z}$ with $N \geq 2$.
Output: Either "N is composite", or "N is possibly prime".

1. $x \xleftarrow{\boxtimes} \{1, \ldots, N-1\}$.
2. $g \leftarrow \gcd(x, N)$. If $g \neq 1$, then Return "N is composite".
3. $y \leftarrow x^{N-1}$ in $\mathbb{Z}_N$.
4. If $y \neq 1$, then Return "N is composite".
5. Return "N is possibly prime".

If N is prime, the Fermat test will correctly answer that N is possibly prime, by Fermat's Little Theorem 15.56. We are looking for primes, and thus there are no "false negatives". The test errs if N is composite but it returns "possibly prime"; this is a "false positive".

Now assume that N is composite. An element x of $\mathbb{Z}_N^\times$ is called a *Fermat witness* for (the compositeness of) N if $x^{N-1} \neq 1$ in $\mathbb{Z}_N$, and a *Fermat liar* otherwise. The set L_N of Fermat liars is a subgroup of $\mathbb{Z}_N^\times$. By Lagrange's Theorem 15.53 (ii), either L_N is equal to $\mathbb{Z}_N^\times$ (there is no Fermat witness) or at most half as large. If there is at least one Fermat witness, then they are abundant, and at most half of all possible x's are liars.

There are composite numbers N without any Fermat witness. They are called *Carmichael numbers*. The Fermat test 3.6 fails miserably for them, returning the wrong answer for all x except the negligibly few for which $g \neq 1$ in step 2. The first Carmichael numbers are $561 = 3 \cdot 11 \cdot 17$, $1105 = 5 \cdot 13 \cdot 17$, and $1729 = 7 \cdot 13 \cdot 19$.

What to do with the challenge posed by the Carmichael numbers? Pomerance (1990) pleads: *"Using the Fermat congruence is so simple that it seems a shame to give up on it just because there are a few counterexamples!"*

We now resolve the shortcoming of the Fermat test 3.6 in a drastic way: the new test not only distinguishes primes from Carmichael numbers, it actually factors these seemingly difficult numbers in random polynomial time. In general, factoring integers seems much harder than testing them for primality, and so these numbers turn out to be quite harmless after all.

If the binary representation of $N - 1$ has e trailing zeros, then the repeated squaring algorithm 15.48 for step 3 of the Fermat test performs those e squarings at the end. The new algorithm performs exactly the same computations, only it looks more closely at the intermediate results. For any odd composite number, either the Fermat test works or we are likely to find a nontrivial square root of 1 in these e steps.

ALGORITHM 3.7. Strong pseudoprimality test.

Input: An odd integer $N \in \mathbb{Z}$ with $N \geq 3$.
Output: Either "composite" or "probably prime".

1. Write $N - 1 = 2^e m$, where e and m are (uniquely determined) positive integers with m odd.
2. $x \xleftarrow{\boxed{\cdot}} \{1, \ldots, N - 1\}$.
3. If $\gcd(x, N) \neq 1$ then return "composite".
4. $y \leftarrow x^m \in \mathbb{Z}_N$.
5. If $y = 1$, then return "probably prime"
6. For i from 0 to $e - 1$ do steps 7–8

7.   If $y = -1$ then return "probably prime"
8.   else $y \leftarrow y^2$ in $\mathbb{Z}_N$.
9. Return "composite".

The correctness of this test is based on the following facts.

○ if N is prime, then $\mathbb{Z}_N$ is a field (Corollary 15.26 (i)).

○ The only elements x of a field with $x^2 = 1$ are $x \in \pm 1$, that is, $\sqrt{1} = \pm 1$ (Corollary 15.47).

Moreover, if we have integers $N \geq 2$ and y with $y^2 = 1$ and $y \notin \pm 1$ in $\mathbb{Z}_N$, then $\gcd(y - 1, N)$ is a proper divisior of N. As an example, $4^2 = 1$ and $4 \notin \pm 1$ in $\mathbb{Z}_{15}$, and $\gcd(4 - 1, 15) = 3$ is a nontrivial factor of 15. This will be a recurrent theme in this and later sections: finding a nontrivial square root of 1, that is, an element of $\sqrt{1} \setminus \pm 1$. Any $s, t \in \mathbb{Z}_N^\times$ with $s^2 = t^2$ and $s \notin \pm t$ yield $s/t \in \sqrt{1} \setminus \pm 1$.

Several, but not all, methods for factoring an integer N rely on a Chinese Remainder decomposition of $\mathbb{Z}_N$. This does not work when N is a perfect power of a prime number. We combine this property with the remarks above in the following lemma, which is at the basis of many factoring methods.

LEMMA 3.8. *Let N be an odd n-bit integer.*

(i) *If $y \in \sqrt{1} \setminus \pm 1 \subseteq \mathbb{Z}_N^\times$, then $\gcd(y - 1, N)$ is a proper factor of N.*

(ii) *One can test in time polynomial in n whether N is a perfect power a^e of an integer a, with $e \geq 2$, and if so, produce such a and e.*

PROOF.   Let $N = q_1 \cdots q_r$ be a factorization of N into powers of distinct primes.

(i) Since $y^2 = 1$ in $\mathbb{Z}_N$, we also have $y^2 = 1$ in each $\mathbb{Z}_{q_i}$, and hence $y \in \pm 1$ in each $\mathbb{Z}_{q_i}$ by Corollary 15.67. Since $y \notin \pm 1$ in $\mathbb{Z}_N$, we have $r \geq 2$, some $I \subseteq \{1, \ldots, r\}$ with $I \neq \emptyset, \{1, \ldots, r\}$ and $y = 1$ in $\mathbb{Z}_{q_i}$ for $i \in I$ and $y = -1$ in $\mathbb{Z}_{q_i}$ for $i \in \{1, \ldots, r\} \setminus I$. Then $\gcd(y - 1, N) = \prod_{i \in I} q_i$ is a proper divisor of N.

(ii) We calculate the real number $r_i = N^{1/i}$ to some, say 5, digits of accuracy after the binary point, for $2 \leq i \leq \log_2 N$. For those r_i which are close to an integer, we take the rounded value $s_i = \lceil r_i \rfloor$, test whether $s_i^i = N$, and return the (s_i, i) with the largest value of i that passed the test. The answer is correct, each test can be performed with $O(n^3)$ operations, and there are $O(n)$ tests.   □

EXAMPLE 3.9. To illustrate the strong pseudoprimality test, we consider the choice $x = 113$ in the three examples of Table 3.1. Step 4 calculates y_0, and $y_1, y_2, \ldots$ denote the values successively computed in Step 8.

N	$N - 1 = 2^e \cdot m$	y_0	y_1	y_2	y_3	y_4
553	$2^3 \cdot 69$	407	302	512	22	
557	$2^2 \cdot 139$	556	1			
$561 = 3 \cdot 11 \cdot 17$	$2^4 \cdot 35$	56	331	116	67	1

Table 3.1: Testing $553, 557$, and 561 for primality.

The test returns "composite" for 553 and 561 since $y_3 \neq 1$ for 553 and $y_3 \neq -1$ for 561, and it returns "probably prime" for 557 since $y_0 = -1$ in $\mathbb{Z}_{557}$.   ◇

In order to study the error probability of the strong pseudoprimality test, we need some notation. An integer is *squarefree* if no square of a prime number divides it. The *order* $k = \mathrm{ord}_G x$ of an element x in a finite group G is the smallest positive integer so that $x^k = 1$. For any integer ℓ, we have

$$(3.10) \qquad\qquad x^\ell = 1 \iff k \mid \ell$$

(Corollary 15.54). By Lagrange's Theorem 15.53, k divides $\#G$.

LEMMA 3.11. *A Carmichael number is squarefree.*

PROOF.   We take a prime number p and suppose that it divides the Carmichael number N exactly $e \geq 2$ times. Since $\gcd(p^e, N/p^e) = 1$, by the Chinese Remainder Theorem 15.44, there exists some $x \in \mathbb{Z}$ so that

$$x = 1 + p^{e-1} \text{ in } \mathbb{Z}_{p^e} \text{ and } x = 1 \text{ in } \mathbb{Z}_{N/p^e}.$$

If $N = p^e$, the second condition is vacuous. Then

$$(1 + p^{e-1})^p = \sum_{0 \leq i \leq p} \binom{p}{i} (p^{e-1})^i = 1 + p \cdot p^{e-1} = 1 \text{ in } \mathbb{Z}_{p^e},$$

since $(e - 1)i \geq e$ for $i, e \geq 2$.

The order of x in $\mathbb{Z}_{p^e}^\times$ divides p by (3.10), and so is either 1 or p. Since $x \neq 1$ in $\mathbb{Z}_{p^e}$, p is indeed the order of x in $\mathbb{Z}_{p^e}^\times$. From the Chinese Remainder Theorem 15.44, it follows that p is the order of x in $\mathbb{Z}_N^\times$. Since $x^{N-1} = 1$ in $\mathbb{Z}_N$, we conclude that p divides $N - 1$. On the other hand, p also divides N, and hence p is a divisor of $N - (N - 1) = 1$. This contradiction proves that no such p exists.   □

One easily obtains a characterization of the Carmichael numbers.

LEMMA 3.12. *An odd composite integer N is a Carmichael number if and only if N is squarefree and $p - 1$ divides $N - 1$ for all prime divisors p of N.*

PROOF. Let N be a Carmichael number and p a prime divisor of it. Then $x^{N-1} = 1$ in $\mathbb{Z}_p$ for all $x \in \mathbb{Z}_p^\times$. Since $\mathbb{Z}_p^\times$ has an element of order $p - 1$ by Theorem 15.66, (3.10) implies that $p - 1$ divides $N - 1$. On the other hand, if all these $p - 1$ divide $N - 1$, then for any such p and $x \in \mathbb{Z}_N^\times$, we have $x^{N-1} = (x^{p-1})^{(N-1)/(p-1)} = 1$ in $\mathbb{Z}_p$, by Fermat's Little Theorem 15.56. By Chinese Remaindering, this implies that $x^{N-1} = 1$ in $\mathbb{Z}_N^\times$. $\square$

A Carmichael number is odd, since otherwise $N - 1$ is odd and not divisible by $p - 1$ for any prime $p \neq 2$. Furthermore, N has at least three prime factors (Exercise 3.3). We will not need this fact. It is useful to consider *Carmichael's lambda function* $\lambda(N)$. For powers 2^e, we set $\lambda(1) = \lambda(2) = 1$, $\lambda(4) = 2$, and $\lambda(2^e) = 2^{e-2}$ for $e \geq 3$. For an integer N with prime factorization $N = 2^e p_1^{e_1} \cdots p_r^{e_r}$ with $e, r \geq 0, e_1, \ldots, e_r \geq 1$, and $p_1, \ldots, p_r$ pairwise distinct odd primes, we define $\lambda(N)$ as

$$
\begin{aligned}
\lambda(N) &= \operatorname{lcm}(\lambda(2^e), p_1^{e_1-1}(p_1 - 1), \ldots, p_r^{e_r-1}(p_r - 1)) \\
&= \operatorname{lcm}(\lambda(2^e), \varphi(p_1^{e_1}), \ldots, \varphi(p_r^{e_r})).
\end{aligned}
$$
(3.13)

Then $\lambda(N)$ divides $\varphi(N)$, and $x^{\lambda(N)} = 1$ for all $x \in \mathbb{Z}_N^\times$, so that the order of x in $\mathbb{Z}_N^\times$ divides $\lambda(N)$. The last condition in Lemma 3.12 is equivalent to $\lambda(N)$ dividing $N - 1$.

EXAMPLE 3.14. The smallest Carmichael number is $N = 561 = 3 \cdot 11 \cdot 17$, for which $N - 1 = 560 = 2^4 \cdot 5 \cdot 7$, $\varphi(N) = 2 \cdot 10 \cdot 16 = 320$ by Lemma 15.64, and $\lambda(N) = \operatorname{lcm}(2, 10, 16) = 80$. For any $x \in \mathbb{Z}_N^\times$, we have $x^2 = 1$ in $\mathbb{Z}_3$, $x^{10} = 1$ in $\mathbb{Z}_{11}$, and $x^{16} = 1$ in $\mathbb{Z}_{17}$, hence $x^{80} = 1$ in $\mathbb{Z}_3, \mathbb{Z}_{11}$, and $\mathbb{Z}_{17}$, and by the Chinese Remainder Theorem 15.44 also in $\mathbb{Z}_N$. Then also $x^{N-1} = x^{560} = (x^{80})^7 = 1$ in $\mathbb{Z}_N^\times$. $\diamond$

LEMMA 3.15. *Let N be odd, not prime or a power of a prime, and $x \xleftarrow{\boxtimes} \mathbb{Z}_N^\times$. Then with probability at least $1/2$, the order k of x is even and $x^{k/2} \notin \pm 1$.*

PROOF. We take the prime factorization $N = p_1^{e_1} \cdots, p_r^{e_r}$ with $r \geq 2$, $p_1, \ldots p_r$ distinct odd primes and each integer $e_i \geq 1$. Then we have the

Chinese Remainder isomorphism (Corollary 15.63)

$$(3.16) \qquad\qquad \mathbb{Z}_N^\times \cong \mathbb{Z}_{p_1^{e_1}}^\times \times \cdots \times \mathbb{Z}_{p_r^{e_r}}^\times .$$

Furthermore, each $\varphi(p_i^{e_i})$ is even and divides $\lambda(N)$ by (3.13). There exists some $i \leq r$ so that 2 divides $\varphi(p_i^{e_i})$ to the same power as it does $\lambda(N)$. We pick such an i, say $i = 1$ after reordering the factors. Then $\lambda(N)/\varphi(p_1^{e_1})$ is odd. We set $c = \lambda(N)/2$ and consider the subgroup

$$G = \{x \in \mathbb{Z}_N^\times : x^c \in \pm 1\}$$

of $\mathbb{Z}_N^\times$. For any $x \in \mathbb{Z}_N^\times$ of odd order k, k divides c and hence $x \in G$. Also, $c_1 = \varphi(p_1^{e_1})/2$ divides c and c/c_1 is odd. By Euler's Theorem, $x_1^{2c_1} = 1$ for all $x_1 \in \mathbb{Z}_{p_1^{e_1}}^\times$ and there exists $x_1 \in \mathbb{Z}_{p_1^{e_1}}^\times$ with $x_1^{c_1} = -1$. We pick such an x_1. Then $x_1^c = (x_1^{c_1})^{c/c_1} = (-1)^{c/c_1} = -1$ in $\mathbb{Z}_{p_1^{e_1}}$. If $y \in \mathbb{Z}_N^\times$ corresponds to $(x_1, 1, \ldots, 1)$ under (3.16), then y^c corresponds to $(-1, 1, \ldots, 1) \notin \pm 1$. Thus $y \notin G$, G is a proper subgroup of $\mathbb{Z}_N^\times$ and has at most $\varphi(N)/2$ elements by Lagrange's Theorem 15.53 (ii). At least half of the elements x of $\mathbb{Z}_N^\times$ are outside of G, and thus have even order k, and $x^{k/2} \notin \pm 1$. $\square$

THEOREM 3.17. *The strong pseudoprimality test 3.7 has the following properties.*

(i) *If N is prime, the test returns "probably prime".*

(ii) *If N is composite, the test returns "composite" with probability at least $1/2$.*

(iii) *For an n-bit input N, the test uses $O(n^3)$ bit operations.* $\square$

PROOF.   When $\gcd(x, N) \neq 1$, step 3 returns correctly. For an RSA modulus N, the probability for this to happen is exponentially small: see (3.3). We now show that the error probability is at most $1/2$ for the $x \in \mathbb{Z}_N^\times$.

(i) When N is prime and $x \in \mathbb{Z}_N^\times$, then $x^{N-1} = 1$ by Fermat's Little Theorem 15.56. Either in step 5, $y = 1$ and the correct answer is returned, or $y^{2^{i+1}} = 1$ for some i with $1 \leq i < e$. We pick the smallest such i. Since $\sqrt{1} = \pm 1$, the value of y is $y^{2^i} = -1$ in the ith iteration, and at that stage, the correct answer is returned in step 7.

(ii) We first assume that N is composite and not a Carmichael number. Thus there exists at least one Fermat witness. The set

$$L_N = \{x \in \mathbb{Z}_N^\times : x^{N-1} = 1\}$$

of Fermat liars is a proper subgroup of $\mathbb{Z}_N^\times$ and hence $\#L_N \leq \#\mathbb{Z}_N^\times/2$ (Lagrange's Theorem 15.53 (ii)). So for a random $x \longleftarrow$ $\mathbb{Z}_N^\times$, $\mathrm{prob}\{x \in L_N\} \leq 1/2$. Hence the test returns "composite" in step 9 with probability at least $1/2$.

Now we suppose that N is a Carmichael number and let k be the order in $\mathbb{Z}_N^\times$ of x from step 2. By Lemmas 3.12 and 3.15, k is even and $x^{k/2} \notin \pm 1$ with probability at least $1/2$. Assuming this, we have $y \neq 1$ in step 5, and one of the repeated squares in the loop of step 6 equals 1. The previous value of y is $x^{k/2}$ and in $\sqrt{1} \smallsetminus \pm 1$, which shows the claimed lower bound on the probability to return correctly from step 9.

(iii) Step 4 can be done with $2\log_2 m$ operations in $\mathbb{Z}_N$ using the repeated squaring Algorithm 15.48, and the loop of step 6 with at most e operations. The total is $2\log_2 m + e \leq 2n$. One operation in $\mathbb{Z}_N$ takes $O(n^2)$ bit operations, for a total of $O(n^3)$. □

For a Carmichael number N, assuming that $x^m \neq 1$ in step 4, the last value of y in the loop of step 6 which is not 1 is actually in $\sqrt{1} \smallsetminus \pm 1$, and $\gcd(y - 1, N)$ is a nontrivial factor of N. Thus we can even find a factor of N from the intermediate results, with probability at least $1/2$. In fact, Carmichael numbers can be factored completely in random polynomial time.

The probability that the strong pseudoprimality test answers incorrectly "probably prime" for a composite N is at most $1/2$. When we use this test t times independently, the error probability is at most 2^{-t}.

THEOREM 3.18. *The strong pseudoprimality test 3.7, repeated t times independently, has the following properties on input N.*

(i) *If it outputs "composite", then N is composite.*

(ii) *If it outputs "probably prime", then N is prime with probability at least $1 - 2^{-t}$.* □

If the test outputs "probably prime", then N is called a *pseudoprime*. What does it mean to be a pseudoprime? Is N then "probably prime"? Of course not; N is either prime or it is not. The "probably" only refers to the random choices we made within the algorithm. (This is not to be confused with the weaker notion of *average case analysis*, where one assumes a probability distribution on the inputs N.) If we have run the test 1000 times, say, then it means the following; if N is not prime, then an event has been witnessed whose probability is at most 2^{-1000}. If you fly in an airplane whose safety depends on the actual primality of such an *industrial-grade pseudoprime*, then this fact should not worry you unduly, since other things are much more likely to fail ☺.

In practice, one first tests whether N has a small prime factor by computing the gcd of N and the first prime numbers, say up to $\log N$. With this amendment, the strong pseudoprimality test is the algorithm of choice for testing the primality of a given number N, unless deterministic security is required.

3.3. Finding prime numbers

A classical method to find all primes below some number x starts with a list $L = (2, 3, 4, \ldots)$ of all integers up to x. Then 2 is declared a prime and all proper multiples of 2 are removed from the list. Continuing like this, the first element of the remaining list is declared a prime, its proper multiples are removed, and the higher values form the new remaining list. This *sieving method* goes back over 2000 years to Eratosthenes (276–194 BC) and gave rise to the word *prime number*: πρῶτος ἄριθμος = pròtos árithmos = first number in Greek. It is old and pretty, but its running time and space are about $x \log x$ and look pretty old in view of cryptographic requirements, namely polynomial in $\log x$. The following algorithm finds a large pseudoprime at the latter cost, in the range prescribed in the RSA Cryptosystem 2.16.

ALGORITHM 3.19. Finding a pseudoprime.

Input: An integer n and a confidence parameter t.
Output: A pseudoprime number N in the range from $2^{(n-1)/2}$ to $2^{n/2}$.

1. $y \longleftarrow 2^{(n-1)/2}$.
2.   Repeat steps 3 and 4 until some N is returned.
3. $N \xleftarrow{\boxdot} \{\lceil y \rceil, \ldots, \lfloor \sqrt{2}\, y \rfloor\}$.
4.   Call the strong pseudoprimality test with input N for t independently chosen $x \xleftarrow{\boxdot} \{1, \ldots, N-1\}$. Return N if and only if all these tests return "probably prime".

In order to find a prime, we choose numbers and test them until we succeed. But if there were, say, no primes at all in the given range, then we would be trapped. Namely, the algorithm would go into an infinite loop (and thus would not be an algorithm in the usual sense). A similar behavior would occur if we just relied on Theorem 3.5 but there were only very few primes: then the algorithm would take a very long time. Fortunately, this assumption is very wrong: there is not only one prime in the desired range, but they are abundant. The famous prime number theorem says approximately how many primes there are up to some bound x. Let $\pi(x)$ denote the number of primes p with $p \leq x$.

Besides $\pi(x)$, some other functions are useful in number theory. Namely, p_n denotes the nth prime number, so that $p_3 = 5$, and

$$(3.20) \qquad \vartheta(x) = \sum_{p \leq x} \ln p = \ln \prod_{p \leq x} p,$$

where p ranges over the prime numbers. Here $\ln$ denotes the natural logarithm. A proof of the following fundamental result is beyond the scope of this text.

PRIME NUMBER THEOREM 3.21. *We have approximately*

$$\pi(x) \approx \frac{x}{\ln x}, \quad \vartheta(x) \approx x, \quad p_n \approx n \ln n,$$

and more precisely

$$\frac{x}{\ln x}\left(1 + \frac{1}{2 \ln x}\right) < \pi(x) < \frac{x}{\ln x}\left(1 + \frac{3}{2 \ln x}\right) \text{ for } x \geq 59,$$

$$\frac{3x}{5 \ln x} < \pi(2x) - \pi(x) < \frac{7x}{5 \ln x} \text{ for } x \geq 21,$$

$$n\left(\ln n + \ln \ln n - \frac{3}{2}\right) < p_n < n\left(\ln n + \ln \ln n - \frac{1}{2}\right) \text{ for } n \geq 20,$$

$$x\left(1 - \frac{1}{2 \ln x}\right) < \vartheta(x) < x\left(1 + \frac{1}{2 \ln x}\right) \text{ for } x \geq 563. \qquad \square$$

We can now estimate the error probability and running time of our pseudoprime finding algorithm.

THEOREM 3.22. *On input $n \geq 23$ and t, the output of Algorithm 3.19 is prime with probability at least $1 - 2^{-t+1} n$. It uses an expected number of $O(tn^4)$ bit operations.*

PROOF. The probability space consists of all random choices in the algorithm. The primes chosen in step 1 of the RSA Cryptosystem 2.16 are in the range from $x = 2^{(n-1)/2}$ to $\sqrt{2}x = 2^{n/2}$. Since $n \geq 23$, we have $x \geq 59$, and the Prime Number Theorem 3.21 says that the probability r for a randomly chosen integer in the range to be prime satisfies

$$r = \frac{\pi(\sqrt{2}x) - \pi(x)}{\sqrt{2}\,x - x} \geq \frac{\frac{\sqrt{2}x}{\ln(\sqrt{2}x)}\left(1 + \frac{1}{2 \ln(\sqrt{2}x)}\right) - \frac{x}{\ln x}\left(1 + \frac{3}{2 \ln x}\right)}{(\sqrt{2} - 1)x}.$$

Ignoring terms of smaller order, this comes to about $1/\ln x$. For an explicit estimate, we choose $a = 1.35$, set $y = \ln x$ and $b = a\ln(\sqrt{2})/(\sqrt{2}-a) < 7.3$. By assumption, we have $y > b$ and hence $\sqrt{2}/(y + \ln\sqrt{2}) \geq a/y$. Thus

$$r > \frac{1}{\sqrt{2}-1}\cdot\left(\frac{\sqrt{2}}{y+\ln\sqrt{2}} - \frac{1}{y}(1+\frac{3}{2y})\right) > \frac{a-(1+3/2b)}{(\sqrt{2}-1)y} > \frac{1}{n},$$

since $y = ((n-1)\ln 2)/2 < (n\ln 2)/2$.

One may be tempted to say that the error probability is at most 2^{-t}, as for the test. But for a valid argument, we have to estimate the conditional probability that the test returns a composite number N. We denote by C the event that N is composite and by T the event that the test accepts N. We use conditional probabilities like the probability $\text{prob}\{C\colon T\}$ that C occurs provided that T does. This is defined as $\text{prob}\{C\cap T\}/\text{prob}\{T\}$, if the denominator does not vanish. By Theorem 3.18, we know that $\text{prob}\{T\colon C\} \leq 2^{-t}$. But we want an estimate on $\text{prob}\{C\colon T\}$. The two are related as $\text{prob}\{T\}\cdot\text{prob}\{C\colon T\} = \text{prob}\{C\cap T\} = \text{prob}\{C\}\cdot\text{prob}\{T\colon C\}$. Since the test answers correctly for any prime, we know $\text{prob}\{N \text{ prime}\} \leq \text{prob}\{T\}$. Putting all this together, we get

$$\frac{1}{2n}\text{prob}\{C\colon T\} \leq \text{prob}\{N \text{ prime}\}\cdot\text{prob}\{C\colon T\}$$

$$\leq \text{prob}\{T\}\cdot\text{prob}\{C\colon T\}$$

$$= \text{prob}\{C\}\cdot\text{prob}\{T\colon C\} \leq \text{prob}\{T\colon C\} \leq 2^{-t}.$$

Hence the probability that our algorithm returns a composite number is at most $n2^{-t+1}$. Since the probability that N is prime is at least $1/n$, we expect to make at most n choices. For each choice, the test takes $O(tn^3)$ bit operations. □

For RSA, with $t = \lceil\log_2(2n)\rceil + s$, the algorithm returns a prime with probability at least $1 - 2^{-s}$ and it uses an expected number of $O((s + \log n)n^4)$ bit operations. With $s = 20 + \log_2 n$ the error probability is at most $0.000001/n$ and we expect $O(n^4 \log n)$ bit operations. In practice, $t = 50$ seems sufficient.

For determining the public exponent e, we are faced with the following question: what is the probability that two random integers are coprime? More precisely, when x gets large and $c_x = \#\{1 \leq a, b \leq x\colon \gcd(a, b) = 1\}$, we are interested in the numerical value of c_x/x^2. Table 3.2 gives c_x/x^2 for some values of x. In fact, a famous theorem of Dirichlet (1849) says that the value is

$$\frac{c_x}{x^2} \in \frac{6}{\pi^2} + O\left(\frac{\log x}{x}\right) \approx 0.60792\,71016 + O\left(\frac{\log x}{x}\right).$$

x	c_x/x^2
10	0.63
100	0.6087
1000	0.608383
10 000	0.60794971
100 000	0.6079301507

Table 3.2: The probabilities that two random positive integers below x are coprime.

Fixing one of the arguments, the average of $\varphi(x)/x$ also equals $6/\pi^2$, and its value is at least $1/3 \ln\ln x$ for $x \geq 31$.

For RSA, it follows that in order to find a suitable e, we have to test an expected number of at most $O(\log n)$ random values, and usually only one or two. Each test for coprimality with $p - 1$ and with $q - 1$ takes $O(n^2)$ bit operations, by Algorithm 15.24.

We can now answer the efficiency questions of Section 2.5, using pseudoprimes instead of primes.

Find $n/2$-bit pseudoprimes at random	$O(n^4 \log n)$,
Find e	$O(n^2 \log n)$,
Calculate N and d	$O(n^2)$,
Calculate powers modulo N	$O(n^3)$.

COROLLARY 3.23. *The key generation in RSA can be done in time* $O(n^4 \log n)$, *and the encryption or decryption of one n-bit plaintext block with* $O(n^3)$ *bit operations.*

3.4. Finding safe prime numbers

The material of this section is only used in Sections 8.5 and 9.9 and the remark after Corollary 13.84, and may be skipped at first reading.

Ideally, we would like to find large primes p so that a random element of $\mathbb{Z}_p^\times$ generates this group with probability close to 1. But this is impossible, since $\varphi(p) = \#\mathbb{Z}_p^\times = p - 1$ is even, the squares form a subgroup consisting of half the elements, and none of them is a generator. The next best thing is to have $\ell = (p - 1)/2$ also prime; then ℓ is called a *Sophie Germain prime*. In this case, a random square generates the group of squares with overwhelming probability $1 - \ell^{-1}$. These primes are used in cryptology, but it is a long-standing open question whether there are infinitely many of them. Thus no cryptosystem using them for arbitrarily large security parameters n can be proven to work.

On the other hand, there is no reason to doubt the *Sophie Germain conjecture* which says that there are about $2Cx/\ln^2 x$ many of them up to x, with some explicitly known constant $C \approx 0.66016$. In practice, Sophie Germain generating algorithms work well.

This section provides a generalization of this notion and an algorithm for generating such primes which can be proven to work in time polynomial in n. This is necessary, e.g., to prove efficiency and the security claims about the HKS cryptosystem in Section 9.12. Historically, the smaller prime ℓ in a pair (p, ℓ) with $\ell = (p-1)/2$ is called a Sophie Germain prime. We have to talk about p and call it a *Marie Germain prime*, after the elder sister Marie Germain of the mathematician Sophie Germain. As a generalization, we allow ℓ to also have two prime factors, both large. More precisely, we let

$$\mathrm{MG}_1 = \{\text{primes } p : (p-1)/2 \text{ prime}\},$$
$$\mathrm{MG}_2 = \{\text{primes } p : (p-1)/2 = \ell_1 \ell_2 \text{ with } \ell_1 < \ell_2 \text{ primes and } \ell_2 \leq \ell_1^3\}.$$

Furthermore, $\mathrm{MG}_1(x)$ and $\mathrm{MG}_2(x)$ are the corresponding subsets of those p with $p \leq x$, and $\pi_1(x)$ and $\pi_2(x)$ denote the respective cardinalities. A prime ℓ is a Sophie Germain prime if and only if $2\ell + 1 \in \mathrm{MG}_1$ is a Marie Germain prime. The last condition in the definition of MG_2 is equivalent to $p^{1/4} \approx (\ell_1 \ell_2)^{1/4} \leq \ell_1$, meaning that both ℓ_1 and ℓ_2 are "large".

The idea for this comes from a result of Heath-Brown (1986) which implies that

$$\pi_1(x) + \pi_2(x) \geq \frac{cx}{\ln^2 x}$$

for some constant $c > 0$. Thus at least one (possibly both) of the alternatives

$$\pi_1(x) \geq \frac{cx}{2\ln^2 x} \quad \text{or} \quad \pi_2(x) \geq \frac{cx}{2\ln^2 x}$$

holds. The following algorithm picks alternatingly random candidates for these two options until success. The smaller values ℓ_1 for MG_2 are chosen at random for each possible bit length individually.

ALGORITHM 3.24. Generating a safe prime.

Input: A positive even integer n.
Output: A prime p with $x/\ln^2 x < p \leq x$, where $x = 2^n$.

1. Compute $y_0 \leftarrow (x - \ln^2 x)/2\ln^2 x$ and $y_1 \leftarrow 2^{n-1} - 1$.
2. Repeat steps 3 and 4 until a prime is returned.
3.   Choose an integer ℓ with $y_0 \leq \ell \leq y_1$ uniformly at random and test ℓ and $2\ell + 1$ for primality. On success, return $p = 2\ell + 1$.

4.   For $j = \lfloor n/4 \rfloor, \ldots, n/2 - 1$, choose integers ℓ_1 and ℓ_2 with $2^j < \ell_1 < 2^{j+1}$ and $y_0/\ell_1 \leq \ell_2 \leq y_1/\ell_1$ uniformly at random and test ℓ_1, ℓ_2, and $2\ell_1\ell_2 + 1$ for primality. On success, return $p = 2\ell_1\ell_2 + 1$.

One can show the following properties.

THEOREM 3.25. *Any p returned by the algorithm satisfies the output specification, $p \equiv 3 \bmod 4$, and $(p-1)/2$ is squarefree with at most two prime divisors, each of them at least $2^{n/4}$. For a sufficiently large input n, the expected number of repetitions made by Algorithm 3.24 until an output is returned is $O(n^3)$, and the expected runtime of the algorithm is polynomial in n.*

A prime p is r-*safe* if $(p-1)/2$ has no prime factor up to r. In order to generate an r-*safe RSA integer* $N = pq$ with r-safe primes p and q, one can first find such primes p and q of appropriate lengths, and then test whether $\gcd((p-1)/2, (q-1)/2) = 1$. For $r \leq 2^{n/8}$, his yields in polynomial time (approximately) uniform safe RSA moduli which are Blum integers (Section 15.14). Furthermore, $\varphi(N)/4$ is squarefree with at most four prime factors, and each of these has at least $n/8$ bits. For uniformly random $a \in \mathbb{Z}_N^\times$, a^2 generates the group of squares in $\mathbb{Z}_N^\times$ with probability at least $2^{-n/8+2}$. Thus a uniformly random element from the group $\square_N^+$ of signed quadratic residues, in the notation of Section 9.9, generates this group with probability at least $1 - 2^{-n/8+2}$. The bounds in von zur Gathen & Shparlinski (2013) imply that

$$(3.26) \qquad\qquad \frac{p+q}{N} \leq 2^{-n/2 + 4 \log_2 n - 4}.$$

We will use the factoring assumption 3.33 for the resulting distribution of values of N in Sections 8.5 and 9.9.

3.5. Security of RSA

We consider as an adversary a (probabilistic) polynomial-time computer $\mathcal{A}$. $\mathcal{A}$ knows $\mathsf{pk} = (N, e)$ and $y = \mathsf{enc}_{\mathsf{pk}}(x)$. There are several notions of "breaking RSA". $\mathcal{A}$ might be able to compute from its knowledge one of the following data.

B_1: the plaintext x,

B_2: the hidden part d of the secret key $\mathsf{sk} = (N, d)$,

B_3: the value $\varphi(N)$ of Euler's totient function,

B_4: a factor p (and q) of N.

These four questions are widely assumed to be hard. The *RSA assumption* is the conjecture that B_1 is hard. To describe the close relationship between these problems concisely, we use the following concept.

DEFINITION 3.27.   *(i) If A and B are two computational problems (given by an input/output specification), then a* random polynomial-time reduction *from A to B is a random polynomial-time algorithm for A which is allowed to make calls to an (unspecified) subroutine for B. The cost of such a call is the combined input plus output length in the call. If such a reduction exists, then A is* random polynomial-time reducible *to B, and we write*

$$A \leq_p B.$$

If such a reduction exists that does not make use of randomization, then A is polynomial-time reducible *to B.*

(ii) If also $B \leq_p A$, we call A and B random polynomial-time equivalent *and write*

$$A \equiv_p B.$$

The existence of a reduction $A \leq_p B$ has two consequences: if B is easy, then also A is easy, and if A is hard, then also B is hard. Please note carefully the direction of these implications; they are easily confused the first time one sees these notions. Such reductions play a major role in several parts of this text. We will later modify our notion of breaking a system by allowing the adversary to not be always successful, only sufficiently often. Example 9.22 remedies a weakness of RSA by introducing randomness.

The secret keys in RSA are assumed to be generated uniformly at random. Implementations sometimes use weak random generators to produce these keys. The ensuing danger is forcefully exhibited by the real-world existence of thousands of pairs of RSA keys N_1 and N_2 with $p = \gcd(N_1, N_2)$ being one of the prime factors of N_1 and N_2.

In Section 2.5, BOB is a probabilistic polynomial-time algorithm. His computation shows that

$$B_1 \leq_p B_2 \leq_p B_3 \leq_p B_4.$$

What about the converse? One reduction is easy to turn around.

LEMMA 3.28. $B_4 \leq_p B_3$.

PROOF. We know $N = p \cdot q$, and with one call to a subroutine for B_3 we find $\varphi(N) = (p-1)(q-1) = pq - (p+q) + 1$. We substitute $q = N/p$ and multiply up the denominator p to find a quadratic equation in p:

$$p^2 - (N + 1 - \varphi(N)) \cdot p + N = 0.$$

We solve it, and have solved B_4.   □

EXAMPLE 3.29. We take $N = 323$ and suppose we receive $\varphi(N) = 288$ on calling B_3. The equation becomes

$$0 = p^2 - (323 + 1 - 288)p + 323 = p^2 - 36p + 323,$$

with the two solutions

$$18 \pm \sqrt{18^2 - 323} = 18 \pm 1,$$

and indeed $N = 17 \cdot 19$. Because of the symmetry in p and q, the two solutions to the quadratic equation will equal p and q.   ◊

If we want to show $B_3 \leq_p B_2$, we may proceed as follows. We keep N fixed, and call the subroutine for B_2 with several random values $e_1, e_2, \ldots$ for e. Each time we get a value $d_1, d_2, \ldots$ with $e_i d_i \equiv 1 \bmod \varphi(N)$. In other words, $\varphi(N)$ is a divisor of each $e_i d_i - 1$, and hence also of $\gcd(e_1 d_1 - 1, e_2 d_2 - 1, \ldots)$.

Now if we could show that our random $e_1, e_2, \ldots$ generate random quotients $(e_1 d_1 - 1)/\varphi(N), (e_2 d_2 - 1)/\varphi(N), \ldots$, then we could conclude that a few choices are already likely to give us $\varphi(N)$ as gcd. This works quite well in practice.

EXAMPLE 3.30. We take again $N = 323$, and choose $e_1 = 5$ and $e_2 = 11$. The subroutine for B_2 returns $d_1 = 173$ and $d_2 = 131$. Then

$$g_1 = e_1 d_1 - 1 = 864,$$
$$g_2 = e_2 d_2 - 1 = 1\,440,$$
$$\gcd(g_1, g_2) = 288.$$

In fact, $g_1 = 3 \cdot 288$ and $g_2 = 5 \cdot 288$. We now take 288 as a test value for $\varphi(N)$, and indeed this is correct. Had we been less lucky, we would have continued with some $e_3, e_4, \ldots$.   ◊

In fact, determining $\varphi(N)$ is deterministic polynomial time equivalent to computing the secret key d. In other words, $B_1 \leq_p B_2 \equiv_p B_3 \equiv_p B_4$.

OPEN QUESTION 3.31. *Does "$B_4 \leq_p B_1$" hold?*

This is an unsolved problem and an active area of research. It asks whether efficient deciphering of RSA encryptions implies efficient factorization. This is equivalent to RSA being secure, provided that factoring RSA moduli is hard. One can set up reasonable but restricted models of computation in which contradictory answers are true. Breaking the HKS cryptosystem in Section 9.9 is equivalent to factoring N.

Our discussion of security so far has been quite superficial. It is clear that if $\mathcal{A}$ can solve B_1, B_2, B_3 or B_4, then the system is broken. But already something much less spectacular might make the system useless.

- $\mathcal{A}$ can find out some information about x, say its least bit (= parity), or whether it encodes a string of English words, or whether it contains the string "2015".

- $\mathcal{A}$ might break the system using some additional information, say the decryption of some other ciphertexts that $\mathcal{A}$ produces.

- The attack might not always work, but only for 1% of the messages. Or, more generally, for a nonnegligible fraction $\varepsilon(n) \geq n^{-k}$ for some $k \geq 1$ (see Section 15.6) of the messages.

Notions of security are discussed in Chapter 9. For the moment, we state the following conjecture.

CONJECTURE 3.32 (RSA assumption). *There is no probabilistic polynomial-time algorithm $\mathcal{A}$ which on input N, e, and y outputs x (problem B_1) with nonnegligible probability.*

The probability space consists of the choices of N, e, y, as prescribed, and the internal randomization of $\mathcal{A}$. For "nonnegligible" and "polynomial-time" to make sense, we need a security parameter n to which they refer. Furthermore, e, x and y might be chosen uniformly at random under the restrictions of Figure 2.15, but it is useful to consider N as being produced by some specific key generation method as in the RSA Cryptosystem 2.16. For any n, this yields a distribution D_n on n-bit integers, and D is the collection of these. The concept that "factoring integers is hard" can then be quantified as follows.

CONJECTURE 3.33 (Distributional factoring assumption). *Let $D = (D_n)_{n \in \mathbb{N}}$ be a collection of distributions D_n on n-bit integers. For all polynomial-time algorithms $\mathcal{A}$, the probability that $\mathcal{A}$ returns a proper factor on input $N \xleftarrow{\boxtimes} D_n$ is negligible.*

Conjecture 3.32 for N distributed according to D implies the factoring assumption for D. The latter is therefore the weaker and hence more desirable assumption to make. When p is very small or $p < q$ are very close to each other, say $q - p \in n^{O(1)}$, then N can be factored efficiently. This leads to a requirement like $1.07 < q/p < 2^{30}$. Similarly, an attack on several messages prohibits the exponent $e = 3$; $e > 2^{16}$ seems reasonable.

3.6. The Wiener attack

According to the RSA rules, the encryption exponent e is chosen at random in $\{2, \ldots, \varphi(N) - 2\}$, and the decryption exponent d so that

$$(3.34) \qquad de = 1 \text{ in } \mathbb{Z}_{\varphi(N)}.$$

The RSA computations are cheaper with small exponents. For example, parts of the German online banking system use 1024-bit RSA. Each user get his own randomly chosen (by the bank) modulus N, but the public exponent is the same for everybody: $e = 2^{16} + 1$, a Fermat prime. As far as we know, this particular choice entails no security risk.

In some scenarios, the party using d has very little computing power, for example a smartcard generating an RSA signature (see Section 7.1). One might then be tempted to first choose a small exponent d, making the smartcard's life comfortable, and then e satisfying (3.34). The following attack by Wiener (1990) shows that this is dangerous and must be avoided.

(3.34) guarantees the existence of an (unknown) positive integer k so that

$$
\begin{aligned}
de - k\varphi(N) &= 1, \\
\frac{e}{\varphi(N)} - \frac{k}{d} &= \frac{1}{d\varphi(N)}.
\end{aligned}
$$

(3.35)

The idea for finding d now is as follows. The two distinct prime factors p and q of N are of the same order of magnitude, namely about $\sqrt{N}$, and $\varphi(N) = N - p - q + 1$ is about N. Thus we have

$$(3.36) \qquad \frac{e}{N} - \frac{k}{d} \approx \frac{1}{dN}.$$

If d is small, then we have an unknown approximation k/d to the known quantity e/N with error about $1/dN$, which is quite small compared to $1/d$. Now there is a marvelous theory—continued fractions—that says that any rational approximation with small denominator and of good quality comes from the Euclidean Algorithm. Thus it is sufficient to run the Extended Euclidean Algorithm 15.19 with inputs N and e—and d will turn up by magic. Theorem 15.39 presents a precise and easily proven result of this flavor.

THEOREM 3.37 (Wiener's attack).   In the RSA notation, suppose that $p < q < 2p$, $1 \le e < \varphi(N)$, and $1 \le d \le N^{1/4}/\sqrt{12}$. Then d can be computed from the public data in time $O(n^2)$.

PROOF.   Using k from (3.35), we define r as

$$r = kN - de = k\varphi(N) - de - k(\varphi(N) - N) = -1 + k(p + q - 1) > 0.$$

Now $k \le d$, since otherwise $1 = de - k\varphi(N) < de - d\varphi(N) < 0$, and also $k > 0$. We have

$$0 < r < k(p + q) < 3kp \le 3dp < 3dN^{1/2}.$$

We furthermore set $a = e$, $b = N$, $s = k$ and $t = d$. Then $4rt \le 4 \cdot 3dN^{1/2} \cdot d \le N = b$ and

$$\left|\frac{a}{b} - \frac{s}{t}\right| = \frac{1}{bt}|de - kN| = \frac{r}{bt} \le \frac{b}{4t \cdot bt} = \frac{1}{4t^2}.$$

The conditions of Theorem 15.39 are satisfied and also $\gcd(s, t) = \gcd(k, d) = 1$. Therefore $d = ut_i$, where $u \in \pm 1$ and t_i is one of the entries in the Extended Euclidean algorithm with inputs N and e. We take some $x \in \mathbb{Z}_N^\times$ with $x \notin \pm 1$, say $x = 2$. If $x^{et_i} = x = x^{-et_i}$, then $x^{2et_i} = x^{\pm 2} = 1$ and $x^2 = 1$. Since $x \notin \pm 1$, this actually gives us the factorization of N, hence d. Otherwise, exactly one of $x^{et_i} = x$ or $x^{-et_i} = x$ holds, and this gives us the value of d.   □

3.7. Chinese remainder computation for RSA

According to the rules, BOB deletes his primes p and q once $N = p \cdot q$ and the exponents e and d are established. We can take the remainder $a_p \in \mathbb{Z}_{p-1} = \{0, \dots, p - 2\}$ of any $a \in \mathbb{Z}_{\phi(N)}$. Then he can simplify his computation of $x = y^d$ in $\mathbb{Z}_N$ by keeping p and q, computing

$$d_p = d \text{ in } \mathbb{Z}_{p-1}, \quad x_p = y^{d_p} \text{ in } \mathbb{Z}_p,$$
$$d_q = d \text{ in } \mathbb{Z}_{q-1}, \quad x_q = y^{d_q} \text{ in } \mathbb{Z}_q,$$

(3.38)
$$x = \begin{cases} x_p \text{ in } \mathbb{Z}_p, \\ x_q \text{ in } \mathbb{Z}_q. \end{cases}$$

Then $x = y^d$ in $\mathbb{Z}_N$ (Exercise 3.13), but the computation becomes cheaper. If N has n bits, then p, q, d_p, d_q, x_p, y_p all have about $n/2$ bits, and we assume that the two exponents have Hamming weight around $n/4$. Then the two exponentiations require only $3n/4$ multiplications each with $n/2$ bits, or $2 \cdot 3n/4 \cdot (n/2)^2 = 3n^3/8$ operations, in contrast to y^d in $\mathbb{Z}_N$ with $3n^3/2$ operations. The Chinese remainder computation has negligible cost $O(n^2)$. Thus we save roughly 75% of the work. The constant factors in the two cost estimates are equal and cancel out in our comparison.

3.8. Fault attacks

For certain implementations of a cryptographic system, say on a smartcard, attacks may be possible that lie completely outside of what we have considered so far.

We suppose that an RSA system with $N = p \cdot q$, d, and e as usual is implemented on a smartcard in the time-saving Chinese remaindering fashion of the previous section, and that we can feed it values $y \in \mathbb{Z}_N$ and get $x = y^d \in \mathbb{Z}_N$ as output. This is the case, for example, in an RSA signature scheme (Section 8.1). We have the card in our possession and may subject it to physical stress which produces errors. The stress might be caused by just bending the card, or by frying it in a microwave oven (take out the cat beforehand!). Furthermore, suppose that we can fine-tune the stress so that one of the x-values in (3.38), say x_p, is incorrect, and the other value x_q is correct.

Now if we start with some $z \in \mathbb{Z}_N$ and use $y = z^e$, then for the correct value of x we have $x = y^d = z^{de} = z$ in $\mathbb{Z}_N$. But for our faulty value of x, calculated by (3.38), we have

$$x - z \equiv \begin{cases} \neq 0 \text{ in } \mathbb{Z}_p, \\ 0 \text{ in } \mathbb{Z}_p. \end{cases}$$

Thus $\gcd(x - z, N) = q$, and this reveals the secret factors of N.

This text usually deals with attacks that read or write messages and perform computations. But here we disturb the legitimate user's calculations. There are other attacks of this nature, such as measuring the time taken for some operations. We do not discuss such *side-channel attacks*; see Anderson (2001).

An amazing piece of acoustic cryptanalysis of RSA is presented in Genkin, Shamir & Tromer (2013). They put a regular mobile phone near a laptop running the GnuPG implementation of RSA, say at a distance of 30 cm, and listen to the high-pitched noise generated by some electronic components. With chosen plaintexts, they can then extract secret 4096-bit RSA keys in an hour.

3.9. Hard core bit for RSA

We assume that the problem B_1 of recovering the plaintest x from the public key $\mathsf{pk} = (N, e)$ and encryption $y = \mathsf{enc_{pk}}(x)$ is hard. But it might be easy to glean some partial information about x. We now show that this is not the case for the least significant bit (or parity) of x. This bit is then called a *hard core bit* for RSA.

In the RSA notation 2.15, we write $x = \sum_{0 \le i < n} x_i 2^i$ in binary, with all $x_i \in \{0, 1\}$. Then x_0 is the *least significant bit* of x. We thus extend the list in Section 3.5 by a new problem:

$$B_0 : \text{compute the least significant bit } x_0 \text{ of } x.$$

Then clearly $B_0 \le_p B_1$.

THEOREM 3.39. $B_0 \equiv_p B_1$.

PROOF.   It is sufficient to show $B_1 \le_p B_0$. As usual, we show the converse: if the least significant bit is insecure, then the whole of RSA is insecure. Suppose we can find x_0. If $x_0 = 0$, then x is even and it is sufficient to deal with the shorter plaintext $x/2$. If $x_0 = 1$, then $N - x$ is even, and again we are reduced to a shorter plaintext. We next turn this observation recursively into an algorithm.

We have the public key $\mathsf{pk} = (N, e)$, with N of bit length n, and x and $y = x^e \in \mathbb{Z}_N$. From an algorithm $\mathcal{A}$ that computes the least bit x_0 from N, e, y, we derive an algorithm $\mathcal{B}$ that computes x.

For $1 \le \ell \le n$, we describe an algorithm $\mathcal{B}_\ell$ which determines x from the input N, e, y, provided that $0 \le x < 2^\ell$. If $\ell = 1$, then $\mathcal{B}_0$ simply outputs $\mathcal{A}(N, e, y)$. If $\ell > 1$, then $\mathcal{B}_\ell$ first computes $x_0 = \mathcal{A}(N, e, y)$. We use an inverse u of 2 in $\mathbb{Z}_N$, so that $2u = 1$ in $\mathbb{Z}_N$ (see Exercise 3.4), and $y/2^e$ is $y \cdot u^e$ in $\mathbb{Z}_N$. We have $2^{ed} = 2$ in $\mathbb{Z}_N$ and distinguish two cases.

If $x_0 = 0$, then $\mathcal{B}_\ell$ calls $\mathcal{B}_{\ell-1}$ with input N, e, yu^e and output z. Then $\mathcal{B}_\ell$ outputs $2z$. If both $\mathcal{A}$ and $\mathcal{B}_{\ell-1}$ return the correct answer, then

$$\frac{x}{2} = \frac{y^d}{2} = \left(\frac{y}{2^e}\right)^d = (yu^e)^d = z \text{ in } \mathbb{Z}_N.$$

Furthermore, $x/2$ is an integer with at most $\ell - 1$ bits. Thus $\mathcal{B}_\ell$ returns the correct answer $x = 2z$ in $\mathbb{Z}_N$.

If $x_0 = 1$, then $\mathcal{B}_\ell$ calls itself with input $N, e, -yu^e$. Now $N - x$ is even and equals $(-yu^e)^d$ in $\mathbb{Z}_N$, since d is odd. The workings of $\mathcal{B}_\ell$ on the new input have been defined in the previous paragraph, and we let z be its output. Now $\mathcal{B}_\ell$ returns $-2z$ for the original input N, e, y. Then

$$-\frac{x}{2} = -\frac{y^d}{2} = \left(-\frac{y}{2^e}\right)^d = (-yu^e)^d = z \in \mathbb{Z}_N.$$

Now $\mathcal{B}_\ell$ returns the correct answer $x = -2z$ in $\mathbb{Z}_N$.

The claimed algorithm $\mathcal{B}$ is simply $\mathcal{B}_n$. $\qquad\qquad\qquad\square$

COROLLARY 3.40. *If RSA is secure, then the least significant bit of the plaintext is also secure.*

The security of this bit means that

$$|\operatorname{prob}\{x_0 = \mathcal{A}(N, e, \operatorname{enc}_{N,e}(x))\} - \frac{1}{2}|$$

is negligible in n for all polynomial-time algorithms $\mathcal{A}$. By just guessing $x_0 \in \{0, 1\}$ uniformly at random, one can achieve success probability $1/2$. The hard core property says that it is impossible to improve on this by any inverse polynomial advantage. Here $\mathcal{A}$ is an algorithm that takes the public key N, e and the encryption $y = \operatorname{enc}_{N,e}(x)$ as input.

3.10. Factoring integers

The remainder of this chapter is devoted to one obvious way of breaking RSA, namely, by factoring N. Some (probabilistic) algorithms to factor an n-bit integer N are listed in Table 3.3.

method	year	time $\exp(O^\sim(t))$
trial division	$-\infty$	$n/2$
Pollard's $p-1$ method	1974	$n/4$
Pollard's ρ method	1975	$n/4$
Pollard's and Strassen's method	1976	$n/4$
Morrison & Brillhart continued fractions	1975	$n^{1/2}$
Dixon's random squares, quadratic sieve	1981	$n^{1/2}$
Lenstra's elliptic curves	1987	$n^{1/2}$
number field sieve	1990	$n^{1/3}$

Table 3.3: Factorization methods for n-bit integers. An entry t in the "time" column means $\exp(O^\sim(t))$ bit operations.

The claims about most of these algorithms are proven, but Pollard's ρ, Lenstra's elliptic curves and the number field sieve rely on heuristic assumptions. Integers with n bits can be tested for primality in (probabilistic) polynomial time, with $O^\sim(n^3)$ operations. If factoring can be achieved within a similar time bound, then systems like RSA are dead.

Before you try to factor an integer, you make sure it is odd, not prime and not a proper power of an integer. You might also remove small factors by taking (repeatedly) the gcd with a product of small primes.

3.11. The birthday paradox

The next section describes a factoring algorithm of Pollard (1975). It generates certain elements $x_0, x_1, x_2, \ldots$ and is successful when a *collision* modulo a prime factor p occurs, that is, when $x_t = x_{t+\ell}$ in $\mathbb{Z}_p$ for some t and ℓ with $\ell \geq 1$.

This section answers the question: how "small" can we expect t and ℓ to be? We certainly have $t + \ell \leq p$, and the following analysis shows that the expected value is only $O(\sqrt{p})$ for a *random* sequence $x_0, x_1, \ldots$. This is known as the *birthday problem*: assuming peoples' birthdays occur randomly, how many people do we need to get together before we have probability at least $1/2$ for at least two people to have the same birthday? Surprising answer: only 23 are sufficient. In fact, with 23 or more people at a party, the probability of two coinciding birthdays is at least 50.7%.

THEOREM 3.41 (Birthday paradox). *We consider random choices, with replacement, among m labeled items. The expected number of choices until a collision occurs is $O(\sqrt{m})$.*

PROOF.   Let s be the number of choices until a collision occurs, that is, two identical items are chosen. This is a random variable. For $j \geq 2$, we have

$$\text{prob}\{s \geq j\} \;=\; \frac{1}{m^{j-1}} \prod_{1 \leq i < j} (m - (i-1)) \;=\; \prod_{1 \leq i < j} \left(1 - \frac{i-1}{m}\right)$$

$$\leq \prod_{1 \leq i < j} e^{-(i-1)/m} = e^{-(j-1)(j-2)/2m} < e^{-(j-2)^2/2m},$$

where we have used $1 - x \leq e^{-x}$. It follows that the expected value of s is

$$\mathcal{E}(s) \;=\; \sum_{j \geq 1} \text{prob}\{s \geq j\} \leq 1 + \sum_{j \geq 0} e^{-j^2/2m} \leq 2 + \int_0^\infty e^{-x^2/2m} dx$$

$$\leq \; 2 + \sqrt{2m} \int_0^\infty e^{-x^2} dx = 2 + \sqrt{\frac{m\pi}{2}},$$

since $\int_0^\infty e^{-x^2} dx = \sqrt{\pi}/2$; see Exercise 3.22.   □

The theorem says that we expect to find a collision with only about $\sqrt{m\pi/2} \approx 1.25\sqrt{m}$ random choices of x, where m is the number of possible values. It provides a generic tool for cryptanalysis, useful in several situations. For example, in Chapter 7 we will apply this to hash functions.

EXAMPLE 3.42. For $m = 2^{20}$, the first collision is expected to occur after about $\sqrt{\pi/2} \cdot 2^{20/2} \approx 1283.4$ samples. Ten experiments with random choices found the first collision after

$$846, 832, 532, 2404, 2270, 973, 1301, 697, 2078, 583$$

steps, with average 1251.6 and standard deviation 688.56. The average matches its prediction reasonably well and the deviation is fairly large. ◊

3.12. Pollard rho with Floyd's trick

The idea is as follows. We choose some function $f\colon \mathbb{Z}_N \to \mathbb{Z}_N$ and a random initial value $x_0 \in \mathbb{Z}_N$, and recursively define $x_i \in \mathbb{Z}_N$ by $x_i = f(x_{i-1})$ for all $i > 0$. We hope that $x_0, x_1, x_2, \ldots$ behave like a sequence of independent random elements in $\mathbb{Z}_N$. If p is an (unknown) prime divisor of N, then we will have a collision modulo p if there are two integers t and ℓ with $\ell > 0$ and $x_t = x_{t+\ell}$ in $\mathbb{Z}_p$. If N is not a prime power and q is a different prime divisor of N, and the x_i's are random residues modulo N, then $x_i \bmod p$ and $x_i \bmod q$ are independent random variables, by the Chinese Remainder Theorem 15.44. Thus it is very likely that $x_t \neq x_{t+\ell}$ in $\mathbb{Z}_q$, and then $p = \gcd(x_t - x_{t+\ell}, N)$ is a nontrivial factor of N.

We now describe Pollard's ρ method for factoring N. It generates a sequence $x_0, x_1, \ldots \in \mathbb{Z}_N$. We pick x_0 at random and define $x_i \in \mathbb{Z}_N$ by $x_i = f(x_{i-1}) = x_{i-1}^2 + 1$ in $\mathbb{Z}_N$. The choice of this iteration function is black magic, but linear functions do not work, and higher degree polynomials are more costly to evaluate, and one cannot prove more about them than about $x^2 + 1$.

Let p be the smallest prime dividing N. Then we have $x_i = x_{i-1}^2 + 1$ in $\mathbb{Z}_p$ for $i \geq 1$. The birthday paradox and heuristic reasoning which says that the x_i's "look" random imply that we can expect a collision in $\mathbb{Z}_p$ after $\mathrm{O}(\sqrt{p})$ steps.

EXAMPLE 3.43. We want to factor $N = 82\,123$. Starting with $x_0 = 631$, we find the following sequence:

i	$x_i \bmod N$	$x_i \bmod 41$	i	$x_i \bmod N$	$x_i \bmod 41$
0	631	16	6	40 816	21
1	69 670	11	7	80 802	32
2	28 986	40	8	20 459	0
3	69 907	2	9	71 874	1
4	13 166	5	10	6 685	2
5	64 027	26	11	14 313	5

The iteration modulo 41 is illustrated in Figure 3.4; the algorithm's name derives from the similarity to the Greek letter ρ (rho). It leads to the factor $\gcd(x_3 - x_{10}, N) = 41$. When we execute the algorithm, only the values modulo N are known; the values modulo 41 are included for our understanding.   ◇

A drawback is that we have to store all values until a collision occur. *Floyd's cycle detection trick* reduces this to a constant number of values. Quite generally, suppose we have finite set Z (here: $Z = \mathbb{Z}_p$), $x_0 \in Z$, a function $f\colon Z \to Z$ and define $x_i = f(x_{i-1}) = f^{(i)}(x_0)$ for $i \geq 1$, where $f^{(i)} = f \circ f^{(i-1)}$ is the ith iterate of f. This is an infinite sequence from a finite set, so that at some point the values repeat. This results in a cycle of some length $\ell \geq 1$ such that $x_i = x_{i+\ell}$ for all $i \geq t$, for some $t \in \mathbb{N}$. In Figure 3.4, we see an example with $t = 3$ and $\ell = 7$.

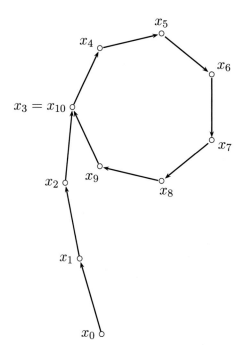

Figure 3.4: Pollard's ρ method.

The idea of Floyd's 1-step/2-step cycle detection method is to use a second sequence $y_0, y_1, \ldots$ that iterates f with double speed, so that $y_i = x_{2i}$ for all i, and to store only the current values of x_i and y_i. The "faster" sequence "overtakes" the slower one for some i, and then we have $x_{2i} = y_i = x_i$.

LEMMA 3.44. *In the above notation, suppose that $x_t = x_{t+\ell}$. Then $x_i = y_i$ for some $i < t + \ell$.*

PROOF. For any $t' \geq t$ and $j \geq 1$, we have $x_{t'} = x_{t'+j\ell}$. When $t = 0$, we have $x_\ell = x_{2\ell} = y_\ell$ and $i = \ell$ is sufficient. Otherwise we set $j = \lceil t/\ell \rceil$ and $i = j\ell$. Then $j \geq 1$, $i \geq t$, and $2i = i + j\ell$. Thus $x_i = x_{2i} = y_i$. Furthermore, $i < (t/\ell + 1) \cdot \ell = t + \ell$. □

ALGORITHM 3.45. Pollard's ρ algorithm.

Input: An odd composite integer N.
Output: A proper divisor of N, or "failure".

1. $i \leftarrow 0$.
2. $x_0 \xleftarrow{\boxdot} \mathbb{Z}_N$.
3. $y_0 \leftarrow x_0$.
4. Repeat steps 5–8
5. $i \leftarrow i + 1$,
6. $x_i \leftarrow x_{i-1}^2 + 1$ in $\mathbb{Z}_N$,
7. $y_i \leftarrow (y_{i-1}^2 + 1)^2 + 1$ in $\mathbb{Z}_N$,
8. $g \leftarrow \gcd(x_i - y_i, N)$.
9. Until $g \neq 1$.
10. If $g < N$ then return g else return "failure".

We identify $\mathbb{Z}_N$ with $\{0, \ldots, N - 1\} \subset \mathbb{Z}$, so that x_i is an element of $\mathbb{Z}_N$ in step 6 and an integer in step 8.

THEOREM 3.46. *Let $N \in \mathbb{N}$ be a composite n-bit integer, p its smallest prime factor, and $f(x) = x^2 + 1$. Under the assumption that the sequence $(f^{(i)}(x_0))_{i \in \mathbb{N}}$ of iterates of x_0 behaves modulo p like a random sequence, the expected number of iterations in Pollard's algorithm for returning a proper factor of N is $O(\sqrt{p}) \subseteq O(N^{1/4})$, using an expected number of $O(N^{1/4} n^2)$ bit operations.*

PROOF. Since p is at most $\sqrt{N}$, we have $\sqrt{p} \leq N^{1/4}$, and Theorem 3.41 implies the first claim. Each step in the algorithm can be performed with $O(n^2)$ bit operations. □

According to Pollard (1975), the expected number of iterations is close to $\sqrt{\pi^5 p / 288} \approx 1.0308 \sqrt{p}$.

EXAMPLE 3.43 CONTINUED. The algorithm calculates in tandem x_i and $y_i = x_{2i}$ and performs the gcd test each time. We have $t = 3$, $\ell = 7$, and in Lemma 3.44, $j = \lceil 3/7 \rceil = 1$ and $i = 7$. In fact, after seven 2-steps the algorithm catches up with the 1-steps:

i	$x_i \bmod N$	$x_i \bmod 41$	$y_i \bmod N$	$y_i \bmod 41$	$\gcd(x_i - y_i, N)$
0	631	16	631	16	82123
1	69 670	11	28 986	40	1
2	28 986	40	13 166	5	1
3	69 907	2	40 816	21	1
4	13 166	5	20 459	0	1
5	64 027	26	6685	2	1
6	40 816	21	75 835	26	1
7	80 802	32	17 539	32	41

The factorization $N = 41 \cdot 2003$ is found as $\gcd(x_7 - y_7, N) = 41$. Of course the sequence also repeats modulo 2003, and in fact $x_{38} = 4430 = x_{143}$ in Z_N, so that $x_{105} = y_{105}$ in $\mathbb{Z}_N$. If this repetition occurred simultaneously with the first one in $\mathbb{Z}_{41}$, we would not find a factor: an "unlucky" x_0.   ◇

The algorithm can be applied recursively to factor N completely; see Exercise 3.24. The iteration $x_i = x_{i-1}^2 + 1$ is clearly not random, although $x_0, x_1, \ldots$ generated in this way "look" quite random in practice. It is an open problem whether the birthday theorem—which assumes true randomness—actually applies to Pollard's rho method. The approach can also be used for discrete logarithms, and in that context the required property is actually proven; see Section 4.4.

3.13. Dixon's random squares

This section presents the first algorithm (Dixon 1981) which beats the $2^{\Omega(n)}$ barrier for factorization. Its main idea, the usage of smooth numbers, is due to Morrison & Brillhart (1975) and a cornerstone of modern factorization methods.

We want to apply Lemma 3.8 and thus are looking for two values x and y with $x^2 = y^2$ in $\mathbb{Z}_N$, but not $x \in \pm y$. This is not an easy task. To illustrate the solution in this section, we begin with an example.

EXAMPLE 3.47. Let $N = 2183$. Suppose that we have found the equations in $\mathbb{Z}_N$

$$
\begin{aligned}
453^2 &= 7, \\
1014^2 &= 3, \\
209^2 &= 21.
\end{aligned}
$$

Then we obtain $(453 \cdot 1014 \cdot 209)^2 = 21^2$, or $687^2 = 21^2$ in $\mathbb{Z}_N$. This yields the factors $37 = \gcd(687 - 21, N)$ and $59 = \gcd(687 + 21, N)$; in fact, $N = 37 \cdot 59$ is the prime factorization of N. $\diamond$

Dixon's factoring algorithm uses this idea in a systematic way. We choose a parameter $B > 0$, and the prime numbers $p_1, \ldots, p_h$ up to B as a *factor base*. Any integer all of whose prime factors are at most p_h is called B-*smooth*. An integer b is called a B-*number* if the integer $b^2 \in \mathbb{Z}_N = \{0, \ldots, N-1\}$ is B-smooth. In Example 3.47, $453, 1014$, and 209 are B-numbers for any $B \geq 7$. The idea is to take a product of such B-numbers such that every prime factor (of these squares) in the factor base occurs an even number of times, so that we get an equation of the type $x^2 = y^2$ in $\mathbb{Z}_N$, which leads (hopefully) to a factorization of N. More formally, we associate to each B-number b with $b^2 = p_1^{\alpha_1} p_2^{\alpha_2} \cdots p_h^{\alpha_h}$ in $\mathbb{Z}_N$ the binary vector $\varepsilon = (\alpha_1, \alpha_2, \ldots, \alpha_h)$ in $\mathbb{Z}_2^h$. For example, the binary vector associated to 453 is $(0, 0, 0, 1)$, and the one associated to 209 is $(0, 1, 0, 1)$. Now the idea, as explained above, is to consider a set of B-numbers $b_1, b_2, \ldots, b_\ell$ so that the associated binary vectors $\varepsilon_1, \ldots, \varepsilon_\ell$ satisfy $\varepsilon_1 + \varepsilon_2 + \cdots + \varepsilon_\ell = 0$ in $\mathbb{Z}_2^h$. Writing α_{ij} for the exponent of p_j in b_i^2, $a_j = \sum_{1 \leq i \leq \ell} \alpha_{ij}$ is the jth component of $\sum_{1 \leq i \leq \ell} \varepsilon_i$ and even, due to the assumption above. We take $\delta_j = a_j/2 \in \mathbb{Z}$, and $x = \prod_{1 \leq i \leq \ell} b_i$ and $y = \prod_{1 \leq i \leq h} p_j^{\delta_j}$ in $\mathbb{Z}_N$. Then

$$
\begin{aligned}
x^2 &= \prod_{1 \leq i \leq \ell} b_i^2 = \prod_{1 \leq i \leq \ell} \prod_{1 \leq j \leq h} p_j^{\alpha_{ij}} = \prod_{1 \leq j \leq h} p_j^{\sum_{1 \leq i \leq \ell} \alpha_{ij}} \\
&= \prod_{1 \leq j \leq h} p_j^{2\delta_j} = \Big(\prod_{1 \leq j \leq h} p_j^{\delta_j} \Big)^2 = y^2 \text{ in } \mathbb{Z}_N.
\end{aligned}
$$

To factor N, we invoke the Euclidean algorithm to compute $\gcd(x - y, N)$, which will exhibit a nontrivial factor of N unless $x \in \pm y$. A subtlety is that b, b^2, x, and y are defined in $\mathbb{Z}_N$ and we compute with them in that domain, but when we factor b^2 or take a gcd, then they are considered as integers under the usual identification of $\mathbb{Z}_N$ with $\{0, \ldots, N-1\}$.

We expect that $b^2 \in \mathbb{Z}_N$ for $b \longleftarrow \boxtimes \mathbb{Z}_N$ behaves approximately like a uniformly random element of $\mathbb{Z}_N$. In fact, one can prove this, but we skip the proof here. Furthermore, a randomly chosen $b \longleftarrow \boxtimes \mathbb{Z}_N$ is a B-number with probability about u^{-u}, where $u = \ln N / \ln B \approx n / \log_2 B$.

ALGORITHM 3.48. Dixon's random squares method.

Input: A composite integer $N \geq 3$, and $B \geq 2$.
Output: Either a proper factor of N, or "failure".

1. Compute all primes $p_1, p_2, \ldots, p_h$ up to B.
2. If p_i divides N for some $i \in \{1, \ldots, h\}$ then return p_i.
3. $A \leftarrow \varnothing$. [Initialize the set of B-numbers.]
4. Repeat steps 5–16
5.   Choose a uniform random number $b \xleftarrow{\boxtimes} \mathbb{Z}_N \setminus \{0\}$.
6. $g \leftarrow \gcd(b, N)$.
7.   If $g > 1$ then return g.
8. $a \leftarrow b^2$ in $\mathbb{Z}_N$. [Factor a over $\{p_1, \ldots, p_h\}$.]
9.   For $i = 1, \ldots, h$ do steps 10–13
                [Determine multiplicity of p_i in a.]
10. $\alpha_i \leftarrow 0$.
11.   While p_i divides a do steps 12–13
12. $a \leftarrow \frac{a}{p_i}$.
13. $\alpha_i \leftarrow \alpha_i + 1$.
14.   If $a = 1$ then
15. $\alpha \leftarrow (\alpha_1, \ldots, \alpha_h)$.
16. $A \leftarrow A \cup \{(b, \alpha)\}$.
17. Until $\#A = h + 1$.
18. Find distinct pairs $(b_1, \alpha^{(1)}), \ldots, (b_\ell, \alpha^{(\ell)}) \in A$ with $\alpha^{(1)} + \cdots + \alpha^{(\ell)} = 0$ in $\mathbb{Z}_2^h$, for some $\ell \geq 1$, by solving an $(h+1) \times h$ system of linear equations over $\mathbb{F}_2$.
19. $(\delta_1, \ldots, \delta_h) \leftarrow \frac{1}{2}(\alpha^{(1)} + \cdots + \alpha^{(\ell)})$.
20. $x \leftarrow \prod_{1 \leq i \leq \ell} b_i$.
21. $y \leftarrow \prod_{1 \leq j \leq h} p_j^{\delta_j}$.
22. $g \leftarrow \gcd(x + y, N)$.
23. If $1 < g < N$ then return g else return "failure".

Figure 3.5: Dixon's integer factorization method.

EXAMPLE 3.47 CONTINUED. We have $B = 7$, factor base $(2, 3, 5, 7)$,

$$b_1 = 453, b_2 = 1014, b_3 = 209,$$
$$\alpha^{(1)} = (0, 0, 0, 1), \alpha^{(2)} = (0, 1, 0, 0), \alpha^{(3)} = (0, 1, 0, 1),$$
$$\alpha^{(1)} + \alpha^{(2)} + \alpha^{(3)} = (0, 2, 0, 2) = (0, 0, 0, 0) \text{ in } \mathbb{Z}_2^4,$$
$$\delta_1 = \delta_3 = 0, \delta_2 = \delta_4 = 1,$$
$$x = 687, y = 21, \text{ and } \gcd(687 - 21, N) = 37.$$

In fact, there are exactly 73 7-numbers in $\mathbb{Z}_N$, excluding 0. Thus we expect $2180/73 \approx 31$ random choices of b in order to find one 7-number. We have $u = \ln 2182 / \ln 7 \approx 3.95108$, $u^{-u} \approx 0.00439$, and $Nu^{-u} \approx 9.58$. This is a serious underestimate, which occurs for small values. However, 7-smoothness is the same as 10-smoothness, and using this value in u, we find $Nu^{-u} \approx 41.4$.   ◇

Since $\text{prob}\{b \text{ is a } B\text{-number}\} \approx u^{-u}$ by (3.49), the expected number of trials to find

○ one B-number is $1/u^{-u} = u^u$,

○ $h + 1$ B-numbers is $(h + 1)u^u$.

The integer arithmetic, like gcd and trial division, can be done with $O(n^2)$ bit operations, if N is an n-bit integer. The algorithm has two stages: collecting the relations, which includes finding $h+1$ B-numbers and factoring them in the factor base (using trial division), for a cost of $hu^u \cdot B \cdot n^2$, and then linear algebra over $\mathbb{Z}_2$ in step 18 which can be done in time $O(B^3)$. Hence the expected total cost, using $h < B$, is

$$O(B^3 + Bu^u \cdot B \cdot n^2).$$

How to choose B? B^3 grows with B while u and $B^2 u^u n^2$ decrease with B. Thus we will choose B such that both terms are approximately equal. We set $m = \ln N \approx \ln 2 \cdot n$ and have

$$B^3 = e^{3 \ln B},$$
$$B^2 u^u n^2 = e^{2 \ln B + m \cdot (\ln m - \ln \ln B)/(\ln B) + 2 \ln n}.$$

Ignoring the small summand $2 \ln n$ and equating exponents approximately yields

$$3 \ln B = 2 \ln B + \frac{m}{\ln B}(\ln m - \ln \ln B),$$
$$\ln^2 B = m(\ln m - \ln \ln B) \approx m \ln m,$$
$$\ln B = \sqrt{m \ln m},$$
$$B = \exp(\sqrt{m \ln m}).$$

With this B, the cost is $O(B^3) = O(\exp(3\sqrt{m \ln m}))$. With a more careful analysis one can do a little better. We use the notation $L(a,b)(n) = \exp_2(b \cdot n^a \cdot (\log n)^{1-a})$.

THEOREM 3.50. *Dixon's random squares method factors an n-bit integer N with an expected number of*

$$O\big(L(1/2, \sqrt{9/2})(n)\big)$$

operations.   □

There are many improvements to this method, including Pomerance's 1982 *quadratic sieve*, see Crandall & Pomerance (2005), Chapter 6. Moreover, the collection of relations can be done in a distributed fashion on parallel machines, networks and by grid computing, whereas the linear algebra phase needs more sophisticated methods. The best time bound that is actually proved is $L(1/2, 1 + o(1))$ for a method based on class groups of binary quadratic forms (Lenstra & Pomerance 1992). Finally, a faster algorithm called the *number field sieve*, relying on an idea of Pollard and described in Lenstra & Lenstra (1993), needs $L(1/3, O(1))$ bit operations, and an improvement by Coppersmith (1993) yields the estimate

(3.51) $\exp_2(1.902 \cdot (n \cdot \log_2^2 n)^{1/3}).$

The record factorization of a 768-bit RSA modulus, completed in 2009 by Kleinjung, Aoki, Franke, Lenstra, Thomé, Bos, Gaudry, Kruppa, Montgomery, Osvik, te Riele, Timofeev & Zimmermann (2010), used this method. The main computational effort was to find relations and took almost two years on many hundreds of machines. This was followed by the linear algebra step, running for about four months on several clusters in three countries.

3.14. Factorization via element order

Chapter 14 presents quantum computers and how they can compute element orders efficiently, without knowing the group order. In particular, for an n-bit integer N, they can calculate orders in $\mathbb{Z}_N^\times$ using $O(n^3)$ operations on $4n$ qubits. This section shows how one can then factor N efficiently. Thus scalable quantum computers can factor integers in polynomial time—if they ever come into existence. The reader might skip this section unless interested in the quantum computations of Chapter 14.

We now define the problem. The notions and facts from group theory in Section 15.13 are assumed throughout this section.

DEFINITION 3.52.   (i) *Let G be a finite group. The problem EO_G (element order in G) has as input some $x \in G$, and as output the order of x in G.*

Continuing our series of RSA problems from Section 3.5, we consider

B_5 : Given N and $x \in \mathbb{Z}_N^\times$, compute the order $\mathrm{ord}(x) = EO_{\mathbb{Z}_N^\times}(x)$.

If we know the factorization of N, we can compute $\varphi(N)$ by Lemma 15.64, and given the factorization of $\varphi(N)$, we can solve B_5 by Corollary 15.59. Thus $B_5 \leq_p B_4$. We can also reduce the factorization problem B_4 to our B_5. In fact, we reduce B_4 to the following possibly easier task.

B_5' : Given $\varepsilon \geq 0$, N, and $x \in \mathbb{Z}_N^\times$, either compute an integer multiple ℓ of $k = \mathrm{ord}(x)$ with bit-size polynomial in that of N, or return "failure"; if k is odd, the latter with probability at most ε.

When k is even, an algorithm may return "failure" all the time; the reason for this relaxation is that the quantum algorithm 14.25 satisfies the specification in B_5', but not the one in B_5. We clearly have $B_5' \leq_p B_5$.

It is sufficient to factor an odd integer which is not a proper power of some integer; see Lemma 3.8. This is achieved by the following reduction. It is a simple adaption of Algorithm 3.7, with $N - 1$ in step 1 replaced by ℓ and the loop suitably modified.

ALGORITHM 3.53. Reduction $\mathcal{A}$ from factoring to B_5'.

Input: An n-bit odd composite integer N, not a proper power of an integer.

Output: A proper factor of N, or "failure".

1. Choose $x \xleftarrow{\;\boxdot\;} \{1, \dots, N - 1\}$.
2. Compute $g \leftarrow \gcd(x, N)$.
3. If $g \neq 1$ then return g.
4. $y \leftarrow x^{2^n}$.
5. Call an oracle for B_5' to either receive a multiple ℓ of the order of y in $\mathbb{Z}_N^\times$ or "failure". In the latter case, return "failure".
6. Write $\ell = 2^e m$ with nonnegative integers e and m, where m is odd.
7. $z \leftarrow y^m$ in $\mathbb{Z}_N$.
8. If $z = 1$ then return "failure".
9. For i from 1 to n do steps 10 through 13.
10.   If $z = -1$ then return "failure".
11. $u \leftarrow z^2$ in $\mathbb{Z}_N$.
12.   If $u = 1$ then compute $r \leftarrow \gcd(z - 1, N)$ and return r.

$\gcd(x, N) \neq 1$	even order						odd order			
	x	y	ℓ	k	z	r	x	y	z	k
3, 6, 7, 9	2	4	3	6	8	7	1	1	1	1
12, 14, 15, 18	5	4	3	6	20	f	4	16	1	3
	8	1	1	2	8	7	16	4	1	3
	10	16	3	6	13	3				
	11	16	3	6	8	7				
	13	1	1	2	13	3				
	17	16	3	6	20	f				
	19	4	3	6	13	3				
	20	1	1	2	20	f				

Table 3.6: The values x and y are from steps 1 and 4, respectively, of Algorithm 3.53 on input $N = 21$, ℓ is the output of the order oracle, assumed to be $\mathrm{ord}(y)$, so that $\ell = m$ in step 5, $k = \mathrm{ord}(x)$, z is from step 7, and r is either the factor of 21 from step 12 or $f = $ "failure".

13. $z \leftarrow u$.

14. Return "failure".

EXAMPLE 3.54. For input $N = 21$, we have $n = 5$ and the 20 choices of x in step 1 of Algorithm 3.53 lead to the values in Table 3.6, where z is the value in step 7. Thus we obtain a proper factor of 21 for $7 + 6 + 2 = 15$ values of x. The two values $x = 5$ and $x = 17$ in the middle column do not lead to a factorization because $x^{k/2} = x^3 = -1$. See Figure 15.2 for squaring in $\mathbb{Z}_{21}$.   ◊

THEOREM 3.55. *If an output is returned in steps 3 or 12, it is correct. The probability of failure is at most $1/2 + \varepsilon$, and for an n-bit input N the reduction uses $O(n^3)$ operations in $\mathbb{Z}_N$ plus one call to B_5' with an argument of odd order.*

PROOF.   An output in step 3 is correct, and we now assume that the condition in step 3 does not hold, so that $x \in \mathbb{Z}_N^\times$. Since $k = \mathrm{ord}(x) \leq \lambda(N) < 2^n$ by Theorem 15.60 (i), $\mathrm{ord}(y)$ is odd. Thus "failure" occurs in step 5 with probability at most ε. On the other hand, by Lemma 3.15, k is even and $x^{k/2} \notin \pm 1$ with probability at least $1/2$. Since $x^{2^n m} = 1$, some u in the loop of step 9 equals 1, its square root z satisfies $z = -1$ in step 10 with probability at most $1/2$ (Lemma 3.15), and step 12 returns a proper factor with probability at least $1/2$. Overall, if step 5 does not fail, a factor is found with probability at least $1/2$.

The most expensive part are the exponentiations in steps 4, 7, and 11, which can be done with $O(n^3)$ bit operations. □

Running the algorithm repeatedly decreases the failure probability exponentially, similar to Theorem 15.94 for decision problems. The quantum algorithm of Theorem 14.26 has failure probability $\varepsilon \leq 1 - 1/30 \ln\ln k$ for odd $k \geq 31$. (We can check for smaller order separately.) Now t repetitions decrease this to ε^t, and $(1 - 1/30 \ln\ln k)^t \leq e^{-2} \leq 1/4$ for $t \geq 60 \ln\ln k$.

COROLLARY 3.56. *Let* $\varepsilon > 0$ *and* $t = \lceil \log_{4/3} \varepsilon^{-1} \rceil$. *If the odd element order subroutine called in step 5 has failure probability at most* ε, *then* t *calls to it are sufficient to find a factor of* N *with probability at least* $1/2$.

Notes 3.2. Hardy's Euclidean quote is in his Chapter 12, page 32. News about the *Great Internet Mersenne Prime Search* GIMPS are available at http://www.mersenne. org/. Alford, Granville & Pomerance (1993) show that there are infinitely many Carmichael numbers. For factoring Carmichael numbers completely, see Bach *et al.* (1986), Theorem 3.

With fast arithmetic, the cost of the strong pseudoprimality test is $O^\sim(n^2)$. The first probabilistic polynomial-time primality test is due to Solovay & Strassen (1977), found in 1974. This revolutionary efficient solution of a classical problem provided a major impetus to popularize probabilistic algorithms. The strong test described above is due to Miller (1976) and Rabin (1976), see also Rabin (1980). These four authors received the ACM Paris Kanellakis Theory and Practice Award. Monier (1980) and Rabin (1980) show that the correctness probability in Theorem 3.17 (ii) is at least $3/4$. Our pseudoprimes are sometimes called *strong pseudoprimes*. The long-standing open problem of a deterministic polynomial-time test was solved in the affirmative by Agrawal, Kayal & Saxena (2004). They received the Gödel Prize and the Fulkerson Prize for this work. The currently fastest version of their method takes $O^\sim(n^6)$ operations using fast arithmetic. See Lenstra & Pomerance (2011). The term *"industrial-grade prime"* was coined by Henri Cohen.

3.3. Eratosthenes' sieve and several improvements to it are discussed in Bach & Shallit (1996), Section 9.8. The bounds of the Prime Number Theorem 3.21 can be found in Rosser & Schoenfeld (1962). Zagier (1997) presents a short proof of the approximate statement, using elementary complex calculus. A proof of a famous conjecture in analytic number theory, the *Riemann Hypothesis* going back to 1859, would yield much better error bounds. The better approximation by the logarithmic integral li$(x) = \int_1^x (\ln t)^{-1} dt$ would satisfy $|\pi(x) - \text{li}(x)| \in O(\sqrt{x} \ln x)$. Similar improvements would hold for the other functions in the Theorem.

Hardy & Wright (1985), Theorem 328, show that for any $\epsilon > 0$, $\varphi(k) \geq (e^{-\gamma} - \epsilon)k/\ln\ln k$ for all but finitely many k. Here γ is Euler's constant, and $e^{-\gamma} \approx 0.56146$. In particular, $\varphi(k) \geq k/(3 \ln\ln k)$ for $k \geq 31$.

For large n, fast arithmetic may be useful in RSA. Key generation then takes $O^\sim(n^2 \, \mathsf{M}(n) \log n)$ or $O^\sim(n^3)$ and encryption $O^\sim(n \, \mathsf{M}(n))$ or $O^\sim(n^2)$ operations; see

Section 15.6 and von zur Gathen & Gerhard (2013), Theorem 18.8. In order to make sure that the pseudoprime output of Algorithm 3.19 is indeed prime, one may run a deterministic primality test; see the Notes for Section 3.2.

3.4. Sophie Germain proved the "first case" of Fermat's Last Theorem for the primes named after her. The conjecture on the number of her primes is stated in Hardy & Littlewood 1923, Conjecture D. (Their conjecture contains a typo: the denominator should be ab instead of just a, in their notation.) The largest one known in 2015 is 18 543 637 900 515 · $2^{666\ 667} - 1$ with 200 701 decimal digits (http://www.primes.utm.edu/). No explicit value for the c from Heath-Brown (1986) is known. The expected number of executions is in $O(n^2)$ for step 3 and in $O(n^3)$ for step 4 of Algorithm 3.24, for a total of $O(n^7)$ bit operations. Details of the proof of Theorem 3.25 are in von zur Gathen & Shparlinski (2013). There is also a more efficient algorithm that is proven to work, uses expected time in $O^\sim(n^5)$, provably beats standard Sophie Germain generation under the conjecture and an analogous conjecture about MG_2 primes, and in practice as well, for example by a factor of 2.7 for $n = 2048$. Given an output of Algorithm 3.24, one may test the primes involved deterministically; see Notes 3.2 and Crandall & Pomerance (2005), §4.5. Under the conjectures, this method can also generate efficiently safe moduli with exactly n bits. See Ziegler & Zollmann (2013).

3.5. Lenstra *et al.* (2012) and Heninger *et al.* (2012) found many RSA moduli which together are easy to factor.

The bound illustrated in Table 3.2 can be generalized to the asymptotic probability that several integers at most x have their gcd equal to 1. For ten integers, this tends to $93555/\pi^{10} \approx 0.99901$ with growing x.

The equivalences $B_2 \equiv_p B_3 \equiv_p B_4$ are already indicated in Rivest *et al.* (1977). May (2004) and Coron & May (2007) show that $B_3 \leq_p B_2$ with a deterministic reduction, provided that $e, d < \varphi(N)$ and that the factors p and q are of the same bit-size. These conditions are satisfied in our description of RSA.

The evidence concerning the question whether $B_1 \equiv_p B_4$ is not yet conclusive. On the one hand, Boneh & Venkatesan (1998) show approximately that any reduction $B_4 \leq_p B_1$ yields a polynomial-time algorithm for factoring the RSA modulus. This means that if B_4 is hard (as we believe), then no such reduction exists. Their restrictions are that the reduction has to be a circuit (or straight-line program), where no branching is allowed, and it may use only a logarithmic number of queries to the variant of B_1 where e is small. On the other hand, Aggarwal & Maurer (2009) provide approximately a reduction from factoring to breaking RSA, that is, $B_4 \leq_p B_1$. Their restriction is that for B_4, only ring circuits (or generic ring algorithms) are allowed, which only use the ring operations $+, -, \cdot, /$. These are the ring analogs of the group circuits (or generic group algorithms) of Section 4.9. We do not go into several subtleties that are needed to obtain technically correct versions of these results. They are explained in the work of Aggarwal & Maurer, where one also finds references to related work. If we believe that factoring is hard, then analogs of these two results without any restrictions would contradict each other.

The continued fraction method for factoring from Morrison & Brillhart (1975) factors N efficiently if q is close to p. Håstad (1988) shows that the public exponent $e = 3$ is insecure; see Exercise 3.20.

3.6. Later work along the lines of Wiener's attack usually employs Coppersmith's method (Section 13.9). Boneh & Durfee (2000) relaxed the upper bound on d in Theorem 3.37 from roughly $N^{1/4}$ to $N^{0.292}$. May (2010) provides further references.

3.9. The material of this section on hard core bits of RSA is from Håstad & Näslund

(1998) and Näslund (1998).

3.10. See Brent (2000) for an overview on integer factorization. Shanks (1971) invented the *baby-step giant-step* method. It takes $O(\sqrt{p})$ steps to find a factor p, but also as much space. This was taken up by Knuth (1973), pages 575–576, and Pollard (1974) and Strassen (1976) present a simple variant. See von zur Gathen & Gerhard (2013), Section 19.3. Pollard (1974) describes his $p - 1$ method, and a generalization which provides the time bound in Table 3.3. Lenstra's powerful method from 1987 is more sensitive to the size of the factor rather than to that of the inputs.

3.12. See Konyagin *et al.* (2013) and the references there for a discussion of the behavior of quadratic (and other) maps over finite fields.

3.13. An early forerunner of the random squares method is suggested in Miller (1975). In step 18 of Algorithm 3.48, $\alpha^{(i)}$ is the vector in $\mathbb{Z}_2^h$ called α in step 15 for b. A proof that b^2 is sufficiently random in $\mathbb{Z}_N$ for random b is due to Schnorr (1982) and described in von zur Gathen & Gerhard (2013), Lemma 19.14. A variation due to Charles Rackoff is in Exercise 19.10 of the cited book.

A theorem of Canfield, Erdős & Pomerance yields asymptotic estimates for the number $\psi(x, y)$ of y-smooth numbers up to x. If $u = u(x, y) = \ln x / \ln y$, the theorem states that

$$(3.49) \qquad \psi(x, y) = x \cdot u^{-u + o(1)},$$

when $x \to \infty$ such that $(\ln x)^\epsilon < u < (\ln x)^{1-\epsilon}$, for some fixed $\epsilon > 0$. *Dickman's rho function* is defined by a differential equation and approximates ψ quite well; u^{-u} is an approximation to this function. Exercises 3.29 and 3.30 ask for experimental data.

Sieving methods were introduced by Schroeppel and improved by Pomerance (1985, 1996).

3.14. In Example 14.30, we found $5^3 = 17^3 = -1$ in $\mathbb{Z}_{21}^\times$. The equation $x^3 - y^3 = (x - y)(x^2 + xy + y^2)$ tempts us to try $\gcd(5 - 17, 21)$, which indeed provides the divisor 3. In general, if $b_1^{k/2} = b_2^{k/2}$, then one may try $\gcd(b_1 - b_2, N)$.

Exercises.

EXERCISE 3.1. *Using the results of the strong pseudoprimality test in Table 3.1, can you factor 553 and 561?*

EXERCISE 3.2 (Fermat liars for prime powers). *Let p be an odd prime, $e \geq 2$, and $N = p^e$. Lemma 15.64 says that $\varphi(N) = (p - 1)p^{e-1}$.*

(i) *Determine the number $\#L_{p^2}$ of Fermat liars for p^2.*

(ii) *Let $g_0 \in \mathbb{Z}_N^\times$ be such that $(g_0 \text{ modulo } p) \in \mathbb{Z}_p^\times$ is a generator of $\mathbb{Z}_p^\times$. Show that $g_1 = g_0^{p^{e-1}}$ and $g_2 = 1 + p$ have orders $p - 1$ and p^{e-1}, respectively, in $\mathbb{Z}_N^\times$.*

(iii) *Let G be any group, and $g_1, g_2 \in G$ of coprime orders d_1, d_2, respectively. Show that $g_1 g_2$ has order $d_1 d_2$. Conclude that $\mathbb{Z}_N^\times$ is cyclic.*

(iv) *Determine $\gcd(N - 1, \varphi(N))$.*

(v) *Show that for $x \in \mathbb{Z}_N^\times$ with $x^{N-1} = 1$, we have $x^{p-1} = 1$ in $\mathbb{Z}_N^\times$.*

(vi) *Let g be a generator of $\mathbb{Z}_N^\times$. Determine the number of exponents a with $1 \leq a < \varphi(N)$ and $(g^a)^{p-1} = 1$. Theorem 15.60 might be useful.*

(vii) *Determine the number $\#L_N$ of Fermat liars for N, and the error probability of the Fermat test on input N.*

EXERCISE 3.3 (RSA numbers are not Carmichael).   Let N be the product of two distinct primes. Without using the statement of Lemma 3.12, show that N is not a Carmichael number.

EXERCISE 3.4 (Inverse of 2).   For an odd integer N, determine explicitly the inverse of 2 in $\mathbb{Z}_N$.

EXERCISE 3.5 (Primality testing).   Which of the two integers $10^{200} + 349$ and $10^{200} + 357$ is probably prime and which is certainly composite? You may not use built-in routines for primality testing.

EXERCISE 3.6 (Primality test via square roots).   Starting with Berlekamp (1970), there are several algorithms to compute square roots modulo an odd prime number. Let $\mathcal{A}$ be one of them, e.g. Algorithm 14.15 from von zur Gathen & Gerhard (2013). On input an odd integer N, choose $b \xleftarrow{\;\Box\;} \mathbb{Z}_N^\times$, set $a = b^2$ in $\mathbb{Z}_N$, and run $\mathcal{A}$ on a. If the output is not in $\pm b$, then return "N composite" else return "N probably prime". Can you prove estimates on the cost and the error probablity of this primality test? It may be useful to first test whether N is a perfect power (Lemma 3.8 (ii)).

EXERCISE 3.7 (Even order).

 (i) In Lemma 3.15, assume that N has $r \geq 2$ distinct prime factors. Show that $\mathrm{prob}\{k \text{ even}\} \geq 1 - 2^{-r} \geq 3/4$.

 (ii) Let p be a prime and $e \geq 1$. Determine the number of elements of $\mathbb{Z}_{p^e}^\times$ with even order.

EXERCISE 3.8 (Prime Number Theorem).

 (i) Verify the precise estimates for $\pi(x)$ in the Prime Number Theorem 3.21 up to $x = 10\,000$.

 (ii) Plot $\pi(x)/(x/\ln x)$ for $x = 1000, 2000, \ldots, 10\,000$.

 (iii) Plot $\pi(x)/(x/\ln x)$ and the corresponding lower and upper bounds in a single figure.

 (iv*) Extend the ranges until a reasonable computing time is reached, say about half an hour.

EXERCISE 3.9 (Find a prime).   Let c be your birth number from Exercise 2.1 and $B = c \cdot 10^6$ The goal is to find the first prime beyond B. Use the Fermat test 3.6 to examine some numbers following B.

 (i) Estimate the number of numbers you need to examine according to the Prime Number Theorem 3.21. That is, determine a small value for k such that the existence of a prime between B and $B + k$ is guaranteed.

 (ii) Which numbers are easy to examine? How can you reduce the number of needed Fermat tests by a factor of $1/2$, or even more?

 (iii) Give a Fermat witness for the first odd composite number following B.

 (iv) For the first possibly prime p following B, give ten numbers $a \in \mathbb{Z}_p^\times$ on whose choice the Fermat test answers "possibly prime".

EXERCISE 3.10 (Finding prime numbers).

(i) Prove that $\pi(2x) - \pi(x) > x/2\ln x$ if $x \geq e^6$.

(ii) Use (i) to give lower bounds on the number of k-bit primes for $k = 16$, $k = 32$ and $k = 64$.

(iii) For $k = 16$, use a computer algebra system to count the primes between 2^{15} and 2^{16}. Compare with (ii).

EXERCISE 3.11 (Keys for the RSA system). *The prime numbers $p = 263$ and $q = 307$ are to be used in RSA. (In practice, these numbers would be much too small.) We choose $e = 54323$ and $N = p \cdot q$ as our public key.*

(i) *Compute the corresponding secret key d.*

(ii) *Encrypt $x = 35698$.*

(iii) *Decrypt $y = 373$.*

If you want to solve this exercise using a computer algebra system, take the following numbers instead of the ones above:

$$p = 2609899,$$
$$q = 3004217,$$
$$e = 54323425121,$$
$$x = 4364863612562,$$
$$y = 850080551629.$$

EXERCISE 3.12 (RSA bad choice). *We claim that two prime numbers close to each other form a bad choice for RSA. We use Fermat's factorization method based on the fact that if $N = pq$, then $N = (\frac{p+q}{2})^2 - (\frac{p-q}{2})^2$.*

(i) *Explain how you can use this fact to find factors of N.*

(ii) *Do it for the 35-bit modulus $N = 23360947609$.*

EXERCISE 3.13 (RSA computation using the Chinese Remainder Theorem 15.44). We consider the RSA notation 2.15 and the CRT implementation of Section 3.7.

(i) Which quantities can be precomputed for decryption, independently of y?

(ii) How many bit operations does the RSA decryption require, using the precomputed values? Compare to the cost of the standard RSA algorithm.

(iii) Prove that x as computed is the correct RSA decryption of y.

(iv) Run 10 computer experiments each with RSA moduli of 100, 200, 400, and 800 bits, with randomly chosen primes. Take the average of these 10 runtimes. How do the precomputation times compare to the rest of the decryption? Normalize the times for the time at 100 bits. Do these normalized times follow an n^2-law? Do the same experiments with the standard RSA decryption in 2.16. Compare the results.

EXERCISE 3.14 (Faulty implementation of RSA). BOB *has set up an RSA system as in 2.15. Consider the following protocol.*

1. ALICE *encrypts a message x using* BOB*'s public exponent and sends him $y = x^e$ in $\mathbb{Z}_N$.*

2. BOB *decrypts $x' = y^d$ in $\mathbb{Z}_N$, eventually making an error. In order to acknowledge receipt of the message, he sends x' to* ALICE, *encrypted with her public key, which is different from (N, e).*

3. ALICE *decrypts x' correctly.*

Suppose that BOB *decrypts y in the CRT fashion of Section 3.7, thus first recovering $x' = (x'_p, x'_q)$.*

(i) *We assume that exactly one of the modular exponentiations of* BOB *in the Chinese Remainder Theorem 15.44 yields an incorrect result, say $x'_p \neq x$ in $\mathbb{Z}_p$, while $x'_q = x$ in $\mathbb{Z}_q$. Show that* ALICE *can recover the secret key of* BOB.

(ii) *Assume that* BOB *is a smartcard which has an error in the implementation of the Chinese Remainder Algorithm 15.45. Suppose that an adversary can activate the card so that it returns decrypted messages. Show that he can easily factor the public modulus on the card.*

EXERCISE 3.15 (Reductions for RSA).

(i) *Check the calculations in Examples 3.29 and 3.30.*

(ii) *Perform the two reductions $B_4 \leq_p B_3 \leq_p B_2$ on the example $N = 391$. The oracle for B_3 returns $\varphi(N) = 352$. For the second reduction, start with $e_1 = 5$ and $e_2 = 11$. Calculate the corresponding d_1 and d_2 (in the reduction, these would just be given to you), and determine $\varphi(N)$ as in the text. Are further values $e_3, e_4, \ldots$ needed?*

(iii) *Suppose that $N = 168149075693$ and $\varphi(N) = 168148245408$. Compute the two prime factors of N.*

EXERCISE 3.16. *(Shamir 1993) Let $p \neq q$ be prime numbers, $N = p \cdot q$, $f = x \in \mathbb{Z}_N[x]$.*

(i) *Show that $p^2 + q^2$ is a unit in $\mathbb{Z}_N^{\times}$.*

(ii) *Let $u \in \mathbb{Z}_N$ be the inverse of $p^2 + q^2$. Show that $f = u(px + q)(qx + p)$.*

(iii) *Prove that the two linear factors in (ii) are irreducible in $\mathbb{Z}_N[x]$.*

(iv) *Conclude that factoring N is polynomial-time reducible to factoring polynomials in $\mathbb{Z}_N[x]$.*

EXERCISE 3.17 (Polynomial-time reduction). *Consider the following two decision problems*

○ Primes: *On input of an integer x, decide whether x is a prime.*

○ Factor: *On input of two integers k and x, decide whether x has a factor at most k.*

(i) *Reduce one problem to the other and use the appropriate notation.*

(ii) *How can you use an efficient algorithm for Factor to actually factor an integer efficiently?*

(iii) *In (i), suppose there was a reduction in the other direction as well. What would that imply? Is such a reduction likely to exist?*

EXERCISE 3.18 (Carmichael's lambda function).   *You are to show that one can use the slightly smaller number $\lambda(N)$ from (3.13) instead of $\varphi(N)$ in the definition of RSA.*

(i)   *Show that $\lambda(MN) = \lambda(M)\lambda(N)$ for coprime integers M and N.*

(ii) *Show that $x^{\lambda(N)} = 1$ for all $x \in \mathbb{Z}_N^\times$, and that $\lambda(N)$ is the smallest positive number with this property.*

(iii) *Show that $\lambda(N)$ divides $\varphi(N)$.*

(iv) *If N is odd and $r \geq 2$ (as in RSA), then $\lambda(N) \leq \varphi(N)/2$.*

(v)   *Give a polynomial-time reduction from factoring N to calculating $\lambda(N)$.*

(vi) *Where does the proof of Theorem 3.55 use the assumption that N is not a proper power of an integer? Is its claim also true when N is a prime power?*

EXERCISE 3.19 (Polynomial Chinese remainder RSA).   *Let $\ell, n \in \mathbb{N}$, $k = 2^\ell$ substantially less than n, say $k \leq n/10$. Furthermore, let $p \neq q$ be randomly chosen $n/2$-bit primes, and $N = pq$. We take the public key $\mathsf{pk} = (N, \ell)$ and secret key $\mathsf{sk} = (p, q)$. We consider message packets $(m_0, \ldots, m_{k-1})$, where each m_i is an integer with bit length less than $n - k$. We append its index j in binary to the jth message to obtain $m'_j = m_j \cdot 2^\ell + j$ in $\mathbb{Z}_N$. The encryption function is*

$$\mathrm{enc}_{\mathsf{pk}}(m_0, \ldots, m_{k-1}) = \text{coefficients } 1, f_{k-1}, f_{k-2}, \ldots, f_0 \text{ of}$$

$$f = \sum_{0 \leq i \leq k} f_i x^i = \prod_{0 \leq j < k} (x - m'_j) \text{ in } \mathbb{Z}_N[x].$$

The decryption procedure factors f modulo p and modulo q, and puts together the corresponding roots via the Chinese Remainder Algorithm 15.45. The correspondence is given by the low-order bits.

(i)   *Verify that the system works correctly.*

(ii) *Determine the cost of encryption.*

(iii) *Determine the cost of decryption.*

(iv) *In order to break the system, it is sufficient to factor polynomials modulo N that have only linear factors. Show that factoring N is reducible to this problem.*

(v)   *Can you conclude that this cryptosystem is secure, assuming that RSA is?*

EXERCISE 3.20 (RSA with small public exponent).   *This exercise shows that small public exponents in RSA may be a real danger. Suppose that in some public key infrastructure, maybe an internet banking system, the exponent $e = 3$ is always used as public exponent. (In fact, $e = 2^{16} + 1$ is often used as a general public exponent in connection with the RSA scheme.) Thus every user chooses a public modulus N such that $\gcd(\phi(N), 3) = 1$ and computes his respective secret exponent d such that $3d = 1$ in $\mathbb{Z}_{\varphi(N)}$. Suppose that the users A_1, A_2, A_3 have the following public moduli:*

$$N_1 = 5000746010773, \quad N_2 = 5000692010527, \quad N_3 = 5000296004107.$$

(i)   ALICE *broadcasts a message x to A_1, A_2, A_3 by encrypting $y_i = x^3$ in $\mathbb{Z}_{N_i}$ for $i \in \{1, 2, 3\}$.* EVE *drops in and captures the following values:*

$$y_1 = 1549725913504, \quad y_2 = 2886199297672, \quad y_3 = 2972130153144.$$

Show that EVE *can recover the value of x without factoring N_i and compute this value. (Hint: Chinese Remainder Theorem 15.44.)*

(ii) *Generalize the method used by* EVE *above for larger public exponents e. How many messages does* EVE *need to intercept in order to recover the plaintext message?*

(iii) *Suppose that RSA is used with a small public exponent e as a signature scheme (Section 8.1). Can a similar attack be used by* EVE *in order to forge signatures? Explain your answer.*

An immediate defense against this attack consists in adding random information to the message x. It was proposed to use a time stamp t. The variation works as follows. If t is the numerical value of the current (computer) time and ℓ its bit length, this will be appended to the initial message x. The resulting encapsulated message will thus be

$$z = 2^{\ell} \cdot x + t.$$

The same message x will receive different encapsulations at different times. Is this protection secure enough? Still using the exponent $e = 3$, we shall show in the following that this is not the case. For this we use the following theorem of Håstad (1988).

Let N_i be seven public RSA moduli and $z_i = a_i \cdot x + b_i \in \mathbb{Z}_{N_i}$, where x is a fixed message and $i = 1, 2, \ldots 7$. Write $z_i^3 = (a_i \cdot x + b_i)^3 = \sum_{0 \le j \le 3} c_{ij} \cdot x^j$ as a polynomial in x over $\mathbb{Z}_{N_i}$, and let $N = \prod_{1 \le i \le 7} N_i$. Then one can find in time polynomial in $\log(N)$ a set of seven multipliers s_i with the following property. If $c_j \in \mathbb{Z}_N$ is such that $c_j = s_i \cdot c_{ij}$ in $\mathbb{Z}_{N_i}$ for $i = 1, 2, \ldots, 7$, then

$$s_i z_i^3 = \sum_{0 \le j \le 3} c_j \cdot x^j \quad \text{and} \quad c_j \cdot x^j < N = \prod_{1 \le i \le 7} N_i.$$

(iv) *Using Håstad's theorem, show that in the above setting, the time stamp randomized message m can be recovered using seven encryptions. Hint: assume that the time stamps t_i are not too far apart and can be found by trial and error. In this case, for each guessed time vector $t = (t_1, t_2, \ldots, t_7)$, the values a_i, b_i, s_i for $i \le 7$ in Håstad's theorem are known, so one can recover the c_j for $j \le 3$.*

(v) *An improved randomization (as essentially used in the current standard PKCS5 consists in requiring that the message length be less than $n/2$ and adding a random pad r of length $n/2$, where the N_i are n-bit numbers. The message z to be encrypted is produced by merging the bits of x and r according to a fixed scheme. Explain why this measure is safe against the previous class of attacks. Explain how the holder of the secret key can recover the message x from the encryption z^e in $\mathbb{Z}_{N_i}$.*

EXERCISE 3.21 (Birthday problem).   *We consider random draws with replacement from m labelled items, as in Theorem 3.41.*

(i) *Show that $1 - x \le e^{-x}$ holds for $x \in \mathbb{R}$.*

(ii) *Let s be the number of rounds until one of the items is drawn for the second time. Show that $\text{prob}\{s \ge j\} = m^{-j+1} \prod_{0 \le i < j} (m - (i - 1))$.*
   Hint: If k_j is the number on the jth item, then

$$\text{prob}\{s \ge j\} = \prod_{0 \le i < j} \text{prob}\{k_i \notin \{k_1, \ldots, k_{i-1}\}: \#\{k_1, \ldots, k_{i-1}\} = i - 1\}.$$

(iii) *In Theorem 3.41 we have shown that the expected number of rounds needed satisfies*

$$\mathcal{E}(s) = \sum_{j \geq 1} \operatorname{prob}\{s \geq j\} \leq 2 + \sqrt{\frac{\pi m}{2}}.$$

Determine a constant c, as large as you can, so that $\mathcal{E}(s) \geq c \cdot \sqrt{m}$.

EXERCISE 3.22 (Integrating the error function). *The Gauß error function* $\operatorname{erf}(t) = c \cdot \int_0^t e^{-x^2}\, dx$, *for a specific constant c, describes the cumulative distribution of the standard normal distribution on $\mathbb{R}$. It is not an elementary function. However, the definite integral occurring in the birthday paradox can be evaluated by elementary calculus.*

(i) *Justify the equations*

$$\left(\int_{-\infty}^{\infty} e^{-x^2}\, dx \right)^2 = \int_{-\infty}^{\infty} \int_{-\infty}^{\infty} e^{-(x^2+y^2)}\, dx\, dy = \int_0^{2\pi} \left(\int_0^{\infty} e^{-r^2} r\, dr \right) d\varphi,$$

where (r, φ) are the polar coordinates of $(x, y) \in \mathbb{R}^2$.

(ii) *Find the indefinite integral $\int e^{-r^2} r\, dr$.*

(iii) *Determine the value of $\int_0^{\infty} e^{-x^2}\, dx$ and that of c so that $\operatorname{erf}(\infty) = 1$.*

EXERCISE 3.23 (A sample run of Pollard's rho).

(i) *Complete the table below, which represents a run of Pollard's rho algorithm for $N = 100181 = 17 \cdot 5893$ and the initial value $x_0 = 399$, up to $i = 13$.*

i	$x_i \operatorname{rem} N$	$x_i \operatorname{rem} 17$	$y_i \operatorname{rem} N$	$y_i \operatorname{rem} 17$	$\gcd(x_i - y_i, N)$
0	399	8	399	8	100181
1	$\cdots$	$\cdots$	$\cdots$	$\cdots$	$\cdots$

(ii) *The smallest prime divisor of N is 17. Describe the idea of the algorithm by looking at x_i and y_i in $\mathbb{Z}_{17}$.*

(iii) *Compute the complete factorization of N with this method.*

EXERCISE 3.24 (Complete factorization with Pollard's rho algorithm). *Let $N = p_1^{e_1} \cdots p_r^{e_r}$ be the prime factorization of N, with primes $p_1 < p_2 < \cdots < p_r$ and positive integers e_i. On input N we test whether N is prime. If not, we run Pollard's rho on N and expect to find p_1. Then we test whether the output is prime. If so, we determine its multiplicity e_1 and run this procedure recursively on $N/p_1^{e_1}$. Show the following.*

(i) *$p_i \leq N^{1/(r-i+1)}$ for all $i \geq 1$.*

(ii) *For $x > 0$, $\sum_{2 \leq i < r} x^{1/i}$ is in $O(x^{1/2})$.*

(iii) *Specifiy in detail the procedure indicated above.*

(iv) *Determine carefully its expected running time.*

EXERCISE 3.25 (Bad luck in Pollard rho algorithm).   *Consider distinct odd primes p and q, $N = pq$, $x_0 \in \mathbb{Z}_N^{\times}$, and the sequence $x_0, x_1, \ldots$ in $\mathbb{Z}_N$ from Pollard's rho Algorithm 3.45.*

(i) *Find p and q so that -3 is a nonzero square in $\mathbb{Z}_p$ and in $\mathbb{Z}_q$. Find x_0 so that the sequence is constant.*

(ii) *For large B, how would you estimate the number of $N \leq B$ for which such an x_0 exists? Your reckoning may be heuristic, but with clearly stated assumptions.*

(iii) *Find p and q so that $x_0 = x_4 \neq x_1$.*

(iv) *What do your examples mean for factoring N?*

EXERCISE 3.26 (Expected distance of a random walk.).

(i) *Prove that $\sum_{0 \leq k < n} \binom{2n}{k} = (4^n - \binom{2n}{n})/2$ and $\sum_{0 \leq k < n} \binom{2n-1}{k} = 4^{n-1}$ for all positive integers n.*

(ii) *Let $n \in \mathbb{N}_{>0}$, X_i for $1 \leq i \leq 2n$ be a collection of independent random variables which take each of the two values 1 and -1 with probability $1/2$, and $X = \sum_{1 \leq i \leq 2n} X_i$ be a random walk of length $2n$. Prove that $\mathrm{prob}\{X = 2(n-k)\} = \mathrm{prob}\{X = -2(n-k)\} = \binom{2n}{k} 4^{-n}$ for $0 \leq k \leq n$.*

(iii) *Show that $\mathcal{E}(X) = 0$ and $\mathcal{E}(|X|) = 2n \binom{2n}{n} 4^{-n}$.*

(iv) *Stirling's formula (see Knuth 1997, 1.2.11.2) says that*

$$(3.57) \qquad n! \approx \sqrt{2\pi n} \left(\frac{n}{e}\right)^n \left(1 + \frac{1}{12n} + \frac{1}{288n^2} + \cdots\right).$$

Use the simplified version $n! \in \sqrt{2\pi n} \, (n/e)^n (1 + O(n^{-1}))$ to show that $\mathcal{E}(|X|) \in 2\pi^{-1/2} n^{1/2} + O(n^{-1/2})$.

(v) *Prove the same formulas as in (iii) when there are $2n - 1$ instead of $2n$ random variables.*

EXERCISE 3.27 (Dixon's random squares).

(i) *Let $N = q_1 q_2 \cdots q_r$ be odd with pairwise distinct prime numbers q_i and $r \geq 2$. Show that there are exactly 2^r square roots of 1 in $\mathbb{Z}_N^{\times}$.*
    Hint: Use the Chinese Remainder Theorem 15.44. Note: The claim is also true if the q_i are pairwise coprime prime powers, since $\sqrt{1} = \pm 1$ in $\mathbb{Z}_q$ also for a prime power q.

(ii) *Show that if s, t are randomly chosen among the elements of $(\mathbb{Z}_N^{\times})^2$ satisfying $s^2 = t^2$ in $\mathbb{Z}_N$, then the probability for $s \notin \pm t$ in $\mathbb{Z}_N$ is at least $1 - 2^{1-r}$.*

EXERCISE 3.28 (Employing Dixon's random squares).

(i) *Find a factor of $N = 1517 = 37 \cdot 41$ using Dixon's random squares method. Choose $B = 5$ and execute the algorithm step by step.*

(ii) *For $N = 184\,53148\,59041$, compute the value $B_1 = \exp(\sqrt{\ln N \ln \ln N})$ used in the text as well as the promised value $B_2 = \exp(\sqrt{\frac{1}{2} \ln N \ln \ln N})$. Try to factor N via Dixon's random squares method and both values for the factor base bound B. No matter whether your factorization attempts succeed, hand in a protocol with all B-numbers found and their exponent vectors. You may use an appropriate concise representation for bit vectors with many zeroes. Comment on what has happened.*

EXERCISE 3.29 (Smooth numbers).   Denote by $\psi(x, B)$ the number of B-smooth integers up to x.

(i) List all 3-smooth and all 7-smooth numbers up to 100 and give the values of $\psi(100, 3)$ and $\psi(100, 7)$.

(ii) Compute $\psi(10\,000, 7)$ and $\psi(10\,000, 100)$.

For fixed x and B consider the set $\{p_1, \ldots, p_h\}$ of all prime numbers up to B and let $\nu = \lfloor \ln(x)/\ln(B) \rfloor$, $(a_1, a_2, \ldots, a_\nu) \in \{p_1, \ldots, p_h\}^\nu$, and $a = a_1 a_2 \cdots a_\nu$.

(iii) Show that a is B-smooth.

(iv) Conclude that $\psi(x, B) \geq h^\nu / \nu!$.

In the text we use $\psi(x, B) \geq h^\nu / \nu! \approx xu^{-u}$ with $u = \ln(x)/\ln(B)$.

(v) Compute the estimates $h^\nu / \nu!$ and xu^{-u} of 7-smooth numbers and of 100-smooth numbers less than $10\,000$. Compare these estimates to the exact values. What do you observe?

(vi) 7-smooth is equivalent to 10-smooth. Compare the estimates as in (v). What do you observe?

EXERCISE 3.30 (Smoothness experiments).   A theorem of Canfield et al. (1983) estimates the probability of smoothness of integers. You are to find empirical data about this probability and compare them to the prediction of (3.49).

(i) Use a computer algebra system for computing the exact values of $\psi(x, y)$ for the following sixteen sets of values: $x = c \cdot 10^6$, with $c \in \{0.5, 1, 1.5, 2\}$ and $y = 10d$, with $d \in \{2, 3, 5, 10\}$.

(ii) Compare the results with the asymptotic prediction (3.49). What is the $o(1)$ term in your results?

(iii) Now we estimate $\psi(x, y)$ probabilistically. (3.49) suggests that on average, out of $x/\psi(x, y)$ random integers up to x, one should be y-smooth. For each pair (x, y) as in (i) do the following.

(a) Draw random integers between 1 and x and test whether they are y-smooth. No complete factorization is required for this. Repeat until $s = 50$ smooth numbers have been found. Let $i(x, y)$ denote the number of random draws which were required for finding these s smooth numbers.

Hint: Use only one loop for each x and all four values of y.

(b) Let $N(x, y) = i(x, y)/s$. Compare $N(x, y)$ to $\rho(x, y) = x \cdot u(x, y)^{-u(x, y)}$ and to the exact value of $\psi(x, y)$ determined in (i). Which of the two values $N(x, y)$ and $\rho(x, y)$ is the better approximation of $\psi(x, y)$?

(iv) The approximation in (iii) is faster then the exact computation of $\psi(x, y)$ in (i); it is thus possible to use larger values of x and y. For each pair (x, y) with $x \in \{10^{10}, 10^{11}, 10^{12}, 10^{13}\}$ and $y \in \{10^4, 2 \cdot 10^4, 5 \cdot 10^4\}$ do the following:

(a) Evaluate $\rho(x, y) = x \cdot u(x, y)^{-u(x, y)}$.

(b) In a loop, draw random integers between 1 and x and test whether they are y-smooth. Repeat until s smooth numbers have been found (set $s = 10$ for the start and try to increase). Let $i(x, y)$ be the number of random draws required for finding s numbers up to x which are y-smooth.

(c) Let $N(x,y) = i(x,y)/s$. Both $N(x,y)$ and $\rho(x,y)$ are approximations to $\psi(x,y)$, which we did not compute exactly. Compare these two values to each other and discuss whether the approximation improves with growing values of x and y as compared to the results in (iii).

(v) Investigate the same problem for values of x near 10^{20} and y near 10^4, and possibly using some more powerful arithmetic system such as NTL, PARI, GMP or even your own long arithmetic—10^{20} is small in this context.

EXERCISE 3.31 (Extrapolating).

(i) Assume that a factoring algorithm requires $\exp\left(\sqrt[2]{n \ln n}\right)$ bit operations to factor an n-bit integer and that it only needs a second to factor a number less than 2^{100}. How large should an RSA modulus be so that this algorithm cannot factor it in less than the lifetime of the universe, which is estimated at about $13.7 \cdot 10^9$ years or about 10^{18} seconds?

(ii) How large should a modulus be if a new algorithm is found that requires only $O\left(\exp(2\sqrt[3]{n \ln^2 n})\right)$ operations? The big-Oh notation hides a constant factor c. How does your result depend on c? For a specific estimate, you may make any reasonable assumption about c.

(iii) How large should a modulus be if the new algorithm is optimized and now requires only $O\left(\exp(\sqrt[3]{n \ln^2 n})\right)$ operations?

(iv) What happens with an algorithm using $O\left(\exp(\sqrt[4]{n \ln^3 n})\right)$? $O(n^{10})$?

EXERCISE 3.32 (Element order modulo pq). Let p and q be distinct odd primes, $N = pq$, $x \longleftarrow^{\boxdot} \mathbb{Z}_N^\times$, and r_p, r_q, r_N be the orders of x in $\mathbb{Z}_p^\times$, $\mathbb{Z}_q^\times$, $\mathbb{Z}_N^\times$, respectively. Furthermore, write $r_p = 2^{e_p} m_p$, $r_q = 2^{e_q} m_q$, and $r_N = 2^{e_N} r_N$ with integers $e_p, e_q, e_N \geq 1$ and m_p, m_q, m_N odd.

(i) Show that $e_N = \max\{e_p, e_q\}$.

(ii) For $i \geq 1$, determine the probability that 2^i divides r_p.

(iii) For $i \geq 1$, determine the probability that 2^i divides r_N.

(iv) Determine the expected value of the maximal i for which 2^i divides e_N.

Cryptography must be secret, swift, and accurate.
Cryptographers must be security conscious and of one mind.
HO CHI MINH (1950)

The invention of the symbol ≡ by Gauss affords a striking
example of the advantage which may be derived
from an appropriate notation, and marks
an epoch in the development of the science of arithmetic.
GEORGE BALLARD MATHEWS (1892)

It is never wise to mix your ciphers,
like mixing your drinks, it may lead to selfbetrayal.
SIR ALFRED EWING (1914)

Kurz, es sind die Arten der Verheimlichung unendlich
und deren Modificationen sind es noch mehr.
Bey allen wird eine Verabredung vorausgesetzt,
und diese ist der Schlüssel zur Geheimpost.[1]
JOHANN SAMUEL HALLE (1786)

Multa patent, sed plura latent, semperque latebunt.[2]
SAMUEL LINDNER (1770)

In the very outset [...] was found a universal clearness
or hardness of belief, almost mathematical in its limitations,
and repellent in its unsympathetic form.
THOMAS EDWARD LAWRENCE (OF ARABIA) (1926)

Those who would give up essential Liberty, to purchase
a little temporary Safety, deserve neither Liberty nor Safety.
BENJAMIN FRANKLIN (1755)

---

[1]In short, the ways of concealing are infinite, and their modifications are even more numerous. All of them assume a prearrangement, and this is the key to the encryption scheme.

[2]Many things lie open, but more are hidden, and will always remain hidden.

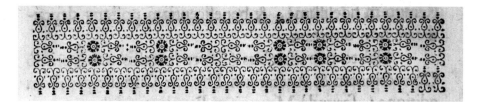

Chapter B

Key addition and modular arithmetic

his chapter presents some historical examples of key addition systems. These are easy to describe with our modern notion of modular arithmetic. Already in 1690, a rather obscure French author, Claude Comiers d'Ambrun, had the right intuition about the arithmetic nature of such systems. But without the proper notions and notations, it is very strenuous to express these things.

B.1. Key addition systems

In a *key addition system*, given a message, one produces a key of the same length and adds the two together, letter by letter, to obtain the encryption. This is then transmitted, and the legitimate receiver only has to subtract the key, again letter by letter, to find the original message. This can be described as

(B.1)
$$\text{ciphertext} = \text{plaintext} + \text{key},$$
$$\text{plaintext} = \text{ciphertext} - \text{key}.$$

More formally, we have letters from a fixed alphabet of some size s (in modern English, $s = 26$), and then the plaintext $x = (x_0, x_1, \ldots)$, the key $k = (k_0, k_1, \ldots)$, and the ciphertext $y = (y_0, y_1, \ldots)$ are related as

$$y_i = x_i + k_i, \quad x_i = y_i - k_i$$

for all i. Here addition and subtraction take place in the additive group $\mathbb{Z}_s = \{0, 1, \ldots, s-1\}$, that is, by doing arithmetic modulo s. The relation between the alphabet and $\mathbb{Z}_s$ is taken in the natural way: A $\longleftrightarrow$ 0, B $\longleftrightarrow$ 1, ..., as in the conversion Table A.1. A simple example is to encrypt VIGENERE CIPHER with the key CAESAR.

plain	V	I	G	E	N	E	R	E	C	I	P	H	E	R
	21	8	6	4	13	4	17	4	2	8	15	7	4	17
key	C	A	E	S	A	R	C	A	E	S	A	R	C	A
	2	0	4	18	0	17	2	0	4	18	0	17	2	0
cipher	23	8	10	22	13	21	19	4	6	0	15	24	6	17
	X	I	K	W	N	V	T	E	G	A	P	Y	G	R

Table B.1: Vigenère encryption.

Thus the ciphertext XIKWN VTEGA PYGR would be transmitted. Many ways of producing the required key have been employed. We have seen the Caesar cipher, where one uses a single letter and repeats it as often as necessary: $k_i = k_0$ for all i. Caesar used $k_0 = 3$, and Augustus $k_0 = 1$ (with Z + k_0 = AA); see Example A.2 (iv). The abbot Johannes Trithemius published in 1518 his *Polygraphia*, the first printed book about cryptography; Section G.3 describes some details. It contains, among other things, his *Recta transpositionis tabula*[1] (Figure G.6) consisting of the 24 Caesar substitutions on his 24-letter alphabet {A, B, C, D, E, F, G, H, I, K, L, M, N, O, P, Q, R, S, T, U, X, Y, Z, W}. Trithemius suggested to use these substitutions one after the other. But together with the idea from Blaise de Vigenère's 1586 *Traicté des Chiffres* of using a keyword-driven alternation between the various Caesar ciphers, this gives the *Vigenère cipher* (Example A.2 (v)), which was famous as the *chiffre indéchiffrable* or *unbreakable cipher* for centuries. Formally, one has a keyword $k_0, \ldots, k_{\ell-1}$ of some length ℓ, and repeats this as necessary: $k_i = k_{i \text{ rem } \ell}$ for all i. This is illustrated in Table B.1 with the keyword $k_0 k_1 k_2 k_3 k_4 k_5$ = CAESAR of $\ell = 6$ letters, and the encryption is $y_i = x_i + k_i$.

Modern variants, usually over the binary alphabet, include the *one-time pad* (Section 9.4) where the key k is a random sequence of the same length as the message, and variations where one has an initial segment of the key (possibly random) and generates the remaining key letters in a pseudorandom fashion (Chapter 11). A systematic method for breaking the Vigenère system was published by Kasiski (1863) and is explained in Chapter C. Charles Babbage had found a solution method earlier, in February or March of 1846, but never published it. His notes were dis-

---

[1]square table of substitutions

covered in the early 1980s in the British Library; Franksen (1984) narrates
the story. The central part of Babbage's success is his discovery of (B.1),
which he writes as

(B.2)
$$\text{Cypher} = \text{Key} + \text{Translation} - 1,$$
$$\text{Translation} = \text{Cypher} - \text{Key} + 1.$$

The ± 1 comes from the fact that he starts his alphabet with $\text{a} = 1$
instead of $\text{a} = 0$, as we do.

Was Babbage the first to discover the key equation (B.1)?

B.2. Claude Comiers d'Ambrun

An interesting little booklet (Figure B.2) appeared in 1690. Its author,
Claude Comiers d'Ambrun, hailed from Embrun in Savoy, was a professor of mathematics at Paris, and wrote about mathematical and cabbalistic subjects and a *Medicine Universelle*. He was blind, and proud to
point out his achievements. His sample message throughout the book is
Comiers d'Ambrun avevgle Roial[2] and the publisher introduces him as *"Le
Sçavant Mr. Comiers [...] tout aveugle & maltraité de la fortune qu'il est,
continuë de travailler, & fait voir toûjours les lumieres de son esprit."*[3]
Comiers presents the first "arithmetization" of cryptography. He uses an
18-letter alphabet, and basically writes the Vigenère system as in (B.2),
computing modulo 18.

In fact, he uses the coding of letters in Figure B.3. Thus his two-digit
number $10 \cdot k$ stands for $9 + k$, and he has a system of representatives
modulo 18. He spells out the same equations (B.2) as Babbage later. They
meant to be modulo 18, but, of course, he has neither the concept nor
the notation at his disposal, and has to make awkward case distinctions
between the first and the second half of his encoding numbers. Modular
arithmetic was put on proper footing and given the modern notation by
Gauß (1801).

Comiers explains his decryption rule for one-digit letters (on his page
45) in the example "ciphertext $8+1-$ key D (equal 4) $= 5 =$ plaintext E":

> *ajoûtez l'unité au nombre envoyé, & de la somme ôtez-en
> le chifre du mot du guet, il restera le nombre qui indiquera
> la quantiéme lettre de mon Alphabet est la lettre du secret.
> Ainsi D.4. étant la lettre du mot du guet, & le chifre envoyé*

---

[2]Comiers d'Ambrun, royal blind man

[3]The scholar Mr. Comiers d'Ambrun, although blind and mistreated by fortune, continues to work and show the lights of his spirit.

L'ART
D'ECRIRE
ET
DE PARLER
OCCULTEMENT
ET SANS SOUPÇON.
Par Mr. COMIERS.

A PARIS,
Chez MICHEL GUEROUT, Cour-neuve du Palais, au Dauphin. 1690.

Se vend A BRUXELLES,
Chez J. LEONARD, Libr. rüe de la Cour.

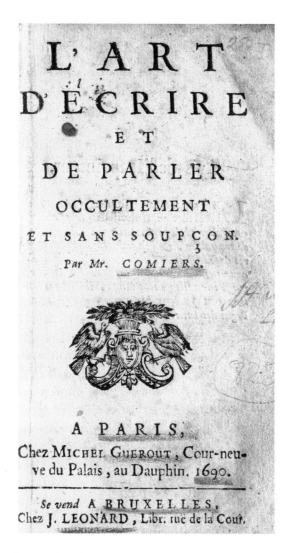

Figure B.2: The title page of Comiers's book.

A.	B.	C.	D.	E.	F.	G.	I.	L.	M.
1.	2.	3.	4.	5.	6.	7.	8.	9.	10.
N.	O.	P.	Q.	R.	S.	T.	V.		
20.	30.	40.	50.	60.	70.	80.	90.		

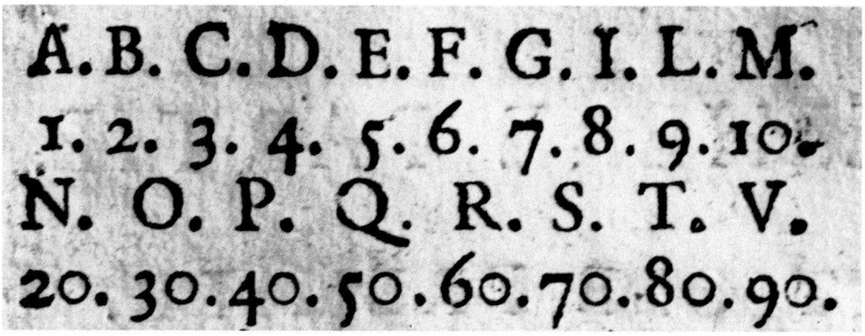

Figure B.3: Comiers's 18-letter alphabet and coding.

> *étant* 8. *dites* 8. *plus* 1. *égale* 9. *que vôtre amy avoit eu pour*
> *la somme du chifre de la lettre D. du mot du guet, & de la*
> *lettre du secret qui vous est encore inconnuë. C'est pourquoy*
> *dites* 9. *moins* 4. *égale* 5. *c'est à dire que la lettre du secret est*
> *la cinquiéme de l'Alphabet qui est la lettre E.*[4]

Comiers also specifies the rules for two-digit letters; they are even more complicated due to the clumsy encoding.

Comiers is well aware of the "cyclic" nature of addition modulo 18: *"Ayez ensuite devant vous l'Alphabet de dix-huit lettres dans leur ordre naturel, que vous devez concevoir comme écrites en chapelet, ou autour de la circonference d'un cercle."*[5]. But because of his unfortunate representation, he expresses our $13 = 9 + 1 - 15$ in $\mathbb{Z}_{18}$ as: *"afin de connoître pour quelle lettre de l'Alphabet le chifre* 9. *a été envoyé par le moyen de la lettre du mot du guet R* 60. *comptez depuis* 60. *inclusivement, disant* 60.70.80.90.1.2.3.4.5.6.7.8.9. *& remarquez que vous avez compté treize nombres ; dont la treiziéme lettre de l'Alphabet est la lettre P. de secret, pour laquelle on a envoyé le chifre* 9.*"*[6].

With hindsight, modular addition is already visible in Alberti's disk (Figure A.14), but Comiers is the first one to relate this to the arithmetic notion. He describes several cryptosystems, for example a simple substitution given by a keyword followed by the letterwise addition of another key word as superencipherment. He also addresses distributed cryptography. In order to share a secret among three recipients, he suggests to give one an encrypted version, and one half of the key each to the other two. Comiers displays the usual inventor's pride in his additive cryptosystem:

> [...] *elle est neanmoins indéchifrable à tout esprit hu-*
> *main. Quand même on donneroit au Déchifreur les lettres*
> *que chaque chifre signifie, il faudroit encore qu'il pust de-*
> *viner le nombre qui sert de premiere clef, & qu'aprés cela il*

---

[4]Add 1 to the number received [ciphertext], and from the sum subtract the key digit. It will remain the number that indicates which letter of my alphabet is the letter of my secret [plaintext]. Say if $D = 4$ is the key letter and 8 is received, then you say 8 plus 1 is 9 which is what your friend had for the sum of the key letter D and the plaintext letter which is yet unknown to you. That is why you say 9 minus 4 is 5, meaning that the plaintext letter is the fifth letter of my alphabet, which is the letter E.

[5]Then take the alphabet of eighteen letters in their natural order, which you must conceive as being written like a rosary, or around the circumference of a circle.

[6]In order to know for which letter the number 9 was sent using the key letter $R = 60$, you count from 60 inclusive up to 9, also inclusive, saying 60, 70, 80, 90, 1, 2, 3, 4, 5, 6, 7, 8, 9. You notice that you have counted thirteen letters, and hence the thirteenth letter P is the plaintext, whose encryption 9 has been sent.

*devinast encore l'ordre de l'Alphabet qui a donné ces lettres
par le moyen du nombre convenu pour clef ; de quoy on peut
faire facilement l'essay avec ceux qui se piquent de pouvoir
déchifrer ; fust-ce avec Mr. Viette, le Pere de nôtre Algebre
specieuse, & le grand Déchifreur de son temps, s'il pouvoit
revenir au monde.*[7]

Comiers first published this work in the journal *Mercure Galant*, in
the three issues of September and October 1684 and February 1685. It
includes a fold-out table shown in Figure B.4. In the centre is a "Vigenère
table" in his alphabet, and just above it we find:

```
2  8  a  d  i  e  F  e  b  r  u  a  r  i  i  1  6  90
4  1  10 20 30 1  6  4  6  50 6  9  1  4  1  8  6  8
c  o  m  i  e  r  a  v  e  u  g  l  e  r  o  y  a  l
```

The first line is the message, the bottom one his usual key, and the center
line the encryption. At the sides we see standards with two alphabets.
The top line at left means that A is to be encrypted as 111, and B as
the same with a dot added, maybe as $11 \cdot 1$. In the lower part is a keyword generated unsorted alphabet, made up from *Profetisandum* and the
remaining letters *bcglq*, a pigpen cipher, and a line cipher. (The reproduction makes reading hard in some parts.)

Comiers d'Ambrun turned blind around 1684 and took the title of
aveugle royal, because he had been accepted in the old-age home of *Les
Quinze-Vingts*. His work remained inconsequential in the history of
cryptography.

B.3. Later work on arithmetic cryptography

In the Hebrew, Greek, and Arab civilizations, letters had a standard numerical value. This allows to encrypt a letter by sending two (or more)
letters whose numerical sum equals the given one. This system was described by al-Kindī (801–876); see Mrayati *et al.* (1987).

Count Josse Maximiliaan van Gronsveld invented during the Thirty
Years' War (1618–1648) the *Gronsfeld cipher*, a Vigenère cipher with a numerically written key.

---

[7] ... [the cipher] is nevertheless indecipherable by any human mind. Even if one gave
to the decipherer the encryption of each letter, he would still have to guess the initial key
element, and then guess the order of the alphabet which provided these encryptions
by means of the number agreed upon as the key. You can easily try this out on those
who pride themselves on being able to decipher; be it with Mr. Viète, the father of our
symbolic algebra and the great decipherer of his age, if he could return to this world.
[See Section D.1]

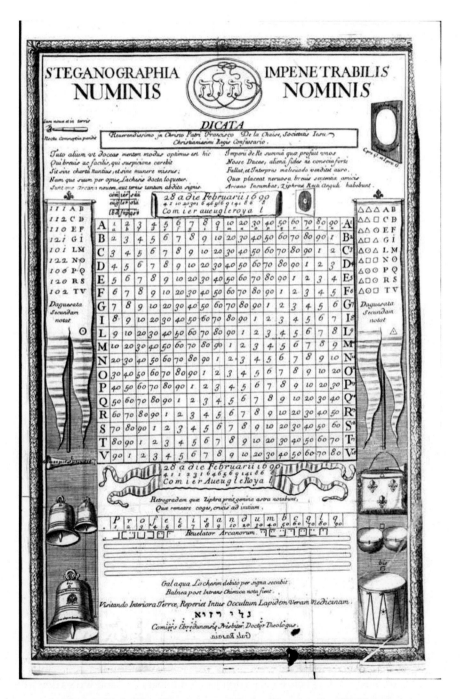

Figure B.4: The Vigenère table in Comiers's 1691 book.

In an anonymous work *Neueröffneter Schauplatz geheimer philosophis-cher Wissenschaften*[9], published in 1770 (Figure B.5), the author proposes an even more general type of arithmetic cryptography. In Chapter 16 *Von der logistischen Steganographia*[10] he adds a fixed number to each letter of the plaintext (Caesar) or subtracts one. But then he also multiplies by a number, or divides by one. Figures B.6 and B.7 illustrate encryption and decryption. He uses a standard alphabet with 24 letters and an offset of 6: $A = 6$, $B = 7, \ldots, Z = 5$. This is *table F* from his *Zahltisch*[11]. The secret key is this "F" and the multiplier 2. The result is taken as an integer, not modulo 24. Since the multiplier 2 is easily revealed as the gcd of a few encrypted letters, this does not offer any more security than a Caesar cipher. Addition and multiplication by a constant are special cases of affine linear transformations. But it is a long way from here to Hill's 1931 general linear transforms (Exercise A.2), requiring the development of the appropriate mathematical machinery over the centuries.

In his 1819 work on *Den hemmelige Skrivekunst*[14], Major Lindenfels from Copenhague presents a Vigenère tableau with a 20-letter alphabet. The 16th letter is "S", and he uses the encryption rule (B.1) with an offset of $S = 16$. On page 177, he encrypts the plaintext ABEL with the key CAIN, obtaining the ciphertext UTIS. To explain the encryption of the last letter, he writes „N" + „L" = „S" . Thus Lindenfels states rule (B.1) in some special cases. On page 181, he says about another table: *Det Ene af dem — efter Omstændighederne, snart det Første, snart det Andet — tjener som Multiplicand, det Andet som Product.*[15] The context is not quite clear to this author, but Lindenfels may have recognized somehow the Vigenère tableau as the table of a group operation. Kasiski (1863) points out explicitly that the cryptosystem is addition modulo 26, and that commutativity of addition corresponds to the (anti-)symmetry of the Vigenère table: plaintext + key = key + plaintext.

Variants of the key addition cipher were rediscovered several times,

---

[9]Newly opened showplace of secret philosophical sciences

[10]On logistic steganography

[11]number table

[12]2) Find in the number table all letters of the plaintext, and write below them the corresponding numbers, as above. [Flihe bald = Flee soon.] 3) Multiply all these numbers by a number of your choosing and which your correspondent knows. We take 2 here, and have the following example:

[13]1) Because he [your correspondent] knows which table is used and by which number has been multiplied, namely here by 2, he divides all numbers by 2 and writes down the quotients in their order:

[14]The art of secret writing

[15]One of them—the first or the second, depending on the circumstances—is a factor, the other one the product.

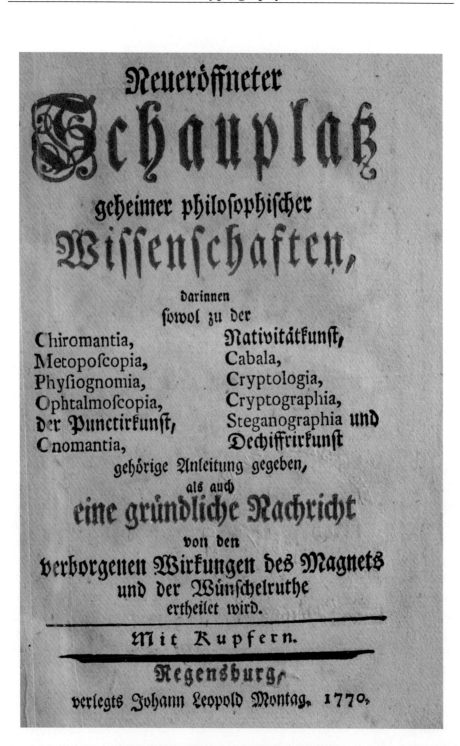

Figure B.5: The *Schauplatz* title page.

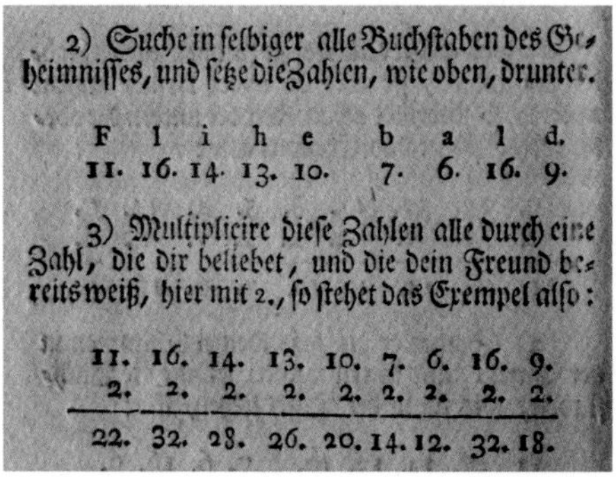

2) Suche in selbiger alle Buchstaben des Geheimnisses, und setze die Zahlen, wie oben, drunter.

F	l	i	h	e	b	a	l	d.
11.	16.	14.	13.	10.	7.	6.	16.	9.

3) Multiplicire diese Zahlen alle durch eine Zahl, die dir beliebet, und die dein Freund bereits weiß, hier mit 2., so stehet das Exempel also:

11.	16.	14.	13.	10.	7.	6.	16.	9.
2.	2.	2.	2.	2.	2.	2.	2.	2.
22.	32.	28.	26.	20.	14.	12.	32.	18.

Figure B.6: Encryption via multiplication by 2.[12]

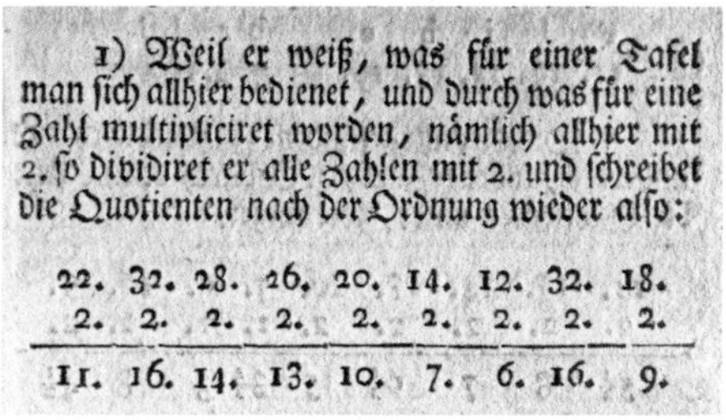

1) Weil er weiß, was für einer Tafel man sich allhier bedienet, und durch was für eine Zahl multipliciret worden, nämlich allhier mit 2. so dividiret er alle Zahlen mit 2. und schreibet die Quotienten nach der Ordnung wieder also:

22.	32.	28.	26.	20.	14.	12.	32.	18.
2.	2.	2.	2.	2.	2.	2.	2.	2.
11.	16.	14.	13.	10.	7.	6.	16.	9.

Figure B.7: First step in the decryption via division by 2.[13]

often without recognizing the similarities with the original idea. Rear Admiral Sir Francis Beaufort (1774–1857) created the currently used scale for wind forces (from *beau temps* to *vent fort*) and also the Beaufort cipher, which is just the Vigenère cipher with subtraction rather than addition of the key. He enciphered parts of his diary, in particular those concerning his incestuous relationship with his sister Harriet. In the US Civil War, the Confederate armies used the Vigenère cipher. One such message fell into Union hands. The enciphering officer had committed the cryptographic sin of writing many words in plaintext. The message was duly broken and even the key recovered. The Confederate General Alexander was aware of the equation "clear + key = cipher", just as Comiers.

Captain Holstein (1917) writes about a new cipher reported in the *Scientific American* that it is *"the old Vinegere [sic]* cipher of the sixteenth century." Lubin (1901) describes Vigenère encryption and its additive character. He then suggests repeated superencipherment with the same method but different keywords whose lengths are pairwise coprime to obtain a long period.

Notes B.1. The author gratefully acknowledges the previous publication of much of this section's material in *Cryptologia* (von zur Gathen 2003).

B.2. Comiers does not give any reference to Trithemius or Vigenère, but attributes the general Caesar substitution to the ancient Hebrews. Claude Comiers d'Ambrun was born in Embrun in the French alps, early in the 17th century. The exact date is not known. He turned blind and died in 1693 in the Hôpital des Quinze-Vingts, which was founded in 1260 with $15 \times 20 = 300$ beds and still is a renowned ophthalmological clinic in Paris. Comiers published in the *Mercure Galant* and the *Journal des Sçavans* and wrote on a wide range of topics: *La nouvelle Science de la nature des comètes; Discours sur les comètes; Instruction pour réunir les églises prétendues réformées à l'Église romaine; Trois discours sur l'art de prolonger la vie; Traité des lunettes; Traité des prophéties, vaticinations, prédictions et pronoçtications, contre le ministre Jurieu; Traité de la parole, des langues et écritures*. That is, about comets, doubling the cube and trisecting angles, church politics, the art of prolonging one's life (with a 400 year old gentleman as a real-life example), mirrors, lenses and solar clocks, mineral waters, machines for raising water in channels, horoscopes and prophecies. Rochas writes in his 1856 *Biographie du Dauphiné* (volume I, page 270) deprecatingly: *"C'était un homme possédé de la manie d'écrire, et qui réussit auprès de ses contemporains à se faire passer pour un savant."*[8]

B.3. The cryptographic sin is in Gallagher (1989), pages 303–304, and Alexander's equation in Gaddy (2004).

---

[8]He was a man possessed by a mania for writing, and succeeded in passing himself off to his contemporaries as a scholar.

Նա, ով չի ճանաչում իր անցյալը,
դատապարտված է ապրելու իր իսկ սերնդի մթության մեջ:[16]
Գարեգին Նժդեհ (1886–1955)

Logarithmorum canonem à me longo tempore elaboratum
superioribus annis edendum curavi, qui rejectis naturalibus
numeris, & operationibus quæ per eos fiunt, difficilioribus,
alios substituit idem præstantes per faciles
additiones, substractiones, bipartitiones, & tripartitiones.[17]
JOHN NAPIER (1617)

Which *Rule* is of Excellent use for finding the
Logarithms of *Prime Numbers*, having the
logarithms of the adjoyning *Numbers* given.
WILLIAM JONES (1706)

The science of calculation also is indispensable as far as the
extraction of the square & cube roots, Algebra as far as the
quadratic equation & the use of logarithms are often of value
in ordinary cases: but all beyond these is but a luxury;
a delicious luxury indeed; but not be indulged in by one
who is to have a profession to follow for his subsistence.
THOMAS JEFFERSON (1799)

The student should not lose any opportunity of exercising himself
in numerical calculation and particularly in the use
of logarithmic tables. His power of applying mathematics
to questions of practical utility is in direct proportion
to the facility which he possesses in computation.
AUGUSTUS DE MORGAN (1902)

---

[16]Garegin Nzhdeh: He who lacks a sense of the past is condemned to live in the narrow darkness of his own generation.

[17]I produced the *canon of logarithms* on which I worked for a long time over the last years. It replaces the natural numbers and the more difficult operations on them by others which yield the same results by simple additions, substractions, and divisions by two and three [in place of square and cube roots].

Chapter 4

Group cryptography and discrete logarithms

Alongside RSA, the most important practical tools for asymmetric (public key) cryptography are protocols that work in groups. Most of them were introduced in the multiplicative group of units in a finite field like $\mathbb{Z}_p^\times$ for some prime p. Today, the elliptic curves of Chapter 5 are often preferred.

We first present the general set-up, study a specific protocol (ElGamal encryption), and then discuss in depth some general attacks on the security of group cryptography, namely algorithms for the discrete logarithm. The upshot of this discussion is a recommendation to use groups of a certain type, namely of large prime order. Special groups like $\mathbb{Z}_p^\times$ allow more efficient attacks. Finally, we consider the natural model of *group circuits* (or *generic algorithms*), where only group operations are performed. There is an exponentially large lower bound on discrete logarithm computations in this model. As of today, no other attacks are known for general elliptic curves. This allows shorter key lengths than in $\mathbb{Z}_p^\times$ and is a central reason to prefer these groups.

4.1. Groups for cryptography

The Diffie-Hellman key exchange, discussed in Section 2.6, was originally proposed for use in the multiplicative group $\mathbb{Z}_p^\times$ modulo a prime p. In 1987, Neal Koblitz and Victor Miller independently suggested that it could also be implemented in the group of an elliptic curve over a finite field. Throughout this text, we discuss such systems in the general context of an arbitrary (cyclic) finite group G. Four examples are of special interest in cryptography:

1. the multiplicative group $G = \mathbb{Z}_p^\times$ of units modulo a prime p,

© Springer-Verlag Berlin Heidelberg 2015
J. von zur Gathen, *CryptoSchool*, DOI 10.1007/978-3-662-48425-8_6

2. the multiplicative group $G = \mathbb{F}_q^\times$ of a finite field $\mathbb{F}_q$,

3. elliptic curves (Chapter 5) and their subgroups,

4. large subgroups of $\mathbb{Z}_p^\times$ for a huge prime p (Sections 4.8 and 8.4).

We recall that in the system, one has a publicly known group element g which generates the finite group G, both parties choose their secret (random) exponents a and b, communicate g^a and g^b, respectively, and then

$$g^{ab} = (g^a)^b = (g^b)^a$$

is their shared secret key. This idea of having two versions of the same key, namely secret a and public g^a, is the foundation for most protocols in this area.

It is actually not necessary that g generate the whole group, but only that the subgroup $\langle g \rangle$ of G generated by g be large enough (more precisely, that its order have a large prime divisor; this is discussed in Section 4.8). For example, elliptic curve groups are often not cyclic, but may have large cyclic subgroups, as required. In the following, we will usually assume that $G = \langle g \rangle$.

We denote as $d = \#G$ the number of elements in G, and we call $n = \lfloor \log_2 d \rfloor + 1$ the *bit size* of G, so that d is an n-bit number. There is a representation for the elements of G with n bits for each element. It will be obvious in the concrete examples. We let

$$(4.1) \qquad\qquad\qquad \mathsf{M}_G$$

be a number of bit operations sufficient to perform the group operations (multiplication, inversion, the unit element 1) in this representation. For a family $(G_n)_{n \in \mathbb{N}}$, we assume that we have a binary representation of size polynomial in n, and that M_{G_n} is polynomial in n. We have $\mathsf{M}_{G_n} \in O(n^2)$ for the usual groups in this text; an exception occurs in Section 8.4.

We consider the map taking an integer a to g^a. Since $g^d = 1$ by Lagrange's Theorem 15.53, multiples of d do not matter in the exponent, that is, $g^{a+kd} = g^a \cdot (g^d)^k = g^a \cdot 1^k = g^a$ for any a and k. Thus we actually have a map $\mathrm{dexp}_g \colon \mathbb{Z}_d \longrightarrow G$ called *discrete exponentiation in base* g, with $\mathrm{dexp}_g a = g^a$. The cost for computing it is $O(n\mathsf{M}_G)$ (Section 15.11). What is its inverse? From calculus we know the answer: the logarithm. So given some $x \in G$, we call $a \in \mathbb{Z}_d$ the *discrete logarithm of* x *in base* g if $g^a = x$, and we write $a = \mathrm{dlog}_g x$. Thus $\mathrm{dexp}_g(\mathrm{dlog}_g x) = x$ and $\mathrm{dlog}_g(\mathrm{dexp}_g a) = a$, for all $x \in G$ and $a \in \mathbb{Z}_d$. Since g has order d, $a \in \mathbb{Z}_d$ is uniquely determined. The "discrete" is not related to the discreet nature of cryptography, but to the fact that these operations

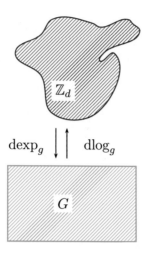

Figure 4.1: The inverse isomorphisms dexp and dlog.

take place in a finite "discrete" domain, in contrast to the continuous exponentiations and logarithms of calculus. The two maps just introduced are of fundamental importance and illustrated in Figure 4.1. They are inverse to each other, and provide an isomorphism between the two groups, $\mathbb{Z}_d$ with addition and G with multiplication. We call this the *exp-log isomorphism*. Furthermore, the map dexp_g is easy to evaluate, but not so dlog_g, in general. Part of this chapter is devoted to the study of algorithms for dlog_g.

We have a phenomenon here that we might call a basic distinction between mathematics and computer science. The two groups G and $\mathbb{Z}_d$ are isomorphic, and any statement in the language of groups that holds for one also holds for the other group. From a mathematical point of view, the two groups are "the same". But in computer science, we have to compute with representations (say, in a computer) of the group elements. And we are aware by now that the conversion from the G-representation to the $\mathbb{Z}_d$-representation of one element may be computationally difficult. It is precisely the discrete logarithm problem. From a computer science point of view, the two groups are quite different.

Thus it is easy to go "down" from $\mathbb{Z}_d$ to G, and hard to go "up". In Chapters 7, 8, and 10, we will see several protocols using this scenario in the context of signatures, identification, etc. The basic idea is that the owner of secret keys can do things at the "secret" level of $\mathbb{Z}_d$, shown in red in Figure 4.1, and everybody has access to the "public" level of G, shown in green.

The owner of the (powerful) secret key, say of a random $a \in \mathbb{Z}_d$, can easily communicate with the world by presenting the (useful, but less powerful) public version $A = \mathrm{dexp}_g\, a = g^a$. And then everybody can calculate in the public world, but not go back to the secret level, provided the discrete logarithm is hard.

The discrete logarithm problem is of central importance in cryptography. We now give some definitions, and the rest of this chapter will look at algorithmic approaches to the problem. It is sometimes convenient to consider integers rather than elements of $\mathbb{Z}_d$, and when we have two elements g and $x = g^a$ of a finite group G, with $a \in \mathbb{Z}$, then we call a a discrete logarithm of x in basis g. It was noted above that we may assume $0 \le a < d$, and then this notion corresponds to the one above under our usual identification of $\mathbb{Z}_d$ with $\{0, \cdots, d-1\}$. If $G = \langle g \rangle$, then a is uniquely determined. The following is a more detailed version of the Discrete Logarithm Problem 2.24.

DEFINITION 4.2. *Let G be a finite group.*

(i) *The (general) discrete logarithm problem in G is, given two elements g and x of G, to decide whether $x \in \langle g \rangle$, and if so, compute some $a \in \mathbb{Z}$ with $x = g^a$.*

(ii) *We are often given additional information:*

  ○ *guaranteed membership: $x \in \langle g \rangle$ is true,*

  ○ *cyclic group: we have $G = \langle g \rangle$,*

  ○ *group order: we are given $\#G$,*

  ○ *inverse logarithm: we are given $b \in \mathbb{Z}$ with $g = x^b$.*

(iii) *When the group order $d = \#G$ is given, we usually require the smallest nonnegative value for a, that is, $a \in \mathbb{Z}_d = \{0, 1, \ldots, d-1\}$.*

(iv) *We write DL_G for the computational version of the most common form of the discrete logarithm problem in the finite group G: a generator g of $G = \langle g \rangle$ and $d = \#G$ are given, the input is some $x \in G$, and the output is the unique integer a with $x = g^a$ and $0 \le a < d$.*

Two related tasks, the *Diffie-Hellman problem DH_G* and the *Decisional Diffie-Hellman problem DDH_G*, are defined in Section 2.6.

EXAMPLE 4.3. Before going further in this chapter, it is worth mentioning that the discrete logarithm problem in the additive group $G = (\mathbb{Z}_d, +)$ can be reduced polynomially to the computation of inverses in $\mathbb{Z}_d$, which in turn is efficiently solvable using the Extended Euclidean Algorithm 15.19. "$x = g^a$" now means $x = a \cdot g$ in $\mathbb{Z}_d$, and for $g \in \mathbb{Z}$ we have by the Extended Euclidean Algorithm:

$$g \text{ generates } \mathbb{Z}_d \iff \gcd(g, d) = 1 \iff \text{there exist } s, t \in \mathbb{Z} \quad sg + td = 1.$$

We claim that for any integer x, the value $a = xs$ in $\mathbb{Z}_d$ is the discrete logarithm of x in basis g. Indeed, $x = x(sg + td) = xsg + xtd = a \cdot g$ in $\mathbb{Z}_d$. This computation can be done in polynomial time, and therefore $\mathbb{Z}_d$ is an *easy group* and useless for group cryptography. ◇

Now any finite cyclic group $(G, \cdot)$ of order d is isomorphic to $(\mathbb{Z}_d, +)$. One might first compute a discrete logarithm in $(\mathbb{Z}_d, +)$ and then deduce the solution in G. But this does not work—the isomorphism is the discrete logarithm and may be hard to compute.

In the rest of the chapter, we first describe ElGamal's encryption system, which works in a group, and then discuss several algorithms for the discrete logarithm problem. Here *time* means number of group operations and *space* denotes the number of memory cells (for one group element each) that are used by the algorithm.

1. *Baby-step giant-step attack:* works for any G and takes time and space $\sqrt{d}$.

2. *Pollard rho:* works for any G and takes time $\sqrt{d}$ and constant space.

3. *Chinese distribution:* reduces from general d to prime power d.

4. *Pohlig-Hellman algorithm:* reduces from prime power d to prime d.

5. *Index calculus:* works for $G = \mathbb{Z}_p^\times$, p a prime, and is more efficient than the methods above. It can be generalized to $\mathbb{F}_q^\times$ with q a prime power.

In the last section, we discuss *arithmetic group circuits*, an abstraction of these algorithms, and convince ourselves that these methods cannot be improved, in a specific sense.

Although we will mainly talk about individual groups in this chapter, for perspective it is useful to mention the asymptotic notion of a family $G = (G_n)_{n \in \mathbb{N}}$ of cyclic groups in which the discrete logarithm is hard. We are also given a generator g_n for G_n and $d_n = \#G_n$ with d_n of bit length polynomial in n and at least n. This leads to the parametrized problem

$\mathrm{DL}_G = (\mathrm{DL}_{G_n})_{n \in \mathbb{N}}$. We consider probabilistic algorithms $\mathcal{A}$ which on input n (in unary) and $x \in G_n$ return either $\mathrm{dlog}_{g_n} x$ or "failure". Then $\sigma_{\mathcal{A}}(n)$ is the probability that on such an input $\mathcal{A}$ returns $\mathrm{dlog}_{g_n} x$. The probability space consists of a uniform choice $x \xleftarrow{\boxtimes} G_n$ and the internal randomization of $\mathcal{A}$. We use the notion of negligible functions from Section 15.16.

DEFINITION 4.4. *In the above notation, DL_G is* hard *if for all polynomial-time $\mathcal{A}$, $\sigma_{\mathcal{A}}$ is negligible in n.*

4.2. The ElGamal encryption scheme

In Section 2.6, we discuss the Diffie-Hellman protocol for exchanging secret keys. It works in a cyclic group. ElGamal (1985a) has adapted this system so that one can securely exchange messages with it. We now describe this public key cryptosystem.

PROTOCOL 4.5. ElGamal encryption scheme.

Set-up:

Input: 1^n.

1. Choose a finite cyclic group $G = \langle g \rangle$ with d elements, where d is an n-bit integer, and a generating element g. These data, including a description of G, are made public.

Key generation:

Output: $(\mathsf{sk}, \mathsf{pk})$.

2. Secret key $\mathsf{sk} = b \xleftarrow{\boxtimes} \mathbb{Z}_d$. Public key $\mathsf{pk} = B \leftarrow g^b \in G$.

Encryption:

Input: Plaintext $x \in G$.
Output: Ciphertext $\mathsf{enc}_{\mathsf{pk}}(x) \in G^2$.

3. Choose a secret session key $a \xleftarrow{\boxtimes} \mathbb{Z}_d$ at random.
4. Public session key $A \leftarrow g^a \in G$, and $k \leftarrow B^a$.
5. $y \leftarrow x \cdot k \in G$.
6. Return $\mathsf{enc}_{\mathsf{pk}}(x) = (y, A)$.

Decryption:

Input: Arbitrary $(y, A) \in G^2$.

Output: Decryption $\mathsf{dec}_\mathsf{sk}(y, A) \in G$.

7. Session key $k \leftarrow A^b$, inverse $k^{-1} \in G$ of the shared session key.
8. $z \leftarrow y \cdot k^{-1}$.
9. Return z.

Knowing sk, one can compute $k = B^a = g^{ab} = A^b$, but presumably not without this knowledge (see below).

EXAMPLE 2.18 CONTINUED. Let $G = \mathbb{Z}_{2579}^\times$ and $g = 2$. BOB has published his public key $B = g^b = 949$. ALICE wants to encrypt the plaintext MY SECRET in ECB mode (Section 2.8) and writes it in the familiar numerical notation $(12, 24, 18, 4, 2, 17, 4, 19)$, where A corresponds to 0, etc. Spaces are often ignored in cryptography. ALICE sends two letters b_1, b_2 as one unit by computing $26b_1 + b_2$. She uses different secret session keys to avoid statistical attacks, and sends (y, A) as her message.

| plaintext | | | encryption by ALICE | | | decryption by BOB | | |
letters x		a	$A = g^a$	$k = B^a$	$y = x \cdot k$	$k = A^b$	k^{-1}	yk^{-1}
MY	336	2111	1617	2430	1516	2430	1402	336
SE	472	367	734	2474	2020	2474	2186	472
CR	69	351	832	203	1112	203	559	69
ET	123	161	1355	1483	1879	1483	873	123

$\diamondsuit$

We discuss the three basic properties: correctness, efficiency, and security. BOB has set up the system, and ALICE sends the message $x \in G$ to him.

Although we have written the input to the decryption algorithm as (y, A), it works for all inputs in G^2, not just encryptions. But for the correctness proof, we assume that the input actually is $\mathsf{enc}_\mathsf{pk}(x)$. As in the Diffie-Hellman key exchange, BOB correctly determines the secret session key as $A^b = (g^a)^b = (g^b)^a = B^a = k$. Because G is a group, BOB can invert any element like $k \in G$. Hence $z = y \cdot k^{-1}$ is well-defined and $z = y \cdot k^{-1} = x \cdot B^a \cdot (B^a)^{-1} = x$. Thus the system works correctly.

ALICE and BOB have to perform the operations as in the Diffie-Hellman key exchange protocol (Section 2.6), plus one division and one multiplication. These can all be done efficiently. Indeed, for $G = \mathbb{Z}_N^\times$ with an n-bit integer N, a multiplication in G can be done with $O(n^2)$ bit operations, and similarly for inversion using the inversion Algorithm 15.24, and hence also division. Most expensive are the four exponentiations, at $O(n^3)$ operations each, using the repeated squaring Algorithm 15.48.

Hence the total cost is $O(n^3)$ in this case. For a general group G, we find a cost of $O(n \mathsf{M}_G)$ bit operations.

For the security, we show that breaking ElGamal's cryptosystem and solving the Diffie-Hellman Problem 2.21 are polynomially equivalent problems, in the sense of Definition 3.27 (ii). We use the strongest notions of breaking and solving, namely for all inputs. See Chapter 9 for more general notions.

THEOREM 4.6. *For any group G, breaking the ElGamal cryptosystem and solving the Diffie-Hellman Problem 2.21 can be reduced to each other in polynomial time, namely, with two group operations.*

PROOF. "$\leq_p$": We are given the public or transmitted values g, B, y, and A in G and have to compute x. From a solver of Diffie-Hellman, we obtain $k = g^{ab}$ from $A = g^a$ and $B = g^b$. Then $x = yk^{-1}$ is immediate.

"$\geq_p$": We are given some $A = g^a$ and $B = g^b$ in G, and need to compute g^{ab}. This quantity is called $k = yx^{-1}$ in the ElGamal system. In order to solve DH_G for A and B, we publish B as BOB's public key, choose some $y \in G$, say, $y = 1$, and send (y, A). Then an ElGamal breaker will return x, and we have $k = yx^{-1}$. □

ElGamal encryption has a homomorphic property similar to (2.4) for RSA. Suppose that $w = \mathsf{enc}_{\mathsf{pk}}(x) = (xB^a, g^a)$ and $w' = \mathsf{enc}_{\mathsf{pk}}(x') = (x'B^{a'}, g^{a'})$. Then with componentwise multiplication,

$$(4.7) \qquad w \cdot w' = (xx'B^{a+a'}, g^{a+a'})$$

is among the possible encryptions of xx', with session key $k = g^{(a+a')b}$.

An important distinction from AES and RSA is that for a fixed secret key, a plaintext has many possible encryptions, one for each session key a. Such randomized encryptions often enjoy better security properties than deterministic ones. See Example 9.22 for RSA and Section 9.6 for further security properties of ElGamal encryption.

4.3. Baby-step giant-step algorithm

In this and the next sections, we study several algorithms for DL_G that all use time about $\sqrt{d}$ for some group G, where $d = \#G$. The space required is initially also $\sqrt{d}$, but gets cut down in several improvement steps to constant. The basic idea in these algorithms is to find a *collision*, namely, a single group element which is computed in two different ways. Tracing the computation then gives us information about the discrete logarithm.

ALGORITHM 4.8. Baby-step giant-step algorithm for the discrete logarithm.

Input: A cyclic group $G = \langle g \rangle$ with d elements, and a group element $x \in G$.

Output: $\mathrm{dlog}_g x$.

1. $m \longleftarrow \lceil \sqrt{d} \rceil$.
2. Baby steps: compute and store $x, xg, xg^2, \ldots, xg^m$ in a table.
3. Giant steps: compute $g^m = xg^m \cdot x^{-1}$, g^{2m}, g^{3m}, ... until one of them, say g^{im}, equals an element in the table, say xg^j.
4. Return $im - j$ in $\mathbb{Z}_d$.

EXAMPLE 4.9. We take a group G with $d = 20$ elements. We might have $G = \mathbb{Z}_{20}$ with addition, or $G = \mathbb{Z}_{25}^{\times}$ with multiplication, since $\varphi(25) = 4 \cdot 5$. Let us take the latter representation. Now $g = 2 \in G = \mathbb{Z}_{25}^{\times}$ is a generator by Corollary 15.59, since $2^{20/2} = 2^{10} = 24 \neq 1$ and $2^{20/5} = 2^4 = 16 \neq 1$ in G. In order to compute the discrete logarithm of $x = 17$, we have $m = \lceil \sqrt{20} \rceil = 5$, and perform the following computations.

k	baby steps xg^k	giant steps g^{km}
0	17	1
1	9	7
2	18	24
3	11	18
4	22	1
5		7
		. . .

In the third giant step, we find the collision $xg^2 = 18 = g^{3 \cdot 5}$, and hence $\mathrm{dlog}_2 17 = 3 \cdot 5 - 2 = 13$. We check that indeed $2^{13} = 17$ in $\mathbb{Z}_{25}^{\times}$.   ◇

THEOREM 4.10. For any group G with d elements, the baby-step giant-step method solves DL_G with at most $2m$ group operations and space for m elements of G, where $m = \lceil \sqrt{d} \rceil$.

PROOF.   For correctness of the algorithm, we have $g^{im} = xg^j$ at the collision in step 3, so that $x = g^{im-j}$, in other words, $\mathrm{dlog}_g x = im - j$ in $\mathbb{Z}$. Thus the result returned is correct.

   To see that step 4 is always reached, we let $x = g^a$ with $0 \leq a < d$. Using division with remainder (Section 15.3), we write $a = bm + c$ with $0 \leq c < m$. Then also $b = (a - c)/m \leq a/m < d/m \leq m$, and the algorithm will find $a = im - j$ when $i = b + 1 \leq m$ and $j = m - c$.

The baby steps take m group operations, and the giant steps at most m operations, since $j = m - c \leq m$. □

Here and in other algorithms, we ignore the cost of control operations like index arithmetic, searching the table, loop control, and intermediate storage during an operation. This is justified by the fact that in practice their cost is at most of the same order of magnitude. For example, if we use a sorted table for the baby results, then a single entry or look-up costs about $O(\log d)$ bit operations, while a single group operation will cost at least as much in terms of bit operations. This will be made more formal in Section 4.9, where we deal with group circuits (or generic algorithms) whose cost is solely measured in terms of group operations.

4.4. The birthday attack

This discrete logarithm algorithm takes its name from the *birthday problem*, discussed in Section 3.11. That observation can be turned into a discrete logarithm computation in an obvious way. We simply keep computing xg^i and g^j for random i and j until a collision $xg^i = g^j$ occurs; then we have $\mathrm{dlog}_g x$, as in the baby-step giant-step method: $\mathrm{dlog}_g x = j - i$ in $\mathbb{Z}_d$.

The algorithm we now describe is not of practical interest in itself. It serves as a preparation for Pollard's rho method in the next section, where we modify the random choices and succeed in reducing the required space to be of constant size. Also, it explains the term $\sqrt{d}$ which is pervasive in discrete logarithm computations.

ALGORITHM 4.11. Birthday algorithm for discrete logarithm.

Input: A cyclic group $G = \langle g \rangle$ with d elements, and a group element $x \in G$.

Output: $\mathrm{dlog}_g x$.

1. $X, Y \longleftarrow \varnothing$.
2. Do step 3 until a collision of X and Y occurs.
3.   Choose uniformly at random a bit $b \longleftarrow^{\boxtimes} \{0, 1\}$ and $i \longleftarrow^{\boxtimes} \{0, \ldots, d-1\}$. Add xg^i to X if $b = 0$ and g^i to Y if $b = 1$, and remember the index i.
4. If $xg^i = g^j$ for some $xg^i \in X$ and $g^j \in Y$, then return $j - i$ in $\mathbb{Z}_d$.

THEOREM 4.12. *The algorithm works correctly as specified. Its expected time is* $O(\sqrt{d} \log d)$ *multiplications in* G, *with expected space for* $O(\sqrt{d})$ *elements of* G.

PROOF. The elements making up $X \cup Y$ are independent and uniformly random in G. So we expect a collision (i, j) after $O(\sqrt{d})$ steps, by Theorem 3.41, where i and j are the indices of the colliding elements u and v in G. If both u and v are in X, or both are in Y, we have made no progress. But otherwise we have, say, $u \in X$ and $v \in Y$, $xg^i = u = v = g^j$, and $\mathrm{dlog}_g x = j - i$ in $\mathbb{Z}_d$. The chances for the two cases of equal and of different membership are equal, so that the expected number of random choices i is at most twice the bound in Theorem 3.41, that is, at most $4 + \sqrt{2\pi d} \in O(\sqrt{d})$. A single xg^i or g^i costs $O(\log d)$ group operations. $\square$

In order to "remember i", one will store (xg^i, i) in X. We have avoided this in the description of the algorithm for simplicity.

The proof would not work for putting each xg^i into X and each g^i into Y, because then the elements of $X \cup Y$ would not be independent. Namely, with each element $z \in X \cup Y$, either xz or $x^{-1}z$ would also be an element, and random sets are highly unlikely to have this property.

4.5. The Pollard rho algorithm

The discrete logarithm algorithm based on the birthday paradox is conceptionally useful. It can be implemented with expected time $O(\sqrt{d} \log d)$ and space $O(\sqrt{d})$ for groups with d elements. In a series of steps, we now transform this method into a practical algorithm, which still uses time $O(\sqrt{d} \log d)$ but only constant space. This is analogous to the development in Sections 3.11 and 3.12 for factoring integers.

In the first intermediate step, we replace the uniformly random choice of the next group element by a uniformly random choice between three options. Thus the amount of random bits used per iteration decreases from $\log d$ to constant. In the second step, we take a partition of the group into three subsets of roughly equal size and replace the random choice by testing in which of the three subsets the previous element lies. Except for the initial choice, there is no more randomness involved.

In the third step, we replace the single iteration for finding collisions by two iterations running in parallel, one twice as fast as the other. This trick, due to Floyd, reduces the storage required from $O(\sqrt{d})$ to constant. The resulting final and glorious version is Pollard's rho algorithm, an important practical tool for discrete logarithm computation and used in the generic discrete logarithm Algorithm 4.28.

We lose one pleasant feature in these transformations: the ability to prove that the algorithm works correctly, with high probability. This issue is discussed at the end of the section. Pollard invented his rho method for factoring integers in 1975 (see Section 3.11), and for discrete logarithms in 1978. We now describe the latter.

We have a cyclic group $G = \langle g \rangle$ with d elements, and an element $x = g^a$ of G. Our task is to calculate $a = \text{dlog}_g\, x$ from g and x.

For the first modification step, we choose a sequence $b_0, b_1, \ldots \xleftarrow{\boxdot}$ $\{0, 1, 2\}$ of uniformly and independently distributed random "trits" b_k, choose $u_0, v_0 \xleftarrow{\boxdot} \mathbb{Z}_d$ at random and start with $y_0 = x^{u_0} g^{v_0}$. Then we calculate $y_1, y_2, \ldots$ in G by

$$(4.13) \qquad y_k = \begin{cases} x \cdot y_{k-1} & \text{if } b_{k-1} = 0, \\ y_{k-1}^2 & \text{if } b_{k-1} = 1, \\ g \cdot y_{k-1} & \text{if } b_{k-1} = 2, \end{cases}$$

until we find a collision $y_i = y_j$ with $i \neq j$. We claim that then we can (probably) calculate $\text{dlog}_g\, x$. We define the *trace exponents* as linear polynomials $\tau_0, \tau_1, \ldots$ in $\mathbb{Z}_d[t]$ by $\tau_0 = u_0 t + v_0$ and for $k \geq 1$

$$(4.14) \qquad \tau_k = \begin{cases} \tau_{k-1} + t & \text{if } b_{k-1} = 0, \\ 2\tau_{k-1} & \text{if } b_{k-1} = 1, \\ \tau_{k-1} + 1 & \text{if } b_{k-1} = 2. \end{cases}$$

Then we check that $y_k = x^{u_k} g^{v_k} = g^{u_k a + v_k} = g^{\tau_k(a)}$ for all k, where $\tau_k = u_k t + v_k$. In a collision, we have

$$x^{u_i} g^{v_i} = y_i = y_j = x^{u_j} g^{v_j},$$

$$g^{(u_i - u_j)a + v_i - v_j} = x^{u_i - u_j} g^{v_i - v_j} = 1,$$

$$(u_i - u_j)a + v_i - v_j = 0 \text{ in } \mathbb{Z}_d.$$

If $\gcd(u_i - u_j, d) = 1$, then $u_i - u_j$ is invertible in $\mathbb{Z}_d$,

$$a = (v_j - v_i) \cdot (u_i - u_j)^{-1} \text{ in } \mathbb{Z}_d,$$

and the discrete logarithm has been computed. We explain in the following remark that, even though the gcd condition need not always be satisfied, the result still is sufficiently strong to consider the discrete logarithm problem to be solved.

REMARK 4.15. *We have the following scenario, which we will also encounter in other situations: we want to compute some discrete logarithm a modulo the (known) group order d, and have integers u and v with*

$$(4.16) \qquad au = v \text{ in } \mathbb{Z}_d.$$

We let $w = \gcd(u, d)$. If $w = 1$, then there exists an inverse u^{-1} of u in $\mathbb{Z}_d$, and

$$(4.17) \qquad a = v \cdot u^{-1} \text{ in } \mathbb{Z}_d.$$

But if $w > 1$, then this does not work. We now make some remarks about this situation.

(i) *There are exactly $\varphi(d)$ many values of $u \in \mathbb{Z}_d$ with $w = 1$, namely the u in $\mathbb{Z}_d^\times$. For an average value of d, we have from Section 3.3*

$$\frac{\varphi(d)}{d} \approx \frac{6}{\pi^2} \approx 0.60792\,71019,$$

so that for about 61% of the possible values of u, (4.17) is applicable. In the worst case, we have $\frac{\varphi(d)}{d} \geq \frac{1}{3\ln\ln d}$; see the notes to Section 3.3. This is still a fairly large fraction of the possible values of u.

(ii) *If an equation like (4.16) can be found for most a, and as long as u is random or depends on the input and there is no general reason why w should be large, then this is sufficient to consider the discrete logarithm problem in G to be broken.*

(iii) *According to the recommendation in Figure 4.6, the group order d should be prime. Then 0 is the only noninvertible element in $\mathbb{Z}_d$.*

(iv) *From (4.16), we have $v = au - kd$ for some integer k, and v is also divisible by w. We let $d' = d/w$, divide the equation by w, and find*

$$a\frac{u}{w} = \frac{v}{w} \text{ in } \mathbb{Z}_{d'}, \quad \gcd\left(\frac{u}{w}, d'\right) = 1.$$

We take an inverse $(u/w)^{-1}$ in $\mathbb{Z}_{d'}$, and find some integer b with

$$b = \frac{v}{w}\left(\frac{u}{w}\right)^{-1} \text{ in } \mathbb{Z}_{d'}.$$

Then

$$a = b + id' \text{ in } \mathbb{Z}_d$$

for some integer i with $0 \leq i < w$. Now if w is a "small" number, we can simply try out whether $x = g^{b+id'}$ for any of these i, and then find $a = \mathrm{dlog}_g x$. This is illustrated below in Example 4.9 continued.

A particular case is when $G = \mathbb{Z}_p^\times$ and $d = p - 1 = 2q$ for a prime q. Then q is called a Sophie Germain prime. There are now four possibilities for w: 1, 2, q, and $2q$. The latter two are improbable, and for $w = 2$ the strategy just described is applicable. This applies similarly to the safe primes of Section 3.4.

The sequence $y_0, y_1, y_2, \ldots$ is not random, since for a random sequence none of the three relations between the y_i's in (4.13) is likely to hold. But in practice, a collision seems to occur among these values as quickly as in a random sequence, namely after $\mathrm{O}(\sqrt{d})$ steps. Assuming that this holds, we now have a probabilistic discrete logarithm computation with expected time and space $O(\sqrt{d})$.

The third and final version replaces these random choices by a deterministic one, and achieves the goal of reducing the storage to a constant amount. The idea is to partition G into three sets S_0, S_1, and S_2 of roughly equal size and determine b_k according to whether y_{k-1} is in S_0, S_1, or S_2.

If p is a prime and G a subgroup of $\mathbb{Z}_p^\times = \{1, \ldots, p-1\}$, including the important case $G = \mathbb{Z}_p^\times$, we might put the x in G with $0 \leq x < p/3$ into S_0, the ones with $p/3 \leq x < 2p/3$ into S_1, and the rest into S_2. The three sets are of roughly equal size.

Then we take random u_0, v_0, and y_0 as above, and calculate $y_k \in G$, $u_k, v_k \in \mathbb{Z}_d$, and τ_k for $k \geq 1$ as in (4.13) and (4.14), using

$$b_k = \begin{cases} 0 & \text{if } y_{k-1} \in S_0, \\ 1 & \text{if } y_{k-1} \in S_1, \\ 2 & \text{if } y_{k-1} \in S_2. \end{cases}$$

When a collision occurs, we are likely to have found $\text{dlog}_g x$, as above.

Furthermore, we use Floyd's trick, as explained in Section 3.11. We recall that the idea is to run two iterations in parallel, the second one twice as fast as the first one. The faster one is likely to catch up with the slower one in the loop after $O(\sqrt{d})$ steps. We do not need to store the intermediate values.

ALGORITHM 4.18. The Pollard rho algorithm for discrete logarithms.

Input: A cyclic group $G = \langle g \rangle$ of order d, a partition $G = S_0 \cup S_1 \cup S_2$ into three disjoint parts of roughly equal size, and $x \in G$.
Output: $\text{dlog}_g x$, or "failure".

1. Define the iteration function $\mathcal{P}$ by $\mathcal{P}(z, \rho) = (z^*, \rho^*)$, where $z, z^* \in G$, $\rho, \rho^* \in \mathbb{Z}_d[t]$, and

$$z^* = \begin{cases} x \cdot z & \text{if } z \in S_0, \\ z^2 & \text{if } z \in S_1, \\ g \cdot z & \text{if } z \in S_2, \end{cases} \qquad \rho^* = \begin{cases} \rho + t & \text{if } z \in S_0, \\ 2\rho & \text{if } z \in S_1, \\ \rho + 1 & \text{if } z \in S_2. \end{cases}$$

2. $k \longleftarrow 0$,
   $u_0, v_0 \xleftarrow{\boxtimes} \mathbb{Z}_d$,
   $x_0, y_0 \longleftarrow x^{u_0} g^{v_0}$,
   $\sigma_0, \tau_0 \longleftarrow u_0 t + v_0$.
3. Do step 4 until $x_k = y_k$.
4. $k \longleftarrow k + 1$.
       Calculate $x_k, y_k \in G$ and $\sigma_k, \tau_k \in \mathbb{Z}_d[t]$ by
       $(x_k, \sigma_k) \leftarrow \mathcal{P}(x_{k-1}, \sigma_{k-1})$, $(y_k, \tau_k) \leftarrow \mathcal{P}(\mathcal{P}(y_{k-1}, \tau_{k-1}))$.

k	x_k	y_k	σ_k	τ_k
0	8	8	$12t + 7$	$12t + 7$
1	11	21	$13t + 7$	$6t + 14$
2	21	14	$6t + 14$	$12t+10$
3	17	17	$6t + 15$	$4t + 1$

Figure 4.2: Pollard rho in $\mathbb{Z}_{25}^{\times}$, as in Example 4.9 continued.

5. Let $\sigma_k = ut + v$ and $\tau_k = u't + v'$, with $u, u', v, v' \in \mathbb{Z}_d$. If $\gcd(u - u', d) = 1$ in $\mathbb{Z}$, then return $(v' - v) \cdot (u - u')^{-1}$ in $\mathbb{Z}_d$, else return "failure".

The coefficients of σ_k and τ_k are reduced modulo d whenever necessary. A basic assumption is that x_0, x_1, x_2, ... behave like independent uniformly random elements in G. Then so do the y_k, and we expect a collision after about $\sqrt{d}$ steps, as in the birthday attack. In fact, one can think of this as a storage-efficient implementation of the birthday attack.

THEOREM 4.19. *Let G be a cyclic group of order d. Then the Pollard rho algorithm 4.18, with Floyd's trick, finds a discrete logarithm in G with an expected number of $\mathrm{O}(\sqrt{d})$ group operations, provided that the sequence $x_0, x_1, x_2, \ldots$ behaves randomly. Space is required for two elements of G and four elements of $\mathbb{Z}_d$.*

EXAMPLE 4.9 CONTINUED. We have $g = 2 \in G = \mathbb{Z}_{25}^{\times}$, $d = 20$, and $x = 17$. As suggested above, we use the partition $S_0 = \{1, 2, 3, 4, 6, 7, 8\}$, $S_1 = \{9, 11, 12, 13, 14, 16, 17\}$, and $S_2 = \{18, 19, 21, 22, 23, 24\}$ of G, with $7, 7$, and 6 elements, respectively. Our random choice is $u_0 = 12$ and $v_0 = 7$, so that $\sigma_0 = 12t + 7$.

Figure 4.2 gives a trace of Algorithm 4.18. We find the collision $x_3 = y_3 = 17$ with $\sigma_3 = 6t+15$ and $\tau_3 = 4t+1$ in $\mathbb{Z}_{20}[t]$. Then $u-u' = 6-4 = 2$ and $w = \gcd(2, 20) = 2 \neq 1$. The algorithm as stated returns "failure". But we persist and compute $a = \mathrm{dlog}_2 17$ according to Remark 4.15 (iv) as a root of $\sigma_3 - \tau_3$. Namely, $d' = d/w = 20/2 = 10$, and dividing $\sigma_3 - \tau_3$ by 2, we have

$$(6t + 15 - (4t + 1))/2 = (2t + 14)/2 = t + 7.$$

The quantity called u in the Remark is now $\tilde{u} = 2$, and $(\tilde{u}/w)^{-1} =$

$(2/2)^{-1} = 1^{-1} = 1$ in $\mathbb{Z}_{d'} = \mathbb{Z}_{10}$. Then

$$b = \frac{-v}{w} \cdot (\frac{u}{w})^{-1} = -\frac{14}{2} \cdot 1^{-1} = -7 = 3 \text{ in } \mathbb{Z}_{10},$$

and the possible values for a are $a = b + id' = 3 + 10i$ in $\mathbb{Z}_{20}$, for $0 \leq i < 2$. Thus $a \in \{3, 13\}$, and a check reveals that $\mathrm{dlog}_2 17 = 13$.

Hopefully this failure does not happen often. But it does! For the nine values $4, 6, 13, 17, 18, 19, 21, 22, 23$ for x in $\mathbb{Z}_{25}^{\times}$, all 400 initial choices of σ_0 lead to failure. For the other eleven values, there are some successful σ_0. But if $x = 3$, then only $\sigma_0 = 11$ works. In groups with large prime order, one does not observe this somewhat pathological behavior. $\Diamond$

Our analysis of Pollard's rho method assumes that the transition $x_{k-1} \mapsto x_k$ is random, so that the birthday theorem can be applied. Since x_k is one of $x_{k-1}x$, $x_{k-1}g$, or x_{k-1}^2, this assumption does not hold. Indeed, it was an open problem for decades to actually prove the claims about the algorithm, finally solved by Miller & Venkatesan (2009) and Kim et al. (2010) with methods from the theory of random walks on graphs and rapidly mixing Markov chains. The corresponding problem for integer factorization (Section 3.12) is still unsolved.

4.6. Chinese distribution of discrete logarithms

In the previous three sections, we have described discrete logarithm computations in a group G with d elements that all use about $\sqrt{d}$ steps.

When d is a prime number, it seems impossible to do better in general. This is explained in Section 4.9 below, and the security parameters in group-based cryptography are based on this assumption.

But in this section and the next, we provide algorithms that show that the difficulty of the discrete logarithm in a group really does at most depend on the largest prime factor of its order. If the order is large but has only small prime factors, then the discrete logarithm is easy to compute.

This proceeds in two steps. In the first step, due to Pollard (1978) and Pohlig & Hellman (1978), we reduce efficiently the discrete logarithm in G to discrete logarithms in groups whose orders are the prime power factors of the group order d. In the second step, due to Pohlig & Hellman (1978), we reduce efficiently the discrete logarithm in a group whose order is a prime power p^e to discrete logarithms in groups of order p. The latter then is the base case, to which we may apply the algorithms from Sections 4.3 through 4.5.

As usual, we have a finite cyclic group $G = \langle g \rangle$ with $d = \#G$ elements. We take a factorization

(4.20) $d = q_1 \cdots q_r$

of d into pairwise coprime $q_1, \ldots, q_r$. For example, the q_i might be the various prime power factors of d. Theorem 15.60 (iv) describes for a divisor q of d the effect of the map $\pi_{d/q} \colon G \to G$, which raises the group elements to the power d/q. In terms of the exp-log correspondence of Figure 4.1, we consider $e = d/q$ and the "multiplication by e" map

$$\mu_e \colon \mathbb{Z}_d \longrightarrow e \cdot \mathbb{Z}_d \cong \mathbb{Z}_{d/e} = \mathbb{Z}_q,$$
$$a \longmapsto ea.$$

This map is a well-defined homomorphism of the additive groups (but not of the rings), since for integers a and b with $a = b$ in $\mathbb{Z}_d$, we have $a = b + td$ for some $t \in \mathbb{Z}$, and

$$ea = eb + t \cdot ed,$$
$$ea = eb \text{ in } e \cdot \mathbb{Z}_d.$$

It is quite intuitive that the factorization (4.20) provides a "Chinese remainder" decomposition of G. An interesting feature is that we can describe the decomposition factors explicitly as subgroups of G. We start with the special case of two factors, showing that discrete logarithms in the small subgroups yield the discrete logarithm in the large group.

LEMMA 4.21. *Suppose that* $d = q_1 q_2$ *with coprime* q_1 *and* q_2, *and that* $a_i = \operatorname{dlog}_{g_i} x_i$ *in* $\mathbb{Z}_{d/q_i}$, *where* $g_i = g^{d/q_i}$ *and* $x_i = x^{d/q_i} \in \pi_{d/q_i}(G)$, *for* $i = 1, 2$. *Then* $\operatorname{dlog}_g x = a_i$ *in* $\mathbb{Z}_{q_i}$ *for* $i = 1, 2$.

PROOF. We set $e_i = d/q_i$, so that $g_i = g^{e_i} = \pi_{e_i}(g)$ for $i = 1, 2$, and consider $\mu_{e_i} \colon \mathbb{Z}_d \longmapsto \mathbb{Z}_{q_i}$ as above. The exp-log isomorphisms in Figure 4.1 lead to Figure 4.3. If $a = \operatorname{dlog}_g x$, then $x = g^a$ and $x_i = x^{e_i} = (g^{e_i})^a = g_i^a$, so that $a_i = \operatorname{dlog}_{g_i} x_i = a$ in $\mathbb{Z}_{q_i}$ for $i = 1, 2$. $\qquad\square$

Thus we can compute $\operatorname{dlog}_g x$ from a_1 and a_2 by the Chinese Remainder Theorem 15.44.

EXAMPLE 4.9 CONTINUED. We have $G = \mathbb{Z}_{25}^\times = \langle 2 \rangle$ with $d = \#G = 20 = 4 \cdot 5$, so that $q_1 = 4$ and $q_2 = 5$, and $x = 17 \in G$. Additively, μ_5 maps $\mathbb{Z}_{20}$ to $5 \cdot \mathbb{Z}_{20} = \{0, 5, 10, 15, 20, \ldots, 95\} = \{0, 5, 10, 15\} \cong \mathbb{Z}_4$ as a subgroup of $\mathbb{Z}_{20}$. Multiplicatively, we have $g_1 = 2^{20/4} = 7$ and $g_2 = 2^{20/5} = 16$, and the two subgroups

$$S_1 = \langle 2^{20/4} \rangle = \{1, 7, 24, 18\} \text{ and } S_2 = \langle 2^{20/5} \rangle = \{1, 16, 6, 21, 11\}$$

have 4 and 5 elements, respectively. See Figure 4.3. The Chinese remainder algorithm for the discrete logarithm of 17 first computes the two

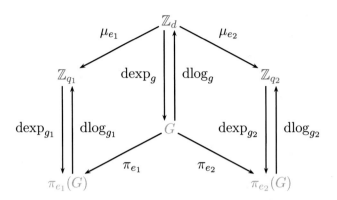

Figure 4.3: Chinese remainder computation for DL_G with $d = q_1 q_2$.

constituents of x in S_1 and S_2: $x_1 = 17^{20/4} = 7$ and $x_2 = 17^{20/5} = 21$. We can read off the discrete logarithms in S_1 and S_2: $a_1 = \mathrm{dlog}_{g_1} x_1 = 1$ and $a_2 = \mathrm{dlog}_{g_2} x_2 = 3$. With the Chinese Remainder Algorithm, we find $a = 13$, which satisfies $a = 1$ in $\mathbb{Z}_4$ and $a = 3$ in $\mathbb{Z}_5$. We are quite happy to have found the same result as with baby and giant steps and Pollard's rho method in Example 4.9. $\Diamond$

We now describe the general situation. This can be applied to the prime power factorization of d.

LEMMA 4.22. *Let* $d = q_1 \cdots q_r$ *be a factorization of* $d = \#G$ *into pairwise coprime factors, with* $G = \langle g \rangle$ *a cyclic group as above, let* $x \in G$ *and for* $i \leq r$, *let* $S_i = \{x^{d/q_i} : x \in G\}$ *and* $T_i = \{x \in G : x^{q_i} = 1\}$. *Then the following hold.*

(i) $S_i = T_i$ *is a subgroup with* q_i *elements, generated by* g^{d/q_i}, *and the map*

$$G \to S_1 \times S_2 \times \cdots \times S_r,$$
$$y \mapsto (y^{d/q_1}, \ldots, y^{d/q_r}),$$

is an isomorphism.

(ii) *If* $x = g^a$, $i \leq r$, *and* $a = a_i$ *in* $\mathbb{Z}_{q_i}$, *then* $x^{d/q_i} = (g^{d/q_i})^{a_i}$.

(iii) *If* $a_i = \mathrm{dlog}_{g^{d/q_i}} x^{d/q_i}$ *and* $a \in \mathbb{Z}_d$ *satisfies* $a = a_i$ *in* $\mathbb{Z}_{q_i}$ *for all* $i \leq r$, *then* $a = \mathrm{dlog}_g x$.

PROOF. The Chinese Remainder Theorem 15.44 says that

$$\mathbb{Z}_d \cong \mathbb{Z}_{q_i} \times \cdots \times \mathbb{Z}_{q_r},$$

since $\gcd(q_i, q_j) = 1$ for $i \neq j$. Theorem 15.60 implies that $S_i = T_i$. For (ii) and (iii), we consider the exp-log correspondence in Figure 4.1 and the corresponding statements at the secret level of $\mathbb{Z}_d$. These are true, again by the Chinese Remainder Theorem, and hence also the claims of the lemma at the public level. □

ALGORITHM 4.23. Chinese remaindering for discrete logarithms.

Input: A cyclic group $G = \langle g \rangle$ of order $d = \#G$, and $x \in G$.
Output: $a = \text{dlog}_g x$.

1. Compute the prime power factorization (4.20) of d.
2. For each $i \leq r$, do steps 3 and 4.
3. Compute $g_i = g^{d/q_i}$ and $x_i = x^{d/q_i}$, with the repeated squaring Algorithm 15.48.
4. Compute the discrete logarithm $a_i = \text{dlog}_{g_i} x_i \in \mathbb{Z}_{q_i}$ in $S_i = \langle g_i \rangle$.
5. Combine these "small" discrete logarithms via the Chinese Remainder Algorithm 15.45 to find the unique $a \in \mathbb{Z}_d$ so that $a = a_i$ in $\mathbb{Z}_{q_i}$ for all $i \leq r$.

THEOREM 4.24. Let G be a cyclic group of n-bit order d. Then Algorithm 4.23 computes a discrete logarithm in G at the following cost:

(i) factoring the integer d,

(ii) one discrete logarithm in each of the groups $S_1, \ldots, S_r$,

(iii) $O(n^2)$ operations in G,

(iv) $O(n^2)$ bit operations.

PROOF. The cost for computing all g_i and x_i is at most $2r \cdot 2n \leq 4n^2$ operations in G. Then come $r \leq n$ discrete logarithm calls in the S_i's, and finally the Chinese remaindering, with a total of $O(n^2)$ bit operations, by Theorem 15.46. □

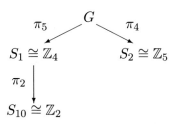

Figure 4.4: Discrete logarithm in a group with 20 elements.

4.7. The Pohlig-Hellman algorithm

We have reduced the discrete logarithm problem in arbitrary groups to groups whose order is a prime power. We now go one step further and reduce it to groups of prime order, corresponding to the lower left arrow in Figure 4.4. Thus our set-up is that $G = \langle g \rangle$ is a cyclic group of order p^e, where p is a prime and $e \geq 2$ an integer. We consider the p^{e-1}th power mapping $\pi_{p^{e-1}} \colon x \mapsto x^{p^{e-1}}$, which maps G to its subgroup $H = \langle g^{p^{e-1}} \rangle = \{x^{p^{e-1}} : x \in G\}$ of order p, according to Theorem 15.60 (iv). Furthermore, an element $y \in G$ is in H if and only if $y^p = 1$.

We will do all our discrete logarithm computations in H. Suppose that we want $a = \operatorname{dlog}_g x$ for some $x \in G$. We compute its representation $a = a_{e-1}p^{e-1} + \cdots + a_0$ in base p, with all $a_i \in \{0, \ldots, p-1\}$, as follows. We abbreviate $h = g^{p^{e-1}}$. The power $x_0 = x^{p^{e-1}}$ is in H, and we compute $a_0 = \operatorname{dlog}_h x_0$, a discrete logarithm in $H = \langle h \rangle$. Now we set $x_1 = x^{p^{e-2}} \cdot g^{-a_0 \cdot p^{e-2}}$. Then

$$x_1^p = x^{p^{e-1}} \cdot g^{-a_0 \cdot p^{e-1}} = x_0 \cdot h^{-a_0} = 1.$$

Thus $x_1 \in H$, and we compute its discrete logarithm $a_1 = \operatorname{dlog}_h x_1$ in H. Continuing in this fashion, we obtain the following algorithm, which uses an unspecified discrete logarithm routine for DL_H.

ALGORITHM 4.25. Pohlig & Hellman.

Input: A cyclic group $G = \langle g \rangle$ with p^e elements, where p is a prime and $e \geq 2$ an integer, and $x \in G$.
Output: $\operatorname{dlog}_g x$.

1. Compute $h = g^{p^{e-1}}$ and set $y_{-1} = 1 \in G$.
2. For i from 0 to $e-1$ do steps steps 3 – 5.
3. $x_i \leftarrow (x \cdot y_{i-1})^{p^{e-i-1}}$. [Then $x_i \in H = \langle h \rangle$.]
4. $a_i \leftarrow \operatorname{dlog}_h x_i$.

5. $y_i \leftarrow y_{i-1} \cdot g^{-a_i p^i}$.

6. Return $a = a_{e-1} p^{e-1} + \cdots + a_0$.

THEOREM 4.26. *The algorithm correctly computes* $\mathrm{dlog}_g\, x$. *It uses* $\mathrm{O}(e^2 \cdot \log p)$ *operations in* G, *plus* e *calls to a subroutine for discrete logarithms in the group* H *with* p *elements.*

PROOF.   Let $b = b_{e-1} p^{e-1} + \cdots + b_0$ be the representation of $\mathrm{dlog}_g\, x$ in base p, and a_i and y_i for $0 \leq i < e$ as computed in the algorithm. We claim the following loop invariants:

$$a_i = b_i, \quad x_i \in H, \quad y_i = g^{-(b_0 + \cdots + b_i p^i)}.$$

We begin with $i = 0$. (The impatient reader may set $x_{-1} = 1$ and $a_{-1} = b_{-1} = 0$, check the invariant for $i = -1$, and then start the inductive step right away.) We have $y_{-1} = 1$, and

$$x_0 = x^{p^{e-1}} = (g^b)^{p^{e-1}} = (g^{p^{e-1}})^b = h^b = h^{b_0 + b_1 p + \cdots} = h^{b_0},$$

since $h^p = 1$. Thus $x_0 \in H$, $a_0 = b_0$ is returned by the discrete logarithm subroutine, and $y_0 = g^{-b_0}$. This shows the claim for $i = 0$.

For the inductive step with $i \geq 1$, we have

$$
\begin{aligned}
x_i &= (x y_{i-1})^{p^{e-i-1}} = (g^{b_0 + b_1 p + \cdots} \cdot g^{-(b_0 + b_1 p + \cdots + b_{i-1} p^{i-1})})^{p^{e-i-1}} \\
&= (g^{b_i p^i + b_{i+1} p^{i+1} + \cdots})^{p^{e-i-1}} = g^{b_i p^{e-1} + b_{i+1} p^e + \cdots} = h^{b_i}.
\end{aligned}
$$

Thus $x_i \in H$, and the discrete logarithm subroutine correctly returns $a_i = b_i$. The claimed property of y_i is clear.

A single jth power in G can be computed with $\mathrm{O}(\log j)$ multiplications, using the repeated squaring Algorithm 15.48, for any positive integer j. Thus all powers in step 1, and each power in steps 3 or 5 can be done with $\mathrm{O}(e \log p)$ operations, which proves the claim about the cost. $\square$

EXAMPLE 4.27. We illustrate Algorithm 4.25 in an example with $p^e = 3^4 = 81$. The group G is the subgroup $G = \langle 4 \rangle \subseteq \mathbb{Z}_{163}^\times$ generated by $g = 4$. We note that 163 is prime and $\#G_{163}^\times = \varphi(163) = 162 = 2 \cdot 81 = 2 \cdot 3^4$. Furthermore, $2^{2 \cdot 27} = 104$ and $2^{81} = -1$ in $\mathbb{Z}_{163}$. Thus 2 is a generator of $\mathbb{Z}_{163}^\times$ by Corollary 15.59, so that the order of 4 in $\mathbb{Z}_{163}^\times$ is 81, by Theorem 15.60 (i).

We have $p = 3$, $e = 4$, and $H = \langle 4^{27} \rangle = \langle h \rangle = \{1, 104, 58\}$ with $h = 104$. We take $x = 60$ and trace the computation of the discrete logarithm $a = \mathrm{dlog}_4\, 60$. Discrete logarithms in H are found by inspection.

$$x_0 = x^{p^3} = x^{27} = 58 \in H, a_0 = \mathrm{dlog}_h\, x_0 = 2, y_0 = y_{-1} \cdot g^{-a_0} = 51,$$

$$x_1 = (xy_0)^9 = 104 \in H, a_1 = \operatorname{dlog}_h x_1 = 1, y_1 = y_0 \cdot g^{-a_1 \cdot 3} = 39,$$

$$x_2 = (xy_1)^3 = 1 \in H, a_2 = \operatorname{dlog}_h x_2 = 0, y_2 = y_1 \cdot g^{-a_2 \cdot 9} = 39,$$

$$x_3 = xy_2 = 58 \in H, a_3 = \operatorname{dlog}_h x_3 = 2.$$

We now have computed $a = a_3 p^3 + a_2 p^2 + a_1 p + a_0 = 2 \cdot 27 + 0 \cdot 9 + 1 \cdot 3 + 2 \cdot 1 = 59$. As a check, we find $4^{59} = 60$ in $\mathbb{Z}_{163}^{\times}$. ◇

Putting together the results so far in this chapter, we have the following algorithm. Suppose that you have a cyclic group G, and you know its order and how to multiply and invert in it, but you know nothing else of relevance about G and you want to compute discrete logarithms in G. Then this is a very reasonable method to use.

ALGORITHM 4.28. Discrete logarithm in a cyclic group.

Input: A finite cyclic group $G = \langle g \rangle$ of order d, and $x \in G$.
Output: $a = \operatorname{dlog}_g x$.

1. Factor $d = p_1^{e_1} \cdots p_r^{e_r}$ into powers of distinct primes, with positive integer exponents $e_1, \ldots, e_r$.
2. For $1 \leq i \leq r$ do the following.
3.   Compute g^{d/q_i} and the image x^{d/q_i} of x in the group $\langle g^{d/q_i} \rangle$ of order q_i, where $q_i = p_i^{e_i}$.
4.   Compute the discrete logarithm a_i of x^{d/q_i} in base g^{d/q_i}, using the Pohlig-Hellman Algorithm 4.25. For the discrete logarithm subroutine in $H_i = \langle g^{d/p_i} \rangle$, with $\#H_i = p_i$, use the Pollard rho Algorithm 4.18.
5. Combine the individual logarithms a_i into a by Chinese remaindering, as in Algorithm 4.23.

THEOREM 4.29. *Let G be a cyclic group with d elements, p the largest prime factor of d, and n the bit length of d. We assume that the sequences in the Pollard rho algorithm behave sufficiently randomly. Then Algorithm 4.28 computes a discrete logarithm in G at the following cost:*

○ *factoring d,*

○ *an expected number of $O(n\sqrt{p} + n^2)$ operations in G,*

○ *$O(n^2)$ bit operations.*

PROOF.   For each $i \leq r$, step 3 takes $O(n)$ operations in G, and step 4 uses $O(e_i^2 \log p_i + e_i \sqrt{p_i})$ operations, according to Theorems 4.26 and 4.19. We have $\sum_{i \leq r} e_i \log p_i = \log d \approx n$. The Chinese remainder phase takes $O(n^2)$ bit operations.   □

The attentive reader may complain that we have reduced the hard problem of discrete logarithms to another hard problem, namely factoring integers. However, the two notions of "hard" are not identical. For a general n-bit group of prime order, no discrete logarithm algorithm better than $2^{n/2}$ is known, while n-bit integers can be factored faster, namely in time $2^{O^{\sim}(n^{1/3})}$; see Section 3.9.

4.8. Index calculus

We now present a method for calculating discrete logarithms in $\mathbb{Z}_p^\times$, where p is a prime, which is more efficient than the one of Theorem 4.29 for general groups. Many group-based protocols were first presented using such special groups. The algorithm borrows ideas from Dixon's random squares method for factoring integers, and we assume some of the notation from Section 3.10. So we have a generator g for the multiplicative group $\mathbb{Z}_p^\times = \langle g \rangle$ of units modulo p, of order $d = p - 1$, and want to compute $\mathrm{dlog}_g x$ for some given $x \in \mathbb{Z}_p^\times$. As in the random squares method, we choose a *factor base* $\{p_1, \ldots, p_h\}$ consisting of the primes up to some bound $B = p_h$.

In a preprocessing step, which does not depend on x, we choose random exponents $e \xleftarrow{\boxtimes} \mathbb{Z}_{p-1}$ and check if g^e in $\mathbb{Z}_p^\times = \{1, \ldots, p-1\}$ is B-smooth. If it is, we find nonnegative integers $\alpha_1, \ldots, \alpha_h$ with

$$
(4.30) \qquad
\begin{aligned}
g^e &= p_1^{\alpha_1} \cdots p_h^{\alpha_h} \text{ in } \mathbb{Z}_p^\times, \\
e &= \alpha_1 \, \mathrm{dlog}_g p_1 + \cdots + \alpha_h \, \mathrm{dlog}_g p_h \text{ in } \mathbb{Z}_{p-1}.
\end{aligned}
$$

We collect enough of such relations until we can solve these linear equations in $\mathbb{Z}_{p-1}$ for the $\mathrm{dlog}_g p_i$. Typically, a little more than h relations (4.30) will be enough, say $h + 20$.

To solve the system of linear equations, we may factor $p - 1$, solve modulo each prime power factor of $p - 1$, and piece the solutions together via the Chinese Remainder Theorem 15.44.

At this point, we know $\mathrm{dlog}_g p_1, \ldots, \mathrm{dlog}_g p_h$. Now on input x, we choose random exponents e until some xg^e in $\mathbb{Z}_p$ is B-smooth, say $xg^e = p_1^{\beta_1} \cdots p_h^{\beta_h}$ in $\mathbb{Z}_p$. Then

$$
(4.31) \quad \mathrm{dlog}_g x = -e + \beta_1 \, \mathrm{dlog}_g p_1 + \cdots + \beta_h \, \mathrm{dlog}_g p_h \text{ in } \mathbb{Z}_{p-1}.
$$

EXAMPLE 4.32. We look at index calculus in $\mathbb{Z}_{163}^\times = \langle 2 \rangle$, as in Example 4.27, and experiment with different factor bases. For each of the six bases in Table 4.5, we ran 1000 experiments to determine $\mathrm{dlog}_2 p$ for all p

in the factor base. Column 3 gives the average number of relations sufficient for a unique solution of the linear equations. Column 4 indicates the
average number of random choices for one relation, and the fifth one is
the product of the two previous columns. It represents the total number
of choices for e which yield all $\operatorname{dlog}_g p_i$ via (4.30). The last two columns
present the theoretical prediction of the expected number of choices to
find one B-smooth number in $\mathbb{Z}_{163}^\times$, exactly (column 6) and approximately
(column 7) as u^{-u} with $u = (\ln 163)/(\ln B)$; see Section 3.13. The last
column is a serious overestimate of column 4; see Example 3.47.

B	h	# rel. unique solution	find one rel.	unique solution	find one rel. (exact)	find one rel. (approx.)
5	3	5.41	3.97	21.47	3.77	38.3
7	4	7.27	2.74	19.9	2.70	12.4
11	5	10.06	2.31	23.2	2.22	4.96
13	6	12.19	1.92	23.4	1.91	3.91
17	7	16.82	1.74	29.23	1.72	2.87
19	8	22.95	1.62	37.2	1.59	2.58

Table 4.5: Index calculus with different factor bases.

For $B = 5$, we find the three discrete logarithms

$$\operatorname{dlog}_2 2 = 1, \operatorname{dlog}_2 3 = 101, \operatorname{dlog}_2 5 = 15,$$

and use them to compute $118 = \operatorname{dlog}_2 60$, consistent with Example 4.27.
This requires only one more relation which is obtained within (on average)
four attempts. We observe that columns 3 and 4 change monotonically
in opposite directions, and that the smallest number of random choices
(column 5) occurs for $B = 7$, which is therefore the best choice in this
example. This estimate ignores the cost of the linear algebra step, which
increases with h.   ◊

The time is, as in the random squares method, driven by estimates
for the cost of finding B-smooth integers, balanced against the cost of
the linear algebra calculations. With an appropriate choice of B, one
calculates the time for this algorithm as $\exp((1 + o(1))\sqrt{n \log n})$ for an
n-bit p, again as for Dixon's random squares method. In particular, the
factorization of $p - 1$ can be computed in this time.

In fact, we do not even need to factor $p - 1$. We simply perform
Gaussian elimination on the matrix of relations (4.30). From time to
time, we have to calculate an inverse in $\mathbb{Z}_{p-1}$ of some integer a. If $w = \gcd(a, p - 1)$ equals 1, there is no problem. If $w \neq 1$, then we factor

$p - 1 = w \cdot \frac{p-1}{w}$. This factorization has to be refined into coprime factors: $w_2 = \gcd(w, p - 1/w)$, $p - 1 = w \cdot w_2 \cdot ((p - 1)/(w \cdot w_2))$, and so on. This task is called *factor refinement*. As the elimination proceeds, we have various factorizations $p - 1 = q_1^{e_1} \cdots q_r^{e_r}$ with pairwise coprime $q_1, \ldots, q_r$ and a solution modulo each $q_i^{e_i}$. At the end, these can be pieced together with the Chinese Remainder Algorithm. The technique of only factoring $p - 1$ when a factor actually turns up is called *lazy factorization*. In record computations, the matrices are huge, with dimensions in the hundreds of millions. Special methods are required to solve them: structured Gaussian elimination and block Wiedemann or Lanczos methods.

An even faster method, the *number field sieve*, was originally invented for factoring integers (see Section 3.10) and adapted to discrete logarithms in $\mathbb{Z}_p^\times$ by Gordon (1993). Its running time is $\exp(O^\sim(n^{1/3}))$ for an n-bit prime p. For finite fields $\mathbb{F}_q^\times$ of small characteristic, say $q = 2^n$, Adleman (1994) introduced the *function field sieve*, obtaining a similar cost. For fields $\mathbb{F}_q$ with $q = p^e$ of "medium characteristic" p, with $\log p$ near e, the running time given above has still not been reached.

Table 3.3 describes some of the major advances in integer factorization, from $\exp(O(n))$ to $\exp(O^\sim(n^{1/2}))$ and $\exp(O^\sim(n^{1/3}))$. Discrete logarithm computations followed with similar improvements, lagging some years behind. Starting as a Christmas gift on 24 December 2012, this situation was reversed: discrete logarithms were computed at bit sizes (then: 1175 bits) which are completely out of reach for the known factorization methods.

These advances were achieved by Antoine Joux and others. The world record in 2015 was set by a new index calculus method of Joux that computed a discrete logarithm in $\mathbb{F}_{2^{6168}}$. Some new index calculus methods work (heuristically) in time $\exp(O^\sim(n^{1/4}))$, others even in quasi-polynomial time $\exp((\log n)^{O(1)})$. These methods apply to certain fields of small characteristic and with field degree of a special form. The area is in a state of flux and more progress can be expected. For prime fields, the record stands at 596 bits, and for fields $\mathbb{F}_{2^n}$ with n prime at $n = 809$.

We now combine the algorithmic insights of this chapter into an overall recommendation about which groups should be used in cryptography. Suppose that we have a cyclic group $G = \langle g \rangle$ of order d and bit size n, and let p be the largest prime factor of d. If p is small, say polynomial in n, then DL_G is easy and G is not useful for cryptography. So suppose that p is large. Then DL_G is about as hard as DL in a group with p elements, as long as we can only use the algorithms of the previous sections. This is made precise in Section 4.9. For any $g \in G$, $h = g^{d/p}$ is either 1 or generates a group $H = \langle h \rangle \subseteq G$ of order p (Theorem 15.60 (ii)). Usually, we like to have $p = d$, but see the Schnorr signatures in Section 8.4 for an example with large p and even larger d. Furthermore, the discrete logarithm

> Choose an elliptic curve G over a finite prime field and with a
> large prime factor p of its order, and then work in the subgroup
> H with p elements for all group-based protocols. Ideally, we can
> take $H = G$, and the prime is chosen randomly among those of
> the desired bit length.

Figure 4.6: Overall recommendation for group-based cryptography.

in $\mathbb{Z}_p^\times$ (or in a finite field) is still hard, but considerably easier than in
general groups of the same bit size. The most popular source of "general"
groups are the elliptic curves of Chapter 5. The *overall recommendation*
for group cryptography is given in Figure 4.6.

4.9. Arithmetic circuits for discrete logarithms

In this section, we shift gears. Instead of providing better and better
algorithms for discrete logarithms, we show that what we have achieved
so far cannot be essentially improved in the usual framework.

We fix the following notation:

(4.33) $G = \langle g \rangle$ is a finite cyclic group, $d = \#G$, p is the largest
           prime divisor of d, and n is the bit length of d.

Our goal is to describe a general type of algorithms for discrete logarithms with two properties:

○ all algorithms in Sections 4.3 through 4.7 are of this type,

○ any algorithm of this type requires at least $\sqrt{p}$ group operations.

According to Theorem 4.29, we can actually compute discrete logarithms with $O(n\sqrt{p} + n^2)$ group operations. Up to the "small" terms n
and n^2, we will then know that the methods incorporated in Theorem 4.29
are optimal: they cannot be essentially improved. In other words, we can
only find better discrete logarithm algorithms if we go beyond the general
type to be discussed. In fact, index calculus shows that this is possible
for special groups, $\mathbb{Z}_p^\times$ in this case. On the other hand, for any class
of groups where no special method is known, we have the comfortable
lower bound of requiring $\sqrt{p}$ operations for discrete logarithms. More
specifically, Theorem 4.42 gives asymptotically matching upper and lower
bounds for discrete logarithm computations.

We have noted in Example 4.3 that different representations of the
"same" group (isomorphic copies of one group) can influence vastly the
complexity of the discrete logarithm. Similarly, the natural representa-
tion of $\mathbb{Z}_p^\times$ as $\{1, 2, \ldots, p-1\}$ allows computations that are not possible

instruction	trace	trace exponent
$y_{-2} \longleftarrow 1$	1	0
$y_{-1} \longleftarrow g$	g	1
$y_0 \longleftarrow x$	x	t
$y_1 \longleftarrow y_0 \cdot y_{-1}$	xg	$t+1$
$y_2 \longleftarrow y_1 \cdot y_{-1}$	xg^2	$t+2$
$y_3 \longleftarrow y_2 \cdot y_{-1}$	xg^3	$t+3$
$y_4 \longleftarrow y_3 \cdot y_{-1}$	xg^4	$t+4$
$y_5 \longleftarrow y_4 \cdot y_{-1}$	xg^5	$t+5$
$y_6 \longleftarrow y_5 \cdot y_0^{-1}$	g^5	5
$y_7 \longleftarrow y_6 \cdot y_6$	g^{10}	10
$y_8 \longleftarrow y_7 \cdot y_6$	g^{15}	15
$y_9 \longleftarrow y_8 \cdot y_6$	g^{20}	20
$y_{10} \longleftarrow y_9 \cdot y_6$	g^{25}	25

Figure 4.7: Baby steps/giant steps in $\mathbb{Z}_{25}^{\times}$ as arithmetic circuit.

in general groups. We can ask whether 2 divides an element of $\mathbb{Z}_p^{\times}$, considered as an integer. But for any elements x and y of any group, $x^{-1}y$ is again a group element, and we might say that "x divides y"; divisibility is an unproductive concept in general groups.

So now we consider "generic" algorithms that do not make use of any special representation. The most natural way is to use only the group operations, starting with three special group elements: 1, the generator g, and x. From these we build further group elements by multiplication and inversion.

EXAMPLE 4.34. Figure 4.7 presents a formulation as an arithmetic circuit of the baby-step giant-step algorithm from Example 4.9 for $\mathbb{Z}_{25}^{\times}$ with $d = 20$. The "trace" gives the group element computed in each step. The "trace exponent" is explained below and has already been used in the Pollard rho Algorithm 4.18.

If $\mathrm{dlog}_g x = 5b + c$, with $0 \le b, c < 5$, as in the proof of Theorem 4.10, then $x = g^{5b+c}$, hence $xg^{5-c} = g^{5(b+1)}$, and if $c > 0$, then both elements appear in the computation. In Example 4.9 on page 177, we have $y_2 = xg^2 = g^{15} = y_8$. ◇

How do we express that the algorithm computes $\mathrm{dlog}_g x$? We are very generous: we say that the algorithm is *successful* if a *collision* $u = v$ occurs for two previously computed results u and v for which "$u = v$ is not trivial". If we computed $y_1 = y_{-1} \cdot y_{-1}^{-1}$, $y_2 = y_0 \cdot y_0^{-1}$, then $y_1 = y_2$ would be trivial. We will make this precise in a minute.

The type of computation shown in the table above is an *arithmetic group circuit* (or *arithmetic circuit* for short, or straight-line program) with inputs 1, g, and x. We abbreviate the assignment $y_k \longleftarrow y_i \cdot y_j^{\pm 1}$ as $(i, j, \pm 1)$, and also trace the exponents of g and x in the circuit. Then we arrive at the following notion.

DEFINITION 4.35.   (i) *An* arithmetic circuit *is a finite sequence* $\mathcal{C} = (I_1, \dots, I_\ell)$ *of instructions* $I_k = (i, j, \epsilon)$ *for* $1 \leq k \leq \ell$, *with* $-2 \leq i, j < k$ *and* $\epsilon \in \{1, -1\}$. *The* size *of* $\mathcal{C}$ *is* ℓ. *Note that* $\mathcal{C}$ *is not connected to any particular group.*

(ii) *If* $\mathcal{C} = (I_1, \dots, I_\ell)$ *is an arithmetic circuit,* G *a group and* $g, x \in G$, *then the* trace *of* $\mathcal{C}$ *on input* (g, x) *is the following sequence* $z_{-2}, z_{-1}, \dots, z_\ell$ *of elements* z_k *of* G:

$$z_{-2} = 1, z_{-1} = g, z_0 = x, z_k = z_i \cdot z_j^\epsilon \text{ for } k \geq 1 \text{ and } I_k = (i, j, \epsilon).$$

(iii) *For an arithmetic circuit* $\mathcal{C} = (I_1, \dots, I_\ell)$, *the* trace exponents *form the following sequence* $\tau_{-2}, \tau_{-1}, \dots, \tau_\ell$ *of linear polynomials* τ_k *in* $\mathbb{Z}[t]$:

$$\tau_{-2} = 0, \tau_{-1} = 1, \tau_0 = t, \tau_k = \tau_i + \epsilon \cdot \tau_j \text{ for } k \geq 1 \text{ and } I_k = (i, j, \epsilon).$$

We think of g *as fixed, and also write* $z_k(x)$ *for the trace elements* z_k *in* (ii).

The connection between the trace and the trace exponents is as follows. If $x = g^a$ and $\tau_k = c \cdot t + b$, then

$$z_k(x) = g^b x^c = g^b \cdot g^{ac} = g^{\tau_k(a)}.$$

Recall that in the exponents, we may calculate modulo the group order d, once we consider a fixed group. We have used the trace exponents in the Pollard rho algorithm; see (4.14) and Figure 4.2.

EXAMPLE 4.36. Here are two more examples of trivial collisions.

(i) We take g and x in a group of order d, and an arithmetic circuit which computes $y_m = g^d$ with an addition chain of some length m, and also $y_{2m} = x^d$. Then $y_m = 1 = y_{2m}$ is a collision, but $\tau_m = d = 0$ and $\tau_{2m} = dt = 0$ in $\mathbb{Z}_d[t]$, and we take the equation $\tau_m - \tau_{2m} = 0$ in $\mathbb{Z}_d[t]$ as an indicator for the triviality of this collision.

(ii) Now let q be an arbitrary prime divisor of d, maybe a small one, and assume that $d \neq q$. Again we calculate some $y_m = g^{d/q}$ and $y_{2m} = x^{d/q}$. Now both results lie in the subgroup $H = \langle g^{d/q} \rangle$ of order q, and we can find a collision with a further q (or even $O(\sqrt{q})$) steps. But we have only calculated a discrete logarithm in H, not in G. If, say, $q = 2$, then $y_m = g^{d/2} \neq 1$ and y_{2m} is either y_m or 1. Thus we have a collision, either $y_{-2} = y_{2m}$ or $y_m = y_{2m}$. $\diamond$

How do we express that "$u = v$ is trivial"? We certainly want to say that "the collision $y_i = y_j$ is trivial" if $\tau_i = \tau_j$, or even if $\tau_i = \tau_j$ in $\mathbb{Z}_d[t]$, but this is not quite enough. We now rule out unpleasant cases like the one in Example 4.36(ii), where a collision occurs but the discrete logarithm is not really computed.

DEFINITION 4.37. Let $\mathcal{C}$ be an arithmetic circuit of size ℓ, $G = \langle g \rangle$, q an arbitrary divisor of the group order $d = \#G$, and $i, j \leq \ell$.

(i) Then (i, j) is said to respect q if and only if $\tau_i - \tau_j \neq 0$ in $\mathbb{Z}_q[t]$.

(ii) If on input of some $g, x \in G$, a collision $y_i = y_j$ occurs, then this collision respects q if and only if (i, j) respects q.

Thus we have the linear polynomial $\tau_i - \tau_j \in \mathbb{Z}[t]$ which is nonzero modulo d, and if a collision occurs for $x = g^a$, then $g^{\tau_i(a)} = z_i(x) = z_j(x) = g^{\tau_j(a)}$, so that $(\tau_i - \tau_j)(a) = 0$ in $\mathbb{Z}_d[t]$. If $q_1 \mid q_2 \mid d$, and (i, j) respects q_2, then it also respects q_1.

EXAMPLE 4.36 CONTINUED. (ii) We have $\tau_m = d/2$, $\tau_{2m} = dt/2$, and $\tau_m - \tau_{2m} = d/2 \cdot (1 - t)$, all in $\mathbb{Z}_d[t]$. We take a prime divisor q of $d/2$. Then $\tau_m - \tau_{2m} = 0$ in $\mathbb{Z}_q[t]$. Thus $(m, 2m)$ does not respect q, and if on some input x, the collision $g^{d/2} = z_m(x) = z_{2m}(x) = x^{d/2}$ occurs, then this does not respect q either. $\diamond$

We do not insist that an arithmetic circuit always compute the right answer, but call the fraction of inputs on which it does its *success rate*.

DEFINITION 4.38. Let $G = \langle g \rangle$ be a finite cyclic group, $\mathcal{C}$ an arithmetic circuit, and q an arbitrary divisor of the group order $d = \#G$. Then the success rate $\sigma_{\mathcal{C}, q}$ of $\mathcal{C}$ over G respecting q is the fraction of group elements for which a collision respecting q occurs:

$$\sigma_{\mathcal{C}, q} = d^{-1} \cdot \#\{x \in G : \text{ on input } x, \text{ a collision respecting } q \text{ occurs in } \mathcal{C}\}.$$

Thus $0 \leq \sigma_{\mathcal{C},q} \leq 1$, and a circuit, for which a collision respecting q occurs for every input x, has $\sigma_{\mathcal{C},q} = 1$. If $q_1 \mid q_2 \mid d$, then $\sigma_{\mathcal{C},q_1} \leq \sigma_{\mathcal{C},q_2}$. Example 4.34 indicates that the baby-step giant-step algorithm gives a circuit of size $O(\sqrt{d})$, where $d = \#G$ and $\sigma_{\mathcal{C},d} = 1$. Our notation does not reflect the dependence of the success rate on the group.

LEMMA 4.39. *Let $d \geq 2$ be an integer, p^e a prime power divisor of d, where p is a prime and $e \geq 1$, and $\tau = c_1 t + c_0 \in \mathbb{Z}_d[t]$ a linear polynomial with $\tau \neq 0$ in $\mathbb{Z}_{p^e}[t]$. Then*

$$\#\{a \in \mathbb{Z}_d \colon \tau(a) = 0 \text{ in } \mathbb{Z}_{p^e}\} \leq d/p.$$

PROOF.   Let $i \geq 0$ be the largest exponent with $\tau = 0$ in $\mathbb{Z}_{p^i}[t]$. Thus $i < e$, and we can write $\tau = p^i \cdot (c_1' t + c_0')$, with $c_0', c_1' \in \mathbb{Z}_{d/p^i}$ and at least one of them nonzero modulo p. If $c_1' = 0$ in $\mathbb{Z}_p$, then there is no $a \in \mathbb{Z}_d$ with $\tau(a) = 0$ in $\mathbb{Z}_{p^{i+1}}$, let alone modulo p^e. Otherwise there is exactly one $a_0 \in \mathbb{Z}_p$ with $c_1' a_0 + c_0' = 0$ in $\mathbb{Z}_p$, namely $a_0 = -c_0' \cdot c_1'^{-1}$ in $\mathbb{Z}_p$. The residue class mapping $\mathbb{Z}_d \longrightarrow \mathbb{Z}_p$ maps any $a \in \mathbb{Z}_d$ to a in $\mathbb{Z}_p$ (Lemma 15.43). Exactly d/p elements of $\mathbb{Z}_d$ are mapped to the same element of $\mathbb{Z}_p$. Now if $p^i(c_1' a + c_0') = \tau(a) = 0$ in $\mathbb{Z}_{p^e}$, then $c_1' a + c_0' = 0$ in $\mathbb{Z}_p$, and hence $a = a_0$ in $\mathbb{Z}_p$. There are exactly d/p such a, and the claim follows.   □

THEOREM 4.40. *Let $G = \langle g \rangle$ be a finite cyclic group, $q = p^e$ a prime power divisor of the group order $d = \#G$, $\mathcal{C}$ an arithmetic circuit of size ℓ, and $\sigma_{\mathcal{C},q}$ its success rate over G respecting q. Then*

$$\ell \geq \sqrt{2\sigma_{\mathcal{C},q} \cdot p} - 3.$$

When $\sigma_{\mathcal{C},q}$ is a constant, then $\ell \in \Omega(\sqrt{p})$.

PROOF.   On some input x, a collision in $\mathcal{C}$ is of the form $z_i(x) = z_j(x)$ with $-2 \leq i < j \leq \ell$. There are $(\ell + 2)(\ell + 3)/2$ such (i,j). Any (i,j) which respects q leads to a collision respecting q for at most d/p values of x, by Lemma 4.39, since the exponents $a \in \mathbb{Z}_d$ correspond bijectively to the group elements $x = g^a$. Thus the total number of possible collisions respecting q is at most $(\ell + 2)(\ell + 3)/2 \cdot d/p$, and hence

$$\begin{aligned} \sigma_{\mathcal{C},q} &\leq (\ell + 2)(\ell + 3)/2p, \\ (\ell + 3)^2 &\geq (\ell + 2)(\ell + 3) \geq 2\sigma_{\mathcal{C},q}p. \end{aligned}$$ □

The main point of Sections 4.3 to 4.7 is the $O(n\sqrt{p} + n^2)$ upper bound for discrete logarithm computations, and we now have an almost matching lower bound $\Omega(\sqrt{p})$ where p is the largest prime divisor of d. We turn to some finer points. The following may be skipped at first reading.

In what follows, we derive upper and lower bounds that differ only by a constant factor, namely 10. We start with a lower bound different from Theorem 4.40, namely $\Omega(n)$. This is not of direct cryptographic relevance, since $n \approx \log_2 d$ is roughly the bit size of G—in contrast to $\sqrt{p}$ which may be exponentially large in n. The interest is a desire to understand the complexity of discrete logarithms as well as possible in the circuit model.

LEMMA 4.41. *Let $\mathcal{C}$ be an arithmetic circuit of size ℓ and $G = \langle g \rangle$ a cyclic group of n-bit order d, with $\sigma_{\mathcal{C},d} = 1$. Then for $d \geq 3$, we have*

$$\ell \geq \frac{n}{2} - 2,$$

and hence $\ell \in \Omega(n)$.

PROOF. In the symmetric system R_d of representatives (15.12), every integer a has exactly one *symmetric representative* $b \in \mathbb{Z}$ with $a = b$ in $\mathbb{Z}_d$ and $-d/2 \leq b < d/2$. For $-2 \leq k \leq \ell$, we write the trace exponent $\tau_k \in \mathbb{Z}_d[t]$ as $\tau_k = c_k t + b_k$, where $c_k, b_k \in \mathbb{Z}$ are symmetric representatives. By induction on k it follows that $|b_k|, |c_k| \leq 2^k$ for $0 \leq k \leq \ell$ (and $|b_k|, |c_k| \leq 1$ for $k = -2, -1$). Now let $a = \lfloor \sqrt{d} \rfloor \in \mathbb{Z}_d$ and $x = g^a \in G$. The assumption $\sigma_{\mathcal{C},d} = 1$ implies that there are $i, j \leq \ell$ with $\tau_i - \tau_j \neq 0$ in $\mathbb{Z}_d$ and $(\tau_i - \tau_j)(a) = 0$ in $\mathbb{Z}_d$. We let

$$u = (c_i - c_j) \cdot a + (b_i - b_j) \in \mathbb{Z}.$$

The above implies that $u = 0$ in $\mathbb{Z}_d$.

If $c_i = c_j$, then $b_i = b_j$ and $\tau_i - \tau_j = 0$ in $\mathbb{Z}_d[t]$, which is ruled out. Thus $c_i \neq c_j$. If $u = 0$, then

$$\sqrt{d} - 1 \leq |a| = \frac{|b_i - b_j|}{|c_i - c_j|} \leq |b_i - b_j| \leq |b_i| + |b_j| \leq 2^{\ell+1}.$$

If $u \neq 0$, then $|u| \geq d$, and

$$
\begin{aligned}
2^{\ell+1}(\sqrt{d} + 1) &= 2^{\ell+1}\sqrt{d} + 2^{\ell+1} \geq |c_i - c_j|a + |b_i - b_j| \\
&\geq |(c_i - c_j)a + (b_i - b_j)| = |u| \geq d, \\
2^{\ell+1} &\geq \frac{d}{\sqrt{d} + 1} \geq \sqrt{d} - 1.
\end{aligned}
$$

Thus $\ell \geq \log(\sqrt{d} - 1) - 1$. The claim now follows from

$$\log(\sqrt{d} - 1) \geq \frac{1}{2}\log d - \frac{1}{2} \geq \frac{1}{2}\lfloor \log d \rfloor - \frac{1}{2} = \frac{n}{2} - 1$$

for $d \geq 4$. (One checks the case $d = 3$ separately.) $\qquad\square$

For an upper bound in our model we take the circuit C that just computes $g^{d/p}$ and $x^{d/p}$, and then performs a baby-step giant-step search in the subgroup $\langle g^{d/p} \rangle$ of p elements. The total cost is $(2 + \epsilon)(n + \sqrt{p})$ for some positive ϵ, and we have the lower bounds of $n/2$ and $\sqrt{2p}$, approximately. Thus the gap is a factor of about 4 or $\sqrt{2}$, depending on whether n or $\sqrt{p}$ is larger. We can obtain a specific estimate as follows.

THEOREM 4.42. *Let G be a cyclic group of n-bit order d, p the largest prime divisor of d, e the multiplicity of p in d,*

$$m = \max\{\sqrt{2p} - 3, n/2 - 2\},$$

and assume that $m \geq 37$. Then there exists an arithmetic circuit C with success rate $\sigma_{C,p^e} = 1$ over G and size at most $10m$. Any circuit with $\sigma_{C,p^e} = 1$ has size at least m.

PROOF.   The last claim follows from Theorems 4.40 and 4.41. For C we take the circuit described above. Then $\sigma_{C,p^e} = 1$, and its size ℓ is at most $2 \cdot 2 \log(d/p) + 2\sqrt{p}$, by Theorems 15.49 and 4.10. Thus

$$\begin{aligned} \ell \leq 4 \log d + 2\sqrt{p} &\leq 8(n/2 - 2) + \sqrt{2} \cdot (\sqrt{2p} - 3) + 17 + 3\sqrt{2} \\ &\leq (8 + \sqrt{2})m + 17 + 3\sqrt{2} \leq 10m. \quad \square \end{aligned}$$

The attentive reader may have observed that Pollard's rho method does not fit into our framework, because it uses random choices. However, the method of this section generalizes easily to the probabilistic scenario.

Notes 4.1. Certain noncommutative groups have also been suggested for cryptography; see Myasnikov *et al.* (2011) for a survey.

4.2. Just as Diffie and Hellman did for their key exchange, ElGamal proposed his system originally for the group of units modulo a prime number. His system is not often used in practice.

4.3. For the baby-step giant-step method in integer factorization, see Table 3.3 and the Notes to Section 3.10.

4.5. The equations in (4.13) can be replaced by other ones, like $y_k = x^{100} y_{k-1}^{1000}$ for the first equation. The choices in (4.13) are the simplest and most efficient ones known. An alternative to the tripartition (S_0, S_1, S_2) in Algorithm 4.18, step 1, is to use a hash value $h(z)$ modulo 3, or a pseudorandom function. Pollard's rho method can be improved in practice, for example by storing appropriate intermediate results.

A first proof of the required randomness properties was given by Horwitz & Venkatesan (2002) for a variant of the Pollard rho method, and by Miller & Venkatesan (2006, 2009) and Kim *et al.* (2010) for the original version. Some of these results assume the group order d to be prime, the order of 2 in $\mathbb{Z}_d^{\times}$ to be large, and an efficient version of the tripartition $G = S_1 \cup S_2 \cup S_3$, where membership of $x \in G$ in some S_i can easily be tested.

4.6. An alternative proof of Lemma 4.22 is by induction on r, using Lemma 4.21.

The estimate $O(n^2)$ in Theorem 4.24 (iii) can be improved to $nH(q_1^{-1}, \ldots, q_r^{-1})(1 + o(1)) \in n \log r \cdot (1 + o(1))$ using Brauer's addition chains and the Huffman method for building balanced trees. Here H denotes the entropy. For details, see von zur Gathen & Gerhard (2013), Exercises 10.5 to 10.7.

4.7. The Pohlig-Hellman Algorithm 4.25 actually works for an arbitrary integer $p \geq 2$. But its most useful application is when p is a prime. It is not clear whether one can reduce the number of group operations, maybe to $O(e \log p)$. The Chinese remainder algorithm of Section 4.6 is in Pollard (1978), and both it and the p-adic method of this section are in Pohlig & Hellman (1978). The whole method is often called the Pohlig-Hellman algorithm.

4.8. The word *index* is a historical synonym for discrete logarithm, going back to Gauß (1801), article 57. In modern terminology, $\gcd(\mathrm{dlog}_g x, d)$ still equals the index of $\langle x \rangle$ in $\langle g \rangle$. For various groups $G = \mathbb{Z}_p^\times = \langle g \rangle$ with primes p, tables like the ones of Jacobi (1839) list (a, g^a) for all $a \in \mathbb{Z}_{p-1}$ and $(x, \mathrm{dlog}_g x)$ for all $x \in \mathbb{Z}_p^\times$. Thus $a = \mathrm{dlog}_g x$ is the index (position) of $x = g^a$ in the first list. These tables were discrete analogs of the pre-electronic marvels of logarithm tables and slide rules, facilitating discrete computations modulo p rather than (approximate) calculations with real numbers.

The runtime $\exp(O^\sim(n^{1/2}))$ of index calculus for discrete logarithms in $\mathbb{F}_q$ with $q = p^e$ and $n = \log q$ is proven for $e \in \{1, 2\}$, and also for "small" p with $\log p \in O(e \log e)$; for the latter, see Bender & Pomerance (1998) and Enge & Gaudry (2002). For other parameter ranges, this is not yet clear, and for the faster algorithms, one has to rely on unproven heuristics. For factor refinement in lazy factorization, see Exercises 15.11 and 15.12.

Joux received the prestigious Gödel Award in 2013 for his work on a one-round implementation of the Diffie-Hellman protocol for three parties, jointly with Boneh and Franklin for their ideas in identity-based cryptography. Both lines of work make essential use of pairings on elliptic curves; see the end of Section 5.1. Their security is partially based on the difficulty of the discrete logarithm in certain finite fields. Quite amusingly, Joux's advances put just this in question, and the security of pairing-based cryptography needs to be re-evaluated.

For discrete logarithms in $\mathbb{F}_{2^{809}}$, see Jeljeli (2014). The recommendation of Figure 4.6 ignores many practical concerns, including compatibility with legacy software and existing hardware.

4.9. Arithmetic circuits are sometimes called *straight-line programs* and were introduced by Strassen (1972b) in great generality. The material of this section is taken from von zur Gathen (2004a), where also probabilistic circuits are discussed. Somewhat different models of "generic" group algorithms were proposed by Nechaev (1994) for deterministic computations, and by Shoup (1997) for the general case of probabilistic algorithms, together with a lower bound $\Omega(\sqrt{p})$. Shoup's model works with "random representations" of the group elements, and can also be applied to the Diffie-Hellman problem. Maurer & Wolf (1999) continued work on this model, in particular relating the two questions of discrete logarithms and the Diffie-Hellman task. See also Boneh & Lipton (1996) and Schnorr (2001). Babai & Szemerédi (1984) introduced a more restricted model, in which a lower bound of $\Omega(p)$ holds. A charm of this section's circuits over other models are the (essentially) matching upper and lower bounds of Theorem 4.42.

Exercises.

EXERCISE 4.1 (Discrete logarithm example). *Let $G = \mathbb{Z}_{73}^{\times}$ and $g = 5$.*

(i) *Verify that $\langle 5 \rangle = G$.*

(ii) *Execute Algorithm 4.28 by hand to determine $\operatorname{dlog}_5 6$ in G. Check your result.*

(iii) *Compute BOB's secret key in Example 2.18 continued on page 175.*

EXERCISE 4.2 (Uniqueness of the discrete logarithms). *Suppose that $g \in G = \langle g \rangle$ has order d, and that $x \in G$. Show that $\operatorname{dlog}_g x \in \mathbb{Z}_d$ is uniquely determined.*

EXERCISE 4.3 (Additive Subgroups).

(i) *Let $N = p \cdot q$ be the product of two distinct primes p and q. Describe two subgroups S and T of the additive group $\mathbb{Z}_N$ of orders p and q, respectively, and a group isomorphism $\mathbb{Z}_N \longrightarrow S \times T$. Is this a ring isomorphism?*

(ii) *More generally, let $q_1, \ldots, q_r$ be pairwise relatively prime, and $N = q_1 \cdots q_r$. Describe subgroups $S_1, \ldots, S_r \subseteq \mathbb{Z}_N$ of the additive group $\mathbb{Z}_N$ with $\#S_i = q_i$ for all $i \leq r$ and a group isomorphism $\mathbb{Z}_N \longrightarrow S_1 \times \cdots \times S_r$. Is this a ring isomorphism?*

EXERCISE 4.4 (Counting ElGamal encryptions). *Let G be a finite cyclic group with d elements and $(\mathsf{sk}, \mathsf{pk}) = (b, B)$ a pair of keys for ElGamal encryption.*

(i) *Let $y \in G$. For how many $A \in G$ is (y, A) the encryption of some plaintext?*

(ii) *Let $A \in G$. For how many $y \in G$ is (y, A) the encryption of some plaintext?*

(iii) *For how many $(y, A) \in G^2$ is (y, A) the encryption of some plaintext?*

EXERCISE 4.5 (ElGamal example). *We take the prime number $p = 20443$ and implement the ElGamal cryptosystem 4.5 using the group $G = \mathbb{Z}_p^{\times}$. A is mapped to 0, B to 1 and so forth, Z is mapped to 25. We combine groups of three letters (a_0, a_1, a_2) to $a_0 + 26 a_1 + 26^2 a_2$. Thus ABC corresponds to the value $0 + 26 \cdot 1 + 2 \cdot 26^2 = 1378$.*

(i) *Show that 2 is a generator of $\mathbb{Z}_{20443}^{\times}$.*

(ii) *Encrypt the word SYSTEM using the ElGamal scheme 4.5 over G with generator $g = 2$ and the ECB mode. The receiver of the message has the public key $\mathsf{pk} = 8224$. Choose the session key to be $A = 2^a$ with $a = 321$.*

(iii) *The following transcript of a conversation was intercepted, which contains a message encrypted with the ElGamal system as above.*

BOB	*has the public key 7189.*
ALICE to BOB:	*message (part 1) (16278, 4151).*
ALICE to BOB:	*message (part 2) (12430, 4151).*
ALICE to BOB:	*message (part 3) (2689, 4151).*

An indiscretion revealed that one part of the message corresponds to the plaintext (value) 8324. Compute the (alphabetic) plaintext of the entire message.

EXERCISE 4.6 (Different generators for Diffie-Hellman). *This exercise shows that the complexity of the Diffie-Hellman is independent of the choice of generator, but the notion of Diffie-Hellman triple is not. Let $G = \langle g \rangle = \langle g' \rangle$ be a cyclic group of size d.*

(i) *Prove that* $\mathrm{dlog}_g\, g'$ *and d are coprime.*

(ii) *Give an example of a Diffie-Hellman triple with respect to g which is not a Diffie-Hellman triple with respect to* g'.

(iii) *Which triples are Diffie-Hellman triples with respect to both g and* g'*?*

(iv) *Reduce the Diffie-Hellman problem* $DH_G(g)$ *with respect to g to* $DH_G(g')$. *Conclude that* $DH_G(g) \equiv_p DH_G(g')$.

EXERCISE 4.7 (Ambiguous Diffie-Hellman triples). *Let* $G = \langle g \rangle$ *be a finite cyclic group and take a Diffie-Hellman triple* (A, B, C) *in G. In the following, consider arbitrary* $A', B', C' \in G$.

(i) *Determine simple conditions that are necessary and sufficient for* (A', B, C) *to be a Diffie-Hellman triple. Give an example with* $A \neq A'$.

(ii) *Same for* (A, B', C) *and* (A, B, C'), *as far as possible.*

(iii) *What can you say about (i) for* (A', B', C), (A', B, C'), *and* (A, B', C')*?*

EXERCISE 4.8 (Variants of the Diffie-Hellman problem). *Let* $G = \langle g \rangle$ *be a cyclic group of order d congruent to* $2 \bmod 4$. *Let* $y = g^{d/2} \in G$, *so that* $y^2 = 1$. *All these data are given. Consider the following problems:*

DH : *Given* g^a *and* g^b, *compute* g^{ab}.
SQ : *Given* g^a, *compute* g^{a^2}.
DH': *Given* g^a *and* g^b, *compute* $\{g^{ab}, yg^{ab}\}$.

Obviously $DH' \leq_p DH$. *Show that* $DH' \leq_p SQ \leq_p DH$. *What can you say about* $DH \leq_p SQ$ *and* $SQ \leq_p DH'$*?*

EXERCISE 4.9 (ElGamal modes). *Encrypt the plaintext from Example 2.18 in other modes from Section 2.8: CBC, CFB, OFB, and autokey.*

EXERCISE 4.10 (Success probability in reduction). *Let* $G = \langle g \rangle$ *be finite. The input values of* DH_G *are* $A = g^a$ *and* $B = g^b$ *in G, and for the ElGamal encryption system they are* (A, B, y), *with* A, B *as in* DH_G. *A randomized algorithm "solves* DH_G *with success probability* ε*" if it returns the correct DH answer with probability at least* ε. *The probability space consists of the algorithm's internal random choices, and uniformly random* (A, B).

(i) *State a similar definition for "plaintext recovery in the ElGamal encryption system with success probability* ε*".*

(ii) *Theorem 4.6 gives two reductions. For each of them, investigate how the success probabilities are related.*

EXERCISE 4.11 (Baby-step giant-step).   *Implement the baby-step giant-step algorithm 4.8. Compute the discrete logarithm of your birth number (Exercise 2.1) in the group $\mathbb{Z}_p^\times$ with $p = 10^6 + 3$ and base $g = 2$. Count the number of group operations and compare with the prediction of Theorem 4.12.*

EXERCISE 4.12 (Pollard rho method).   *Implement the Pollard rho algorithm 4.18 for the discrete logarithm. Compute the discrete logarithm of your birth number (Exercise 2.1) in the group $\mathbb{Z}_p^\times$ with $p = 10^6 + 3$ and base $g = 2$. Count the number of group operations needed and compare with the prediction from Theorem 4.19.*

EXERCISE 4.13 (Chinese remaindering and Pohlig-Hellman).   *The numbers in this exercise are deliberately small and the use of a computer algebra system is discouraged.*

(i) *Let $G = \mathbb{Z}_p^\times = \langle g \rangle$ with $p = 2 \cdot 3 \cdot 5 \cdot 7 + 1$ and $g = 2$, and $x = 10$. Verify that p is prime and $\mathbb{Z}_p^\times = \langle g \rangle$. Compute $\mathrm{dlog}_g\, x$ using Algorithm 4.23.*

(ii) *Let $G = \mathbb{Z}_p^\times = \langle g \rangle$ with $p = 2^4 + 1$, $g = 3$, and $x = 7$. Verify that $\mathbb{Z}_p^\times = \langle g \rangle$. Compute $\mathrm{dlog}_g\, x$ using Algorithm 4.25.*

EXERCISE 4.14 (Pohlig-Hellman). *Let $G = \mathbb{Z}_p^\times$ with $p = 1000771201$, $d = 2^7 3^7 5^2 \cdot 11 \cdot 13$, $g = 7$, and $x = 6$.*

(i) *Verify that p is prime, $\#\mathbb{Z}_p^\times = d$, and $\mathbb{Z}_p^\times = \langle g \rangle$.*

(ii) *Let $a = \mathrm{dlog}_g\, x$. Compute $(a_1 = a \text{ in } \mathbb{Z}_{2^7})$, ..., $(a_5 = a \text{ in } \mathbb{Z}_{13})$ using Algorithm 4.28.*

(iii) *Now compute a in $\mathbb{Z}_d$.*

EXERCISE 4.15 (Easy discrete logarithms).   *The discrete logarithm problem can be exceptionally easy. Take $p = 436916347656251$ and use a computer algebra system to solve the following tasks.*

(i) *Show that p is prime and find the prime factorization of $p - 1$.*

(ii) *Show that 2 generates $\mathbb{Z}_p^\times$.*

(iii) *Let $x = 194471139368208$. Compute $\mathrm{dlog}_2\, x$ using Algorithm 4.28.*

The prime in the example above is too small for cryptographic purposes. You are to generalize your experience from this example to cryptographic proportions.

(iv) *Describe how you would choose a prime number p with 1024 bits so that discrete logarithms in $\mathbb{Z}_p^\times$ are fairly easy to compute, as was the case in the example.*

(v) *Estimate the probability that a randomly chosen prime number p has the desired property, so that discrete logarithms modulo p can be computed with fewer than about 10^{15} operations. (Hint: You may use estimates about smooth numbers from Exercise 3.30.)*

(vi) *Use a computer algebra system to find a prime number p as in (iv). Next find a generator g of $\mathbb{Z}_p^\times$ and a random $x \xleftarrow{\;\boxtimes\;} \mathbb{Z}_p$. Compute $\mathrm{dlog}_g\, x$ using Algorithm 4.28.*

EXERCISE 4.16 (Index calculus). *Consider the 24-bit prime* $p = 10000223$, $g = 5$, $x = 42$, *and the factor base* $\{p_1, p_2, \cdots, p_{46}\}$ *up to* $B = 200$, *which is less than* $\exp(\sqrt{\ln p \ln \ln p}) \approx 807$.

(i) *Compute the discrete logarithms* e_i *of the elements* p_i *of the factor base by factoring many* g^e *in* $\mathbb{Z}_p$, *considered as integers, over the factor base. After about 50 successful factorizations you should be able to find the wanted logarithms.*

(ii) *Now compute* $\mathrm{dlog}_g x$ *via the index calculus.*

(iii) *Give a rough estimate of how many operations in* $\mathbb{Z}_p^\times$ *Algorithm 4.28 would use. You do not have to implement the algorithm.*

You should use a computer algebra system for this exercise. If you have to calculate by hand, you may work with the following smaller example: $p = 107$, $g = 2$, *and* $x = 42$.

EXERCISE 4.17 (Product of cyclic groups). *You are to generalize Theorem 4.40 to a product of two cyclic groups. This is studied in Shoup (1997). So let* $H_1 = \langle g_1 \rangle$ *and* $H_2 = \langle g_2 \rangle$ *be two cyclic groups of orders* d_1 *and* d_2, *respectively, and* $G = H_1 \times H_2$. *Thus* G *consists of all* (x_1, x_2) *with* $x_1 \in H_1$ *and* $x_2 \in H_2$, *using componentwise multiplication.*

(i) *Show that there are cyclic groups* K_1 *and* K_2 *of orders* e_1 *and* e_2 *respectively, with* e_1 *dividing* e_2 *and* $G \cong K_1 \times K_2$.

Hint: The Chinese Remainder Theorem 15.44 may come in handy. See the Notes to Section 5.1 for elliptic curves being of this form.

(ii) *Assume that the prime factorizations of* d_1 *and* d_2 *are given. Show how to find* K_1 *and* K_2 *as in (i) as products of subgroups of* H_1 *and* H_2. *Hint: a construction as in Lemma 4.22 (i) may be useful.*

Let p *and* q *be the largest prime divisor of* $\gcd(d_1, d_2)$ *and* d_2, *respectively, and* $\mathcal{C}$ *an arithmetic circuit of size* ℓ, *which on input* g_1, g_2, *and any* $x \in G$ *produces a nontrivial collision.*

(iii) *Show that* $\ell \geq \sqrt{2q}$.

(iv) *Show that* $\ell \geq \sqrt{2p}$.

Hint: Assign two trace polynomials to each step of the computation.

In all our Reasonings about Infinity, there are certain
Bounds set to our finite and limited Capacities,
beyond which all is Darkness and Confusion [...]
The Usefulness and Excellence of this Method
is shewn by a successful Application of it to the making
of several Improvements in the Geometry of Curve-Lines.

ISAAC NEWTON (1737)

It is a serious question whether America,
following England's lead, has not gone into problem-solving
altogether too extensively. Certain it is that we are producing no
text-books in which the theory is presented in the delightful style
which characterizes many of the French works [...],
or those of the recent Italian school [...],
or, indeed, those of the continental writers in general.

DAVID EUGENE SMITH (1900)

Vous ne sçauriez accorder si peu de chose à un Amant,
que bientôt après il ne faille lui en accorder davantage,
& à la fin cela va loin. De même accordez à un Mathematicien
le moindre principe, il va vous en tirer une consequence,
qu'il faudra que vous lui accordiez aussi,
& de cette consequence encore un autre; & malgré vous-même,
il vous mene si loin, qu'à peine le pouvez-vous croire;[1]

BERNARD LE BOUYER DE FONTENELLE (1686)

---

[1]You would not know to grant such a small trifle to a lover that soon afterwards you would have to concede more, and in the end this may go far. Similarly, you grant the smallest principle to a mathematician, he will then draw a consequence from this which you will also have to grant, and from this consequence yet another one, and no matter what you intend he will take you so far that you can hardly believe it yourself.

Chapter 5

Elliptic curves

This chapter discusses a class of groups called *elliptic curves*, which arise from algebraic geometry. We can compute efficiently in them and, in general, only the slow generic algorithms for discrete logarithms (Section 4.9) are known. This allows smaller key sizes than RSA or groups like $\mathbb{Z}_p^{\times}$ at comparable security levels and makes them attractive for public-key cryptography. They are particularly useful in environments with limited communication bandwidth.

Elliptic curves find many uses in standardized form. We describe the *NIST curves* in Section 5.6. The *ECC Brainpool* curves are available in OpenSSL and BouncyCastle, used in some European passports, and in TLS and IPsec for internet security.

5.1. Elliptic curves as groups

The following presents our operational definition of elliptic curves. Some of the geometric background is explained in Section 5.2.

DEFINITION 5.1. *Let F be a field of characteristic different from 2 and 3, and $a, b \in F$ with $4a^3 + 27b^2 \neq 0$. Then*

$$E = \{(u, v) \in F^2 : v^2 = u^3 + au + b\} \cup \{\mathcal{O}\} \subseteq F^2 \cup \{\mathcal{O}\}$$

is an elliptic curve over F. Here $\mathcal{O}$ denotes the "point at infinity" on E.
   The Weierstraß equation for E is

$$(5.2) \qquad\qquad y^2 - (x^3 + ax + b) = 0,$$

E consists of its roots (u, v) plus $\mathcal{O}$, and a and b are the Weierstraß coefficients of E.

© Springer-Verlag Berlin Heidelberg 2015
J. von zur Gathen, *CryptoSchool*, DOI 10.1007/978-3-662-48425-8_7

There are other—equivalent—ways of defining and presenting elliptic curves, but they all lead to curves that are isomorphic to one in Weierstraß form. In characteristic 2 or 3, elliptic curves can be described by similar equations, stated in (5.8).

EXAMPLE 5.3. Taking $a = -1$, $b = 0$, we have $4a^3 + 27b^2 = -4 \neq 0$ if char $F \neq 2$. The corresponding elliptic curve given by $y^2 = x^3 - x$, together with other examples of elliptic curves, is drawn in Figure 5.1 for $F = \mathbb{R}$. Over $\mathbb{F}_7$, this equation gives a curve E with eight points:

$$(0,0),\ (1,0),\ (-3,2),\ (-3,-2),\ (-2,1),\ (-2,-1),\ (-1,0),\ \mathcal{O}.$$

It is illustrated in Figure 5.2. (The dashed lines are explained below.)

Another example is the curve E^* over $\mathbb{F}_7$ with the equation $y^2 = x^3 + x$, comprising the eight points

$$(0,0),\ (1,3),\ (1,-3),\ (3,3),\ (3,-3),\ (-2,2),\ (-2,-2),\ \mathcal{O}. \qquad \Diamond$$

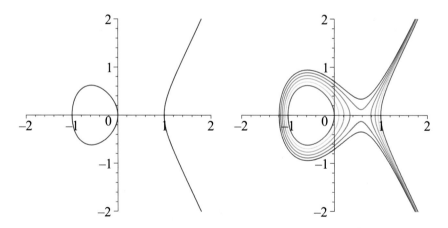

Figure 5.1: The elliptic curve $y^2 = x^3 - x$ over the real numbers (left diagram), and the elliptic curves $y^2 = x^3 - x + b$ for $b = 0, 1/10, 2/10, 3/10, 4/10, 5/10$.

The reader should imagine that $\mathcal{O}$ lies beyond the horizon in the direction of the y-axis (up *and* down), and that any two vertical lines "intersect" at $\mathcal{O}$. Projective geometry provides a rigorous framework for these notions, and Figure 5.3 an illustration.

Let $L = \{(u,v) \in F^2 : v = ru + s\}$ be a line, for some $r, s \in F$. Then

$$(5.4) \qquad L \cap E = \{(u,v) \in F^2 : (ru + s)^2 = u^3 + au + b\}.$$

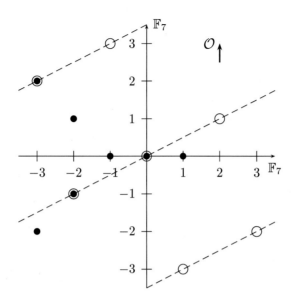

Figure 5.2: The elliptic curve E for $y^2 = x^3 - x$ over $\mathbb{F}_7$ (bold points) and the (dashed) line $y = -3x$ containing seven (circled) points, three of them on the curve.

Since a, b, r, and s are all fixed, this is a cubic equation for u. In the case of a vertical line $L = \{(u, v): v \in F\}$, where $u \in F$ is fixed, one of the points is $\mathcal{O}$.

In general, a cubic polynomial $g = x^3 + cx^2 + dx + e$ can have 0, 1, or 3 roots in F. Certainly not more than 3, because the product of $x - x_0$ for all roots x_0 divides g. If there are four or more, then this product has degree at least 4, and thus cannot divide g. But why not 2 roots? If x_1 and x_2 are distinct roots, then $(x - x_1)(x - x_2)$ divides g, and the result is a linear polynomial. It has a root, and this is the third root of g. Multiple roots require special consideration, but when this is done properly, the statement is again true.

The fundamental property that makes elliptic curves interesting for cryptography is that they carry a (commutative) group structure in a natural way. We define the group operation "+" as follows. The neutral element is $\mathcal{O}$. The negative of a point $P = (u, v) \in E$ is its mirror image $-P = (u, -v)$ upon reflection at the x-axis, and $-\mathcal{O} = \mathcal{O}$. Consider the line through P and Q. When we intersect it with E, we get three collinear points, say P, Q, and a third one, say S. Then

(5.5) $P + Q = -S$

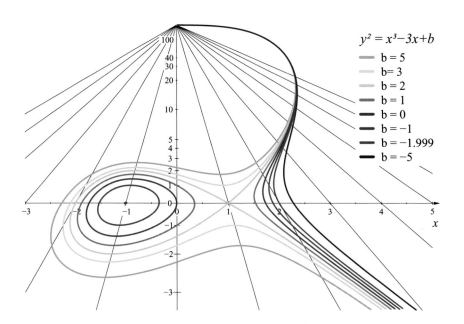

Figure 5.3: A family of curves over $\mathbb{R}$ with the point at infinity.

is the sum of P and Q (Figure 5.4). In other words, the three collinear points on E satisfy $(P+Q) + S = 0$. We have the following special cases.

1. $Q = P$. We take the tangent line at P. Since E is nonsingular (Section 5.2), the tangent is always well defined.

2. $Q = \mathcal{O}$. We take the vertical line through P:

$$P + \mathcal{O} = -(-P) = P.$$

3. $Q = -P$. We take again the vertical line through P and Q and obtain

$$P + (-P) = -\mathcal{O} = \mathcal{O}.$$

It turns out that these definitions make E into a commutative group. The second special case above shows that $\mathcal{O}$ is indeed the neutral element of E, and the third case says that the inverse of a point P is its negative $-P$. As usual, for $k \in \mathbb{Z}$ and $P \in E$, we write kP for adding P exactly k times. In particular, $0P = \mathcal{O}$ and $(-k)P = -(kP)$ for a negative multiple, with $k > 0$.

We now derive rational expressions for addition on an elliptic curve E. Suppose that $P = (x_1, y_1)$, $Q = (x_2, y_2)$, and $x_1 \neq x_2$. Then $R = (x_3, y_3) = P + Q \in E \setminus \{\mathcal{O}\}$. The line through P and Q has the equation

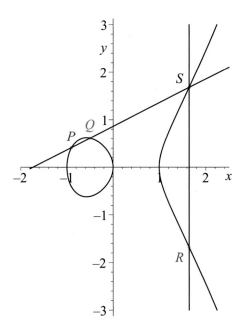

Figure 5.4: Adding two points P with $x = -0.9$ (red) and Q with $x = -0.5$ (green) on the elliptic curve $y^2 = x^3 - x$. The point $R = P + Q$ (blue) is the negative of the intersection point S (black) of the two lines with the curve.

$y = \alpha x + \beta$, where $\alpha = (y_2 - y_1)/(x_2 - x_1)$ and $\beta = y_1 - \alpha x_1$. Let $S = (x_3, -y_3)$ be the third intersection point of this line with the curve. Then $(-(\alpha x_3 + \beta))^2 = x_3^3 + a x_3 + b$. Since x_1, x_2 are the two other roots of the cubic equation $(u^3 + au + b) - (\alpha u + \beta)^2 = 0$ in u, we have $x_1 + x_2 + x_3 = \alpha^2$, by Viète's rule. It follows that

$$(5.6) \quad x_3 = \left(\frac{y_2 - y_1}{x_2 - x_1} \right)^2 - x_1 - x_2, \quad y_3 = \frac{y_2 - y_1}{x_2 - x_1} \cdot (x_1 - x_3) - y_1.$$

Thus, the coefficients of the sum of two distinct points are given by rational functions of the input coefficients. Interestingly, these formulas do not explicitly use the Weierstraß coefficients of E, which are, in fact, determined by the two points on it. A similar formula holds for doubling a point $P = (x_1, y_1)$. Writing $(x_3, y_3) = R = 2P = 2 \cdot (x_1, y_1)$, we obtain

$$(5.7) \quad x_3 = \left(\frac{3x_1^2 + a}{2y_1} \right)^2 - 2x_1, \quad y_3 = -y_1 + \frac{3x_1^2 + a}{2y_1} \cdot (x_1 - x_3)$$

if $y_1 \neq 0$, and $2P = \mathcal{O}$ if $y_1 = 0$; see Exercise 5.7.

The curve E with this operation is a commutative group. It is easy to check the properties (ii) and (iii) in the additive version of Definition 15.51, but not associativity. The latter can be verified on a computer algebra system (Exercise 5.5).

A single addition or doubling on E takes a constant number of operations in $\mathbb{Z}_p$, or $O(n^2)$ bit operations for an n-bit p. An "exponentiation" like z^k in a multiplicative group now becomes a multiplication $k \cdot P$ in the additive group E and costs about $\log_2 d \approx n$ additions on E, or $O(n^3)$ bit operations.

EXAMPLE 5.3 CONTINUED. The curve E over $\mathbb{F}_7$ given by $y^2 = x^3 - x$ has eight points, as determined above. The group E is generated by the point $(-3, 2)$ of order 4 plus the point $(0, 0)$ of order 2, and hence is isomorphic to $\mathbb{Z}_4 \times \mathbb{Z}_2$. The points $(0, 0)$, $(-3, 2)$ and $(-2, -1)$ lie on the line $y = -3x$, drawn as a dashed line in Figure 5.2. Thus

$$(0,0) + (-3, 2) = -(-2, -1) = (-2, 1),$$

$$(0,0) + (-2, -1) = -(-3, 2) = (-3, -2),$$

$$(-3, 2) + (-2, -1) = (0, 0).$$

In fact, we already noted that if you take any two distinct points P and Q in Figure 5.2, the line through them will contain exactly one other point, namely $-(P + Q)$. The four other points on our line, but not on E, are drawn as white circles. There is an eighth point on the line, at infinity, but it is not on E.

As another example, we had the curve E^* with the equation $y^2 = x^3 + x$, also comprising eight points. E^* is generated, for example, by $(3, 3)$. Figure 5.5 illustrates the group structures of E and E^*. The "cyclic" nature of E^* is well visible.   ◇

As a group, an elliptic curve over a finite field $\mathbb{F}_q$ is isomorphic to $\mathbb{Z}_k \times \mathbb{Z}_\ell$ with integers k and ℓ, and ℓ dividing $\gcd(k, q-1)$. In Example 5.3 we have $(4, 2)$ and $(8, 1)$, respectively, for (k, ℓ).

In the following, we mainly discuss elliptic curves over a finite field $\mathbb{F}_q$ where q is not a power of 2 or 3. But arithmetic in $\mathbb{F}_{2^n}$ can be implemented particularly efficiently, so that such fields are actually quite popular. In this case the Weierstraß form of most elliptic curves is given by

$$(5.8) \qquad\qquad y^2 + xy = x^3 + ax^2 + b,$$

with $a, b \in \mathbb{F}_{2^n}$ and $b \neq 0$. This means that the formulas for addition and doubling also look slightly different from the ones above, but they can be derived with the same elementary method. The description (5.8) is valid unless the characteric p divides $\#E - 1$ (in which case E is called *supersingular*).

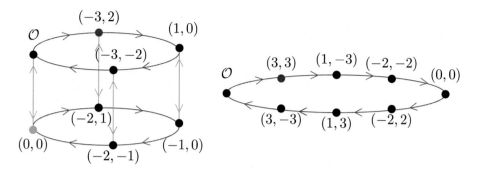

Figure 5.5: Structure of the elliptic curve groups $E = \{y^2 = x^3 - x\} \cup \{\mathcal{O}\}$ (left) and $E^* = \{y^2 = x^3 + x\} \cup \{\mathcal{O}\}$ (right) from Example 5.3. E is generated by $(-3, 2)$ (red) and $(0, 0)$ (green), and E^* is generated by $(3, 3)$ (red). There is a colored arrow from a point P to a point Q if $Q = P$ + the generator of that color.

So far in this text, we have written the operation in most groups as multiplication. The additive notation for elliptic curves is a historical standard, derived from general considerations (see the end of Section 5.2). This is just a naming convention and makes no material difference. The exp-log isomorphisms of Figure 4.1 become $\mathrm{dexp}_P(m) = mP$ and $\mathrm{dlog}_P(Q) = m$ if $Q = mP$. Example 4.3 shows that the additive group of integers modulo some number is not useful for cryptography.

Elliptic curves possess an additional structure called *pairing* which maps pairs of points to elements of an extension of the ground field. It can be used to compute discrete logarithms on special types of curves, and also in *identity-based cryptography*. Here keys are set up so that the public identity of participants, maybe an email address, can be used as a public key. However, the dramatic recent advances by Joux and others on discrete logarithm computations in finite fields of small characteristic (Section 4.8) have obliterated some proposals in this direction and it is not clear (in 2015) in which form this approach will survive. Therefore we skip this topic.

5.2. The geometric background

Elliptic curves come from *algebraic geometry*, one of the most important and deepest areas of mathematics. It deals with the solutions of polynomial equations. The material of this section is not required in the following and may well be skipped. For our purposes, a very special case of equations is sufficient: a single equation $f(x, y) = 0$ in two variables x and y.

If F is our ground field, so that $f \in F[x, y]$, then

$$X = \{(u, v) \in F^2 : f(u, v) = 0\} \subseteq F^2$$

is the *curve defined by f*. We assume that f is not a constant polynomial, and write $X = \{f = 0\}$. If f is irreducible, we also call X *irreducible*. For this and some other notions, it is useful to assume F to be algebraically closed.

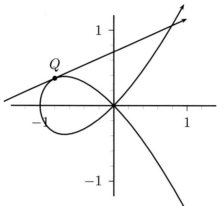

Figure 5.6: Descartes' foil $y^2 = x^3 + x^2$ over the real numbers.

A point $P = (u, v) \in X$ is *nonsingular* (or *smooth*) on X if the tangent to X at P is well-defined, that is, if

$$\left(\frac{\partial f}{\partial x}(u, v), \frac{\partial f}{\partial y}(u, v) \right) \neq (0, 0).$$

X is *nonsingular* if all points on it are nonsingular.

EXAMPLE 5.9. *Descartes' foil* is given by the polynomial $f = y^2 - (x^3 + x^2)$; see Figure 5.6. We have $(\partial f / \partial x, \partial f / \partial y) = (-3x^2 - 2x, 2y)$, and this evaluates to $(0, 0)$ at only one point on the curve, namely at $P = (0, 0)$. This is the only singular point; the curve is not nonsingular, it is *singular*. At $Q = (-3/4, 3/8) \in X$, the partial derivatives are $(-3/16, 3/4) \neq (0, 0)$. Therefore Q is a nonsingular point on X, and in fact the slope of the tangent to X at Q is $-3/16 : 3/4 = -1/4$.   ◊

To any curve X is associated a nonnegative integer, the *genus $g(X)$* of X. If X is a projective nonsingular irreducible plane curve, defined by an irreducible polynomialof degree d, then $g(X) = (d-1)(d-2)/2$. Thus the parabola $\{y = x^2\}$ and the hyperbola $\{xy = 1\}$ have genus 0. An *elliptic curve* is a nonsingular irreducible curve of genus 1.

The polynomial $x^3 + ax + b$ is squarefree, that is, without multiple roots, if and only if the "discriminant" condition $\Delta = 4a^3 + 27b^2 \neq 0$ from Definition 5.1 holds (Exercise 5.2).

We convince ourselves that an elliptic curve is nonsingular. Let $f = y^2 - (x^3 + ax + b) \in F[x,y]$ define the elliptic curve $E = \{f = 0\} \cup \{\mathcal{O}\}$, assuming char $F \neq 2, 3$. For $(u, v) \in E \setminus \{\mathcal{O}\}$, we have

$$\left(\frac{\partial f}{\partial x}(u, v), \frac{\partial f}{\partial y}(u, v) \right) = (-3u^2 - a, 2v).$$

If this is $(0, 0)$, meaning that (u, v) is singular on E, then $v = 0$ (since $2 \neq 0$ in F) and $0 = 3 \cdot (u^3 + au + b) + u \cdot (-3u^2 - a) = 2au + 3b$. Eliminating u and squaring yields

$$-\frac{a}{3} = u^2 = \left(\frac{-3b}{2a} \right)^2 = \frac{9b^2}{4a^2}.$$

This implies that $\Delta = 0$, which contradicts our choice of E. On the other hand, if $\Delta = 0$, then $(-3b/2a, 0)$ is singular on the curve $\{f = 0\}$ if $a \neq 0$. For $a = 0$, we have $b = 0$ and $(0, 0)$ singular.

Besides the points of X as explained above, X also has *points at infinity*, namely, the points of the *projective plane* that satisfy the projective version of the polynomial f. This is visualized in Figure 5.3 and best understood in the framework of projective geometry. The *projective plane* $\mathbb{P}^2$ over F consists of all triples $(u : v : w)$ with $u, v, w \in F$, not all zero, where we identify two such triples if they are multiples of each other: $(u : v : w) = (\lambda u : \lambda v : \lambda w)$ for all nonzero $\lambda \in F$. We may also regard $(u : v : w)$ as the line in F^3 through (u, v, w) and the origin. We can embed the *affine plane* F^2 in $\mathbb{P}^2$ by mapping $(u, v) \in F^2$ to $(u : v : 1)$. If we denote by x, y, and z the projective coordinates on $\mathbb{P}^2$, then the image of this map is $\{z \neq 0\} \cong F^2$, and the inverse of this embedding maps $(u : v : w)$ with $w \neq 0$ to $(u/w, v/w) \in F^2$.

The corresponding transformation on polynomials associates to $f \in F[x, y]$ its homogeneous version $\tilde{f} \in F[x, y, z]$, where all summands are multiplied by an appropriate power of z so that all degrees become equal. Thus $y^2 - (x^3 + ax + b)$ is mapped to $y^2z - (x^3 + axz^2 + bz^3)$, which is homogeneous of degree 3. It does not make sense to "evaluate" $\tilde{f}$ at $(u : v : w)$, but it does make sense to say that "$\tilde{f}(u : v : w) = 0$", since $\tilde{f}(u, v, w) = 0$ if and only if $\tilde{f}(\lambda u, \lambda v, \lambda w) = \lambda^{\deg \tilde{f}} \tilde{f}(u, v, w) = 0$ for all nonzero $\lambda \in F$. Now the projective curve in $\mathbb{P}^2$ corresponding to an elliptic curve E given by $y^2 = x^3 + ax + b$ is

$$\tilde{E} = \{(u : v : w) \in \mathbb{P}^2 : v^2w = u^3 + auw^2 + bw^3\},$$

and $E \cap F^2 = \{(u,v) \in F^2 : v^2 = u^3 + au + b\}$ is in bijective correspondence with the *affine part* $\tilde{E} \cap \{z \neq 0\}$ via the above substitution. The intersection with the *line at infinity* is

$$\tilde{E} \cap \{z = 0\} = \{(u : v : w) \in \mathbb{P}^2 : w = u = 0\} = \{(0 : 1 : 0)\} = \{\mathcal{O}\}.$$

The choice of 1 for the second coordinate is arbitrary.

Over the complex numbers, an elliptic curve is a one-dimensional curve and hence a two-dimensional real surface. It looks like a torus. You can think of the real picture as being the intersection of such a torus with a plane; this requires a four-dimensional imagination in $\mathbb{C}^2 \cong \mathbb{R}^4$.

Elliptic curves were instrumental in proving one of the most celebrated mathematical results of the end of the last century: Fermat's Last Theorem. It says that $a^n + b^n = c^n$ is impossible with positive integers a, b, c, n and $n \geq 3$. It had been stated by Pierre de Fermat (1601/1607/1608–1665) in 1640 *"Cuius rei demonstrationem mirabilem sane detexi. Hanc marginis exiguitas non caperet."*[1] A proof remained a major challenge until Wiles (1995) and Taylor & Wiles (1995).

To put elliptic curves into perspective, we note that all curves of genus 0 are "isomorphic" to the line, via rational mappings. On the line, we have the addition as a group operation. This actually only works on the affine line, but not on the projective line $\mathbb{P}^1$. On the affine hyperbola $\{xy = 1\}$, we can define a group operation by $(u,v) \cdot (u',v') = (uu', vv')$. On any curve of genus 1 (that is, an elliptic curve), we have an addition as defined above. And, in fact, on any curve of higher genus it is impossible to define a group law by rational functions.

But one can define a natural group for any irreducible nonsingular projective curve X. We take as G the group of all formal finite sums $\sum_i a_i P_i$, where all $a_i \in \mathbb{Z}$, all $P_i \in X$, and $\sum_i a_i = 0$. Let f be a rational function that is the quotient of two homogeneous polynomials of the same degree. One can associate such a sum to f; this is the *divisor* $\mathrm{div}(f)$ of f. The P_i that occur are the zeroes and poles of f on X, and the a_i are the corresponding multiplicities. If H denotes the subgroup of all these $\mathrm{div}(f)$, then $J = G/H$ is the group mentioned. It is called the *Jacobian* of X and is an irreducible nonsingular projective variety, with dimension equal to the genus of X. When X is an elliptic curve, the dimension of J is 1, so it is also a curve and in fact isomorphic to X. Three collinear points P, Q, S on X, as in (5.5), satisfy $P + Q + S = 0$. Then the formal sum $s = P + Q + S - 3\mathcal{O}$ is an element of H, and H is generated by all

---

[1] I have discovered a truly marvelous demonstration of this proposition that this margin is too narrow to contain. [This is not admissible for a student to write on an exam.]

these expressions. If $\ell \in F[x, y, z]$ is the equation of the projective line through the three points, then $s = \mathrm{div}(\ell/z)$ and ℓ/z has a threefold pole at 0.

Jacobians of higher genus curves can be substituted in cryptography for elliptic curves. Their advantage is an even smaller key size, their disadvantage a more complicated calculation for the addition. A popular choice for genus $g \geq 2$ are the *hyperelliptic curves* given by $y^2 = f$ with $f \in F[x]$ squarefree of degree $2g + 1$ (in odd characteristic); for $g = 1$ we recover elliptic curves. We do not treat these more general curves in this text.

5.3. The size of an elliptic curve

In the remainder of this chapter, we consider elliptic curves over finite fields. Our first task is to determine the size of such an elliptic curve, that is, to estimate the number of points on it. The following estimate is easy and crude.

THEOREM 5.10. *Let E be an elliptic curve over the finite field $\mathbb{F}_q$ of characteristic greater than three. Then $\#E \leq 2q + 1$.*

PROOF. For each of the q possible values for u, there are at most two possible values for v such that $v^2 = u^3 + au + b$, corresponding to the two possible square roots of $u^3 + au + b$. Adding the point at infinity gives the required estimate. □

As an example, it follows that the size s of any elliptic curve over the field $\mathbb{F}_{25\,013}$ is between 1 and $50\,027$. But when we actually count those curves, we find that s always lies in a tiny subinterval, as shown in Figure 5.7. See Figure 5.9 for an enlargement of the critical zone.

One reason to think that the estimate is crude is the following. Pretending that the value of $u^3 + au + b$ varies randomly as u ranges over $\mathbb{F}_q$, we should expect that for about half of the u's, there would be two solutions v for the equation, and no solution for the other half. In other words, $u^3 + au + b$ should be a square about half of the time. Random elements have this property; see Theorem 15.68 (i). More formally, we consider the *quadratic character* $\chi\colon \mathbb{F}_q \to \{1, 0, -1\}$ defined by

$$\chi(c) = \begin{cases} 1 & \text{if } c \text{ is a square,} \\ 0 & \text{if } c = 0, \\ -1 & \text{otherwise.} \end{cases}$$

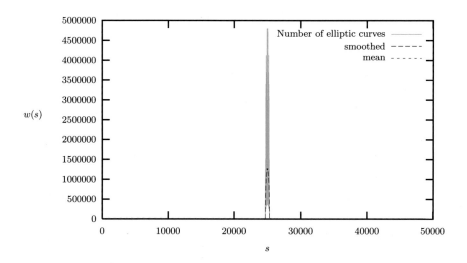

Figure 5.7: The number $w(s)$ of Weierstraß parameters of elliptic curves over $\mathbb{F}_{25\,013}$ with s points in the range 0 to 50027.

For q prime, $\chi(c) = \left(\frac{c}{q}\right)$ is the *Legendre symbol* (Section 15.14), and the number of square roots of any $c \in \mathbb{F}_q$ is

$$\#\{v \in \mathbb{F}_q : v^2 = c\} = 1 + \chi(c).$$

From this we conclude that

$$(5.11) \quad \#E = 1 + \sum_{u \in \mathbb{F}_q} (1 + \chi(u^3 + au + b)) = q + 1 + \sum_{u \in \mathbb{F}_q} \chi(u^3 + au + b).$$

If $\chi(u^3 + au + b)$ was a uniformly distributed random variable, then the sum would behave like a random walk on the line. After q steps of such a random walk, we expect to be about $\sqrt{q}$ steps away from the origin; see Exercise 3.26. Of course this is not at all a random process, but the analogy provides some motivation for the following result, which we state without proof.

THEOREM 5.12 (Hasse's bound). *If E is an elliptic curve over the finite field $\mathbb{F}_q$, then $|\#E - (q+1)| \leq 2\sqrt{q}$.*

EXAMPLE 5.13. Let $q = 7$. By the Hasse bound, each elliptic curve E over $\mathbb{F}_7$ has $|\#E - 8| \leq 2\sqrt{7} \simeq 5.3$, so that $3 \leq \#E \leq 13$. We have seen two curves in Example 5.3, both of size 8. Figure 5.8 represents the sizes of all 42 elliptic curves in Weierstraß form over $\mathbb{F}_7$.   ◇

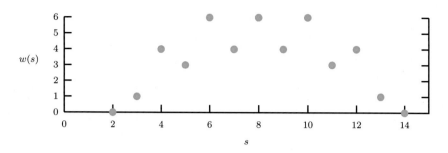

Figure 5.8: The number $w(s)$ of Weierstraß parameters of elliptic curves over $\mathbb{F}_7$ with s points.

Hasse's bound implies that roughly the top halves of the decimal (or binary) representations of q and of $\#E$ agree, except for strings of 0's or 9's. The NIST curve of Table 5.14 provides an example.

We now write $\#E = q + 1 - t$, with $t \in \mathbb{Z}$. By Hasse's Theorem, $|t| \leq 2\sqrt{q}$. Which values of t do occur? If q is prime, then actually all values permitted by Hasse's bound occur, and in the "middle half" $|t| \leq \sqrt{q}$, the values of t are fairly evenly distributed.

EXAMPLE 5.14. We take the prime $q = 25013$. Hasse's bound in Theorem 5.12 says that for any elliptic curve with s points over $\mathbb{F}_q$ we have $24698 \leq s \leq 25340$. Figure 5.9, a blow-up of the central part of Figure 5.7, shows for each such s the number $w(s)$ of Weierstraß parameters (a, b) whose curves have exactly s points.   ◊

Any elliptic curve can be represented by some Weierstraß parameters a and b. But different parameters may give *isomorphic* elliptic curves, as considered in Exercise 5.11. The frequency statements above are in terms of Weierstraß parameters.

5.4. Elliptic curve cryptography

All group-based cryptographic systems that we have discussed can be implemented with elliptic curves, among them:

- Diffie-Hellman key exchange (Protocol 2.20),

- ElGamal encryption system (Protocol 4.5),

- ElGamal signature scheme (Protocol 8.1),

- Schnorr identification scheme (Section 10.2),

- Okamoto identification scheme (Section 10.3).

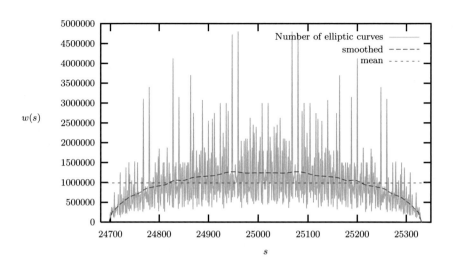

Figure 5.9: The number $w(s)$ of Weierstraß parameters of elliptic curves over $\mathbb{F}_{25013}$ with s points.

A point (u, v) on an elliptic curve over an n-bit finite field $\mathbb{F}_q$ is given by its two coordinates, of total length $2n$. Given just u, it is easy to calculate $w = u^3 + au + b$. Now v is one of the two square roots of w (if $w \neq 0$), and can easily be calculated from w, for example, by factoring $y^2 - w$. The other root is $-v$. If q is prime and the elements of $\mathbb{F}_q$ are written in the symmetric representation (15.12), then these roots are $\pm v$, one negative and one positive. Thus, the point can be specified as "$(u, +)$" or "$(u, -)$". This is called *point compression* and we illustrate it in the following. It only uses $n + 1$ bits.

EXAMPLE 5.15. We perform the Diffie-Hellman key exchange 2.20 on the elliptic curve E^* with Weierstraß equation $y^2 = x^3 + x$ over $\mathbb{F}_7$ from Example 5.3. $E^* = \langle P \rangle$ is generated by $g = P = (3, 3)$ and has $d = 8$ elements. The global set-up produces all these values. The following calculations are easy to follow by looking at the right part of Figure 5.5, but of course the example is too small to convey any sense of security.

2. ALICE chooses her secret key $a = 3 \xleftarrow{\boxtimes} \mathbb{Z}_8$. She computes her public key $A \leftarrow 3P = (-2, -2) = (-2, -) \in G$. The multiplicative assignment in Protocol 2.20 now becomes additive: $A \leftarrow g^a$ becomes $A \leftarrow a \cdot P$.

3. BOB chooses his secret key $b = 5 \xleftarrow{\boxtimes} \mathbb{Z}_8$. He computes his public key $B \leftarrow 5P = (-2, 2) = (-2, +) \in G$.

4. ALICE and BOB exchange their public keys A and B.

5. ALICE computes the common key $k_A = 3B = 3 \cdot (-2, 2) = (3, -3) = (3, -)$.

6. BOB computes the common key $k_B = 5A = 5 \cdot (-2, -2) = (3, -3) = (3, -)$.

Thus $k_A = k_B = (3, -3) = (3, -) = 15P = 7P$ is the secret key shared by ALICE and BOB. $\Diamond$

A concrete application is a secure Diffie-Hellman key exchange on a NIST curve over a 192-bit prime field (Section 5.6) via just two short messages (SMS) on GSM mobile phones (Berndt 2010, Schröder 2012). This cannot be achieved in groups like $\mathbb{Z}_p^\times$.

EXAMPLE 5.16. Suppose that ALICE wants to encrypt the message $1 \in \mathbb{Z}_p$ for BOB, using the ElGamal encryption scheme 4.5. She turns the plaintext into a point on $G = E^*$ by choosing one of the possible second coordinates in ± 3, say $x = (1, -3) \in G$. The global set-up is the same as for the previous example. The rest of the protocol runs as follows.

2. BOB chooses his secret key $\mathsf{sk} = b = 3 \xleftarrow{\boxtimes} \mathbb{Z}_8$ at random. He computes his public key $\mathsf{pk} = B = bP = 3(3,3) = (-2, -2) \in G$ and publishes it.

3. ALICE chooses a secret session key $a = 4 \xleftarrow{\boxtimes} \mathbb{Z}_8$ at random.

4. Public session key $A \xleftarrow{\boxtimes} aP = 4P = (0, 0) \in G$, and shared session key $k = aB = 4(-2, -2) = (0, 0)$.

5. $y \leftarrow x + k = (1, -3) + (0, 0) = (1, 3) \in G$.

6. Return $\mathsf{enc}_{\mathsf{pk}}(x) = (y, A) = ((1, 3), (0, 0))$.

7. BOB calculates the shared session key $k \leftarrow bA = 3(0, 0) = (0, 0)$ and the inverse $-k = -(0, 0) = (0, 0) \in G$ of the shared key.

8. $z \leftarrow y + (-k) = (1, 3) + (0, 0) = (1, -3)$.

9. Return $z = (1, -3)$. Indeed, this equals the plaintext x. $\Diamond$

In order to generate a group whose size is a prime number with a specified number n of bits, say $n = 256$, one chooses a prime p with n digits and n-bit integers a and b, and computes the size d of the elliptic curve $\{y^2 = x^3 + ax + b\}$ over $\mathbb{Z}_p$ as described in Section 5.7 below. The

NIST procedure of Section 5.6 actually takes a special p and $a = -3$. We want d to be prime, or not far from being prime. A typical choice is to test the integer values among d, $d/2$, $d/3$, and $d/4$ for primality and accept the curve if one of them is indeed prime. Since the sizes of elliptic curves are reasonably equally distributed in (half of) the Hasse range (Section 5.3), one expects to test about $256 \cdot \log_e 2 \approx 162$ many Weierstraß pairs a, b before finding an appropriate curve E of prime order. A point P on E is easy to find by testing for random $u \in \mathbb{Z}_p$ whether $u^3 + au + b$ is a square in $\mathbb{Z}_p$ and, if so, extracting a square root v to obtain $P = (u, v) \in E$. If E has prime order, any point $P \neq \mathcal{O}$ on it generates the whole group. (In fact, it is more convenient to generate u, v, and a, maybe at random, and then compute b.) The NIST, the European *ECC Brainpool* and others have published lists of appropriate (p, a, b, d, P); see Section 5.6.

What about discrete logarithms in elliptic curves? No general method is known except generic algorithms (Sections 4.3 through 4.6). If the size of E is an n-bit prime, then these take time about $2^{n/2}$. This lack of other methods makes elliptic curves attractive for cryptography. Will a better method be found? Some special types of curves are known to have an easy discrete logarithm problem.

At the time of writing (2015), the record in elliptic curve discrete logarithm computations over a prime field is by Bos *et al.* (2009). They took a 112-bit prime p and an elliptic curve $E = \langle P \rangle$ over $\mathbb{Z}_p$ recommended in the two standards SEC and SEC2. Furthermore, they took a point Q on E whose x-coordinate agreed with the decimal expansion of π (after the decimal point) to 34 digits. And then they started playing, from January to July 2009. Namely, they house in a cool basement a lab of over 200 PlayStation 3 game consoles. They ran their discrete logarithm software, and eventually the cool playstations coughed up $\mathrm{dlog}_P Q$, after an effort totalling about 2^{63} operations. Wenger & Wolfger (2014) computed a discrete logarithm in an elliptic curve over $\mathbb{F}_{2^{113}}$ on a small FPGA cluster.

5.5. Empirical cryptographic security

Two types of asymmetric cryptosystems are prevalent today: RSA and group-based cryptography. These are broken when we can factor integers or compute discrete logarithms efficiently. It is also conjectured—but not proven—that they are secure when these two number-theoretic problems are hard. This security property is proven for the HKS cryptosystem of Section 9.9.

We now describe some practical considerations for the security of these systems. Such requirements increase continuously but slowly as computing power evolves according to Moore's law and in jumps when unexpected algorithmic progress is made. The basic question is: how large a computation can one perform in a real-world cryptanalytic attack? The current assumption is that 2^{80} operations may be feasible, and 2^{128} are not; in between is a gray area. A system resisting known attacks using 2^{128} operations is said to have "128-bit security". The 2^{63} factorization success of Bos *et al.* (2009) showed that the earlier assumption of only 2^{60} attacks being feasible is obsolete.

There are two complementary notions of security. In *empirical security*, the system parameters are chosen so that known attacks do not work, but new attack methods cannot be ruled out. Often this is the best one can achieve, for example, for symmetric systems. The second approach, *axiomatic security*, is discussed in Chapter 9.

DEFINITION 5.17. *A cryptographic system has s-bit empirical security if it withstands the known attacks when 2^s operations are allowed.*

	s	cycles	time at 100 GHz	time at 10^7 GHz
insecure	60	10^{18}	4 months	
legacy	80	10^{24}	300 000 years	3 years
current	100	10^{30}	$3 \cdot 10^{11}$ years	3 million years
near-term	128	$4 \cdot 10^{38}$	10^{20} years	80,000 LU
long-term	256	10^{77}	$3 \cdot 10^{58}$ years	$3 \cdot 10^{43}$ LU

Table 5.10: 60- to 256-bit security.

Table 5.10 gives a rough perspective on various values of s. We just account for 2^s cycles, while the "operations" in Definition 5.17 may be more costly, say, computing an encryption. The "times" are properly calculated but should not be taken too seriously. The last "astronomical" line goes to show where such calculations can take you. LU is the lifetime of the universe, which is estimated at around $13.7 \cdot 10^9$ years. The fourth column represents the computing power of a small university laboratory, and for the last column, we think of grid computing or large clusters.

Near-term refers to the next ten years, say until around 2025, and long-term to the next 30 to 50 years. As long as Moore's law continues to hold, security needs to be increased by about one bit every 1.5 years. The slow but methodical progress on cryptanalytic algorithms over the past decades has contributed about as much as hardware advances according to Moore, so that one bit every nine months should be added to security requirements. Of course, the long-term considerations are pretty much guesswork and may well be ruined by unexpected advances; see Section 4.8 for one that actually happened for special discrete logarithms.

method	security	$s = 80$	$s = 128$
AES	≥ 128	✓	✓
RSA, DL in finite fields	$1.902 \cdot \sqrt[3]{n \log_2^2 n}$	$n \approx 780$	$n \approx 2352$
DL in elliptic curves	$n/2$	$n = 160$	$n = 256$

Table 5.11: Empirical security (in 2015) of AES, RSA, and elliptic curve cryptography.

Standardization bodies like the NIST (2012b) and the German *Bundesamt für Sicherheit in der Informationstechnologie* (BSI 2014) extrapolate minimal key size recommendations in Table 5.12 by comparing the cost of currently (and publicly) known attacks to the security requirements. The "hash" column presents the minimal bit size of hash values for use in signatures. In other scenarios such as HMACs (Section 10.5), shorter values are deemed sufficient, since only one-wayness rather than collision resistance is required. For Schnorr signatures (or DSA) one should combine the RSA and ECC values, thus taking $\ell \geq 256$ and $n \geq 3000$ in Protocol 8.5 for near-term security.

How does one arrive at such recommendations? For RSA, we have to factor n-bit integers, and for the DL, to compute discrete logarithms in an n-bit group G. There are two fundamental computational approaches: index calculus for factoring integers and discrete logarithm in finite fields, and Pollard's rho method for elliptic curves; nothing better is known in general. The running times for these approaches are (heuristically) expected to be $\exp_2((1.902 + o(1)) \sqrt[3]{n \log_2^2 n})$ and $\exp_2(n/2)$, respectively, at key length n. Here $\exp_2(x) = 2^x$ and $o(1)$ tends to zero as n grows; it is customary to ignore this term in practical considerations. Table 5.11 gives the numerical solution of the corresponding equations. The recommendations roughly match the values in Table 5.11, with an additional security margin.

	s	AES	RSA	ECC	hash
legacy	80		1024	160	160
near-term	128	128	3000	256	256
long-term	256	256	15360	512	512

Table 5.12: Key-size recommendations in bits.

Such recommendations, often made by the relevant government authority of a country, have far-reaching consequences. Systems to be used in government contracts or for legally binding digital signatures, will have to implement them. This is often difficult to impossible, since legacy software or hardware is hard to change. Publications such as those of NIST and the BSI discuss these issues and give advice on how to deal with them.

The security levels of RSA vs. ECC compare roughly as $\sqrt[3]{n}$ vs. $n/2$. The consequence is visible in Tables 5.11 and 5.12: much smaller key lengths are required for ECC than for RSA to provide comparable security.

All such recommendations make the explicit assumption of no unexpected major advances in cryptanalysis, such as the surprising progress on discrete logarithms in special finite fields (Section 4.8) or scalable quantum computers (Chapter 14). It would be naive to assume that the best methods are always publicly known.

5.6. The NIST curves

The US government agency NIST (1999b) has set standards for elliptic curves. These NIST curves must be used for US government applications, but are deployed much more widely. For five values of the security parameter s, they specify bit lengths $\ell = \lceil \log_2 p \rceil$ for prime fields $\mathbb{F}_p$ and m for fields $\mathbb{F}_{2^m}$. Table 5.13 exhibits the four larger ones. We note

s	ℓ	m
112	224	233
128	256	283
192	384	409
256	521	571

Table 5.13: Bit lengths for NIST curves.

that $\ell \approx 2s$ and m is somewhat larger, and always prime. The three larger values of s equal the AES key lengths. In the prime field case, for each ℓ an ℓ-bit prime p_ℓ is chosen as a *generalized Mersenne prime*, which is the sum of terms $\pm 2^{32k}$ for at most five different values of k, except that $p_{521} = M_{521} = 2^{521} - 1$ is a Mersenne prime (Section 3.2). An integer can be reduced modulo such a prime with fairly simple arithmetic on 32-bit words, faster than modulo an arbitrary prime of the same size. Furthermore, some $b_\ell \in \mathbb{F}_{p_\ell}$ is chosen, as explained below, and a generator $P_\ell = (x_\ell, y_\ell)$ of the elliptic curve E_ℓ with Weierstraß equation $y^2 = x^3 - 3x + b_\ell$. The choices of p_ℓ and -3 as coefficient are motivated by computational advantages; one hopes that this does not essentially affect the security. E_ℓ is chosen so that $\#E_\ell = p_\ell + 1 - t_\ell$ is prime. In view of the reproach of intransparency in the choice of DES in 1971 (see

$$p_{256} = 2^{256} - 2^{224} + 2^{192} + 2^{96} - 1$$

$$= \quad 11579208\ 9210356248\ 7626974469\ 4940757353\backslash$$
$$0086143415\ 2903141955\ 3363130886\ 7097853951$$

$$\#E_{256} = \quad 11579208\ 9210356248\ 7626974469\ 4940757352\backslash$$
$$9996955224\ 1357603424\ 2225906106\ 8512044369$$

$$b_{256} = \quad 4105836\ 3725152142\ 1293261297\ 8004726840\backslash$$
$$9114441015\ 9937255548\ 3525631403\ 9467401291$$

$$t_{256} = \quad 89188191\ 1545538531\ 1137224779\ 8585809583$$

$$x_{256} = \quad 4843956\ 1293906451\ 7590525852\ 5279791420\backslash$$
$$2762949526\ 0417479958\ 4408071708\ 2404635286$$

$$y_{256} = \quad 3613425\ 0956749795\ 7985851279\ 1958788195\backslash$$
$$6611106672\ 9850150718\ 7719825356\ 8414405109$$

Table 5.14: The NIST curve for $n = 128$ (and $\ell = 256$).

Section 2.2), NIST explains the choices it made. For $\ell = 256$, this looks as follows: The prime p_ℓ has been motivated above. For b_{256}, they start with a certain 160-bit string s, obtain c via the concatenation of SHA-1(s) and SHA-1($s + 1$) (see Section 7.5) and take b_{256} so that $b_{256}^2 = -27/c$ in $\mathbb{F}_{p256}$ and $\#E_{256}$ has prime order. This is repeated with various choices of s until the requirements are met. The generator P_{256} is included only for convenience; users may take other generators. For the other values of $\ell \neq 256$, the procedure is similar. Given s, it is easy to check that the values in the above table were actually produced in this way; see Exercise 15.6 for verifying $\#E_\ell$.

The choice of s is not motivated in the standard. Under the assumption that either no-one has special values of b_ℓ for which discrete logarithms on E_ℓ are particularly easy, or that no-one knew how to invert SHA-1 in 1999 (see Section 7.5 for the state of the art in 2015), all is well. The opposite assumption leads to an implausible conspiracy theory.

Discrete logarithm computations can be sped up somewhat by precomputations that are independent of the point whose dlog is sought. Conceivably, security agencies or organized criminality might gain an advantage if many users take the same curve.

The European ECC Brainpool (2010) proposes a method to generate elliptic curves of a desired bit length starting from a 160-bit seed and using SHA-1 (Section 7.5) to produce random-looking numbers. The primes are random rather than special as with the NIST curves. They also suggest some specific curves, taking parts of the binary expansion of $e = \exp(1)$ as seed. Security-minded organizations may want to produce their own seeds—which may be made public—and generate their own curves. It is required that the parameter generation be random and well documented.

5.7. Computing the size of an elliptic curve

For cyptographic applications, it is usually required to know exactly how many points there are on a given elliptic curve. We now describe a polynomial-time algorithm by Schoof (1985) which counts such a curve. Its running time is $O(\log^8 q)$ over $\mathbb{F}_q$, where q is a prime power. This is polynomial in the input size. The original version presented here is fairly slow, but there are clever variations of it that are eminently practical.

Hasse's bound in Theorem 5.12 is equivalent to

$$(5.18) \qquad\qquad |t| \leq 2\sqrt{q}$$

for the integer $t = q + 1 - \#E$. In the following, we describe an algorithm to compute t, from which $\#E$ is immediately calculated. The *Frobenius automorphism* is defined as

$$\varphi: \begin{array}{ccc} E & \longrightarrow & E, \\ (u,v) & \longmapsto & (u^q, v^q), \end{array}$$

with $\varphi(\mathcal{O}) = \mathcal{O}$. Fermat's Little Theorem 15.56 says that $\varphi(P) = P$ for all P, so that φ is quite boring. But in the next section, we apply the following ideas to an extension field K of $\mathbb{F}_q$. A fundamental relation is

$$(5.19) \qquad\qquad \varphi^2 - t\varphi + q = 0,$$

that is, for all $P \in E$

$$P^{q^2} - tP^q + qP = \mathcal{O}.$$

A first attempt to calculate the "trace" t would be to take an arbitrary point P on E, compute qP, $\varphi(P) = P^q$, and $\varphi^2(P) = P^{q^2}$, and try out all possible values of t until one finds one with

$$(5.20) \qquad\qquad t\varphi(P) = \varphi^2(P) + qP.$$

According to (5.18), we may have to try up to $4\sqrt{q} + 1$ values of t, which is exponential in the bit size $\log q$ of $\mathbb{F}_q$. In fact, computing $t = \mathrm{dlog}_{\varphi(P)}(\varphi^2(P) + qP)$ is a discrete logarithm problem in E. As in any finite group, $(\#E) \cdot P = \mathcal{O}$ for any $P \in E$, by Lagrange's Theorem 15.53. The order d of P is the smallest positive integer so that $dP = \mathcal{O}$; d is a divisor of $\#E$. These are the additive translations of the multiplicative statements in Section 15.13.

Now we take some integer ℓ and the set

$$E[\ell] = \{P \in E : \ell P = \mathcal{O}\}$$

of *ℓ-torsion points* in E. It is a subgroup of E, since $\ell(P+Q) = \ell P + \ell Q = \mathcal{O} + \mathcal{O} = \mathcal{O}$ for any $P, Q \in E[\ell]$. We have $E[1] = \{\mathcal{O}\}$, $E[m] \subseteq E[\ell]$ if m divides ℓ, and $P \in E[\ell]$ if and only if the order of P in E divides ℓ, by Corollary 15.54. But this order divides $\#E$, so that for a prime number ℓ, we have $E[\ell] = \{\mathcal{O}\}$ unless ℓ divides $\#E$. There are few such primes.

A brilliant idea of Schoof (1985) leads to a polynomial-time algorithm for t. It has several ingredients.

○ Apply the Chinese Remainder Theorem 15.44 by computing t modulo several small primes ℓ.

○ Each ℓ is so small that we can search exhaustively for $t_\ell = t$ in $\mathbb{Z}_\ell$, satisfying (5.20) in $E[\ell]$.

A problem is that most $E[\ell]$ have just one element, except when ℓ is a divisor of $\#E$. These are few, perhaps large, and we do not know them. But we do not give up: if $E[\ell]$ is too small by nature, just make it bigger artificially!

○ Consider the elliptic curve $E(K)$ given by the same equation as E, but over an extension field K of $\mathbb{F}_q$.

In fact, there is a standard tool for doing this: instead of our field $\mathbb{F}_q$, we take an algebraic closure K of $\mathbb{F}_q$. This is obtained by "adjoining" to $\mathbb{F}_q$ roots of any irreducible polynomial in $\mathbb{F}_q[x]$. It is an infinite field of the same characteristic as $\mathbb{F}_q$, and any polynomial in $K[x]$ has a root in K. As an example, the complex numbers form an algebraic closure of the field of real numbers.

The polynomial $f = y^2 - (x^3 + ax + b) \in \mathbb{F}_q[x, y]$ from (5.2) also defines a plane curve over K, denoted as

$$E(K) = \{(u, v) \in K^2 : f(u, v) = 0\} \cup \{\mathcal{O}\}.$$

The Frobenius automorphism φ also acts on $E(K)$, since for $P = (u, v) \in E(K)$ we have from (15.4)

(5.21)
$$\begin{aligned} f(\varphi(P)) &= (v^q)^2 - ((u^q)^3 + au^q + b) \\ &= (v^2 - (u^3 + au + b))^q = 0. \end{aligned}$$

Now φ is more interesting, since $\varphi(P) \neq P$ for "most" $P \in E(K)$. We can apply φ to the addition and doubling formulas (5.6) and (5.7). Since the Frobenius map is an automorphism of K (that is, a ring isomorphism from K to K), we have $\varphi(P) + \varphi(Q) = \varphi(P + Q)$. We also find $\varphi(2P) = 2\varphi(P)$, where we use the additional fact that $a^q = a$, by Fermat's Little Theorem 15.56. Overall, we find that $\varphi(kP) = k\varphi(P)$ for all $k \in \mathbb{Z}$ and $P \in E(K)$. A fundamental fact is that the equation (5.19) also holds for φ over $E(K)$. For a prime ℓ, the ℓ-torsion points $E(K)[\ell]$ in $E(K)$ form a group isomorphic to $\mathbb{Z}_\ell^2$. Now a new problem arises: how do we find a "random" point in P in $E(K)$? Or in $E(K)[\ell]$? Computing with general elements of K quickly leads to intermediate results of unmanageable size. Again, there is an easy solution.

○ Find the defining polynomial of $E(K)[\ell]$, called the *division polynomial* ψ_ℓ, and compute modulo ψ_ℓ.

We start with the first ingredients, leaving the division polynomials to the next section.

So we have a uniformly random ℓ-torsion point $P \xleftarrow{\;\boxtimes\;} E(K)[\ell] \setminus \mathcal{O}$ for some prime ℓ different from the characteristic p of $\mathbb{F}_q$. Then all points Q under consideration, namely $Q \in \{qP, \varphi(P), t\varphi(P), \varphi^2(P), \varphi^2(P) + qP\}$, are also in $E(K)[\ell]$, and $\ell Q = \mathcal{O}$ for any of them. Thus we may work modulo ℓ in this group, and only have to test the values in the symmetric residue system $R_\ell = \{-\lfloor \ell/2 \rfloor, \ldots, \lfloor (\ell-1)/2 \rfloor\}$ from (15.12) for t. If we let t_ℓ and q_ℓ be the remainders of t and q, respectively, in R_ℓ, then we can compute t_ℓ in this way, trying out at most ℓ values for t_ℓ.

Now if we do this for sufficiently many primes ℓ, namely such that

(5.22) $$u = \prod \ell > 2\sqrt{q},$$

then by the Chinese Remainder Theorem 15.44 we have t modulo u, from which we immediately get t (see the end of Section 15.11). If we take all primes from 2 to some bound B, then by the Prime Number Theorem 3.21 we have

$$\ln \prod_{\substack{2 \leq \ell \leq B \\ \ell \text{ prime}}} \ell = \vartheta(B) \approx B,$$

and it is sufficient to take $B \approx \ln(2\sqrt{q}) = \left(1 + \frac{1}{2}\log_2 q\right)\ln 2$. If m is the number of primes used, then $m \approx B/\ln B$, and both B and m are polynomial in the bit size of the field. We have to test fewer than $mB \in O(\log^2 q)$ values of t; this is a polynomial number of tests.

5.8. Division polynomials

We now discuss the *division polynomials* ψ_ℓ which describe the ℓ-torsion points in $E(K)[\ell]$. It will turn out that there are always such points, so that we can implement the algorithm of the previous section over K. Furthermore, rather than computing points in $E(K)[\ell]$ explicitly, we can work with them symbolically over $\mathbb{F}_q$, using the polynomials ψ_ℓ. We assume $\ell \neq p$ in this section.

Since addition and doubling on an elliptic curve are described by rational functions in the coordinates, for any ℓ also the two coordinates of an arbitrary multiple ℓP of $P = (u, v) \in E$ are given by rational functions, provided no intermediate result equals $\mathcal{O}$. These functions can be written in the special form

$$(5.23) \qquad \ell P = \left(x - \frac{\psi_{\ell-1}\psi_{\ell+1}}{\psi_\ell^2}, \frac{\psi_{\ell+2}\psi_{\ell-1}^2 - \psi_{\ell-2}\psi_{\ell+1}^2}{4y\psi_\ell^3}\right)(u, v)$$

with $\psi_\ell \in \mathbb{F}_q[x, y]$ as described below. Equation (5.23) is valid if $\ell P \neq \mathcal{O}$, and in fact

$$\ell P = \mathcal{O} \Longleftrightarrow P = \mathcal{O} \text{ or } \psi_\ell(P) = 0.$$

We give recursive formulas for these polynomials in the case $p > 3$:

$$\psi_0 = 0,$$
$$\psi_1 = 1,$$
$$\psi_2 = 2y,$$
$$\psi_3 = 3x^4 + 6ax^2 + 12bx - a^2,$$
$$\psi_4 = 4y(x^6 + 5ax^4 + 20bx^3 - 5a^2x^2 - 4abx - a^3 - 8b^2),$$
$$\psi_{2\ell+1} = \psi_{\ell+2}\psi_\ell^3 - \psi_{\ell-1}\psi_{\ell+1}^3 \qquad \text{for } \ell \geq 2,$$
$$\psi_{2\ell} = (\psi_{\ell+2}\psi_{\ell-1}^2 - \psi_{\ell-2}\psi_{\ell+1}^2)\psi_\ell/2y \qquad \text{for } \ell \geq 3.$$

Higher powers y^n of y, with $n \geq 2$, are reduced modulo the equation $f = y^2 - (x^3 + ax + b)$.

THEOREM 5.24. *The above recursive equations, taken modulo f, yield polynomials in $\mathbb{F}_q[x, y]$ satisfying (5.23), and we have the following bounds on the degree of ψ_ℓ in its two variables:*

$$\deg_x \psi_\ell \leq \begin{cases} (\ell^2 - 1)/2 & \text{if } \ell \text{ is odd,} \\ (\ell^2 - 4)/2 & \text{if } \ell \text{ is even,} \end{cases}$$

$$\psi_\ell \in \begin{cases} \mathbb{F}_q[x] & \text{if } \ell \text{ is odd,} \\ y \cdot \mathbb{F}_q[x] & \text{if } \ell \text{ is even.} \end{cases}$$

PROOF.   We claim that for even ℓ, ψ_ℓ is divisible by y exactly once. This is visible when ℓ is 2 or 4. In the formula for $\psi_{2\ell}$ with $\ell \geq 3$, one distinguishes the cases ℓ even, with $\psi_{\ell-2}$, ψ_ℓ, $\psi_{\ell+2}$ divisible by y, and ℓ odd, with $\psi_{\ell-1}$ and $\psi_{\ell+1}$ divisible by y. In either case, after division by y one factor y is left. In particular, $\psi_{2\ell}$ is a polynomial in $y \cdot \mathbb{F}_q[x]$. It follows that the y in one denominator of (5.23) can be cancelled. Similarly, one shows that $\psi_{2\ell+1} \in \mathbb{F}_q[x]$, since the factor y^4 in $\psi_{\ell-1}\psi_{\ell+1}^3$ is replaced by $(x^3 + ax + b)^2$ modulo f.

The degree bounds are proven by induction on ℓ. We only execute this for $\psi_{2\ell+1}$ with even ℓ. By the above, ψ_ℓ and $\psi_{\ell+2}$ are each divisible once by y. Thus $\psi_{\ell+2}\psi_\ell^3$ has a factor of y^4, which by the reduction modulo f becomes x^6 plus lower terms. Furthermore, $\psi_{\ell-1}$ and $\psi_{\ell+1}$ do not contain y. Thus

$$\deg_x(\psi_{\ell+2}\psi_\ell^3) \leq 6 + \frac{(\ell+2)^2 - 4}{2} + 3\frac{\ell^2 - 4}{2} = \frac{(2\ell+1)^2 - 1}{2},$$

$$\deg_x(\psi_{\ell-1}\psi_{\ell+1}^3) \leq \frac{(\ell-1)^2 - 1}{2} + 3\frac{(\ell+1)^2 - 1}{2} = \frac{(2\ell+1)^2 - 1}{2}.$$

The claimed bound on $\deg_x \psi_{2\ell+1}$ follows.

The fact that the division polynomials satisfy (5.23) is shown in Theorem 9.31 of Washington (2008).   □

Actually these degree bounds are equalities. The most costly operations are the computations of x^q, y^q, x^{q^2}, y^{q^2}, as required for (5.20). These are done modulo the curve equation f and modulo the division polynomial ψ_ℓ. The first reduces the degree in y to at most 1, and the second the degree in x to less than ℓ^2. For an n-bit q, the total cost is roughly

(number of ℓ's) $\cdot$ (multiplications per exponentiation) $\cdot$ (cost of arithmetic modulo division polynomials) $\cdot$ (cost of field arithmetic)

$$\approx n \cdot n \cdot n^4 \cdot n^2 = n^8.$$

EXAMPLE 5.25. We take the elliptic curve $E = \{y^2 = x^3 - 3x + 1\}$ over $\mathbb{F}_{61}$, with $a = -3$ and $b = 1$. One checks that $4a^3 + 27b^2 = -20 \neq 0$ in $\mathbb{F}_{61}$. Thus $\#E = 62 - t$ with $|t| \leq 2\sqrt{61}$ or $|t| \leq 15$. It is sufficient to compute $t \bmod \ell$ for $\ell = 2, 3$, and 5, unless we are in the unlucky case that t equals 15 in $\mathbb{Z}_{30}$, since $2 \cdot 15 \leq 2 \cdot 3 \cdot 5$. But we are lucky ... We simply state that t is -1 in $\mathbb{Z}_2$ and 2 in $\mathbb{Z}_5$, and look at the computation modulo 3. From the recursive formulas, we have

$$\psi_3 = 3x^4 - 18x^2 + 12x - 9.$$

The roots of ψ_3 are precisely the x-coordinates of the (as yet undetermined) $P = (x, y) \in E[3]$, that is, with $3P = \mathcal{O}$. For the left hand side of (5.20), we first need $qP = 61P$. But since $3P = \mathcal{O}$ and $61 \equiv 1 \bmod 3$, we have $qP = P = (x, y)$. For $\varphi^2 P = (x^{61^2}, y^{61^2})$, we reduce these powers of x and y modulo the curve equation and ψ_3 and find

$$\varphi^2 P = (4x^3 + 25x^2 - 5x - 2, \, y \cdot (-8x^3 + 11x^2 + 10x)),$$

for which we write (x_2, y_2). Now we apply the first formula in (5.6) to calculate the x-coordinate of $P + \varphi^2 P$:

$$
\begin{aligned}
x_3 &= \left(\frac{y_2 - y}{x_2 - x}\right)^2 - x - x_2 \\
&= \left(\frac{y \cdot (-8x^3 + 11x^2 + 10x - 1)}{4x^3 + 25x^2 - 6x - 2}\right)^2 - 4x^3 - 25x^2 + 4x + 2.
\end{aligned}
$$

Since $P + \varphi^2 P$ is a point in the 3-torsion, we may reduce x_3 modulo ψ_3. If the dominator $d = 4x^3 + 25x^2 - 6x - 2$ were coprime to ψ_3, then we could take an inverse of d modulo ψ_3 and x_3 would become a polynomial in x after this reduction.

However, $\gcd(d, \psi_3) = x - 28$ in $\mathbb{F}_{61}[x]$ and there is no such inverse. On the other hand, $28^3 - 3 \cdot 28 + 1 = 31$ is not a square in $\mathbb{F}_{61}$, so that there is no point on E over $\mathbb{F}_{61}$ with x-coordinate 28. Since ψ_3 vanishes on all points of $E(K)[3]$ and $x - 28$ on at most two points of $E(K)$, also $\psi_3/(x - 28) = 3x^3 + 23x^2 + 16x - 28$ has some root in $E(K)[3] \setminus \{\mathcal{O}\}$. Reducing x_3 modulo this polynomial (and modulo f, as always) yields

$$x_3 = 26x^2 + 5x + 18.$$

We have to find the $t \in \mathbb{Z}_3 = \{-1, 0, 1\}$ for which $(x_3, y_3) = \varphi(tP)$. Now

$$\varphi(1 \cdot P) = (x^{61}, y^{61}) = (26x^2 + 5x + 18, \, y \cdot (-4x^2 + 18x + 24)).$$

The x-coordinate equals x_3, so that $(x_3, y_3) \in \pm\varphi(P)$ and $t \bmod 3 \in \{1, -1\}$. Inspecting y_3 determines the sign: $t \bmod 3 = 1$. Together with the values modulo 2 and 5 and the Hasse bound, this implies $t = 7$ and $\#E = 55$. ◇

Notes 5.1. The dashed line in Figure 5.2 has one point at infinity, which is not shown.

Washington (2008), Section 2.4, presents a (rather lengthy) proof of associativity of addition on elliptic curves.

In practice, it is common to use Karatsuba's fast multiplication at n about 200 or more, so that $O(n^{1.59})$ bit operations (Section 15.6) is more realistic for one field multiplication. There are many suggestions for computing efficiently with elliptic curves; see Cohen & Frey (2006).

Equation (5.8) does not produce all elliptic curves in characteristic 2; see Washington (2008), Section 2.8. It is useful to distinguish between variables x, y, z and values u, v, w in F for them. In Example 5.25, we take the liberty of denoting an undetermined point on E as $P = (x, y)$.

5.2. General references on elliptic curves are Silverman (1986), Koblitz (1987a), and Washington (2008). Blake, Seroussi & Smart (1999) and the handbook of Cohen & Frey (2006) also deal with many cryptographic aspects.

5.3. Characters are group homomorphisms $\mathbb{F}_q^\times \to \mathbb{C}$. We extend this concept slightly by allowing 0 as an argument. Hasse proved his famous theorem in 1933. An elementary proof is presented in Washington (2008), Theorem 4.2. Waterhouse (1969) shows that the whole Hasse interval is taken up by the sizes of elliptic curves, and Lenstra (1987) describes the distribution in the middle half.

5.4. Elliptic curves were introduced into this area by Lenstra (1987), who used them to factor integers. Since then they have served in primality tests and prime generation. Miller (1986) and Koblitz (1987b) suggested their use in cryptosystems.

See von zur Gathen & Gerhard (2013), Section 14.5, for root-finding over finite fields. Over a finite field of characteristic p, the discrete logarithm is easy for *supersingular elliptic curves*, where $\#E = q + 1 - t$ and p divides t. (Menezes, Okamoto & Vanstone 1993, Frey & Rück 1994). This includes all curves with an odd number of points for $p = 2$. Similarly, *anomalous elliptic curves*, where $t = 1$, are insecure (Satoh & Araki 1997, Semaev 1998, Smart 1999); see Blake *et al.* (1999) for details.

5.5. We do not specify what an "operation" is in a cryptanalytic attack. Although there is a difference between an evaluation of AES and a multiple on an elliptic curve, it seems best to ignore this factor for the time being. The issues in key size recommendations are explained in great detail in Lenstra (2006) and BSI (2014).

5.6. The special form of NIST primes speeds up user software, and even more so an adversary's heavier computations. The ECC Brainpool curves can be found at http://www.ecc-brainpool.org.

5.7. The characteristic equation (5.19) of the Frobenius automorphism is proven in Theorem 4.10 of Washington (2008). It implies the Hasse bound (5.18). Namely, when we consider φ as a variable and factor (5.19) over the complex numbers as

$$\varphi^2 - t\varphi + q = (\varphi - r_0)(\varphi - r_1),$$

then $r_0 r_1 = q$ and $r_0 + r_1 = t$. One can show that r_1 and r_2 are either equal or nonreal. In the latter case, being nonreal roots of the same polynomial with integer coefficients, r_0 and r_1 are conjugate and have the same absolute value. Thus $|r_0| = |r_1| = \sqrt{q}$ and $|t| = |r_0 + r_1| \leq 2\sqrt{q}$. For the computation of t_ℓ we might also use any discrete logarithm algorithm such as Pollard's rho method (Theorem 4.19), reducing the number of group operations from $O(\ell)$ to $O(\sqrt{\ell})$. Our estimate of B must be amended if $\ell = p = \operatorname{char} \mathbb{F}_q$ occurs in (5.22).

5.8. Insights of Elkies (1998) and Atkin (1992) have greatly improved the running time of Schoof's method, and this is now called the SEA algorithm. Its running time is $O(n^6)$, and $O^\sim(n^4)$ with fast arithmetic. This improves the original $O(n^8)$ and $O^\sim(n^5)$, respectively. It is eminently practical for elliptic curves at key sizes dictated by current security requirements.

Schoof's point-counting algorithm is now considered a seminal advance, but the time was not ripe for it and it was difficult to publish it. He applied his ideas to computing square roots, as in his title, and did not even state his counting algorithm explicitly. Atkin and Elkies sent their improvements to the experts in the area, but did not formally publish them and left this task to Schoof (1995), where the complete SEA algorithm is described.

Exercises.

EXERCISE 5.1 (Easy computations).

(i) Check that the Weierstraß equation $y^2 = x^3 + 3x + 2$ defines an elliptic curve E over $\mathbb{F}_{13}$.

(ii) Are the points $P = (4, 2)$ and $P_1 = (3, 5)$ on E?

(iii) What are the negatives of the points $P = (4, 2)$, $Q = (5, -5)$, $R = (3, -1)$ on E?

(iv) Compute $(5, -5) + (3, -1)$ on E.

(v) Compute $2 \cdot (5, -5)$ on E.

EXERCISE 5.2 (Smoothness of elliptic curves).   Let F be a field of characteristic neither 2 nor 3. We consider elements $a, b \in F$, the polynomial $g = x^3 + ax + b \in F[x]$, the discriminant $\Delta = 4a^3 + 27b^2$ and the curve $K = \{(u, v) \in F^2 \colon v^2 = u^3 + au + b\}$. Show that the following statements (i) through (iv) are equivalent.

(i) The curve K is smooth.

(ii) The greatest common divisor $\gcd(g, g')$ of g and the formal derivative g' of g is 1.

(iii) The polynomial g is squarefree, that is, there exists no nonconstant $h \in F[x]$ with h^2 dividing g. [Here equivalent: g has no multiple roots.]

(iv) $\Delta \neq 0$.

(v) For which values of b does $y^2 = x^3 - x + b$ not define an elliptic curve over $F = \mathbb{R}$? Plot the curves for all these values.

(vi) Assuming the above statements, verify that the elliptic curve $E = K \cup \{\mathcal{O}\}$ is also smooth at $\mathcal{O}$. Look at the curve $\{y^2 z = x^3 + axz^2 + bz^3\}$ in the projective plane and $\mathcal{O} = (0 : 1 : 0)$ on E. The usual partition of the curve is

$$\underbrace{\{(u : v : 1) \in \mathbb{P}^2 \colon v^2 = u^3 + au + b\}}_{w \neq 0} \cup \underbrace{\{(0 : 1 : 0)\}}_{w = 0}.$$

In the partition

$$\underbrace{\{(u : 1 : w) \in \mathbb{P}^2 \colon w = u^3 + auw^2 + bw^3\}}_{v \neq 0} \cup \underbrace{\{(u : 0 : 1) \in \mathbb{P}^2 \colon 0 = u^3 + au + b\}}_{v = 0}$$

the point $\mathcal{O} = (0 : 1 : 0)$ is not isolated any more. Check that $(0 : 1 : 0)$ is nonsingular on E, and that in the second part "$v = 0$" we may indeed assume that $w = 1$.

EXERCISE 5.3 (Elliptic curves?).   Which of the curves in Figure 5.3 are elliptic? Determine the discriminant and its roots $b \in \mathbb{R}$. Explain the blob at $(-1, 0)$ and draw a blow-up of it.

EXERCISE 5.4 (Addition on elliptic curves).   We consider the elliptic curve E with Weierstraß equation $y^2 = x^3 + x - 1$ over $\mathbb{F}_{11}$. It consists of the ten points

$$(1,1),\ (6,1),\ (1,10),\ (2,3),\ (2,8),\ (4,1),\ (4,10),\ (6,10),\ (9,0),\ \mathcal{O}.$$

We take $P = (1,1)$. Determine $3P$ and $5P$ by hand. Show your intermediate results.

EXERCISE 5.5 (Elliptic curves: associativity).   Using a computer algebra system, show that the addition on an elliptic curves is in fact associative. As a simplification, you may assume that no doubling or addition of two inverse points occurs.
    [Hint: You need to reduce modulo the curve equation after every tiny calculation.]

EXERCISE 5.6 (Alternative addition?). For two points P, Q on an elliptic curve E, define $P \oplus Q = S$, where S is the third intersection point of the line through P and Q with E, so that $S = -(P + Q)$. Explain why $\oplus$ almost never generates a group structure on E.

EXERCISE 5.7 (Doubling on an elliptic curve).   Let $P = (x_1, y_1) \neq 0$ be a point on an elliptic curve E with Weierstraß equation $y^2 = x^3 + ax + b$ over $\mathbb{R}$.

(i) Determine the equation for the tangent line at P to E.

(ii) Show that $Q = (x_3, y_3) = P + P = 2P$ can be computed using the following formula if $y_1 \neq 0$:

$$\alpha = \frac{3x_1^2 + a}{2y_1}, \quad x_3 = \alpha^2 - 2x_1, \quad y_3 = \alpha(x_1 - x_3) - y_1.$$

(iii) What happens if $y_1 = 0$?

(iv) Verify that over $\mathbb{R}$ the formula for doubling a point is the limit of the formula for addition of two points $P = (x_1, y_1)$ and $Q = (x_2, y_2) = (x_1 + \varepsilon_x, y_1 + \varepsilon_y)$, if Q (on the curve) converges to P. In order to do this, show that in this case the value $\alpha = (y_2 - y_1)/(x_2 - x_1)$ of the formula (5.6) for addition converges to the α from the formula of doubling above. You may want to use the fact that both P and Q are on the curve, that is, that (x_1, y_1) and $(x_1 + \varepsilon_x, y_1 + \varepsilon_y)$ satisfy the curve equation.

EXERCISE 5.8 (Points of order 2).

(i) Consider the two examples for elliptic curves E (on the left) and E^* (on the right) over $\mathbb{F}_7$ from Figure 5.5. Which are the elements of order 2?

(ii) Show that an elliptic curve E has at most three points P of order 2. (Those are the points P with $P \neq \mathcal{O}$ and $2P = \mathcal{O}$.) In other words, $\#E[2] \leq 4$.

EXERCISE 5.9 (Halving on elliptic curves). Consider a prime number $p \geq 5$ and an elliptic curve E over $\mathbb{F}_p$.

(i) Let Q be a point on E and $R = 2Q$. How can one compute Q from R? [Hint: Letting $Q = (u, v)$ find an equation for u of degree 4 and a further equation that is linear in v.]

(ii) What are possible numbers of solutions?

(iii) What can you say about $\#E$ or the group structure of E, respectively, if there is a point R with 2 (or with 4) solutions Q for $R = 2Q$?

EXERCISE 5.10 (Point tripling on elliptic curves).

(i) *Use the expressions from (5.6) and (5.7) to design a formula that gives the coordinates of $3 \cdot (u, v)$, for $(u, v) \in E$ with $v \neq 0$ and E in Weierstraß form.*

(ii) *Can you simplify your formula so that its computation uses fewer field operations?*

EXERCISE 5.11 (Isomorphisms of elliptic curves). *Let E_1 and E_2, with Weierstraß equations $y^2 = x^3 + a_1 x + b_1$ and $y^2 = x^3 + a_2 x + b_2$, respectively, be two elliptic curves over a field F with $\operatorname{char} F \neq 2, 3$. A rational map $\varphi \colon E_1 \to E_2$ is given by a pair of rational functions $f, g \in F(x, y)$, so that $(f(u, v), g(u, v))$ is well-defined and on E_2 for all $(u, v) \in E_1$. Such a φ is an isomorphism if there exists a rational map $\psi \colon E_2 \to E_1$ so that $\psi \circ \varphi = \operatorname{id}_{E_1}$ and $\varphi \circ \psi = \operatorname{id}_{E_2}$.*

(i) *Let $z \in F$. Show that*

$$\varphi \colon \quad \begin{array}{ccc} E_1 & \longrightarrow & E_2, \\ (u, v) & \longmapsto & (z^{-2} u, z^{-3} v) \end{array},$$

$$\psi \colon \quad \begin{array}{ccc} E_2 & \longrightarrow & E_1, \\ (u, v) & \longmapsto & (z^2 u, z^3 v) \end{array}$$

define inverse mappings if and only if $z^4 a_2 = a_1$ and $z^6 b_2 = b_1$.

(ii) *Let F be algebraically closed. Then a transformation as in (i) exists if and only if $(a_1/a_2)^3 = (b_1/b_2)^2$.*

(iii) *The quantity $j(E) = 1728 \cdot 4a^3 / \Delta$ with $\Delta = 4a^3 + 27b^2$ is called the j-invariant of E_1. Show that two isomorphic curves as in (i) have the same j-invariant.*

(iv) *Illustrate the transformation for the case $F = \mathbb{R}$ by drawing some curves in the (a, b)-plane on which a^3/b^2 is constant.*

Note: *In fact, all isomorphisms of elliptic curves in Weierstraß form arise in this way.*

EXERCISE 5.12 (Number of elliptic curves).

(i) *Let $p = 5$. For each pair $(a, b) \in \mathbb{F}_p^2$, count the number of points on the elliptic curve over $\mathbb{F}_p$ with Weierstraß equation $y^2 = x^3 + ax + b$. You first have to check whether this actually describes an elliptic curve. Make a table of the results: possible number N of points and the number of Weierstraß coefficients resulting in a curve with exactly N points. Compare with the Hasse bound and see Figure 5.8.*

(ii) *Same for $p = 37$. [A computer algebra system will be useful for this task.]*

EXERCISE 5.13 (Torsion points). *Let q be a prime power, not divisible by 2 or 3, and E the elliptic curve over $\mathbb{F}_q$ given by $y^2 = x^3 + ax + b$, with $a, b \in \mathbb{F}_q$. Furthermore, for an integer $\ell \geq 1$, the ℓ-torsion group $E[\ell]$ occurs in Schoof's method for calculating $\#E$.*

(i) *Show that $E[\ell]$ is a subgroup of E. If ℓ is prime, show that $E[\ell]$ is an $\mathbb{Z}_\ell$ vector space in a natural way, and, in particular, $\#E[\ell]$ is a power of ℓ.*

(ii) *Show that $P = (u, v) \in E[2]$ if and only if $u^3 + au + b = 0$. Conclude that $\#E[2] \leq 4$.*

(iii) We have $P = (u, v) \in E[3]$ if and only if $P + P = -P$. Show that this holds if and only if $3u^4 + 6au^2 + 12bu - a^2 = 0$. Conclude that $\#E[3] \leq 9$.

In fact, if ℓ is a prime not dividing q, then over an algebraic closure K of $\mathbb{F}_q$, $E(K)[\ell] \cong \mathbb{Z}_\ell \times \mathbb{Z}_\ell$, with ℓ^2 elements. In particular, $E[\ell]$ over $\mathbb{F}_q$ is a subgroup of $E(K)[\ell]$, and $\#E[\ell]$ divides ℓ^2.

EXERCISE 5.14 (Curve order from prime multiple). Let $\mathbb{F}_q$ be a finite field, P a point on an elliptic curve over $\mathbb{F}_q$, and ℓ a prime number with $\ell P = O$. Determine a bound B depending only on q and as small as possible so that just from these facts you can determine $\#E$, provided that $\ell \geq B$.

EXERCISE 5.15 (Weil's Theorem). This theorem says the following. Let E be an elliptic curve over $\mathbb{F}_q$ with Weierstraß equation $y^2 = x^3 + ax + b$ for some $a, b \in \mathbb{F}_q$, and set $t = q + 1 - \#E$. For $k \geq 1$, we consider the curve $E(\mathbb{F}_{q^k})$, given by treating a and b as elements of $\mathbb{F}_{q^k}$. Then $\#E(\mathbb{F}_{q^k}) = q^k + 1 - \alpha^k - \beta^k$, where α and β are the complex numbers in the factorization $1 - tz + qz^2 = (1 - \alpha z)(1 - \beta z)$ in $\mathbb{C}[z]$. Use some computer algebra system to compute the corresponding values t, α and β for $q = 5$, all $a, b \in \mathbb{F}_q$ and $1 \leq k \leq 7$, and compare $\#E(\mathbb{F}_{q^k})$ with $q^k + 1 - \alpha^k - \beta^k$.

EXERCISE 5.16 (ElGamal encryption on elliptic curves). Example 5.16 presents a special case of ElGamal encryption on an elliptic curve, and you are to describe the general situation. Thus ALICE and BOB have agreed on an elliptic curve E with Weierstraß equation $y^2 = x^3 + ax + b$ over $\mathbb{F}_q$ and a point P on E. They use the ElGamal cryptosystem in the cyclic subgroup of E generated by P to exchange secret messages. In the first step BOB chooses a secret key sk_B and publishes $Q = \mathsf{sk}_B \cdot P$. ALICE translates her messages into a point R on the curve E. Deviating from the original ElGamal cryptosystem, R need not necessarily be an element of the subgroup generated by P.

To handle yes-or-no questions, the two have agreed that a point R with odd x-coordinate means 'Yes'.

(i) How does ALICE encrypt?

(ii) How does BOB decrypt?

(iii) Determine the cost of encryption and decryption.

(iv) Let the curve and the point be given by $q = 23$, $a = 4$, $b = 13$, $P = (4, 1)$. BOB chooses $\mathsf{sk}_B = 17$ and publishes $Q = (2, 12)$. He asks ALICE to tell him whether every elliptic curve has a commutative group structure. ALICE answers with the message

$$((1, 8), (13, 13)).$$

What is the plaintext R? Is ALICE's answer correct?

EXERCISE 5.17 (Elliptic curves in characteristic two). The Weierstraß form of an elliptic curve over a finite field $\mathbb{F}_{2^n}$ of characteristic 2 is given by (5.8).

(i) Find the condition on a and b that corresponds to smoothness of the curve.

(ii) Derive the formulas for addition and doubling on the curve.

EXERCISE 5.18 (Generating elliptic curves of prime order). *Let q be a prime power and for $a, b \in \mathbb{F}_q$, let $d_{a,b} = \#E$, where E is given by $y^2 = x^3 + ax + b$. Furthermore, let $H = \{d : |q + 1 - d| \leq \sqrt{q}\}$ be the middle half of the Hasse interval from Theorem 5.12. As a and b range through $\mathbb{F}_q$, assume that more values $d_{a,b}$ lie in H than outside of it, and that those in H are uniformly distributed. The latter is not true but almost; see Lenstra (1987) for proven bounds. We do not assume anything about the distribution of $d_{a,b}$ outside of H. When we choose a and b uniformly at random, what is the expected number of choices until $d_{a,b}$ is prime? Is the Prime Number Theorem 3.21 sufficient to prove your claim?*

Rien n'est si dangereux qu'un ignorant ami;
Mieux vaudrait un sage ennemi.[2]
JEAN DE LA FONTAINE (1678)

Genie, gesunder Witz, lebhafte Einbildungskraft,
treues Gedächtniß, philosophisch-grammaticalische
Sprachenkenntniß, und vornehmlich eiserne Geduld,
sind die Eigenschaften, die der Dechiffrirer haben muß.[3]
FRIEDRICH AUGUST LEO (1797)

Il faut, pour opérer avec certitude, toute la lucidité d'esprit
et l'ensemble des aptitudes qui distinguent le cryptologue.[4]
PAUL LOUIS EUGÈNE VALÈRIO (1896)

L'Art de déchiffrer des lettres peut avoir son usage particulier;
puisque la pratique en conduit au but, que j'ai déjà indiqué,
& qui m'a fait recommander l'étude de l'Algèbre.[5]
WILLEM JACOB 'S GRAVESANDE (1748)

There was an orgy of ciphering and deciphering
going on in the Intelligence Office, and the others
which had been received really seemed to mean business.
FILSON YOUNG (1922)

Et ideo insanus est qui aliquod secretum scribit nisi ut a vulgo
celetur, et ut vix a studiosissimis et sapientibus possit intelligi.[6]
ROGER BACON (13th century)

---

[2]Nothing is as dangerous as an ignorant friend; better a wise enemy.

[3]Genius, healthy wit, vivid imagination, reliable memory, philosophical and grammatical knowledge of languages, and above all iron patience are the characteristics that the cryptanalyst must have.

[4]In order to proceed with certainty, you need all the mental lucidity and the whole of the abilities that distinguish the cryptanalyst.

[5]The art of deciphering letters can have its particular uses; as practice leads to success, as I have already indicated, and which makes me recommend the study of algebra.

[6]A man is crazy who writes a secret unless it remains concealed from the vulgar, and it can be understood quickly by studious people in the know.

Chapter C

Breaking the unbreakable

Esteemed as "le chiffre indéchiffrable", the Vigenère cipher was considered unbreakable for over three centuries. Its workings and arithmetic nature have been explained in Example A.2 (v) and Chapter B. We now present an attack from 1863 which brings the system to its knees. However, it did not really diminish the system's popularity, and the cipher was reinvented again and again by people unaware of this attack.

In fact, the British scientist Charles Babbage (1791–1871), inventor of the mechanical computer, broke the Vigenère system even earlier, but his work was kept secret (Section B.1). The first published attack on the Vigenère was a 95-page booklet written by the Prussian officer Wilhelm Kasiski (1805–1881). We present his cryptanalysis, and also a tool later developed by the US cryptographer William Frederick Friedman (1891–1969): the index of coincidence.

C.1. Kasiski's attack on Vigenère

In this section, we discuss the attack by Kasiski (1863) on the Vigenère cryptosystem. We have an alphabet A of s letters and a secret key $k = (\pi_0, \ldots, \pi_{m-1})$ consisting of m permutations $\pi_0, \ldots, \pi_{m-1}$ of A. The encryption is by applying $\pi_0, \ldots, \pi_{m-1}, \pi_0, \ldots, \pi_{m-1}, \pi_0, \ldots$ to the consecutive letters of the plaintext $x = (x_0, x_1, \ldots)$, so that the ciphertext is

(C.1)
$$y = (y_0, y_1, \ldots)$$
$$= (\pi_0(x_0), \ldots, \pi_{m-1}(x_{m-1}), \pi_0(x_m), \ldots, \pi_{m-1}(x_{2m-1}), \pi_0(x_{2m}), \ldots).$$

In other words, we have $y_j = \pi_i(x_j)$ for each j congruent to $i \in \mathbb{Z}_m$ modulo m. The Vigenère system is the special case where an alphabetical key $(k_0, \ldots, k_{m-1}) \in A^m$ is given and

$$\pi_i(x) = x + k_i,$$

under the usual identification $A \leftrightarrow \mathbb{Z}_s = \{0, \ldots, s - 1\}$ as in Table A.1, and addition modulo s. Thus each π_i is the Caesar cipher with shift by k_i. An example is given in Table B.1.

Kasiski's cryptanalysis, that is, finding the message x and key k from y, proceeds in two stages:

o find the key length m,

o determine each permutation $\pi_0, \ldots, \pi_{m-1}$.

As our running example, we take the following ciphertext of 348 letters, generated by a Vigenère system:

	0	5	10	15	20	25
0	KODGD	UCXEM	XGMFQ	PUEUX	DDOVA	ZXLOE
30	HSMVY	YEJRV	YPAMC	LWGAQ	YXYSK	CFOKI
60	VKYIN	CSLAC	BLJGW	HDQXN	GMMGA	NJRVM
90	FQRNC	GNYDE	CSTXF	MNPIV	UWFHN	RWVIN
120	UCRGM	RULUC	GNYDE	MISWZ	GTHSM	TPQTX
150	FWVSF	DXAFT	JUVNE	FWWAU	AFGPC	XSCST
180	XRMKN	RGNRM	NMFMK	LFBNJ	GKCKO	DVXTA
210	QYXYJ	ACMDR	WLHZQ	SNZWK	CPFAS	ERMGR
240	KSVRY	ZDHSM	KZADH	XGUCP	IEMVX	BUNCS
270	XHSDQ	DEJMC	DSJRV	MFMTH	SMKFQ	AMEFW
300	OGAAX	WKQNE	MMKIM	EEMSX	PFQRN	LALKM
330	JNWLR	QTAUP	LAGZK	OML		

In any classical cryptanalysis, the first thing is to count how often each encrypting symbol occurs (Table C.1), as was done in Figure A.23 for the Gold-Bug cryptogram. Figure C.2 shows the frequency-ordered graphs for English and for our cryptogram. The two curves differ sufficiently for us to conclude that we are not dealing with a simple substitution. No wonder—we set it up as a Vigenère encryption.

M	N	X	S	A	G	F	C	R	K	D	E	W
8.62	5.46	5.17	5.17	5.17	4.89	4.89	4.89	4.6	4.6	4.31	4.02	3.74
V	U	Q	L	Y	T	P	J	H	Z	O	I	B
3.74	3.45	3.45	3.45	3.16	2.59	2.59	2.59	2.59	2.01	2.01	2.01	0.86

Table C.1: Frequency table for the cryptogram, in percent.

For the first step in his cryptanalysis, Kasiski looks at all polygrams (sequences of two or more letters) that occur repeatedly in the ciphertext, then factors the differences of their initial positions, and determines m via the most conspicuous factors. He writes:

> *Jetzt sucht man zuerst zu ermitteln, aus wieviel Buch-staben der Schlüssel besteht. Zu diesem Zweck sucht man in der aufgeschriebenen Chiffre=Schrift alle Wiederholungen von zwei und mehreren Chiffern auf, zählt dann die Entfernung der gleichen Wiederholungen von einander, schreibt diese mit der Zahl ihrer Entfernung von einander unter die Chiffre=Schrift und sucht diese Zahl in ihre Faktoren zu zerlegen.*[1]

The idea is that, with sufficiently long plaintext and short key, there will a repeated polygram like ...you ...you ... in the plaintext which happens to be encrypted by the same piece of the key:

```
...Y  O  U  ...  Y  O  U  ...
...M  A  J  ...  M  A  J  ...
...K  O  D  ...  K  O  D  ...
```

In fact, this is precisely what takes place at positions 0 and 203 of our example. Of course, it may also occur that unrelated pieces of plaintext and key happen to add up to the same ciphertext. In the example, the repeated TXF at positions 102 and 148 is of this nature. But these accidents are not a serious obstacle.

We now turn to Kasiski's suggestion: Look at repeated polygrams! The polygrams of length at least three that occur repeatedly are listed in Table C.3, together with the factorization of the difference in positions of occurrences. The column "rep" gives the number of repetitions; a polygram repeated four times gives rise to six pairwise differences. The

---

[1] Now one first tries to determine of how many letters the key consists. To this end, one finds all repetitions in the ciphertext of two or more letters, counts the relative distances of repetitions of the same polygram, writes these with their distance below the ciphertext, and tries to factor these distances.

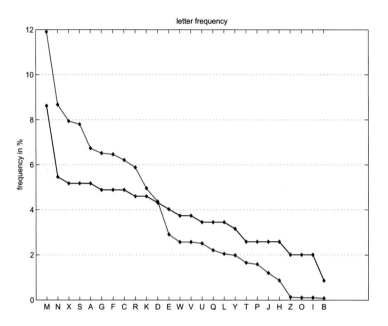

Figure C.2: English frequencies (COCA, in red) and ciphertext (in black, letters labelling the horizontal axis), both ordered by frequency.

polygram	rep	first	distance	polygram	rep	first	distance
CGNYDE	2	94	$5 \cdot 7$	JRV	3	37	$7^2, 5 \cdot 7^2,$
AQYXY	2	48	$7 \cdot 23$				$2^2 \cdot 7^2$
JRVMF	2	86	$2^2 \cdot 7^2$	AQY	2	48	$7 \cdot 23$
CGNYD	2	94	$5 \cdot 7$	QYX	2	49	$7 \cdot 23$
GNYDE	2	95	$5 \cdot 7$	YXY	2	50	$7 \cdot 23$
AQYX	2	48	$7 \cdot 23$	NCS	2	64	$7 \cdot 29$
QYXY	2	49	$7 \cdot 23$	RVM	2	87	$2^2 \cdot 7^2$
JRVM	2	86	$2^2 \cdot 7^2$	VMF	2	88	$2^2 \cdot 7^2$
RVMF	2	87	$2^2 \cdot 7^2$	FQR	2	90	$3 \cdot 7 \cdot 11$
FQRN	2	90	$3 \cdot 7 \cdot 11$	QRN	2	91	$3 \cdot 7 \cdot 11$
CGNY	2	94	$5 \cdot 7$	CGN	2	94	$5 \cdot 7$
GNYD	2	95	$5 \cdot 7$	GNY	2	95	$5 \cdot 7$
NYDE	2	96	$5 \cdot 7$	NYD	2	96	$5 \cdot 7$
CSTX	2	100	$7 \cdot 11$	YDE	2	97	$5 \cdot 7$
THSM	2	141	$3 \cdot 7^2$	CST	2	100	$7 \cdot 11$
HSMK	2	247	$2 \cdot 3 \cdot 7$	STX	2	101	$7 \cdot 11$
KOD	2	0	$7 \cdot 29$	TXF	2	102	$2 \cdot 23$
MFQ	2	12	$7 \cdot 11$	THS	2	141	$3 \cdot 7^2$
HSM	4	30	$2^4 \cdot 7, 7 \cdot 31,$	EFW	2	164	$7 \cdot 19$
			$7 \cdot 37, 3 \cdot 5 \cdot 7,$	MFM	2	191	$2 \cdot 47$
			$3 \cdot 7^2, 2 \cdot 3 \cdot 7$	SMK	2	248	$2 \cdot 3 \cdot 7$

Table C.3: Repeated polygrams of length at least three.

prime power	total	6	5	4	3	2	prime power	total	6	5	4	3	2
2	72		1	3	8	60	13	10					10
2^2	42		1	2	4	35	19	5				1	4
2^3	21				1	20	23	18		1	2	4	11
2^4	10				1	9	29	10				2	8
3	53			3	7	43	31	6				1	5
3^2	6					6	37	4				1	3
5	40	1	2	3	6	28	43	7					7
5^2	2					2	47	6				1	5
7	129	1	4	11	28	85	59	2					2
7^2	28		1	3	7	17	109	2					2
11	26			2	5	19	127	4					4

Table C.4: The number of times that prime powers divide distances between repetitions.

column "first" is the first occurrence. Thus the C of CGNYDE occurs in positions 94 and $94 + 35 = 129$.

Furthermore, there are 73 repeated digrams, three of them five times, five of them four times, fifteen thrice and fifty twice. Their statistics are included in Table C.4 which shows for each prime power how many positional differences it divides. Thus the factor 4 for the pentagram JRVMF gives a contribution of one for the prime powers 2 and 2^2. The prime powers $2^5, 2^6, 2^7, 13^2, 17, 41, 62, 71, 73, 89, 113, 137, 179, 197, 229$, and 241 divide exactly one digram difference and are not shown. We also ignore repeated monograms (individual letters). This table strongly indicates a key length of seven, and we take seven as our first guess for the key length. We split y into seven blocks $z_0, \ldots, z_6$, consisting of each seventh letter, so that

$$\text{(C.2)} \qquad z_i = (y_i, y_{i+7}, y_{i+14}, \ldots).$$

z_0 : KXQDOYAQFIBQAQDMFNUDGQFUAXMMFKQMQPGZAPUDDMQGNEQMTZ

z_1 : OEPOEEMYONLXNRENHULETTDVUSKNBOYDSFRDDINQSTAAEERJAK

z_2 : DMUVHJCXKCJNJNCPNCUMHXXNACNMNDXRNAKHHECDJHMAMMNNUO

z_3 : GXEASRLYISGGRCSIRRCISFAEFSRFJVYWZSSSXMSERSEXMSLWPM

z_4 : DGUZMVWSVLWMVGTVWGGSMWFFGTGMGXJLWEVMGVXJVMFWKXALLL

z_5 : UMXXVYGKKAHMMNXUVMNWTVTWPXNKKTAHKRRKUXHMMKWKIPLRA

z_6 : CFDLYPACYCDGFYFWIRYZPSJWCRRLCACZCMYZCBSCFFOQMFKQG

The method described so far applies to any system where each block z_i is enciphered by some arbitrary simple substitution π_i. We now assume that each π_i is a Caesar substitution, corresponding to the Vigenère system, and which is indeed used in our example.

The unknown plaintext corresponding to a block z_i is made up of each seventh letter of some English text. Thus it does not consist of English words, but can still be expected to follow the frequency distribution of English letters. The same holds for z_i, except that it is translated by a Caesar shift, and we can expect to solve it by frequency analysis. An inconvenience is that the available ciphertext is much shorter, namely, only one seventh of the original length, which comes to about 50 letters in our case.

We set up the seven frequency tables:

$$z_0 : \frac{\text{Q M D F A U G Z X P N K Y T O I E B}}{9\ 6\ 5\ 4\ 4\ 3\ 3\ 2\ 2\ 2\ 2\ 1\ 1\ 1\ 1\ 1\ 1\ 1} \rightsquigarrow \text{M}$$

$$z_1 : \frac{\text{E N O D T S R A Y U L K X V Q P M J I H F B}}{7\ 5\ 4\ 4\ 3\ 3\ 3\ 3\ 2\ 2\ 2\ 2\ 1\ 1\ 1\ 1\ 1\ 1\ 1\ 1\ 1\ 1} \rightsquigarrow \text{A}$$

$$z_2 : \frac{\text{N M C H X J U D A K V R P O E}}{9\ 6\ 6\ 5\ 4\ 4\ 3\ 3\ 3\ 2\ 1\ 1\ 1\ 1\ 1} \rightsquigarrow \text{J}$$

$$z_3 : \frac{\text{S R E X M I G F Y W L C A Z V P J}}{11\ 6\ 4\ 3\ 3\ 3\ 3\ 3\ 2\ 2\ 2\ 2\ 1\ 1\ 1\ 1} \rightsquigarrow \text{O}$$

$$z_4 : \frac{\text{G V W M L X F T S J Z U K E D A}}{8\ 7\ 6\ 6\ 5\ 3\ 3\ 2\ 2\ 2\ 1\ 1\ 1\ 1\ 1\ 1} \rightsquigarrow \text{C}$$

$$z_5 : \frac{\text{K M X W V U T R N H A P Y L I G}}{8\ 6\ 5\ 3\ 3\ 3\ 3\ 3\ 3\ 3\ 2\ 1\ 1\ 1\ 1} \rightsquigarrow \text{G}$$

$$z_6 : \frac{\text{C F Y Z R W S Q P M L G D A O K J I B}}{9\ 6\ 5\ 3\ 3\ 2\ 2\ 2\ 2\ 2\ 2\ 2\ 2\ 1\ 1\ 1\ 1\ 1} \rightsquigarrow \text{Y}$$

The letters that do not occur in z_i are not shown. Now the easiest approach is to assume that the most frequent letter represents E. For z_0 we take Q and obtain the key as $k_0 = $ Q - E = M. The seven key letters obtained by this crude guess are given at the end of each row. But Kasiski warns that only in rare cases will one be so lucky as to determine all key letters correctly by this table, given a ciphertext of a few lines:

> *Man wird jedoch nur in seltenen Fällen so glücklich sein, aus einigen Zeilen alle richtigen Buchstaben des Schlüssels durch die Schlüssel-Tabelle zu ermitteln, [. . .] weil die Buchstaben der Schrift zwar im Allgemeinen in dem [üblichen] Verhältniß vorkommen; in kürzern Schriften jedoch sehr auf-*

fallende Abweichungen statt-finden können.[2]

And indeed ours is not one of those rare cases. When we decipher with the key MAJOCGY, we find:

```
YOUSB   OELED   JEGHE   PLQSR   FROMM   XRNCE
YEKPA   MEADT   SROMT   XUACE   YOKQE   ETOBU
TEAWN   TEJUE   PLASU   BFEXE   SKGIO   NADTG
HEREO   EHARE   TERRH   ANGUT   OYTHE   DUPKB
UTDEG   TILLO   EHARE   DUQQB   UTYEK   NRETO
RUPUT   DOMDN   LIVEQ   DQYOU   RREJE   LSTER
RTAKE   DEHTA   NDRKE   NTBEV   EEEYO   UHVNC
EYOKH   UEADI   IJBBE   SELUE   EDFRE   CLOUR
BETLA   NDYEK   EBODY   JEOED   IVYTR   DINTE
VBURQ   UQHGE   RSADT   GHATY   EKEHE   ADQDQ
QUARJ   UESBE   DYICO   SEDEV   JHERE   XYFMA
JEIJL   SHALB   JUINK   FYJ
```

We recognize some English-looking pieces of text, but clearly we have not deciphered the message.

A more successful method is not to rely just on E as occurring most frequently, but to try to match visually the English frequencies with a shift of the ciphertext frequencies. We only do this for the fourth block z_3. According to the categories in Table A.21, the most frequent ciphertext letter S (= 18) is likely to stand for E, T, A, or O (= 4, 19, 0, 14). This corresponds to shifts by O, Z, S, or E (= 14, 25, 18, 4), which sum modulo 26 pairwise to $18 = 4 + 14 = 19 + 25 = \cdots$.

Table C.5 gives a systematic calculation. For any of the 26 letters x in the English alphabet A, we let diff_x be the difference between its frequency in English (first column in Table A.19) and in z_3. The second and fifth rows in the table show $\sigma_1 = \sum_{x \in A} |\text{diff}_x|$, corresponding to the statistical distance (15.85) between these two distributions. The third and sixth rows present $\sigma_2 = \sum_{x \in A} \text{diff}_x^2$. These quantities are scaled so that the smallest value equals 1.

Figure C.6 shows a visual representation. The horizontal axis contains all 26 possible key guesses (or shifts). The blue curve gives the values of σ_1 in Table C.5 and the red curve those for σ_2. Clearly, the

---

[2]Only in rare cases will one be so lucky as to determine all key letters correctly by this table, given a ciphertext of a few lines. The reason is that the plaintext letters occur in general with the usual frequencies, but that there can be considerable fluctuations in short texts.

	0	1	2	3	4	5	6	7	8	9	10	11	12
σ_1	1.40	1.65	1.85	1.77	1.00	1.48	1.69	1.93	1.68	1.70	1.51	1.49	1.60
σ_2	1.30	1.43	1.64	1.58	1.00	1.38	1.56	1.57	1.55	1.60	1.39	1.33	1.49
	13	14	15	16	17	18	19	20	21	22	23	24	25
σ_1	1.35	1.38	1.52	1.70	1.36	1.51	1.45	1.48	1.71	1.92	1.53	1.32	1.55
σ_2	1.37	1.17	1.45	1.60	1.40	1.31	1.50	1.51	1.63	1.61	1.55	1.30	1.33

Table C.5: The values of σ_1 and σ_2.

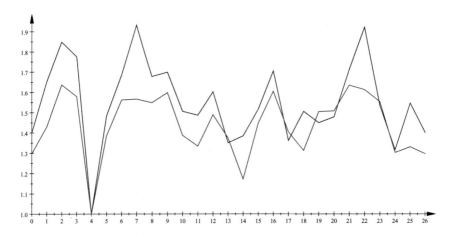

Figure C.6: Two measures for the distance between English frequencies and all Caesar shifts of z_3 as in Table C.5; σ_1 in blue and σ_2 in red.

ciphertext z_3 shifted by $-$ E comes closest to the English distribution. Indeed, this is the right key guess!

In Section C.2, we learn a computational method, called the index of coincidence, that implements such a visual approach quite reliably.

This tabular or visual analysis, applied to all seven subtexts, reveals the correct keyword MAJESTY. The plaintext is the sentence of the British conspirator Layer in 1722 (page 564), and you can check that the "English-looking" parts of the incomplete decipherment on page 247 agree with the plaintext in the four consecutive positions $-1, 0, 1, 2$, corresponding to the correct key letters YMAJ. Somewhat informally, we can describe this cryptanalytic method as follows.

ALGORITHM C.3. Kasiski attack on Vigenère cipher.

Input: ciphertext y, assumed to be Vigenère-encrypted.
Output: key length ℓ, key k and plaintext x, hopefully. Otherwise "no Vigenère".

1. Set up the table of repeated polygrams and their factored positional differences, as in Table C.3.
2. For each prime power, determine how many positional differences it divides, as in Table C.4.
3. Guess ℓ as the product of some of the most frequently occurring prime powers in step 2.
4. Form ℓ ciphertexts $z_0, \ldots, z_{\ell-1}$ by taking each ℓth letter from y, as in (C.2).
5. Assume that each z_i is a Caesar substitution, and cryptanalyze it. If one of these returns "no Caesar substitution", then go to step 3.
6. Try to match the various answers returned in step 5.

The Vigenère cipher was considered unbreakable for several centuries. Even Kasiski's successful attack in 1863 took quite some time to become widely known. But the basic idea of Kasiski's method had already been glimpsed in the Renaissance!

Section A.2 describes Porta's cipher disk (Figures A.16 and A.15). In the second edition of his book, from 1602, Porta proudly starts his Chapter 17 with the words "*NVNC rem arduam & magnam molimur*"[3]. He explains how a message prepared with a key may be solved and read without the key. He deciphers the following message of 77 letters, which he has set up himself:

```
0        5        10       15       20       25       30       35       40
mmmbtxco pxb dfbv gst inrgtn gtc cc ctg amhcm ahto
    45       50       55       60       65       70       75
xtmoq slqpr mmmbtth mhv, aceohg lll li nxiog.
```

Porta's original text shows some word divisions, but not the position numbers that we put on top. Porta makes several observations, most of which are not useful in general. But he points to the repetitions of mmm in positions 0 and 51, and the llll in position 67. And then he says:

> Since there are 17 letters between the 3 letters **MMM** and the 4 letters **LLLL** and 51 between the first 3 **MMM** and the same 3 letters repeated in the thirteenth word, I conclude that the key has been given 3 times, and decide correctly that it consists of 17 letters.

For the repeated mmm, this is Kasiski's argument! Porta fails to say that he has to take the second l of llll. He does not look at arbitrary repeated polygrams, as Kasiski does, but only at consecutive repetitions of the same letter in the ciphertext. These arise, for example, when there

---

[3] We will NOW undertake a difficult and great enterprise.

are arithmetic progressions in the plaintext and the key, one with the negative increment of the other.

He then guesses the 17-letter keyword, first *studens sic deficio* and *studium sic deficio* incorrectly, then *studium hic deficit*[4] correctly, to find the plaintext

```
0        5          10          15         20         25         30         35
pontiane, est uxor tua mortua, vix ut sit nomen
   40         45        50          55        60         65           70
suum, nihil manet, pontius cur studet non me
   75
latet.
```
[5]

We note that Porta has taken great care to include the arithmetic sequences `cdef`, `pon` and `[r]stu` for plaintext and key. He has encrypted four letters incorrectly. Porta's observations do not yield a general method for breaking Vigenère systems. The key ingredient of Kasiski's approach is present: the key length is likely to divide positional differences of repeated polygrams. But neither Porta nor any other cryptanalyst took up this insight at the time—as far as we know—and the Vigenère remained secure for another 250 years.

C.2. Friedman's index of coincidence

We describe a method developed by William Friedman in the 1920s for analyzing cryptosystems involving several substitutions.

We have a message $y \in A^n$ of n letters from some alphabet A with $s = \#A$ letters. Each letter $a \in A$ occurs some number t_a of times in y, so that $\sum_{a \in A} t_a = n$. Note that here a is not the first letter of the alphabet, but a variable denoting an arbitrary one. The *index of coincidence* $c(y)$ of y is defined as the probability with which two letters in y are equal. There are $t_a(t_a - 1)/2$ many pairs of positions in y with the same letter $a \in A$, and $n(n - 1)/2$ many pairs of positions in y. Thus

$$c(y) = \frac{\sum_{a \in A} \frac{t_a(t_a-1)}{2}}{\frac{n(n-1)}{2}} = \frac{\sum_{a \in A} t_a(t_a - 1)}{n(n - 1)}.$$

In other words, there are

$$\frac{n(n - 1) \cdot c(y)}{2}$$

---

[4] eagerness is missing here

[5] Pontiane is your wife, recently deceased; let her name be [praised], nothing remains; it is hidden [not understandable] to me why Pontius does not grieve.

many coincidences in y. If we consider $y\colon \{1,\dots,n\} \longrightarrow A$ as a hash function, then coincidences are collisions in the sense of Section 7.1.

We now look at the situations where y is either random or produced by a Vigenère cipher, and will find a noticeable difference between the values of c in these two cases. When y is random and n is large enough, we will have

$$t_a \approx \frac{n}{s}$$

for each letter a. Then

$$c(y) \approx \frac{\sum_{a \in A} \frac{n}{s}\left(\frac{n}{s}-1\right)}{n(n-1)} = s \cdot \frac{\frac{n}{s}}{n} \cdot \frac{\frac{n}{s}-1}{n-1} = \frac{n-s}{s(n-1)} \approx \frac{1}{s} = c_{\text{rand}}.$$

This is the simplest case of repetitions, namely where $z = a$ is a monogram of length $\ell = 1$.

If y is generated with the frequencies p_a for each letter $a \in A$ of some language, then $t_a \approx np_a$, and thus

$$c(y) \approx \frac{\sum_{a \in A} np_a \cdot (np_a - 1)}{n(n-1)} = \sum_{a \in A} p_a(p_a - \frac{1-p_a}{n-1}) \approx \sum_{a \in A} p_a^2.$$

The latter value depends only on the language, and a simple substitution does not change this value. According to Table A.19, we have for English

$$c_{\text{Eng}} \approx 6.36\% \gg 3.85\% = \frac{1}{26} = c_{\text{rand}}.$$

Now assume that y is the encryption of an English plaintext $x = (x_0, \dots, x_{n-1})$ as in (C.1). We know neither m nor the substitutions, but now calculate m via the index of coincidence $c(y)$. Namely, the monogram repetitions (coincidences) $y_k = y_\ell$ in y are of two types: generated by the same permutation, that is $y_k = \pi_i(x_k) = \pi_i(x_\ell) = y_\ell$ for some $i < m$ with $x_k = x_\ell$ and $\ell = k$ in $\mathbb{Z}_m$, or by two different permutations, that is, $y_k = \pi_i(x_k) = \pi_j(x_\ell) = y_\ell$ with $i \neq j$ (and then $k \neq \ell$ in $\mathbb{Z}_m$). (If $i \neq j$, $x_k = x_\ell$, and $\pi_i(x_k) = \pi_j(x_\ell)$, then this contributes to the second type.)

For the first type, we consider some substitution π_i with $i < m$. There are about n/m letters enciphered with π_i, and about

$$\frac{n}{m}\left(\frac{n}{m} - 1\right)/2 = \frac{n(n-m)}{2m^2}$$

many pairs of letters enciphered under π_i. The repetition probability is c_{Eng}, and since there are m substitutions, we expect

$$m \cdot c_{\text{Eng}} \cdot \frac{n(n-m)}{2m^2} = \frac{c_{\text{Eng}} \cdot n(n-m)}{2m}$$

repetitions of the first type.

For the second type, we take two substitutions π_i and π_j with $0 \leq i < j < m$. The number of pairs of letters enciphered with π_i and π_j is about $(n/m)^2$, and we expect $c_{\mathrm{rand}} \cdot (n/m)^2$ many coincidences, if π_i and π_j distribute letters in an independent fashion. Since there are $m(m-1)/2$ choices for i and j, we expect

$$\frac{m(m-1)}{2} \cdot \frac{c_{\mathrm{rand}} \cdot n^2}{m^2} = \frac{c_{\mathrm{rand}} \cdot n^2(m-1)}{2m}$$

many coincidental repetitions of the second type. The index of coincidence for y is the sum of these two values, and so we expect

$$
\begin{aligned}
c(y) &\approx \frac{2}{n(n-1)} \left(\frac{c_{\mathrm{Eng}} \cdot n(n-m)}{2m} + \frac{c_{\mathrm{rand}} \cdot n^2(m-1)}{2m} \right) \\
&= \frac{c_{\mathrm{Eng}} \cdot (n-m) + c_{\mathrm{rand}} \cdot n(m-1)}{m(n-1)} \\
&= \frac{(c_{\mathrm{Eng}} - c_{\mathrm{rand}}) \cdot n}{m(n-1)} + \frac{c_{\mathrm{rand}} \cdot n - c_{\mathrm{Eng}}}{n-1}.
\end{aligned}
$$

Solving for m, we find

$$m \approx \frac{(c_{\mathrm{Eng}} - c_{\mathrm{rand}}) \cdot n}{(n-1)c(y) - nc_{\mathrm{rand}} + c_{\mathrm{Eng}}}.$$

When n is large, we may approximate $n/(n-1)$ by 1, and ignore the summand c_{Eng} in the denominator. Then

(C.4) $$m \approx \frac{c_{\mathrm{Eng}} - c_{\mathrm{rand}}}{c(y) - c_{\mathrm{rand}}}.$$

This is the main step in the cryptanalysis of this kind of cipher: estimating the number m of simple substitutions used. In practice, we will determine m_0 as the rounded value of (C.4), and then also try $m_0 \pm 1$, and maybe even $m_0 \pm 2$. For each guessed value of m, we split the ciphertext y into m blocks $z_i = (y_i, y_{i+m}, \ldots)$ for $0 \leq i < m$ as in Section C.1 and calculate the index of coincidence for each block z_i; these indices should all be near c_{Eng}. If they are, we have presumably found the correct key length, and otherwise we have to continue with other guesses of m. And after trying a reasonable range of values, we may abandon our hypothesis that the ciphertext was formed in the assumed system. Once we have m correctly, each z_i is a simple substitution of a text with English frequencies. This can be handled with the frequency analysis as in Section A.3. We have two additional challenges: the ciphertext is only of length n/m, and the plaintext consists of each mth letter in an English text, thus displaying English frequency but no hints via letter contacts. One can get such hints by solving the z_i in parallel, but this is a more difficult task than Legrand's in Section A.3.

When the cryptosystem is a Vigenère cipher, so that each π_i is a Caesar substitution with $\pi_i(a) = a + k_i$ for some key letter k_i, then we can even use the coincidences to determine the secret key $k = (k_0, \ldots, k_{m-1})$. Here we identify the alphabet A with $\mathbb{Z}_s$, so that $a + b$ is a letter for any two letters a and b, with addition modulo s.

Thus we have the correct key length m, and the division of the ciphertext y into m blocks $z_0, \ldots, z_{m-1}$, as above. We could now prepare a frequency diagram for each z_i as in Section C.1. It should resemble the frequency diagram for English, shifted by k_i. If we put it onto a transparency, slide it across the standard English diagram, and visually observe the best match, then this should give away k_i. In fact, this works reasonably well, as we saw in Section C.1.

A computational version of this visual method goes as follows. For an arbitrary text $z = (z_0, z_1, \ldots, z_{n-1})$ of length n and a letter $a \in A$, we let $t_a(z)$ be the number of a's in z, let $q_a(z) = t_a(z)/n$ be the frequency of a, and denote by $z \oplus b$ the text z with each letter shifted by some $b \in A$:

$$z \oplus b = (z_0 + b, z_1 + b, \ldots).$$

This is just the Caesar encryption of z with key b, and $t_a(z \oplus b) = t_{a-b}(z)$. If z is a random text, then each $t_a(z)$ is near n/s, and $q_a(z)$ near $1/s$, and if it is an English plaintext, then $q_a(z)$ is expected to be near p_a, the standard frequency of a, as in Table A.19. We define the *mutual index of coincidence* $mc(z)$ of z and English to be

$$mc(z) = \sum_{a \in A} q_a(z) p_a.$$

For a random z, we have

$$mc(z) \approx \sum_{a \in A} \frac{1}{s} p_a = \frac{1}{s},$$

and for an English text z,

$$mc(z) \approx \sum_{a \in A} p_a \cdot p_a = c_{\text{Eng}}.$$

If z is English and $b \in A$ a letter, then the shifted version $z \oplus b$ has

$$mc(z \oplus b) \approx \sum_{a \in A} p_{a-b} p_a = mc_b.$$

The following is the first half of the values of mc_b, as defined here:

b	A	B	C	D	E	F	G	H	I	J	K	L	M	N
mc_b	6.36	3.89	3.13	3.65	4.31	3.28	3.65	3.99	3.31	3.34	3.90	4.35	3.85	4.32

These 14 values are sufficient to know, since

$$mc_b = \sum_{a \in A} p_{a-b}p_a = \sum_{a \in A} p_a p_{a+b} = mc_{-b}.$$

Thus the next 12 values for OPQ . . . Z are the same as those for MLK . . . B. (Note that $-N = N$ in A.)

Back to our cryptanalysis! Each block z_i of our Vigenère ciphertext, for $i < m$, is a shift of some text with English frequencies. We calculate $mc(z_i \oplus b)$ for each $i < m$ and each $b \in A$. Then the large value $c_{Eng} \approx 6.36\%$ should show up as $mc(z_i \oplus (-k_i))$, and the other values should follow the table above, in particular, they should not be much larger than the maximal value of 4.35.

C.3. Polygram substitutions

This and the next section do not deal directly with breaking the Vigenère cipher, but rather with historical attempts to harden encryption schemes against frequency attacks.

One defense against frequency analysis is to encrypt not single letters, but pairs (or triples, . . .) of letters. Thus we have a bijective encryption function enc: $A^\ell \to A^\ell$ for some $\ell \geq 1$. These are just simple substitutions on A^ℓ, but for $\ell \geq 2$ they are also called *polygram substitutions*. Porta invented in 1563 the 20×20 table shown in Figure C.7 of strange symbols for a digram substitution.

An interesting example is the *Playfair cipher*. It was invented in 1854 by Charles Wheatstone (1802–1875)—who also invented the *Wheatstone bridge* for measuring electrical resistance—and so eagerly promoted by his friend Lyon Playfair that the latter's name got stuck to it, although he never claimed its invention. It is handy, and was still used in the Second World War.

The system has two stages: from a keyword you derive a 5×5 square of 25 letters, and then perform a digram substitution with this square. The first stage uses the Argenti mnemonics from the end of Section A.3. You select a key word k, remove repetitions, and append the remaining letters of the alphabet, leaving out J. In the first step, you write this row-wise into a (partial) rectangle of predetermined width. Figure C.8 shows Wheatstone's original example, with $k = $ MAGNETIC (and no repetitions) and rectangle width 8. Then you read this columnwise and enter it row-wise into a 5×5 square; this gives the right-hand array in Figure C.8.

Figure C.7: Porta's digram table.

For the second stage, we use Figure C.8 as an example. In order to encrypt a pair of letters like IR that stand in different rows and columns, you take the two letters that form the other two corners of the corresponding rectangle, first the one in I's row and then the one in R's row. In this example, enc$_k$(IR) = UE, as shown in red and yellow in Fig-

```
M A G N E T I C          M  B  P  Y  A
                         D  Q  Z  G  F
B D F H J K L O          R  N  H  S
P Q R S U V W X          U  T  K  V  I
Y Z                      L  W  C  O  X
```

Figure C.8: Wheatstone's Playfair key mnemonics at left. At right the magnetic Playfair square; red is coded by yellow, and blue by green, as explained in the text.

th	he	in	er	an	re	on	ti	at	en
2.09	1.68	1.62	1.33	1.29	1.29	1.25	1.10	1.09	1.06

Table C.9: The most popular digrams in English and their frequencies in percent.

ure C.8. If both letters are in the same row or column, their encryption is the pair of letters immediately to the right or below them, respectively. Thus $enc_k(\text{FA}) = \text{EF}$ and $enc_k(\text{AY}) = \text{MA}$. Rows and columns wrap around cyclically, with $enc_k(\text{AY}) = \text{MA}$ and $enc_k(\text{SO}) = \text{VY}$. A dummy X is inserted between a pair of double letters (but not if they fall into adjacent pairs), and a dummy Z is added after a message of odd length, so that BALL and ALLY are encrypted as $enc_k(\text{BALXLZ}) = \text{PMWLCD}$ and $enc_k(\text{ALLY}) = \text{XMDM}$. We have $enc_k(\text{PLAYFAIR}) = \text{CMMAEFUE}$.

Such a system can also be attacked by frequency analysis. But now we have $26^2 = 676$ pairs, and the peaks and valleys in their distribution are much less pronounced. The most frequent digrams in English are given in Table C.9.

In her 1932 detective story *Have His Carcase*, Dorothy Sayers gives her sleuth Lord Peter Wimsey a Playfair cipher to solve. He comments:

> *If you ask me, it's either one of those devilish codes founded on a book — in which case it must be one of the books in the dead man's possession, and we only have to go through them — or it's a different kind of code altogether — the kind I was thinking about last night, when we saw those marked words in the dictionary."—"What kind's that, my lord?"—"It's a good code", said Wimsey, "and pretty baffling if you don't know the key-word. It was used during the War. I used it myself, as a matter of fact, during a brief interval of detecting under a German alias. But it isn't the exclusive property of the War Office. In fact, I met it not so long ago in a detective story. It's just —" He paused, and the*

> *policemen waited expectantly. "I was going to say, it's just*
> *the thing an amateur English plotter might readily get hold*
> *of and cotton on to. It's not obvious, but it's accessible and*
> *very simple to work [...] probably Playfair.*

The first word of the ciphertext is `xnatnx`, whose first and last pairs of letters are identical, except that they are reversed. The circumstances lead Wimsey, a *connaisseur in mysteries*, to guess the plaintext `warsaw`, and also to guess the next word, which is a date, and soon the Playfair unravels.

C.4. Polyalphabetic substitutions

Substitutions enc: $A \longrightarrow \mathbb{A}$ with a small alphabet A are susceptible to frequency analysis (and some to other attacks as well). *Polyalphabetic substitutions* provide higher security. Instead of a single encryption function enc, we have several of them: $\text{enc}_1, \ldots, \text{enc}_k \colon A \longrightarrow A$. In a certain rhythm, encryption rotates between the various enc_i's. There are many such systems; each of them can be described by the sequence $E = (\text{enc}_1, \ldots, \text{enc}_k)$ of substitutions, the rule for rotation among them, and possibly a key for initialization. If the rotation is just repetition, we can also view E as a simple substitution on A^k.

In Section B.1, we have seen Trithemius' progression through the Caesar shifts $\text{enc}_0, \text{enc}_1, \text{enc}_2, \ldots$ in order, and Vigenère's idea of moving through a fixed sequence of Caesar shifts determined by a key word. Now Vigenère (1587) proposed as well two cipher chaining methods (see Section 2.8), also called *autokey methods*. In his autokey method, we take as the key sk a single letter, and the rotation rule is as follows. The first letter is encrypted by enc_{sk}. If xy are consecutive letters in the plaintext, then y is encrypted as $\text{enc}_x(y)$. Thus *Traicté des chiffres*, with key sk $= v = 21$, is encrypted as follows.

plaintext	T	R	A	I	C	T	E	D	E	S	C	H	I	F	F	R	E	S
number	19	17	0	8	2	19	4	3	4	18	2	7	8	5	5	17	4	18
substitution	21	19	17	0	8	2	19	4	3	4	18	2	7	8	5	5	17	4
encryption	14	10	17	8	10	21	23	7	7	22	20	9	15	13	10	22	21	22
	O	K	R	I	K	V	X	H	H	W	U	J	P	N	K	W	V	W

Thus $\text{enc}_v(\texttt{TRAICTE DES CHIFFRES})=\texttt{OKRIK VXHHW UJPNK WVW}$.

The first three steps of the decryption procedure are as follows:

$$O - V = 14 - 21 = 19 = T,$$
$$K - T = 10 - 19 = 17 = R,$$
$$R - R = 17 - 17 = 0 = A.$$

Such a cryptogram is, of course, trivial to analyze even by hand: one simply has to try the 26 possibilities for the key. Guessing the first five letters a, b, c, d, e as keys gives the following first four letters:

A: OWVN   B: NXUO   C: MYTP   D: LZSQ   E: KARR

Most of the guesses can be abandoned after a short initial segment, as is visible in these five examples.

Vigenère also proposed a variation where not the plaintext but the ciphertext drives the rotation. This has the unpleasant feature that a single mistake makes the following encryption undecryptable.

If we consider the numerical letter values $x_0 x_1 x_2 \cdots$ and $y_0 y_1 y_2 \cdots$ of the plaintext x and the ciphertext y, then $y_i = x_{i-1} + x_i$ for $i \geq 1$. This might be called a *linear shift register code* over the alphabet.

Instead of using a single letter as key, a longer one makes things more difficult. Say, a Vigenère with key sk = rotate would mean that the first six letters of a message $x = (x_0, x_1, \ldots)$ are encoded by adding rotate:

$$\text{enc}_{\text{sk}}(x) = (x_0 + \text{r}, x_1 + \text{o}, x_2 + \text{t}, x_3 + \text{a}, x_4 + \text{t}, x_5 + \text{e}, \ldots),$$

and then the next six letters by adding $x_0, \ldots, x_5$ (or the first six letters of $\text{enc}_{\text{sk}}(x)$), and so on.

On the other hand, Vigenère had really proposed a more general cryptosystem, of which only the simplest variant bears his name today. Namely, if $\sigma \in \text{Sym}_A$ is the cyclic shift that moves each letter in the alphabet A to the next one (with cyclic wraparound $\sigma(z) = \text{a}$), then $E = \{\sigma^i : 0 \leq i \leq 25\}$ is the set of substitutions possible in the usual Vigenère system. But Vigenère had suggested to also use an arbitrary substitution $\tau \in \text{Sym}_A$ and $E_\tau = \{\tau\sigma^i : 0 \leq i \leq 25\}$. The system described above is the special case where τ is the identity mapping, with $\tau(x) = x$ for all letters x in A. But even this goes further back, together with hardware implementations, shown in Figures A.14 through A.16.

Notes. For Babbage's cryptanalysis of the Vigenère cipher, see Franksen (1984, 1993).

C.1. In the 1602 edition of Porta's *De Furtivis Literarum Notis*[6], the disk is on page 93 and the Vigenère cipher with solution in Chapter 17 of Book 4, *Quomodo scriptum cum*

---

[6]On secret writing of letters.

clavi, sine clavi aperiri, & interpretari possit[7], on pages 122–126. This part is not in the 1563 edition. Mendelsohn (1939) was the first to point out the whiff of a Kasiski analysis in Porta's work. Porta writes the unusual SI for 51, and makes several mistakes in counting the words, for example, calling the thirteenth word the twelfth. Porta's four encryption errors are r for r+d=x at position 20, g for t+c=a at 31, o for m+m=a at 40, and n for l+i=u at 72. Presumably he had a hard time making up a phrase that displays his arithmetic progressions in the right places, and it is not easy to make sense of the plaintext.

The Vigenère cipher also has its place in literature. Jules Verne tells in his adventure story *Jangada*—a large raft drifting down the Amazonas—how an innocent is saved from hanging at the very last moment by the decipherment of a Vigenère ciphertext. Judge Jarriquez, the fictitious cryptanalyst, knows a little bit about cryptography, enough to assume he deals with an "indecipherable" Vigenère cipher, but his solution depends substantially on guesswork. A crib, that is, a probable word in the plaintext, is brought to him at the last minute, and the cryptogram unravels by magic.

C.2. Friedman (1921) introduced the index of coincidence in his analysis of two encryption systems by the American E. J. Vogel from 1917 and by the French Commandant L. Schneider from 1912, both involving several permutations of the alphabet.

The assumption that π_i and π_j (with $i \neq j$) distribute letters independently is realistic when they are independent "random" permutations, but not when they are both Caesar ciphers. In that case, the table of the values of mc_b applies. For two arbitrary messages z and z', one can define the mutual index of coincidence as $\sum_{a \in A} q_a(z) q_a(z')$. We do not need this here, only the special case where z' is like a (very long) text with English frequencies.

C.3. For a homophonic polygram substitution, we have a subset $S \subseteq A^k \times A^\ell$, with positive integers $k \leq \ell$, where A is our alphabet. Then for a k-tuple $a = (a_1, \ldots, a_k) \in A^k$, the encryptor selects one of the $b \in A^\ell$ with $(a, b) \in S$ as encryption enc(a) of a. It is required that such a b always exist, and also that for each $b \in A^\ell$ there is at most one $a \in A^k$ with $(a, b) \in S$; then $a = \text{dec}(b)$ is the unique decryption of b.

The Playfair description lacks a rule on how to encrypt XX. The Australian coastwatch using the Playfair cipher in 1943 is in Kahn (1967), Chapter 17, page 592. This happened in connection with the Japanese sinking of a US patrol torpedo boat commanded by John F. Kennedy. The Playfair ciphertext and Wimsey's comments are at the beginning of Chapter 26 in *Have His Carcase*.

Exercises.

EXERCISE C.1 (Cipher composition). *You are to compare the two possibilities for composition outlined in Example A.5 (iii). Thus $A = \{a, \ldots, z\}$ is the standard alphabet, x the plaintext of Layer's sentence on page 565, σ the Vigenère with key* `majesty`, *and τ the 3×2 transposition in Example A.3 (ii).*

(i) *Determine the ciphertext y_1 obtained by applying $\tau_{A^7} \circ \sigma^6$ to consecutive blocks of 42 letters of x, padded with 30 x's at the end.*

(ii) *Do the same using $\tau_A^7 \circ \sigma^6$ to find y_2.*

(iii) *Perform a Kasiski analysis on y_1 and y_2, that is, find repeated polygrams and the divisors of their distance. Compare the two.*

----

[7] How it is possible to understand something written with a secret key, without having the key.

(iv) A "flaw" in τ is that $\tau(0) = 0$ and $\tau(5) = 5$, so that bigrams at the "borders" $6i, 6i + 1$ in (ii) are left intact. Compute the ciphertext as in (ii) and the Kasiski analysis with τ' which equals τ except that $\tau'(3) = 5$ and $\tau'(5) = 4$.

EXERCISE C.2 (General index of coincidence). *Friedman's index of coincidence exploits the statistical irregularities in multiple occurrences of single letters. The present exercise generalizes this to arbitrary polygrams. As usual, we have a text $y \in A^n$ of length n, and an integer $\ell \geq 1$. For any ℓ-gram $z \in A^\ell$, we let t_z be the number of occurrences of z in y, so that $\sum_{z \in A^\ell} t_z = n - \ell + 1$. The index of coincidence $c^{(\ell)}(y)$ of order ℓ is the probability with which two ℓ-grams in y are equal. There are $t_z(t_z - 1)/2$ many pairs of occurrences of z in y, and $(n - \ell + 1)(n - \ell)/2$ many pairs of ℓ-grams in y. Thus*

$$c^{(\ell)}(y) = \frac{\sum_{z \in A^\ell} \frac{t_z(t_z - 1)}{2}}{\frac{(n-\ell+1)(n-\ell)}{2}} = \frac{\sum_{z \in A^\ell} t_z(t_z - 1)}{(n - \ell + 1)(n - \ell)}.$$

This generalizes the index of coincidence $c(y) = c^{(1)}(y)$ in a natural way.

(i) *When y is random and n is large enough compared to s^ℓ, we may make the approximation $t_z \approx n/s^\ell$ for each ℓ-gram z. Show that then*

$$c^{(\ell)}(y) \approx s^{-\ell}.$$

(ii) *Next assume that y is a text in some language, say English. Just as a single letter, also an ℓ-gram z has its probability p_z of occurring. Then $t_z \approx np_z$. Show that*

$$c^{(\ell)}(y) \approx \sum_{z \in A^\ell} p_z^2.$$

(iii) *The last value in (ii) only depends on the language, and from the COCA corpus used in Table A.19, we find the following values for $c_{Eng}^{(\ell)} = \sum_{z \in A^\ell} p_z^2$:*

ℓ	1	2	3	4	5
$26^\ell \cdot c_{Eng,\ell}$	1.65	4.83	23.65	135.17	1238.02

As before, we now take y as encryption via m simple substitutions $\pi_0, \ldots, \pi_{m-1}$ of an English text. Let $0 \leq i < m$. There are about n/m ℓ-grams z enciphered with $(\pi_i, \pi_{i+1}, \ldots, \pi_{i+\ell-1})$ (with wrap-around) as $(\pi_i(z_0), \pi_{i+1}(z_1), \ldots, \pi_{i+\ell-1}(z_{\ell-1}))$, and about $n(n - m)/2m^2$ pairs of ℓ-grams enciphered in this way. Therefore we expect about

$$m \cdot c_{Eng}^{(\ell)} \cdot \frac{n(n - m)}{2m^2} = \frac{c_{Eng}^{(\ell)} \cdot n(n - m)}{2m}$$

coincidences of the first type.
For the second type, we take $0 \leq i < j < m$ and assume that $\pi_i, \ldots, \pi_{i+\ell-1}$ and $\pi_j, \ldots, \pi_{j+\ell-1}$ distribute ℓ-grams in an independent fashion. Thus we expect

$$\frac{m(m - 1)}{2} \cdot \frac{c_{rand}^{(\ell)} \cdot n^2}{m^2} = \frac{c_{rand}^{(\ell)} \cdot n^2(m - 1)}{2m}$$

many coincidences of the second type. Calculate an approximation for $c^\ell(y)$. Solve this for m to find

$$m \approx \frac{c_{Eng}^{(\ell)} - c_{rand}^{(\ell)}}{c^{(\ell)}(y) - c_{rand}^{(\ell)}}.$$

(iv) *Apply this estimate with $\ell = 2, 3, 4$ to the ciphertext in Section C.1. Is this a useful extension of the index of coincidence?*

Hand me down a solution. Pass it on.

THE WHO (1972)

Ceux qui se vantent de dèchiffrer une lettre sans être instruits des
affaires qu'on y traite & sans avoir des secours préliminaires,
sont de plus grands charlatans que ceux qui se vanteraient
d'entendre une langue qu'ils n'ont point apprise.[8]

VOLTAIRE (1764)

Ich blicke jetzt auf eine zehnjährige Praxis als Dechiffreur zurück.
Im Jahre 1921 habe ich [...] eine Vorlesung über die „De-
chiffrierung von Geheimschriften" gehalten. Was mir damals nur
instinktiv zum Bewußtsein kam, das weiß ich jetzt aus Erfahrung:
daß man das Dechiffrieren weder lehren noch lernen kann,
ebensowenig wie das Dichten und Komponieren. Dechiffrieren
ist eine Sache der analytischen Begabung, und wer analytisch
unbegabt ist, lernt und kann es nie, mag er sonst ein Genie sein.
Man kann lediglich Anhaltspunkte und Hilfsmittel geben.[9]

LOTHAR PHILIPP (1931)

Das Entziffern erfordert dabey einen offenen Kopf. Wer eine
Sache langsam begreift, wird zwar, wenn er sonst die Sprache
kennt, und die gemeinen Regeln weiß, entziffern können;
aber es wird ihm dieses eine sehr mühsame Arbeit seyn,
und hoch wird ers in der Kunst gewiß nicht bringen. Ein wahrer
Dummkopf aber bleibe lieber ganz von der Arbeit weg;
denn er und die Entzifferungskunst sind ein Widerspruch.[10]

CARL ARNOLD KORTUM (1782)

---

[8]Those who boast about deciphering a letter without being acquainted with the
subject that it treats and without preliminary assistance, are greater charlatans than
those who boast to understand a language which they have never learned.

[9]I am now looking back on ten years of practice as a decipherer. In the year 1921,
I [...] taught a course on "deciphering secret writing". Something that then entered
my mind only instinctively, I now know from experience: that you can neither teach
nor learn deciphering, just as with poetry or musical composition. Deciphering is a
question of analytic talent, and he who is analytically untalented, will never learn or
perform it, be he a genius otherwise. One can only provide pointers and tools.

[10]Deciphering demands an open mind for it. He who understands things slowly may,
if he knows about language and the general rules, be able to decipher; but it will be
tiresome work for him and he will not get far in this art. But a real moron had better
stay away from this work, since he and the art of deciphering are a contradiction.

Chapter 6

Differential and linear cryptanalysis

In this chapter, we discuss two attacks on "Boolean" cryptosystems, that is, on systems which employ sequences of operations on bits and bytes, but do not work in a "large" algebraic structure, as RSA and group cryptography do. *Differential cryptanalysis* takes two messages that are close to each other, say, that differ only in a single bit, and traces this small difference through several rounds of an encryption system. *Linear cryptanalysis* looks for linear relations between the input and output bits of several successive rounds. Both attacks are probabilistic and recover the secret key given sufficiently many plaintext-ciphertext pairs, using the same secret key. Any Boolean cryptosystem must avoid falling prey to them.

These attacks were invented in the context of DES, an encryption standard from 1977 to 2005; see Section 2.2. It was later revealed that the designers of DES knew about differential cryptanalysis—published only 13 years later—and incorporated that knowledge. On the other hand, there is no evidence that resistance to linear cryptanalysis was an objective in the design of DES.

We first describe the two attacks in the context of the Boolean encryption system AES. Fortunately for the system but unfortunately for our exposition, AES is specially hardened against these attacks (see Section 6.4). Its authors, Daemen & Rijmen, put forth general theories of these attacks and countermeasures against them, and designed AES according to these principles.

So we will first describe a reduced variant of AES, called baby-AES, in Section 6.1 and then discuss differential cryptanalysis (Section 6.2) and linear cryptanalysis (Section 6.3) in its context.

© Springer-Verlag Berlin Heidelberg 2015
J. von zur Gathen, *CryptoSchool*, DOI 10.1007/978-3-662-48425-8_9

6.1. Baby-AES

The basic unit in AES is a byte (8 bits) and each state consists of $4 \times 4 = 16$ bytes, or 128 bits. For baby-AES, we choose a nibble (4 bits) as the basic unit and deal with a block-size of $2 \times 2 = 4$ nibbles. More precisely, baby-AES works with

- 4-bit *nibbles*, realized as elements of $\mathbb{F}_{16} = \mathbb{F}_2[t]/(t^4 + t + 1)$, and

- 16-bit *states*, realized as 2×2 arrays of nibbles.

By convention, the states of (baby-)AES are written and read "column-before-row", as in AES. In other words, we identify a 16-bit message $x = (x_1, x_2, x_3, x_4) \in \mathbb{F}_{16}^4$ with the array

$$
(6.1) \qquad \begin{pmatrix} x_1 & x_3 \\ x_2 & x_4 \end{pmatrix} \in \mathbb{F}_{16}^{2 \times 2}.
$$

In the spirit of Boolean cryptography, we will focus on the description of states as 16-bit strings, but for the moment keep the representation (6.1) by state arrays in mind when describing SUBBYTES, SHIFTROWS, MIXCOLUMNS, and ADDROUNDKEY for baby-AES. As an example, we process the 16-bit message
(6.2)
$$
u = 0001\,0000\,1010\,1011 = \begin{pmatrix} 0001 & 1010 \\ 0000 & 1011 \end{pmatrix} = \begin{pmatrix} 1 & t^3 + t \\ 0 & t^3 + t + 1 \end{pmatrix} \in \mathbb{F}_{16}^{2 \times 2}.
$$

The nonlinear core of the S-box in baby-AES is the "patched inverse" in $\mathbb{F}_{16}$ (mapping 0 to 0). This is followed by the affine transformation $b \mapsto t_1 \cdot b + t_0 \bmod t^4 + 1$ on $\mathbb{F}_{16}$, with $t_1 = t^3 + t^2 + 1$ and $t_0 = t^2 + t$. Thus the S-box of baby-AES maps a nibble $x \in \mathbb{F}_{16}$ to $t_1 \cdot \mathrm{inv}_{\mathbb{F}_{16}}(x) + t_0 \bmod t^4 + 1$. Table 6.1 presents the corresponding lookup-table of 4-bit strings. SUBBYTES applies this S-box simultaneously to the four nibbles of a state (6.1). For u as in (6.2), this yields

$$
u = 0001\,0000\,1010\,1011
$$

$$
(6.3) \qquad v = \text{SUBBYTES}\,(u) = 1011\,0110\,1111\,1100.
$$

SHIFTROWS of baby-AES leaves the first row of a state array unchanged and moves the two nibbles in the second row one position to the

x	S-box(x)	x	S-box(x)
0000	0110	1000	1001
0001	1011	1001	1101
0010	0101	1010	1111
0011	0100	1011	1100
0100	0010	1100	0011
0101	1110	1101	0001
0110	0111	1110	0000
0111	1010	1111	1000

Table 6.1: The S-box of baby-AES as a lookup-table of 4-bit strings.

right (with wrap-around), effectively exchanging them. MIXCOLUMNS of baby-AES reads a column

$$\begin{pmatrix} x_i \\ x_{i+1} \end{pmatrix} \in \mathbb{F}_{16}^{2\times 1}, i \in \{1,3\},$$

as $x_{i+1}s + x_i \in \mathbb{F}_{16}[s]/(s^2+1)$ and multiplies it with $c = t^2 s + 1$ in $\mathbb{F}_{16}[s]/(s^2+1)$. In other words, the state array (6.1) is multiplied from the left by the 2×2-array

$$(6.4) \qquad\qquad C = \begin{pmatrix} t+1 & t \\ t & t+1 \end{pmatrix} \in \mathbb{F}_{16}^{2\times 2}.$$

We continue with v as in (6.3), and obtain

$$v = 1011\,0110\,1111\,1100$$

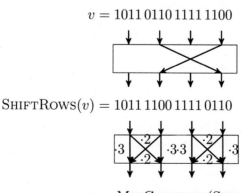

$$\text{SHIFTROWS}(v) = 1011\,1100\,1111\,0110$$

$$(6.5) \qquad \begin{aligned} w &= \text{MIXCOLUMNS}\,(\text{SHIFTROWS}\,(v)) \\ &= 0101\,0010\,1110\,0111. \end{aligned}$$

Finally, ADDROUNDKEY of baby-AES XORs a state with a 16-bit round key derived from the 16-bit secret key. Taking u of (6.2) itself as round key, we continue with w as in (6.5) for

$$w = 0101\,0010\,1110\,0111$$

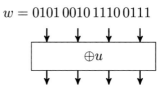

$$\textsc{AddRoundKey}_u(w) = 0100\,0010\,0100\,1100$$

as the state after one complete round of baby-AES.

This concludes the description of a single round of baby-AES. For R rounds of baby-AES, we require $R + 1$ round keys of 16 bits each. These are generated by the key schedule from a single 16-bit secret key. The key schedule of Baby-AES is the key schedule of AES as described in Algorithm 2.14 with $\ell_k = 2$, $\ell_r = R$, and K_i and E_i as 2-nibble words instead of 4-byte words. As in the case of AES, the R-round baby-AES starts with an initial ADDROUNDKEY, followed by $R - 1$ rounds of SUBBYTES, SHIFTROWS, MIXCOLUMNS, ADDROUNDKEY, and concludes with a single round of SUBBYTES, SHIFTROWS, and ADDROUNDKEY. Figure 2.1 describes the overall structure and Figure 6.2 in more detail the 3-round baby-AES version which we attack. We use the following notation with the abbreviations SB, SR, MC, and AK for the four steps.

- The initial state equals the plaintext $x = w^0$,

- round 0 maps w^0 to $u^1 = \text{AK}_0(w^0)$,

- each successive round $r = 1, 2, 3$ maps its input u^r to its output u^{r+1} through the following steps

$$u^r \xrightarrow{\text{SB}} v^r \xrightarrow{\text{MC} \circ \text{SR}} w^r \xrightarrow{\text{AK}_r} u^{r+1},$$

  where the last round omits MC,

- the ciphertext equals the final state $y = u^4$.

Generally, rounds are noted by superscripts $r = 0, 1, 2, 3$, nibbles by subscripts $i = 1, 2, 3, 4$, and the ith S-box in round r by S_i^r.

We conclude with a few comments on the construction.

- AES and baby-AES are examples of substitution-permutation networks, one of two popular designs for block ciphers. The other one is the Feistel structure with DES as a prime example.

- For the choice of MIXCOLUMNS as an MDS matrix and its omission in the last step, see Section 2.2.

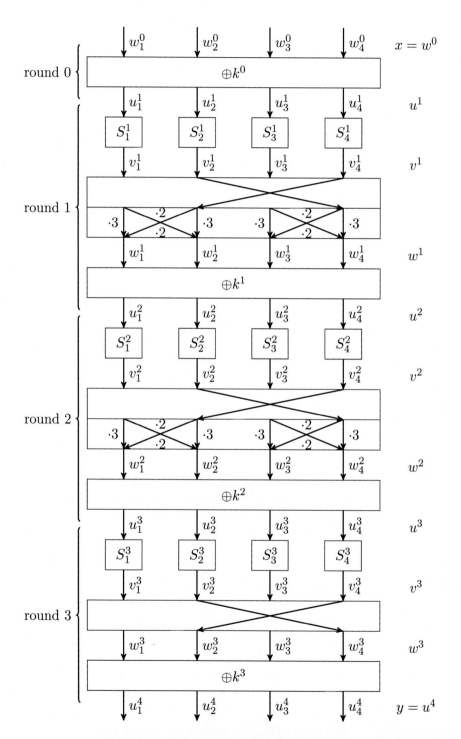

Figure 6.2: Three rounds of baby-AES.

○ Having ADDROUNDKEY as first and last operation is called *whitening*.

○ The key schedule of (baby-)AES is invertible. This is not necessary, but reduces the memory requirement and yields uniformly distributed individual round keys from a uniformly distributed secret key. On the other hand, recovery of a round key immediately yields the secret key. Differential and linear cryptanalysis take this approach by aiming at the last round key. We encounter them in the following two sections.

6.2. Differential cryptanalysis

Differential cryptanalysis recovers the round key of the last round using several well-chosen plaintext-ciphertext pairs. Since the key schedule is invertible, this is Key Recovery under Chosen Ciphertext Attack in the language of Section 9.7. It was invented by Biham & Shamir (1991b) with DES in mind. Surprisingly, DES proved particularly resistant against this attack. This was no accident! Don Coppersmith and IBM were already aware of this method and designed DES to resist it as well as possible. Coppersmith (1992) writes:

> We (IBM crypto group) knew about differential cryptanalysis in 1974. This is why DES stood up to this line of attack; we designed the S-boxes and the permutation in such a way as to defeat it. [...]
>
> We kept quiet about this for all these years because we knew that differential cryptanalysis was such a potent form of cryptanalysis, and we wished to avoid its discovery and use (either for designing or for attacking) on the outside.

The basic idea is to pick two plaintexts that differ only in a few bits and trace this difference through the first $R - 1$ rounds. If chosen with care, there is a high probability that the difference in the input states will propagate to a predetermined difference in the state at the beginning of round R. The ciphertexts of the two chosen plaintexts are then run backwards through the last round testing several candidates for the last round key. Only with the correct one can the decryptions be expected to match the given difference in the state at the beginning of round R. We repeat for several such pairs of plaintexts and record the number of matches. Given enough samples, the correct round key stands out.

We now walk through the three steps of differential cryptanalysis, discuss the costs, and conclude with possible countermeasures.

x	x^*	y	y^*	Δy	x	x^*	y	y^*	Δy
0000	1000	0110	1001	1111	1000	0000	1001	0110	1111
0001	1001	1011	1101	0110	1001	0001	1101	1011	0110
0010	1010	0101	1111	1010	1010	0010	1111	0101	1010
0011	1011	0100	1100	1000	1011	0011	1100	0100	1000
0100	1100	0010	0011	0001	1100	0100	0011	0010	0001
0101	1101	1110	0001	1111	1101	0101	0001	1110	1111
0110	1110	0111	0000	0111	1110	0110	0000	0111	0111
0111	1111	1010	1000	0010	1111	0111	1000	1010	0010

Table 6.3: Output pairs and output differences for $(x, x^*) \in D(1000)$.

Step 1: Build the differential distribution table of the S-box.
Before we trace differences of states through several rounds, let us trace
differences of nibbles through a single S-box. Following the description as
a Boolean cryptosystem, we view the S-box of baby-AES as a permutation
$S \colon \mathbb{F}_2^4 \to \mathbb{F}_2^4$ of 4-bit strings, see Table 6.1.

For two 4-bit strings $x, x^* \in \mathbb{F}_2^4$, we write Δx for their *difference*
$x \oplus x^* \in \mathbb{F}_2^4$ and for any $\Delta x \in \mathbb{F}_2^4$, we denote by

$$D(\Delta x) = \{(x, x^*) \in \mathbb{F}_2^4 \times \mathbb{F}_2^4 \colon x \oplus x^* = \Delta x\}$$
$$= \{(x, x \oplus \Delta x) \in \mathbb{F}_2^4 \times \mathbb{F}_2^4 \colon x \in \mathbb{F}_2^4\}$$

the set of the 16 pairs with difference Δx. For example,

$$\begin{aligned}
D(1000) = \{&(0000, 1000), (0001, 1001), (0010, 1010), (0011, 1011), \\
&(0100, 1100), (0101, 1101), (0110, 1110), (0111, 1111), \\
&(1000, 0000), (1001, 0001), (1010, 0010), (1011, 0011), \\
&(1100, 0100), (1101, 0101), (1110, 0110), (1111, 0111)\}
\end{aligned}$$

consists of the 16 pairs with difference 1000.

Let $(x, x^*) \in D(\Delta x)$, $y = S(x)$, and $y^* = S(x^*)$. What are the
possible values for $\Delta y = y \oplus y^*$? Table 6.3 lists the values of y, y^*, and
Δy for all $(x, x^*) \in D(1000)$. The symmetry $x \oplus x^* = x^* \oplus x$ of XOR
explains the symmetry in the values of Δy. Hence, every value for Δy
appears an even number of times. Furthermore, only 7 out of 16 possible
values occur, and in general at most 8 values of Δy are possible.

A pair $(\Delta x, \Delta y) \in \mathbb{F}_2^4 \times \mathbb{F}_2^4$ of 4-bit strings is called a (nibble) *differential*. We denote it by $\Delta x \to \Delta y$ and call Δx its *input difference* and Δy
the *output difference*. We take a closer look at the frequency with which a
certain output difference Δy occurs for a given input difference Δx, when
the messages $(x, x^*) \in D(\Delta x)$ are processed by the S-box.

DEFINITION 6.6. *For a differential* $\Delta x \to \Delta y \in \mathbb{F}_2^4 \times \mathbb{F}_2^4$, *we define its absolute frequency (in S) as*

$$N_S(\Delta x \to \Delta y) = \#\{(x, x^*) \in D(\Delta x): (S(x), S(x^*)) \in D(\Delta y)\}$$
$$= \#\{x \in \mathbb{F}_2^4: S(x) \oplus S(x \oplus \Delta x) = \Delta y\},$$

and its relative frequency *(or propagation ratio) (in S) as*

$$\text{freq}_S(\Delta x \to \Delta y) = \frac{N_S(\Delta x \to \Delta y)}{\#D(\Delta x)} = \frac{N_S(\Delta x \to \Delta y)}{16}.$$

For input difference $\Delta x = 1000$, we find from Table 6.3 the following absolute frequencies for differentials $1000 \to \Delta y$.

Δy	0000	0001	0010	0011	0100	0101	0110	0111
$N_S(1000 \to \Delta y)$	0	2	2	0	0	0	2	2
Δy	1000	1001	1010	1011	1100	1101	1110	1111
$N_S(1000 \to \Delta y)$	2	0	2	0	0	0	0	4

For all 256 differentials, this information is stored in the *differential distribution table* (or *difference distribution table*) containing $N_S(\Delta x \to \Delta y)$ for $\Delta x = 0000..1111$ and $\Delta y = 0000..1111$. For the S-box of baby-AES, this is shown in Table 6.4, and the absolute frequencies of the differentials $1000 \to \Delta y$ form the ninth row (in italics). All entries are even due to the symmetry of XOR and the values in each row and in each column sum up to 16. Since S is deterministic, the output difference is 0000 if the input difference is 0000. This explains the first row. Since S is invertible, the converse is also true and this explains the first column. Therefore, $0000 \to 0000$ has absolute frequency 16 and is called the *trivial* differential. We are looking for nontrivial differentials with large absolute frequency and thus ignore the first row and column.

The differential distribution table shows the absolute frequencies and it is common to use their maximal value over all nontrivial differentials as a quality measure for the security of an S-box, see Exercise 6.2. For the description of the differential attack it is more convenient to work with the relative frequency due to the following probabilistic reasoning. The relative frequency of $\Delta x \to \Delta y$ is the conditional probability that a uniformly chosen pair $(x, x^*) \xleftarrow{\boxtimes} \mathbb{F}_2^4 \times \mathbb{F}_2^4$ has the output difference Δy under the condition that it has the input difference Δx, that is

$$\text{freq}_S(\Delta x \to \Delta y) = \text{prob}\{S(x) \oplus S(x^*) = \Delta y: (x, x^*) \xleftarrow{\boxtimes} D(\Delta x)\}$$
$$= \text{prob}\{S(x) \oplus S(x + \Delta x) = \Delta y: x \xleftarrow{\boxtimes} \mathbb{F}_2^4\}.$$

This motivates the alternative term *propagation ratio* which we use from now on.

$\Delta x \backslash \Delta y$	0000	0001	0010	0011	0100	0101	0110	0111	1000	1001	1010	1011	1100	1101	1110	1111
0000	16	0	0	0	0	0	0	0	0	0	0	0	0	0	0	0
0001	0	2	2	2	2	0	0	0	2	0	0	0	2	4	0	0
0010	0	2	0	4	2	2	2	0	0	2	0	0	0	0	0	2
0011	0	2	4	0	0	2	0	0	2	2	0	2	0	0	2	0
0100	0	0	2	0	4	2	0	0	0	0	2	0	2	0	2	2
0101	0	0	0	2	0	0	0	2	4	2	0	0	2	0	2	2
0110	0	4	0	0	0	2	0	2	0	2	2	0	2	2	0	0
0111	0	2	0	0	0	0	2	0	0	0	0	2	4	2	2	2
1000	*0*	*2*	*2*	*0*	*0*	*0*	*2*	*2*	*2*	*0*	*2*	*0*	*0*	*0*	*0*	*4*
1001	0	0	2	2	0	0	0	0	0	2	2	4	0	2	0	2
1010	0	0	2	0	2	0	2	2	0	4	0	2	2	0	0	0
1011	0	0	0	0	2	0	2	0	2	2	4	0	0	2	2	0
1100	0	0	0	2	0	4	2	0	2	0	2	2	2	0	0	0
1101	0	0	0	0	2	2	0	4	2	0	0	2	0	2	0	2
1110	0	0	2	2	0	2	4	2	0	0	0	0	0	2	2	0
1111	0	2	0	2	2	0	0	2	0	0	2	2	0	0	4	0

Table 6.4: Differential distribution table of the S-box of baby-AES.

For $\Delta x = 1000$, there is a 25% chance that the S-box will turn this input difference into an output difference of 1111, assuming a uniform sampling over all possible pairs of inputs. This is substantially above the value for a random mapping, where all output differences are equally likely (6.25%). We will try and trace such "above average" differentials through several rounds.

Step 2: Find a differential trail through $R - 1$ rounds. The first step focusses on differences of nibbles processed by a single application of the S-box. The second step extends this to differences of states processed by the round functions. We are again looking for differentials with large propagation ratio—just on a bigger scale.

We start with SUBBYTES and examine a state differential $\Delta x \to \Delta y \in \mathbb{F}_2^{16} \times \mathbb{F}_2^{16}$ with input difference $\Delta x = (\Delta x_1, \Delta x_2, \Delta x_3, \Delta x_4) \in (\mathbb{F}_2^4)^4$ and output difference $\Delta y = (\Delta y_1, \Delta y_2, \Delta y_3, \Delta y_4) \in (\mathbb{F}_2^4)^4$. The differential distribution Table 6.4 of the S-box gives the propagation ratio for each nibble differential $\Delta x_i \to \Delta y_i$ for $1 \le i \le 4$. Since the nibbles are processed independently, we find as propagation ratio in SUBBYTES

$$(6.7) \qquad \mathrm{freq}_{\mathrm{SB}}(\Delta x \to \Delta y) = \prod_{1 \le i \le 4} \mathrm{freq}_S(\Delta x_i \to \Delta y_i).$$

For example, the differential $0110\,0000\,0000\,1000 \to 0001\,0000\,0000\,1111$ has a propagation ratio of

$$\mathrm{freq}_{\mathrm{SB}}(0110\,0000\,0000\,1000 \to 0001\,0000\,0000\,1111)$$
$$= \mathrm{freq}_S(0110 \to 0001) \cdot \mathrm{freq}_S(0000 \to 0000)^2 \cdot \mathrm{freq}_S(1000 \to 1111)$$

$$(6.8)$$
$$= \frac{1}{4} \cdot 1^2 \cdot \frac{1}{4} = \frac{1}{16}$$

from the seventh, first (twice), and ninth row of the difference distribution Table 6.4, respectively. This is also the most likely differential with input difference $\Delta x = 0110\,0000\,0000\,1000$. Some of the involved nibble differentials $\Delta x_i \to \Delta y_i$ may be trivial and the corresponding S-box S_i is then called *inactive*. Otherwise, it is called *active*, so that S_1 and S_4 are active in our example. We can rewrite (6.7) as

$$\mathrm{freq}_{\mathrm{SB}}(\Delta x \to \Delta y) = \prod_{S_i \text{ active}} \mathrm{freq}_S(\Delta x_i \to \Delta y_i).$$

The propagation ratio of differentials through SHIFTROWS and MIX-COLUMNS is even easier to compute. SHIFTROWS permutes the nibbles unmodified within in a row and MIXCOLUMNS applies a linear map to the nibbles in a column. Both operations commute with XOR. More precisely, $\Delta y = y \oplus y^* = \mathrm{SR}(x) \oplus \mathrm{SR}(x^*) = \mathrm{SR}(x \oplus x^*) = \mathrm{SR}(\Delta x)$ and $\Delta y = y \oplus y^* = \mathrm{MC}(x) \oplus \mathrm{MC}(x^*) = \mathrm{MC}(x \oplus x^*) = \mathrm{MC}(\Delta x)$, respectively. We find as propagation ratios

$$\mathrm{freq}_{\mathrm{SR}}(\Delta x \to \Delta y) = \begin{cases} 1 & \text{if } \Delta y = \mathrm{SR}(\Delta x), \\ 0 & \text{otherwise,} \end{cases}$$

$$\mathrm{freq}_{\mathrm{MC}}(\Delta x \to \Delta y) = \begin{cases} 1 & \text{if } \Delta y = \mathrm{MC}(\Delta x), \\ 0 & \text{otherwise.} \end{cases}$$

Finally, ADDROUNDKEY leaves a given input difference unchanged, since x and x^* with $x \oplus x^* = \Delta x$ are both XORed with the same round key k and thus $\Delta y = y \oplus y^* = (x \oplus k) \oplus (x^* \oplus k) = x \oplus x^* = \Delta x$. In other words,

$$\mathrm{freq}_{\mathrm{AK}}(\Delta x \to \Delta y) = \begin{cases} 1 & \text{if } \Delta y = \Delta x, \\ 0 & \text{otherwise.} \end{cases}$$

To summarize, all round functions but SUBBYTES propagate a difference Δx uniquely and independent of the choice of messages $(x, x^*) \in D(\Delta x)$. Thus, it is sufficient to specify the input difference to SUBBYTES and the output difference after ADDROUNDKEY to uniquely describe a path of state differences through a single round. We find as the propagation ratio of a state differential $\Delta u^r \to \Delta u^{r+1}$ through round r of baby-AES the propagation ratio of the corresponding SUBBYTES step

$$\text{freq}_{\text{round } r}(\Delta u^r \to \Delta u^{r+1}) = \text{freq}_{\text{SB}}(\Delta u^r \to \Delta v^r)$$

$$= \prod_{S_i^r \text{ active}} \text{freq}_S(\Delta u_i^r \to \Delta v_i^r),$$

where $\Delta v^r = (\text{MC} \circ \text{SR})^{-1}(\Delta u^{r+1})$, for $r = 1, 2$, and $\Delta v^3 = \text{SR}^{-1}(\Delta u^4)$.

DEFINITION 6.9. *A differential trail (or differential characteristic) through the first $R - 1$ rounds of baby-AES is a sequence $(\Delta u^1 \to \Delta u^2 \to \cdots \to \Delta u^R) \in (\mathbb{F}_2^{16})^R$ of R state differences Δu^r for the beginning of each round r for $1 \le r \le R$.*

The corresponding differential $\Delta u^1 \to \Delta u^R$ is the crucial ingredient for the forthcoming third step. The "intermediate differences" Δu^r for $1 < r < R$ allow us to approximate its propagation ratio — an essential information for the analysis in the final fourth step, where we determine the number of required plaintext-ciphertext pairs.

Assuming independent round keys (*Markov assumption*), we find as lower bound

$$\text{freq}_{\text{rounds 1 to } R-1}(\Delta u^1 \to \Delta u^R) \ge \prod_{1 \le r \le R-1} \text{freq}_{\text{round } r}(\Delta u^r \to \Delta u^{r+1})$$

$$(6.10) \qquad\qquad\qquad = \prod_{S_i^r \text{ active}} \text{freq}_S(\Delta u_i^r \to \Delta v_i^r).$$

Other differential trails $(\Delta u^1 \to \Delta u^{2'} \to \cdots \to \Delta u^{R'} \to \Delta u^R)$ with the same differential $\Delta u^1 \to \Delta u^R$ may be possible and they also contribute to $\text{freq}_{\text{rounds 1 to } R-1}(\Delta u^1 \to \Delta u^R)$. We usually single out the one with the largest contribution (6.10). Let us try and hike a differential trail through the first two rounds of our 3-round baby-AES, see Figure 6.5.

We start with the input difference

$$\Delta x = 0110\,0000\,0000\,1000 = \Delta w^0.$$

ADDROUNDKEY in round 0 leaves this difference unchanged and we begin the first round with $\Delta u^1 = \Delta w^0$. The most likely state after SUBBYTES is then

$$\Delta v^1 = 0001\,0000\,0000\,1111,$$

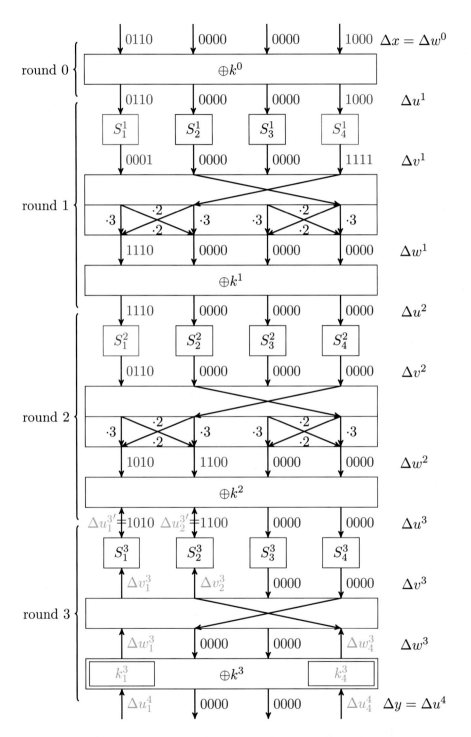

Figure 6.5: Differential trail 0110 0000 0000 1000 → 1110 0000 0000 0000 → 1010 1100 0000 0000 through two rounds of baby-AES.

as computed in (6.8). Application of SHIFTROWS and MIXCOLUMNS turns this into

$$\Delta w^1 = 1110\,0000\,0000\,0000,$$

which remains unchanged under ADDROUNDKEY and we have $\Delta u^2 = \Delta w^1$ at the beginning of the second round.

We use the most likely differential $1110 \to 0110$ for the only active S-box S_1^2 in the second round and find ourselves after application of SHIFTROWS, MIXCOLUMNS, and ADDROUNDKEY with the state difference

$$\Delta u^3 = 1010\,1100\,0000\,0000.$$

This completes our differential trail $\Delta u^1 \to \Delta u^2 \to \Delta u^3$ through the first two rounds. What is its propagation ratio? We have three active S-boxes and a propagation ratio of $1/4$ for the nibble differential in each of them. Therefore, by (6.10),

$$(6.11) \qquad \mathrm{freq}_{\text{rounds 1 to 2}}(\Delta u^1 \to \Delta u^3) \geq \left(\frac{1}{4}\right)^3 = 2^{-6}.$$

In other words, we have

$$\mathrm{prob}\{\Delta u^3 = 1010\,1100\,0000\,0000 \mid (x, x^*) \in D(0110\,0000\,0000\,1000)\}$$
$$\geq 2^{-6}.$$

(This lower bound is quite accurate. The only other differential trail for $\Delta u^1 \to \Delta u^3$ is along $\Delta u^{2'} = 1010\,0000\,0000\,0000$ with a propagation ratio of only 2^{-9}.) For a random mapping, we could only expect a lower bound of 2^{-16}.

Here Δu^3 is the input difference to round three. In the following third step, we will approach this position running backwards from two ciphertexts y and y^* corresponding to plaintexts $(x, x^*) \in D(0110\,0000\,0000\,1000)$ encrypted with an unknown secret key.

Step 3: Nibbles of the last round key. After the preparation of a 2-round differential $\Delta u^1 \to \Delta u^3$ with high propagation ratio it is time to launch our attack on the last round key. More precisely, we will recover the two nibbles k_1^3 and k_4^3 of the last round key k^3 that are influenced by the nonzero nibbles Δu_1^3 and Δu_2^3 of the output difference. The remaining two nibbles can then be found by brute force.

This key recovery requires chosen plaintexts (UBK-CPA model in Section 9.7). More precisely, for an unknown secret key k, the adversary is allowed to choose some x and request its encryption $y = \mathrm{AES}_k(x)$ along with the encryption y^* of $x^* = x + \Delta u^1$. Such samples (x, x^*, y, y^*) are collected in the set $\mathcal{S}$ and part of the input to Algorithm 6.12.

ALGORITHM 6.12. Differential attack on 3-round baby-AES.

Input: 2-round differential $\Delta u^1 \to \Delta u^3$ with propagation ratio $p = \text{freq}_{\text{rounds 1 to 2}}(\Delta u^1 \to \Delta u^3)$ [Note: $\Delta u_3^3 = \Delta u_4^3 = 0000$], set of samples $\mathcal{S} \subseteq \{(x, x^*, y, y^*) \in (\mathbb{F}_2^{16})^4 : (x, x^*) \in D(\Delta u^1)\}$.

Output: Partial round key $(k_1^3, k_4^3) \in \mathbb{F}_2^4 \times \mathbb{F}_2^4$.

1. Initialize a counter for every *candidate key* $(k_1^3, k_4^3) \in \mathbb{F}_2^4 \times \mathbb{F}_2^4$ with 0.
2. For $(x, x^*, y, y^*) \in \mathcal{S}$ do steps 3–13
3.   If $y_2 \neq y_2^*$ or $y_3 \neq y_3^*$ then
4.   discard sample and skip iteration [*filter*].
5.   For $(k_1^3, k_4^3) \in \mathbb{F}_2^4 \times \mathbb{F}_2^4$ do steps 6–13
6.   [Run the last round backwards on ciphertext nibbles y_1 and y_4.]
7. $(w_1^3, w_4^3) \leftarrow (y_1, y_4) \oplus (k_1^3, k_4^3)$ [Undo AK on nibbles 1 and 4].
8. $(v_1^3, v_2^3) \leftarrow (w_1^3, w_4^3)$ [Undo SR on nibbles 1 and 4].
9. $(u_1^3, u_2^3) \leftarrow (S^{-1}(v_1^3), S^{-1}(v_2^3))$ [Undo SB on nibbles 1 and 2].
10.   Similarly (u_1^{*3}, u_2^{*3}) from (y_1^*, y_4^*).
11. $(\Delta u_1^3, \Delta u_2^3)' \leftarrow (u_1^3 \oplus u_1^{*3}, u_2^3 \oplus u_2^{*3})$.
12.   If $(\Delta u_1^3, \Delta u_2^3)' = (\Delta u_1^3, \Delta u_2^3)$ then
13.   increase the counter for (k_1^3, k_4^3) by 1.
14. Return candidate key with highest counter.

How many samples should we request? Elements of $(y, y^*) \in D(\Delta u^1)$ with difference Δu^3 after two rounds as predicted are called *compliant*. These are (mostly) elements that follow the differential path laid out by $\Delta u^1 \to \Delta u^2 \to \Delta u^3$. Otherwise, we call an element *noncompliant*.

If the condition in step 12 holds and the counter is increased, we say that the pair (y, y^*) *suggests* the key (k_1^3, k_4^3). A compliant pair suggests the correct round key, but also some wrong subkeys, for example, the key $k + y + y^*$ that interchanges y and y^*. A noncompliant pair usually suggests wrong subkeys. If $p(\Delta u^1 \to \Delta u^3) \gg p(\Delta u^1 \to (\Delta u^3)')$ for all $(\Delta u^3)' \neq \Delta u^3$, we can assume these suggestions to be uniformly distributed. The attack is successful if the correct subkey is suggested significantly more often than any other subkey. If the noise from wrong suggestions is sufficiently low, already a small number c of compliant pairs suffices. As a rule of thumb, we ask for

(6.13) $$\#\mathcal{S} = c/p.$$

For our example with propagation ratio $p = \text{freq}_{\text{rounds 1 to 2}}(\Delta u^1 \to \Delta u^3) = 2^{-6}$ from (6.11), we choose $c = 3$.

What is the cost of this attack? Ignoring the filtering in step 3 for the moment, we loop over all 2^8 candidate keys and $\#\mathcal{S}$ pairs (y, y^*). In

steps 6–10, we invert the last round of baby-AES on two nibbles of y and
y^*, respectively. Each execution of the inner loop requires two partial
(that is, half an array) 1-round decryptions. Furthermore, two XORs of
nibbles in step 11 and two comparisons of nibbles in step 12 at negligible
costs. This yields a total of $2^8 \cdot \#\mathcal{S}$ partial (inverse) rounds of baby-AES
to recover 8 bits of the last round key. Recovering the remaining 8 bits
by brute force requires $2^8 \cdot 3$ rounds of baby-AES, for a total cost of

$$2^8 \cdot \#\mathcal{S} + 2^8 \cdot 3 = 3 \cdot 2^{14} + 3 \cdot 2^8 \approx 3 \cdot 2^{14}$$

rounds with $\#\mathcal{S} = 3 \cdot 2^6$ as above. This is already a (small) improvement
over the $3 \cdot 2^{16}$ rounds required for a direct brute force attack on the 16-bit
secret key.

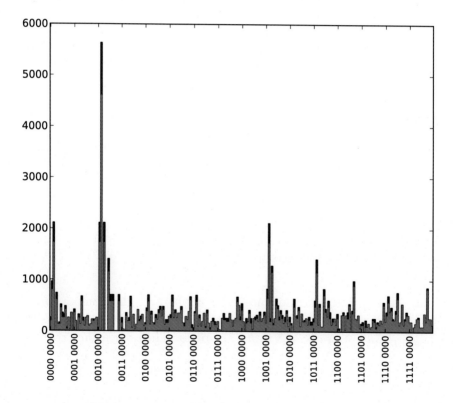

Figure 6.6: Counters for all 8-bit partial round keys for all possible samples
without (red) and with (green) filtering.

The "filtering" in step 3 widens this gap by discarding a large proportion of noncompliant pairs early, but passing all the compliant ones.
This merely reduces the noise—and the cost. Compliant elements have
$\Delta u_3^3 = \Delta u_4^3 = 0000$ and hence $\Delta y_3 = \Delta y_4 = 0000$. Elements that

subkey	unfiltered	filtered	subkey	unfiltered	filtered
$(0000, 0000)$	256	192	$(0001, 0100)$	128	128
$(0000, 0001)$	960	800	$(0001, 0101)$	672	576
$(0000, 0010)$	2112	1728	$(0001, 0110)$	256	192
$(0000, 0011)$	0	0	$(0001, 0111)$	96	96
$(0000, 0100)$	736	608	$(0001, 1000)$	256	256
$(0000, 0101)$	128	128	$(0001, 1001)$	288	288
$(0000, 0110)$	160	128	$(0001, 1010)$	96	96
$(0000, 0111)$	512	448	$(0001, 1011)$	128	128
$(0000, 1000)$	352	256	$(0001, 1100)$	224	224
$(0000, 1001)$	320	256	$(0001, 1101)$	224	192
$(0000, 1010)$	480	416	$(0001, 1110)$	224	224
$(0000, 1011)$	64	32	$(0001, 1111)$	256	256
$(0000, 1100)$	256	256	$(0010, 0000)$	0	0
$(0000, 1101)$	32	32	$(0010, 0001)$	2112	1728
$(0000, 1110)$	352	352	$\mathbf{(0010, 0010)}$	**5632**	**4608**
$(0000, 1111)$	0	0	$(0010, 0011)$	0	0
$(0001, 0000)$	416	352	$(0010, 0100)$	2112	1728
$(0001, 0001)$	192	192	$(0010, 0101)$	0	0
$(0001, 0010)$	0	0	$(0010, 0110)$	0	0
$(0001, 0011)$	320	256	$(0010, 0111)$	1408	1152

Table 6.7: Table with the 40 leftmost 8-bit partial round keys in Figure 6.6 and associated counters when sampling over all 2^{16} message-ciphertext pairs; for the counter of the correct subkey (in boldface), we expect—ignoring the noise from noncompliant pairs—at least $p \cdot \#\mathcal{S} = 2^{10} = 1024$.

do not satisfy this condition are noncompliant. The comparisons for filtering are essentially free and we eventually expect to process only $\text{freq}_{\text{round } 1}(\Delta u^1 \to \Delta u^2) \cdot \#\mathcal{S} = \#\mathcal{S}/16$ pairs. Then our attack requires

- $3/p = 3 \cdot 2^6$ chosen plaintext-ciphertext pairs,

- $2^8 \#\mathcal{S}/16 + 2^8 \cdot 3 = 3 \cdot 2^{10} + 3 \cdot 2^8 \approx 2^{12}$ rounds of baby-AES.

This is better than the $3 \cdot 2^{16}$ rounds for a brute-force attack.

To conclude our example, we choose for 3-round baby-AES the secret key $k = 0010\,0100\,0011\,1111$ (nothing up my sleeve: these are the binary digits of the number $\pi = 3.14\ldots$ after the binary point) and as 2-round differential $0110\,0000\,0000\,1000 \to 1010\,1100\,0000\,0000$ of Step 2. We then sample all 2^{16} possible pairs and run the differential attack Algorithm 6.12.

The resulting counters for each candidate key are visualized in Figure 6.6 and the exact numbers for the 40 leftmost candidate keys in Table 6.7. This took 4 hours and 43 minutes on a single core 3.4 GHz using Sage 6.3. The propagation ratio of 2^{-6} guarantees at least $2^{16-6} = 2^{10}$ suggestions by compliant elements for the correct candidate key. In an actual attack, one would take only a small random sample of all pairs as determined by (6.13).

6.3. Linear cryptanalysis

As the differential cryptanalysis of the previous section, linear cryptanalysis targets the round key of the last round. In contrast to differential cryptanalysis, it requires only *known* plaintext-ciphertext pairs and is therefore a key recovery under known ciphertext attack in the terminology of Section 9.7. Linear cryptanalysis, due to Matsui (1994b), is the best theoretical attack on the (obsolete) DES with 2^{43} operations requiring 2^{43} plaintext-ciphertext pairs. If such a large number of pairs (with known plaintexts) is not available, a brute-force attack on the 56-bit secret key is also feasible today.

The basic idea of linear cryptanalysis is to find sums of input and output bits that are equal (or unequal) with probability sufficiently different from the average value of $1/2$. Combining such linear relations from single S-boxes over all but the last round, we find a relation between bits of a known plaintext, unknown key bits (constant for all plaintexts), and input bits to the last round. As in differential cryptanalysis, we decrypt the corresponding ciphertext under all possible candidate keys for the last round and increase the key's counter when the expected relation at the beginning of the last round holds. Finally, we choose the candidate key whose counter is furthest away from the expected average which equals half the number of samples. We describe the attack in three steps.

Step 1: Linear approximations to the S-box. Recall the S-box $S \colon \mathbb{F}_2^4 \to \mathbb{F}_2^4$ of baby-AES as specified in Table 6.1. For a random mapping we would expect that any output bit flips with probability $1/2$ whenever an input bit flips. We generalize this comparison of single bits to sums of selected input bits and sums of selected output bits. In other words, we are interested in the correlation of linear functions on the input bits and linear functions on the resulting output bits.

A 4-bit string $a = a_3 a_2 a_1 a_0 \in \mathbb{F}_2^4$ defines an $\mathbb{F}_2$-linear Boolean function

$$a^* \colon \mathbb{F}_2^4 \to \mathbb{F}_2, \quad x = x_3 x_2 x_1 x_0 \mapsto \sum_{0 \leq j \leq 3} a_j x_j = a \star x,$$

also known as the *inner product* with a. All $\mathbb{F}_2$-linear Boolean functions on the $\mathbb{F}_2$-vector space $\mathbb{F}_2^4$ are of this form, see Exercise 6.9. Their set $\mathbb{F}_2^{4*} = \{\mathbb{F}_2\text{-linear functions } \ell\colon \mathbb{F}_2^4 \to \mathbb{F}_2\}$, called the *dual space* of $\mathbb{F}_2^4$, is isomorphic to $\mathbb{F}_2^4$.

We try to approximate a linear function b^* on the output $y = S(x)$ by a linear function a^* on the input x. The quality of the approximation can be measured by the number of "agreements", that is the number of $x \in \mathbb{F}_2^4$ such that the linear relation

$$(6.14) \qquad a \star x = a^*(x) = b^*(y) = b \star y$$

holds for $y = S(x)$. We call the pair $(a, b) \in \mathbb{F}_2^4 \times \mathbb{F}_2^4$ the *selection mask* (or *selection pattern*) of the *linear approximation* (6.14) (of the S-box) and denote it by $a \parallel b$. Instead of working with the number of agreements for a selection mask $(a \parallel b) \in \mathbb{F}_2^4 \times \mathbb{F}_2^4$, it is more convenient to use its *signed agreement*

$$
\begin{aligned}
C_S(a \parallel b) &= \#\{x \in \mathbb{F}_2^4\colon a \star x = b \star S(x)\} - \#\{x \in \mathbb{F}_2^4\colon a \star x \neq b \star S(x)\} \\
(6.15) \qquad &= 2 \cdot \#\{x \in \mathbb{F}_2^4\colon a \star x = b \star S(x)\} - 16,
\end{aligned}
$$

namely the number of agreements minus the number of disagreements.

For example, we find $C_S(0010 \parallel 0011) = -8$ from the following list of comparisons, looking up $S(x)$ in Table 6.1.

x	0000	0001	0010	0011	0100	0101	0110	0111
$0010 \star x$	0	0	1	1	0	0	1	1
$0011 \star S(x)$	1	0	1	0	1	1	0	1
"agreement"	-	+	+	-	-	-	-	+
x	1000	1001	1010	1011	1100	1101	1110	1111
$0010 \star x$	0	0	1	1	0	0	1	1
$0011 \star S(x)$	1	1	0	0	0	1	0	0
"agreement"	-	-	-	-	+	-	-	-

We tabulate $C_S(a \parallel b)$ for $a = 0000..1111$ and $b = 0000..1111$ in the *signed agreement table* of the S-box, as shown in Table 6.8. We find $C_S(0100 \parallel 0011)$ from above as the fourth entry in the third row (in boldface). Trivially, $C_S(0000 \parallel 0000) = 16$ and $C_S(a \parallel 0000) = C_S(0000 \parallel a) = 0$ for all nonzero $a \in \mathbb{F}_2^4$. All entries are divisible by 4, see Exercise 6.11, and the entries in each row and each column sum up to ± 16, as we will show below.

For a linear f, we have $C_f(a \parallel b) \in \pm 16$ for all nonzero $a, b \in \mathbb{F}_2^4$. For a random mapping f, we would expect (6.14) to hold for half of all inputs and hence $C_f(a \parallel b) = 0$ for all nonzero $a, b \in \mathbb{F}_2^4$. The S-box of baby-AES

$a\backslash b$	0000	0001	0010	0011	0100	0101	0110	0111	1000	1001	1010	1011	1100	1101	1110	1111
0000	16	0	0	0	0	0	0	0	0	0	0	0	0	0	0	0
0001	0	−4	−4	0	0	4	4	0	8	4	−4	0	0	4	−4	8
0010	0	−4	−4	−8	4	−8	0	4	0	4	−4	0	4	0	0	−4
0011	0	8	0	0	4	−4	−4	−4	0	0	−8	0	−4	−4	−4	4
0100	0	−4	4	0	−8	−4	−4	0	−4	8	0	−4	−4	0	0	4
0101	0	0	0	−8	0	0	8	0	−4	−4	4	−4	−4	−4	−4	4
0110	0	0	0	0	4	4	4	4	−4	4	−4	4	−8	0	8	0
0111	0	4	4	0	−4	−8	8	−4	4	0	0	4	0	4	4	0
1000	0	4	−8	−4	−4	0	−4	0	4	0	4	0	−8	4	0	−4
1001	0	0	4	−4	−4	4	0	0	4	−4	−8	−8	0	0	4	−4
1010	0	8	−4	4	0	0	4	4	−4	4	0	−8	4	4	0	0
1011	0	4	0	−4	0	4	0	−4	4	8	4	0	4	−8	4	0
1100	0	0	4	4	4	−4	0	8	8	0	4	−4	−4	−4	0	0
1101	0	4	0	−4	−4	0	−4	8	0	−4	0	4	4	0	4	8
1110	0	−4	−8	4	0	−4	0	−4	0	−4	0	−4	0	−4	8	4
1111	0	0	−4	4	−8	0	4	4	0	0	−4	4	0	−8	−4	−4

Table 6.8: Signed agreement table of the S-box of baby-AES.

lies between these two extremes and we are looking for nonzero selection masks with large absolute value for the signed agreement.

To put this approach into a larger context, we define the *signed agreement* of two Boolean functions $f, g \colon \mathbb{F}_2^n \to \mathbb{F}_2$ as

$$(6.16) \quad C(f,g) = \#\{x \in \mathbb{F}_2^n \colon f(x) = g(x)\} - \#\{x \in \mathbb{F}_2^n \colon f(x) \neq g(x)\}$$
$$= 2\#\{x \in \mathbb{F}_2^n \colon f(x) = g(x)\} - 2^n.$$

Then $C(f,g)$ is even and $-2^n \leq C(f,g) \leq 2^n$. A mask $(a \parallel b) \in \mathbb{F}_2^4 \times \mathbb{F}_2^4$ on the S-box S of baby-AES defines the two Boolean functions a^* and $b^* \circ S$, and we find

$$C_S(a \parallel b) = C(a^*, b^* \circ S).$$

To turn these quantities into probabilities, we define the *correlation* of two Boolean functions $f, g \colon \mathbb{F}_2^n \to \mathbb{F}_2$ as

$$\mathrm{corr}(f,g) = C(f,g)/2^n$$
$$(6.17) \qquad = \mathrm{prob}\{f(x) = g(x) \colon x \xleftarrow{\boxtimes} \mathbb{F}_2^n\}$$
$$- \mathrm{prob}\{f(x) \neq g(x) \colon x \xleftarrow{\boxtimes} \mathbb{F}_2^n\}$$
$$= 2\,\mathrm{prob}\{f(x) = g(x) \colon x \xleftarrow{\boxtimes} \mathbb{F}_2^n\} - 1.$$

Then $-1 \leq \mathrm{corr}(f, g) \leq 1$. Furthermore, if $\mathrm{corr}(f, g) = 1$, then f and g are equal, if $\mathrm{corr}(f, g) = -1$, then they are each other's complement, and if $\mathrm{corr}(f, g) = 0$, we call them *uncorrelated*.

Let us return to our S-box S of baby-AES. For a selection mask $(a \parallel b) \in \mathbb{F}_2^4 \times \mathbb{F}_2^4$, we have the Boolean functions $f = a^*$ and $g = b^* \circ S$ and define the *correlation* of the mask $a \parallel b$ as

$$\mathrm{corr}_S(a \parallel b) = \mathrm{corr}(a^*, b^* \circ S) = C_S(a \parallel b)/16,$$

and find from (6.15)

$$\mathrm{prob}\{a \star x = b \star S(x)\colon x \xleftarrow{\;\boxtimes\;} \mathbb{F}_2^n\} = \frac{\mathrm{corr}_S(a \parallel b) + 1}{2}.$$

For $n = 4$ and the mask $0010 \parallel 0011$, this means

$$\mathrm{prob}\{0010 \star x = 0011 \star S(x)\colon x \xleftarrow{\;\boxtimes\;} \mathbb{F}_2^4\} = \frac{-8/16 + 1}{2} = 1/4$$

from the boldface entry in Table 6.8. This is sufficiently different from $1/2$, and we will combine and exploit such deviations in the following.

Step 2: Linear approximations through $R - 1$ rounds. The first step focusses on masks for a single S-box and their correlations. The second step now extends this to masks for the round functions and we are looking for masks that have a correlation with large amplitude.

We begin with SUBBYTES. Let $a = a_1 a_2 a_3 a_4, b = b_1 b_2 b_3 b_4 \in (\mathbb{F}_2^4)^4$ be two 16-bit (state) selection masks. The nibbles of a state are processed by independent S-boxes and the correlations multiply as the following lemma shows.

LEMMA 6.18. *Let $f_1, f_2, g_1, g_2\colon \mathbb{F}_2^n \to \mathbb{F}_2$ be Boolean functions on n bits. On $2n$ bits, we define $f, g\colon \mathbb{F}_2^n \times \mathbb{F}_2^n \to \mathbb{F}_2$ by $f(x_1, x_2) = f_1(x_1) + f_2(x_2)$ and $g(x_1, x_2) = g_1(x_1) + g_2(x_2)$, respectively, and find for the correlation*

$$\mathrm{corr}(f, g) = \mathrm{corr}(f_1, g_1) \cdot \mathrm{corr}(f_2, g_2).$$

PROOF.   For $i, j \in \{0.1\}$, we write $p_{i,j} = \mathrm{prob}\{f_i(x_i) + g_i(x_i) = j\colon$

$x_i \xleftarrow{\boxed{\square}} \mathbb{F}_2^n\}$. Then $\operatorname{corr}(f_i, g_i) = p_{i,0} - p_{i,1}$ and we compute directly

$$
\begin{aligned}
\operatorname{corr}&(f, g) \\
&= \operatorname{prob}\{f(x) + g(x) = 0 : x \in \mathbb{F}_2^{2n}\} - \operatorname{prob}\{f(x) + g(x) = 1 : x \in \mathbb{F}_2^{2n}\} \\
&= \operatorname{prob}\{f_1(x_1) + f_2(x_2) + g_1(x_1) + g_2(x_2) = 0 : (x_1, x_2) \in \mathbb{F}_2^n \times \mathbb{F}_2^n\} \\
&\quad - \operatorname{prob}\{f_1(x_1) + f_2(x_2) + g_1(x_1) + g_2(x_2) = 1 : (x_1, x_2) \in \mathbb{F}_2^n \times \mathbb{F}_2^n\} \\
&= \operatorname{prob}\{f_1(x_1) = g_1(x_1) : x_1 \in \mathbb{F}_2^n\} \cdot \operatorname{prob}\{f_2(x_2) = g_2(x_2) : x_2 \in \mathbb{F}_2^n\} \\
&\quad + \operatorname{prob}\{f_1(x_1) \neq g_1(x_1) : x_1 \in \mathbb{F}_2^n\} \cdot \operatorname{prob}\{f_2(x_2) \neq g_2(x_2) : x_2 \in \mathbb{F}_2^n\} \\
&\quad - \operatorname{prob}\{f_1(x_1) = g_1(x_1) : x_1 \in \mathbb{F}_2^n\} \cdot \operatorname{prob}\{f_2(x_2) \neq g_2(x_2) : x_2 \in \mathbb{F}_2^n\} \\
&\quad - \operatorname{prob}\{f_1(x_1) \neq g_1(x_1) : x_1 \in \mathbb{F}_2^n\} \cdot \operatorname{prob}\{f_2(x_2) = g_2(x_2) : x_2 \in \mathbb{F}_2^n\} \\
&= (p_{1,0}p_{2,0} + p_{1,1}p_{2,1}) - (p_{1,0}p_{2,1} + p_{1,1}p_{2,0}) = \operatorname{corr}(f_1, g_1) \cdot \operatorname{corr}(f_2, g_2),
\end{aligned}
$$

as claimed. $\qquad\square$

Repeated application shows for the correlation of a^* and $b^* \circ \text{SUBBYTES}$

$$(6.19) \qquad \operatorname{corr}_{\text{SB}}(a \parallel b) = \prod_{1 \leq i \leq 4} \operatorname{corr}_S(a_i \parallel b_i).$$

If $a_i = b_i = 0000$ for some nibble i, then the ith S-box is ignored by the mask, and we call it *inactive*. Otherwise, we call it *active*. Since $\operatorname{corr}_S(0000 \parallel 0000) = 1$, we can rewrite (6.19) as

$$\operatorname{corr}_{\text{SB}}(a \parallel b) = \prod_{S_i \text{ active}} \operatorname{corr}_S(a_i \parallel b_i).$$

For example, with $a = 0010\,0000\,0000\,1000$ and $b = 0101\,0000\,000\,1100$, we have as correlation for SUBBYTES

$$(6.20) \qquad \operatorname{corr}_{\text{SB}}(a \parallel b) = -\frac{1}{2} \cdot \left(-\frac{1}{2}\right) = 2^{-2}$$

from the third and ninth row of the correlation Table 6.8. This absolute value is maximal for selection masks on SUBBYTES with input mask a, but in contrast to the propagation of differentials, there are several output masks b for a given input mask a that achieve this maximal value.

SHIFTROWS exchanges two nibbles, so that $b \star \text{SR}(x) = \text{SR}^{-1}(b) \star x$ and

$$\operatorname{corr}_{\text{SR}}(a \parallel b) = \begin{cases} 1 & \text{if } a = \text{SR}^{-1}(b), \\ 0 & \text{otherwise.} \end{cases}$$

MIXCOLUMNS multiplies each column $(x_i, x_{i+1})^T$ for $i = 1, 3$ from the left with the matrix C of (6.4). At the bit level, this is multiplication of the 8-bit string $(x_{i,3}x_{i,2}x_{i,1}x_{i,0}x_{i+1,3}x_{i+1,2}x_{i+1,1}x_{i+1,0})^T$ with the matrix

$$
C_2 = \begin{pmatrix}
1 & 1 & 0 & 0 & 0 & 1 & 0 & 0 \\
0 & 1 & 1 & 0 & 0 & 0 & 1 & 0 \\
1 & 0 & 1 & 1 & 1 & 0 & 0 & 1 \\
1 & 0 & 0 & 1 & 1 & 0 & 0 & 0 \\
0 & 1 & 0 & 0 & 1 & 1 & 0 & 0 \\
0 & 0 & 1 & 0 & 0 & 1 & 1 & 0 \\
1 & 0 & 0 & 1 & 1 & 0 & 1 & 1 \\
1 & 0 & 0 & 0 & 1 & 0 & 0 & 1
\end{pmatrix} \in \mathbb{F}_2^{8 \times 8}.
$$

We denote the multiplication with its transpose C_2^T as MC' and have $b \star \mathrm{MC}(x) = \mathrm{MC}'(b) \star x$, see Exercise 6.8. Therefore,

$$
\mathrm{corr}_{\mathrm{MC}}(a \parallel b) = \begin{cases} 1 & \text{if } a = \mathrm{MC}'(b), \\ 0 & \text{otherwise.} \end{cases}
$$

Finally, ADDROUNDKEY is bitwise addition of a constant and we have $b \star (x + k) = b \star x + b \star k$, so that

$$
\mathrm{corr}_{\mathrm{AK}}(a \parallel b) = \begin{cases} (-1)^{b \star k} & \text{if } b = a, \\ 0 & \text{otherwise.} \end{cases}
$$

In total, all round functions but SUBBYTES allow only one output mask b for a given input mask a to obtain a linear approximation $a \parallel b$ with nonzero correlation. Moreover, the absolute value of the correlation for a state selection mask $a^r \parallel a^{r+1}$ for round r of baby-AES depends only on the absolute values of the employed masks on SUBBYTES, and we find

$$
\mathrm{corr}_{\text{round } r}(a^r \parallel a^{r+1}) = (-1)^{a^{r+1} \star k^r} \mathrm{corr}_{\mathrm{SB}}(a^r \parallel b^r)
$$
$$
= (-1)^{a^{r+1} \star k^r} \prod_{S_i^r \text{ active}} \mathrm{corr}_S(a_i^r \parallel b_i^r),
$$

where k^r is the round key, $b^r = \mathrm{SR}^{-1}(\mathrm{MC}'(a^{r+1}))$ for $r = 1, 2$, and $b^3 = \mathrm{SR}^{-1}(a^4)$.

DEFINITION 6.21. *A linear trail (or linear characteristic) through the first R rounds of baby-AES is a sequence $(a^1 \parallel a^2 \parallel \cdots \parallel a^{R+1}) \in (\mathbb{F}_2^{16})^{R+1}$ of $R+1$ masks a^r for the beginning of each round r for $1 \leq r \leq R+1$.*

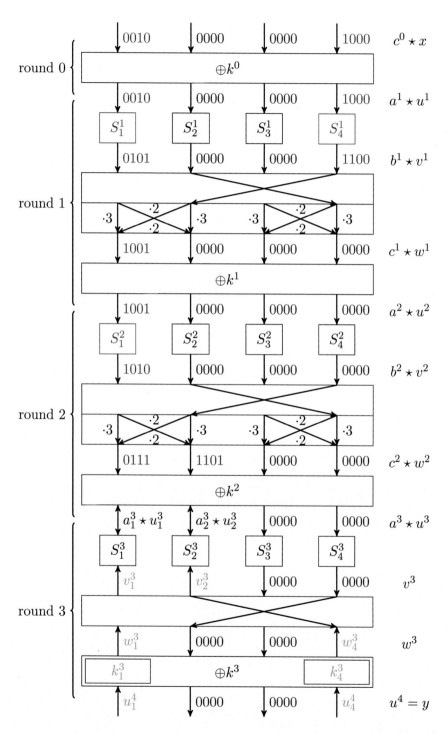

Figure 6.9: Linear trail $a^1 \parallel a^2 \parallel a^3 = 0100\,0000\,0000\,1000 \parallel 1001\,0000\,0000\,0000 \parallel 0111\,1101\,0000\,0000$ through two rounds of baby-AES.

The *(correlation) contribution* of a trail $a^1 \parallel a^2 \parallel \cdots \parallel a^{R+1}$ is

(6.22)
$$
\mathrm{corr}_{\text{rounds 1 to } R}(a^1 \parallel a^2 \parallel \cdots \parallel a^{R+1})
$$
$$
= \prod_{1 \leq r \leq R} \mathrm{corr}_{\text{round } r}(a^r \parallel a^{r+1})
$$
$$
= (-1)^{\sum_{1 \leq r \leq R} a^{r+1} \star k^r} \prod_{\substack{S_i^r \text{ active} \\ 1 \leq r \leq R}} \mathrm{corr}_S(a_i^r \parallel b_i^r),
$$

assuming independent round keys $k_1, \ldots, k_R$. The resulting mask $a^1 \parallel a^{R+1}$ on the input of the first round and the output of the Rth round is the crucial ingredient for a linear attack in the next step. We sum over all intermediate masks a^r for $1 < r < R + 1$ to determine its correlation, and have

$$
\mathrm{corr}_{\text{rounds 1 to } R}(a^1 \parallel a^{R+1})
$$
$$
= \sum_{\substack{\text{linear trails} \\ a^1 \parallel a^2 \parallel \cdots \parallel a^{R+1}}} \mathrm{corr}_{\text{rounds 1 to } R}(a^1 \parallel a^2 \parallel \cdots \parallel a^{R+1}).
$$

The contribution of each trail is signed. One assumes that the contributions outside a "main path" are small and roughly cancel each other.

Let us try and hike a linear trail through the first two rounds of our 3-round baby-AES; see Figure 6.9. We begin with the input mask

$$
c^0 = 0010\,0000\,0000\,1000 = a^1.
$$

Then $\mathrm{corr}_{\text{AddRoundKey}}(c^0 \parallel a^1) = (-1)^{c^0 \star k^0}$. For the two active S-boxes S_1^1 and S_4^1, we choose as output masks 0101 and 1100, respectively. In other words,

$$
b^1 = 0101\,0000\,0000\,1100
$$

with $\mathrm{corr}_{\text{SubBytes}}(a^1 \parallel b^1) = 2^{-2}$ as computed in (6.20). After ShiftRows, MixColumns, and AddRoundKey, we have

$$
a^2 = 1001\,0000\,0000\,0000
$$

and $\mathrm{corr}_{\text{round 1}}(a^1 \parallel a^2) = (-1)^{a^2 \star k^1} 2^{-2}$. We choose the output mask 1010 for the first S-box S_1^2 of the second round. This yields a correlation $\mathrm{corr}_{\text{SubBytes}}(a^2 \parallel b^2) = -2^{-1}$ for

$$
b^2 = 1010\,0000\,0000\,0000.
$$

After ShiftRows, MixColumns, and AddRoundKey, we arrive at

$$
a^3 = 0111\,1101\,0000\,0000
$$

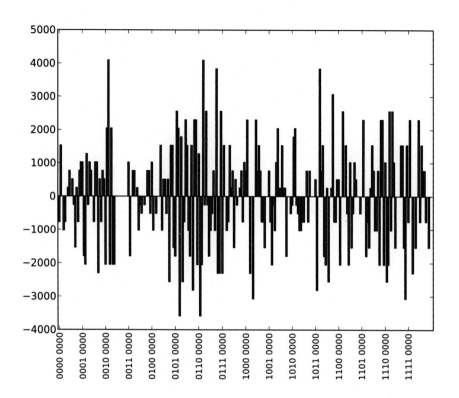

Figure 6.10: Counters for all 8-bit partial round keys for all possible samples.

with $\text{corr}_{\text{round }2}(a^2 \parallel a^3) = (-1)^{1+a^3 \star k^2} 2^{-1}$. This completes our linear trail $a^1 \parallel a^2 \parallel a^3$. Its contribution to the mask $a^1 \parallel a^3$ on the input to the first round and output of the second round has an absolute value of

$$2^{-2} \cdot 2^{-1} = 2^{-3},$$

according to (6.10). Adding the 0th round to our discussion, we find

(6.23) $\qquad \text{corr}_{\text{rounds 0 to 2}}(a^1 \parallel a^3) \approx (-1)^{\sum_{0 \leq r \leq 2} a^{r+1} \star k^r} 2^{-3}.$

For a fixed secret key, the sum in the exponent is unknown but constant, and the absolute value of the correlation is $1/8$.

As in differential cryptanalysis, we will now approach the nonzero nibbles of a^3 by running known ciphertext backwards decrypting with all possible candidate keys for the last round.

Step 3: Nibbles of the last round key. After the preparation of a 2-round linear trail $a^1 \parallel a^2 \parallel a^3$ with large absolute value of correlation,

subkey	counter	subkey	counter
(0000, 0000)	-768	(0001, 0100)	-256
(0000, 0001)	1536	(0001, 0101)	1024
(0000, 0010)	0	(0001, 0110)	768
(0000, 0011)	-1024	(0001, 0111)	0
(0000, 0100)	-768	(0001, 1000)	-768
(0000, 0101)	0	(0001, 1001)	1024
(0000, 0110)	256	(0001, 1010)	1024
(0000, 0111)	768	(0001, 1011)	-2304
(0000, 1000)	0	(0001, 1100)	512
(0000, 1001)	512	(0001, 1101)	-768
(0000, 1010)	-256	(0001, 1110)	768
(0000, 1011)	-1536	(0001, 1111)	512
(0000, 1100)	256	(0010, 0000)	-2048
(0000, 1101)	-768	(0010, 0001)	2048
(0000, 1110)	768	**(0010, 0010)**	**4096**
(0000, 1111)	1024	(0010, 0011)	-2048
(0001, 0000)	1024	(0010, 0100)	2048
(0001, 0001)	-1792	(0010, 0101)	-2048
(0001, 0010)	-2048	(0010, 0110)	-2048
(0001, 0011)	1280	(0010, 0111)	0

Table 6.11: The 40 leftmost 8-bit partial round keys in Figure 6.10 and associated counters when sampling over all 2^{16} message-ciphertext pairs; for the correct subkey (in boldface), we expect a magnitude of $2^{16} \cdot |\text{corr}|/2 = 4096$.

we launch our attack on the last round key. More precisely, we will recover the two nibbles k_1^3 and k_4^3 of the last round key k^3 that are influenced by the nonzero nibbles a_1^3 and a_2^3 of the output mask. The remaining two nibbles of k^3 can then be found by brute force.

This key recovery requires known plaintexts (UBK-KPA model in Section 9.7). For example, with an unknown secret key k, the adversary is allowed to ask for the encryptions $y = \text{baby-AES}_k(x)$ of plaintexts x that he chooses uniformly at random. These pairs (x, y) are collected in the set $\mathcal{S}$ and form the second input (apart from the mask $a^1 \parallel a^3$) to Algorithm 6.24.

ALGORITHM 6.24. Linear attack on 3-round baby-AES.

Input: selection mask $(a^1 \parallel a^3)$ with correlation corr,
         set of samples $\mathcal{S} = \{(x, y) \in \mathbb{F}_2^{16} \times \mathbb{F}_2^{16} : y = \text{baby-AES}_k(x)\}$.

Output: Partial round key $(k_1^3, k_4^3) \in \mathbb{F}_2^4 \times \mathbb{F}_2^4$.

1. Initialize a counter for each *candidate key* $(k_1^3, k_4^3) \in \mathbb{F}_2^4$ with 0.
2. For $(x, y) \in \mathcal{S}$ do steps 3–11
3.   For $(k_1^3, k_4^3) \in \mathbb{F}_2^4 \times \mathbb{F}_2^4$ do steps 4–11
4.   [Run the last round backwards on (y_1, y_4).]
5. $(w_1^3, w_4^3) \leftarrow (y_1, y_4) \oplus (k_1^3, k_4^3)$ [Undo AK on nibbles 1 and 4].
6. $(v_1^3, v_2^3) \leftarrow (w_1^3, w_4^3)$ [Undo SR on nibbles 1 and 4].
7. $(u_1^3, u_2^3) \leftarrow (S^{-1}(v_1^3), S^{-1}(v_2^3))$ [Undo SB on nibbles 1 and 2].
8.   If $a^1 \star x = a_1^3 \star u_1^3 \oplus a_2^3 \star u_2^3 (= a^3 \star u^3)$ then
9.   increase the counter for (k_1^3, k_4^3) by 1.
10.   Else
11.   decrease the counter for (k_1^3, k_4^3) by 1.
12. Return candidate key with counter of largest magnitude.

The else clause can be omitted by initializing the counters to $-\#\mathcal{S}/2$ instead of 0. The attack is successful, if the correct subkey's counter has sufficiently large magnitude. As a rule of thumb, we ask for

$$(6.25) \qquad\qquad \#\mathcal{S} = c/\operatorname{corr}^2$$

with a small number c. For our example with correlation of magnitude $|\operatorname{corr}| = |\operatorname{corr}_{\text{rounds 0 to 2}}(a^1 \parallel a^3)| \approx 2^{-3}$ from (6.23), we choose $c = 8$. Then our attack requires

- $\#\mathcal{S} = 8/\operatorname{corr}^2 = 2^9$ known plaintext-ciphertext pairs,

- $2^8 \#\mathcal{S} + 2^8 \cdot 3 \approx 2^{17}$ rounds of baby-AES with similar arguments as for Algorithm 6.12.

This is only slightly better than the $3 \cdot 2^{16}$ rounds for a brute-force attack. Choosing $c = 1$ leads to about 2^{14} rounds.

To conclude our example, we choose for 3-round baby-AES the secret key $k = 0010\,0100\,0011\,1111$ as in Section 6.2. We use the selection mask $0100\,0000\,0000\,1000 \parallel 0111\,1101\,0000\,0000$ of Step 2 and sample all 2^{16} message-ciphertext pairs. The resulting counters for each candidate key are visualized in Figure 6.10 and the exact numbers for the 40 leftmost candidate keys are in Table 6.11. This computation took 2 hours and 26 minutes on a single core at 3.4 GHz using Sage 6.3. For a correlation of magnitude $|\operatorname{corr}| = 1/8$, we have a probability of $(|\operatorname{corr}| + 1)/2 = 9/16$ for agreement or disagreement. Ignoring interference, we expect a counter of magnitude $2^{16} \cdot |\operatorname{corr}|/2 = 4096$. In an actual attack, one would take only a small random sample of all pairs as determined by (6.25).

6.4. Countermeasures and near-optimality of SUBBYTES

AES is well prepared against the attacks of the previous sections. Indeed, Daemen & Rijmen (2002a) write on their page 108: "The main crypto-graphic criterion in the design of Rijndael has been its resistance against differential and linear cryptanalysis." The workload of such an attack is determined by the employed trail's propagation ratio freq or magnitude of correlation |corr|, respectively. The corresponding formulae (6.10) and (6.22), respectively, suggest countermeasures at two levels.

1. (Global) Ensure a large number of active S-boxes in every differential or linear trail.

2. (Local) Use an S-box with small maximal propagation ratio and small maximal magnitude of correlation.

In Shannon's terminology (Section A.1), the first point addresses *diffusion* and the second one *confusion*. We now discuss how they are realized in the design of AES.

The diffusion property of a map $f\colon \mathbb{F}_q^n \to \mathbb{F}_q^n$ is measured by the *differential* and *linear branch numbers*

$$b_{\mathrm{diff}}(f) = \min\{\mathrm{wt}(x + y) + \mathrm{wt}(f(x) + f(y))\colon x, y \in \mathbb{F}_q^n \text{ with } x \neq y\},$$
$$b_{\mathrm{lin}}(f) = \min\{\mathrm{wt}(a) + \mathrm{wt}(b)\colon a, b \in \mathbb{F}_q^n \text{ with } \mathrm{corr}_f(a \parallel b) \neq 0\}.$$

For a *linear* f represented by the matrix $M \in \mathbb{F}_q^{n \times n}$, we have

$$b_{\mathrm{diff}}(f) = \min\{\mathrm{wt}(\Delta x) + \mathrm{wt}(M \cdot \Delta x)\colon \Delta x \in \mathbb{F}_q^n \setminus \{(0, \ldots, 0)\}\},$$
$$b_{\mathrm{lin}}(f) = \min\{\mathrm{wt}(a) + \mathrm{wt}(M^T \cdot a)\colon a \in \mathbb{F}_q^n \setminus \{(0, \ldots, 0)\}\}.$$

The MDS property of MIXCOLUMNS (Section 2.2) gives us

$$b = b_{\mathrm{diff}}(\mathrm{MIXCOLUMNS}) = b_{\mathrm{lin}}(\mathrm{MIXCOLUMNS}) = n + 1,$$

where the vector space dimension n is the number of rows, and therefore 2 for baby-AES and 4 for AES. Together with SHIFTROWS' property to distribute every single column over all columns, this yields the *Four-Round Propagation Theorem*: Every 4-round differential or linear trail has at least b^2 active S-boxes.

At the summer school crypt@b-it 2009, Bart Preneel summarized the effect of a good diffusion layer as follows: "*Every substitution-permutation network is secure after sufficiently many rounds. Every substitution-per-mutation network is also unusable after sufficiently many rounds.*"

The rest of this section deals with local countermeasures. We consider an invertible map $S\colon \mathbb{F}_2^n \to \mathbb{F}_2^n$. Its *maximal absolute frequency* is

$$(6.26) \quad N_{\max}(S) = \max\{N_S(\Delta x \to \Delta y)\colon \Delta x, \Delta y \in \mathbb{F}_2^n \setminus \{(0\ldots 0)\}\}.$$

A small value indicates a high resistance against differential cryptanalysis. We have as lower bound $N_{\max}(S) \geq 2$, see the comment following Table 6.4. We consider the n-bit generalization of inv (2.7) in SUBBYTES, namely

$$\text{inv}\colon \quad \begin{array}{ccc} F & \longrightarrow & \mathbb{F}_{2^n}, \\ a & \longmapsto & \begin{cases} a^{-1} & \text{if } a \neq 0, \\ 0 & \text{if } a = 0. \end{cases} \end{array}$$

Then $N_{\max}(\text{inv}) \leq 4$ and thus inv is optimal up to a factor of 2 with respect to the lower bound from above.

We turn to the resistance against linear cryptanalysis. The *maximal magnitude of signed agreement* of an invertible map $S\colon \mathbb{F}_2^n \to \mathbb{F}_2^n$ is

$$C_{\max}(S) = \max\{|C_S(a \parallel b)|\colon a, b \in \mathbb{F}_2^n \setminus \{0\ldots 0\}\}.$$

A small value indicates a high resistance against linear cryptanalysis. In the rest of this section, we prove the lower bound $C_{\max}(S) \geq 2^{n/2}$ in (6.30), and then use elliptic curves to show that inv is nearly optimal in Theorem 6.31.

The signed agreements $C_S(a \parallel b) = C(a^*, b^* \circ S)$ under investigation are special cases of $C(f, g)$ as defined in (6.16). We investigate the latter for linear functions $f \in \mathbb{F}_2^{n*}$ and fixed second argument $g\colon \mathbb{F}_2^n \to \mathbb{F}_2$. This is called the *Walsh transform* g^W of g:

$$(6.27) \quad g^W\colon \quad \begin{array}{ccc} \mathbb{F}_2^{n*} & \longrightarrow & \mathbb{Z}, \\ \ell & \longmapsto & C(\ell, g) = \sum_{x \in \mathbb{F}_2^n} (-1)^{\ell(x)+g(x)} \end{array} \cdot$$

This is an important tool in the theory of discrete Fourier transforms and the sum is called a *Kloosterman sum*. Its value indicates the amount of agreement between g and a linear function ℓ. For linear g, we have $g^W(\ell) = 2^n$ if $\ell = g$ and $g^W(\ell) = 0$ otherwise.

A linear cryptanalyst is looking for large absolute values $|g^W|$ of the signed agreement. We prove some general lower bounds for $\max_{\ell \in \mathbb{F}_2^{n*}} |g^W(\ell)|$ to establish a baseline for the evaluation of the inv-map. As a first step, we determine the average value of g^W, since "not everybody can be below average". We take a nonzero entry $x_i = 1$ of any nonzero $x \in \mathbb{F}_2^n$ and denote by $e_i \in \mathbb{F}_2^n$ the bit string with ith bit 1 and all other bits 0. Arranging the linear functions in $\mathbb{F}_2^{n*}$ into (unordered) pairs $\{\ell, \ell + e_i^*\}$, we find $(-1)^{\ell(x)} + (-1)^{\ell(x)+e_i^*(x)} = 0$. For any $x \in \mathbb{F}_2^n$, this yields

$$(6.28) \qquad \sum_{\ell \in \mathbb{F}_2^{n*}} (-1)^{\ell(x)} = \begin{cases} 2^n & \text{if } x = 0, \\ 0 & \text{otherwise,} \end{cases}$$

and we have for every Boolean function $g \colon \mathbb{F}_2^n \to \mathbb{F}_2$

$$\sum_{\ell \in \mathbb{F}_2^{n*}} g^W(\ell) = \sum_{x \in \mathbb{F}_2^n} \sum_{\ell \in \mathbb{F}_2^{n*}} (-1)^{\ell(x)+g(x)}$$

$$= \sum_{x \in \mathbb{F}_2^n} (-1)^{g(x)} \sum_{\ell \in \mathbb{F}_2^{n*}} (-1)^{\ell(x)} = 2^n (-1)^{g(0)}.$$

This shows that the sum of every single column in Table 6.8 has absolute value 2^n and sign $(-1)^{(b^* \circ S)(0000)} = (-1)^{b \star S(0000)} = (-1)^{b \star 0110}$. Taking uniformly random $\ell \xleftarrow{\boxdot} \mathbb{F}_2^{n*}$, we find the expectation (see Section 15.16) to be

$$\mathcal{E}(g^W(\ell)) = 2^{-n} \sum_{\ell \in \mathbb{F}_2^{n*}} g^W(\ell) = (-1)^{g(0)} \in \pm 1,$$

and thus $\max_{\ell \in \mathbb{F}_2^{n*}} |g^W(\ell)| \geq 1$.

We can do better with *Parseval's identity*

$$\sum_{\ell \in \mathbb{F}_2^{n*}} (g^W(\ell))^2 = \sum_{\ell \in \mathbb{F}_2^{n*}} \sum_{x \in \mathbb{F}_2^n} (-1)^{\ell(x)+g(x)} \sum_{y \in \mathbb{F}_2^n} (-1)^{\ell(x+y)+g(x+y)}$$

$$= \sum_{x,y \in \mathbb{F}_2^n} (-1)^{g(x)+g(x+y)} \sum_{\ell \in \mathbb{F}_2^{n*}} (-1)^{\ell(y)}$$

$$= 2^n \sum_{x \in \mathbb{F}_2^n} (-1)^{2g(x)} = 2^{2n},$$

using (6.28). For uniformly distributed $\ell \xleftarrow{\boxdot} \mathbb{F}_2^{n*}$ as above, this yields

$$\mathcal{E}((g^W(\ell))^2) = 2^{2n}/2^n = 2^n,$$

$$(6.29) \qquad \mathcal{E}(|g^W(\ell)|) = 2^{n/2}.$$

In particular, we have

$$(6.30) \qquad \max_{\ell \in \mathbb{F}_2^{n*}} |g^W(\ell)| \geq 2^{n/2}$$

for all Boolean functions $g \colon \mathbb{F}_2^n \to \mathbb{F}_2$.

The best resistance against linear cryptanalysis that we can hope for is that the maximal absolute value of signed agreement with linear functions equals the average (6.29). In other words, the Walsh transform has constant absolute value and we have equality in (6.30). If this holds, then g is called a *bent function*. Such functions were first studied by Rothaus (1976). If n is odd, then no bent functions exist. If n is even, then $x = x_{n-1}x_{n-2}\dots x_1x_0 \mapsto x_{n-1} \cdot x_{n-2} + \dots + x_1x_0$ is bent, see Exercise 6.7. Several other constructions for bent functions are known, but no general classification or enumeration.

For a bent g, we have $|g^W(0)| = 2^{n/2}$. If, say, $g^W(0) = 2^{n/2}$, then g takes the value 0 exactly $2^{n-1} + 2^{n/2-1}$ times and the value 1 only $2^{n-1} - 2^{n/2-1}$ times. Functions with such discrepancies are called *unbalanced*. If employed in stream ciphers, they are vulnerable to *correlation attacks*. Furthermore, they are not suitable for *invertible* S-boxes.

Thus we are looking for balanced g with $\max_{\ell \in \mathbb{F}_2^{n*}}|g^W(\ell)|$ close to $2^{n/2}$. We now show that the patched inverse inv in a finite field $\mathbb{F}_{2^n}$ yields, for any nonzero $b \in \mathbb{F}_2^n$, a Boolean function $g = b^* \circ \mathrm{inv}$ that satisfies this requirement. This is based on a rather surprising connection between two cryptographic topics, elliptic curves and AES, that at first sight have nothing in common.

THEOREM 6.31. *For the n-bit patched inversion map* $\mathrm{inv} \colon \mathbb{F}_{2^n} \to \mathbb{F}_{2^n}$, *we have*

$$C_{\max}(\mathrm{inv}) = \max\{|C(a^*, b^* \circ \mathrm{inv})| \colon a, b \in \mathbb{F}_2^n \setminus \{0\dots0\}\} \leq 2 \cdot 2^{n/2} + 1.$$

PROOF. We abbreviate $F = \mathbb{F}_{2^n}$ and have the *trace function* on F:

$$\mathrm{tr} \colon \begin{array}{ccc} F & \longrightarrow & \mathbb{F}_2, \\ a & \longmapsto & a + a^2 + a^4 + \dots + a^{2^{n-1}} = \sum_{0 \leq i < n} a^{2^i}. \end{array}$$

Every $\mathbb{F}_2$-linear function $\ell \colon F \to \mathbb{F}_2$ is of the form

$$\mu_c \colon \begin{array}{ccc} F & \longrightarrow & \mathbb{F}_2, \\ a & \longmapsto & \mathrm{tr}(ca) \end{array}$$

for a unique $c \in F$. Conversely, every μ_c for $c \in F$ is $\mathbb{F}_2$-linear, in particular $\mu_1 = \mathrm{tr}$. For all $a^*, b^* \in F^*$ there are $c, d \in F$ such that $\mathrm{corr}(a^*, b^* \circ \mathrm{inv}) = \mathrm{corr}(\mu_c, \mu_d \circ \mathrm{inv}) = \mathrm{corr}(\mu_{cd^{-1}}, \mathrm{tr} \circ \mathrm{inv})$. It is therefore sufficient to bound the Walsh transform of $f = \mathrm{tr} \circ \mathrm{inv}$.

We consider an arbitrary $\mathbb{F}_2$-linear function ℓ and will eventually show that $|f^W(\ell)| \leq 2 \cdot 2^{n/2} + 1$. We take $c \in F$ with $\ell = \mu_c$ and have the Walsh transform (6.27)

$$f^W(\ell) = \sum_{u \in F}(-1)^{f(u)+\ell(u)} = \sum_{u \in F}(-1)^{\mathrm{tr}(\mathrm{inv}(u))+\mathrm{tr}(cu)} = \sum_{u \in F}(-1)^{\mathrm{tr}(cu+\mathrm{inv}(u))}.$$

Furthermore, we take the set

$$X = \{(u, v) \in F^2 : v^2 + uv = cu^3 + u\}.$$

For $(u, v) \in F^2$ with $u \neq 0$, we set $w = v/u$ and have

$$(u, v) \in X \iff (uw)^2 + u(uw) = cu^3 + u$$
$$\iff w^2 + w = cu + u^{-1} = cu + \mathrm{inv}(u).$$

A famous result called Hilbert's Theorem 90 says that for $a \in F^\times$ we have

$$\mathrm{tr}(a) = 0 \iff \text{there is } b \in F \text{ with } a = b^2 + b$$
$$\iff \text{there are exactly two } b_1 \neq b_2 \in F$$
$$\text{with } a = b_i^2 + b_i \text{ for } i = 1, 2.$$

It follows that for nonzero $u \in F$

$$\#\{v \colon (u, v) \in X\} = \begin{cases} 2 & \text{if } \mathrm{tr}(cu + \mathrm{inv}(u)) = 0, \\ 0 & \text{otherwise.} \end{cases}$$

Furthermore, $(0, 0) \in X$ is the only point $(0, v)$ on X. We set

$$U = \{u \in F \colon u \neq 0, \mathrm{tr}(cu + u^{-1}) = 0\}.$$

Except for $u = 0$, U is the projection of X on the first coordinate, and this projection always maps two points to one. Thus $\#X = 2\#U + 1$, and

$$f^W(\ell)$$
$$= (-1)^{\mathrm{tr}(c \cdot 0 + \mathrm{inv}(0))} + \sum_{u \in U} (-1)^{\mathrm{tr}(cu + \mathrm{inv}(u))} + \sum_{u \in F \setminus (U \cup \{0\})} (-1)^{\mathrm{tr}(cu + \mathrm{inv}(u))}$$
$$= 1 + \sum_{u \in U} 1 + \sum_{u \in F \setminus (U \cup \{0\})} (-1) = 1 + \#U - (2^n - (\#U + 1))$$
$$= 2\#U - 2^n + 2 = \#X - 2^n + 1.$$

How to get an upper bound on $|f^W(\ell)|$, as desired? Lo and behold! For $c \neq 0$, $E = X \cup \{\mathcal{O}\}$ is an elliptic curve over F. It is not quite in Weierstraß form as in (5.8), but we check that it is nonsingular. The curve is given by the polynomial $g = y^2 + xy - (cx^3 + x)$, and the partial derivatives

$$\frac{\partial g}{\partial x} = y + cx^2 + 1 \quad \text{and} \quad \frac{\partial g}{\partial y} = x$$

have no common root on X. One also checks, by a transformation as in Exercise 5.2, that $\mathcal{O}$ is nonsingular on E if $c \neq 0$.

For $c = 0$, we have

$$f^W(0) = \sum_{u \in F}(-1)^{\mathrm{tr}(\mathrm{inv}(u))} = \sum_{u \in F}(-1)^{\mathrm{tr}(u)} = 0 \leq 2 \cdot 2^{n/2} + 1.$$

For $c \neq 0$, we use Hasse's bound (Theorem 5.12) for $q = 2^n$. Thus

$$|f^W(\ell)| = |\#E - 1 - 2^n + 1| = |\#E - 2^n| \leq 2 \cdot 2^{n/2} + 1. \qquad \square$$

Thus inversion, and hence SUBBYTES, is optimally nonlinear, except for the extra factor of 2 and the addition of 1.

Notes. This chapter was largely written by Konstantin Ziegler, using some of the author's notes. AES is a Boolean cryptosystem as we understand it here. It uses arithmetic in the field $\mathbb{F}_{256}$ for SUBBYTES, its only nonlinear operation, and some of the other steps are also most easily described in algebraic terms. But $\mathbb{F}_{256}$ is not a "large" algebraic structure as used in RSA and group cryptography. DES was reaffirmed as FIPS PUB 46-1/2/3 in 1983, 1988, and 1993, and in 1999 as 3-DES allowing DES only for legacy systems. In 2005, NIST withdrew FIPS 46-3.

Heys (2001) gives a tutorial for differential and linear cryptanalysis using his own Boolean (toy) cipher. Differential and linear cryptanalysis have been developed independently, but recent attacks exploit links between the two, see Blondeau & Nyberg (2013).

6.1. Our baby-AES is an instance of the small-scale variants defined by Cid *et al.* (2005) and available as `mq.SR(3, 2, 2, 4, allow_zero_inversions=True, star=True)` in Sage. Other suggestions for such "toy" versions of AES are due to Phan (2002) and Musa *et al.* (2003).

6.2. For the early work on differential cryptanalysis, see also Biham & Shamir (1991a, 1992, 1993). Don Coppersmith's quote is from an email to Adi Shamir, dated 19 February 1992, and quoted from Biham (2006). Some explanations about the DES S-boxes were later given by Coppersmith (1994a). Differentials with positive probability are sometimes called *possible differentials*, and those with probability 0 *impossible differentials*. These can also be used in the aptly-named *impossible differential cryptanalysis*.

The size of the subgroup generated by the round functions as permutations of the state space is important, because of its connection to the Markov cipher approach to differential cryptanalysis. Hornauer *et al.* (1994) showed that if the round functions of an substitution-permutation network generate the alternating or the symmetric group, then after sufficiently many rounds all differences become equally probable and the cipher is secure against differential cryptanalysis. Wernsdorf (2002) showed that the round functions of AES generate the alternating group.

Let γ be the average number of suggested keys per pair (we have $\gamma \approx 1.12$). The signal-to-noise ratio of our path is $S - N = \gamma^{-1} \cdot (\#\text{candidate keys}) \cdot p$. In our example, this is 4. This value is crucial for the number of required pairs. Biham & Shamir (1991b) motivate this by experimental results: $S - N$ between 1 and 2 require 20–40 compliant pairs. For larger values of $S - N$ even fewer may suffice and for $S - N$ below 1 there may be too much noise, rendering a differential attack infeasible.

6.3. Matsui (1994a) writes: "We carried out, the first, experimental attack of the full 16-round DES using twelve computers (HP9735/PA-RISC 99MHz) to confirm this scenario. The program, described in C and assembly languages consisting of a total of 1000 lines, was designed to solve two equations while generating 2^{43} random plaintexts and enciphering them. We finally reached all of the 56 secret key bits in fifty days, out of which forty days were spent for generating plaintexts and their ciphertexts and only ten days were spent for the actual key search." Junod (2001) presents experimental data that suggests 2^{41} evaluations suffice for the same level of confidence.

The Electronic Frontier Foundation (1998) built a special purpose machine to attack DES within eighteen months and for "well under US\$ 250,000". The EFF DES Cracker is designed to find DES keys in an average of 4.5 days. The COPACOBANA (2006) and its succesor RIVYERA (2008) are FPGA-based special-purpose hardware for brute force attacks with even fewer *dollardays*.

Originally, Matsui (1994b) considered the random variable $z = a \star x + b \star S(x)$ with bias $\epsilon = \text{prob}\{z = 0\} - 1/2$. We study the correlation corr of the Boolean functions a^* and $b^* \circ S$. This is equivalent, since $\text{corr} = 2 \cdot \epsilon$ by (6.17). Daemen & Rijmen (1999), pages 108-109, advise us that "the usage of probabilities [...] requires the application of the so-called piling-up lemma in the computation of probabilities of composed transformations. When working with correlations, no such tricks are required: correlations can be simply multiplied." We speak of *constructive interference* in the case of two linear trails with the same sign and of *destructive interference* if they have opposing signs. For a fixed key, constructive interference will occur for certain selection masks. But "the strong round round key dependence of interference makes locating the input and output selection patterns for which high correlations occur practically infeasible. This is true if the key is known, and even more so if it is unknown." (Daemen & Rijmen 1999, Section 9.1.1).

Algorithm 6.24 is Algorithm 2 of Matsui (1994a). In Lemma 5, he gives a formula for the success probability of the attack for a given number of samples under reasonable assumptions. He also provides numerical data in his Table 3 recommending $8/\text{corr}^2$, $16/\text{corr}^2$, and $32/\text{corr}^2$ samples for success probabilities of 48.6%, 78.5%, and 96.7%, respectively.

6.4. (Baby-)AES's permutation layer (MixColumns ∘ ShiftRows) is designed according to the *wide trail strategy*, see Daemen & Rijmen (1999), Section 9.2, for a description. The Four-Round Propagation Theorem is Theorem 9.5.1 of Daemen & Rijmen (1999). Preneel's quote combines remarks of Luke O'Connor and James L. Massey, as cited in Preneel *et al.* (1998).

A Boolean function S with $N_{\max}(S) \leq \delta$ is called *differentially δ-uniform*. We have $N_{\max}(S) \geq 2$ for all Boolean functions S, and differentially 2-uniform functions are called *almost perfectly nonlinear*; see Nyberg & Knudsen (1993), Section 3, for examples. Nyberg (1994) shows $N_{\max}(\text{inv}) = 2$ for odd n and $N_{\max}(\text{inv}) = 4$ otherwise. We define the Walsh transform over $\mathbb{F}_2$. For any prime p, one can define a corresponding mapping over $\mathbb{F}_p$ by using a complex primitive pth root of unity $\exp(2\pi i/p)$ instead of -1 in (6.27).

The *nonlinearity* of g is $2^{n-1} - \max_{\ell \in \mathbb{F}_2^n *} |g^W(\ell)|/2$. A bent function has nonlinearity $2^{n-1} - 2^{n/2-1}$. For odd n, the maximal nonlinearity lies between $2^{n-1} - 2^{(n-1)/2}$ and $2^{n-1} - 2^{n/2-1}$. For balanced functions within these bounds, see Carlet (2010), Section 6, with constructions and further pointers to the literature. The maximal nonlinearity of balanced functions is unknown for $n \geq 8$. Oscar Rothaus worked until 1966 for a subcontractor of the NSA and produced for them an internal report *On Bent Functions*, (see Joyner 2013). Its publication in 1976, together with Dillon (1974), created an area of research that is still active today. For an excellent survey, see Mesnager (2015).

The connection between inv and elliptic curves is based on private conversations with Hans Dobbertin, around 1999. Every Boolean function $f\colon \mathbb{F}_{2^n} \to \mathbb{F}_2$ has a *trace representation* $f = \mathrm{tr} \circ h$ for some $h\colon \mathbb{F}_{2^n} \to \mathbb{F}_{2^n}$. The unique representation of g by μ_c in the proof of Theorem 6.31 is easy to see. The linear map $\mathrm{tr}\colon \mathbb{F}_{2^n} \to \mathbb{F}_2$ has at most $\deg \mathrm{tr} = 2^{n-1}$ roots. Thus, there exists some $w \in \mathbb{F}_{2^n}$ with $\mathrm{tr}(w) \neq 0$. The map $c \mapsto \mu_c$ from $F = \mathbb{F}_{2^n}$ to its dual vector space $F^* = \{g\colon F \to \mathbb{F}_2 \text{ linear}\}$ is $\mathbb{F}_2$-linear, and $\mu_c \neq 0$ if $c \neq 0$, since $\mu_c(wc^{-1}) = \mathrm{tr}(w) \neq 0$. Since $\dim F = \dim F^*$, it follows that this map is an isomorphism. Hilbert's Theorem 90 is from his *Zahlbericht* (Hilbert 1897). We give a short proof over $\mathbb{F}_2$. In the expression for $\mathrm{tr}(b^2 + b)$, the ith summand is $(b^2 + b)^{2^i} = b^{2^i} + b^{2^{i+1}}$, and the second term cancels with the first term of the $(i+1)$st summand. Hence $\mathrm{tr}(b^2 + b) = b + b^{2^n} = 0$, by Fermat's Little Theorem 15.56. Hence $\mathrm{im}(b \mapsto b^2 + b) \subseteq \ker \mathrm{tr}$. The linear map $b \mapsto b^2 + b$ has $\mathbb{F}_2$ as its kernel, and hence its image equals $\ker \mathrm{tr}$. The claim that there exist exactly two $b_1 \neq b_2$ follows from the fact that the latter kernel equals $\mathbb{F}_2$.

Exercises.

EXERCISE 6.1 (Inverting AES). *(Baby-)AES has the following structure:*

○ ADDROUNDKEY$_0$.

○ *For rounds* $i = 1, \ldots, R$ *do:*

— SUBBYTES,

— SHIFTROWS,

— MIXCOLUMNS *(omit in round R)*,

— ADDROUNDKEY$_i$.

(i) *Show that decryption of (baby-)AES can be achieved with a similar structure using modifications* ADDROUNDKEY', SUBBYTES', SHIFTROWS', MIXCOLUMNS' *instead of their "originals". (Hint:* SHIFTROWS *and* MIXCOLUMNS *of baby-AES are self-inverse and the key schedule is invertible.)*

(ii) *Look up the structure of DES in FIPS PUB 46 and repeat the exercise for this Feistel network, where the flip in the last round is omitted.*

EXERCISE 6.2 (Propagation ratio). *What is the expected maximal absolute frequency of a nontrivial differential for a randomly chosen S-box? More precisely, do the following.*

(i) *Randomly generate S-boxes* $\mathbb{F}_2^4 \to \mathbb{F}_2^4$, $\mathbb{F}_2^8 \to \mathbb{F}_2^8$, *and* $\mathbb{F}_2^6 \to \mathbb{F}_2^4$.

(ii) *Compute the differential distribution table.*

(iii) *Derive the maximal absolute frequency (for a nontrivial differential).*

(iv) *Plot the distribution and mark the values for baby-AES, AES, and DES within them.*

EXERCISE 6.3 (Differential trail). *Find a better differential trail for the first 2 rounds of 3-round baby-AES. Find a feasible differential trail for 4-round baby-AES.*

EXERCISE 6.4 (Correlation examples).

(i) *Use a computer algebra system of your choice to compute the correlations* $\mathrm{corr}(\ell_i \circ f_j, \ell_k)$ *of the following functions* $\mathbb{F}_2^8 \longrightarrow \mathbb{F}_2$. *Please hand in a little matrix for each of the* f_j.

- $f_0(a) = \mathrm{inv}(a)$,
- $f_1(a) = a$,
- $f_2(a) = a^2$,
- $f_3(a) = a^3$,
- $f_4(a) = (a_7 + a_6)x^7 + (a_5 + a_3)x^6 + (a_6 + a_5)x^5 + (a_7 + a_4 + a_2)x^4 + (a_7 + a_6 + a_5 + a_4)x^3 + (a_5 + a_1)x^2 + (a_7 + a_6 + a_4)x + a_6 + a_4 + a_0$.

- $\ell_0(a) = a_0$,
- $\ell_1(a) = a_7 + a_6 + a_5 + a_4 + a_3 + a_2 + a_1 + a_0$,
- $\ell_2(a) = a_7 + a_4 + a_0$,
- $\ell_3(a) = a_7 + a_5 + 1$,
- $\ell_4(a) = a_5 + a_7$.

(ii) *Compute all possible values of* $\mathrm{corr}(f, \ell)$ *if* f *and* ℓ *are linear. Hint: Show first that you may assume* f *to be the zero function.*

EXERCISE 6.5 (Different correlation). *What happens if you just count agreement instead of difference for the correlation?*

EXERCISE 6.6 (Coincidence). *How is the correlation related to Friedman's index of coincidence (Section C.2)?*

EXERCISE 6.7 (Walsh example). *For even* n *compute the Walsh transform of*

$$g \colon \mathbb{F}_2^n \to \mathbb{F}_2, \quad (x_1, x_2, x_3, x_4, \ldots, x_{n-1}, x_n) \mapsto x_1 x_2 + x_3 x_4 + \cdots + x_{n-1} x_n.$$

EXERCISE 6.8 (Linearity of the scalar product). *For all vectors* $a, b \in \mathbb{F}_2^n$ *and matrices* $M \in \mathbb{F}_2^{n \times n}$ *with transpose* M^T, *show that*

$$a \star (Mb) = (M^T a) \star b.$$

EXERCISE 6.9 (Linear functions). *Let* $n > 0$ *and* $f \colon \mathbb{F}_2^n \to \mathbb{F}_2$ *be an* n-*ary Boolean function.*

(i) *Show that the following two properties are equivalent.*

- $f(x + y) = f(x) + f(y)$ *for all* $x, y \in \mathbb{F}_2^n$.
- *There is an* $a \in \mathbb{F}_2^n$ *such that* $f(x) = a \star x$ *for all* $x \in \mathbb{F}_2^n$.

If these hold, the function f *is called linear.*

(ii) *Show that for a linear* f, *the* a *from above is uniquely determined.*

(iii) *Derive the fraction of linear* n-*ary Boolean functions among all* n-*ary Boolean functions.*

EXERCISE 6.10 (Nonlinear examples). *Let $a \in \mathbb{F}_2^n$ be nonzero. Show that the following two functions $f: \mathbb{F}_2^n \to \mathbb{F}_2$ are nonlinear.*

(i) $f(x) = a \star x + 1$.

(ii)

$$f(x) = \begin{cases} 1 & \text{if } x = 0, \\ a \star x & \text{otherwise.} \end{cases}$$

EXERCISE 6.11 (Correlation). *Let $n \geq 2$, $a, b \in \mathbb{F}_2^n$, and $S: \mathbb{F}_2^n \to \mathbb{F}_2^n$ be a permutation on n bits.*

(i) *Show that $\sum_{x \in \mathbb{F}_2^n} a \star x + b \star S(x) = 0$.*

(ii) *Show that the set of x for which $a \star x + b \star S(x) = 1$ and the set where $a \star x + b \star S(x) = 0$ both have even cardinality.*

(iii) *Conclude that the absolute correlation, generalized from (6.15),*

$$C_S(a \parallel b) = 2 \cdot \#\{x \in \mathbb{F}_2^n : a \star x = b \star S(x)\} - 2^n$$

is divisible by 4.

It is impossible to predict the future, and all attempts to do so
in any detail appear ludicrous within a very few years.
ARTHUR CHARLES CLARKE (1958)

There is a remarkably close parallel between the problems of the
physicist and those of the cryptographer. The system on which a
message is enciphered corresponds to the laws of the universe,
the intercepted messages to the evidence available,
the keys for a day or a message to important constants
which have to be determined. The correspondence is very close,
but the subject matter of cryptography is very easily dealt
with by discrete machinery, physics not so easily.
ALAN MATHESON TURING (1948)

Yu, shall I teach you what knowledge is? When you know a
thing, to hold that you know it; and when you do not know a
thing, to allow that you do not know it;— this is knowledge.
CONFUCIUS (551–479 BC)

Applications programming is a race between software engineers,
who strive to build idiot-proof programs,
and the Universe which strives to produce bigger idiots.
So far the Universe is winning.
RICK COOK (1989)

C'est ce que nous appelons un cryptogramme, disait-il,
dans lequel le sens est caché sous des lettres brouillées
à dessein, et qui convenablement disposées
formeraient une phrase intelligible.[1]
JULES VERNE (1867)

The discussion of ciphers is second-hand and erroneous.
The British are not only behind in their secret service,
but also in books about it.
DAVID SHULMAN (1976)

---

[1]This is what we call a cryptogram, he said, where the meaning is hidden under letters purposely jumbled and which properly arranged would form an intelligible phrase.

Chapter 7

Hash functions

The purpose of a hash function is to distill a small amount of information out of large messages. But the amount has to be large enough so that it (usually) identifies the message uniquely. One requirement for cryptography is that it should be computationally hard for an adversary to generate two different messages with the same hash value.

In cryptography, hash functions are used, among other areas, in connection with signature schemes (Chapter 8), where only the (short) hash values of (long) messages are signed. In the standard SHA-256 of Section 7.6, this looks as follows:

length	arbitrary	256 bits	512 bits
data	message $x \longrightarrow$ hash value $h(x) \longrightarrow$ signature $\mathrm{sig}(h(x))$		

This is more efficient and, above all, more secure than breaking x into small pieces of 256 bits and signing each piece separately.

Further applications include *key derivation*, where a master key K and a counter i yield a new key $K_i = h(K|i)$. In Section 10.5 we will see *message authentications* of the form $h(K|m)$.

7.1. Hash functions

Hash functions are used in many areas of computer science. An example is *bucket sort*, where integers are sorted by putting them into different "buckets" according to their leading digit(s) which are the hash values in this case, first prefixing zeros to make them of uniform length.

A more interesting example is the following. Suppose that one maintains a large data base in North America and a mirror image in Europe, by performing all updates on both. Each night, one wants to check whether the data bases indeed are identical. Sending the whole data base would take too long. So one considers the data base as a string of 64-bit words,

© Springer-Verlag Berlin Heidelberg 2015
J. von zur Gathen, *CryptoSchool*, DOI 10.1007/978-3-662-48425-8_10

many gigabytes long, and the (large) number x whose 2^{64}-ary representation this is. Then one chooses a prime p, computes the remainder of x
modulo p and sends this to the mirror site. The corresponding calculation
is performed on the other data base, and the two results are compared. If
they disagree, then the two data bases differ. If they agree, then probably the two data bases are identical, provided p was chosen appropriately.
This can be set up so that the size of the transmitted message is only logarithmic in the size of the data bases. This simple hash function h with
$h(x) = x \bmod p$ has many uses, but is too simplistic for cryptographic
purposes, which we now consider.

DEFINITION 7.1. *Let* $h\colon X \longrightarrow Z$ *be a mapping between two finite sets*
X *and* Z. *If* $x \neq y$ *are messages in* X *with* $h(x) = h(y)$, *then* x *and* y
collide, and (x, y) *is a collision.*

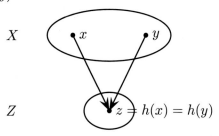

Often $h(x)$ is called the *digest* of x: after hashing the substantial
item x, digestion leaves only the small item $h(x)$. Hash functions map
large sets (or long messages) to small sets (or short digests), so that X
is much larger than Z, and it is unavoidable that many collisions exists.
For concreteness, we assume throughout this chapter that

(7.2) $$\#X \geq 2 \cdot \#Z.$$

A further assumption is that h be polynomial-time computable.

We consider three types of adversaries $\mathcal{A}$ against a hash function h.

   (i) A *collision finder* $\mathcal{A}$ takes no input and outputs either a collision
        (x, y) or "failure".

  (ii) A *second-preimage finder* $\mathcal{A}$ takes an input x and outputs either
        some $y \in X$ that collides with x, or "failure".

 (iii) An *inverter* $\mathcal{A}$ takes an input $z \in Z$ and outputs either some $x \in X$
        with $h(x) = z$, or "failure".

In any of these situation, we define the *success probability* $\sigma_{\mathcal{A}}$ of $\mathcal{A}$ as

$$\sigma_{\mathcal{A}} = \mathrm{prob}\{\mathcal{A} \text{ does not return "failure"}\}.$$

The probability is taken over the internal random choices of $\mathcal{A}$ plus uniformly random choices $x \xleftarrow{\boxtimes} X$ in (ii) and $z \xleftarrow{\boxtimes} h(X) \subseteq Z$ in (iii).

It would be nice if we could do without the parameter of success probability, by simply requiring that $\sigma_{\mathcal{A}} = 1$. After all, it is a common requirement for an algorithm to return correctly for all inputs (ignoring the internal random choices). As often in cryptology, this would be asking too much. Namely, we can modify any hash function—even a cryptographically weak one—by adding a new message x_0 to X and its new value $h(x_0)$ to Z. Then there would be no second preimage for x_0, therefore no second-preimage algorithm $\mathcal{A}$ with $\sigma_{\mathcal{A}} = 1$, but for an irrelevant reason.

A further trivial example is the identity function $h \colon X \to Z = X$, which is easy to invert but for which no collision or second-preimages exist. However, this is ruled out by (7.2).

When we allocate more computing time to an algorithm or allow it to encode some values $(x, h(x))$, then its success probability may increase. To take this into account, we denote as $T(\mathcal{A})$ the cost of an algorithm $\mathcal{A}$, being the sum of its expected runtime plus the bitlength to encode it, fixing some standard way of doing this.

We assume locality of variables, so that when $\mathcal{A}$ is used as a subroutine in some other algorithm, a "new" variable can be given a name of constant length. We also take a constant c so that in time c, we can generate $x \xleftarrow{\boxtimes} X$, compute $h(x)$ for some $x \in X$, or test equality of elements. For a hash function h and $\varepsilon > 0$, we denote as $\mathrm{coll}_\varepsilon$, sec_ε, and inv_ε the minimal value of $T(\mathcal{A})$ over all algorithms $\mathcal{A}$ as in (i), (ii) and (iii), respectively, with $\sigma_{\mathcal{A}} \geq \varepsilon$. Then we have the following inequalities between these three measures. Statement (i), for example, says that if it is expensive to find a collision, so that $\mathrm{coll}_\varepsilon$ is large, then this also holds for finding second preimages.

THEOREM 7.3. *For any hash function $h \colon X \to Z$ and $\varepsilon > 0$ we have the following.*

(i) $\mathrm{coll}_\varepsilon \leq \mathrm{sec}_\varepsilon + c$.

(ii) $\mathrm{sec}_\delta \leq \mathrm{inv}_\varepsilon + 3c$, *where* $\delta = \varepsilon \cdot (1 - \#Z/\#X) - 2\#Z/\#X$.

PROOF. (i) We take a second-preimage finder $\mathcal{B}$ with $\sigma_{\mathcal{B}} \geq \varepsilon$ and $T(\mathcal{B}) = \mathrm{sec}_\varepsilon$ and consider the following collision finder $\mathcal{C}$.

1. $x \xleftarrow{\boxtimes} X$.

2. Call $\mathcal{B}$ with input x.

3. Return $\mathcal{B}(x)$.

$\mathcal{C}$ is clearly a collision-finder with success probability $\sigma_\mathcal{C} = \sigma_\mathcal{B} \geq \varepsilon$, and thus $\mathrm{coll}_\varepsilon \leq T(\mathcal{C}) \leq T(\mathcal{B}) + c = \mathrm{sec}_\varepsilon + c$.

(ii) It is also easy to obtain the following second preimage finder $\mathcal{B}$ from an inverter $\mathcal{A}$ with $\sigma_\mathcal{A} \geq \varepsilon$ and $T(\mathcal{A}) = \mathrm{inv}_\varepsilon$.

1. $x \xleftarrow{\;\boxdot\;} X$.

2. Call $\mathcal{A}$ with input $z = h(x)$.

3. If $\mathcal{A}(z) = $ "failure" or $\mathcal{A}(z) = x$ then return "failure".

4. Else return $\mathcal{A}(z)$.

$\mathcal{B}$ is a second-preimage finder and $T(\mathcal{B}) \leq T(\mathcal{A}) + 3c = \mathrm{inv}_\varepsilon + 3c$. It remains to calculate its success probability $\sigma_\mathcal{B}$. We first assume that the call to $\mathcal{A}$ in step 2 on any input x does not return "failure". By the probabilistic nature of $\mathcal{A}$, the value of $\mathcal{A}(z)$ is not fixed, but a probability distribution on the set

$$h^{-1}(z) = \{x \in X : h(x) = z\}$$

of preimages of z. For $x \in X$, we denote by $\pi(x)$ the probability that $\mathcal{A}$ returns x on input $h(x)$. Then for any $z \in h(X)$, we have

$$\sum_{x \in h^{-1}(z)} \pi(x) = \sum_{x \in h^{-1}(z)} \mathrm{prob}\{(\mathcal{A}(z) = x)\} = 1,$$

and the sets $h^{-1}(z)$ for $z \in h(X) \subseteq Z$ form a partition of X.

Let $a = \#X$ be the number of possible messages, $b = \#h(X) \leq \#Z$ the number of possible hash values, and let

$$E = \{x \in X : h(x) \neq h(y) \text{ for all } y \in X \setminus \{x\}\}$$

be the set of x that do not participate in any collision. Then $\#E < b$ and

(7.4) $\mathrm{prob}\{x \in E : x \xleftarrow{\;\boxdot\;} X\} < b/a$.

The inverter $\mathcal{A}$ might always be successful on input $z \in h(E)$, but its success probability on input $z = h(x)$ with $x \xleftarrow{\;\boxdot\;} (X \setminus E)$ is still at least $\sigma_\mathcal{A} - b/a$. Now further assume that $x \notin E$, so that $\#h^{-1}(z) \geq 2$. Since x is uniformly random in X, it is also uniformly random in $h^{-1}(z)$ and, no matter how $\mathcal{A}(z)$ is distributed, $\mathrm{prob}\{x = \mathcal{A}(z)\} = 1/\#h^{-1}(z)$. Thus on the $x \notin E$, the success probability $\sigma_\mathcal{B}$ of $\mathcal{B}$ is at least

$$\frac{\sigma_\mathcal{A} - \frac{b}{a}}{\#X} \sum_{x \in X} (1 - \pi(x)) = \frac{\sigma_\mathcal{A} - \frac{b}{a}}{a} \left(a - \sum_{z \in Z} \sum_{x \in h^{-1}(z)} \pi(x) \right)$$

$$\geq \left(\varepsilon - \frac{b}{a} \right) \left(1 - \frac{1}{a} \sum_{z \in h(X)} 1 \right) = \left(\varepsilon - \frac{b}{a} \right) \left(1 - \frac{b}{a} \right).$$

Taking into account the probability that $x \in E$, we find

$$\sigma_{\mathcal{B}} \geq (\sigma_{\mathcal{A}} - \frac{b}{a}) \cdot (1 - \frac{b}{a}) - \frac{b}{a} > \varepsilon \cdot (1 - \frac{b}{a}) - \frac{2b}{a}. \qquad \square$$

If $\#X = 2\#Z$, then the conclusion of Theorem 7.3 (ii) is rather useless. For SHA-256 (Section 7.6), we have $\#Z/\#X = 2^{256}/2^{512} = 2^{-256}$, and the subtractions in the conclusion are not worrisome.

We have kept the exposition simple and concentrated on a single hash function in order not to drown in notation. For a more general (and appropriate) notion, we have for each (or at least infinitely many) positive integer n three finite sets, namely, a key space K_n, a message space X_n, and a hash value space Z_n. One may think of these spaces as consisting of all bit strings of a certain length. The hash function family then is $h = (h_n)_{n \in \mathbb{N}}$. For $n \geq 1$ and a key $k \in K_n$, we have a *keyed hash function* $(h_n)_k \colon X_n \longrightarrow Z_n$. An example is in Section 7.2, and we use them for the MACs of Section 10.5.

The runtime of algorithms and the success probabilities depend in a natural way on n. Fixing n, the success probability $\sigma_{\mathcal{A}}$ of an inverter $\mathcal{A}$ is the probability with which $(h_n)_k(\mathcal{A}(k,z)) = z$ for $(k,z) \xleftarrow{\boxed{\cdot}} K_n \times Z_n$.

Now we want to say that h is *collision-resistant* if no polynomial-time collision-finder exists for h. Since $\#X_n > \#Z_n$, collisions always exist. We might find one "offline" at arbitrary cost and just "hard-wire" it into a collision-finder $\mathcal{A}$, which would then have maximal success probability $\sigma_{\mathcal{A}} = 1$. However, this does not work for an infinite family h_n of hash functions as above. An algorithm $\mathcal{A}$ as discussed will then receive n (in unary) plus the usual inputs, and has to work for all n. In particular, one can only hard-wire finitely many collisions. We allow some positive success probability but insist that it be negligible (Definition 15.30).

We can now describe some properties that are relevant for hash functions in cryptography.

DEFINITION 7.5. *Let* $h = (h_n)_{n \in \mathbb{N}}$ *be a family of hash functions with both* $\log_2(\#X_n)$ *and the time to evaluate* h_n *polynomial in* n. *We call* h

> collision resistant,
>
> second-preimage resistant,
>
> inversion resistant *(or one-way)*,

if for all probabilistic polynomial-time

> collision finders $\mathcal{A}$,
>
> second-preimage finders $\mathcal{A}$,
>
> inverters $\mathcal{A}$,

respectively, for h the success probability σ_A is a negligible function of n.

The following consequence of Theorem 7.3 is immediate. It shows that our three requirements for hash functions have decreasing strength. The "constant" c now depends on n, but only polynomially.

COROLLARY 7.6. *For a family $h = (h_n)_{n\in\mathbb{N}}$ of hash functions where the fraction $\#Z_n/\#X_n$ is negligible in n, we have*

> *h collision resistant $\Longrightarrow$ h second-preimage resistant $\Longrightarrow$ h one-way.*

In view of this result, we concentrate on collision resistance in the following. Some applications require only one of the weaker properties. This may allow shorter lengths of hash values than required for collision resistance.

The birthday paradox of Theorem 3.41 says that we expect to find a collision with only about $\sqrt{b\pi/2} \approx 1.25\sqrt{b}$ random choices of x, where b is the number of hash values. Hence the set of hash values must be large enough to withstand this attack. Thus $b \approx 2^{40}$ is much too small, requiring only 1.3 million attempts on average to find a collision, while $b = 2^{256}$, as in SHA-256, provides 128-bit security (Section 5.5) and is presumably safe against this attack.

If Z_n is very small, say of some constant size like 1000, then we can find a collision for h_n by just evaluating h_n at no more than 1001 messages. Thus collision-resistance implies, besides the lower bound (7.2) on $\#X_n$, also an upper bound; $\log(\#X_n) \in (\log(\#Z_n))^{O(1)}$ is sufficient.

For one-wayness to be of interest, $\#h(X_n)$ should be reasonably close to $\#Z_n$. Otherwise we can take a weak hash function h, extend Z_n by 100 bits to Z'_n and duplicate the last 100 bits of each $h_n(x)$ at the end. Then the fraction of new hash values is at most 2^{-100}, and random values in Z'_n will rarely have an inverse.

7.2. A discrete logarithm hash function

Can we find a hash function h using a finite cyclic group $G = \langle g \rangle$, as in Chapter 4, so that collisions correspond to discrete logarithms? Let $d = \#G$. Then $h\colon \mathbb{Z}_d \longrightarrow G$ with $h(a) = g^a$ for $a \in \mathbb{Z}_d$ is collision-resistant (in fact, collision-free), but not a hash function since $\#\mathbb{Z}_d = \#G$; we require $\#X \geq 2 \cdot \#Z$. What about $h\colon \mathbb{Z}_d \times \mathbb{Z}_d \longrightarrow G$ with $h(a_1, a_2) = g^{a_1+a_2}$? Now it is easy to find a collision: $h(a_1 + i, a_2 - i) = h(a_1, a_2)$ for any $i \in \mathbb{Z}_d$. But here is a solution.

We take another generator z of G and the function

$$(7.7) \qquad\qquad h_z\colon \begin{array}{ccc} \mathbb{Z}_d \times \mathbb{Z}_d & \longrightarrow & G, \\ (a,b) & \longmapsto & g^a z^b \end{array} .$$

This is a keyed hash function h consisting of hash functions h_z, one for each key $z \in G$ which is a generator. Choosing $e \xleftarrow{\quad} \mathbb{Z}_d^\times$ yields a uniformly random key $z = g^e$ (Theorem 15.60 (ii)). A collision finder $\mathcal{A}$ for h gets the key z as input and returns a collision for h_z, or "failure". It may perform computations and tests in G and in $\mathbb{Z}_d$. The success probability $\sigma_{\mathcal{A}}$ is taken over uniformly random inputs z and the internal randomization of $\mathcal{A}$, and for $\varepsilon > 0$, $\mathrm{coll}_\varepsilon$ is the minimal $T(\mathcal{A})$ over all $\mathcal{A}$ with $\sigma_{\mathcal{A}} \geq \varepsilon$.

Since we are now dealing with algorithms that do not necessarily succeed on all inputs, we have to modify the discrete logarithm problem DL_G in Definition 4.2 accordingly. Namely, for $\varepsilon > 0$ we let

$$\mathrm{dl}_{G,\varepsilon} = \min\{T(\mathcal{A}) : \mathcal{A} \text{ solves } \mathrm{DL}_G \text{ with success probability } \sigma_{\mathcal{A}} \geq \varepsilon\}.$$

The probability is taken over uniform random choices of an input in $G = \langle g \rangle$ to DL_G and the internal randomization of the algorithm. The generator g of G is fixed and, for simplicity, not included in the notation. The timing constant c now includes the cost of arithmetic in $\mathbb{Z}_d$, with exponentiation.

THEOREM 7.8. *Suppose that $d = \#G$ is prime, and h as in (7.7). Then there is a constant c so that for all $\varepsilon > 0$ we have $\mathrm{coll}_\varepsilon \leq \mathrm{dl}_{G,\varepsilon} \leq \mathrm{coll}_\varepsilon + 6c$.*

PROOF. For the left-hand inequality, it is sufficient to observe that if $y = \mathrm{dlog}_g z$ and $a, b, a', b' \in \mathbb{Z}_d$ are such that $a' = a + y(b - b')$, then $h_z(a, b) = g^a z^b = g^{a+yb} = g^{a'+yb'} = g^{a'} z^{b'} = h_z(a', b')$. We can take $a = b = 1$, $a' = 1 + y$, $b' = 0$, for example.

For the right-hand inequality, we take a collision finder $\mathcal{C}$ with $\sigma_{\mathcal{C}} \geq \varepsilon$ and $T(\mathcal{C}) = \mathrm{coll}_\varepsilon$. The following algorithm $\mathcal{A}$ takes an element $x \in G$ as input and returns either $\mathrm{dlog}_g x$ or "failure".

1. $e \xleftarrow{\quad} \mathbb{Z}_d^\times$.

2. $z \longleftarrow x^e$.

3. Call $\mathcal{C}$ with input z.

4. If $\mathcal{C}$ returns "failure", then return "failure".

5. If $\mathcal{C}$ returns $\{(a_1, b_1), (a_2, b_2)\}$, then return

$$\frac{a_1 - a_2}{e(b_2 - b_1)} \text{ in } \mathbb{Z}_d.$$

The division in step 5 is well-defined. Namely, the choice of e implies that it is invertible in $\mathbb{Z}_d$, and it is sufficient to see that $b_2 - b_1 \neq 0$ in $\mathbb{Z}_d$, since d is prime and hence every nonzero element is invertible in $\mathbb{Z}_d$. The collision in step 5 means that

$$g^{a_1} z^{b_1} = g^{a_2} z^{b_2},$$

(7.9) $$g^{a_1 - a_2} = z^{b_2 - b_1} = x^{e(b_2 - b_1)} = g^{\operatorname{dlog}_g x \cdot e(b_2 - b_1)}.$$

If $b_2 - b_1 = 0$, it follows that $g^{a_1 - a_2} = 1$ and $a_1 = a_2$ in $\mathbb{Z}_d$. But then $(a_1, b_1) = (a_2, b_2)$, contradicting the collision property. Now (7.9) implies that in step 5, indeed the output equals $\operatorname{dlog}_g x$.

Since z is a uniformly random generator of G, we have $\sigma_\mathcal{C} = \sigma_\mathcal{A} \geq \varepsilon$. The runtime of $\mathcal{A}$ is at most $T(\mathcal{C}) + 6c$. It follows that $\operatorname{dl}_{G,\varepsilon} \leq T(\mathcal{A}) \leq T(\mathcal{C}) + 6c = \operatorname{coll}_\varepsilon + 6c$. $\square$

When d is not prime, then $\mathcal{A}$ has success probability at least $\sigma_\mathcal{C} \cdot \operatorname{prob}\{b_2 - b_1 \in \mathbb{Z}_d^\times\}$. If the collisions returned by $\mathcal{C}$ are sufficiently random, one may use the estimate of Remark 4.15 (i) to bound the success probability of $\mathcal{A}$.

We take a family $(G_n)_{n \in \mathbb{N}}$ of groups, each with a fixed generator $g_n \in G_n$ and, for simplicity, with $d_n = \#G_n$ being an n-bit prime. Then the "constant" c of Theorem 7.8 is in $O(n^3)$, assuming that a multiplication in G_n or in $\mathbb{Z}_{d_n}$ can be done with $O(n^2)$ operations. We then have the following consequence.

COROLLARY 7.10. *Discrete logarithms in $(G_n)_{n \in \mathbb{N}}$ are hard to compute if and only if h in (7.7) is collision-resistant.*

In Chapter 4, we studied *probabilistic algorithms* $\mathcal{A}$ for DL_G. For any input, such an $\mathcal{A}$ returns the correct answer with probability at least $2/3$. Presently, we have a weaker type of algorithm, namely *approximate algorithms* with success probability taken over uniformly randomly chosen inputs. This is a less powerful form of algorithm; their study is related to *average-case analysis*. For example, there may exist certain inputs on which $\mathcal{A}$ always returns "failure"; this is disallowed for a probabilistic algorithm.

For discrete logarithms, we can actually obtain a probabilistic algorithm $\mathcal{B}$ from an approximate one $\mathcal{A}$. The data $G = \langle g \rangle$ and $d = \#G$ is fixed in the following.

ALGORITHM 7.11. Probabilistic algorithm $\mathcal{B}$ for DL_G from approximate $\mathcal{A}$ with success probability $\sigma_\mathcal{A}$.

Input: $x \in G$.

Output: Either $\mathrm{dlog}_g x$ or "failure".

1. Compute $k = \lceil 2/\sigma_{\mathcal{A}} \rceil$.
2. Do k times steps 3–6.
3. $\quad c \xleftarrow{\boxdot} \mathbb{Z}_d^\times, b \xleftarrow{\boxdot} \mathbb{Z}_d$.
4. $\quad g' \longleftarrow g^c, x' \longleftarrow x^b$.
5. $\quad$ Call $\mathcal{A}$ with input (g', x').
6. $\quad$ If $b \in \mathbb{Z}_d^\times$ and $\mathcal{A}$ returns $a \in \mathbb{Z}$, then return $c \cdot a/b$ in $\mathbb{Z}_d$.
7. Return "failure".

THEOREM 7.12. *Assume that $d \geq 7$ is prime. Then there is a constant c so that for any approximate algorithm $\mathcal{A}$ for DL_G with runtime t, $\mathcal{B}$ as above is a probabilistic algorithm for DL_G with runtime $2\sigma_{\mathcal{A}}^{-1} t + c$ which for any input $x \in G$ returns $\mathrm{dlog}_g x$ with probability at least $2/3$.*

PROOF. If in some execution of the loop, $\mathcal{A}$ does not return "failure", then the output of $\mathcal{B}$ is correct, since in step 6 we then have

$$a = \mathrm{dlog}_{g'} x' = \mathrm{dlog}_{g^c} x^b = b \cdot \mathrm{dlog}_{g^c} x = (b \,\mathrm{dlog}_g x)/c \text{ in } \mathbb{Z}_d.$$

In step 4, g' and x' are a uniformly random generator and element, respectively, of G. We abbreviate $\varepsilon = \sigma_{\mathcal{A}}$. Then in step 5, a single call to $\mathcal{A}$ returns "failure" with probability at most $1 - \varepsilon$, and all k independent calls fail with probability at most

$$(1 - \varepsilon)^k \leq e^{-k\varepsilon} \leq e^{-2}$$

by (15.92). In step 6, we have $b \notin \mathbb{Z}_d^\times$ only for $b = 0$, which happens with probability $1/d$. Thus $\mathcal{B}$ returns $\mathrm{dlog}_g x$ with probability at least $1 - e^{-2} - d^{-1} \geq 2/3$. $\qquad \square$

For an asymptotic notion, we have a family $G = (G_n)_{n \in \mathbb{N}}$ of cyclic groups $G_n = \langle g_n \rangle$ with $d_n = \#G_n$ elements for which DL_G is hard (Definition 4.4).

COROLLARY 7.13. *Let $G = (G_n)_{n \in \mathbb{N}}$ be a family of cyclic groups for which DL_G is hard. The following two statements are equivalent.*

(i) *For any nonnegligible function ε of n, $\mathrm{dl}_{G_n, \varepsilon(n)}$ is not in $\mathrm{poly}(n)$.*

(ii) *The hash function (7.7) based on G is collision resistant.*

PROOF. (i) The "constant" c in Theorem 7.8 is in $\mathrm{O}(d_n^3) \subseteq \mathrm{poly}(n)$. $\quad \square$

7.3. Hashing long messages

For cryptographic purposes, one has to hash (and then sign) long messages, maybe contracts, software, audio or video files of mega- or gigabyte size. In practice, this is performed by a single short hash function such as SHA-256 (Section 7.6) from 512 to 256 bits, plus a general strategy to hash long messages using a short hash function. We now explain this strategy: the Merkle-Damgård construction, invented by Damgård (1990) and Merkle (1990).

We take bit strings as messages and hash values, a hash function $h\colon \mathbb{B}^{\ell} = \{0,1\}^{\ell} \longrightarrow \mathbb{B}^n$, and assume $\ell \geq n + 2$. In order to hash a long message x of $k > \ell$ bits, the basic idea is quite simple. We apply h to some initial ℓ-bit value, obtaining n bits, and then continue to apply h to the previous hash value concatenated with the next $\ell - n$ bits of x, until all of x has been digested. Thus each step digests $\ell - n$ bits of x, and we need about $k/(\ell - n)$ steps. This does not quite work, since the hash values of x and the slightly shorter message obtained in the first step are identical. Thus collisions are trivial to find.

But the following modification does work. We set $i = \lceil k/(\ell - n - 1) \rceil$, so that x fits into i blocks of $\ell - n - 1$ bits each, and $d = i(\ell - n - 1) - k < \ell - n - 1$ is the excess. We now pad x by d zeros, so that the length of this new word $\bar{x}$ is the multiple $i(\ell - n - 1)$ of $\ell - n - 1$. We split it as follows:

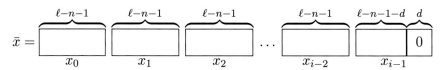

where each x_j has $\ell - n - 1$ bits. We attach another word x_i of $\ell - n - 1$ bits containing the binary representation of d, filled with leading zeroes.

Since $d < \ell - n - 1$, the binary representation of d has fewer than $\ell - n - 1$ bits. Hashing x now goes as described in Figure 7.1. Each internal state y_j of the procedure consists of ℓ bits, initialized at the top. The state y_j following y_{j-1} is the concatenation of $h(y_{j-1})$, the block x_j, and a 1. This last bit is 0 only in the initial state.

Thus we obtain a hash function $h^*\colon \bigcup_{k \geq 0} \mathbb{B}^k \to \mathbb{B}^n$, where $h^*(x) = h(y_i)$ is the hash value of x. We have to evaluate h exactly $1 + \lceil k/(\ell - n - 1) \rceil$ times to compute $h^*(x)$.

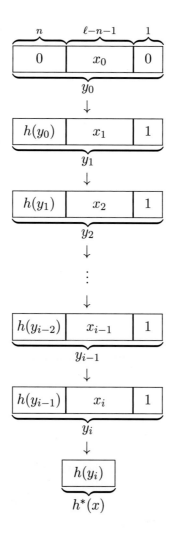

Figure 7.1: Hashing a long message x.

THEOREM 7.14. *A collision for h^* yields a collision for h.*

PROOF. Given a collision (x, x') for h^*, we perform the deterministic steps described in the following and prove that they find a collision for h.

So we have $x \neq x'$ and $h^*(x) = h^*(x')$. We denote by k and k' the lengths of x and x', by $x_0, \ldots, x_i$ and $x'_0, \ldots, x'_{i'}$ the blocks of length $\ell - n - 1$ obtained from x and x', by d and d' the padding lengths for x and x', and by $y_0, \ldots, y_i$ and $y'_0, \ldots, y'_{i'}$ the values computed in the algorithm, respectively.

We distinguish three cases.

Case 1. $k \neq k'$ modulo $\ell - n - 1$.

Then $d \neq d'$, $x_i \neq x'_{i'}$, $y_i \neq y'_{i'}$, and

$$h(y_i) = h^*(x) = h^*(x') = h(y'_{i'}),$$

so that $(y_i, y'_{i'})$ is a collision for h.

Case 2. $k = k'$.

Then $i = i'$ and $x_i = x'_i$. As in case 1, we have

$$h(y_i) = h(h(y_{i-1}), x_i, 1) = h(h(y'_{i-1}), x'_i, 1) = h(y'_i).$$

If $h(y_{i-1}) \neq h(y'_{i-1})$, we have a collision for h. If $h(y_{i-1}) = h(y'_{i-1})$, we either have $y_{i-1} \neq y'_{i-1}$ and a collision for h, or else $y_{i-1} = y'_{i-1}$. In the latter case, we continue this argument for $i - 2$, $i - 3$, $\ldots$, and either find a collision for h or $x_j = x'_j$ for all j. But then $x = x'$, a contradiction.

Case 3. $k = k'$ modulo $\ell - n - 1$ and $k \neq k'$.

We may assume that $k < k'$, and proceed as in case 2. We now have $x_i = x'_{i'}$ and $i < i'$. We either find a collision (by having $x_{i-j} \neq x'_{i'-j}$ for some j), or we eventually reach the top y_0 of Figure 7.1 for x, and

$$h(0, \ldots, 0, x_0, 0) = h(y_0) = h(y'_{i'-i}) = h(y'_{i'-i-1}, x'_{i'-i}, 1).$$

Since the last bits of these two inputs to h are different, we again have a collision for h. This collision is the raison d'être of the last bit in the state. □

The cost of this procedure is, given (x, x'), the computation of $h^*(x)$ and $h^*(x')$ with the intermediate results of Figure 7.1 plus some comparisons and index arithmetic.

Intuitively, Theorem 7.14 says that if h is collision-resistant, then so is h^*. But this asymptotic concept does not apply to the single hash function h. We would consider h^* as a family $(h^*_k)_{k \in \mathbb{N}}$ with h^*_k being h^* as in Figure 7.1 restricted to k-bit messages. This family is not collision-resistant, since for $k = 2^\ell$, we can build a table of all values of h, find a collision for h, and then one for h^*. (Of course, such exponentially large input sizes are outside of the realm of practicality.) See also the end of Section 7.1.

The first value y_0 in Figure 7.1 begins with $z = 0^n$. If we allow an arbitrary *initial value* $z \in \mathbb{B}^n$ in its stead, we obtain a keyed hash function with key space $\mathbb{B}^n$.

7.4. Time stamps

Suppose that in a signature scheme, EVE somehow obtains BOB's secret key. Then she can forge BOB's signature on any message. Also, it is not clear which previous messages of BOB's were signed properly or with a forged signature. In fact, all of BOB's signatures may be considered invalid.

A malicious BOB might turn this around to his advantage if he later regrets signing a message. He simply releases his secret key. Now everyone can forge BOB's signatures, and BOB may claim that his previous signature was forged.

Time stamping provides a safeguard against this kind of misuse. For this, one uses a piece *pub* of public information that was presumably not known to anybody before a specific date. Examples: the complete New York Stock Exchange listing, or a precise weather picture, with hundreds of temperature and pressure measurements.

Now BOB timestamps a message x with a publicly known hash function h as follows: He computes $z = h(x)$ and $z' = h(z, \text{pub})$, signs this as $y = \text{sig}_K(z')$ with his secret key K, and publishes (z, pub, y) in next day's newspaper. This fixes the date of the signature to within one day.

It is also easy to timestamp using a trusted authority.

7.5. The MD and SHA hash function families

The recent history of practically used hash functions and their cryptanalysis is interesting and led to a temporary *hash crisis* which only subsided in 2012.

For a while, the most popular hash functions came from the *message digest* (MD) family, designed by Ron Rivest. MD4 from 1990, with hash values of 128 bits, was used to hash passwords on some Windows systems, attacked by Dobbertin (1998), and is completely broken: collisions can be found in microseconds. MD5 from 1992 has been an internet standard, for SSL certificates and signatures. After preliminary attacks by Dobbertin, Wang *et al.* (2004), and others, Stevens, Sotirov, Appelbaum, Lenstra, Molnar, Osvik & de Weger (2009) put it completely to rest by forging a security certificate provided under MD5. The latest version MD6 was submitted to the NIST competition (see below), but did not make it into the second round of selections. Some security providers were not much worried by these attacks, but government authorities were. After a preliminary version called SHA-0, the NIST proposed SHA-1 with 160-bit hash values in 1995, soon adopted as a standard. It is closely modelled after MD4.

The Merkle-Damgård construction of Section 7.3 is used to hash long messages. But the hash functions of this section use this principle also internally, disecting the input into manageable chunks and digesting them in several rounds, chunk by chunk. A first cryptanalysis of SHA-1 by Rijmen & Oswald broke 53 out of the 80 rounds. Then came Wang *et al.* (2005b) with an attack on the full SHA-1 with 2^{69} operations, later improved to 2^{63} operations. Some preprocessing is needed, and an operation is just the evaluation of SHA-1 on some specific input. This was improved by Cannière & Rechberger (2006). The attacks carry the differential cryptanalysis of Section 6.2 much further and are beyond the scope of this text.

In view of concerns that also the successor SHA-2, discussed in the next section, might come under attack, the NIST started a hash function competition in 2007, considered now as successful as its AES competition (Section 2.2). Entries could be submitted until 31 October 2008, and a new standard SHA-3 was announced on 3 October 2012. Named *Keccak* on its submission by Bertoni *et al.* (2013), it enjoys excellent security and efficiency properties. A major novel feature is its design principle, called *sponge functions* and quite different from those used previously. One of its designers, Joan Daemen, was also a co-winner of the AES competition.

7.6. The SHA-2 hash functions

SHA-2 comprises four similar hash functions, with outputs of 224, 256, 384, and 512 bits, respectively. It continues being a NIST standard since 2001, now in co-existence with SHA-3. We describe SHA-256, omitting some of the more tedious details.

After some preprocessing, a message of $512 = 16 \cdot 32$ bits is hashed to 256 bits. The computation operates on 32-bit words. The 16 inputs words are first extended to 64 words $W_0, \ldots, W_{63}$. An internal state consists of eight words a, b, c, d, e, f, g, h. Initially, these are some constants. These are then modified in 64 rounds, with W_i being digested in the ith round, as presented in Figure 7.2.

The top row consists of the eight input words of round i, wich are constants if $i = 0$ and the output of the $(i - 1)$st round otherwise. The bottom row is the output. W_i is the ith part of the extended input, and K_i some specified constant. The various functions are explained in Figure 7.3; Ch, +, and Maj are nonlinear.

Notes. The word *to hash* comes, according to Webster's Dictionary, from the Old French *hacher* = to mince or chop, and is related to the German *hacken* = to hack or

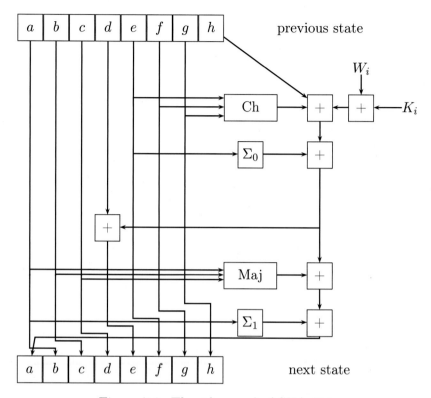

Figure 7.2: The ith round of SHA-256.

function	semantic
$+$	The sum modulo 2^{32} of its two input integers
$\mathrm{Ch}(e, f, g)$	*choice*: if e then f else g; equivalently $(e$ and $f)$ or $($not e and $g)$
$\Sigma_0(a)$	$\mathrm{ROTR}^2(a) \oplus \mathrm{ROTR}^{13}(a) \oplus \mathrm{ROTR}^{22}(a)$
$\mathrm{Maj}(a, b, c)$	*majority*: $(a$ and $b)$ or $(a$ and $c)$ or $(b$ and $c)$
$\Sigma_1(e)$	$\mathrm{ROTR}^6(e) \oplus \mathrm{ROTR}^{11}(e) \oplus \mathrm{ROTR}^{25}(e)$

Figure 7.3: The internal SHA-256 functions. All except $+$ are executed bitwise, and ROTR^k is a rotation (cyclic shift) to the right by k places.

cut. To hash means to chop (meat and vegetables) into small pieces (for cooking) and a *hacker*'s work has just that effect on a computer system.

7.1. A simple example of hashing modulo a prime is your birth number in Exercise 2.1. As mentioned in the text, it seems necessary to include the program length of $\mathcal{A}$ in $T(\mathcal{A})$. If we did not assume locality of variables, then a "new" variable outside of $\mathcal{A}$ might require a long name, with up to about $\log T(\mathcal{A}) \log\log T(\mathcal{A})$ bits. This is spurious to our concerns.

See Rogaway & Shrimpton (2004) for a discussion of concepts for hash function security and references. A second-preimage resistant hash function is sometimes called *weakly collision-free*, and for collision resistant, one also finds *strongly collision-free*. See the Notes 13.13 for a brief mention of lattice-based collision resistant hash functions.

7.2. The hash function of (7.7) was proposed by Chaum, van Heijst & Pfitzmann (1992).

7.3. An alternative to the Merkle-Damgård construction of Section 7.3 are hash trees (Merkle 1988). These are used in BitCoin, BitTorrent, and other protocols. A small modification makes the Merkle-Damgård construction also work when $\ell = n + 1$.

7.5. Further attacks on MD4 and MD5 are in Wang & Yu (2005), Lenstra *et al.* (2005), Wang *et al.* (2005a) and Klíma (2005). The standardization of SHA-1 is in FIPS PUB 180-1 (NIST 1993) and the draft for SHA-3 is FIPS PUB 202 (NIST 2014).

7.6. The standardization of SHA-2 is in FIPS PUB 180-2 through 180-4 (NIST 2012a). The shortest output size of 224 bits may be too small for some applications.

Exercises.

EXERCISE 7.1 (Birthday collision). *Suppose we have a class of 30 students, all born in the year 1990 and with birthdays uniformly random among the 365 possible values, ignoring leap years.*

(i) *Compute the probability that some different students $x \neq y$ have the same birthday.*

(ii) *Referring to Exercise 2.1, (i) asks for the probability of $d(x) = d(y)$ and $x \neq y$. Determine the three probabilities that $a(x) = a(y)$, $b(x) = b(y)$, and $c(x) = c(y)$, always assuming $d(x) \neq d(y)$.*

(iii) *Is the function $x \mapsto c(x)$ a hash function? If so, a good one?*

EXERCISE 7.2 (sec $\leq$ inv). *We use the notation in the proof of Theorem 7.3 (ii). The elements $z \in h(E)$ have only one preimage and finding it with an inverter does not help in finding second preimages. The goal of this exercise is to strengthen the Theorem's conclusion under suitable assumptions. Assume that $b = \#Z$, set $e = \#E = \varepsilon b$ with $0 \leq \varepsilon < 1$, and show the following.*

(i) *The upper bound b/a in (7.4) can be replaced by e/a. Where can you make the same replacement in the later calculations?*

(ii) *Show that for some constant c, $\mathrm{sec}_\delta \leq \mathrm{inv}_\varepsilon + 3c$, where $\delta = \varepsilon(1 - b/a) + e/a^2 \cdot (a(2 - \varepsilon) - b + e)$.*

(iii) *Find conditions on ε, a, b, and e under which $\delta \leq \varepsilon/2$.*

EXERCISE 7.3 (Concatenating hash functions). *Let h_1 and h_2 be hash functions and $h = h_1 | h_2$ their concatenation.*

(i) *Prove that if at least one of h_1 or h_2 is collision-resistant, then so is h.*

(ii) *Determine whether the analogous claim holds for second-preimage resistance or one-wayness, respectively.*

EXERCISE 7.4 (Building new hash functions). Let $h_0\colon \mathbb{B}^{2n} \to \mathbb{B}^n$ be a collision-resistant hash function.

(i) We construct a hash function $h_1\colon \mathbb{B}^{4n} \to \mathbb{B}^n$ as follows: parse the bit string $x \in \mathbb{B}^{4n}$ as $x = (x_1|x_2)$, where both $x_1, x_2 \in \mathbb{B}^{2n}$ are $2n$-bit words and $|$ is the concatenation of bit strings. Then compute the hash value $h_1(x)$ as

$$h_1(x) = h_0(h_0(x_1)|h_0(x_2)).$$

Show that h_1 ist collision resistant.

(ii) Let $i \geq 1$. We define a hash function $h_i\colon \mathbb{B}^{2^{i+1}n} \to \mathbb{B}^n$ recursively using h_{i-1} as h in (i): parse the bit string $x \in \mathbb{B}^{2^{i+1}n}$ as $x = (x_1|x_2)$, where both $x_1, x_2 \in \mathbb{B}^{2^i n}$ are $2^i n$-bit words. Then

$$h_i(x) = h_0(h_{i-1}(x_1)|h_{i-1}(x_2)).$$

Show that h_i is collision resistant.

(iii) The number $p = 2017$ is prime. Show that 5 and 19 are generators of $\mathbb{Z}_p^{\times}$. Now define $h_0\colon \mathbb{B}^{22} \to \mathbb{B}^{11}$ as in Section 7.2: Let $x = (b_{21}, \ldots, b_0) \in \mathbb{B}^{22}$. We set $x_1 = \sum_{0 \leq i \leq 10} b_{11+i} 2^i$ and $x_2 = \sum_{0 \leq i \leq 10} b_i 2^i$ in $\mathbb{Z}_{p-1}$. Now compute $y = 5^{x_1} \cdot 19^{x_2}$ in $\mathbb{Z}_p^{\times}$ and let $h_0(x) = (y_{10}, \ldots, y_0)$ be the binary representation of y, so that $y = \sum_{0 \leq i \leq 10} y_i 2^i$. Use a computer algebra system and the birthday attack to find a collision of h_0 and one for h_1 from (i). Can you describe succinctly the hash function $h_2\colon \mathbb{B}^{88} \to \mathbb{B}^{11}$ as in (ii)?

EXERCISE 7.5 (Shortening hash functions). Let $h\colon \mathbb{B}^{\ell} \longrightarrow \mathbb{B}^n$, $0 \leq m < 2^n$, and $g\colon \mathbb{B}^{\ell} \longrightarrow \mathbb{Z}_m$ with $g(x) = h_0(x) \operatorname{rem} m$, where $h_0(x) \in \{0, \ldots, 2^n - 1\}$ is the integer with binary representation $h(x)$.

(i) Determine functions ε_1 and t_1 of ε and t so that g is (ε, t)-inversion resistant (one-way) if h is $(\varepsilon_1(\varepsilon, t), t_1(\varepsilon, t))$-inversion resistant. The functions may also depend on ℓ, n and m. You may make further reasonable assumptions about h; state them clearly.

(ii) Can you do the same for second-preimage and collision resistance?

(iii) Now take a family $h = (h_k)_{k \in \mathbb{N}}$ of functions $h_n\colon \mathbb{B}^{\ell(k)} \longrightarrow \mathbb{B}^{n(k)}$, a family $m = (m(n))_{n \in \mathbb{N}}$ of integers, and consider $g = (g_k)_{k \in \mathbb{N}}$ with $g_n\colon \mathbb{B}^{\ell(k)} \longrightarrow \mathbb{Z}_{m(k)}$ as above. Find conditions on h and m so that g_k is inversion-resistant if h is. Can you do the same for second-preimage and collision resistance?.

EXERCISE 7.6 (Collisions and one-way functions). We want to compare collision-resistant hash functions $h\colon X \to Z$ to one-way functions, allowing the probabilistic algorithms to make mistakes. We suppose additionally that every element $z \in Z$ is the hash value of at least $\#X/(2\#Z)$ elements of X. (On average, an element $z \in Z$ is the hash value of $\#X/\#Z$ elements of X. Therefore our assumption means that the images of the hash function are approximately evenly distributed.) A probabilistic algorithm $\mathcal{A}$ is called (δ, ε)-reliable (for a certain problem) if for at least a fraction δ of the inputs it returns the correct result with probability at least ε. For all other inputs it may return arbitrary nonsense. A mistake need not be recognized by $\mathcal{A}$.

We consider (δ, ε)-reliable probabilistic algorithms $\mathcal{A}$ for some $\delta, \varepsilon > 0$. Adapt the definitions to this kind of algorithm. A hash function $h\colon X \to Z$ is called

○ approximately *collision resistant* if there is no efficient $\mathcal{A}$ that finds a collision from scratch.

○ approximately *second preimage resistant* if there is no efficient $\mathcal{A}$ that given a message $x \in X$ finds a second message y with the same hash value.

○ an *approximate one-way function* if there is no efficient $\mathcal{A}$ that, given a hash value $z \in Z$, finds a message $x \in X$ with $h(x) = z$.

The goal now is to derive quantitative estimates for the (δ, ε)'s so that the analog of Theorem 7.3 holds: If h is approximately collision resistant, then h is an approximate one-way function. You may want to start with $\delta = 1$ and the following steps may be helpful for the solution.

(i) For an approximate inverter $\mathcal{A}$, compute the error probability

$$\text{prob}\left(\mathcal{A}(h(x)) = x \text{ or } h(\mathcal{A}(h(x))) \neq h(x)\right).$$

(ii) For any two events U and V, we have $\text{prob}(U \text{ or } V) \leq \text{prob}(U) + \text{prob}(V)$. Deduce that

$$\text{prob}(U \text{ and } V) \geq \text{prob}(U) + \text{prob}(V) - 1.$$

If only $\text{prob}(V) \geq 0$ is known, we still have $\text{prob}(U \text{ and } V) \geq 0$, even if $\text{prob}(U) - 1 < 0$.

(iii) There is a set $Z_0 \subseteq Z$ of inputs with $\#Z_0 \geq \delta \#Z$ and such that for each input $z \in Z_0$, we have $\text{prob}(h(\mathcal{A}(z)) = z) \geq \varepsilon$.

(iv) Let $X_0 = h^{-1}(Z_0)$ be the set of elements $x \in X$ whose hash value is in Z_0. Show that $\#X_0 \geq \delta \cdot \#X/2$.

(v) Show that for each $z \in Z$ we have $\sum_{x \in h^{-1}(z)} \text{prob}(\mathcal{A}(z) = x) \leq 1$.

(vi) Show that the error probability is at most

$$\frac{\#Z}{\#X} + 1 - \frac{1}{2}\varepsilon\delta.$$

(vii) Accomplish the goal stated above.

(viii) Does a similar claim hold if we replace approximately collision-resistant with approximately second-preimage-resistant?

EXERCISE 7.7 (Double DH). Let $G = \langle g \rangle$ be a finite group and let x, y, z, and u be elements of G. Let 2DH_G be the problem to compute from the input (x, y, z, u) the output $x^{\text{dlog}_g y} z^{\text{dlog}_g u} \in G$. Here G and g are publicly known. Show that

$$2\text{DH}_G \equiv_p \text{DH}_G.$$

EXERCISE 7.8 (DL and hash functions).   The numbers $q = 7541$ and $p = 15083 = 2q + 1$ are prime. Take the group $G = \{z \in \mathbb{Z}_p^\times : \text{ord } z \text{ divides } q\} \subset \mathbb{Z}_p^\times$, and $x = 604$ and $y = 3791 \in \mathbb{Z}_p^\times$.

(i) Show that both elements x and y have order q in $\mathbb{Z}_p^\times$ and thus each generates G.

(ii) Consider the hash function

$$h: \mathbb{Z}_q \times \mathbb{Z}_q \to G, \text{ with } (a, b) \mapsto x^a y^b.$$

Compute $h(7431, 5564)$ and $h(1459, 954)$.

(iii) Find $\text{dlog}_x y$.

EXERCISE 7.9. *We consider hash functions h and h^* as in Figure 7.1.*

  (i) *Given a collision for h, can you find one for h^*?*

 (ii) *Describe a weakness for h under which you can find collisions for h^*. Phrase your result in the language used in this section.*

EXERCISE 7.10. *Modify h^* in Figure 7.1 as follows.*

  (i) *Omit the last bit.*

 (ii) *Omit x_i.*

Take $\ell = 10$ and $n = 4$ and show how to construct a collision for the modified h^* in (i) and in (ii), respectively, without having one for h.

EXERCISE 7.11. *Section 7.3 derives a "long" hash function h^* from a "short" h. Suppose that we leave out the last step in Figure 7.1 and define $h'(x) = h(y_{i-1})$. Show how to find the hash value $h'(x, z)$ of x extended by a (nonempty) message z of your choice, just knowing $h'(x)$ but not x. (Such a length-extension weakness actually occurs in SHA-1 and SHA-2.)*

Systematisch ausgebaut aber wurde das Geheimschriftenwesen
erst mit der Einführung der ständigen Gesandtschaften.
Der diplomatische Verkehr, wie er sich im 15. Jhdt.
durch die Sforza in Mailand und bald im übrigen Italien
und Europa eingebürgert hatte, führte notgedrungen
zur weiteren Ausgestaltung der Kryptographie.[1]

FRANZ STIX (1936)

Car ce ne sont point lettres appostees qui ie represente. J'en tiens
et garde soigneusement les originaux, que ie recongnois en
bonne forme, et bien sellez et signez, lesquels ie representeray
tousiours auecques mes traductions, et les Alphabets et
Dictionaires que i'ay compris pour y paruenir, à qui et quand de
par vous il me sera ordonné. Et ne doit esmouuoir, SIRE, que
cela sera occasion à voz ennemis de changer leur chiffres, et se
tenir couuerts, et à nous voz officiers plus enemies empeschez à
vous y servir. Ils en ont chãgé et rechangé, et neantmoins
ont esté et seront tousiours surpris en leurs finesses.
Car votre cause est iuste et la leur inique.[2]

FRANÇOIS VIÈTE (1589)

Codes and ciphers appear at first sight to be such complicated
and difficult affairs, and so completely wrapped in mystery,
that all but the boldest hesitate to tackle them.

ALEXANDER D'AGAPEYEFF (1939)

---

[1]Cryptographic chancelleries were only expanded systematically with the introduction of permanent legations. The diplomatic traffic, as established by the Sforza in Milan in the 15th century and soon after in the rest of Italy and Europe, led inevitably to a further development of cryptography.

[2]For I do not attach the letters. I hold and keep carefully their originals, which I know to be in good shape, well sealed and signed, and which I can always present with my translations and the alphabets and dictionaries that I compiled to achieve [my decipherment] to whomever and whenever you will order me to do so. And do not worry, SIR, that this will be an occasion for your enemies to change their ciphers and to keep themselves better protected, and to hinder us, your officers, to serve you. They have changed them and changed them again, but nevertheless they have been and will be caught in their tricks. For your cause is just and theirs is unjust.

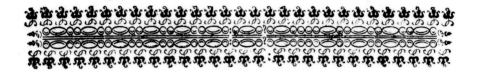

Chapter D

Codebooks

 imple substitutions generalize the Caesar cipher. One step further are the nomenclators and codebooks, which we present in this chapter. They work like simple substitions, except that they have much larger alphabets: not just letters, but also digrams, syllables, words, and names of people and places. Examples exist already from the 14th century, and a century later we find code factories at work that output series of codebooks by minor variation of a general template. In the First World War, top secret diplomatic messages were encrypted in this way, for example the Zimmermann telegram discussed in Chapter I. These nomenclators encode many frequently occurring terms with individual encryptions. We will see examples of their use by British and by Cuban conspirators (Sections H.4 and H.5) and in private correspondence (Sections H.2 and H.3). The idea was employed in a different way when the telegraph came into use, namely in the form of commercial codebooks for reduced telegraph fees. If secrecy was needed, they could be superenciphered.

D.1. Nomenclators 14th century

A *codebook* (or *code*) is a list of frequently used terms (plus individual letters and, sometimes, syllables) and a codeword for each of them. They have been used since before the Renaissance and had their own renaissance with the rise of telegraphic correspondence, in particular the trans-Atlantic cable in 1866.

© Springer-Verlag Berlin Heidelberg 2015
J. von zur Gathen, *CryptoSchool*, DOI 10.1007/978-3-662-48425-8_11

Historically, they were called *nomenclators*. This was originally the designation of the ushers who called out (*calamare*) the name (*nomen*) of a dignitary entering a party, and carried over to those secret books that contained the names of many dignitaries.

We do not know when codeboooks of substantial size came into use, but an example from 1377 claims to be an original invention by the King of Navarra, and seems to be the oldest surviving sample. During the hundred-year war, the Spaniards were allied with the English against the French. King Carlos II of Navarra (1332–1387) used a codebook to communicate with his agent Pierre du Tertre at Bernay in Normandy, and with his English allies. Both the complete codebook and the story of its invention have survived in the *Chronique Normande*, written by Pierre Cochon (c. 1360–c. 1434).

> *En l'an mil .CCC. LXXvij., en Karesme, fu aprocheue une soutille maniere de faire du roy de Navarre devant dit contre le roy de France, en maniere de traison, d'escripre couvertement et muer les nomz des prinches, des chastiax et bonnes villes en aultres nomz que les euz propres, si comme il aperra cy après, et fais par la sutilité mestre Pierre du Tuetre, conseillier du dit roy de Navarre.*[1]

When Charles V (1338–1380), King of France, captured the city of Bernay, du Tertre was caught, and he and another councillor *"ourent les colz trenchez"*[2] on 28 June 1378. Part of the codebook of 124 words is shown in Figure D.1.

The Navarran King's two teenage sons were held hostages by their uncle, King Charles V of France. A sample letter from the King of Navarra, written on 1 May 1378 at Pamplona, begins as follows:

[1]During Lent 1377, a subtle method of acting secretly against the King of France was devised by the King of Navarra. This was by writing covertly and moving the names of princes, castles and larger cities to other names, not their own, as apparent below, and it was made by the subtlety of Master Pierre du Tertre, councillor to the King of Navarra.

[2]had their necks cut

[3]Charles le Roy de navarre: S'il est ainsi que *nommularius* ne lessat partir de liy *Repertum*, il est de neccessité que *Vexatus* pense et ymagine aucune voie comment *Repertus* peust venir en *Bispartie* vers *Capitolium*. Car si ... In English: Charles, the King of Navarra: If the King of France will not release my son Charles, it is necessary that du Tertre think and imagine a way of how my son can come to Cherbourg in Normandy.

...

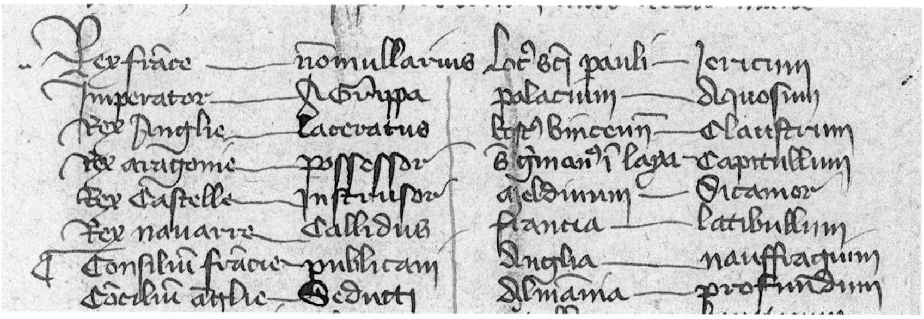

plaintext	ciphertext	meaning
Rex france	nommullarius	King of France
Imperator	Agrippa	German Emperor
Rex Anglie	Laceratus	King of England
Rex arragonie	possessor	King of Aragon
Rex Castelle	Instrusor	King of Castille
Rex navarre	Callidus	King of Navarra
⋮	⋮	⋮
Dominus Karolus Navarre infans	Repertus	Charles Prince of Navarra (1361–1425)
Dominus Petrus	Restaurator	Peter Prince of Navarra (1366–1412)
⋮	⋮	⋮
Cesarisburgum	Capitolium	Cherbourg
Mare	Planicies	sea
Naves	Aquatice	ships
Monspessulanus	Bipennis	Montpellier
Burdegalis	Ambrosia	Bordeaux
⋮	⋮	⋮
Burgundia	Detenta	Burgundy
Normannia	Bispartita	Normandy
Britannia	Vulnerata	Brittany
⋮		

Figure D.1: The codebook of Carlos II of Navarra from 1377.

The King of France later did release the two sons and gave them back their lands in Normandy, now as fief of the King of France.

D.2. Codebooks 15th century

One century later, the new invention has become routine business. Figure D.2 exhibits an example from 1463. It comes from the records of the Milanese *Cancelleria segreta*[4] which were mainly produced by Francesco Tranchedino (1411–c. 1496). They show an early Renaissance code factory at work. Since 1450, Francesco "Cicco" Simonetta (1410–beheaded 1480) had been First Secretary of the Secret Chancellery at the court of the Sforza Dukes in Milan. He wrote in 1474 the oldest Western text on cryptanalysis that has been conserved. (The Arab cryptographers like Al-Kindi had been centuries earlier; see Section G.1.) See Section G.2 for his cryptanalytic achievements.

Tranchedino worked for Cicco Simonetta and produced in 1475 a catalog of 159 Milanese codebooks up to that time. This forms the nucleus of a manuscript which was continued by other officials and gives 297 such ciphers in total and is edited with beautiful illustrations as Tranchedino (1970). The cipher in Figure D.2 is quite typical. It starts with the date *23 August 1463* and the recipient D. [Dominus = Mr.] Antonio de Besana. Then the cipher begins with either two or three encryptions of the 21 letters plus &, con, and ex. The letters A, e, h, l, and q get three possibilities, the others get two. Then come 12 dummies (*Nulle*) and 12 signs for doubled letters (*Duplicate*), from bb to tt. The center part has 63 signs for digrams of the form 'vowel plus consonant'. The last part is the nomenclator proper and has encryptions of 31 codewords:

> Pope, King of France, René d'Anjou (titular King of Naples), King Ferrante (Ferdinand) I of Naples, Duke Philip the Beautiful of Burgundy, Duke Johannes, Duke of Milan, Venetians, Florentinians, Saône, Genova, Genovese, Santa Liga federatore, your government (La S.^{ria} V$\bar{r}$a = La Signoria Vostra), soldiers, cavalry, footsoldiers, money, ships, galleons, King Alfonso VI of Aragon, Count Iacobo Picinino, Italy, Germany, Duke of Savoy, council of cardinals, France, D. Philip of Savoy, that, because, not.

The total comes to 165 cipher symbols for the 130 plaintext terms. Some of the encrypting symbols resemble letters or digits, but most are phantasy signs. It takes a careful and patient hand, experienced in this kind of

---

[4]Italian for MSA = Municipal Security Agency

Figure D.2: One of Tranchedino's nomenclators, from 1463.

cryptTEX, to put down long messages with such contrived symbols. The
difficulty in reading them may have suggested a false sense of security,

but in fact, a legitimate user faced the same problem, at least initially. Most of the codebooks in Tranchedino's compilation are dated, between 1450 and 1496. The longest one has 283 symbols.

The various codebooks all follow the same structure but use varying symbols, with plenty of room for the designer's fancy. These records form an impressive display of the power of Northern Italian cryptography in the early Renaissance.

D.3. Codebooks 16th century

We now jump another hundred years ahead. Henry III (1551–1589), a Calvinist King of France, had as powerful enemies the family of Guise. They formed the Catholic *Holy League* in 1576, with the goal of putting one of their bloodline on the throne. Henry III had the two leading brothers murdered in 1588; the narrow passage in the Blois castle on the Loire, where Henri de Guise was assassinated at 8 am on 23 December 1588, is now a favorite tourist sight. A third brother, Charles de Mayenne, Duke of Mayenne, took over leadership of the *League*. After the murder of Henri III by Jacques Clément, a Catholic priest, in 1589, the Protestant King Henri IV of France (1553–1610) quickly gained the upper hand militarily. The Duke of Mayenne's ambition still was to become King himself. When it became apparent that the League's military power was not sufficient, he schemed to involve Philip II (1527–1598), King of Spain and Portugal, in his plans. Besides invoking their common religion—then as now a major excuse for killing the others—he offered a substantial prize: large parts of France, namely the Roussillon in the South and Picardy bordering on the Spanish Netherlands. Their possession had been a Spanish goal for some time.

Commander Juan de Moreo was delegated in 1589 to the Spanish army ready to aide the League. For his communication with the Spanish court, he had a codebook of 423 terms, plus dummies and signs for doubling letters and for numbers, and used it 1589–1597. Figure D.3 shows its initial part in modern type. The codebook presumably also contained encryptions of letters, digrams, and frequent words like *and*, *with*, etc. The original seems to have been lost, but the Spanish archives at Simancas contain another codebook with striking similarities. This was issued for use with Juan Baptista de Taxis around 1590. Its beginning is shown in Figure D.4; in the original, this is just one column (out of seven in total) which we have split into two for the reproduction. The plaintext words are the same in both codebooks, but the encryptions are different. For Moreo's book, they are the underlined numbers from $\underline{0}$ on (and going to $\underline{99}$, not shown here), while in Taxis' cipher they are three-letter syllables "consonant + vowel + m". The consonants are used in descending order: s, r, qu, p, n, m, l, j, h, g, f, d (and continuing to b on another page). For each consonant (except s) the five vowels are used, for example towards the end: fum, fom, fim, fem, fam. In other parts, Moreo's cipher has such two- and three-letter syllables, and Taxis' has underlined numerals.

We see a well-organized cipher factory at work. They have a list of plaintext words, which may be copied for the different ciphers, and standardized (but not identical) types of cipher equivalents, mainly certain two- and three-letter syllables and over/underlined or dotted numerals. These are inserted in several sections, with an alphabetical or numerical order (or reversed order) in each section. The use of standard signs is progress over the contrived symbols in Tranchedino's codebooks.

Both codebooks contain provisions for dummies, double letters, and numbers. In Taxis' cipher, this reads in the bottom lines of Figure D.4: *Las Nullas tendran una raya enzima exemplo $\overline{19}$, y las dupplicis un 0, como esto $\overset{0}{46}\,\overset{0}{25}$ y todos los que fueron num.^{os} tendran una cruz encima $\overset{+}{10}\,\overset{+}{20}$.*[5]

Henri IV, King of France and Philip's adversary, had in his services the lawyer François Viète (1540–1603), Seigneur de la Bigotière, who also happened to be one of the leading mathematicians of his times. He introduced the use of letters for known quantities in algebra, and expressed by *Viète's formula* the coefficients of a polynomial in terms of its roots; we use this for the elliptic curve addition rules in Section 5.1. Viète deciphered Moreo's codebook; this was a major cryptanalytic achievement. See Comiers' praise in Section B.2. After such a success, one usually keeps mum about it, expecting the enemy to continue using it and so to provide more secret messages which can then be deciphered. But here something unusual happened: Viète published a booklet of 14 pages, containing a lengthy letter, sent by Moreo from Anvers (Amveres, Antwerp) to Madrid and which he had deciphered. Figure D.5 shows the title page.

Decipherment of a letter written by Commander Moreo to his chief-in-command, the King of Spain, on 28 October 1589

---

[5]The dummies will have an overlining bar, for example $\overline{19}$, and double letters a 0, as $\overset{0}{46}$ and $\overset{0}{25}$, and those that signify numbers have a cross above them: $\overset{+}{10}$, $\overset{+}{20}$.

Cifra particular [1589-1597].

A			
		Bruselas	35
Aca	0	bueno	36
adelante	1	**C**	
advertimiento	2	camino	37
Africa	3	campo	38
agora	4	capitan	39
Aleman	5	capitulo	40
alla	6	cardenal	41
Alteracion	7	cargo	42
amigo	8	carta	43
amistad	9	caso	44
andamiento	10	Castellano	45
año	11	castigo	46
antes	12	castillo	47
Amveres	13	catolico	48
apparencia	14	cavallo	49
aqui	15	causa	50
arcabuz	16	cautela	51
Argel	17	christiandad	52
armada	18	cifra	53
armas	19	ciudad	54
artilleria	20	color	55
assi	21	comissario	56
assistencia	22	comission	57
authoridad	23	comodidad	58
aviso	24	como	59
aún	25	comunicacion	60
aúnque	26	compañero	61
		compania	62
B		concierto	63
bastimento	27	confederacion	64
batalla	28	confederado	65
bateria	29	conclusion	66
beneficio	30	concordia	67
Berberia	31	concurso	68
Bohemia	32	condicion	69
bondad	33	confianza	70
Brabante	34	conformidad	71

Figure D.3: The initial part of the Spanish codebook for Juan de Moreo.

Figure D.4: The initial part of the original codebook for Taxis.

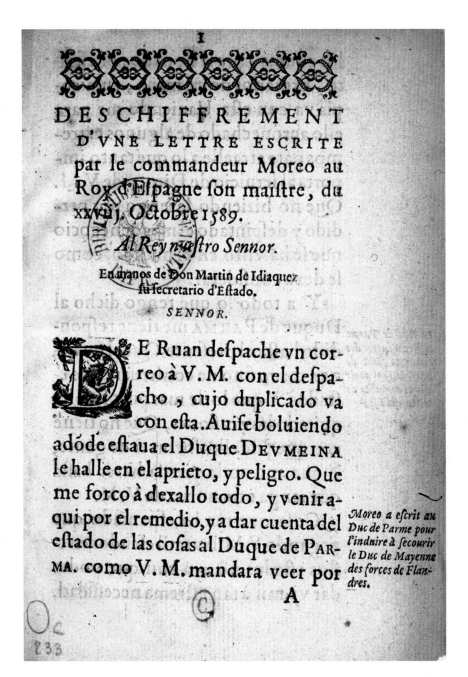

Figure D.5: Viète's decipherment of a Spanish missive.

> *To the King our Lord*
> *in the hands of Don Martin de Idiaquez, his secretary of state.*
> *Sir. From Rouen I sent a letter to Y. M. with the message*
> *whose duplicate goes with the present one. I mentioned that,*
> *after returning to where the Duke of Mayenne was, I found*
> *him in open and dangerous territory. This obliged me to drop*
> *everything and to come here for help, and to give the Duke*
> *of Parma an account of the state of affairs. . . .*

Throughout the Spanish text, Viète gave his marginal précis in French, as on the title page:

> *Moreo has written to the Duke of Parma to induce him to*
> *relieve, with his forces from Flanders, the Duke of Mayenne.*

We can only speculate about the reason for going public, but it had the effect of rallying the French nobility around Henri IV, enraged about the Duke's proposed betrayal of French territories. Henri IV converted pragmatically to the Catholic faith: *"Paris vaut bien une messe."*[6] The Duke of Mayenne gave up his fight for the crown and Henri treated him generously, praising him for *not having permitted, in good or bad luck, the dismembering of France.*

Viète was a successful cryptanalyst, but his vanity was counter-productive. He bragged in front of Giovanni Mocenigo, the Venetian ambassador in France, about his abilities in code-breaking. The wily diplomat teased him into admitting that he also solved Venetian codes, and even into exhibiting an example. When the Council of Ten, back home in Venice, learnt about this, they immediately changed their codes. Philip II also became aware of Viète's bragging. He had been so convinced of the security of his nomenclator that he alleged black magic publicly—a dangerous accusation in those days. Viète's biography mocks the Spanish

> *qui ad odium & invidiam nihil non comminiscuntur, ma-*
> *gicis artibus, nam aliter fieri non potuisse, à Rege id factum,*
> *passim & Romæ præcipuè non sine risu & indignatione rec-*
> *tiùs sentientium per emissarios suos publicabant.*[7]

---

[6]Paris is well worth a mass.

[7]who out of hate and envy do not refrain from inventing anything, had their envoys state publicly everywhere, and in Rome particularly, that this [cryptanalysis] had been achieved by the King through black magic because it was not otherwise possible - not without incurring the derision and indignation of those in the know.

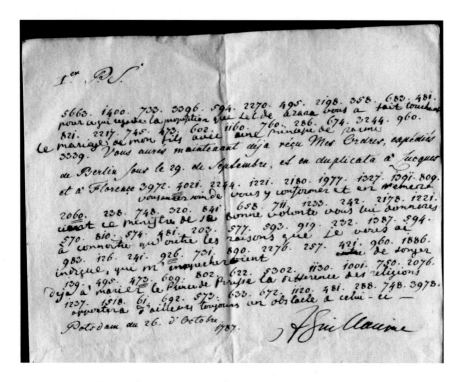

Figure D.6: A letter from King Frederick William II of Prussia to his envoy Girolamo Lucchesini, dated 26 October 1787.

D.4. Codebooks 18th century

Two hundred years later, we see royal correspondence routinely encrypted at the Prussian court. King Frederick William II (1744–1797) employed Marchese Girolamo Lucchesini (1751–1825), originally from the Italian town of Lucca in Tuscany, for delicate diplomatic missions. Presumably this polyglot man of the world was more apt at subtle negotiations than the King's Prussian underlings, largely influenced by military thinking.

Lucchesini had been a confident of Frederick the Great of Prussia (1712–1786), Frederick William's uncle, and continued his position under the latter's son, Frederick William III (1744–1797). This ended only in 1806, when Lucchesini signed an armistice with Napoleon after the battle of Jena and Auerstädt and the King refused to ratify it. In 1787, Lucchesini was on a diplomatic mission in Italy, where he met Johann Wolfgang von Goethe (1749–1832).

Figure D.6 shows a letter dated 26 October 1787 from the King to his envoy concerning a delicate matter, namely the rejection of an Italian

1er, P.S.

5663 1400	733	3396	594 2270	495	2198	358 683	481
pour ce qui	regardes la	proposition que le C.[onte] de Azana	vous	a	fait	touchant
821 2217	745	473	602 1160	760	286	674 3244	960
le	mariage de	mon	fils avec	une	princesse de	Parme

3339 Vous aurés maintenant déja réçu Mes Orders, expédiés

de Berlin sous le 29. de Septembre, et en duplicata à Lucques

et à Florençe	3972	4021 2244	1221	2180	1977	1327	1391 809
	vous aurez soin de	vous	y	conformer et		en	remer-
2060	238	748	320 841	658	711	1233	242	2178 1221
ciant	ce	ministre	de sa	bonne	volonte vous	lui	donnerres
570	810	574	481 203	577	593	919	232	1387 594
à	connoitre que	outre les	raisons	que	je	vous ai
983	126	241	926 731	890	2276	257	421	960 1886
indique,	qui	m'empecheroient				de	songer
139	495	473	609 802	622	5302	1130	1001	750 2076
deja	à	marier	le Prince de	Prusse	la	difference des religions
1237	1518	61	692 573	633	672	1120	481	288 748 3978
apportera d'ailleurs toujours	un	obstacle a	celui-	ci	-

Potsdam au 26. d'Octobre 1787.   F. Guillaume

Table D.7: Transcription of Frederick William's letter in Figure D.6.

marriage proposal to the Prince of Prussia.

> *Regarding the proposal that the Count of Azana made to you concerning the marriage of my son with a Princess of Parma, you will by now already have reveived My Orders, despatched from Berlin on 29 September, duplicate to Lucca and to Florence. You will take care to follow them and, thanking this Minister for his good intentions, you will let him know that, apart from the reasons that I indicated to you and that already would prevent me from considering to marry the Prince of Prussia, the difference of religions will in addition always present an obstacle to this.*

The secretary penning the missive has encrypted most of the words using a numerical codebook, with some text in clear and the King's own signature at the bottom. On receipt, Lucchesini consulted his copy of the same codebook and wrote the plaintext under the same lines of numerical ciphertext. This is fortunate for us, because the letter itself cannot be deciphered. Its ciphertext consists of 88 numbers between 61 and 5663, four of them underlined. The six ciphertext numbers 473, 495, 594, 748, 960, and 1221 occur twice, and 481 three times.

Lucchesini's plaintext makes perfect sense, but does not seem to match the ciphertext. 481 is variously decrypted as *touchant* (line 03), *outre* (line 13), and some word from *un obstacle à celui-ci* (line 19). Possibly 481 stands for *a* or *à* and the plaintext is not aligned with the ciphertext. Some mystery. In any case, these repetitions would not be enough for solving this cipher.

However, in reality there would be a stream of such missives and a *black chamber* of some other power might intercept them and eventually have enough material with repetitions to start a cryptanalysis.

D.5. Commercial codebooks

The introduction of the telegraph and its rate structure made it desirable to shorten message. Commercial codebooks catered to this need. Words and whole phrases are replaced by short codewords, regulated by the International Telecommunications Union in 1932 to be at most five letters long. (For the younger reader: once upon a time there was neither email nor SMS, Chat, or Twitter, and people had to rely on primitive forerunners called *telegram* and *telex*.)

The first telegraphic codebook was published in 1845, just one year after the start of commercial telegraphic operations. These codes reduce the cost and safeguard against accidental reading, but provide no security. As an example, Lieber's 1896 *Standard Telegraphic Code* presents on its 800 pages about 75 000 entries. Each entry associates both a 5-digit number and a (phantasy) word of at most ten letters, beginning with a letter from A to F, to a phrase. The words and the phrases are sorted alphabetically, the latter by keywords. In this code, the message

(D.1)   27556   03529   09715   00029   24695   04305   22454   28909

of eight words is synonymous with

> Babishly, Acerquen, Aggiunsero, Aalkasten, Atortolar, Acontiadae, Arrozzendo, Barbarizo.

Both encode the less than cheerful (fictitious) message of 70 words:

> A great panic prevails here, caused by the news [that] | there has been a very heavy bank failure here to-day which will seriously affect our market. | Present acting officers of this corporation | have absconded; [we] are on their track, utmost secrecy necessary. | Money market is in a panic. | Bonds are depressed on rumors that they will default on the interest. | [We] have suffered heavy losses. | Send immediately for best physician.

The words in brackets have been added, and the vertical strokes separate phrases.

These codebooks serve no cryptographic purpose, being publicly available. A certain level of secrecy can be gained through superencipherment, by choosing a secret key and using it in a (carryless) key addition scheme. This was proposed (in a slightly different context) by Alberti in 1467 and by the German cipher bureau during the First World War (Section I.1). With their key 718, the message (D.1) would be superenciphered as

$$98327 \quad 80606 \quad 17423 \quad 71890 \quad 01772 \quad 12013 \quad 93225 \quad 05086.$$

D.6. Unicity distance for codebooks

This section adapts Shannon's unicity distance (Definition A.13) to codebooks. The results do not aid in solving codes, but allow to assess whether a claimed solution is valid.

In our usual language, the key for a codebook is just the codebook itself. We start with the codebook designer's point of view and model codebooks in the language of Section A.5. Thus we have an alphabetically ordered list A of s plaintext terms, a (possibly ordered) list B of s codewords, and a set K of possible codebooks. For each codebook $k \in K$ and plaintext term $a \in A$, $\text{enc}_k(a) \in B$ is the codeword encoding a in k. According to Kerckhoffs' Principle (A.1), A should be assumed to be known to a cryptanalyst, who also sees in samples of ciphertexts parts of B and makes a guess at the method, embodied by K. Thus A, B and K are known to him.

We distinguish three types of codebooks. In a *one-part codebook*, both A and B are ordered and encryption preserves the order. Thus there is no secret. In reality, A may not be completely known to an adversary. We discuss this below. As examples, we can take the German naval codebook from 1913 in Figure I.1. B consists of either 5-digit numbers or 3-letter codewords. Both encodings have a natural order. In Figure D.4, the "natural order" of the codes is s, r, q, p, n, m, l, j, h, g, f, d, that is, the reverse of the alphabetic order. Then the syllables are completed by appending -um, -om, -im, -em, and -am.

This construction provides a great help to the cryptanalyst. In Figure D.3, if catolico=48 and christiandad=52 are already known, then the code for cavallo[8] must be 49, 50 or 51. On the other hand,

---

[8] Catholic, christianity, horse

type	abbr	$I_0(K)$
one-part	1	0
mixed	1.5 (b)	$\log_2(b!)$
two-part	2	$\log_2(s!)$

Table D.8: Key entropy $I_0(K)$ for codebooks with known plaintext list.

if he encounters the unknown ciphertext 50, then its plaintext is guaranteed to lie between católico and christiandad in any (contemporary) dictionary. The advantage for the legitimate user is that a single list permits an *alphabetic search* both for encryption and decryption.

In a *mixed codebook* with b blocks, A is split into b blocks of about s/b terms each, each block is mapped to contiguous block of terms in B, but the order of these (non-overlapping) blocks is random. Thus there are $b!$ codebooks. The German diplomatic codebook 13040 (Figures I.6 and I.5), in which the Zimmermann telegram was sent in 1917, was of this type. It consists of pages of alphabetically ordered words, numbered from 0 to 99, but where the pages themselves are randomly shuffled.

In a *two-part codebook*, the order is random and there are $s!$ such codebooks. If A is a list of persons' names and not alphabetically arranged, then the codebook is of this type. This provides more security, because now encryptions cannot be inferred from neighboring words, but has the disadvantage of requiring two separate lists for easy encryption and decryption. Using fantasy symbols like the funny characters in the Porta table (Figure C.7) without a natural order falls into this category.

The entropy $I_0(K)$ of the set of all codebooks is given in Table D.8. Its second column introduces abbreviations. We make the (usually unrealistic) assumption that A and B are known to the cryptanalyst.

For the unicity distance, we need the *word entropy* $H(p)$ of the probability distribution p on words. This is even harder to determine than letter entropy. Based on von zur Gathen & Loebenberger (2015), we use

$$\text{(D.2)} \qquad\qquad H(p) = 11$$

as our operational value for English, and due to the lack of data also for other languages. For short codebooks of $s < 1000$ terms, we use $H(p) = 2.7$, somewhat more than the English letter entropy 1.7 from (A.15).

Example A.2 (vi) explains homophonic substitutions, where some plaintext terms have several possible encryptions. How does this affect the language entropy $H(p)$, where p_x is the probability of $x \in X$

	type	$s, (t)$	$I_0(K)$	$H(p)$	d
Carlos II, Figure D.1	2	124	7	2.7	1
Tranchedino, Figure D.2	2, h	139, 165	7.1	2.7 + 1	1
Moreo, Figure D.3	1.5 (6)	423	55.4	2.7	5
Signalbuch	1	10 000	0	11	0
13040	1.5 (110), h	11 000	6.6	11 + 1	1
0075	2, h	11 000	13.3	11 + 1	1

Table D.9: Fictitious unicity distance (D.4) for some codebooks assuming a known plaintext list.

(Definition A.9)? In a simple case, we have m homophones for each term $x \in X$. This is equivalent to having m copies of x, each encrypted with its own symbol. If the homophonic choices are applied randomly (as they should), then each new letter has probability p_x/m. The new entropy is

$$\sum_{x \in X} m \cdot \left(\frac{p_x}{m} \cdot \log_2(\frac{p_x}{m})^{-1} \right) = H(p) + \log_2 m.$$

Homophony flattens the distribution and thus encreases the entropy. Reliable values for general homophony are hard to come by, and we take

(D.3) $\log_2 2 = 1$

as our operational value for the entropy increase. The "h" in the "type" column of Table D.9 indicates homophony, and (6) in the "Moreo" row indicates six blocks; the values of $I_0(K)$ are from Table D.8. Just using these considerations, Table D.9 presents the unicity distance

(D.4) $d = \lceil I_0(K)/(\log_2 s - H(p)) \rceil$

from Definition A.13 for some of the codebooks described. The ridiculously small distance values indicate that this approach misses an essential point.

According to Kerckhoffs' Principle (A.1), we should assume that the plaintext list A of a codebook is known to the cryptanalyst. In reality, he supposes that the ciphertext comes from a codebook with a list A of relevant plaintext terms (knowing the language and the style, say military, naval, diplomatic, or so), but only knows a slightly larger list A' from which most terms of A are supposed to come. We put ourselves into his shoes, trying to convince some higher authority (the King in Viète's case of Section D.3, the diplomat Bell for the Zimmermann telegram of Section I.3) that he has deciphered messages and is beyond the unicity bound.

Thus we have $s' = \#A' \geq \#A = s$ and some (small) fraction $\varepsilon s = \#(A \setminus A')$ of plaintext terms is not in his guessed superset A', with $0 \leq \varepsilon \leq 1$. One (not infrequent) case is when previous codebooks with the same plaintext list A have already been deciphered; then $s' = s$ and $\varepsilon = 0$. Another extreme case is when not much is known about A, except a general vocabulary A'' with $\#A'' = s''$ terms from which it comes. Thus from the cryptanalyst's point of view, the codebook designer may choose $(1 - \varepsilon)s$ of his plaintext terms from A', and εs from A''. This amounts to

$$(D.5) \qquad\qquad u = \binom{s'}{s - \varepsilon s} \binom{s''}{\varepsilon s}$$

many choices. The choice of A now becomes part of the secret key and adds $\log_2(u)$ to its entropy $I_0(K)$ from Table D.8, so that $I(K) = I_0(K) + \log_2 u$, and the unicity distance d becomes

$$(D.6) \qquad\qquad d = \left\lceil \frac{I(K)}{\log_2 s - H(p)} \right\rceil.$$

Among the few historical data points available to calculate s', s'', and ε, we take some of the 297 codebooks of Tranchedino from Section D.2. Figure D.2 shows the codebook, say T_0, on leaf 37^v of Tranchedino (1970). We now compare it to three other codebooks, say, T_1, T_2, and T_3, all from the same collection. The following six categories of terms form their stock of cleartexts, with their usual numbers in parentheses:

C_1 :   letters and standard abbreviations (24 to 25),
C_2 :   short words (10 to 12),
C_3 :   dummies (10 to 12),
C_4 :   digrams of the form vowel + consonant (60 to 70),
C_5 :   doubled consonants (12),
C_6 :   words and names of people and places (30 to 60).

A codebook need not contain all categories; for example, T_0 does not include C_2. When they do appear, categories C_1 through C_5 are fairly uniform, while C_6 is adapted to the mission for which the codebook is intended and varies considerably. The letters in C_1 have two homophones, and the more frequent ones three. Besides the many dummies, all other plaintext terms have one encryption. T_1 differs from T_0 in category C_6 by 35 of 50 plaintext terms. Augmenting T_0 by C_2 to obtain a fictitious

T_0' with $\#A' = 130 + 11 = 141$ plaintext terms, $\varepsilon = 35/110 \approx 31.82\%$ of all plaintexts in T_1 are not in T_0'. We assume that the missing plaintexts are taken from a list of 1000 words and names of leading dignitaries. The number t of ciphertext symbols in T_0' and T_1 is $165 + 11 = 176$ and 140, respectively. These are phantasy symbols with little or no visible order, and we assume $\log_2(t!)$ for their entropy $I_0(K)$.

Taking $H(p) = 2.7$ for the word entropy, adding 1 for homophony (D.3) and with d as in (D.6), we arrive at the first line in Table D.10. The two lines below it refer to the other two codebooks T_2 and T_3 from the same source and also compared to T_0'.

	s	εs	s'	s''	$\log u$	t	$I_0(K)$	$H(p)$	d
T_1	110	35	141	1000	351.7	140	7.1	$2.7 + 1$	117
T_2	163	28	141	1000	213.7	192	7.6	$2.7 + 1$	61
T_3	149	20	141	1000	194.1	177	7.5	$2.7 + 1$	58
S_1	139	6	423	1000	425.6	139	9.5	2.7	99
S_2	175	31	423	1000	582.3	175	9.5	2.7	125
S_3	423	1	423	1000	18.7	423	9.5	2.7	5

Table D.10: Unicity distance d for some historical codebooks.

We recall what the first line means. The cryptanalyst has established a successful break of codebook T_0' containing 141 plaintext and 176 ciphertext terms. Then he claims an analysis of texts in codebook T_1, finding 110 plaintext and 140 ciphertext items. All but 35 plaintexts are already in T_0', and the remaining ones in a large list of 1000 items. The value $\log u$ represents the uncertainty in determining the plaintext list. Then deciphering texts with at least 38 codewords will convince his boss that the work is correct—assuming that all our assumptions hold. Thus 38 can only be taken as a ballpark figure and slightly more than that will be more convincing. Note the large values of $\log u$ compared to $I_0(K)$, as used in Table D.9.

As a second example, we take for the lower three lines in Table D.10 four of the 66 Spanish codebooks in Devos (1950) mentioned in Section D.3. We pick the one with 423 terms in Figure D.3 as a base, call it S_0, and compare the three others to it. The ciphertexts are made up from a few templates. In Figure D.3, we see underlined one- and two-digit numbers, and in Figure D.4, the pattern "consonant + vowel + m", starting with som, sim, sem, sam, rum, ... In other codebooks, we find numbers with one or two dots above them, and sequences ba, be, bi, bo, bu, ca, ... and bal, bel, bil, bol, bul, cal, ... This makes for six such sequences, and we have a mixed codebook with six blocks and key entropy $I_0(K) = \log_2(6!) \approx 9.5$. This value is larger than the $\log s$ that we would have for a two-part codebook, reflecting the fact that the ciphertext words do not come from s choices, but many more. The relation is reversed as soon as the codebook has more than $6! = 720$ terms.

The first codebook, say S_1 and used 1589–1592, contains $s = 139$ codewords, 6 of which are not in S_0. Thus $\varepsilon s = 6$, $s' = 423$, and we assume that the six missing terms come from a known stock of $s'' = 1000$ words. This value is also assumed for the next two codebooks.

What does this mean? First of all: this theory does not give advice on how to decipher, but a tool for justifying a claimed decipherment. Suppose an able cryptanalyst like Viète has completely and convincingly broken the codebook S_0. Now he wants to show off to his boss, King Henri IV, with his new break of S_1. To this end, he has to provide the data yielding ε and s', justify the stock of 1 000 words containing all plaintexts that are expected in his context, and show consistent decipherments of at least 49 encrypted codewords. Bingo!

Of course, this is highly ahistorical, since this theory was out of reach at the time. But it does apply to modern historians and cryptanalysts breaking historical codebooks.

Table D.11 amends Table D.9 to take into account the fact that the plaintext list is only partially known, with uncertainty $\log u$. We take $s' = 1.1 \cdot s$ and $s'' = 2 \cdot s$. For the first three "small" codebooks we assume $\varepsilon = 30\%$, and for the last three "large" ones $\varepsilon = 10\%$. Thus the codebreaker has a list A' of words, 10% larger than the actual list A, from which all but 30% (or 10%) of the actual plaintext terms in A are chosen. The remaining ones are chosen from a stock A'' of plaintext terms which is twice as large as A.

	$s, (t)$	$\log u$	$I_0(K)$	d
Carlos II	124	272.6	7	29
Tranchedino	139, (165)	306.3	7.1	29
Moreo	423	946.5	9.5	88
Signalbuch	10 000	13 239.4	0	524
13040	11 000	14 564.5	6.6	574
0075	11 000	14 564.5	13.3	574

Table D.11: Unicity distances (D.6) for some codebooks with partially known plaintext list.

Corresponding rows of Tables D.9 and D.11 show the importance of the uncertainty $\log u$ about the cleartext. Still, there is much imprecision

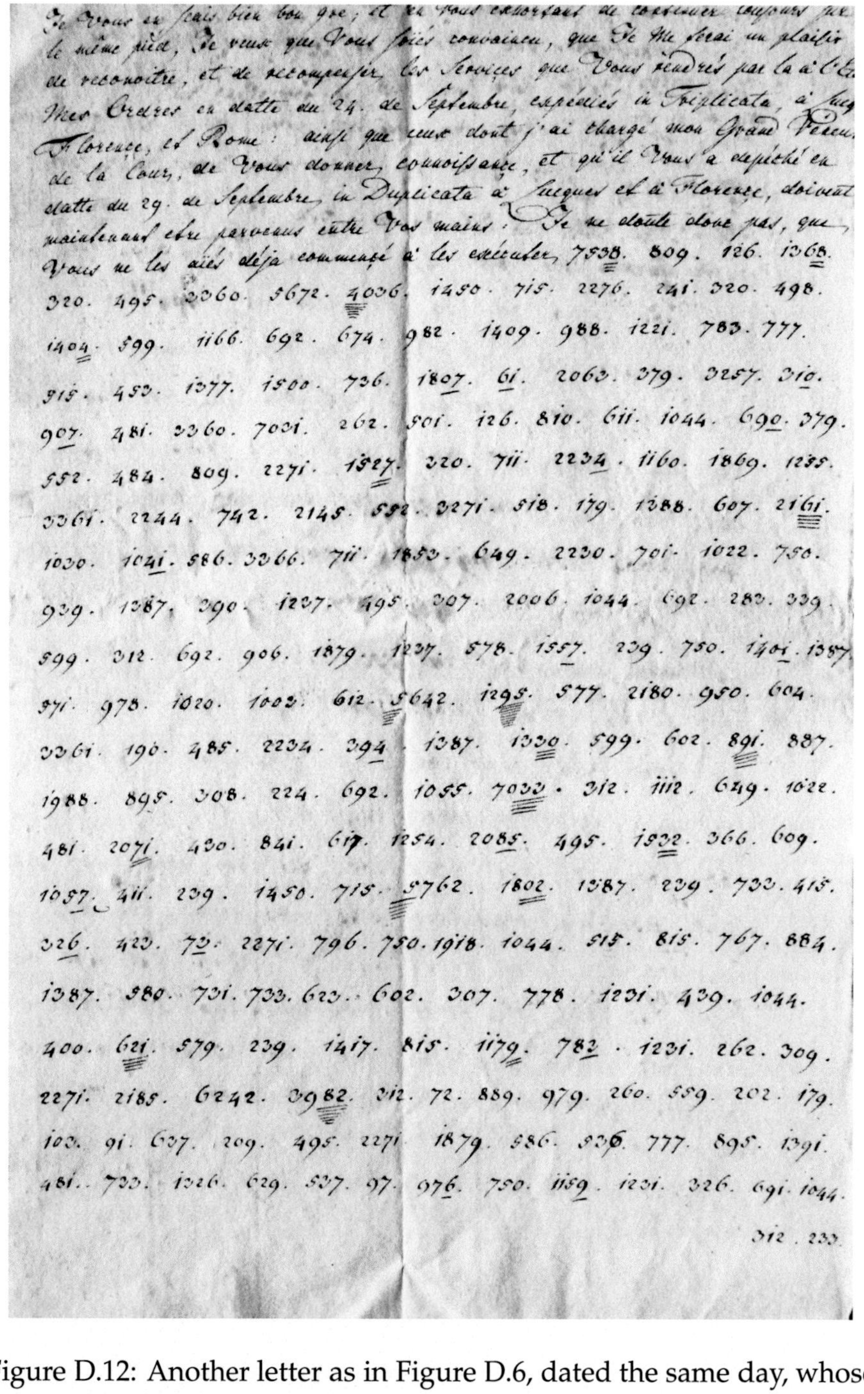
Je Vous en sçais bien bon gré; et en Vous exhortant de continuer toujours sur
le même pied, Je veux que Vous soiés convaincu, que Je me ferai un plaisir
de reconoitre, et de recompenser les Services que Vous rendrés par la [illegible]
Mes Ordres en datte du 24. de Septembre, expediés in Triplicata, à Lucq
Florence, et Rome: ainsi que ceux dont j'ai chargé mon Grand Veneur
de la Cour, de Vous donner connoissance, et qu'il Vous a depeché en
datte du 29. de Septembre in Duplicata à Lucques et à Florence, doivent
maintenant etre parvenus entre Vos mains: Je ne doute donc pas, que
Vous ne les aiés déja commencé à les executer 7538. 809. 126. 1368.
320. 495. 2360. 5672. 4036. 1450. 715. 2276. 241. 320. 498.
1404. 599. 1166. 692. 674. 982. 1409. 988. 1221. 783. 777.
915. 453. 1377. 1500. 736. 1807. 61. 2063. 379. 3257. 310.
907. 481. 2360. 7031. 262. 501. 126. 810. 611. 1044. 690. 379.
552. 484. 809. 2271. 1527. 320. 711. 2234. 1160. 1869. 1255.
3361. 2244. 742. 2145. 552. 3271. 518. 179. 1388. 607. 2161.
1030. 1041. 586. 3366. 711. 1853. 649. 2230. 701. 1022. 750.
939. 1387. 390. 1237. 495. 307. 2006. 1044. 692. 283. 339.
599. 312. 692. 906. 1879. 1237. 578. 1557. 239. 750. 1401. 1387.
971. 978. 1020. 1003. 612. 5642. 1295. 577. 2180. 950. 604.
3361. 196. 485. 2234. 394. 1387. 1330. 599. 602. 891. 887.
1988. 895. 308. 224. 692. 1055. 7033. 312. 1112. 649. 1022.
481. 2071. 430. 841. 617. 1254. 2085. 495. 1532. 366. 609.
1057. 411. 239. 1450. 715. 5762. 1802. 1387. 239. 733. 415.
326. 423. 73. 2271. 796. 750. 1918. 1044. 515. 815. 767. 884.
1387. 580. 731. 733. 623. 602. 307. 778. 1231. 439. 1044.
400. 621. 579. 239. 1417. 815. 1179. 782. 1231. 262. 309.
2271. 2185. 6242. 3982. 312. 72. 889. 979. 260. 559. 202. 179.
103. 91. 637. 209. 495. 2271. 1879. 586. 536. 777. 895. 1391.
481. 733. 1326. 629. 537. 97. 976. 750. 1159. 1231. 326. 691. 1044.
312. 233.

Figure D.12: Another letter as in Figure D.6, dated the same day, whose ciphertext is as yet undeciphered.

inherent in the guesswork involved. However, a successful cryptanalyst will have precise data to convince his audience.

Notes D.1. Bernay is in the Département de l'Eure, about 50 km south-east of Le Havre, a port on the French Channel Coast. Cochon's life dates are not known. He was *prêtre notaire en cour ecclésiastique* (notary at a Church court). The codebook and sample letter are in Secousse (1755), pages 414–417, Cochon (1870), pages 148–157, and Moranvillé (1891), pages 348–357; the latter work, written in Latin, calls the codebook a *kalendarium*.

D.2. See Perret (1890) for a discussion of Simonetta's work. Tranchedino's records are conserved as *Codex Vindobonensis 2398* at the Österreichische Nationalbibliothek in Vienna. They had purchased it from the Fugger family in 1656. A facsimile was published in 1970. The codebook in Figure D.2 is on leaf 37^v. The 283-symbol codebook is on leaf 88^v of Tranchedino (1970).

D.3. Taxis' codebook, part of which is shown in Figure D.4, is in the Spanish *Archivo General* at Simancas, code E 2846. The typeset version of Moreo's cipher in Figure D.3 is from Devos (1950), pages 328–334. Devos writes that it was presumably in use from 1589 to about 1597, and that he did not find this cipher at Simancas; possibly this means that they are from the *Archives Générales du Royaume* at Bruxelles (see his page 11). He also shows Taxis' codebook. His version agrees verbatim with our copy from Simancas (Figure D.4), except that the last lines about dummies etc. are slightly different. We may suppose that two almost but not quite identical copies of the codebook are in the archives.

The use of letters or names for unknowns in algebraic equations dates from before Viète. His contribution was to introduce symbolic notation also for known quantities, so that one can say "a is a root of $x - a$" instead of repeating this for $a = 1, 2, 3, \ldots$. We use elsewhere his formula for the coefficients of a polynomial in terms of its roots; see (15.40).

Spanish ciphers of the time were usually valid for four years, but Moreo's was for 1589 to 1598; see Devos (1946). The story of Moreo and Viète is also told in Kahn (1967), pages 117–118. Henri IV. wrote his gracious words about the Duke of Mayenne in the 1596 *Edit de Folembray*. The description of the Spanish is the last sentence of Viète's *Vita*, on the first pages of his *Opera Mathematica* (1646). The story of Viète's cryptographic achievements is told at length in a contemporary account of Pierre de Segusson Longlée (c. 1540–c. 1598), published as Longlée (1912).

D.4. The letter in Figure D.6 measures 32.9 cm × 20 cm. Only the top half is shown, the bottom half has been left empty. As was customary at the time, the number 1 has a dot above it. These dots and those after the ciphertext numbers are not shown in the transcription. Some cipher numbers are underlined, with 2, or 3, or 5 lines. These are also not shown. The association of the interline plaintext to the ciphertext above it is not always clear and this gives rise to empty fields in Table D.7. The spelling is reproduced literally.

D.5. Francis O. J. Smith, the lawyer of Samuel Morse, inventor of the Morse code, published *The Secret Corresponding Vocabulary; Adapted for the Use to Morse's Electro-Magnetic Telegraph* in 1845.

D.6. Writing $s' = c's$ and $s'' = c''s$, the contribution $\log_2 u$ from (D.5) to $I(K)$ is

$$\log_2 u \approx s \cdot \log_2 \left(\frac{(c')^{c'} \, (c'')^{c''}}{(1 - \varepsilon)^{1-\varepsilon} \, (c' - 1 + \varepsilon)^{c'-1+\varepsilon} \, (c'' - \varepsilon)^{c''-\varepsilon} \, \varepsilon^{\varepsilon}} \right)$$

by Stirling's formula (3.57), where the second factor does not depend on s.

The first two entries 130 and 13042 in Figure I.4 are actually not encryptions, but indicate the code used. Figure I.5 shows one page of the British solution of 13040, with 35 out of 240 encryptions deciphered. The total number of deciphered words is roughly 1800, meaning that messages with well above 574 ciphertext terms have been deciphered. Showing all deciphered messages would have convinced a skeptic who knows this theory, but de Grey's deciphering demonstration (Section I.4) would not.

In the Spanish codebooks, we do not take the two possibilities of straight and reverse order for the ciphertext sequences into account.

The three codebooks from Tranchedino (1970) are: T_1 on page 36^v, 9 June 1465, Cum Azone Vicecomite Locumtenente Ducatus Bari; T_2 on page 38^r, 18 October 1463, with P^{mo} D Sr de Nardinis ..., and T_3 is on page 38^v, 19 November 1468, Cum Nicolosio de Spedio. Besides the Spanish codebook S_0 from Figure D.3, the following ones from Devos (1950) are used: S_1 is codebook 34 on pages 304–309, active 1589–1592; S_2 is codebook 33 on pages 300–308, used with Juan Baptista de Taxis between 1580 and 1590; S_3 is codebook 37, on pages 319–327, also with Taxis in 1590–1592, item Estado 2846 in the Spanish Archives at Simancas.

Exercises.

EXERCISE D.1 (Lucchesini).

(i) *The ciphertext and plaintext in Figure D.6 do not seem to match. Can you explain or rectify this?*

(ii) *Figure D.12 shows a letter from King Frederick William II of Prussia to Lucchesini, sent the same day as that in Figure D.6. Possibly they both use the same codebook. A transcription of the text at the top is available on the book's webpage. The letter has not been deciphered. Can you do this?*

EXERCISE D.2 (Unicity distance for 13040 and 0075). *As reported in Section I.4, the British cryptographer de Grey explained to a US diplomat his break of a 153-word message in the German mixed codebook 13040 of $s = 11\,000$ terms.*

(i) *Assume that the list of plaintexts in 13040 is known and broken into 110 alphabetically sorted blocks of 100 terms each, and that the blocks are randomly arranged. Show detailed calculations for $I(K)$ and d.*

(ii) *Assume that the new two-part codebook 0075 in this affair also has 11 000 plaintext terms, all but 5% of which are contained in the broken 13040, and the rest in a list of 20 000 terms. Show detailed calculations for $I(K)$ and d.*

(iii) *Discuss your findings.*

EXERCISE D.3 (Unicity distance parameters). *Consider a large one-part codebook with $s = 10\,000$ terms, $I_0(K) = 0$, $H(p) = 2.7$, $s'' = 20\,000$, and s' one of 11 000, 12 000, and 15 000. Plot the unicity distance for $0 \leq \varepsilon \leq 0.2$. Stirling's simplification in the Notes D.6 may be helpful.*

EXERCISE D.4 (Unicity).   *You are to investigate the quantities $v' \approx (\log_2 u)/s$ and d as in Section D.6.*

  (i) *Based on the examples in the text, describe some scenarios of codebook design and reasonable ranges for s, c', c'', ε for them; see Notes D.6. Plot values of v and d in such ranges.*

  (ii) *Let $0 < \alpha < 1$, s be an integer, and $b = \binom{s}{\alpha s}$. (The denominator is rounded to the nearest integer.) Using Stirling's formula (3.57), derive approximations for large s:*

$$b \approx \frac{(1-\alpha)^{-(1-\alpha)s}}{\sqrt{2\pi(\alpha - \alpha^2)s}\, \alpha^{\alpha s}},$$

$$\frac{\log_2 b}{s} \approx -(1-\alpha)\log_2(1-\alpha) - \alpha\log_2\alpha \approx \alpha\log_2((\alpha^{-1}-1)e).$$

  (iii) *Derive an approximation to v' that depends on c', c'', and ε, but not on s. Then approximate d. In some ranges, plot the quotient of the true values and their approximations. Are the latter helpful?*

> A signature always reveals a man's character—
> and sometimes even his name.
> ANONYMOUS

> Certains déchiffreurs exercés sont capables de faire des prodiges
> dans ce genre de recherche en opérant, ou par un calcul de
> probabilités, ou par un travail de tâtonnements. Rien qu'en se
> basant sur les lettres que leur emploi plus fréquent fait répéter un
> plus grand nombre de fois dans le cryptogramme, — e dans les
> langues française, anglaise et allemande, o en espagnol, a en
> russe, e et i en italien, — ils parviennent à restituer aux lettres
> du texte cryptographié la signification qu'elles ont dans le texte
> clair. Aussi est-il peu de dépêches, établies d'après ces méthodes,
> qui puissent résister à leurs sagaces déductions.[9]
> JULES VERNE (1885)

> There was no letter frequency ratio,
> nothing a cryptanalyst could hook onto in the way
> of a regular substitution or transposition cipher.
> TOM CLANCY AND STEVE PIECZENIK (1995)

> We have heard much about the poetry of mathematics,
> but very little of it has as yet been sung.
> The ancients had a juster notion of their poetic value than we.
> The most distinct and beautiful statements of any truth
> must take at last the mathematical form.
> We might so simplify the rules
> of moral philosophy, as well as of arithmetic,
> that one formula would express them both.
> HENRY DAVID THOREAU (1849)

---

[9]Certain skilled cryptanalysts are able to perform wonders in this type of research, by operating either by a probabilistic approach or by trial and error. Basing themselves on the letters whose more frequent usage makes them appear more often in the plaintext — e in French, English, and German, o in Spanish, a in Russian, e and i in Italian — they succeed in giving back to the cyphertext letters the meaning they have in the plaintext. Thus there are few messages, composed according to these methods, that can resist their clever deductions.

Chapter 8

Signatures

For electronic business transactions over the internet, it is important to have a system for legally binding digital contracts. In a traditional paper-bound agreement between two parties, both get a copy of the agreement signed by both parties. If a dispute arises later, there is a finely tuned legal system to deal with this. Courts will hold both signatories to the terms of the agreement. At least, that is the principle; sharp lawyers may argue this way or the other.

An electronic version of this procedure has to enjoy similar properties: the signature is tightly bound to the document, the identity of the signer is clear, only that person (or other entity) should be able to sign in his own name, he should not be able to deny his signature, and a signed agreement should hold up in court. That is, a third party should be able to check whether a claimed signature is valid.

Electronic documents are easy to edit, copy, and manipulate in many ways. Signed documents should resist such manipulations. For example, a scanned hand-written signature does not work, since it might be cut and pasted into other documents.

This chapter discusses the basic task and two signature schemes: by ElGamal and by Schnorr; the latter is used in the US *Digital Signature Algorithm* (DSA). Chapter 9 presents security notions for signatures and an example of a security reduction.

8.1. Digital signatures

The following are important requirements for digital signatures.

○ The signature must be tightly attached to the signed document.

○ It should be easy to *sign* for the legitimate signer, easy to *verify* the signature for the recipient, and hard to *forge* a signature.

© Springer-Verlag Berlin Heidelberg 2015
J. von zur Gathen, *CryptoSchool*, DOI 10.1007/978-3-662-48425-8_12

o The signer should not be able to deny that he signed the document.

o Sometimes it is important that a signed document can only be used once for its legitimate purpose, not several times (say, a cheque).

A signature scheme consists of four functions. The first one, called *set-up*, establishes the basic parameters of the scheme, driven by a security parameter n. The second one, called *key generation*, produces a pair $(\mathsf{sk}, \mathsf{pk})$ consisting of a secret key sk and a public key pk. The third one is the *signing function* $\mathsf{sig}_{\mathsf{sk}}$ and produces, on input a message m, its signature $\mathsf{sig}_{\mathsf{sk}}(m)$. The signing step may produce additional secret and public session keys. The fourth one is the Boolean *verifying function* $\mathsf{ver}_{\mathsf{pk}}$. It is given a message m and an arbitrary string z, and outputs

$$\mathsf{ver}_{\mathsf{pk}}(m, z) = \begin{cases} \text{``true''} & \text{if } z \text{ is } \mathsf{sig}_{\mathsf{sk}}(m), \\ \text{``false''} & \text{otherwise.} \end{cases}$$

Thus sk is the signer's secret key, and pk is the public key for verification. The verifier has no access to sk or $\mathsf{sig}_{\mathsf{sk}}(m)$ (unless this equals z) and has to perform the verification using only pk. A message m may have several valid signatures $\mathsf{sig}_{\mathsf{sk}}(m)$. We can use any public key encryption scheme (set-up, keygen, enc, dec) (Definition 2.25) to obtain a signature scheme, by simply reversing encryption and decryption.

BOB m, x ALICE
$\bullet$ $\xrightarrow{\hspace{6cm}}$ $\bullet$
key sk key pk

$x = \mathsf{dec}_{\mathsf{sk}}(m)$ $y = \mathsf{enc}_{\mathsf{pk}}(x)$; verify: $m \stackrel{?}{=} y$

$$\mathsf{ver}_{\mathsf{pk}}(m, z) = \text{``true''} \iff m = \mathsf{enc}_{\mathsf{pk}}(z).$$

We may use a system like RSA (Section 2.5) or ElGamal (Section 4.2). Then BOB's sk is his secret key, used for decryption in the communication mode, and ALICE verifies BOB's signature using his public key pk. The system is based on a counter-intuitive way of using encryption "the wrong way". The plaintext message m is decrypted, although it is (usually) not the encryption of any message, and then this decryption is encrypted to reveal the original $m = \mathsf{enc}_{\mathsf{pk}}(\mathsf{dec}_{\mathsf{sk}}(m))$.

Digital signatures are at least as long as the document signed; in ElGamal's scheme, the signature is twice as long. This makes it necessary to reduce long documents to a short string (the "message digest") whose signing is as secure as signing the whole thing. This is achieved by the cryptographic hash functions discussed in Chapter 7.

8.2. ElGamal signatures

This section presents a digital signature scheme proposed by ElGamal (1985b). A drawback is that there are several forgeries known for it; see Section 8.3 and Exercise 8.9. These attacks are not serious in a practical sense, but they rule out certain security reductions, as discussed in Chapter 9. The scheme is described in the framework of group cryptography from Chapter 4 and related to ElGamal encryption (Section 4.2).

PROTOCOL 8.1. ElGamal signature scheme.

Set-up.

Input: a security parameter n given in unary.
Output: as below.

1. A cyclic group $G = \langle g \rangle$ with $d = \#G$ elements, where d is an n-bit number. We also have an injective encoding function $G \to \mathbb{Z}_d$, denoted as $x \mapsto x^*$, which is easy to compute but otherwise has no particular properties. All these data are published.

Key generation.

Output: secret and public keys.

2. Secret key $\mathsf{sk} = a \xleftarrow{\square} \mathbb{Z}_d$ and public key $\mathsf{pk} = A \longleftarrow g^a \in G$.

Signature generation.

Input: a message $m \in \mathbb{Z}_d$.
Output: a signature $\mathsf{sig}_{\mathsf{sk}}(m) \in G \times \mathbb{Z}_d$ of m.

3. Choose a secret session key $k \xleftarrow{\square} \mathbb{Z}_d^\times$.
4. $K \longleftarrow g^k \in G$.
5. $b \longleftarrow k^{-1} \cdot (m - aK^*)$ in $\mathbb{Z}_d$, where k^{-1} is the inverse of k in $\mathbb{Z}_d$.
6. Transmit m and its signature $\mathsf{sig}_{\mathsf{sk}}(m) = (K, b)$.

Verification.

Input: message m and a pair $(z, c) \in G \times \mathbb{Z}_d$.
Output: $\mathsf{ver}_{\mathsf{pk}}(m, z, c)$, which is either "true" or "false".

7. Compute $u \longleftarrow g^m$ and $v \longleftarrow A^{z^*} z^c$ in G.
8. If $z \neq 1$ and $u = v$ then return "true" else return "false".

$$A = g^a$$

BOB $m, (K, b)$ ALICE

• ———————————————————→ •

$$K = g^k$$ $$g^m \overset{?}{=} A^{K^*} K^b$$
$$b = k^{-1}(m - aK^*)$$

The quantity K plays a curious double role as an element of G (in K^b) and one of $\mathbb{Z}_d$ (in A^{K^*}). In ElGamal's original proposal, he took $G = \mathbb{Z}_p^\times$, and simply identified $G = \{1, \ldots, p-1\}$ and $\mathbb{Z}_d = \mathbb{Z}_{p-1} = \{0, \ldots, p-2\}$ using the "patched identity" mapping with $K^* = K$, except that $(p-1)^* = 0$. In general, this map $G \longrightarrow \mathbb{Z}_d$ must be easy to compute. In particular, it is not the discrete logarithm from the exp-log isomorphism in Figure 4.1, which is assumed to be hard.

EXAMPLE 8.2. BOB has set up the publicly known group $G = \mathbb{Z}_{17}^\times$, with order $d = 16$, $g = 3$, and the "identity" mapping $*$ as above. Now he chooses his secret key, say $\mathsf{sk} = a = 9$, and publishes $\mathsf{pk} = A = 3^9 = 14$ in $\mathbb{Z}_{17}^\times$. Suppose that he wants to sign $m = 11$ and chooses $k = 5$ as secret session key. Then indeed $\gcd(5, 17-1) = 1$, and $k^{-1} = -3 = 13$ in $\mathbb{Z}_{16}$. He calculates

$$K = 3^5 = 5 \text{ in } \mathbb{Z}_{17},$$

$$b = 13 \cdot (11 - 9 \cdot 5) = 6 \text{ in } \mathbb{Z}_{16},$$

since $K^* = K = 5$. BOB sends the message 11 together with its signature $\mathrm{sig}_9(11) = (5, 6)$. ALICE checks that $5 \neq 1$ and computes

$$u = g^m = 3^{11} = 7 \text{ in } \mathbb{Z}_{17},$$
$$v = A^{K^*} K^b = 14^5 \cdot 5^6 = 7 \text{ in } \mathbb{Z}_{17},$$

and accepts the message as properly signed.   ◇

We now check two of the usual requirements: correctness and efficiency. Security is discussed in the next section. According to the recommendation of Figure 4.6, we assume d to be prime.

THEOREM 8.3. *Let d be a prime number. Then the verification procedure works correctly as specified, and the signature scheme can be implemented efficiently.*

PROOF.   All operations in the scheme are exponentiations in G or arithmetic in $\mathbb{Z}_d$, so that efficiency is clear.   Now first assume that

$(z, c) = \text{sig}_{\text{sk}}(m)$ is a correct signature, so that $(z, c) = (K, b)$ for some session key k. Then $K \neq 1$ and

$$aK^* + kb = m \quad \text{in } \mathbb{Z}_d,$$
(8.4) $$A^{K^*} K^b = g^{aK^*}(g^k)^b = g^{aK^*+kb} = g^m \quad \text{in } G,$$

so that $\text{ver}_{\text{pk}}(m, z, c) = $ "true".

Now assume that $\text{ver}_{\text{pk}}(m, z, c) = $ "true", set $k = \text{dlog}_g z$, and let $\text{sig}_{\text{sk}}(m) = (K, b)$ be the correct signature produced in step 6 with secret session key k. Then $K = z$. Since $z \neq 1$, we have $k \neq 0$ and hence $k \in \mathbb{Z}_d^\times$, because d is prime. Using (8.4), we find

$$g^{aK^*+kb} = g^m = u = v = A^{z^*} z^c = A^{K^*} K^c = g^{aK^*+kc} \quad \text{in } G,$$
$$k(b - c) = 0 \quad \text{in } \mathbb{Z}_d,$$
$$b = c \quad \text{in } \mathbb{Z}_d.$$

Thus indeed (z, c) is a correct signature, produced with session key k. $\square$

If $z = 1$, then $z^* = 1$ in our example and $k = \text{dlog}_g z = 0$ in the proof above. Thus $u = v$ would hold for $m = a$ and any claimed signature (z, c), no matter whether $c = b$ or not. We do not expect to stumble into this scenario, since then $m = a = \text{dlog}_g A$ also solves a discrete logarithm problem. The situation for composite d is discussed in Exercises 8.4 and 8.9.

The random choice of k means that there are many valid signatures for a single message m. This is a desirable feature.

8.3. Forging ElGamal signatures

Suppose that EVE wants to forge a signature for a message m. She does not know a. What to do? BOB's secret keys consist of $\text{sk} = a = \text{dlog}_g A$ and $k = \text{dlog}_g K$. Clearly we have to assume that discrete logarithms in G are hard to compute. Now EVE might choose K and would have to find the corresponding $b = \text{dlog}_K(g^m A^{-K^*})$, but this is again a discrete logarithm problem in G.

EVE might choose K and b, and then would have to find the corresponding $m = \text{dlog}_g(A^{K^*} K^b)$, again a discrete logarithm problem. (This m would be determined by the choice of K and b.)

If EVE wants to forge a signature for a specific message m, she might set $A = g^m$ and try to find K and b with

$$A^{K^*} K^b = A.$$

This strange equation in two unknowns, where K appears as a group element and also as an exponent (called K^*), looks hard to solve. But there does not seem to be any theory applicable to it, so that we cannot classify it with respect to any of the usual problems on which we base cryptographic security.

It would be a protocol failure to reveal the secret session key k, since then a can be calculated as

$$a = (m - bk)(K^*)^{-1} \text{ in } \mathbb{Z}_d,$$

assuming that K^* is invertible in $\mathbb{Z}_d$. Another protocol failure is to use the same value of k in two different signatures. Suppose that (K, b_1) and (K, b_2) are valid signatures of two messages m_1 and m_2, respectively. Then

$$\begin{aligned} A^{K^*} &= g^{m_1} K^{-b_1} = g^{m_2} K^{-b_2}, \\ g^{m_1 - m_2} &= K^{b_1 - b_2} = g^{k(b_1 - b_2)}. \end{aligned}$$

Now g has order d, and the last equation is equivalent to

$$m_1 - m_2 = k(b_1 - b_2) \quad \text{in } \mathbb{Z}_d.$$

If $\gcd(b_1 - b_2, d) = 1$, then we have $k = (b_1 - b_2)^{-1}(m_1 - m_2)$ in $\mathbb{Z}_d$; see Remark 4.15 for a discussion of the gcd condition.

A further danger is the *woman-in-the-middle attack*: EVE can intercept a signed message and send it to ALICE. We will discuss identification schemes in Chapter 10 that prevent EVE from impersonating BOB.

However, there are clever ways to select all three parameters m, K, and b, and actually forge a signature. EVE chooses integers $i, j \in \mathbb{Z}_d$ with $j \in \mathbb{Z}_d^\times$. Then she sets

$$\begin{aligned} K &= g^i A^j, \\ b &= -K^* j^{-1} \text{ in } \mathbb{Z}_d, \\ m &= -K^* i j^{-1} \text{ in } \mathbb{Z}_d, \end{aligned}$$

where j^{-1} is the inverse of j in $\mathbb{Z}_d$. Then (K, b) is a valid signature for m:

$$\begin{aligned} A^{K^*} K^b &= A^{K^*} (g^i A^j)^{-K^* j^{-1}} \\ &= A^{K^*} A^{-K^* \cdot j \cdot j^{-1}} \cdot g^{-K^* i j^{-1}} = g^{-K^* i j^{-1}} = g^m. \end{aligned}$$

Another attack goes as follows. Here, EVE has intercepted a message m and its signature (K, b) and can now sign further messages. She picks integers h, i, and $j \in \mathbb{Z}_d$ with $hK^* - jb \in \mathbb{Z}_d^\times$. Then she computes

$$\begin{aligned} z &= K^h g^i A^j, \\ c &= bz^*(hK^* - jb)^{-1} \text{ in } \mathbb{Z}_d, \\ \ell &= z^*(mh + ib)(hK^* - jb)^{-1} \text{ in } \mathbb{Z}_d. \end{aligned}$$

Then (z, c) is a valid signature for ℓ. To verify this claim, we begin with

$$A^{K^*} K^b = g^m,$$

which holds by assumption. We raise to the power h and manipulate:

$$A^{hK^* - jb} K^{hb} g^{ib} A^{jb} = A^{K^* h} K^{bh} g^{ib} = (A^{K^*} K^b)^h g^{ib} = g^{mh + ib}.$$

We raise to the power $(hK^* - jb)^{-1}$ in $\mathbb{Z}_d$ and manipulate:

$$\begin{aligned} A \cdot z^{b(hK^* - jb)^{-1}} &= A \cdot (K^h g^i A^j)^{b(hK^* - jb)^{-1}} \\ &= (A^{hK^* - jb} K^{hb} g^{ib} A^{jb})^{(hK^* - jb)^{-1}} \\ &= g^{(mh + ib)(hK^* - jb)^{-1}}. \end{aligned}$$

Raising to the power z^*:

$$v = A^{z^*} z^c = A^{z^*} z^{bz^*(hK^* - jb)^{-1}} = (A \cdot z^{b(hK^* - jb)^{-1}})^{z^*}$$
$$= g^{z^*(mh + ib)(hK^* - jb)^{-1}} = g^\ell = u,$$

and ALICE will accept (z, c) as a valid signature of ℓ.

These two attacks allow EVE to sign messages. But she cannot choose the message to be signed, and the attacks are not considered to be an essential threat to the security of the system.

8.4. Schnorr signatures and the *Digital Signature Algorithm*

Smartcards, RFIDs, or NFCs often implement protocols in which signatures play an important role. It is therefore of great interest to reduce the computing effort and transmission requirements for such weak computing devices. Schnorr proposed an idea which allows much shorter signatures, in a system based on ElGamal's. His method is now implemented in the DSA, a US NIST standard.

From Chapter 4, we know two types of discrete logarithm computations: generic ones and special ones for special groups. Schnorr's idea is to take a fairly large special group and then a fairly small subgroup G of it. Then we have two attacks on G: a generic computation in G, or a special one in the larger group. The trick is to choose the parameters so that the costs for both attacks are roughly equal, and both beyond feasibility.

Schnorr (1991a) takes a large prime p and the large group $\mathbb{Z}_p^\times$; this is what ElGamal used. But now Schnorr chooses a fairly small prime divisor d of $p - 1 = \#\mathbb{Z}_p^\times$ with ℓ bits, and a subgroup $G \subseteq \mathbb{Z}_p^\times$ of order d. Then we have index calculus attacks on $\mathbb{Z}_p^\times$ (Section 4.8) or generic methods for G; nothing better is known. The latter take time $\Omega(\sqrt{d})$, according to Section 4.9.

Thus we now have two security parameters $\ell < n$, and the assumption is that discrete logarithms in $\mathbb{Z}_p^\times$ for an n-bit prime p are hard, and so are discrete logarithms in the ℓ-bit group G. Specific values chosen by Schnorr and in the DSA are $\ell = 160$ and $n = 1024$. These should be increased at least to $\ell = 250$ and $n = 3000$ for current security requirements, and we now work with these numbers.

So we choose an ℓ-bit prime d (Section 3.2), and an n-bit prime p which is congruent to 1 modulo d; that is, d divides $p - 1$, as required. A famous theorem by Dirichlet (1837) on primes in arithmetic progressions says that there are many such p. In practice, one finds a prime among random numbers congruent to 1 modulo d as fast as among general random numbers of the same size.

We now need an element $g \in \mathbb{Z}_p^\times = H$ of order d, so that $G = \langle g \rangle$ has order d. Following Theorem 15.60 (iv), we simply take a random element $h \in \mathbb{Z}_p^\times$ and $g = h^{(p-1)/d}$, unless the latter equals 1. Then the order of g divides d. But d is prime, so either the order is d (and g is a generator) or 1 (then $g = 1$ and we have to try another h).

If we now just used ElGamal's scheme, the signature (K, b) would have $n + \ell = 3000 + 250$ bits, but our goal is a length of only of $2 \cdot \ell = 2 \cdot 250$ bits. We remedy this by taking our function $K \longmapsto K^*$ now to map $\mathbb{Z}_p^\times$ to $\mathbb{Z}_d$, namely by identifying $\mathbb{Z}_p^\times$ with the set $\{1, \dots, p - 1\}$ of integers, and taking $K^* \in \mathbb{Z}_d$ to be the remainder of K on division by d. Again, this is just a funny mapping with no nice algebraic properties. It is unlikely to be injective on G, although $\#G = d = \#\mathbb{Z}_d$. Then we take (K^*, b) as the signature, of the desired length 2ℓ.

Putting this together, we have the following modification of the ElGamal signature scheme. Schnorr's original proposal and the DSA standard are slightly different and discussed in Exercise 8.8. One may think of $H = \mathbb{Z}_p^\times$ for an appropriate prime p.

PROTOCOL 8.5. Schnorr signature scheme (DSA).

Set-up.

Input: two security parameters $\ell < n$ in unary.

1. Choose an ℓ-bit prime d and a cyclic group H whose n-bit order is a multiple of d, a generator g of a group $G = \langle g \rangle \subseteq H$ of order d, and a map $*$ from H to $\mathbb{Z}_d$.

Key generation.

Output: secret and public keys.

2. Secret key $\mathsf{sk} = a \xleftarrow{\square} \mathbb{Z}_d$ and public key $\mathsf{pk} = A \longleftarrow g^a \in G$.

Signature generation.

Input: a message $m \in \mathbb{Z}_d$.
Output: a signature $\mathrm{sig}_{\mathsf{sk}}(m) \in \mathbb{Z}_d^2$ of m.

3. Choose a secret session key $k \xleftarrow{\square} \mathbb{Z}_d^{\times}$.
4. $K \longleftarrow g^k \in G$.
5. $b \longleftarrow k - a(K^* + m)$ in $\mathbb{Z}_d$.
6. Transmit m and its signature $\mathrm{sig}_{\mathsf{sk}}(m) = (K^*, b) \in \mathbb{Z}_d^2$.

Verification.

Input: message m and a pair $(z, c) \in \mathbb{Z}_d^2$.
Output: $\mathrm{ver}_{\mathsf{pk}}(m, z, c)$, which is either "true" or "false".

7. $u \longleftarrow g^c A^{z+m} \in G$.
8. If $u^* = z$ then return "true" else return "false".

THEOREM 8.6. *Any correct signature is verified as "true". The cost of steps 2 through 8 is polynomial in n and dominated by the cost of three exponentiations in G with ℓ-bit exponents. The signature length is 2ℓ bits.*

PROOF. We take sk and pk from step 2 and $\mathrm{sig}_{\mathsf{sk}}(m) = (K^*, b)$ from step 6 as input $(z, c) = (K^*, b)$ to verification. Then

$$u = g^b A^{K^* + m} = g^{k - a(K^* + m) + a(K^* + m)} = g^k = K,$$

so that $u^* = K^*$ and "true" is returned in step 8.

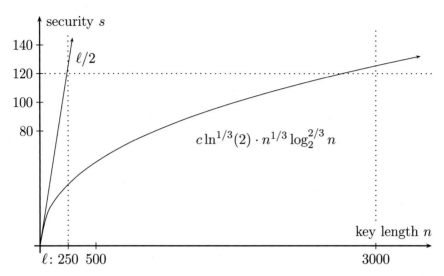

Figure 8.1: Tradeoff between ℓ-bit generic discrete logarithms and n-bit index calculus, with $c = 1.902$.

One exponentiation in step 4 and two in step 7 all have ℓ-bit exponents in $\mathbb{Z}_d$, and each of them takes about $1.5\,\ell$ operations in G, which are actually executed in H. This dominates the cost of arithmetic in $\mathbb{Z}_d$ in steps 5 and 7 and the multiplication in H in step 7.   □

In contrast to Theorem 8.3 for ElGamal signatures, it is not clear whether "true" in step 8 implies that (z, c) is indeed a signature of m. In fact, this is unlikely to be always correct. A successful forger is given the set-up and $\mathsf{pk} = A$ and faces the problem of generating a meaningful message $m \in \mathbb{Z}_d$ (thus excluding variations of the ElGamal examples from Section 8.3) and $(K^*, b) \in \mathbb{Z}_d^2$ so that $(g^b A^{K^*+m})^* = K^*$. This problem is poorly understood and no general solution is known. For $K_1, K_2 \in G$, $K_1^* = K_2^*$ holds if and only if $A^{K_1^*} = A^{K_2^*}$.

Figure 8.1 and Table 8.2 compare $\ell = 250$ and $n = 3000$ with ElGamal, also using a 3000-bit prime; see also Tables 5.11 and 5.12. The cost of arithmetic is explained in Section 15.6; we have left out the constant factor in $O(n^2)$ since it cancels out in the comparison. The gain is the much shorter signature of 500 bits and the faster signing and verification procedures, because only 250-bit exponents are required. It does not matter that the message is also shorter, since we will use hashing anyway and thus have short messages. Thus we have shorter signatures but apparently the same security as if we worked in $\mathbb{Z}_p^\times$ directly, since the only way known to attack a subgroup like G are either generic algorithms for G or index calculus for the whole of $H = \mathbb{Z}_p^\times$.

	ElGamal	Schnorr / DSA
cost A of arithmetic	$(3000)^2$	$(3000)^2$
cost of g^k	$An \approx A \cdot 3000$	$A\ell \approx A \cdot 250$
message length	3000 bits	250 bits
signature length	6000 bits	500 bits (!)
attack	DL in $\mathbb{Z}_p^\times$	DL in $\mathbb{Z}_p^\times$ or in G

Table 8.2: Signature lengths and costs for ElGamal and Schnorr's signatures.

The DSA was developed by the NSA, and the process by which NIST made it a standard happened behind closed doors and was heavily criticized in the US cryptography community. Furthermore, elliptic curves (Chapter 5) provide essentially the same parameters as G does in this section, but with cheaper arithmetic. The resulting Elliptic Curve Digital Signature Algorithm (ECDSA) is now also a standard.

8.5. The Gennaro-Halevi-Rabin signature (GHR) scheme

We describe the GHR signature scheme, due to Gennaro, Halevi & Rabin (1999), with which one can sign short messages, say of ℓ bits. The reader may think of $\ell = 30$; the computation time will be a multiple of 2^ℓ, so that this quantity should be polynomial in the key size n. We use this scheme in Section 15.7 to illustrate security notions for signatures.

We take the set P of the first 2^ℓ odd primes and enumerate them in binary, so that we have a bijection $h\colon \{0,1\}^\ell \to P$, which is easy to compute, and so is its inverse. We denote by h_m the prime $h(m)$ corresponding to $m \in \{0,1\}^\ell$, and by r the largest prime in P. If $\ell = 5$, then $h_{00000} = 3$, $h_{00001} = 5$, $\ldots$, and $h_{m+1} = \text{nextprime}(h_m)$, using the binary representation. By the Prime Number Theorem 3.21, $r = p_{2^\ell+1}$ is near $\ln 2 \cdot \ell \cdot 2^\ell$.

An RSA modulus $N = pq$ is called *r-safe* if neither $(p-1)/2$ nor $(q-1)/2$ have a prime factor up to r. Section 3.5 shows that r-safe n-bit moduli can be efficiently generated even for an exponentially large r, say $r = 2^{n/8}$. We denote the resulting distribution on RSA moduli by $D_{n,\ell}$.

SIGNATURE SCHEME 8.7. Gennaro-Halevi-Rabin signature scheme GHR.

Key generation.

Input: Security parameter n, and ℓ and $h\colon \{0,1\}^\ell \to P$ as above.
Output: N, public key pk, secret key sk, and $t \in \mathbb{Z}_N^\times$.

1. Choose uniformly random r-safe primes p and q with $2^{(n-1)/2} \leq p, q < 2^{n/2}$, where r is the largest prime in P.
2. RSA modulus $N \leftarrow p \cdot q$.
3. $t \xleftarrow{\boxdot} \mathbb{Z}_N^\times$.
4. $\mathsf{sk} \leftarrow (N, \varphi(N))$, $\mathsf{pk} \leftarrow (N, t)$.

Signature generation.

Input: $m \in \{0, 1\}^\ell$.
Output: $s = \mathsf{sig}_{\mathsf{sk}}(m) \in \mathbb{Z}_N^\times$.

5. Compute the inverse v of h_m in $\mathbb{Z}_{\varphi(N)}^\times$.
6. Compute $s \longleftarrow t^v$ in $\mathbb{Z}_N^\times$. [Then $s = t^{1/h_m}$ in $\mathbb{Z}_N^\times$.]

Verification.

Input: $m \in \{0, 1\}^\ell$ and $u \in \mathbb{Z}_N^\times$.
Output: "true" or "false".

7. If $u^{h_m} = t$ in $\mathbb{Z}_N^\times$ then return "true" else return "false".

Although the message space is small enough for exhaustive search, we only know how to calculate the inverse in step 5 (and thus a signature) with the help of sk. We have the usual three tasks of discussing correctness, efficiency, and security of GHR.

THEOREM 8.8. *In the Gennaro-Halevi-Rabin Signature Scheme 8.7, the verification step works correctly. If 2^ℓ is polynomial in n, then the operations can be executed in time polynomial in n.*

PROOF.   For the efficiency, we can set up a table of P (Eratosthenes' sieve, Section 3.3) and, given m, look up h_m in time polynomial in 2^ℓ. We claim that $s = t^{1/h_m}$ in step 6. Since $3 \leq h_m \leq r$, $\varphi(N) = (p-1)(q-1)$, and neither $p - 1$ nor $q - 1$ have an odd prime factor up to r, we have $\gcd(h_m, \varphi(N)) = 1$. Since $t^{\varphi(N)} = 1$ in $\mathbb{Z}_N^\times$ by Euler's Theorem and $v \cdot h_m = 1$ in $\mathbb{Z}_{\varphi(N)}^\times$, we have

$$s^{h_m} = t^{v \cdot h_m} = t^1 = t \text{ in } \mathbb{Z}_N^\times,$$

which shows the claim. The modular inverse v is easy to calculate when $\varphi(N)$, a part of the secret key, is known. For the correctness, it is sufficient to check that for $u \in \mathbb{Z}_N^\times$ we have

$$\mathsf{ver}_{\mathsf{pk}}(m, u) = \text{"true"} \iff u = \mathsf{sig}_{\mathsf{sk}}(m).$$

The direction "$\Longleftarrow$" is clear. For the other direction, we know from the above or from Corollary 15.62 that the exponentiation map $x \mapsto x^{hm}$ is a permutation of $\mathbb{Z}_N^\times$. Therefore $u^{hm} = s$ implies that $u = s^{1/hm} = \mathsf{sig}_{\mathsf{sk}}(m)$.

$\square$

Notes 8.1. Obtaining signatures from asymmetric encryption schemes is suggested in Rivest *et al.* (1977).

8.2. ElGamal proposed his signature scheme in 1985 using the particular type of group $G = \mathbb{Z}_p^\times$ for a prime p. He later became Chief Scientist at Netscape. His original verification procedure does not perform the test whether $z \neq 1$.

8.3. Both ElGamal forgeries in Section 8.3 are from Mitchell *et al.* (1992).

8.4. Schnorr patented his method (see Schnorr 1991b for the US patent), but the NIST used it without obtaining a license. A legal fight ensued. Schnorr (1997) explains how he presented his idea at the Eurocrypt '89 meeting (Schnorr 1989) and that *"the inventor of the DSA, David W. Kravitz, attended this presentation"*. The original 1992 NIST proposal came under heavy fire also for its short keys—originally 512 bits, amended in response by allowing any length between 512 and 1024 that is a multiple of 64—and its opaque selection process: *"the NIST proposal presents an attempt to install weak cryptography [...] in order to please the NSA"* (Rivest 1992), and *"NIST's actions give strong indications of favoring protection of NSA's espionage mission"* (Hellman 1992). DSA is detailed in NIST (2000) and also called the *Digital Signature Standard* DSS.

8.5. For r polynomial in n, n-bit r-safe moduli can also be generated efficiently by testing random primes p and q for the required property.

Exercises.

EXERCISE 8.1 (ElGamal signatures). *We choose the prime number $p = 12347$ and the group $G = \mathbb{Z}_p^\times$ in the ElGamal signature scheme 8.1. We use $\mathsf{sk} = a = 9876 \in \mathbb{Z}_{12346}$ as the secret key. The message m to be signed consists of the four last digits of your birth number; see Exercise 2.1. In the example of Exercise 2.1, we have $m = 3116$.*

(i) *Show that $g = 2$ generates G. Compute $\mathsf{pk} = A = g^a \in G$.*

(ii) *Compute the signature $\mathsf{sig}_{\mathsf{sk}}(m) = (K, b)$ using $k = 399$ as your secret session key from $\mathbb{Z}_{12346}^\times$. Verify your signature.*

EXERCISE 8.2 (Insecure signatures). *Let p be a prime and g a generator of $\mathbb{Z}_p^\times$. BOB signs his messages with the ElGamal signature scheme 8.1 in the group $G = \mathbb{Z}_p^\times$. EVE has intercepted BOB's signed messages $(m_1, (K_1, b_1))$ and $(m_2, (K_2, b_2))$ that use the same session key k. She observes that, by chance, $K_1 = K_2$.*

(i) *Can EVE compute the secret session key k easily?*

(ii) *If EVE knows k, can she generate a valid signature for an arbitrary $m \in \mathbb{Z}_{p-1}$?*

EXERCISE 8.3 (Faulty use of ElGamal signatures). BOB *sets up the ElGamal signature scheme 8.1 by choosing the prime number* $p = 1\,334\,537$ *and the generator* $g = 3$ *of* $G = \mathbb{Z}_p^\times$. *His public key is* $\mathsf{pk} = A = 143\,401$.

(i) BOB *sends the message* $m = 7\,654$ *with signature* $\mathrm{sig}_{\mathsf{sk}}(m) = (335\,037, 820\,465)$. EVE *somehow gets to know the secret session key* k, *which is 377. Compute the secret key* a.

(ii) BOB *changes his secret key and now has the public key* $A = 568\,267$. *But his random number generator was jammed and has output the same value for* k *twice in a row. This was the case for the messages* $2\,001$ *and 243 with signatures* $(576\,885, 1\,323\,376)$ *and* $(576\,885, 1\,161\,723)$, *respectively. Now compute the secret key* a.

EXERCISE 8.4 (ElGamal verification). *Theorem 8.3 shows correctness of the ElGamal verification procedure when the group order* d *is prime. Now assume that* d *is composite.*

(i) *Does the proof still work?*

(ii) *Would further tests beyond* $z \neq 1$ *help?*

(iii) *Without such tests, can you forge signatures?*

(iv) *Suppose we augment the protocol by including in the public key* pk *a list* L *of all prime factors of* d. *In the verification, we also test whether the elements of* L *are indeed prime,* d *is a power product of them, and whether* $z^{d/\ell} \neq 1$ *for all* $\ell \in L$. *What would the outcome of this test be if* d *were prime? Do these additional tests provide further security?*

EXERCISE 8.5 (Number of ElGamal signatures). *In the ElGamal signature scheme, we fix* G, d, g, *and* pk. *How many signatures are possible for a given message* $m \in \mathbb{Z}_d$?

EXERCISE 8.6 (ElGamal and Schnorr signatures). *You are to compare two versions of these signature schemes.*

(i) *In the ElGamal signature scheme, replace* b *in step 5 by* $b \longleftarrow k - a(K^* + m)$. *Design a verification procedure for which you can prove the analog of Theorem 8.3, including the fact that an input* $(z, c) \in G \times \mathbb{Z}_d$ *verified as "true" is a correct signature of* m.

(ii) *Compare signature length and computing time for (i) to Schnorr signatures 8.5.*

(iii) *In Schnorr's signature scheme 8.5, replace step 6 by* $\mathrm{sig}_{\mathsf{sk}}(m) \longleftarrow (K, b)$. *Execute tasks (i) and (ii) for these signatures.*

EXERCISE 8.7 (Forging Schnorr signatures). *Can you adapt the ElGamal forgeries of Section 8.3 to the Schnorr signatures 8.5? Discuss your result.*

EXERCISE 8.8 (Short Schnorr signatures).   Schnorr (1991a) actually proposes even shorter signatures than those in Scheme 8.5. In the notation of that scheme, he takes a one-way hash function $h \colon G \times \mathbb{Z}_d \longrightarrow \mathbb{Z}_{2^t}$ with $t \approx \ell/2$, considering $\mathbb{Z}_{2^t} = \{0, \dots, 2^t - 1\}$ as a subset of $\mathbb{Z}_d = \{0, \dots, d - 1\}$. The following replacements are made in Scheme 8.5.

> Step 5: $e \longleftarrow h(K, m) \in \mathbb{Z}_{2^t}$, $b \longleftarrow k + ae \in \mathbb{Z}_d$.
> Step 6: $\mathrm{sig}_{\mathrm{sk}}(m) = (e, b) \in \mathbb{Z}_{2^t} \times \mathbb{Z}_d$.
> Verification input: $(z, c) \in \mathbb{Z}_{2^t} \times \mathbb{Z}_d$.
> Step 7: $u \longleftarrow g^c A^z \in G$.
> Step 8: If $z = h(u, m)$ then return "true" else return "false".

(i) Show the analog of Theorem 8.6 for this scheme.

(ii) Compare the length and cost of these signatures to those of the Scheme 8.5, as in Table 8.2.

(iii) Compare also to ElGamal signatures on an elliptic curve over $\mathbb{F}_p$ with a 250-bit prime p.

(iv) Suppose that for $e \in \mathbb{Z}_{2^t}$ and $u \in G$ one can efficiently compute $m \in \mathbb{Z}_d$ with $e = h(u, m)$. Show how to forge signatures. [Thus h must be one-way in the argument m for unforgeability.]

(v) Consider the following attack on this signature scheme. Suppose that for some $m \in \mathbb{Z}_d$ and $e \in \mathbb{Z}_{2^t}$,

$$\mathrm{prob}\{h(K, m) = e \colon K \xleftarrow{\;\boxtimes\;} G\} \geq \varepsilon.$$

Then pick $y \xleftarrow{\;\boxtimes\;} \mathbb{Z}_d$ until $e = h(u, m)$ with $u = g^y A^e$. Show that such a y allows to forge signatures and that the expected number of trials is ε^{-1}. Give a condition on the distribution of $h(K, m)$ for $K \xleftarrow{\;\boxtimes\;} G$ that renders this attack infeasible.

(vi) For a second attack, we assume to have about $2^{t/2}$ pairwise distinct $(y_i, z_i) \in \mathbb{Z}_{2^t} \times \mathbb{Z}_d$ so that $h(u_i, m) = h(u_j, m)$ for all i, j and all $m \in \mathbb{Z}_d$, where $u_i = g^{z_i} A^{y_i}$. Then generate $m_1, \dots, m_{2^{t/2}}$ and find i and j with $h(u_i, m_j) = z_i$. Show that $\mathrm{sig}_{\mathrm{sk}}(m_j) = (y_i, z_i)$. Discuss a condition on h that rules out this attack.

(vii) Replace the target range $\mathbb{Z}_{2^t}$ of h by $\mathbb{Z}_d$ and determine the function $h_1 \colon G \times \mathbb{Z}_d \longrightarrow \mathbb{Z}_d$ whose substitution for h in the above yields Scheme 8.5. What can you say about h_1? Is it one-way?

EXERCISE 8.9 (Bleichenbacher's method for forging ElGamal signatures).   The security of ElGamal signatures relies on the difficulty to compute the discrete logarithm in subgroups of $\mathbb{Z}_p^\times$. In Theorem 8.3, we assume $d = \#G$ to be prime, which is not the case for $G = \mathbb{Z}_p^\times$, since then $d = p - 1$. This exercise discusses a forgery for certain composite d; compare Exercise 8.4.

Let p be a prime such that $g = 2$ generates $G = \mathbb{Z}_p^\times$. Let $A \in G$ be BOB's public key and $m \in \mathbb{Z}_{p-1}$ some message to be signed à la ElGamal. For $w, z \in \mathbb{Z}_{p-1}$, we consider the condition

(8.9) $$g^{w \cdot z} = g^w.$$

Show that EVE can sign m on behalf of BOB, without knowing his secret key, as follows.

(i) Let $w = (p - 1)/2$. Show that EVE can easily find some z satisfying (8.9).

(ii) Let e be a divisor of $p-1$, $K = (p-1)/e$, $t = (p-3)/2$, and $b = t \cdot (m - wz)$ in $\mathbb{Z}_{p-1}$. Show that (K, b) is a valid ElGamal signature of m.

(iii) Which values in the above signature should arouse the suspicion of the verifier, even though the validity of the signature is incontestable?

Generalizing this attack, we assume that $p-1 = y \cdot w$, with y a sufficiently smooth integer, and that for some c with $0 < c < y$, also $h = cw \in \mathbb{Z}_p^\times$ generates $\mathbb{Z}_p^\times$. Thus $y = 2$ in (i).

(iv) Show how to generate a signature of m on behalf of BOB, as above, without knowledge of his secret key. Prove that the signature is valid.

(v) Let $s = \mathrm{sig}_{sk}(m) = (K, b)$ be a signature of m and $t = (p-1)/K \in \mathbb{Q}$. If t is a smooth integer, one may suspect the attack above. An indication can be obtained from $u = \gcd(K, p-1)$. If a random value was used for k, then u will usually not be large. Thus $u > 2^{160}$, say, is a good indication that a forgery may have happened. Show that by adding this constraint to ElGamal signatures, the above attack becomes impossible.

(vi) Adapt the attack to the DSA signature system. [As a reaction to the publication of this attack, NIST enforced the verification that $(p-1)/K$ is non-smooth for the DSA signature protocol.]

EXERCISE 8.10 (Rabin signatures).   In a signature scheme suggested by Rabin (1979), we have an RSA modulus $N = pq$ and a, $b \in \mathbb{Z}_N^\times$ with Legendre symbols $\left(\frac{a}{p}\right) = -\left(\frac{a}{q}\right) = 1$ and $\left(\frac{b}{p}\right) = -\left(\frac{b}{q}\right) = -1$, as explained in Section 15.14.

For any $m \in \mathbb{Z}_N^\times$, exactly one of the four values m, am, bm, and abm is a square in $\mathbb{Z}_N^\times$; see Sections 9.9 and 11.8 and Figure 15.3. The set-up is to choose $N = pq$ with random primes p and q of about $n/2$ bits for security parameter n, and $sk = (p, q)$ and $pk = N$.

The signer determines the unique $i, j \in \{0, 1\}$ for which $a^i b^j m$ is a square in $\mathbb{Z}_N^\times$, computes a square root y of it (the four roots are easy to find using sk), and signs with $\mathrm{sig}_{sk}(m) = (y, i, j)$. The verifier simply checks that $y^2 = a^i b^j m$.

(i) Verify that the scheme works efficiently and correctly.

(ii) Show that the scheme is secure in the sense that any signature forgery using only pk yields an efficient way to factor N. We assume this to be impossible. You may want to use Lemma 3.8 (i).

(iii) If you know pk and $x \xleftarrow{\boxtimes} \mathbb{Z}_N^\times$, and receive a square root of x^2, what is the probability that you can factor N?

(iv) Show that the signature scheme is insecure under a chosen message attack, where an adversary can submit messages and ask for their signatures.

(v) Does the attack in (iv) work when the messages are chosen uniformly at random and known to the adversary? When some other party chooses them as it likes and gives them to the adversary?

EXERCISE 8.11 (Message Recovery).   The ElGamal type signature schemes share the need for separate transmission of the message m which was signed. A variant called signatures with message recovery allows to compute m from the signature. This is due to Nyberg & Rueppel (1996). In their scheme, the public data $G = \langle g \rangle$, $d = \#G$, $A = g^a$ are defined as for the ElGamal signature scheme 8.1. We also have $sk = a \in \mathbb{Z}_d$ and $* \colon G \to \mathbb{Z}_d$. BOB produces his signature on a message $m \in G$ as follows.

1. $k \xleftarrow{\boxdot} \mathbb{Z}_d$ and $K \leftarrow g^{-k} \in G$.

2. $c \leftarrow m \cdot K \in G$ and $b \leftarrow ac^* + k \in \mathbb{Z}_d$.

3. $\mathrm{sig}_{\mathsf{sk}}(m) \leftarrow (c, b) \in G \times \mathbb{Z}_d$.

Prove the following properties of the Nyberg-Rueppel signature:

(i) For $v = g^b \cdot A^{-c^*} \in G$, we have $m = v \cdot c \in G$. Thus ALICE can recover the message from the signature.

(ii) Show that by recovering m, ALICE also verifies that (c, b) is a valid signature of BOB on the message m.

(iii) Given one valid signature, one can forge further ones as follows. Let $\mathrm{sig}_{\mathsf{sk}}(m) = (c, b) \in G \times \mathbb{Z}_d$ be a valid signature of $m \in G$ by BOB, $0 < \ell < d$, $m' = g^\ell \cdot m \in G$, and $b' = b + \ell$ in $\mathbb{Z}_d$. Show that (c, b') is a valid signature of m' by BOB. [Such forgeries can be prevented by appropriate padding, not discussed here.]

(iv) For an n-bit group G, compare the number of bits required to send a message and its signature in the ElGamal scheme and in the Nyberg & Rueppel scheme. Discuss the implications when BOB uses a hash function h to sign long messages $\overline{m}$ as $\mathrm{sig}_{\mathsf{sk}}(h(\overline{m}))$.

EXERCISE 8.12 (Subgroup key exchange). Section 8.4 has explained Schnorr's idea to obtain shorter signatures in subgroups of $H = \mathbb{Z}_p^\times$, apparently without sacrificing security. Can this also be usefully employed in the Diffie-Hellman key exchange 2.20?

EXERCISE 8.13 (Schnorr practice). In this exercise you will make practical computations with Schnorr signatures 8.5 in $H = \mathbb{Z}_p^\times$, using real-life key sizes.

(i) Generate a uniformly random prime number d with exactly 250 bits.

(ii) Generate a uniformly random prime p with exactly 3000 bits such that d divides $p - 1$.

(iii) Discuss (but do not execute) the following task: generate a 3000-bit uniformly random prime p and then a 250-bit prime divisor of $p - 1$.

(iv) Find a $g \in \mathbb{Z}_p^\times$ with order d. Let $G = \langle g \rangle \subset \mathbb{Z}_p^\times$ be the cyclic group with d elements generated by g. The data d, p, and g are publicly known. Choose $\mathsf{sk} = a \xleftarrow{\boxdot} \mathbb{Z}_d$ and $\mathsf{pk} = A = g^a \in G$.

(v) Let $m \in \mathbb{Z}_d$ be the integer value of the ASCII text: DSA for real (note the two blanks in the text). Using a random session key $k \xleftarrow{\boxdot} \mathbb{Z}_d^\times$ and sk as above, produce BOB's signature $\mathrm{sig}_{\mathsf{sk}}(m) = (K^*, b)$ on m.

(vi) Verify the signature in (v).

(vii) Let m' be the integer value of the ASCII text: The Lord of the Rings has no secrets. Can you produce a signature of this text using the same setting as above? If not, what additional steps are required?

(viii) The signature scheme can be attacked in two different ways:

(a) By solving the discrete logarithm problem in the group G with d elements. Only group circuits seem available in general, at cost $\Omega(\sqrt{d})$.

(b) By solving the general discrete logarithm problem in $\mathbb{Z}_p^\times$ with the number field sieve. As in (3.51), the cost of this method for an n-bit p is estimated as

$$\exp_2\left(1.902 \cdot \left(n \cdot (\log_2 n)^2\right)^{1/3}\right).$$

The text presents a comparison when $p \approx 2^{3000}$ and $d \approx 2^{250}$. Now suppose that only $p \approx 2^{1024}$ is required. What is the corresponding bit length of d? What if $p \approx 2^{15360}$ (Table 5.11)?

EXERCISE 8.14 (RSA blind signatures).   It is sometimes required that a signature protocol between two parties ALICE and BOB runs in such way that BOB signs implicitly a message m on behalf of ALICE, but does not know explicitly the message he is signing. Thus BOB cannot associate the signature to the user ALICE. Such protocols are called blind signatures and play a key role in electronic cash schemes.

We describe a blinding protocol based on the RSA signature scheme (Section 8.1). Let BOB have the secret and public RSA keys $\mathsf{sk} = (N, d)$ and $\mathsf{pk} = (N, e)$. In order to receive blind signatures from BOB, ALICE uses her own blinding key $k \in \mathbb{Z}_N^\times$.

(i) Suppose that ALICE wants to have BOB sign the message $m \in \mathbb{Z}_N$ so that the signature can be verified, but BOB cannot recover the value of m. Show that the following protocol fulfills the requirements for a blind signature scheme.

    1. ALICE sends $M = m \cdot k^e \in \mathbb{Z}_N$ to BOB.

    2. BOB produces the signature $\mathsf{sig}_{\mathsf{sk}}(M) = M^d \in \mathbb{Z}_N$ and sends it to ALICE.

    3. ALICE recovers $\mathsf{sig}_{\mathsf{sk}}(m) = k^{-1} \cdot \mathsf{sig}_{\mathsf{sk}}(M) \in \mathbb{Z}_N$. Then $\mathsf{sig}_{\mathsf{sk}}(m)$ is a valid signature of m by BOB.

(ii) Let $N = p \cdot q$ where $p = 1\,000\,000\,000\,039$, $q = 10\,000\,001\,000\,029$ and $e = 2^{16} + 1 = 65537$. Compute the secret exponent d of BOB. Let $k \xleftarrow{\;\boxtimes\;} \mathbb{Z}_N$ and $m \in \mathbb{Z}_N$ be the integer value of the ASCII text: blinded.

    (a) Compute the blinded message M.

    (b) Compute BOB's blinded signature $\mathsf{sig}_{\mathsf{sk}}(M)$ and also BOB's plaintext signature $\mathsf{sig}_{\mathsf{sk}}(m)$, using BOB's secret key d.

    (c) Compute the plaintext signature $\mathsf{sig}_{\mathsf{sk}}(m)$ as ALICE recovers it using k. Compare it to the value $\mathsf{sig}_{\mathsf{sk}}(m)$ in (b).

(iii) Design a blind signature scheme based on the ElGamal signature scheme 8.1 in a group $\mathbb{Z}_p^\times$ and explain why it has the properties of a blinding scheme. With a secret random blinding key k, ALICE builds a function $f_k : \mathbb{Z}_{p-1} \to \mathbb{Z}_{p-1}$ which transforms a message m into $M = f_k(m)$ and a function g_k such that $g_k(\mathsf{sig}_{\mathsf{sk}}(M)) = \mathsf{sig}_{\mathsf{sk}}(m)$, where $\mathsf{sig}_{\mathsf{sk}}$ is BOB's signing algorithm.

(iv) Let $p = 10^{20} + 39$ and $g = 3 \in \mathbb{Z}_p^\times$ a generator. Let BOB's secret exponent be $a = 10^{10}$. With the same message as in (ii), use your ElGamal blind signature scheme to compute the blinded message M, its blind signature $\mathsf{sig}_{\mathsf{sk}}(M)$ and the recovered cleartext signature $\mathsf{sig}_{\mathsf{sk}}(m)$.

EXERCISE 8.15 (Blind signature based on the discrete logarithm problem).   Read Camenisch, Piveteau & Stadler (1995). In their Section 2.1, the authors define a modification of the DSA signature scheme. In their Section 3.1, they use this modified signature scheme and define a blind signature protocol.

(i) In what way does the signature scheme in Section 2.1 differ from the DSA signature scheme?

(ii) The DSA scheme is a modification of ElGamal working in the cyclic subgroup $G \subset \mathbb{Z}_p^\times$ with d elements, where d divides $p - 1$ and is prime. In what other points does the DSA scheme differ from the ElGamal scheme?

(iii) *Adapt the protocol of the paper's Section 3.1 to a blinding protocol based on El-Gamal. Describe your protocol in pseudocode, so that you can easily implement it in some computer algebra system.*

(iv) *Generate a 3000-bit prime and perform your complete blind signature protocol from (iii).*

EXERCISE 8.16 (Time stamps). *An electronic notary is offering the service to create trusted time stamps. This works as follows.*

A. *A customer sends a message m. The notary computes the hash value $h(m)$ under some publicly known hash functions h and signs it using his own secret key sk obtaining $\text{sig}_{\text{sk}}(h(m))$.*

B. *Customer entries of the form $c_i = h(m)|\,\text{sig}_{\text{sk}}(h(m))$ accumulate during the day. At the end of the day all c_i are concatenated to obtain the day list $d = c_1|c_2|\ldots|c_n$. Then the day hash $h(d)$ is computed, which together with the signature $\text{sig}_{\text{sk}}(h(d))$ is published as the day stamp $D = h(d)|\,\text{sig}_{\text{sk}}(h(d))$ the next morning in a daily newspaper.*

Show that this scheme has the following advantages and disadvantages.

(i) *Presenting his own message m, a customer can always prove that it was included in the day hash. The day stamp D from the newspaper and the day list d from the notary's archive are needed. Describe a protocol that can be used to verify that m was on the day list.*

(ii) *Explain the purpose of $\text{sig}_{\text{sk}}(h(d))$.*

(iii) *A disadvantage for the customer is that he cannot argue convincingly that his message was received. This can be rectified by the notary handing out time-stamped receipts.*

Find an example of how such a receipt can be generated using the time stamp D' of the previous day and the public key of the notary. Show how, for this example, the customer can use the receipt to prove that the text was not sent on the previous day or earlier.

EXERCISE 8.17 (Hidden message). *Once again a new mission is waiting for her Majesty's finest agent. Old Q has received an assignment from M to find a way how 007 may send a secret message to the London headquarters unnoticed.*

In the guise of a broker James Bond has easy access to the Internet. Q has learned that at the stock market, buyers' and sellers' orders are signed using the ElGamal signature scheme. The mastermind of the Q-Branch starts from there:

Q: *Here is the solution, 007. Naturally you are well acquainted with the signing of electronic messages using the ElGamal scheme.*

007: *I have read the Russian translation of the article, Q.*

Q: *Splendid! We will use this scheme to hide the message you want to send to M. The present system uses the*

prime number $p = 311303$ and the group $\mathbb{Z}_p^\times$. The element $g = 5$ is the generator of $\mathbb{Z}_p^\times$ that was adopted. The secret part of the key is $a = 45678$. Is everything quite clear so far, 007?

007: *Yes, Q. Everything quite standard. So where is the trick?*

Q: *007, for the first time you are showing*

some interest in my work! Instead
of the random number k used for
signing the message m you will use
your secret message s. This is the
date (formatted **DDMMYY**) on which
we — how would you put this —
must be prepared for a surprise.
Good luck, 007!

(i) What is the "conventional" purpose of a randomly chosen component for a digital
signature, for example the k in the ElGamal scheme?

(ii) Explain why Q assumes that the transmission of s is secure.

(iii) After some time Q receives the following message and signature:
$(654321, (214027, 41635))$. Check whether this message originates with 007.
What is the date that 007 predicts for the surprise?

(iv) Which conditions (with respect to the variables) must be met so that this computation works?

EXERCISE 8.18 (Signed key exchange). ALICE and BOB want to exchange messages
using a symmetric cryptosystem. To do this they need to agree on a shared session key
K. They choose the Diffie-Hellman key exchange and safeguard the exchange using
ElGamal signatures. The basis of all computations is a group $G = \langle g \rangle$. ALICE signs
her public key $A = g^a \in G$ using her secret key $\mathsf{sk} = a$. BOB does the same thing with
b and $B = g^b \in G$. To compute the common session key ALICE chooses e_A and BOB
chooses e_B. Those parts of the protocol that are not specified by the instructions of
this exercise should be chosen by you with ample comments.

(i) Describe the individual steps of the corresponding protocol that allows ALICE
and BOB to agree on their common session key $g^{e_A e_B}$.

(ii) Execute the computations needed for the individual steps using the group $G =
\mathbb{Z}_{27011}^\times$. Furthermore let $g = 101$, $a = 21768$, $b = 9898$, $e_A = 54321$, and
$e_B = 13551$.

(iii) Explain why the protocol from part (i) is secure with respect to a "man in the
middle" attack.

(iv) Would the claim of (iii) still be correct if ALICE and BOB had not signed their
public keys and instead exchanged them at the beginning of the protocol?

EXERCISE 8.19 (Authenticated key exchange). In this exercise you are to simulate
the start of an electronic conversation of ALICE and BOB, using RSA authentication
and subsequent key exchange.

(i) Create 768-bit RSA keys for both ALICE and BOB.

(ii) Compose a text and sign it first using ALICE's key and verify (by BOB), and
then sign it using BOB's secret key and verify (by ALICE).

(iii) Generate a random 128 bit key, encrypt it using ALICE's public key and subsequently decrypt it using her secret key. Now ALICE and BOB share a common
secret key.

(iv) Compose a short text (two to three sentences) and encrypt it in ALICE's stead
using AES and the shared secret key. Afterwards, decrypt the text in the name
of BOB.

Note: Those parts of the protocols that are not fully specified in the instructions of
this exercise are to be chosen by you, with ample comments.

EXERCISE 8.20 (Signatures with elliptic curves). *Adapt the ElGamal signature scheme to elliptic curves. The public system parameters are a prime power q, a curve E over $\mathbb{F}_q$, a point $P \in E$ and its order $d \in \mathbb{N}$, which has a large prime divisor.*

  (i) What are the appropriate domains for key and message?

  (ii) How does BOB verify a signature? [In other words: what is considered a valid signature?]

  (iii) How does ALICE sign? [Note: According to Section 8.2, an injective function $\langle P \rangle \to \mathbb{Z}_d$ is needed that can be computed efficiently. How can this demand be reasonably weakened without security risks?]

If I were a secret, Lord, I never would be told.
CREEDENCE CLEARWATER REVIVAL (1969)

By what course of calculation can these results
be arrived at by the machine in the *shortest time*?
CHARLES BABBAGE (1854)

There is a growing conviction among many computer scientists
that the analysis of algorithms, or the study of
"computational complexity", is the most important
and fundamental part of computer science.
GEORGE EDWIN COLLINS (1973)

Malcolm said. "It's a paleontological mystery.
But I believe that complexity theory has a lot to tell us about it."
JOHN MICHAEL CRICHTON (1996)

En effet, il n'est pas nécessaire de se créer
des fantômes imaginaires et de mettre en suspicion
l'incorruptibilité des employés ou agents subalternes,
pour comprendre que, si un système exigeant le secret
se trouvait entre les mains d'un trop grand nombre d'individus,
il pourrait être compromis à chaque engagement
auquel l'un ou l'autre d'entre eux prendrait part.[1]
AUGUSTE KERCKHOFFS (1883)

There is scarce any thing writ in Cypher,
however ingeniously contrived, which in these days,
can lie long hidden for want of a Decypherer,
tho' perhaps there are too few who have made this their study.
PHILIP THICKNESSE (1772)

---

[1]In fact, it is not necessary to conjure up imaginary phantoms and to suspect the incorruptibility of employees or subalterns to understand that, if a system requiring secrecy were in the hands of too large a number of individuals, it could be compromised at each engagement in which one or another of them took part.

Chapter 9

Security and reductions

Chapter 6 presents two general attacks on Boolean cryptosystems. Once such attacks are known, cryptographers have to design systems that resist them.

This chapter provides a different approach to security. Instead of a certain attack methodology, only the interaction of an adversary with the system is specified. Security now means resisting any such attack, no matter which method it uses, known or unknown. The tool is called *security reductions*, of which we have seen examples in previous chapters. One starts by assuming that a certain computational problem X is hard to solve. Then one shows for a certain cryptosystem S that breaking S implies solving X. Since we postulate the axiom "X is hard", no such break is possible and S is secure. We call this notion *axiomatic security*. It complements the *empirical security* discussed in Section 5.5 in that now arbitrary attacks, also currently unknown ones, are ruled out—under the axiom, of course.

There are several types of goals that an adversary may have, and various types of tools at his disposal. This leads to a plethora of notions for such reductions, which we illustrate in this chapter.

9.1. When can we consider a system "secure"?

The history of cryptography is full of confident inventors and users, certain of the security of their cryptosystems, who faced a rude awakening and unexpected punishment when someone succeeded in breaking their system. Mary Queen of Scots was beheaded in 1587, due to the cryptographic nonchalance of her secretaries and the skills of the cryptographers serving Queen Elizabeth I. The Jacobite conspirator Christopher Layer was hanged, drawn and quartered in 1722 (Section H.4), a cryptographic and diplomatic blunder of the German Foreign Office in 1917 hastened

© Springer-Verlag Berlin Heidelberg 2015
J. von zur Gathen, *CryptoSchool*, DOI 10.1007/978-3-662-48425-8_13

the end of World War I (Chapter I), and cryptographic negligence of the German military shortened World War II, presumably by a year or two (Chapter J).

How can we avoid this? A convincing solution would be a rigorous mathematical proof of the security of some system. The 1976 Diffie-Hellman revolution (Section 2.4) initiated a development where—for the first time in its long history—this central question of cryptography could be stated with mathematical precision: is a given system secure? Namely: is no probabilistic polynomial-time algorithm able to break it?

Alas, the answer is elusive. The question begs for the methods of computational complexity, a subarea of theoretical computer science. Indeed, a central goal in that theory is to prove *lower bounds* on the complexity of some problem, implying that any algorithm that solves the problem takes at least that much time (or memory, communication, or some other cost measure). The central notion of "polynomial time = feasible = efficient = easy" emerged, and the goal of proving super polynomial lower bounds became a well-defined task, that is, that of finding "hard" problems. It was achieved in some cases, but mainly found to be a difficult challenge. In his seminal paper from 1971, Cook introduced the notions of P for polynomial time, that is, feasible problems and NP for a larger class. The million dollar question here still flummoxes computer scientists: to prove *Cook's hypothesis* that P ≠ NP.

At this sorry state of affairs, there are no tools in sight to obtain systems secure in the above sense. In the lamentable absence of such "absolute" bounds, there is one highly successful tool, namely that of "relative" bounds.

Cook (1971) adapted the notion of *reduction* from the theory of (un)-computability to this situation. A reduction from a problem X to a problem Y is an efficient algorithm solving X that may call an unspecified subroutine for Y; see Definition 3.27. It follows that if Y is easy, then so is X, and if X is hard, then so is Y. (Non-experts beware: it is easy to confuse the directions.) This leads to the notions of NP-hard and NP-complete problems.

The goal now is to find secure cryptosystems, but instead of actually proving their security, the task is made easier by allowing to assume that some computational problem is hard, say that of factoring integers. Technically, we have to reduce integer factorization to a break of the cryptosystem under consideration.

Thus we postulate the axiom that a certain problem X is "hard"—although maybe not NP-hard—and reduce solving X to breaking the cryptosystem at hand. We call this *axiomatic security* and it forms the topic of this chapter. A number of restrictions and options turn up, and lead

scalable	fixed-size
asymmetric (public key)	symmetric (secret key)
axiomatic security	empirical security

Figure 9.1: Three dichotomies in cryptography

to a rather congested notation. Mathematical rigor requires to state the hypotheses, but like an engineer or physicist, in practice we do not worry unduly about such assumed *"laws of nature"*, provided the evidence for them is strong enough. The complementary *empirical security* of Section 5.5 is based on the axiom that no attacks will be found that are substantially stronger than currently known ones.

One of the basic tasks in cryptography is that of transmitting a message x securely over an open channel. *Stream ciphers* work naturally with messages of arbitrary lengths, while a *block cipher* has a fixed finite set of possible messages, say $\{0,1\}^{128}$ in AES. For arbitrary messages, one then uses a chaining method (Section 2.8). Now in a block cipher, the sender has a key K and uses an encryption algorithm enc to produce the encrypted message $y = \text{enc}_K(x)$. The receiver has a key S and uses a decryption algorithm dec to obtain $x^* = \text{dec}_S(y)$.

Three conditions are of central importance:

○ correctness: we must have $x^* = x$.

○ efficiency: the functions enc and dec should be *easy* to execute.

○ security: just using the publicly available information like y (and maybe other items, depending on the system) it should be *hard* to compute x.

We consider three fundamental distinctions in cryptographic systems, as shown in Figure 9.1. In a *scalable cryptosystem*, there is a security parameter n which can take arbitrarily large values. Usually the key length is directly determined by n. In the RSA cryptosystem, say, the public key $\text{pk} = (N = pq, e)$ has length $2n$; see Figure 2.15. For such systems, the standard notions of theoretical computer science suggest the following interpretations: "easy" should mean "(random) polynomial time $\text{poly}(n)$", and "hard" its complement "superpolynomial time". Usually, one likes to have this even more precise, say cubic, quadratic, or even "softly linear" time $n \log^{O(1)} n$ for "easy", and exponential time 2^{cn} or "frexponential" time 2^{n^c}, with some $c > 0$, for "hard". Such questions have their natural abode in the theory of computational complexity.

In a *fixed-size cryptosystem*, the size of all relevant items (message and key lengths) is fixed. AES allows three key sizes, but this still counts as "fixed". "Easy" now means that we should be able to build inexpensive hardware or software systems with high throughput, say for secure satellite video transission. For security, around 2010 "easy" meant up to 2^{60} computations (or a bit more), and "hard" meant 2^{100} operations (called "100-bit security"), up from 2^{80} just before. These values have to be taken with a grain of salt, there is a gray area between them, and they change slowly with time; see Section 5.5.

The second dichotomy, between *asymmetric* (or *public key*) and *symmetric* (or *secret key*) cryptosystems, should be clear to the reader of Section 2.4. In the first type of system, each player A has a public key pk_A, which anybody has access to (say, over the internet), for use in enc, and also a secret key sk_A, used in dec. Signature schemes also use both types of keys. In a symmetric system, the two keys are the same: $\mathsf{pk} = \mathsf{sk}$, or at least each is easy to compute from the other (and this computation might be incorporated into enc and dec). In particular, pk had better not be publicly available.

The third dichotomy, of *axiomatic* and *empirical* security, has been explained above. A distinction is that the empirical view is usually driven by current technology, while the axiomatic approach tends to have a more timeless outlook. It is useful to keep theoretical and practical aspects apart. For example, security results based on integer factorization should be valid for any bit length. But when it comes to practical use, we must take the current power of factorization methods into account. Thus axiomatic results deal with scalable systems, while empirical ones talk about fixed-sized systems, including concrete versions of scalable ones.

In complexity theory, a useful approach is to not consider *general models* of computation such as Turing machines or Boolean circuits, but rather *structured models* in which only specific types of operations can be executed. This has led to success, namely to asymptotically matching upper and lower bounds, for sorting with comparisons and for arithmetic problems like interpolation and the Euclidean algorithm. We have seen in Section 4.9 a lower bound for discrete logarithm computations, essentially matching an upper bound in such a structured model.

Things are slightly different in Boolean cryptography. General attack methods are known, including the linear and differential cryptanalyses of Chapter 6 and their many variations. For the security of such systems, there are design strategies meant to thwart such attacks, and these should always be considered and implemented, as appropriate.

However, basically we have to work with the algorithms that we know, that is, determine key lengths so that our cryptosystems are safe from the known attacks. Of course, someone might come up (or already have come up, but not told us) with a more powerful attack. As an example, there were persistent rumors—never substantiated—that the developers of the DES had built in a "trapdoor" with which the system could be broken.

More importantly, we can at best hope to have security for some basic building blocks of cryptography, such as encryption and signatures. Real-life systems are built by cementing together many such blocks, with numerous interactions that are nigh impossible to model exactly. But at least we have secure blocks and only need to worry about the cement.

Furthermore, there are successful attacks which are outside of our considerations: fault and side channel attacks, spoofing and phishing, key loggers, social engineering and human negligence, and many more. For such attacks, there does not seem to exist a consistent theory that will last—in contrast to many topics in this text.

9.2. Security of encryption schemes

We consider a cryptographic transmission protocol as above, with $\text{enc}_{\text{pk}}(x) = y$ and $\text{dec}_{\text{sk}}(y) = x$. When can we consider it secure? Or rather: when is it insecure? We list in decreasing order of success some things that an adversary might achieve, knowing y and the public information (which includes pk in the case of an asymmetric system):

○ for any x, compute sk,

○ for any x, compute x,

○ for most x, compute x,

○ for some x, compute x,

○ for some x, find out something nontrivial about x,

○ distinguish somehow between encryptions and random values.

○ for some x, possibly find out something nontrivial about x.

The latter is called *semantic insecurity* and means that for some Boolean predicate f on the message space with average value δ for inputs from some distribution X, there is a probabilistic polynomial-time algorithm computing f with success probability at least $\delta + \epsilon$ for some nonnegligible $\epsilon > 0$ for inputs according to X. It turns out that semantic insecurity is equivalent to distinguishability. In the following, we use the latter rather than semantic (in)security.

9.3. One-way and trapdoor functions

Some of the basic properties that we desire for cryptosystems can be succinctly expressed via the concepts of one-way functions and of trapdoor functions. Intuitively, a one-way function is easy to evaluate but hard to invert. A trapdoor function has the additional property that with a certain private key, inversion becomes easy.

DEFINITION 9.1. Let $f\colon \{0,1\}^* \to \{0,1\}^*$ provide a mapping $\{0,1\}^n \to \{0,1\}^{k(n)}$ for each $n \in \mathbb{N}$, with a polynomially bounded function k of n. An inversion algorithm $\mathcal{A}$ for f with success probability $\epsilon(n)$ is such that for all $n \in \mathbb{N}$,

$$\mathrm{prob}\{f(\mathcal{A}(z)) = z\} = \epsilon(n),$$

where $z \xleftarrow{\boxtimes} f(\{0,1\}^n)$ is chosen uniformly at random in the image of $\{0,1\}^n$ under f. Then f is a one-way function if

(i) f can be computed in random polynomial time,

(ii) all polynomial-time inversion algorithms for f have negligible success probability.

Furthermore, f is a trapdoor function if there is some secret key sk_n for each n, of polynomially bounded length, and a probabilistic polynomial time inversion algorithm for f which takes both $z \xleftarrow{\boxtimes} f(\mathbb{B}^n)$ and sk_n as inputs and has nonnegligible success probability.

The last definition sounds quite intuitive, but requires some care in a nonuniform model of algorithms, because sk_n could simply be built into the algorithm for inputs from $f(\{0,1\}^n)$, and we would have an algorithm contradicting property (ii).

EXAMPLE 9.2. The product of two $n/2$-bit integers, possibly with leading zeros, is an n-bit integer, again allowing leading zeros. Thus we get the multiplication function $f_n\colon \{0,1\}^n = \{0,1\}^{n/2} \times \{0,1\}^{n/2} \longrightarrow \{0,1\}^n$. It is easy to compute. Inverting it means factoring the product into two $n/2$-bit numbers. More closely related to RSA is the function f_n' which returns the product if both inputs are prime numbers in the appropriate range, and 0 otherwise. It is conjectured that factoring such products is hard; then $f' = (f_n')_{n\in\mathbb{N}}$ would be a one-way function. ◇

9.4. Perfect security: the one-time pad

We consider the bit string $x = 100110$ of length six. BOB and ALICE have previously established the secret key $\mathsf{sk} = 010011$ which was randomly chosen among the strings of six bits. Then BOB forms the bitwise *exclusive or (XOR)* $\oplus$, so that $y = x \oplus \mathsf{sk} = 100110 \oplus 010011 = 110101$. We have $(a \oplus b) \oplus b = a \oplus (b \oplus b) = a \oplus 0 = a$ for any bits a, b, and so $y \oplus \mathsf{sk} = (x \oplus \mathsf{sk}) \oplus \mathsf{sk} = x \oplus (\mathsf{sk} \oplus \mathsf{sk}) = x \oplus (0 \cdots 0) = x$. We define $\mathsf{enc}_{\mathsf{sk}}(x) = x \oplus \mathsf{sk}$, $\mathsf{dec}_{\mathsf{sk}}(y) = y \oplus \mathsf{sk}$. Then $\mathsf{dec}_{\mathsf{sk}}(\mathsf{enc}_{\mathsf{sk}}(x)) = x$.

THESIS 9.3. *This system is perfectly secure.*

What does this mean? Can we make this precise?

We have some $n \in \mathbb{N}$. The message space is $\{0,1\}^n$, so that each message x is a string of n bits. The key sk is chosen *uniformly at random* among the 2^n possibilities in $\{0,1\}^n$. For each string $z \in \{0,1\}^n$, we have

$$\mathrm{prob}\{\mathsf{sk} = z\} = 2^{-n}.$$

Now x and sk are chosen. Then $y = x \oplus \mathsf{sk}$ is determined, and $x = y \oplus \mathsf{sk}$. Furthermore, the eavesdropper EVE sees y, does not know sk, and wants to compute x.

Concerning security, we can ask for how many pairs (x', sk') is $\mathsf{enc}_{\mathsf{sk}'}(x') = \mathsf{enc}_{\mathsf{sk}}(x) = y$? For each $x' \in \{0,1\}^n$, there is precisely one sk' with this property, namely $\mathsf{sk}' = x' \oplus y$, since $x' \oplus \mathsf{sk}' = x' \oplus (x' \oplus y) = (x' \oplus x') \oplus y = y$. Therefore, just given y, each x' is equally likely to have been the message. EVE has learnt nothing from y about x.

A more detailed argument goes as follows. Each $x \in \{0,1\}^n$ occurs with some probability p_x as a message. So we have $p_x \geq 0$ and $\sum_{x \in \{0,1\}^n} p_x = 1$. The next statement uses conditional probabilities.

THEOREM 9.4. *For all $x, y \in \{0,1\}^n$, we have:*

$$\mathrm{prob}\{message = x \colon encryption = y\} = p_x.$$

PROOF. We let m run through the messages, and $c = x \oplus \mathsf{sk}$ be a (randomized) encryption of x. Then

$$\mathrm{prob}\{m = x \colon c = y\} = \frac{\mathrm{prob}\{m = x \text{ and } c = y\}}{\mathrm{prob}\{c = y\}},$$
$$c = y \iff \mathsf{sk} = x \oplus y.$$

For the numerator, we have

$$
\begin{aligned}
\mathrm{prob}\{m = x \text{ and } c = y\} &= \mathrm{prob}\{m = x \text{ and } \mathsf{sk} = x \oplus y\} \\
&= \mathrm{prob}\{m = x\} \cdot \mathrm{prob}\{\mathsf{sk} = x \oplus y\} \\
&= p_x \cdot 2^{-n},
\end{aligned}
$$

where the second equality holds since sk is independent of m, and the last equation since sk is chosen uniformly. For the denominator, we calculate

$$
\mathrm{prob}\{c = y\} = \mathrm{prob}\{\mathsf{sk} = x \oplus y\} = 2^{-n},
$$

where we use again that sk is chosen independently of m. We find

$$
\mathrm{prob}\{m = x \colon c = y\} = \frac{p_x \cdot 2^{-n}}{2^{-n}} = p_x. \qquad \square
$$

In particular, each encryption is equally likely, independent of the message. This property depends critically on the uniform random choice of the keys sk. Otherwise, it is "false".

REMARK 9.5. *What about two-time pads? Suppose that two messages x and x' are encrypted with the same key sk:*

$$
y = x \oplus \mathsf{sk}, y' = x' \oplus \mathsf{sk}.
$$

Then

$$
\begin{aligned}
y \oplus y' &= (x \oplus \mathsf{sk}) \oplus (x' \oplus \mathsf{sk}) \\
&= x \oplus x' \oplus \mathsf{sk} \oplus \mathsf{sk} = x \oplus x'.
\end{aligned}
$$

With further general information, say that x and x' are English text, correlation methods can (usually) recover x and x'.

You are not convinced? So consider images, coded as strings of bits. The exclusive OR of two images x and x' clearly still contains a lot of information:

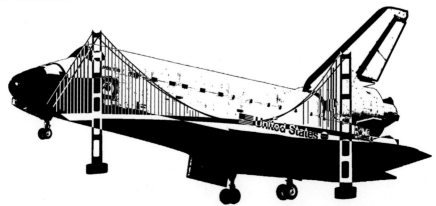

Now the conclusion is clear . . .

The invention of the one-time pad is usually credited to Gilbert Sandford Vernam. His 1919 patent describes an electromechanical implementation of addition in $\mathbb{Z}_2$ (or Boolean XOR) and applies it to the bitwise addition of a plaintext and a key tape. Both tapes use the 5-bit *Baudot code* of the English alphabet. No condition on the key tape is stipulated, but in Vernam (1926) he requires a random tape as long as the message. He also suggests to use the sum of two random key tapes of lengths k and $k+1$, which will in general only repeat after $n = k(k+1)$ steps. However, this combined string of n bits has entropy only about $2\sqrt{n}$ and is *"not, strictly speaking, a purely random key"* in Vernam's words, but more like a two-time pad. Friedman suggested ways of pausing the addition of the key stream. Vernam had proposed this system to the US Signal School during the First World War, but got little interest from them.

The idea of key addition, usually over a larger alphabet, say $\mathbb{Z}_{26}$, goes back to the 15th century (Chapter B) and Vernam's new idea is to use a key as long as the message. The one-time pad was used in practice by East bloc secret services since the 1920s, but its deployment is limited by the enormous key length required, which must be larger than all messages that will eventually be encrypted with it. The idea survives todays in the stream ciphers that add pseudorandom strings to messages, as already in the Siemens Geheimschreiber from 1932. The Soviet KGB reused by mistake (or lack of suitable paper) some one-time keys in the 1940s and 1950s. The US NSA found this out and uncovered several Soviet spy networks in their VENONA operation (Benson & Warner 1996).

9.5. Taxonomy of security reductions for signatures

We now consider the standard framework of modern cryptographic security, which we call *axiomatic security*. One proves rigorously the security of some cryptoscheme assuming the *axiom* that a certain problem is hard to solve. We have seen several examples of this, and now present the general approach.

Thus we consider security reductions "$X \leq_p$ breaking S", in analogy with Definition 3.27, or $X \leq_p S$ for short, where X is a computational problem and S a cryptoscheme. For the moment, we discuss signature schemes S only.

What does it mean to "break S"? There are several such notions, and we classify them by two parameters: the *attack goal* and the *attack resources*. We start by listing four attack goals.

Existential forgery: The adversary produces a signature s of some message m over which he has no control.

Selective forgery: The adversary chooses a message m and produces a
signature s of m.

Universal forgery : The adversary receives a message m and produces a
signature s of m.

Key recovery: The adversary produces the secret key for signing.

These four goals are sorted by difficulty:

$$(9.6) \qquad \begin{aligned} \text{existential forgery} &\leq \text{selective forgery} \\ &\leq \text{universal forgery} \leq \text{key recovery}. \end{aligned}$$

An inequality $X \leq Y$ in (9.6) means that the goal X is easy to attain
if one is allowed to use Y. For simplicity, we ignore selective forgery in
what follows.

Next we list some of the resources at the adversary's disposal. This
may include the possibility of submitting a message to a "signing oracle"
which returns a signature of it. In all cases, the public key for verification
is known to the adversary.

(Public) Key only: The adversary knows the public key, as everybody
does.

Known messages: There is a set M of messages, and a signature of each
$m \in M$ is known.

Nonadaptively chosen messages: The adversary chooses and submits a
set M of messages and receives a signature for each $m \in M$.

Chosen message: The adversary submits during his computation messages m and receives a signature of each of them.

These resources are ordered in a natural way:

$$\begin{aligned} \text{chosen messages} &\leq \text{nonadaptively chosen messages} \\ &\leq \text{known messages} \leq \text{key only}. \end{aligned}$$

Here an inequality $X \leq Y$ means that a successful attack with resources Y implies one with resources X. In all cases, the forged signature
returned by the adversary must be of a message that was not submitted
to the signing oracle. The known messages attackplays only a minor role
in this business, and we leave it out in what follows.

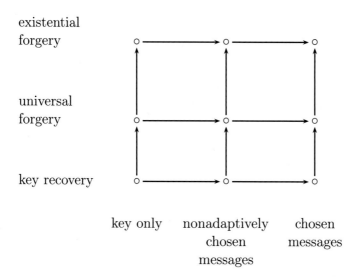

existential
forgery

universal
forgery

key recovery

key only   nonadaptively   chosen
                 chosen   messages
               messages

Figure 9.2: Relations between attack goals and resources.

As an example, Section 8.3 provides two existential forgeries with the key only resource for the ElGamal Signature Scheme 8.1.

We can combine the two lines of reductions into the 2D-picture of Figure 9.2. An arrow $\longleftarrow$ stands for $\leq$.

Now back to the reduction "$X \leq_p$ breaking S". X is a computational problem, S a signature scheme, and we can now specify the second term as "$S(\text{goal}, \text{resource})$", with one of the goals and resources discussed above. The intention of such a reduction is to have a theorem saying that if X actually is hard, then there is no attack on S with the resources and achieving the goal, both as specified. In other words, S is then secure against the stated attacks.

The security of S means that no polynomial-time algorithm with the stated resources can achieve the stated goal with nonnegligible success probability. The impossibility of attaining the three goals in Figure 9.2 is called existential unforgeability, universal unforgeability, , respectively. These are just Boolean values, either "true" or "false". S either has this property, or does not. The relations of (9.6) now become implications between these truth values, in the opposite direction:

$$\text{existential unforgeability} \Longrightarrow \text{universal unforgeability}$$
$$\Longrightarrow \text{impossible key recovery.}$$

The relations beween resources given above are similarly reversed. This turns Figure 9.2 for reductions into Figure 9.3 for security notions. The notation we use in this book deviates somewhat from the standard

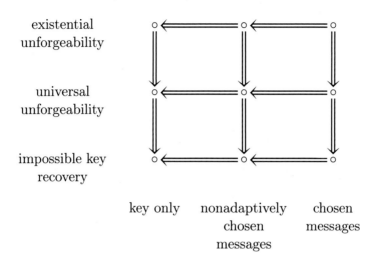

existential unforgeability		
universal unforgeability		
impossible key recovery		
key only	nonadaptively chosen messages	chosen messages

Figure 9.3: Implications between security notions.

one; see the Notes.

9.6. Security of the GHR signature scheme

We reduce a variation—stated below—of the RSA problem to breaking the *Gennaro-Halevi-Rabin signature* (GHR) from Section 8.5. It follows that if this special RSA problem is hard to solve, then it is hard to forge a GHR signature. There is, of course, no proof of the hardness of RSA, nor is there any indication to the contrary.

In the RSA problem, we have publicly known N, e, y and secret x and d, all of bit length n, with the property that $y = x^e$ and $x = y^d$ in $\mathbb{Z}_N^\times$. The RSA problem is to compute x from the public data. The RSA assumption is that this is hard to solve, that is, it cannot be done in time polynomial in n. We now make the task easier by not requiring an x that fits with the public RSA exponent e, but with any old exponent.

PROBLEM 9.7. Weak RSA problem.
Input: an RSA modulus $N = pq$ and $y \in \mathbb{Z}_N^\times$.
Output: $x \in \mathbb{Z}_N^\times$ and an integer $c \geq 2$ with $y = x^c$ in $\mathbb{Z}_N^\times$.

The (unique) solution to the RSA problem provides a solution to this problem. In other words, we have a trivial reduction

$$\text{weak RSA problem} \leq_p \text{RSA problem}.$$

In the weak RSA problem, we are given N as stated and $y \in \mathbb{Z}_N^\times$. The following reduction 9.8 solves the problem, using its access to an algorithm

$\mathcal{F}$ that forges GHR signatures with the same modulus N, requesting signatures of chosen messages.

The idea of the reduction is as follows. It chooses randomly a message i and will solve the weak RSA problem if it receives a signature of i from $\mathcal{F}$. The random value t in the key generation step is chosen judiciously, namely so that $\mathcal{A}$ can answer all requests from $\mathcal{F}$ for sample signatures of messages m, provided that $m \neq i$. All is well if $\mathcal{F}$ actually returns a signature of i.

ALGORITHM 9.8. Reduction $\mathcal{A}$ from weak RSA to forging GHR signatures.

Input: RSA modulus N, $y \in \mathbb{Z}_N^\times$.
Output: x, c with $x^c = y$ in $\mathbb{Z}_N^\times$ and $c \geq 2$, or "failure".

1. Choose $i \xleftarrow{\boxdot} \mathbb{B}^\ell$.
2. For each $m \in \mathbb{B}^\ell$, compute h_m. Compute the integer

$$u = \prod_{\substack{m \in \mathbb{B}^\ell \\ m \neq i}} h_m.$$

3. Choose $r \xleftarrow{\boxdot} \mathbb{Z}_N^\times$. Compute $t = (r^{h_i}y)^u$ in $\mathbb{Z}_N^\times$, and send the GHR public key $\mathsf{pk} = (N, t)$ to the forger $\mathcal{F}$.
4. When $\mathcal{F}$ requests a signature for some message $m \in \mathbb{B}^\ell$, $\mathcal{A}$ replies with $s = (r^{h_i}y)^{u/h_m}$ if $m \neq i$; note that u/h_m is an integer. If $m = i$, $\mathcal{A}$ reports "failure".
5. $\mathcal{F}$ returns some (m, s) at the end to $\mathcal{A}$, which is a validly signed message with probability at least ε. If $m \neq i$, $\mathcal{A}$ reports "failure". If $m = i$, $\mathcal{A}$ checks $\mathrm{ver}_{N,t}(m, s)$. If the signature is valid, it computes integers v and w so that $v \cdot h_i + w \cdot u = 1$ via the Extended Euclidean Algorithm in Section 15.5, using that $\gcd(h_i, u) = 1$. Then $\mathcal{A}$ returns $x = s^w r^{-uw} y^v$ and $c = h_i$.

THEOREM 9.9. Let N be an n-bit RSA modulus, $2^\ell \in \mathrm{poly}(n)$, and $\mathcal{F}$ be an existential (τ, ε)-forger of GHR signatures with modulus N, making at most $q < 2^\ell$ signature queries with chosen messages. Then using $\mathcal{F}$ one can solve the weak RSA problem for an n-bit input $y \xleftarrow{\boxdot} \mathbb{Z}_N^\times$ in time $\tau + (q+1) \cdot O(n^3 + \ell 2^\ell)$ with success probability at least $2^{-\ell}\varepsilon$.

PROOF. We will show the following properties of algorithm $\mathcal{A}$:

○ $\mathcal{A}$ can be executed in time polynomial in n,

○ $\mathcal{A}$ satisfies $\mathcal{F}$'s key and signature requests properly, or fails,

○ $\mathcal{A}$ returns a solution to the weak RSA problem with probability at least $\varepsilon \cdot 2^{-\ell}$.

Except for step 2, $\mathcal{A}$ only executes exponentiations in $\mathbb{Z}_N^\times$ and the Extended Euclidean algorithm, all of which can be done with $O(n^3)$ bit operations. In step 2, we compute the first 2^ℓ odd primes. This can be done with Eratosthenes sieve in about $\ell \cdot 2^\ell$ steps (see Section 3.3). By assumption, this is polynomial in n.

In step 3, the input $t \in \mathbb{Z}_N^\times$ to the GHR scheme is required to be uniformly random. Since $\gcd(h_i, \varphi(N)) = 1$, the h_ith power map $r \mapsto r^{h_i}$ on $\mathbb{Z}_N^\times$ is a bijection (Corollary 15.61), also the multiplication by $y \in \mathbb{Z}_N^\times$, and finally the uth power, since $\gcd(u, \phi(N)) = 1$. Thus $r \mapsto t$ as in step 3 is a bijection of $\mathbb{Z}_N^\times$, and therefore t is uniformly random, since r is. In step 4, any signature (m, s) sent from $\mathcal{A}$ to $\mathcal{F}$ is valid, since

$$s^{h_m} = (r^{h_i} y)^u = t$$

and hence $\mathrm{ver}_{N,s}(m, s) = $ "true".

It remains to bound the success probability in step 5. The algorithm does not even reach that step if ever a signature for $m = i$ is requested in step 4. We denote by q the number of signature requests made by $\mathcal{F}$ in a particular run. Since i is uniformly random and $\mathcal{F}$'s signature requests are made independently of it, the probability of failure is at most $q \cdot 2^{-\ell}$, and that of reaching step 5 is at least $1 - q \cdot 2^{-\ell}$. In the signed message (m, s) delivered by $\mathcal{F}$ to $\mathcal{A}$, m is not among the q messages whose signature was previously requested. Therefore we have $m = i$ with probability $1/(2^\ell - q)$ and $\mathcal{A}$ fails at this point with probability $(2^\ell - q)^{-1}$. If it succeeds, then s is a valid signature of m with probability at least ε, by the assumption on $\mathcal{F}$'s forging capabilities. In that case, $s^{h_i} = t$ and $\mathcal{A}$ indeed outputs a solution to the weak RSA problem:

$$x^c = (s^w r^{-uw} y^v)^{h_i} = (s^{h_i})^w r^{-uwh_i} y^{vh_i} = t^w r^{-uwh_i} y^{vh_i}$$
$$= r^{h_i uw} \cdot y^{uw} r^{-uwh_i} y^{vh_i} = y^{wu+vh_i} = y.$$

Since the three events discussed above are independent, the success probability of $\mathcal{A}$ is at least

$$(1 - q2^{-\ell}) \cdot (2^\ell - q)^{-1} \cdot \varepsilon = 2^{-\ell}\varepsilon. \qquad \square$$

If we allowed $q = 2^\ell$ queries to $\mathcal{F}$, then the reduction might always fail.

COROLLARY 9.10. *The weak RSA Problem 9.7 with modulus N is polynomial-time reducible to existentially forging GHR signatures with modulus N using chosen messages.*

The estimates in the reduction are precise enough to allow quantitative comparisons. Forgery probabilities around $2^{-30} \approx 10^{-9}$, one in a billion, are certainly dangerous, and maybe even ones around 2^{-40}. As to the value of ℓ, we will have to store the first 2^ℓ odd primes, requiring space about $\ell 2^\ell$. If we store only every thousandth one and recompute the intermediate ones whenever necessary, then we need $30 \cdot 2^{30-10}$ bits or 4 MB at $\ell = 30$, $40 \cdot 2^{40-10}$ bits or 5 GB at $\ell = 40$, and 6 TB at $\ell = 50$. Even if we start with a fairly strong forger $\mathcal{F}$, say with success probability $\varepsilon = 2^{-30}$, and use only a fairly small set of messages, say 2^{40} of them, then the success probability $\varepsilon \cdot 2^{-\ell} = 2^{-70}$ against the weak RSA problem is rather small. On the other hand, if even a tiny success probability worries us, then the information given by the reduction is that the GHR key length should be at least so that we consider weak RSA hard at the same key length. For standard RSA, $n = 3000$ bits are currently suggested (Table 5.12).

We now state a stronger assumption than the one that breaking RSA is hard, namely, that even the weak version is hard.

ASSUMPTION 9.11 (Strong RSA assumption). *The weak RSA Problem 9.7 is hard, that is, there is no probabilistic polynomial-time algorithm that solves it with nonnegligible success rate.*

COROLLARY 9.12. *Under the strong RSA assumption, GHR signatures are existentially unforgeable with chosen messages.*

In Section 8.5, we denote by $D_{n,\ell}$ the distribution of the moduli N used in GHR. More generally, we define the following.

DEFINITION 9.13. *(Distributional weak RSA problem). Given a distribution D on RSA moduli, an RSA modulus $N \xleftarrow{\boxtimes} D$, and $y \in \mathbb{Z}_N^\times$, compute $x \in \mathbb{Z}_N^\times$ and an integer $c \geq 2$ with $x^c = y$ in $\mathbb{Z}_N^\times$.*

The general reduction of average case to worst case from Section 15.7 yields

distributional weak RSA problem $\leq$ weak RSA problem.

An analog of the distributional factoring assumption 3.33 goes as follows. We assume a function $\ell \colon \mathbb{N} \longrightarrow \mathbb{N}$ with $2^{\ell(n)} \in \text{poly}(n)$.

CONJECTURE 9.14 (Distributional strong RSA assumption). *For ℓ as above and $D = D_{n,\ell(n)}$, the distributional weak RSA problem Definition 9.13 is hard to solve.*

As usual, hard means that no probabilistic polynomial-time algorithm has nonnegligible success probability. The reduction of Theorem 9.9 shows the following.

COROLLARY 9.15. *If the distributional strong RSA assumption 9.14 holds, then GHR signatures are existentially unforgeable with chosen messages attacks.*

This is a pleasant result in that we have identified a computational problem and proven rigorously that the signatures are secure, provided that the problem is hard.

Is this a satisfactory reduction? We would like to have a signature scheme with the property that under any reasonable notion, a forging algorithm implies that we can solve the (standard) RSA problem or, even better, factor the modulus. If we make the plausible but unproven assumption that the RSA problem is hard, then it follows that forging signatures is hard. In our reduction we do not use the standard but the weak RSA problem, and we assume a somewhat special distribution $D_{n,\ell(n)}$ on the RSA moduli.

9.7. Security reductions for encryption

We adapt the taxonomy of Section 9.5 from signatures to encryption, by defining various *attack goals* and *attack resources*. We consider probabilistic public-key block encryption schemes $S = $ (set-up, keygen, enc, dec), as in Definition 2.25, and start with four attack goals.

Distinguishing: The adversary presents two plaintexts of equal length. Then one of the two is chosen uniformly randomly and its encryption is given as the challenge ciphertext to the adversary, whose task it is to determine which one was chosen.

Malleability: The adversary is given a challenge ciphertext and transforms it into another one so that the corresponding plaintexts are distinct but meaningfully related.

Deciphering: The adversary is given a challenge ciphertext and finds the corresponding plaintext.

Key recovery: The adversary computes the secret key.

There are equivalent ways of describing the distinguishing goal. For one, the second plaintext is chosen uniformly at random from the message space; the rest of the task is unchanged. The second one is called *semantic insecurity* (and resistance against it is *semantic security*); see Section 9.2.

For malleability, we are given a polynomial-time decidable relation R on pairs (x, y) of plaintexts, and x and y are "meaningfully related" if $R(x, y)$ holds. An example is given by the homomorphic property (3.4) of RSA, so that $\text{enc}_{N,e}(2x) = 2^e \cdot \text{enc}_{N,e}(x)$ in $\mathbb{Z}_N$, and encryptions are malleable with respect to the relation $R(x, y) = "y = 2x$ in $\mathbb{Z}_N"$. Resistance against deciphering attacks is also called one-wayness.

We use *decryption* for the function $\text{dec}_{\text{sk}} \colon C \longrightarrow M$ as in Definition 2.25, and *deciphering* (or *cryptanalysis*) for computing this function without knowledge of the secret key sk. These four goals are sorted by difficulty:

$$(9.16) \qquad \text{distinguishing} \leq \text{malleability} \leq \text{deciphering} \leq \text{key recovery.}$$

What exactly does "distinguishing" mean? For a distribution X on a finite set S and an algorithm $\mathcal{D}$ with inputs from S and output 0 or 1, the expectation of $\mathcal{D}$ on inputs from X is, as in (15.83),

$$\mathcal{E}_\mathcal{D}(X) = \text{prob}\{\mathcal{D}(x) = 1 \colon x \xleftarrow{\boxtimes} X\}.$$

Definition 15.86 translates into the following.

DEFINITION 9.17. *(i) The* advantage *(or distinguishing power)* $\text{adv}_\mathcal{D}$ *of $\mathcal{D}$ between* two distributions X_0 and X_1 is

$$\text{adv}_\mathcal{D}(X_0, X_1) = |\mathcal{E}_\mathcal{D}(X_0) - \mathcal{E}_\mathcal{D}(X_1)|.$$

If $\text{adv}_\mathcal{D}(X_0, X_1) \geq \varepsilon$ *and $\mathcal{D}$ takes time at most τ, it is a (τ, ε)-distinguisher between X_0 and X_1. If no such distinguisher exists, then X_0 and X_1 are (τ, ε)-indistinguishable. When X_0 is a suitable uniform distribution, we also call $\text{adv}_\mathcal{D}$ the advantage of $\mathcal{D}$ on X_1.*

(ii) X_0 and X_1 are computationally indistinguishable *if* $\text{adv}_\mathcal{D}(X_0, X_1)$ *is negligible for all polynomial-time distinguishers $\mathcal{D}$.*

Part (ii) assumes that X_0 and X_1 are part of an infinite family of pairs of distributions.

If the difference inside the absolute value for $\text{adv}_\mathcal{D}(X_0, X_1) \geq \varepsilon$ is negative, then we may swap the outputs 0 and 1 of $\mathcal{D}$ and find for this new distinguisher $\mathcal{D}'$

$$(9.18) \qquad \text{adv}_{\mathcal{D}'}(X_0, X_1) = \mathcal{E}_{\mathcal{D}'}(X_1) - \mathcal{E}_{\mathcal{D}'}(X_0) \geq \varepsilon.$$

Then an output 1 of $\mathcal{D}'$ indicates that $\mathcal{D}'$ found its input to be more likely to come from X_1 rather than from X_0. Now we specify the work of a distinguisher $\mathcal{D}$ between encryptions in more detail.

DEFINITION 9.19. *Encryption distinguisher $\mathcal{D}$.*

   (i) *$\mathcal{D}$ generates two plaintexts x_0 and x_1.*

   (ii) *$\mathcal{D}$ receives a challenge ciphertext y, namely $y = \text{enc}_{\text{pk}}(x_i)$ for a uniformly random $i \xleftarrow{\;\boxtimes\;} \{0,1\}$ which is unknown to $\mathcal{D}$.*

  (iii) *$\mathcal{D}$ outputs $i^* \in \{0,1\}$.*

We denote by Y_0 and Y_1 the distributions of $\text{enc}_{\text{pk}}(x_0)$ and $\text{enc}_{\text{pk}}(x_1)$, respectively. The advantage *(or distinguishing power) of $\mathcal{D}$ is $\text{adv}_{\mathcal{D}}(Y_0, Y_1)$.*

Thus we view $\mathcal{D}$ as successful if and only if $i = i^*$. $\mathcal{D}$ may choose its plaintexts from any distribution it likes. For example, it might pick the all-zero string for x_0. If the encryption function enc is deterministic, as in RSA, then $\mathcal{D}$ can calculate the encryptions of x_0 and x_1, and easily make the distinction. Only probabilistic encryption schemes, such as ElGamal, stand a chance of being indistinguishable.

    For the success probability $\sigma_{\mathcal{D}}$ of $\mathcal{D}$ as above, we have

$$
\begin{aligned}
\sigma_{\mathcal{D}} &= \text{prob}\{\mathcal{D}(y) = 0 \colon y \leftarrow Y_0\} \cdot \text{prob}\{i = 0\} \\
&\quad + \text{prob}\{\mathcal{D}(y) = 1 \colon y \leftarrow Y_1\} \cdot \text{prob}\{i = 1\} \\
&= (1 - \mathcal{E}_{\mathcal{D}}(Y_0)) \cdot \frac{1}{2} + \mathcal{E}_{\mathcal{D}}(Y_1) \cdot \frac{1}{2} \\
&= \frac{1}{2}(1 + \text{adv}_{\mathcal{D}}(Y_1, Y_0)).
\end{aligned}
$$

(9.20)

    Next we list some resources potentially at the adversary's disposal. This may include the possibility of submitting a ciphertext to a "decryption oracle" which returns the corresponding plaintext.

Public key only: The adversary knows the public key, as everybody does.

Non-adaptively chosen ciphertexts: The adversary may submit ciphertexts of his choice to a decryption oracle, and after this phase receives the challenge ciphertext.

(Adaptively) chosen ciphertexts: The adversary submits ciphertexts to the decryption oracle, then receives the challenge ciphertext, and then may submit further ciphertexts (distinct from the challenge) to a decryption oracle. We usually leave out the "adaptively".

The adversary may submit invalid strings as ciphertexts and then receives a special reply "failure".

These resources are ordered in a natural way:

(9.21)
$$\text{public key only} \leq \text{non-adaptively chosen ciphertexts}$$
$$\leq \text{chosen ciphertexts.}$$

If an attack can be performed successfully with the resources on the left-hand side of an inequality, then that is also true for the right-hand side.

EXAMPLE 9.22. As an example, RSA can be deciphered with chosen ciphertexts due to its deterministic nature and the homomorphic property (3.4). The adversary knows the public key $\mathsf{pk} = (N, e)$ in the RSA notation of Figure 2.15. On receipt of the challenge ciphertext $y \in \mathbb{Z}_N^\times$, he submits $y_1 = y^{-1} \in \mathbb{Z}_N^\times$ to the decryption oracle and receives $x_1 = \mathsf{dec}_{\mathsf{sk}}(y_1) = y_1^d$. Then $y \neq y_1$ unless $y \in \pm 1$, and we have $y \cdot y_1 = 1$, so that $x \cdot x_1 = y^d \cdot y_1^d = 1^d = 1$ for $x = \mathsf{dec}_{\mathsf{sk}}(y)$. Thus he can successfully return $x = x_1^{-1} \in \mathbb{Z}_N^\times$. If the challenge y is not a unit in $\mathbb{Z}_N$, then $y = 0$ (and $x = 0$) or $\gcd(y, N)$ is a proper divisor of N, in which case he can even recover the secret key.

The following modified system called *randomized RSA* does not suffer from this drawback and may be secure. With N, e, and d as in the RSA notation of Figure 2.15 and even n, we encrypt an $n/2$-bit string $x \in \mathbb{B}^{n/2} = \{0, 1\}^{n/2}$ as follows. We choose $r \xleftarrow{\;\varnothing\;} \mathbb{Z}_N$, let $\ell(r) \in \mathbb{B}^{n/2}$ be the string of its $n/2$ lowest bits, and $\mathsf{enc}_{N,e}(x) = (r^e, \ell(r) \oplus x) \in \mathbb{Z}_N \times \mathbb{B}^{n/2}$. For any $(s, y) \in \mathbb{Z}_N \times \mathbb{B}^{n/2}$, we take $\mathsf{dec}_{N,d}(s, y) = \ell(s^d) \oplus y \in \mathbb{B}^{n/2}$. This is secure under a reasonable assumption, stronger than the RSA assumption 3.32. ◇

In an asymmetric cryptosystem, anyone can encrypt using the public key. This is not the case for symmetric systems, and obtaining encryptions may be an additional resource for an adversary. These might be for plaintexts that come from the outside (known plaintexts), of plaintexts that the adversary chooses himself (chosen plaintexts), or of random plaintexts (Chapter 6), either once at the start of the attack or dynamically during its course.

Combining the two lines of relations (9.16) and (9.21) yields the 2D-picture of Figure 9.4, where an arrow ⟵ stands for $\leq$.

We now adapt the end of Section 9.5 to the current question of encryption. So we consider a reduction "$X \leq_p$ breaking S", where X is a computational problem and S an encryption scheme. We can now specify the second term as "$S(\text{goal}, \text{resource})$", with one of the goals and resources discussed above. The intention of such a reduction is to have a theorem saying that if X actually is hard, then there is no polynomial-time attack on S with the resources and achieving the goal, both as specified, with

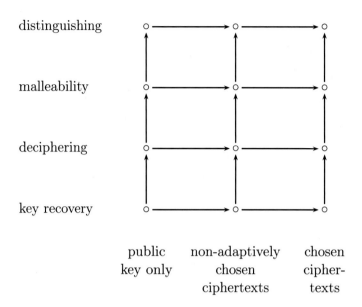

Figure 9.4: Relations between attack goals and resources.

nonnegligible probability. In other words, S is then secure against the stated attacks.

The impossibility of attaining the four goals in (9.16) is called non-malleability, indistinguishability, unbreakability, and impossible key recovery, respectively. These are just Boolean values, either "true" or "false". As in Section 9.5, the relations of (9.16) now become implications between these truth values, in the opposite direction, and the relations beween resources given above are similarly reversed. This turns Figure 9.4 for reductions into Figure 9.5 for security notions.

We see that indistinguishability under chosen ciphertexts is the strongest of these concepts, and hence the most desirable one for secure systems.

EXAMPLE 9.23. We have shown in Section 9.4 that the one-time pad, used with a random key sk, is perfectly secure. The public key is empty, and under a "public key only" attack, the system is malleable but indistinguishable. It is easily broken with a single chosen plaintext attack, since one pair (x, y) with $enc_{pk}(x) = x \oplus sk = y$ reveals the secret key $sk = x \oplus y$. Of course, this contradicts the spirit of the one-time pad, where the same sk is used only once.   ◇

This example points to two issues. Firstly, the security notions discussed here give a power to the adversary which is absent from more naive notions (as in Section 9.4) and provide a better insight into the security

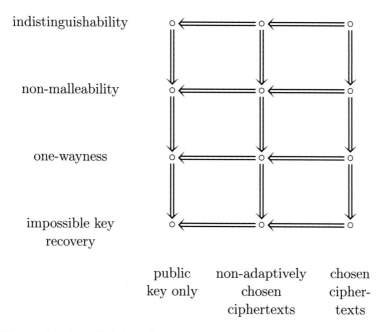

Figure 9.5: Implications between security notions for encryption.

against realistic attacks.

More importantly, it shows that these notions make sense for block ciphers, but possibly not for stream ciphers such as the one-time pad or a block cipher employed under some chaining mode (Section 2.8). A satisfactory theory for such encryption schemes is more difficult to obtain.

9.8. ElGamal encryption and decisional Diffie-Hellman

We present a reduction from the Decisional Diffie-Hellman Problem 2.23 (DDH) to distinguishing ElGamal encryptions with key only. We have a group $G = \langle g \rangle$ with d elements and consider DDH_G and ElGamal over G. The input to DDH_G is a triple $(A, B, C) \in G^3$. Let (A, B, C_0) be the corresponding Diffie-Hellman triple (Definition 2.22). Then either $C = C_0$ or $C \xleftarrow{\boxdot} G$. (In the latter case, we have $\text{prob}\{C = C_0\} = d^{-1}$.) The reduction $\mathcal{R}$ below returns 1 in the former and 0 in the latter case, with nonnegligible distinguishing power. We let $\mathcal{D}$ be a public key only ElGamal distinguisher.

REDUCTION 9.24. $\mathcal{R}$ FROM DDH_G TO $\mathcal{D}$.

Input: $A, B, C \in G$.

Output: 0 or 1.

1. Set-up: d, g, description of G.

2. Public key pk $\longleftarrow B$.
3. $\mathcal{D}$ chooses two distinct plaintexts $x_0, x_1 \in G$ and gives them to $\mathcal{R}$.
4. $i \xleftarrow{\boxed{\cdot}} \{0,1\}$.
5. $e \leftarrow (x_i \cdot C, A) \in G^2$.
6. $\mathcal{R}$ sends e to $\mathcal{D}$ and receives i^*.
7. If $i = i^*$ then return 1 else return 0.

THEOREM 9.25. *Let $\tau_\mathcal{D}$ and $\mathrm{adv}_\mathcal{D}$ denote the running time and advantage, respectively, of $\mathcal{D}$. Then the reduction takes time $\tau_\mathcal{D}$ plus one multiplication in G and distinguishes DH triples from random triples in G^3 with distinguishing power at least $\mathrm{adv}_\mathcal{D}/2$.*

PROOF.   With C_0 as above, we set $z_j = x_j C C_0^{-1}$ and $e_j = \mathrm{enc}_{\mathsf{pk}}(z_j) = (x_j C C_0^{-1} k, A) = (x_j C, A)$ for $j \in \{0,1\}$, both with secret session key $a = \mathrm{dlog}_g A$ and common session key $k = A^b = B^a = C_0$. Then $e = e_i$. We distinguish two cases. If $C = C_0$, then e is distributed as specified for the distinguisher $\mathcal{D}$, namely as $\mathrm{enc}_{\mathsf{pk}}(x_i)$. According to (9.20), the success probability of having $i = i^*$ is $(1 + \mathrm{adv}_\mathcal{D})/2$.

In the second case, C is uniformly random in G. Then both z_0 and z_1 are uniformly random in G, and e_0 and e_1 have the same distribution. For the output $i^* = \mathcal{D}(e)$, we have

$$\mathrm{prob}\{\mathcal{D}(e) = 1 \colon e = e_0\} = \mathrm{prob}\{\mathcal{D}(e) = 1 \colon e = e_1\}.$$

Since $i \xleftarrow{\boxed{\cdot}} \{0,1\}$, it follows that

$$\mathrm{prob}\{i = i^*\} = \frac{1}{2}$$

for both choices of i. Together we find

$$\begin{aligned}
\mathrm{adv}_\mathcal{R}(e_0, e_1) &= |\,\mathrm{prob}\{\mathcal{R}(A, B, C) = 1 \colon C = C_0\} \\
&\quad - \mathrm{prob}\{\mathcal{R}(A, B, C) = 1 \colon C \xleftarrow{\boxed{\cdot}} G\}| \\
&= |\,\mathrm{prob}\{i = i^* \colon C = C_0\} - \mathrm{prob}\{i = i^* \colon C \xleftarrow{\boxed{\cdot}} G\}| \\
&= \frac{1 + \mathrm{adv}_\mathcal{D}}{2} - \frac{1}{2} = \frac{\mathrm{adv}_\mathcal{D}}{2}.
\end{aligned}$$

Besides the call to $\mathcal{D}$, the time that the reduction takes is dominated by one multiplication in step 5.   □

We obtain the usual security consequence.

non-malleable	-	-
indistinguishable	DDH	-
undecipherable	DH	-
impossible key recovery	DL	GDH
	public key only	chosen ciphertexts

Figure 9.6: The security profile of ElGamal encryption.

COROLLARY 9.26. *If DDH_G is hard, then ElGamal encryption in G is indistinguishable by public key only attacks.*

This is a worst case result in that A and B are arbitrary inputs to the reduction and C either completes a DH triple or is uniformly random. In turn, $\mathcal{D}$ is assumed to distinguish between $\mathrm{enc}_{\mathsf{pk}}(x_0)$ and a random pair from G^2, for any A (session key) and B (public key). By the general considerations of Section 15.7, this also leads to a reduction between the average-case versions.

Exercise 9.3 asks to show that $DH_G \leq$ key only deciphering ElGamal over G, and it is clear that $DL_G \leq$ key recovery with key only attacks on ElGamal. The Gap Diffie-Hellman Problem GDH_G is defined as solving DH_G with access to a DDH_G oracle, and one can show that $GDH_G \leq$ key recovery in ElGamal with chosen ciphertexts. Thus we have the following security results on ElGamal encryption, assuming that the problems in the body of Figure 9.6 are hard. The three problems in the middle column are sorted in increasing difficulty:

$$\mathrm{DDH} \leq \mathrm{DH} \leq \mathrm{DL},$$
$$\mathrm{DDH\ hard} \implies \mathrm{DH\ hard} \implies \mathrm{DL\ hard}.$$

The results in Figure 9.6 are pleasantly consistent with this. Under the strongest assumption, hardness of DDH, we get the strongest security, namely indistinguishability. Under the weakest assumption, hardness of DL, we get the weakest security, namely resistance against key recovery.

9.9. Hofheinz-Kiltz-Shoup (HKS) encryption

According to Figure 9.5, indistinguishability under chosen ciphertexts (IND-CCA2 in crypto slang) implies all security notions we consider.

Among the possible hardness axioms on which one can base such security properties, factoring integers is arguably the most natural one. Thus IND-CCA2 under the factoring assumption is a highly desirable security property. We now present a system, due to Hofheinz, Kiltz & Shoup (2013) and here called HKS, which enjoys the desired security.

The HKS cryptosystem uses *safe primes* and *safe moduli* as in Section 3.5. In particular, $(p-1)/2$ is odd with at most two (large) prime factors and hence $p = 3$ in $\mathbb{Z}_4$. The same holds for q. We use a key generator sgen which on input n in unary produces such (approximately uniformly random) p and q of about $n/2$ bits and outputs $N = pq$ of n bits. Thus N is a Blum integer. $\mathbb{Z}_N$ is identified with the symmetric system of representatives (15.12)

$$(9.27) \qquad R_N = \{-(N-1)/2, \dots, (N-1)/2\},$$

and for $x \in \mathbb{Z}_N$, $|x|$ denotes its absolute value under this identification. Furthermore, this encryption scheme makes essential use of the Blum-Blum-Shub pseudorandom generator. This generator and its security are discussed in Section 11.8, and the required background in Section 15.14. We assume familiarity with this material. For output length ℓ and on input a random square $u \in \mathbb{Z}_N$, the generator produces

$$\mathrm{BBS}_N^{(\ell)}(u) = (\mathrm{lsb}(u), \mathrm{lsb}(u^2), \mathrm{lsb}(u^4), \dots, \mathrm{lsb}(u^{2^{\ell-1}})) \in \mathbb{B}^\ell,$$

where all powers of u are taken in $\mathbb{Z}_N$, and $\mathrm{lsb}(x)$ denotes the least significant bit of $x \in R_N$ in the symmetric (or signed) binary representation. Thus $\mathrm{lsb}(-1) = 1$, although $-1 = N-1$ in the usual system $\{0, \dots, N-1\}$ would be even.

$\square_N$ denotes the group of squares in $\mathbb{Z}_N^\times$. Since N is Blum, $-1 \notin \square_N$ and the absolute value function is injective on $\square_N$. We can transfer the group structure from $\square_N$ to its image $\square_N^+ = \{|x| : x \in \square_N\}$ and make $\square_N^+$ into a group, isomorphic to $\square_N$, with $x * y = |x \cdot y|$, where $\cdot$ is the standard multiplication in $\mathbb{Z}_N$. Then $\square_N^+ \subset \mathbb{Z}_N^\times$ is a group, but not a subgroup of $\mathbb{Z}_N^\times$, since the group operations are different. We have $\#\square_N^+ = \varphi(N)/4$. If $\square_N$ is cyclic, then so is $\square_N^+$. We recall the set $\boxtimes_N = -\square_N$ of antisquares and the subgroup

$$J_N = \left\{x \in \mathbb{Z}_N^\times : \left(\frac{x}{N}\right) = 1\right\} = \square_N \cup \boxtimes_N$$

of elements with Jacobi symbol 1. For $x \in \mathbb{Z}_N^\times$, we have

$$x \in J_N \iff -x \in J_N.$$

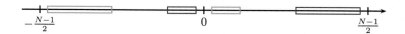

Figure 9.7: $\square_N$ and $\boxtimes_N$; $\square_N^+$ in the positive half.

For a positive $x \in R_N$, we have

$$x \in \square_N^+ \iff x \in J_N.$$

Thus membership in $\square_N^+$ can be decided in polynomial time. For $x \xleftarrow{\boxdot} \mathbb{Z}_N^\times$, $|x^2|$ is a uniformly random element of $\square_N$. In Figure 15.4 we have the cyclic group $\square_{21} = \{-5, 1, 4\} = \langle 4 \rangle$, $\boxtimes_{21} = \{-4, -1, 5\}$ and $\square_{21}^+ = \{1, 4, 5\}$ in $\mathbb{Z}_{21}^\times$. In particular, $4^2 = -5$ in $\mathbb{Z}_{21}^\times$ but $4^2 = 5$ in $\square_{21}^+$. The group operations $*$ in HKS take place in $\square_N^+$. But a sequence of multiplications or an exponentiation in $\square_N^+$ may be first performed in $\mathbb{Z}_N^\times$, then taking the absolute value of the result: $x_1 * x_2 * \cdots * x_m = |x_1 \cdot x_2 \cdots x_m|$. Thus we have $x^a * x^b = |x^c|$ if $c = a + b$ in $\mathbb{Z}_{\varphi(N)/4}$.

In Figure 9.7, the set of $\square_N$ of squares is represented by two green rectangles. (In reality, it will not consist of two intervals, but look rather randomly.) Then $\boxtimes_N = -\square_N$ and $\square_N^+$ consists of one green and one red rectangle to the right of 0.

KEY GENERATION 9.28. keygen.

Input: $n \in \mathbb{N}$.
Output: $\mathsf{pk} = (N, g, A)$, $\mathsf{sk} = (N, g, a)$.

1. Choose a random safe n-bit RSA modulus $N = pq$.
2. $g_0 \xleftarrow{\boxdot} \mathbb{Z}_N^\times$, $g \leftarrow |g_0^2| \in \square_N^+$.
3. $a \xleftarrow{\boxdot} \{1, \ldots, (N-1)/4\}$.
4. $A \leftarrow g^{a2^{4n}}$ in $\square_N^+$.
5. $\mathsf{pk} \leftarrow (N, g, A)$.
6. $\mathsf{sk} \leftarrow (N, g, a)$.
7. Return $(\mathsf{pk}, \mathsf{sk})$.

To encrypt an n-bit message $m \in \mathbb{B}^n$, we pick an irreducible polynomial $f \in \mathbb{F}_2[t]$, say the lexicographically first one, and take $\mathbb{F}_{2^n} = \mathbb{F}_2[t]/(f)$, a field with 2^n elements. For any $c = (c_0, \ldots, c_{n-1}) \in \mathbb{B}^n$, $\tilde{c} = (\sum_{0 \le i < n} c_i t^i \mod f) \in \mathbb{F}_{2^n}$ is the field element represented by c. Now BBS generates three n-bit session keys k_0, k_1 and k_2. Then k_0 is added to m bitwise like a one-time pad, and a MAC of the result is calculated, using the elements $\tilde{k}_1$ and $\tilde{k}_2$ of $\mathbb{F}_{2^n}$. The quantities A, R, S, and mac serve to check authenticity.

ENCRYPTION 9.29. enc.

Input: $\mathsf{pk} = (N, g, A), m \in \mathbb{B}^n$.
Output: $\mathsf{enc}_{\mathsf{pk}}(m) = C \in \square_N^+ \times \square_N^+ \times \mathbb{B}^n \times \mathbb{F}_{2^n}$.

8. $r \xleftarrow{\square} \{1, \dots, (N-1)/4\}$.
9. $R \leftarrow g^{r2^{4n}}$ in $\square_N^+$.
10. $S \leftarrow (g^R * A)^r$ in $\square_N^+$.
11. $u \leftarrow g^{r2^n}$ in $\square_N^+$.
12. $k_0 || k_1 || k_2 \leftarrow \mathrm{BBS}_N^{(3n)}(u)$.
13. $y \leftarrow m \oplus k_0 \in \mathbb{B}^n$.
14. mac $\leftarrow \tilde{k}_1 \cdot \tilde{y} + \tilde{k}_2 \in \mathbb{F}_{2^n}$.
15. $C \leftarrow (R, S, y, \mathrm{mac})$.
16. Return C.

DECRYPTION 9.30. dec.

Input: $\mathsf{sk} = (N, g, a), C = (R, S, y, \mathrm{mac}) \in \square_N^+ \times \square_N^+ \times \mathbb{B}^n \times \mathbb{F}_{2^n}$.
Output: $\mathsf{dec}_{\mathsf{sk}}(C) = m \in \mathbb{B}^n$.

17. If $R^{R+a2^{4n}} \neq S^{2^{4n}}$ then return "failure".
18. Using the Extended Euclidean Algorithm 15.19, find integers s, t, w
    such that $2^w = \gcd(R, 2^{4n}) = sR + t2^{4n}$.
19. $u \leftarrow (S^s * R^{t-sa})^{2^{n-w}}$ in $\square_N^+$.
20. $k_0 || k_1 || k_2 \leftarrow \mathrm{BBS}_N^{(3n)}(u)$.
21. mac$' \leftarrow \tilde{k}_1 \cdot \tilde{y} + \tilde{k}_2 \in \mathbb{F}_{2^n}$.
22. If mac $\neq$ mac$'$ then return "failure".
23. $m \leftarrow y \oplus k_0 \in \mathbb{B}^n$.
24. Return m.

When the conditions in steps 17 and 22 do not hold — so that "failure" is not returned — we call C *consistent*. We now check correctness and efficiency of the HKS scheme, and deal with security in the next section.

THEOREM 9.31. *The HKS Scheme can be implemented efficiently and works correctly.*

PROOF.   For security parameter n, most of the operations are exponentiations in $\mathbb{Z}_N$ with exponents of $O(n)$ bits. In step 9, $R \in \square_N^+$ is a positive integer less than $N/2$ and used as an exponent in step 10. In step 18, we have $R < N$, hence $w < n$ and the exponent 2^{n-w} in step 19 is an integer. These exponentiations can be implemented efficiently, in polynomial time. This also holds for the squarings in $\mathbb{Z}_N$ in BBS and the Euclidean algorithm in step 18.

Based on Theorem 3.25, also the safe modulus generation of step 1 can be done efficiently. The assignments in steps 12 and 20 mean that the $3n$-bit string on the right is parsed as a sequence of three n-bit strings.

For correctness, we assume the key pair $(\mathsf{sk}, \mathsf{pk})$ and an encryption C of a message $m \in \mathbb{Z}_N$ as input to decryption, and claim that the decryption equals m. For the condition in step 17, we calculate in $\square_N^+$

$$S^{2^{4n}} = (g^R * A)^{r2^{4n}} = g^{r2^{4n} \cdot R} * g^{a2^{4n} \cdot r2^{4n}}$$
$$= R^R * R^{a2^{4n}} = R^{R + a2^{4n}},$$

so that "failure" is not returned in step 17. The values of u calculated in steps 11 and 19 coincide, because

$$(S^s * R^{t-sa})^{2^{n-w}} = (g^{srR} * A^{sr} * g^{r2^{4n}(t-sa)})^{2^{n-w}}$$
$$= g^{r(sR + sa2^{4n} + t2^{4n} - sa2^{4n}) \cdot 2^{n-w}} = g^{r \cdot 2^w \cdot 2^{n-w}} = g^{r2^n}.$$

It follows that k_0, k_1, and k_2 have the same values in encryption and decryption, mac $=$ mac$'$, and step 24 returns the original plaintext m. $\square$

As noted in Section 3.4, g generates $\square_N^+$ with probability at least $1 - 2^{-n/8+2}$. When it does not generate the group, even in the extreme case $g = 1$, the system still works correctly. However, this is about as likely as a random guess of m being correct.

9.10. Security of HKS

We now show that HKS is indistinguishable under chosen ciphertext attacks provided the factoring assumption holds. Theorem 11.48 shows the corresponding property for the Blum-Blum-Shub generator and the following reduction is sufficient.

ALGORITHM 9.32. BBS-distinguisher $\mathcal{B}$ from HKS-distinguisher $\mathcal{D}$.

Input: A Blum integer N, $z \in \square_N^+$, $v \in \mathbb{B}^{3n}$.
Output: $b^* \in \mathbb{B}$ or "failure".

1. $g_0 \xleftarrow{\boxdot} \mathbb{Z}_N^\times$, $g \leftarrow |g_0^2| \in \square_N^+$.
2. $c \xleftarrow{\boxdot} \{1, \ldots, (N-1)/4\}$.
3. $R^* \leftarrow z$.
4. $A^* \leftarrow g^{c2^{4n} - R^*}$.
5. $S^* \leftarrow (R^*)^c$.
6. $\mathsf{pk} \leftarrow (N, g, A^*)$.
7. Send pk to $\mathcal{D}$, receive messages m_0, m_1 from $\mathcal{D}$.

8. $\hat{b} \xleftarrow{\boxtimes} \mathbb{B}$.

9. $k_0 \| k_1 \| k_2 \leftarrow v$.

10. $y^* \leftarrow m_{\hat{b}} \oplus k_0$.

11. $\mathrm{mac}^* \leftarrow \tilde{k}_1 \cdot \tilde{y}^* + \tilde{k}_2 \in \mathbb{F}_{2^n}$.

12. $C^* \leftarrow (R^*, S^*, y^*, \mathrm{mac}^*)$.

13. Pass the challenge C^* to $\mathcal{D}$. Decryption requests $C = (R, S, y, \mathrm{mac})$ from $\mathcal{D}$ are handled by Algorithm 9.33.

14. Receive "failure" or $\hat{b}^*$ from $\mathcal{D}$.

15. Return "failure" or $b^* \leftarrow \hat{b}^* \oplus \hat{b}$, respectively.

ALGORITHM 9.33. Handling decryption requests from $\mathcal{D}$ within $\mathcal{B}$.

Input: Decryption request $C = (R, S, y, \mathrm{mac})$.

Output: $m \in \mathbb{B}^n$ or "failure".

1. If $C = (R, S, y, \mathrm{mac}) \notin \square_N^+ \times \square_N^+ \times \mathbb{B}^n \times \mathbb{F}_{2^n}$, $C = C^*$, or $R^{R-R^*+c2^{4n}} \neq S^{2^{4n}}$ then return "failure".

2. If $R \neq R^*$ then

3.   Using the Extended Euclidean Algorithm 15.19, find s, t, w such that
$$2^w = \gcd(R - R^*, 2^{4n}) = s(R - R^*) + t2^{4n}. \quad \text{[Here } R \text{ and } R^* \text{ are}$$
considered as positive integers less than $N/2$.]

4. $u \leftarrow (S^s * R^{t-sc})^{2^{n-w}}$ in $\square_N^+$.

5. $k_0 \| k_1 \| k_2 \leftarrow \mathrm{BBS}_N^{(3n)}(u)$.

6. Else $k_0 \| k_1 \| k_2 \leftarrow v$.

7. If $\tilde{k}_1 \cdot \tilde{y} + \tilde{k}_2 \neq \mathrm{mac}$ in $\mathbb{F}_{2^n}$ then return "failure".

8. $m \leftarrow y \oplus k_0 \in \mathbb{B}^k$.

9. Return m.

The distinguisher $\mathcal{D}$ requests only a polynomial number q of decryptions. To be specific, we make the weak assumption that $q \leq 2^{n/2}$.

THEOREM 9.34. Let $\mathcal{D}$ be a distinguisher for HKS with advantage ε and at most $2^{n/2}$ decyption requests on n-bit moduli, and $n \geq 165$. Then the BBS-distinguisher $\mathcal{B}$ in Algorithm 9.32 has advantage at least $\epsilon - 2^{-n/8+3}$.

PROOF.   The distinguisher $\mathcal{B}$ executes Algorithm 9.33 and knows C^*, c, and v, but $\mathcal{D}$ only knows m_0, m_1, C^* and the public values.

To prove that $\mathcal{B}$ is in fact a BBS-distinguisher with the claimed advantage, it is sufficient to show the following claims:

   o REQUESTS. Algorithm 9.33 answers the decryption requests from $\mathcal{D}$ correctly.

○ DISTRIBUTIONS. The distributions of the challenge ciphertext C^* and the public key pk in the distinguisher $\mathcal{B}$ are identical to their distribution in the HKS distinguishing experiment, up to an error of less than $2^{-n/8+3}$.

○ FAILURE. The probability for the reduction to fail is less than $2^{-n/8+3}$.

○ ADVANTAGE. The distinguishing advantages of $\mathcal{B}$ and $\mathcal{D}$ differ at most by the failure probability.

We denote as $d = \#\square_N^+ = \varphi(N)/4$ the group order.

REQUESTS. Since $\mathcal{D}$ is a distinguisher using chosen ciphertexts, we have to make sure that $\mathcal{B}$ can handle decryption requests correctly. That is, on input $C = (R, S, y, \text{mac}) = \text{enc}_{\text{pk}}(m)$, Algorithm 9.33 should return $m = \text{dec}_{\text{sk}}(C)$ as long as C is consistent with the public key pk published by $\mathcal{B}$. The verification steps 1 and 7 correspond to those in the original decryption and guarantee that the input $C = (R, S, y, \text{mac})$ is consistent.

HKS encryption uses secret keys a and r which are not known to the decryption handler 9.33. We take the "fictitious" values

$$(9.35) \qquad \begin{aligned} a &= c - R^* 2^{-4n} \text{ in } \mathbb{Z}_d, \\ r &= (\text{dlog}_g R) \cdot 2^{-4n} \text{ in } \mathbb{Z}_d, \end{aligned}$$

both unknown to the decryption handler, and show that its output in step 9 equals $m = \text{dec}_{\text{sk}}(C)$ if C is $\text{enc}_{\text{pk}}(m)$ with $\text{sk} = (N, g, a)$ and r as session key in encryption step 8. For r to be defined, we assume that g generates $\square_N^+$. The two values in steps 4 of HKS Key generation 9.28 and of Algorithm 9.32 agree, since

$$A = g^{a2^{4n}} = g^{c4^n - R^*} = A^*.$$

Furthermore, we have $R^{R+a2^{4n}} = R^{R-R^*+c2^{4n}}$, so that "failure" occurs in HKS step 17 if and only if it does in step 1 of Algorithm 9.33 (assuming that the first two conditions there are not satisfied). In HKS step 10, we have $S = (g^R A)^r$, dropping from now on the $*$ for multiplication in $\square_N^+$. If $R \neq R^*$, the decryption handler with its values of s, t, w computes

$$\begin{aligned} u &= (S^s R^{t-sc})^{2^{n-w}} = ((g^R A)^{rs} g^{r2^{4n}(t-sc)})^{2^{n-w}} \\ &= (g^{sR} g^{s(c2^{4n}-R^*)} g^{t2^{4n}-sc2^{4n}})^{r2^{n-w}} \\ &= (g^{s(R-R^*)+sc2^{4n}+t2^{4n}-sc2^{4n}})^{r2^{n-w}} \\ &= (g^{s(R-R^*)+t2^{4n}})^{r2^{n-w}} = (g^{2^w})^{r2^{n-w}} = g^{r2^n}. \end{aligned}$$

This equals the BBS seed in step 11 of encryption. Hence the decryption request is answered correctly if $R \neq R^*$. The case $R = R^*$ is discussed at the end of this proof.

DISTRIBUTIONS. The distributions of pk and C^* within our distinguisher $\mathcal{B}$ differ slightly from the distributions that HKS produces. The value $g \in \square_N^+$ is generated in HKS step 2 and step 1 of $\mathcal{B}$ in the same way, and it generates $\square_N^+$ with probability at least $1 - 2^{-n/8+2}$. When it does not generate the group, HKS still works correctly but $\mathcal{B}$ possibly does not, because there may not exist a value r as in (9.35).

In order to show that the encryption C^* in step 12 of $\mathcal{B}$ is distributed almost like actual HKS encryptions of $m_{\hat{b}}$, we note that in keygen, a is uniformly distributed in $B = \{1, \ldots, (N-1)/4\}$. Is that also true of $c - R^* 2^{-4n}$ in (9.35)? No, it is not, but almost.

Since a is always positive, it is convenient to take the representatives $\{1, \ldots, d\}$ for $\mathbb{Z}_d$ in (9.35). Thus the values of a from (9.35) are all in $\mathbb{Z}_d \subseteq B$. Suppose first that the shift $R^* 2^{-4n}$ in (9.35) is zero. Since $c \xleftarrow{\boxtimes} B$ is taken modulo $d < (N-1)/4$ in $\mathbb{Z}_d$, a "small" value $a \leq (N-1)/4 - d$ may come from $c = a$ or from $c = a + d$, and the others come from the single value $c = a$. Thus the latter have probability $p = 1/((N-1)/4)$ under both distributions, and the statistical distance (15.85) is

$$\sum_{1 \leq a \leq \frac{N-1}{4} - d} |p - 2p| + \sum_{d < a \leq \frac{N-1}{4}} |p - p| = \left(\frac{N-1}{4} - d\right) \cdot p$$

$$= \frac{4(N - 1 - \varphi(N))}{N-1} = \frac{4(pq - 1 - (p-1)(q-1))}{N-1}$$

$$< \frac{4(p+q)}{N} \leq 2^{-n/2 + 4\log_2 n - 2} < 2^{-n/3},$$

using (3.26) and $n \geq 165$.

A nonzero shift $R^* 2^{-4n}$ just means shifting the correspondence between a and c and does not change the statistical distance.

Next we check that C^* is a correct encryption of $m_{\hat{b}}$, with pk $= (N, g, A^*)$. The session key in step 8 of HKS is, in analogy with (9.35),

$$r^* = (\mathrm{dlog}_g R^*) \cdot 2^{-4n} \text{ in } \mathbb{Z}_d,$$

so that $R^* = z = g^{r^* 2^{4n}}$ in step 3 of $\mathcal{B}$.

The values in steps 9 and 10 in HKS encryption with session key r^* in step 8 match those in steps 3 and 5 of $\mathcal{B}$, respectively, since

$$R = g^{r^* 2^{4n}} = R^*,$$

$$S = (g^{R^*} * A)^{r^*} = g^{(R^* + c2^{4n} - R^*)r^*} = g^{cr^* 2^{4n}} = (R^*)^c = S^*.$$

Step 11 of encryption uses

$$u = g^{r^*2^n} = (g^{r^*2^{4n}})^{2^{-3n}} = z^{2^{-3n}},$$

consistent with the BBS condition $z = u^{2^{3n}}$.

ADVANTAGE. The relation between the outputs of $\mathcal{B}$ and $\mathcal{D}$ is:

$$\mathcal{D} \text{ answers correctly} \iff \hat{b}^* = \hat{b} \iff \mathcal{B} \text{ returns } 1.$$

We denote the events where v is a sequence produced by the BBS generator and where it is a uniform random $3n$-bit sequence as real$_v$ and rand$_v$, respectively. In the event rand$_v$, k_0 is a uniform random element of $\mathbb{B}^n$ and the challenge ciphertext C^* is independent of $m_{\hat{b}}$. This implies that the guess $\hat{b}^*$ of $\mathcal{D}$ is independent of the choice of $\hat{b}$, and since the latter is uniformly random, we have

$$(9.36) \qquad \mathcal{E}_{\mathcal{B}}(N, z, U^{3n}) = \text{prob}\{\hat{b}^* = \hat{b}: \text{rand}_v\} = 1/2.$$

Next we assume real$_v$. Then $v = \text{BBS}^{(3n)}(u)$ for some $u \xleftarrow{\;\boxdot\;} \square_N^+$ and $z = u^{2^{3n}}$. On such a legitimate input, $\mathcal{D}$ answers correctly with probability $1/2 + \varepsilon$. By the arguments above, C^* is an encryption of $m_{\hat{b}}$ consistent with the public key $\mathsf{pk} = (N, g, A^*)$, and we have

$$(9.37) \qquad \text{prob}\{\hat{b}^* = \hat{b}: \text{real}_v\} = 1/2 + \varepsilon.$$

Thus $\mathcal{E}_{\mathcal{B}}(N, z, \text{BBS}^{(3n)}) = 1/2 + \varepsilon$, and the advantage of $\mathcal{B}$, as in Definition 9.17 (i), is

$$(9.38) \qquad \begin{aligned} \text{adv}_{\mathcal{B}} &= \left| \mathcal{E}_{\mathcal{B}}(N, z, \text{BBS}^{(3n)}) - \mathcal{E}_{\mathcal{B}}(N, z, U^{3n}) \right| \\ &= \frac{1}{2} + \varepsilon - \frac{1}{2} = \varepsilon. \end{aligned}$$

Additionally, we have to consider the cases where the reduction fails. We already discussed the cases where the distributions of public or secret keys are wrong in the reduction, resulting in an error probability of at most $2^{-n/8+2} + 2^{-n/3}$.

The distinguisher $\mathcal{D}$ may not ask for a decryption of C^*, but it might request that of some $C = (R, S, y, \text{mac})$ with $R = R^*$. Then also $S = S^*$ unless already rejected in step 1 of Algorithm 9.33. In step 3, we would have $w = 4n$ and step 4 could not be executed, since we do not know how to compute a 2^{3n}th root $x^{2^{-3n}}$ without the factorization of N. However, in that case step 6 of Algorithm 9.33 uses the challenge sequence v as the session key (k_0, k_1, k_2), yielding the encryption $(R^*, S^*, y, \text{mac})$. In the case of real$_v$, this will yield a correct decryption. In the case of rand$_v$,

v is a uniform random element of $\mathbb{B}^{3n}$. $\mathcal{D}$ can guess k_0 with chance $1/2$ due to step 10 of $\mathcal{B}$, but has no information about k_1 and k_2. Whatever value he chooses for mac in his request, it equals the uniformly random $\tilde{k}_1\tilde{y}+\tilde{k}_2 \in \mathbb{F}_{2^n}$ only with probability 2^{-n}. Hence the probability that step 7 of Algorithm 9.33 fails on a single legitimate encryption request is at most 2^{-n}. The failure probability on some of the q requests of $\mathcal{D}$ is at most $q2^{-n} \leq 2^{n/2} \cdot 2^{-n} = 2^{-n/2}$.

Including the probability that $\mathcal{B}$ fails, the distinguishing advantage of $\mathcal{B}$ is

$$\mathrm{adv}_{\mathcal{B}} \geq \varepsilon - 2^{-n/8+2} - 2^{-n/3} - 2^{-n/2} - 2^{-n/2+1} > \varepsilon - 2^{-n/8+3}. \qquad \square$$

Notes 9.1. For the cryptographic blunders of Mary Queen of Scots' secretaries, see Kahn (1967), Chapter 2. An example of a proven lower bound concerns *Gröbner bases*, a particularly useful way of presenting a multivariate polynomial ideal. They have been used in cryptanalysis. Ernst Mayr and his collaborators proved that the computation of such a basis requires, in general, exponential space, and thus also exponential time. See Mayr (1997) and von zur Gathen & Gerhard (2013), Section 21.7, for details.

The class P of polynomial-time decidable problems was introduced by Cobham (1965) and, with a hippie title, by Edmonds (1965). Levin (1973) described NP, independently of Cook. The Clay Foundation selected seven mathematical problems for its Millennium Prize, with a prize of one million US dollars for each of them. One has been solved (the Poincaré Conjecture, by Grigori Yakovlevich Perelman), and three interact with the topic of this book: P vs. NP, the Riemann Hypothesis (see the Prime Number Theorem 3.21), and the Birch and Swinnerton-Dyer conjecture on elliptic curves (Chapter 5). See http://www.claymath.org/millenium.

We do not even get secure cryptography under Cook's hypothesis. What seems to be needed is a trapdoor function (see Section 2.1) that is hard to invert even on average, not just in the worst case as in the definition of NP. Furthermore, problems of cryptographic interest such as factoring integers or discrete logarithms in specific groups are in NP, but are considered unlikely to be NP-complete. It is an open question whether the assumption $P \neq NP$ is sufficient for cryptographic security; see Goldreich & Goldwasser (1998).

Axiomatic security is also called *reductionist security*. A popular name in the literature is *provable security*, but that makes a promise that no-one can keep: proving security without assumptions. Actually, it means provable equivalence to a problem of unproven hardness (Lenstra 2006). In analogy, it would be considered pretentious to claim *provable hardness* for NP-hard problems.

Complexity theory manages a whole zoo of "hard" problems and reductions, but polynomial-time reductions from NP-complete problems have an outstanding importance. For security reductions, there is no such standard and many supposedly "hard" problems are used as the basis for the reductions. I heard the "frexponential" (exponential in n^{fraction}) from Volker Strassen.

It was alleged that the NSA could break DES at its inception; see Section 2.2. Government agencies worldwide command huge computing resources and manpower. They might be ahead of published research somewhat. By how much? Can they break AES? Factor large integers? In polynomial time? This is a great area for conspiracy theories.

9.2. Goldreich (2004), Section 5.2.3, shows the equivalence of semantic security and indistinguishability.

9.3. Diffie & Hellman (1976) introduced one-way and trapdoor functions. They are discussed extensively in Goldreich (2001), Chapter 2.

The German translation of *one-way function* is *Einbahnfunktion*, evoking the pictorial meaning of one-way street = Einbahnstraße. However, in German texts, one often sees *Einwegfunktion*. But Einweg refers to a one-time use, such as for glass bottles or session keys.

9.4. Vernam (1926) gives an account of his invention. Klein (2006) describes Vernam's unsuccessful contact with the military and the influence of his *"perhaps one of the most important [patents]"* in the history of cryptography. Kahn (1967), Chapter 13, explains lucidly the history of Vernam's discovery and Mauborgne's role in its development. According to Bellovin (2011), a Californian banker named Frank Miller had proposed a similar system in 1882. The basic idea of adding a short secret key repeatedly to the plaintext was known to Renaissance cryptographers like Alberti, Trithemius, and Vigenère, and explicitly described as modular addition (in clumsy notation) by Comiers d'Ambrun; see Section B.2.

9.5. One finds the further resource "single occurrence chosen messages" in the literature. This is the restriction of "chosen messages" where each message is allowed to be submitted only once. This makes sense if the signing oracle is required to behave exactly as the specified signing algorithm and the latter is probabilistic. But as long as one requires the oracle to just return a signature, as we do, there is no difference between the two resources, since the oracle might keep track of submitted messages and always return the same signature.

Readers who go to the original research literature will find some differences in notation. The standard is to use acronyms like EUF, UUF, and UB for existential and universal unforgeabilityand impossible key recovery ("unbreakability"), respectively. Similarly, KOA, NCMA, and CMA stand for the resources: key only attack, nonadaptively chosen messages attack (also DCMA = directed chosen messages), and chosen messages attack. This allows succinct statements like "S is EUF-CMA-secure". People find this useful in written communications, maybe even in an oral dialogue, but it poses an unnecessary difficulty for a beginner. The author's attitude towards such abbreviations is captured in PCMCIA = people cannot memorize computer industry acronyms.

A less superficial deviation from the standard presentation is in our way of proceeding. We state a reduction $X \leq_p Y$, prove its required properties, and conclude as a corollary that if X is hard, then so is Y (and our system described in Y is secure). It is more common to mix the two arguments and talk about the hardness of X already inside the reduction. As an example, one does not use "EF-CMA-break" for Y in a reduction, but only EUF-CMA-security, as in Corollary 9.12. If done properly, this is ok. But it is cleaner and technically easier to separate the reduction and the security conclusion. In my experience, a major source of confusion for beginners (and maybe for others ☺) is a perceived change of direction in

$$X \leq_p Y \Longrightarrow (X \text{ hard} \Longrightarrow Y \text{ hard}).$$

We try to avoid this confusion here.

9.6. The weak RSA problem and the strong RSA assumption were introduced in Barić & Pfitzmann (1997).

Satisfactory reductions are hard to come by. Many cryptosystems involve a specific hash function. Then a popular relaxation is to replace the hash values by the outputs of a *random oracle* under the reduction's control. Several systems S admit random oracle reductions "$X \leq$ breaking S" but no standard reduction is known. However, this notion has the drawback that even if X is hard, we cannot draw conclusions about the security of S for any particular hash function. In fact, there are provably insecure systems with a random oracle reduction (Canetti *et al.* (2004)). We do not pursue this approach here, but the reader can find more information and an example on the book's website.

9.7. The notions of distinguishability and semantic security come from Goldwasser & Micali (1984), who also proved their equivalence; see also Yao (1982). Malleability was introduced by Dolev *et al.* (1991). In its origin from materials science, the word refers to the ability to deform under compressive stress, as in hammering steel into thin plates. The non-adaptively chosen ciphertexts scenario was introduced by Naor & Yung (1990), and the adaptive one by Rackoff & Simon (1992).

In Definition 9.17 (ii), we have a family $(X_n, Y_n)_{n \in \mathbb{N}}$ of distributions on sets A_n and a distinguisher $\mathcal{D}_n$ for (X_n, Y_n) on A_n for all n, and the notions of polynomial time and negligible are with respect to n. This corresponds to "nonuniform" circuits $\mathcal{D}_n$. For the corresponding "uniform" notion, $\mathcal{D}$ is a Turing machine that takes inputs from $\bigcup_{n \in \mathbb{N}} A_n$. For randomized RSA and its security, see Goldreich (2004), Section 5.3.4.

Some of the Notes to 9.5 apply here *mutatis mutandis*. The usual acronyms are NM for non-malleability, IND for indistinguishablility, OW (one-way, see Section 9.3) for non-decipherability, UBK (unbreakability) for impossible key recovery, CPA (chosen plaintext attack) for key only, CCA1 for non-adaptively chosen ciphertexts, and CCA2 or CCA for adaptively chosen ciphertexts. According to Figure 9.5, IND-CCA is the strongest notion of security and hence the most desirable security property in this area. For symmetric systems, we have the known message attack (KMA) for encryptions of known plaintexts, CPA1 for statically chosen plaintexts, and CPA2 or CPA for dynamically chosen plaintexts.

In his lecture notes, Charles Rackoff (2012) considers the *counter mode* of encrypting a sequence $x_0, x_1, \ldots$ of n-bit message blocks into $y_0, y_1, \ldots$ as $y_i = x_i \oplus f_K(i)$ for $i \geq 0$. Here $f_K \colon \{0,1\}^n \to \{0,1\}^n$ is a pseudorandom function driven by a secret key K shared between sender and receiver and its argument is the n-bit representation of the block number i. Under suitable definitions and assumptions, this provides *privacy*, so that an adversary learns nothing about the message. It does not provide message integrity, since an adversary can send anything she wishes over the line. See also Goldwasser & Bellare (2008).

9.9. A precursor of HKS by Blum & Goldwasser (1985) achieved indistinguishability under chosen plaintext attacks. In the HKS system as published, four problems prevent a proof of the claimed security property:

- The use of Sophie Germain primes; it is not proven that there exist arbitrarily large ones. Hofheinz *et al.* (2013) circumvent this problem in an alternative version using a construction called the Goldreich-Levin predicate, but this yields a less elegant system.

- The use of a hash function in order to work with smaller exponents and thus more efficienctly; no hash function as required is known to be secure under the factoring assumption.

- The system is described as a key encapsulation mechanism (KEM) A general approach of Cramer & Shoup (2004) describes how, together with a secure data encapsulation mechanism (DEM) one can obtain a secure encryption scheme. However, no suitably secure DEM seems to be known.

- The security of a non-standard version of the BBS generator is assumed, where additional information—the value of u^{2^ℓ} —is publicly provided.

The description in this text avoids these problems by appealing to Section 3.5 for safe primes, dropping the use of a hash function, and describing HKS directly as an encryption scheme. The standard security reduction for the Blum-Blum-Shub generator can easily be adapted to the current requirement, and this is done in Section 11.8.

9.10. Many thanks go to Dennis Hofheinz for correcting an oversight in this presentation. The alternative HKS decryption of Exercise 9.7 was suggested by Christoph Peters, a student in one of the author's crypto courses.

Exercises.

EXERCISE 9.1 (Perfect secrecy).   *You are to examine how the distribution of keys affects the distribution of output bits in a one-time pad encryption system.*

We consider the encryption system $y = \mathrm{enc}_k(x)$, where the encryption key k is in the key set $K = \{1, 2, 3, 4\}$. The plaintext x and the ciphertext y belong to same set $P = C = \{a, b, c, d\}$. The mapping enc is given by the following table:

	a	b	c	d
1	b	c	d	a
2	c	d	a	b
3	d	a	b	c
4	a	b	c	d

This means for example that $\mathrm{enc}_1(a) = b$, and enc_4 does not change the data.

Let $\mathcal{P}$ be the plaintext distribution, so that $\mathrm{prob}\{a \xleftarrow{\boxdot} \mathcal{P}\} = 1/2$ means that the character a appears in a plaintext with probability $1/2$. Suppose further that $\mathrm{prob}\{b \xleftarrow{\boxdot} \mathcal{P}\} = 1/4$, $\mathrm{prob}\{c \xleftarrow{\boxdot} \mathcal{P}\} = \mathrm{prob}\{d \xleftarrow{\boxdot} \mathcal{P}\} = 1/8$. Similarly, denote as $\mathcal{K}$ and $\mathcal{C}$ the distributions of keys and ciphertexts.

(i) *Show the identity*

$$\mathrm{prob}\{y \xleftarrow{\boxdot} \mathcal{C}\} = \sum_{\mathrm{enc}_k(x)=y} \mathrm{prob}\{x \xleftarrow{\boxdot} \mathcal{P}\} \cdot \mathrm{prob}\{k \xleftarrow{\boxdot} \mathcal{K}\}.$$

(ii) *Suppose that $\mathrm{prob}\{4 \xleftarrow{\boxdot} \mathcal{K}\} = 0$ and the other three keys each occur with probability $1/3$.*

   (a) *For a, b, c, d compute the probability of observing them as output.*

   (b) *For each $x, y \in P$ compute the conditional probability $\mathrm{prob}\{x \xleftarrow{\boxdot} \mathcal{C} : y \xleftarrow{\boxdot} \mathcal{C}\}$ that the plaintext is x if we observe the ciphertext y.*

(iii) *Now suppose that the keys are uniformly chosen in K. Do the same as in (ii).*

(iv) *Which of these key schedules is better for a one-time pad system? Can you find an even bettter one?*

EXERCISE 9.2 (Randomized RSA).   *We have seen in Section 9.7 that RSA is malleable, and in Example 9.22 that it is even decipherable with chosen ciphertexts, and also an attempt at a repair. As an alternative, consider encrypting $x \in \mathbb{Z}_N$ by choosing $r \xleftarrow{\boxdot} \mathbb{Z}_N$ and computing $\mathrm{enc}_{\mathrm{pk},r}(x) = (r^e, (x+r)^e) \in \mathbb{Z}_N^2$. Decryption of $(s, y) \in \mathbb{Z}_N^2$ yields $\mathrm{dec}_{\mathrm{sk}}((s, y)) = (s^d, y^d)$, so that $\mathrm{dec}_{\mathrm{sk}}(\mathrm{enc}_{\mathrm{pk},r}(x)) = (r, x+r)$, from which x is easily computed. Is this cryptosystem malleable with chosen ciphertexts? Distinguishable? Decipherable? Present the assumptions (if any) and the reasoning for your answer.*

EXERCISE 9.3 (Diffie-Hellman and ElGamal).   Given a group $G = \langle g \rangle$, you are to reduce the Diffie-Hellman Problem 2.21 to deciphering ElGamal encryptions with key only. The idea is to use the inputs A and B to Problem 2.21 in the ElGamal system (Protocol 4.5), choose $y \longleftarrow G$, and submit (y, A) to the adversary $\mathcal{A}$.

   (i) If $\mathcal{A}$ correctly returns the decipherment x, how do you determine g^{ab} from it?

  (ii) State the reduction in detail, and show that the distribution of the submissions to $\mathcal{A}$ equals the distribution of ElGamal encryptions.

 (iii) Letting $\tau_{\mathcal{A}}$ and $\sigma_{\mathcal{A}}$ denote the running time and success probability of $\mathcal{A}$, derive bounds on the corresponding quantities for the reduction.

 (iv) Conclude that if DH_G is hard, then ElGamal encryptions are secure against deciphering with key only.

EXERCISE 9.4 (Worst-case DL $\leq$ average-case DL). Let G be a finite cyclic group of order $d = \#G$ and bit size $n \approx \log_2 d$, and DL_G the discrete logarithm problem from Definition 4.2 (iv). We think of G as being the nth member in a family of groups, so that it makes sense to talk about "polynomial in n".

   (i) In the spirit of Sections 7.2 and 15.7, define carefully worst-case DL_G (with a success rate between 0 and 1) and average-case DL_G.

  (ii) Show

$$\text{worst-case } DL_G \leq_p \text{average-case } DL_G.$$

EXERCISE 9.5 (Rabin encryption).   The following asymmetric encryption scheme is from Rabin (1979). Let $N = pq$ be an n-bit Blum integer, $\mathsf{sk} = (p, q)$, and $\mathsf{pk} = N$, and for a plaintext $x \in \mathbb{Z}_N^\times$, let $\mathsf{enc}_{\mathsf{pk}}(x) = x^2 \in \mathbb{Z}_N^\times$. Square roots modulo a prime can be computed efficiently; see von zur Gathen & Gerhard (2013), Chapter 14.

   (i) State a "list decryption" algorithm that returns a short list of values with x among them.

  (ii) Suppose that x itself is a square. Knowing sk, can you identify it uniquely in the short list?

 (iii) The elements of $\mathbb{Z}_N$ are represented as n-bit strings. Suppose that the plaintexts x to be sent are 8-bit ASCII encoded English sentences, that the entropy of English is 1.7 bits per letter (see (A.15)), and that about 2^{cn} of all n-bit strings represent English text, with $c = 1.7/8$. Furthermore, pretend that all entries (except the plaintext) in the decryption list are random. What is the probability that the list contains more than one English text?

 (iv) Let k be a small number, say $k = 32$, suppose that the binary representation of x has only $n - k$ bits, and pad x (at the low end) with a random k-bit string. What are the chances that this new number x' is a square? What can you say about different choices of k?

  (v) Devise a new encryption algorithm based on the above that allows unique decryption for most plaintexts x.

 (vi) Show that if one can recover x efficiently from $\mathsf{enc}_{\mathsf{pk}}(x)$ with key only, then one can factor N efficiently. This means that the system is secure if factoring is hard.

(vii) *Show that an efficient adversary who can ask for decryptions of ciphertexts of his choice (chosen ciphertext attack) can factor N. This means that the system is insecure.*

(viii) *Evaluate the conclusions of (vi) and (vii).*

You may want to consult Exercises 3.16, 8.10, 15.20, and 15.21.

EXERCISE 9.6 (McEliece encryption scheme).

(i) *Learn the basic theory of error-correcting codes, for example in MacWilliams & Sloane (1977), including Goppa codes and their decoding.*

The system works as follows. Given publicly known parameters d and t, set $n = 2^d$ and $k = n - dt$, produce a generator matrix $G \in \mathbb{F}_2^{k \times n}$ of a Goppa code of dimension k and length n, the matrix $P \in \mathbb{F}_2^{n \times n}$ of a random permutation in Sym_n, a random invertible matrix $S \in \mathbb{F}_2^{k \times k}$, and $H = SGP \in \mathbb{F}_2^{k \times n}$. Then $\mathsf{pk} = H$ and $\mathsf{sk} = (G, P, S)$. For encrypting $x \in \mathbb{F}_2^k$, choose uniformly random $e \in \mathbb{F}_2^n$ of Hamming weight t. Then $\mathsf{enc}_{\mathsf{pk}}(x) = xH + e$.

(ii) *Use sk and a Goppa decoding algorithm to determine xS from yP^{-1}, and then x.*

(iii) *Devise a distinguishing attack on the system using chosen ciphertexts.*

(iv) *Read Berlekamp et al. (1978) on the difficulty of decoding random linear codes, McEliece (1978) for his encryption scheme, and Bernstein et al. (2008) on currently suggested parameters and key sizes.*

The promise of this cryptosystem is that so far it resists number-theoretic attacks as for discrete logarithms or factoring integers. Its drawback is the large key size.

EXERCISE 9.7 (Alternative HKS decryption).

(i) *Show that the following algorithm alt-dec for decrypting HKS works correctly. We use $d = \# \square_N^+$.*

ALGORITHM. Alt-dec.

Input: As for Decryption 9.30.
Output: $m \in \mathbb{B}^\ell$.

1. $e \longleftarrow 2^{-4n}$ in $\mathbb{Z}_d$.
2. $S' \longleftarrow R^{(R + a2^{4n}) \cdot e}$ in $\square_N^+$.
3. Execute steps 18, 19, 20, 23, and 24 using S' for S.

(ii) *The algorithm alt-dec from (i) does not process the value S produced in step 10 of HKS encryption enc 9.29. Omitting the output of S from enc yields enc-without-S. For which keys do (enc, dec), (enc, alt-dec), and (enc-without-S, alt-dec) provide bijections between plaintexts and ciphertexts?*

(iii) *According to Section 8.1, an asymmetric cryptosystem yields a signature scheme. For which of the schemes in (ii) does this work?*

(iv) *Study the reduction from BBS-distinguishing to HKS-distinguishing. Does it work for (enc-without-S, alt-dec)? If not, where does it fail?*

EXERCISE 9.8. *Consider the unit group $\mathbb{Z}_N^\times$ and R_N as in Section 9.9.*

(i) *Show that $R_N^+ = \{x \in R_N : x \geq 1, x \in \mathbb{Z}_N^\times\}$ with the usual multiplication is not a group.*

(ii) *Describe a natural operation on R_N^+ that makes it into a group.*

(iii) *Let $x_1, \ldots, x_m \in \mathbb{Z}_N^\times$. Show that $|x_1| * \cdots * |x_m| = |x_1 \cdots x_m|$.*

(iv) *Let $x \in \square_N^+$, $e \in \mathbb{N}$, and $y = x^e$ in $\mathbb{Z}_N^\times$. Show that x^e in $\square_N^+$ equals $|y|$.*

EXERCISE 9.9 (Inverse of 2). *Let $d \geq 3$ be an odd integer and $\ell \geq 1$. Determine explicitly the inverses of 2 and of 2^ℓ in $\mathbb{Z}_d$.*

EXERCISE 9.10 (HKS security). *We consider some arguments in the proof of Theorem 9.34.*

(i) *Why is 2^{4n} invertible in $\mathbb{Z}_d$?*

(ii) *$\mathcal{B}$ knows C^*, c, and v. What is meant by saying that it does not know $a = c - R^* 2^{-4n}$ in (9.35)?*

Les vraies conquêtes, les seules qui ne donnent aucun regret,
sont celles que l'on fait sur l'ignorance.[1]
NAPOLÉON BONAPARTE (1797)

These devices, taken as a whole, have added little,
if at all, to the security of the straight-alphabet ciphers,
though, for the most part, they have succeeded admirably
in rendering their ciphers totally unfit for general purposes.
HELEN FOUCHÉ GAINES (1939)

THERE is another Art or Method of Writing which has been of
very antient Usage, and tho' it is not now much in use, yet we
have the Equivalent to it now, which we call a Cypher, and they
are indeed the same thing that the Antients call'd *Steganography*,
as above. I shall give a brief Account of this Way of Writing here,
and dismiss it at once, for it has no great matter in it to make it
worth while to dwell on, for any long time.
DANIEL DEFOE (1726)

Elle me dit qu'elle ne l'enfermait pas sous clef, parce qu'il
était écrit en chiffres et qu'elle seule en avait la clef.—
Vous ne croyez donc pas, Madame, à la stéganographie?
[...] Je la quittai emportant avec moi son âme,
son cœur, son esprit et tout ce qui lui restait de bon sens.[2]
JACQUES CASANOVA DE SEINGALT (1832)

Da die Steganographie hauptsächlich in dem Krieg
ihre Verdienste hat, so setze ich hier mein Augenmerk
sonderheitlich auf die Herren Befehlshaber bey Armeen;
im übrigen wird es niemand schwehr fallen,
diese Regeln auf seinen Zustand zu appliciren.[3]
C. W. P. (1764)

---

[1]The real conquests, those that leave no regret, are those that you make on ignorance.

[2]She told me that she did not keep it [her secret recipe for making gold] under lock, because it was written in cipher and only she had the key. You do not believe, Madame, in steganography? [...] I left her, taking with me her soul, her heart, her mind, and all the good sense that she still had.

[3]Since steganography enjoys its merits mainly in war, I focus here particularly on the gentlemen that are commanders of armies; in addition, nobody will find it difficult to apply these rules to his situation.

Chapter E

Steganography

ncrypted messages usually contain funny symbols or jumbled letters which give away their nature. In steganography, one tries to hide even the fact that the message carries hidden information. Secret inks were popular through the centuries, and today we have steganographic techniques that try to hide information in digital files.

E.1. Invisible ink

The Greek word στεγανός (steganos, meaning *covered*) forms together with γραφειν (graphein, *to write*) the word steganography. It refers to the art of hiding a message in some other text (or picture) in a way that disguises its pure existence.

Invisible (or *sympathetic*) *ink* has been used throughout history. It can be made from a variety of substances, say fresh milk, lemon juice, urine, and others. Each has their own recipe for becoming visible after some treatment—with a chemical agent or by heating. However, modern-day censors have devised general methods that detect any such substance, by discerning the slight disturbance caused in a smooth paper surface when the invisible ink is applied.

Donatien Alphonse François Marquis de Sade (1740–1814) spent 27

© Springer-Verlag Berlin Heidelberg 2015
J. von zur Gathen, *CryptoSchool*, DOI 10.1007/978-3-662-48425-8_14

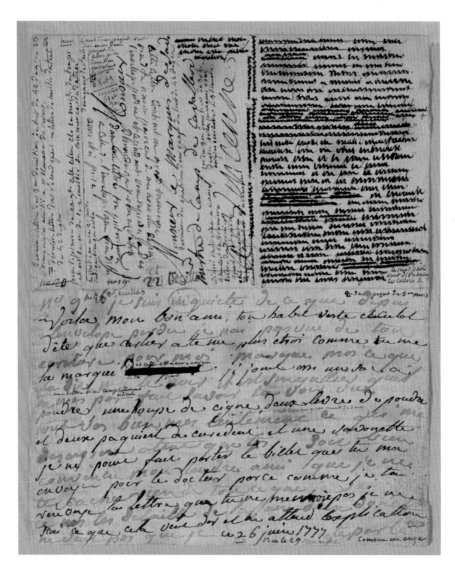

Figure E.1: Letter to the Marquis de Sade in prison from his wife, dated 26 June 1777. Plaintext is in black, sympathetic ink writing in light brown, and prison censorship at top right.

years in various jails, for his "sadistic" sexual practices. His extensive correspondence involved a constant fight against prison censorship. Figure E.1 shows a letter from his wife (black writing) with interline messages in secret ink (light brown) after heating. The brown spotting presumably comes from candles held too close. The top right part has been censored and is unreadable. The letter says the following:

[In brown ink:] *Je suis inquiète de ce que, depuis l'enveloppe perdue, je n'ai pas vu de ton écriture pour moi. Marque-moi ce que t'a dit M. Le Noir [Sade's lawyer]. Il est singulier qu'il ne m'ait pas fait savoir t'avoir vu. Tout va bien, mais lentement, ce qui me désespère autant que toi. Sois bien convaincu, mon tendre ami, que je ne te cache rien de tout ce que je sais. Ce n'est que les détails de l'affaire d'Aix dont on ne veut pas que je te parle.*

[In black ink:] *Voilà, mon bon ami, ton habit, veste et culotte d'été, que Carlier a tenu plus étroit, comme tu me l'as marqué. J'y joins aussi un sac à poudre, une houppe de cygne, deux livres de poudre et deux paquets de curedents et une savonnette. Je ne peux faire partir le billet que tu m'as envoyé pour le docteur, parce que, comme je t'ai renvoyé la lettre que tu ne me renvoies pas, je ne sais ce que cela veut dire et en attends l'explication.*

[In the top left part an annotation by de Sade:] *Cette lettre contient un grand mensonge; vous dites n'avoir pas reçu de mes nouvelles depuis l'enveloppe perdue, et cependant vous répondez à des demandes que je vous ai faites dans une lettre postérieure à celle de l'enveloppe que vous dites avoir été prise.*[1]

Besides invisible ink, a plethora of other steganographic techniques have been used since ancient times. Hiding papers in hollowed-out shoe

---

[1] I am worried about the fact that, since an envelope got lost, I have not seen anything in writing from you. Note well what Mr. Black told you. It is strange that he did not tell me that he saw you. All goes well but slowly, which dispairs me as it does you. Rest quite assured, my dearest friend, that I do not hide anything from you that I know. It is only the details of the Aix affair that they do not want me to talk about.

Here are, my dear friend, your jacket, vest and summer pants which Carlier made tighter as you told me. I also add a bag for powder, a swan powder-puff, two pounds of powder, two packs of tooth-picks, and a piece of soap. I cannot pass to the doctor the note that you sent me, since I returned your letter and you do not send it back to me. I do not know what it means and wait for an explanation of it.

This letter contains a big lie; you say that you did not receive any news from me since the lost envelope, and still you reply to inquiries that I made to you in a letter posted after the one that you say was confiscated.

Figure E.2: Friderici's steganography

soles, books, pieces of furniture, cigarette lighters and the like achieves this purpose (well, until they are discovered), but are not really cryptographical. The Russian spy Richard Sorge in Japan during the Second World War used photographic reduction onto *microdots*, about as large as this printed period ⟶ ·, to send his hidden messages. This was major progress over the *macrodot* of Histiaeus, the Greek tyrant of Milet in Asia Minor during an uprising against the Persians. He shaved the head of his slave Spiridon, tattooed a message asking for support from his son-

Figure E.3: Friderici's steganography

in-law Aristagoras in Greece on it, and sent him off when the hair had grown back. At his destination, he received another total haircut. Poor Spiridon knew nothing about the message and was told that the double shaving would cure his bad eyes. If Histiaeus' message consisted of 50 letters of 5 bits each, transmitted within a month, it would amount to about 100 μbits/sec. The message can't have been really urgent, at least by today's standards.

Steganography is used today, for example, for watermarking files to protect intellectual property. Most of the techniques are not secure in the sense of Chapter 9, and many have been broken. Some exceptions are in Kiayias *et al.* (2005), Hopper *et al.* (2002), and Cachin (2004). We do not treat them in this text.

E.2. Steganographic images

Friderici (1685) is the first book in German language on cryptography. It is a large compendium of cryptographic recipes, mainly simple substitutions and some transpositions. Each method is illustrated by a complete example, and followed by observations on decryption and possible variations. Most of the methods are from the literature, but no references are given. Codebooks are not mentioned at all. He describes an autokey system, where plaintext $x_0, x_1, \ldots$ is encrypted as $y_0 = x_0, y_1 = x_1 - x_0$, $y_2 = x_2 - x_1, \ldots$. He gives clear instructions for "circular" counting modulo 24, his alphabet size, but without using arithmetic terms like addition.

Figure E.2 shows a steganographic image from his book. You solve it by drawing (or imagining) vertical lines through the alphabetic marks at the bottom. Wherever a special mark, namely an eye of a person, animal or fruit lies on a vertical line, that letter is counted. The whole image is then read from left to right and top to bottom. It decrypts line by line as: u ns er co m m en d ant ist t o d = Unser Commendant ist tod = Our commander is dead. Such images are hard to produce. They are elegant examples of *salon cryptography*, meant to entertain, and to show some intellectual possibilities, but not for practical use.

E.3. Bacon's biliteral cipher

Sir Francis Bacon (1561–1626), Baron Verulam, Viscount St. Albans and Lord Chancellor to Queen Elizabeth I. was one of the great scientists of the late Renaissance. He advocated the "new" science of experiment and deduction over the "old" science of interpreting the writings of the ancients. In his *Instauratio Magna* or *Advancement of Learning* from 1640, he laid down his system.

In 1605, he proposed a binary encoding of letters—a far-sighted forerunner of Baudot and ASCII codes, and Bacon was quite proud of it:

> *We will annexe an other invention, which, in truth, we*
> *devised in our youth, when we were at Paris: and is a thing*

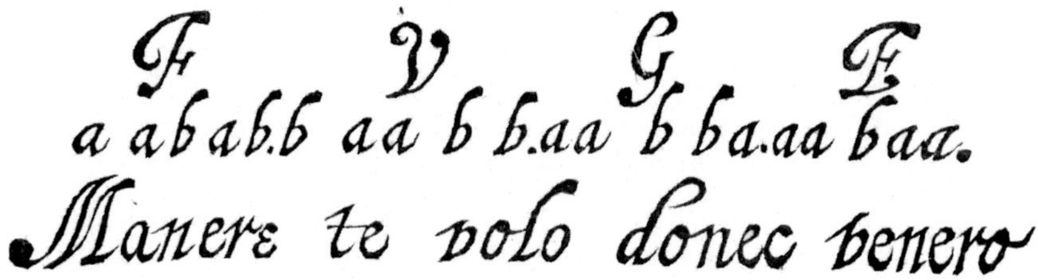

Figure E.4: Bacon's sample message in his biliteral cipher.

that yet seemeth to us not worthy to be lost. It containeth the highest degree of Cypher.

In BSCII[2], five bits are used to encode the 24 letters of his alphabet (without j and u) from 00000 for a to 10111 for z. He actually writes a for the digit 0 and b for 1, so that a is aaaaa and z is babbb. (The full power of the binary notation, also for numbers, was realized by Leibniz 1679, 1703.) In Bacon's *biliteral cipher*, one then takes a harmless cover text and writes it in two fonts. Then each letter in the 0-font is converted to a 0, and in the 1-font to a 1, so that a series of zeroes and ones results, and this encodes the true message in binary. Figure E.4 shows an example of his, with the cover text ma*nere* te volo *do*nec *v*enero (= I want you to stay, so I will come), to be read as 00101 10011 00110 00100 = 5, 19, 6, 4, hiding the message f*u*ge (= flee!). Due to the resulting message expansion by a factor of five, this is quite impractical.

Bacon is a victim in the pseudo-cryptographic *Bacon-Shakespeare controversy*. Beginning in the 19th century, some people claimed that Bacon was the author of Shakespeare's works. Their main argument is a "biliteral" reading of the *First Folio*, the main edition of Shakespeare's plays, into which they read various claims of authorship by Bacon. The US cryptographer Elizebeth Friedman and her husband (see Section C.2) have put such claims to rest in the eyes of all rational people. But some of the others continue to write B-S books. More difficult to refute is the theory of P. G. Wodehouse that William Shakespeare's plays were not written by him, but by another chap of the same name.

There are other ways of secretly marking text: raising or lowering letters, puncturing the paper, placing points—possibly in invisible ink—above letters, and using up- and downstrokes in handwriting.

At the beginning of the English Civil War 1642–1651, the Royalist Sir John Trevanion was languishing in a cell at Colchester Castle and facing

---

[2]Bacon's Standard Code for the Information Industry, pronounced bess-key.

execution when a letter from his friend "R.T." saved his life.

> *Worthie Sir John: —Hope, that is ye best comfort of ye*
> *afflictyd, cannot much, I fear me, help you now. That I wolde*
> *says to you, is this only: if ever I may be able to requite that*
> *I do owe you, stand not upon asking of me. 'Tis not much*
> *I can do; but what I can do, bee verie sure I wille. I knowe*
> *that, if dethe comes, if ordínary men fear it, it frights not you,*
> *accounting it for a high honour, to have such a rewarde of*
> *your loyalty. Pray yet that you may be spared this soe bitter,*
> *cup. I fear not that you will grudge any sufferings; only if bie*
> *submission you can turn them away, 'tis the part of a wise*
> *man. Tell me, an if you can, to do for you any thinge that*
> *you would have done. The general goes back on Wednesday.*
> *Restinge your servant to command. R.T.*

According to a previously agreed system, the third letters after punctuation marks form the steganogram *Panel at East end of chapel slides*, and he vanished during his next hour of prayer. Not for a long respite—he fell in 1643 during the siege of Bristol.

Already Trithemius' *Steganographia* had suggested such methods (Section G.3).

Notes E.1. The *deck*, and *tectonics* via Latin, derive from στεγανός. The German words *Decke* (blanket, ceiling) and *Dach* (roof) have the same root as the Latin *tectum* (roof) and its derivates like *toît, tetto, techo* in French, Italian, Spanish.

E.2. Friderici's autokey system is on pages 29–33 of his *Cryptographia*.

E.3. Bacon's biliteral cipher is in Bacon (1605), and also in posthumous works such as Bacon (1640), Book VI, Chapter 1 on cyphars, pages 264–271, with "Paris" on page 265. The letter to Trevanion is from Bombaugh (1961), Chapter *Puzzles*, Section *Curiosities of Cipher*, without a reference.

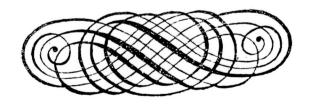

'Identity?' said Jack, comfortably pouring out more coffee.
'Is not identity something you are born with?'
'The identity I am thinking of is something that hovers
between a man and the rest of the world: a mid-point
between his view of himself and theirs of him — for each,
of course, affects the other continually. A reciprocal fluxion, sir.
There is nothing absolute about this identity of mine.'
PATRICK O'BRIAN (1969)

Bertrand Russell har skrevet,
at ren matematik er det emneområde,
i hvilket vi ikke véd, hvad vi taler om,
eller hvorvidt det vi siger er sandt eller falsk.
Det er på den måde jeg har det med madlavning.[3]
PETER HØEG (1992)

Agrippa [...] dice: *Saluberrimum in administratione
magnarum rerum est, summam imperii apud unum esse.*[4]
NICCOLÒ MACHIAVELLI (1513)

Le Predicateur qui voudra se seruir des Mathematiques
pour éclarcir les difficultez de nos mysteres,
& de tout ce qu'il traitera dans ses sermons,
aura de tres-grands auantages;
car s'il est question de comparer le finy à l'infiny
pour expliquer l'vnion des deux natures
dans vne mesme personne diuine,
il monstrera aisément qu'vn mouuement infiny
se peut faire sur vn espace finy.[5]
MARIN MERSENNE (1637)

---

[3]Bertrand Russell wrote that pure mathematics is the field in which we don't know
what we're talking about or to what extent what we say is true or false. That's the
way I feel about cooking.

[4]Agrippa says: In the conduct of great enterprises it is most advantageous if the
supreme command rests in one person.

[5]The preacher who wants to use mathematics to clarify the difficulties of our mysteries and all that he treats in his sermons will have a very great advantage. For when
the question arises to compare the finite to the infinite in order to explain the two
natures in a single divine being, he will easily show that an infinite movement can be
performed in a finite space.

Chapter 10

Identification and authentication

When we meet people face-to-face, we usually either know who they are or they are introduced to us by someone whom we know. Things are different in the e-world. If BOB gets a message from ALICE claiming that she has been robbed in a foreign country and urgently needs some money for a hotel bill, how can he tell it was really ALICE who sent this? Many such situations occur regularly: a website asking for a credit card number, a computer asking for a VPN tunnel to another computer, an RFID reader at immigration requesting the electronic data on a passport, a mobile phone base station establishing a connection to a mobile device, and so on.

This chapter first discusses methods of *identifying* someone over an open line. Then comes *message authentication* which ensures that a message actually comes from whomever claims to be the sender.

Such methods use a *trusted authority TA* to whom each user identifies himself once and for all. The TA issues a certificate which can be used in all later transactions, both for identification and for message authentication. The initial identification can be performed "classically", say, by presenting a passport at some office, by a trusted sysadmin passing data of a known member of his organization to the TA, or by similar means.

10.1. Identification schemes

An identification protocol must safeguard against the following misuses, when ALICE identifies herself to BOB:

- If EVE intercepts the transmission, she should not be able to impersonate ALICE later.

© Springer-Verlag Berlin Heidelberg 2015
J. von zur Gathen, *CryptoSchool*, DOI 10.1007/978-3-662-48425-8_15

○ BOB should not be able to impersonate ALICE.

These requirements preclude the possibility of just ALICE sending a message to BOB. Most identification schemes incorporate three steps:

1. ALICE sends information to BOB.

2. BOB issues a *challenge* to ALICE.

3. ALICE responds.

If ALICE's response is satisfactory, then BOB will assume that it is really ALICE who is talking to him.

A simple system, without the first step, is possible if ALICE and BOB have previously agreed on a secret key sk for a symmetric cryptosystem, with encryption $\mathsf{enc_{sk}}$ and decryption $\mathsf{dec_{sk}}$. The system might be AES.

2. BOB sends ALICE a random message x as challenge. In AES, this might be a 128-bit string.

3. ALICE sends $y = \mathsf{enc_{sk}}(x)$ to BOB.

Then BOB checks whether $\mathsf{dec_{sk}}(y) = x$. If ALICE actually wants to prove that some message z originates from her, then she can send $(y, z, h(y, z))$ to BOB, with some public secure hash function h.

10.2. Schnorr identification scheme

This scheme works with a trusted authority TA and a publicly known cyclic group $G = \langle g \rangle$ with $d = \#G$ of n bits; see Chapter 4. The TA also has a secure signature scheme, with a signing algorithm $\mathsf{sig_{TA}}$ using a secret key and public verification $\mathsf{ver_{TA}}$; see Chapter 8. We write $\mathsf{ver_{TA}}(m, s)$ for verifying TA's signature $s = \mathsf{sig_{TA}}(m)$ of the message m.

ALICE has her identity information ID(ALICE), a secret exponent $a \xleftarrow{\boxtimes} \mathbb{Z}_d^\times$, and sends its public version $A = g^a \in G$ to TA. It computes $s = \mathsf{sig_{TA}}(\text{ID}(\text{ALICE}), A)$ and sends the certificate $C_{\text{ALICE}} = (\text{ID}(\text{ALICE}), A, s)$ to ALICE. These steps are performed once when ALICE enters the protocol. Now each time that ALICE wants to send a message to BOB identifying herself as the sender, she does the following.

ALGORITHM 10.1. Schnorr identification scheme.

Input: Publicly known $G = \langle g \rangle$, d, ID(ALICE), and A, as above.
       Known to ALICE: C_{ALICE}, a.

1. ALICE chooses a secret session key $b \xleftarrow{\boxtimes} \mathbb{Z}_d$ and sends C_{ALICE} and the public version $B = g^b \in G$ to BOB.

2. BOB checks that $\text{ver}_{\text{TA}}((\text{ID}(\text{ALICE}), A), s) = $ "true", chooses $c \xleftarrow{\boxtimes} \mathbb{Z}_d$, and sends c to ALICE.
3. ALICE computes $r = b + ac$ in $\mathbb{Z}_d$ and sends it to BOB.
4. BOB accepts ALICE as herself if $BA^c = g^r$, and not if otherwise.

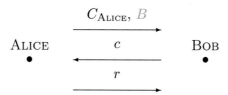

We assume that two group elements can be multiplied with $O(n^2)$ operations; see Section 4.1.

THEOREM 10.2. *The protocol works correctly, that is, if it is properly executed, then* BOB *will accept* ALICE*'s identification. The total computing time is* $O(n^3)$ *bit operations, and the number of bits communicated is*

$$\text{length}(\text{ID}(\text{ALICE})) + \text{length}(\text{TA } signature) + 4n.$$

PROOF.   For the correctness, we have

$$BA^c = g^b(g^a)^c = g^{b+ac} = g^r.$$

The claims about time and communication are clear.   □

EXAMPLE 10.3. We arrive at the following concrete numbers. We assume that ID(ALICE) is a string of 512 bits. If we use the 8-bit extended ASCII coding of $2^8 = 256$ characters, then this corresponds to $512/8 = 64$ letters. The DSA on elliptic curves provides signatures of $2 \cdot 256 = 512$ bits for the recommended bit size of 256 for group cryptography (Table 5.12). A, B, c, and r have 256 bits each. In total, ALICE's certificate comprises $512 + 256 + 512 = 1280$ bits and $1280 + 3 \cdot 256 = 2048$ bits are transmitted.

ALICE's computation is the choice of b, the exponentiation g^b in G in step 1, and $b + ac$ in $\mathbb{Z}_d$ in step 3. The latter is mainly one multiplication modulo d, but the former are on average about $256 \cdot 1.5 = 384$ multiplications in the group G.

If this is to run on a *smart card* with limited computing capability, then the computation of g^b is a bottleneck. But it can be performed *off-line*: after each identification, the smart card already starts computing b and g^b and has them ready to dispatch at the next identification process.

$\diamond$

Next we discuss the protocol's security. Since all communication takes place over public lines, we must assume that EVE knows C_{ALICE}. She also sees B, and then c. Her goal is to find an r that satisfies BOB.

The sequence: *first B, then c* is important. For if EVE can guess c beforehand, then she chooses any old r, sends $B = g^r A^{-c}$ to BOB first, and then r in reply to his challenge.

ALICE's secret session key b has the same function as the randomness in the ElGamal cryptosystem: it protects the secret key. But she does not send it to BOB, only an encrypted version $B = g^b$. The group must be chosen so that discrete logarithms are difficult, and EVE cannot compute b from B, and similarly she cannot compute a from $A = g^a$.

The following statement about security is quite strong. It makes a weak assumption about EVE's power of impersonating ALICE, and draws the strong conclusion that then EVE can actually compute a. *Double impersonation* means that an adversary has two values of c for which she can impersonate ALICE, using the same session key b.

THEOREM 10.4. *The discrete logarithm problem* DL_G *can be reduced to double impersonation in the Schnorr identification scheme.*

PROOF. Suppose that (c_1, r_1) and (c_2, r_2) are two lucky values for an adversary EVE, with all elements in $\mathbb{Z}_d$. If $c_1 = c_2$, then $r_1 = r_2$, since a and b are the same for both executions. But then $(c_1, r_1) = (c_2, r_2)$, which is assumed not to be the case. Thus $c_1 \neq c_2$ and also $r_1 \neq r_2$. From

$$BA^{c_1} = g^{r_1}, \qquad BA^{c_2} = g^{r_2},$$

we conclude that

$$A^{c_1 - c_2} = g^{a(c_1 - c_2)} = g^{r_1 - r_2} \text{ in } G,$$
$$\text{dlog}_g A = a = (r_1 - r_2)(c_1 - c_2)^{-1} \text{ in } \mathbb{Z}_d,$$

provided that $c_1 - c_2$ is invertible in $\mathbb{Z}_d$, that is, $\gcd(c_1 - c_2, d) = 1$. This is not a serious restriction; see Remark 4.15. $\qquad\square$

10.3. Okamoto identification scheme

We next discuss a modification to Schnorr's scheme that was proposed by Okamoto (1993). It strengthens the security of Schnorr's protocol in the following way. Even if ALICE collaborates with EVE, they cannot cause damage unless they solve a hard discrete logarithm problem. Such a property is not known for Schnorr's protocol. It is possible—albeit unlikely—that ALICE might identify herself several times to EVE and

EVE would glean enough knowledge about ALICE's secret a that enables her to impersonate ALICE.

In any cryptographic protocol, it is prudent to consider each player as potentially unreliable. The theorem about the security of Schnorr's scheme says that under certain assumptions EVE can compute ALICE's secret a. Now if ALICE has somehow given her a away, this theorem provides no comfort. We will prove a stronger security property about the following protocol: if EVE can break it, then, with ALICE's help, they can compute a discrete logarithm that the TA keeps secret.

We use the same notation G, d, ID, $\mathrm{sig_{TA}}$, and $\mathrm{ver_{TA}}$ as in Schnorr's scheme. But now the TA chooses two $g_1, g_2 \in G$ of order d. If g_1 has order d, then $g_2 = g_1^k$ has order d if and only if $\gcd(k, d) = 1$ (Theorem 15.60). The hard DL problem is to compute $k = \mathrm{dlog}_{g_1} g_2$.

ALICE now chooses two secret exponents $a_1, a_2 \in \mathbb{Z}_d^{\times}$, so that her secret key is $\mathsf{sk_{ALICE}} = (a_1, a_2)$, and sends $A = g_1^{a_1} g_2^{a_2}$ to TA. Then TA sends ID(ALICE), A and $s = \mathrm{sig_{TA}}(\mathrm{ID(ALICE)}, A)$ to ALICE. The identification protocol goes as follows.

ALGORITHM 10.5. Okamoto identification scheme.
Input: Publicly known $G = \langle g \rangle$, d, ID(ALICE), and A, as above.
   Known to ALICE: C_{ALICE}, (a_1, a_2).

1. ALICE chooses $b_1, b_2 \in \mathbb{Z}_d$ at random and sends her certificate $C_A = (\mathrm{ID(ALICE)}, A, s)$ and $B = g_1^{b_1} g_2^{b_2}$ to BOB.
2. BOB verifies that $\mathrm{ver_{TA}}((\mathrm{ID(ALICE)}, A), s) = \text{"true"}$ and chooses $c \xleftarrow{\;\boxdot\;} \mathbb{Z}_d$.
3. ALICE sends

$$r_1 = b_1 + a_1 c \quad \text{and} \quad r_2 = b_2 + a_2 c \text{ in } \mathbb{Z}_d$$

to BOB.
4. BOB verifies that $B A^c = g_1^{r_1} g_2^{r_2}$.

Correctness is again easy:

$$B A^c = g_1^{b_1 + a_1 c} g_2^{b_2 + a_2 c} = g_1^{r_1} g_2^{r_2}.$$

With the same method as before, we can show the following. We now assume that d is prime, so that every nonzero element in $\mathbb{Z}_d$ is invertible.

LEMMA 10.6. If EVE has a value B for which she can impersonate ALICE for at least two values of c, then EVE can easily compute $e_1, e_2 \in \mathbb{Z}_d$ such that $A = g_1^{e_1} g_2^{e_2}$.

PROOF.   By assumption, EVE can compute r_1, r_2, r'_1, r'_2, c, and c' with $c \neq c'$ and

(10.7) $$BA^c = g_1^{r_1} g_2^{r_2}, \qquad BA^{c'} = g_1^{r'_1} g_2^{r'_2},$$

which implies

$$A^{c-c'} = g_1^{r_1 - r'_1} g_2^{r_2 - r'_2}.$$

Then $c - c'$ is invertible in $\mathbb{Z}_d$, and EVE calculates its inverse $(c - c')^{-1}$, $e_1 = (r_1 - r'_1)(c - c')^{-1}$, and $e_2 = (r_2 - r'_2)(c - c')^{-1}$, all in $\mathbb{Z}_d$. Then $A = g_1^{e_1} g_2^{e_2}$ in G.   □

Of course, ALICE can "impersonate" herself. This corresponds to taking ALICE's own results for the four values of r and leads to $(e_1, e_2) = (a_1, a_2) = \mathsf{sk}_{\text{ALICE}}$ in Lemma 10.6. We now assume that this is not the case, but that EVE's values are chosen uniformly at random among the r that make (10.7) true. We say that such an impersonation is independent of $\mathsf{sk}_{\text{ALICE}}$, and have the following security result, again for prime order.

THEOREM 10.8. If EVE has a value B for which she can impersonate ALICE for at least two values of c, independently of $\mathsf{sk}_{\text{ALICE}}$, then ALICE and EVE together can, with probability at least $1 - 1/d$, easily compute $k = \mathrm{dlog}_{g_1} g_2$.

PROOF.   By the lemma, EVE can compute e_1 and e_2 so that $A = g_1^{e_1} g_2^{e_2}$. Now if ALICE tells EVE a_1 and a_2, then $A = g_1^{e_1} g_2^{e_2} = g_1^{a_1} g_2^{a_2}$ and $g_1^{e_1 - a_1} = g_2^{a_2 - e_2}$. If $(a_1, a_2) \neq (e_1, e_2)$, then $a_2 \neq e_2$, $e_2 - a_2$ is invertible in $\mathbb{Z}_d$, $g_2 = g_1^{(e_1 - a_1)(e_2 - a_2)^{-1}}$, and $\mathrm{dlog}_{g_1} g_2 = (e_1 - a_1)(e_2 - a_2)^{-1}$ in $\mathbb{Z}_d$.

So now we have to consider the possibility that $(a_1, a_2) = (e_1, e_2)$. Then the computation above does not work. But we will show that this happens only with small probability. Otherwise, by taking ALICE for EVE, it would follow that ALICE can compute the discrete logarithm by herself. We consider the set

$$E = \{(a'_1, a'_2) \in \mathbb{Z}_d^2 \colon A = g_1^{a'_1} g_2^{a'_2}\}$$

of possible exponents of ALICE that lead to A. Then

$$(a'_1, a'_2) \in E$$
$$\Longleftrightarrow \quad 1 = g_1^{a_1 - a'_1} g_2^{a_2 - a'_2} = g_1^{a_1 - a'_1} g_1^{k \cdot (a_2 - a'_2)} = g_1^{a_1 - a'_1 + k \cdot (a_2 - a'_2)}$$
$$\Longleftrightarrow \quad a_1 - a'_1 + k \cdot (a_2 - a'_2) = 0 \text{ in } \mathbb{Z}_d.$$

Thus we can give an arbitrary value to a'_2, and then a'_1 is determined. Since k is invertible in $\mathbb{Z}_d$, different values of a'_2 give rise to different values of a_1'. It follows that E consists of exactly d pairs. We have $(a_1, a_2), (e_1, e_2) \in E$.

Since $\mathsf{sk}_{\text{ALICE}} = (a_1, a_2) \in E$ is chosen uniformly in $\mathbb{Z}_d^2$ and hence also uniformly in E, EVE's $(e_1, e_2) \in E$ equals it only with probability d^{-1}. $\square$

10.4. RSA-based identification

We apply Schnorr's method, using the RSA system. The TA chooses two different large primes p and q and an exponent e, and publishes the n-bit $N = pq$ and e, as in Figure 2.15. We assume that $e \geq 2^{n/4}$ is prime. The TA also chooses a signature scheme sig_{TA} and ver_{TA}, possibly also based on RSA (Section 8.1).

For the set-up, the TA produces ALICE's identification ID(ALICE). ALICE chooses $a \xleftarrow{\boxdot} G = \mathbb{Z}_N^\times$, and sends $A = a^e \in G$ to the TA. The TA computes $s = \text{sig}_{\text{TA}}(\text{ID}(\text{ALICE}), A)$ and sends the certificate $C_{\text{ALICE}} = (\text{ID}(\text{ALICE}), A, s)$ to ALICE. Now if ALICE wants to identify herself to BOB, the following protocol is executed.

ALGORITHM 10.9. RSA identification scheme.

Input: Publicly known N, prime e, ID(ALICE), and A, as above.
    Known to ALICE: C_{ALICE}.

1. ALICE chooses $b \in \mathbb{Z}_N^\times$ at random. She does this by choosing $b \in \mathbb{Z}_N$ at random and repeating if $\gcd(b, N) \neq 1$. (This is very unlikely to happen; in particular, ALICE would have broken the TA's RSA scheme.) She sends C_{ALICE} and $B = b^e \in G$ to BOB.
2. BOB verifies the signature by checking that $\text{ver}_{TA}((\text{ID}(\text{ALICE}), A), s) = $ "true", chooses $c \xleftarrow{\boxdot} \{1, \dots, e-1\}$, and sends c to ALICE.
3. ALICE computes $r = ba^{-c} \in G$ and sends it to BOB.
4. BOB accepts ALICE's identification if and only if $B = A^c r^e$.

The proof of correctness is as before:

$$A^c r^e = (a^e)^c (ba^{-c})^e = b^e = B.$$

For the security, we have the following.

THEOREM 10.10. If EVE knows a value of B and two values of c for which she can impersonate ALICE, then EVE can compute a.

PROOF. EVE is assumed to have r_1, r_2, c_1, c_2 with $c_1 \neq c_2$ and

$$B = A^{c_1} r_1^e = A^{c_2} r_2^e.$$

We may assume that $c_1 > c_2$. Then

$$A^{c_1-c_2} = (r_2/r_1)^e.$$

Since $0 < c_1 - c_2 < e$ and e is prime, we have $\gcd(c_1 - c_2, e) = 1$, and EVE can compute $t = (c_1 - c_2)^{-1}$ in $\mathbb{Z}_e$. Then $(c_1 - c_2)t = 1 + fe$ for some $f \in \mathbb{N}$, and

$$
\begin{aligned}
A^{1+fe} &= A^{(c_1-c_2)t} = (r_2/r_1)^{et}, \\
A &= (r_2/r_1)^{et} A^{-fe}.
\end{aligned}
$$

We now raise to the dth power, where d is the TA's secret RSA exponent, with $ed = 1$ in $\mathbb{Z}_{\varphi(N)}$. Then

$$a = a^{ed} = A^d = (r_2/r_1)^{edt} A^{-fedt} = (r_2/r_1)^t \cdot A^{-ft}.$$

Without knowing d, EVE can determine f, evaluate the last expression, and compute a. $\square$

Thus breaking this scheme in the sense of the theorem implies breaking the underlying RSA scheme, which is hard under Conjecture 3.32.

10.5. Message authentication codes

ALICE wants to send a message x to BOB over an insecure channel, allowing BOB to check whether the message has arrived unchanged (*message integrity*) and that it originated from her (*message authentication*). The tool to achieve this is a *Message Authentication Code* (MAC). It requires a secret key sk shared between the two and is also called a *keyed hash function*. Such a mac is used in the HKS encryption scheme of Section 9.9.

Why MACs? The 2G mobile phone standard GSM does not use a MAC. This allows an adversary to set up a fake base station with a strong signal, so that mobile phones in its vicinity will route their traffic through it. For lack of authentication, the mobile phone and its user do not detect the ruse, and the fake station can read these calls. The newer mobile phone standards UMTS and LTE have implemented MACs. The shared secret key is generated and stored by the service provider, and also transferred to the SIM card that the user buys. His phone will then only connect to base stations in phone networks that have an agreement with the service provider to receive its secret key.

In general, a MAC is a publicly known algorithm that takes a secret key sk and a message x as input and outputs a value $\text{mac}_{\text{sk}}(x)$. ALICE sends x together with $\text{mac}_{\text{sk}}(x)$ to BOB. He performs the same calculation, using the shared secret key, and accepts the message if his value equals the $\mathrm{mac}_{\mathsf{sk}}(x)$ he received, and rejects otherwise. Security of a MAC means existential unforgeability under chosen message attacks (Section 9.5).

A concrete implementation of this idea is via a keyed hash message authentication code (HMAC). It uses a hash function h and essentially sets $\mathrm{mac}_{\mathsf{sk}}(x) = h(\mathsf{sk}, h(\mathsf{sk}, x))$. Such HMACs are used in the internet protocols IPsec and TLS.

Some care has to be taken so that the lengths of these items are appropriate. We illustrate this with the hash function SHA-256: $\{0,1\}^{512} \to \{0,1\}^{256}$ (Section 7.6), which can be applied to strings of arbitrary length (Section 7.3).

First, if x has more than 256 bits, then it is replaced by SHA-256(x). Otherwise, it is padded to exactly 256 bits. In either case, we obtain a new message y of 256 bits. The secret key has some number $\ell \leq 256$ of bits and is padded by $256 - \ell$ bits on the right. Furthermore, we have two fixed and publicly known padding strings of 256 bits each, the inner padding *InPad* and the outer padding *OutPad*.

ALGORITHM 10.11. HMAC-SHA-256.

Input: sk, y, InPad, OutPad.
Output: $\mathrm{mac}_{\mathsf{sk}}(x)$.

1. InKey $\leftarrow$ (sk$||0^{256-\ell}$) $\oplus$ InPad,
2. OutKey $\leftarrow$ (sk$||0^{256-\ell}$) $\oplus$ OutPad,
3. $z \leftarrow$ SHA-256(InKey$||y$),
4. $\mathrm{mac}_{\mathsf{sk}}(x) \leftarrow$ SHA-256(OutKey$||z$).

As usual, $||$ denotes the concatenation of bit strings. Thus the secret key is added to both paddings giving an inner and an outer key. The inner key is hashed, together with the message, and the result is hashed together with the outer key. All strings computed have 256 bits. Figure 10.1 illustrates the algorithm, assuming that sk has 128 bits, just like a key for AES-128.

The paddings InPad and OutPad must differ in at least one bit. It is suggested to choose them so that they both have large Hamming weight and large Hamming distance from each other, say all three around 128. But this choice does not seem crucial for the security of the HMAC.

The double hashing in steps 3 and 4 makes HMAC-h stronger than the underlying hash function h, and it seems that even a weak hash function like $h =$ MD5 (Section 7.5) can be used to obtain a secure HMAC.

The new NIST standard SHA-3 (Section 7.5) does not suffer from some of the weakness of previously used hash functions (Exercise 10.6), and HMAC-SHA-3$_{\mathsf{sk}}(x) =$ SHA-3(sk$||x$) provides a secure MAC.

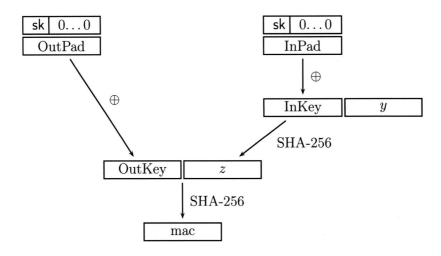

Figure 10.1: $\mathrm{mac}_{\mathsf{sk}}(y) = \mathrm{HMAC\text{-}SHA\text{-}256}_{\mathsf{sk}}(y)$. Each box holds 256 bits, and two adjacent boxes together 512 bits.

10.6. Authenticated key agreement

We recall the Diffie-Hellman key exchange 2.20, one of the great achievements of asymmetric cryptography. The woman-in-the-middle attack of Figure 2.17 is a dangerous threat, and we now see how to protect against it. This is an important tool in the symbiosis of symmetric and asymmetric cryptography, described in Section 2.4.

As in Protocol 2.20 and Figure 2.16, we have a group $G = \langle g \rangle$ with $d = \#G$ of n bits, ALICE's secret and public keys $a \xleftarrow{\boxed{\ }} \mathbb{Z}_d$ and $A = g^a \in G$, and similarly b and $B = g^b$ for BOB. All participants use a signature scheme, for example the ElGamal signatures of Section 8.2. But different users might even use different schemes. Furthermore, as in Section 10.2 there is a trusted authority TA, and ALICE has her certificate $C_{\mathrm{ALICE}} = (\mathrm{ID}(\mathrm{ALICE}), A, \mathrm{sig}_{\mathrm{TA}}(\mathrm{ID}(\mathrm{ALICE}), A))$, and similarly for other users. The following *authenticated key agreement* establishes a secret key k shared between ALICE and BOB.

1. ALICE sends A to BOB.

2. BOB computes $k = A^b \in G$, $s_{\mathrm{BOB}} = \mathrm{sig}_{\mathrm{BOB}}(B, A)$ and sends $(C_{\mathrm{BOB}}, B, s_{\mathrm{BOB}})$ to ALICE.

3. ALICE computes $k = B^a$, verifies s_{BOB} using BOB's verification routine, and verifies C_{BOB} using the TA's verification procedure. She computes $s_{\mathrm{ALICE}} = \mathrm{sig}_{\mathrm{ALICE}}(A, B)$ and sends $(C_{\mathrm{ALICE}}, s_{\mathrm{ALICE}})$ to BOB.

4. BOB verifies s_{ALICE} and C_{ALICE} using the verification routines of ALICE and TA, respectively.

ALICE and BOB then share the secret key k and have fended off the woman-in-the-middle attack. Namely, if EVE wanted to pretend to ALICE that she is BOB, she would have to compute some value for s_{BOB} that ALICE would verify to be $\text{sig}_{\text{BOB}}(B, A)$; this is hard by assumption.

Notes 10.2. In a public key infrastructure (PKI) with *key escrow*, there is a central authority that can uncover the secret key of any participant. This would give governments and their law enforcement agencies access to encrypted data, possibly with legal restrictions similar to those for search warrants. It can only work if potential offenders have no access to strong cryptography outside of this infrastructure. In the 1990s, the US government prohibited the installation of strong cryptography in software products for export, by classifying it as war material. Phil Zimmermann's *Pretty Good Privacy* (PGP) software, a forerunner of GnuPG, could not be exported except as printouts of the source code, and he was sued by the government. Diffie & Landau (1998) present their view on this controversy, which ended with a resounding court victory of the liberal side. Some people might harbor similar reservations about the *trusted authority* (TA) in identification schemes.

10.5. The attack on GSM phones using fake base stations is described in Biryukov *et al.* (2001) and extended by Meyer & Wetzel (2004) to UMTS via downgrade to GSM. HMACs were first defined and analyzed in Bellare *et al.* (1996); see also Bellare (2014).

10.6. The authenticated key agreement, called *station-to-station protocol*, was proposed by Diffie *et al.* (1992) and can be used in the ISO 9798-3 protocols. Versions of this agreement protocol are used in IPsec/IKE, TLS, and SSH; see Krawczyk *et al.* (2013) for their security. At the end of the agreement, ALICE does not know whether BOB received her last message. This is only implicitly confirmed when the correct keys are used.

Exercises.

EXERCISE 10.1 (Identification, chronological order). *We simplify the Schnorr identification scheme:* BOB *sends the challenge r and* ALICE *answers with C_{ALICE}, B, and c. Is this protocol secure?*

EXERCISE 10.2 (Example of Schnorr identification). *As in the Schnorr signature scheme 8.5, we use a subgroup $G \subseteq \mathbb{Z}_p^{\times}$ of small order d inside the much larger group $\mathbb{Z}_p^{\times}$. Specifically, we take $d = 1201$, $p = 122503$, and $g = 11538$.* ALICE *uses the Schnorr identification scheme in G.*

(i) *Verify that $g \in \mathbb{Z}_p^{\times}$ has order d.*

(ii) ALICE's *secret exponent is $a = 357$. Compute her public key A.*

(iii) ALICE *chooses $b = 868$. Compute B.*

(iv) BOB *issues the challenge $c = 501$. Compute* ALICE's *response r.*

(v) *Perform* BOB's *calculations to verify r.*

EXERCISE 10.3 (Attack on Schnorr identification). EVE *has intercepted two Schnorr identifications by* ALICE *and now knows* (B_1, r_1, c_1) *and* (B_2, r_2, c_2). *Furthermore,* EVE *somehow knows a value* $k \in \mathbb{Z}_d$ *and* $\mathrm{dlog}_g(B_1^k B_2^{-1})$.

(i) *Show that* EVE *can easily compute* ALICE*'s secret exponent* a. *[Hint: Look at the case $k = 1$ first.]*

(ii) EVE *knows* ALICE*'s software dealer and has purchased the same identification software from him. This way she learned that* ALICE *uses a linear congruential generator to generate her random secret numbers* b. *Therefore* $b_2 = sb_1 + t$ *in* $\mathbb{Z}_q$ *for known values of* q, $s \in \mathbb{Z}_q^\times$, *and* $t \in \mathbb{Z}_q$ *(Section 11.2). (The programmer has used* q *as the modulus for the random generator so that the numbers* b_i *are automatically in the correct range.) Show how* EVE *can compute* $\mathrm{dlog}_g(B_1^k B_2^{-1})$ *for a specific value of* k *and by (i) also* ALICE*'s secret exponent* a.

EXERCISE 10.4 (Okamoto identification).   ALICE *uses Okamoto's identification scheme with* G, g, d *as in Exercise 10.2,* $g_1 = 60497$ *and* $g_2 = 17163$.

(i) ALICE*'s secret exponents are* $a_1 = 432$ *and* $a_2 = 423$. *Compute* A.

(ii) ALICE *chooses* $b_1 = 389$ *and* $b_2 = 191$. *Compute* B.

(iii) BOB *issues the challenge* $r = 21$. *Compute* ALICE*'s response* $c = (c_1, c_2)$.

(iv) *Perform* BOB*'s calculations to verify* c.

EXERCISE 10.5 (Attack on Okamoto identification).   ALICE *uses Okamoto's identification scheme with the same parameters as in Exercise 10.4. Furthermore, let* $A = 119504$.

(i) EVE *has discovered the equality*

$$g_1^{70} g_2^{1033} A^{877} = g_1^{248} g_2^{883} A^{992} \text{ in } \mathbb{Z}_p^\times.$$

Verify this.

(ii) *Use this information to find numbers* e_1 *and* e_2 *satisfying*

$$g_1^{e_1} g_2^{e_2} = A \text{ in } \mathbb{Z}_p^\times.$$

(iii) ALICE *gives her secret exponents* $a_1 = 717$ *and* $a_2 = 266$ *to* EVE. *Show how* ALICE *and* EVE *together can compute* $\mathrm{dlog}_{g_1} g_2$.

EXERCISE 10.6 (Length extension).   *Consider the weak "long" hash function* h' *from Exercise 7.11 and the alternative HMAC* $\mathrm{mac}_{\mathsf{sk}}(x) = h'(\mathsf{sk}\|x)$. *Show how to generate* $\mathrm{mac}_{\mathsf{sk}}(x\|z)$ *for strings* z *of your choice, just knowing* $\mathrm{mac}_{\mathsf{sk}}(x)$, *but not* sk *or* x. *Conclude that this HMAC is insecure.*

EXERCISE 10.7 (Naive HMAC). *Suppose that* ALICE *wants an HMAC without the nuisance of a shared secret key, and simply sends* x *together with* $h(x)$, *where* h *is a collision-resistant hash function.*

(i) *Does this ensure message integrity?*

(ii) *Does this ensure message authentication?*

Wherefore let us come to CYPHARS. Their kinds are many, as
Cyphars simple; Cyphars intermixt with Nulloes, or non-significant
Characters; *Cyphers of double Letters under one Character; Wheele-*
cyphars; Key-cyphars; Cyphars of words; others. But the virtues of
them [cyphars] whereby they are to be preferr'd are Three;
That they be ready, and not laborious to write;
That they be sure, and lie not open to Deciphering; And lastly,
if it be possible, that they may be managed without suspition.
FRANCIS BACON (1640)

Nos cryptographes, qui se sont acquis une prestigieuse habileté
dans l'art cabalistique et passionnant de traduire ou,
comme on disait jadis, de «perlustrer» la correspondance chiffrée
des ambassades étrangères, s'évertuent depuis trois jours
à découvrir la signification d'un logogriphe,
qui va peut-être nous livrer tout le secret de l'affaire Dreyfus.[1]
MAURICE PALÉOLOGUE (1955)

ἑάλω δὲ καὶ γράμματα λακωνικῶς φράζοντα τοῖς Ἐφόροις τὴν
γεγενημένην ἀτυχίαν: 'ἔρρει τὰ κᾶλα: Μίνδαρος ἀπεσσούα:
πεινῶντι τὤνδρες: ἀπορίομες τί χρὴ δρᾶν.'[2]
Πλούταρχος (PLUTARCH) (c. 75)

Εἰς τὰς συμμαχίδας πόλεις σκυτάλας διέπεμπον.[3]
Ξενοφῶν (XENOPHON) (c. 430–354 BC)

21. Nur genaueste Beachtung dieser Regeln schützen
den Schlüsseltext vor unberufener Entzifferung.
Bei voller Beachtung aber bietet das
„Wehrmacht- Handschlüsselverfahren" hohe Sicherheit.[4]
REICHSKRIEGSMINISTER (1940)

---

[1] Our cryptographers, who have acquired a prestigious mastery of the cabbalistic and exciting art of deciphering or, as they used to say, of "screening" the encrypted correspondence of foreign embassies, strive since three days to discover the meaning of a cryptogram which might deliver us the secret of the Dreyfus affair.

[2] A dispatch was captured announcing the desaster to the ephors in true laconic style: Our ships are lost; Mindarus is slain; our men are starving; we know not what to do.

[3] To the allied cities they sent skytales.

[4] Only meticulous observation of these rules protects the ciphertext against unwarranted decipherment. But under strict observation the "Army hand cipher" provides high security.

Chapter F

Transposition ciphers

etters, words, and other pieces of text are changed by a substitution into a different piece. This creates *confusion*. A completely different effect is obtained by transpositions, which move the pieces around in a text without changing them individually; this creates *diffusion*. Suitably combined and generalized, these two operations form the basis of almost any strong cryptosystem, as postulated by Claude Shannon. We discuss three types of transpositions in this chapter: the Greek skytale, columnar transpositions and grilles.

F.1. The skytale tale

Our civilization owes much to the classical culture of the Greeks. Among them, the Spartans contributed little to improving human existence; their forte was warfare. It is not surprising that one of their few novelties was a military cryptosystem, based on transposition and called a σκυτάλη (*skytale*, rhymes with *Italy*). The historian Plutarch (c. 46–c. 120)

Figure F.1: Porta's fictitious skytale.

© Springer-Verlag Berlin Heidelberg 2015
J. von zur Gathen, *CryptoSchool*, DOI 10.1007/978-3-662-48425-8_16

describes in his *Parallel Lives* the unscrupuolous Spartan general Lysandros (c. 454–395 BC) whose motto was: *You cheat children with dice, and men with oaths.* When Lysandros' brutal and corrupt reign over the Greek cities that he had subdued became too much for the rulers of Sparta, they sent him an encrypted message ordering him back to Sparta. Plutarch writes:

> *When the ephores, Sparta's rulers, send out a military ex-pedition, they have two round wooden sticks made, exactly equal in length and thickness and whose ends fit together. One of them they keep, the other they give to the expedition leader. They call this wooden piece a skytale. If they have a secret important message, they prepare a long strip of pa-pyrus or leather like a belt and wind it around their skytale. They leave no spaces, but the surface is covered everywhere with the strip. When this is done, they write their message on the strip wound around the skytale. After writing, they remove the strip and send it without the piece of wood to the expedition leader. When he receives it, he cannot read any-thing, because the letters are not connected but torn apart. So he takes his own skytale and winds the strip around it. If this is done properly as before, the eye can detect the connection of the letters.*

Back home in Sparta, Lysandros was able to appease the rulers, went on a pilgrimage, later became a general again and fell in battle some years later.

Porta imagined a skytale to look as in Figure F.1. Although it may have provided some security in its time, the skytale is a weak form of cryptography, and a few trials with the *strip of papyrus* give away the secret.

In fact, the story is weak as well. Besides Plutarch, several authors from the third century BC or later, including Gellius, mention the sky-tale's use in the fifth century or before. But in the older writings, up to the fifth century BC, the skytale usually plays the role of a *message stick*, around which a (plaintext) message is wound for convenient long-dis-tance transportation, but no cryptographic purpose is ever mentioned. Thus it is quite possible that the cryptographic use of the skytale is a figment of the imagination of later ancient writers, which has been per-petuated in many cryptographic writings to this day. However, there is no final proof one way or the other. The famous cryptosystems of Cae-sar and Augustus (Example A.2 (iv)) are in a similar state of limbo. The

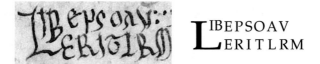

LIBEPSOAV
LERITLRM

Figure F.2: The *Liber epistolarum* of Hieronymus Stridonius.

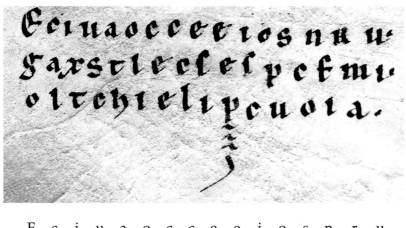

```
E  c  i  u  a  o  c  c  e  e  i  o  s  n  r  u
g  a  x  s  t  l  e  c  s  e  s  p  c  f  m  i
o  l  t  c  h  i  e  l  i  p  c  u  o  i  a
```

Figure F.3: Pope Calixtus II.'s confirmation of the Cistercian rules.

later writers tell us profusely about them, but we have no contemporary documents exhibiting their actual use.

F.2. Columnar transpositions

These transpositions were briefly described in Example A.3 (ii): one writes message in rows which are then read columnwise. Such ciphers were used in the Layer conspiracy (Section H.4, see page 571). In fact, there exist examples of text written in columns already from the Middle Ages. Figure F.2 shows at left a columnar text from a 9th century manuscript of letters, *"liber epistolarum"*, of Hieronymus Stridonius.

A document called the *Exordium parvum*, first written in 1112, explains the early history and the rules of the Cistercian order. In the transposition of Figure F.3 from a manuscript dated around 1260, Pope Calixtus II (reigned 1119–1124) signs his *confirmation* of those rules. Reading it columnwise, it translates as *"I, Calixtus, bishop [pope] of the Catholic church, confirmed [this]."*

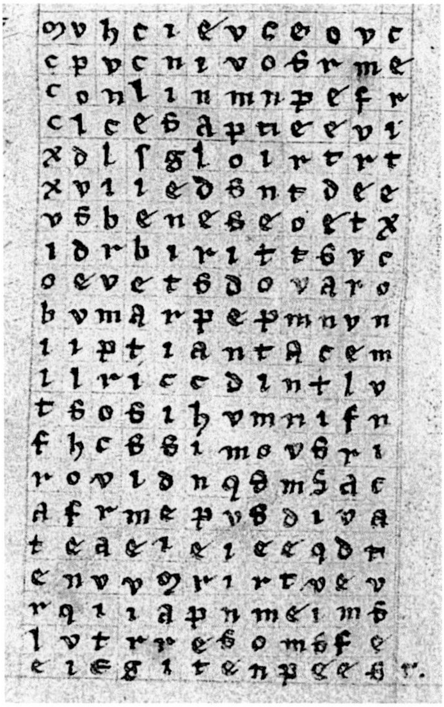

```
m u h c i e u c e o u r
c p u c n i u o s r m e
c o n l i n m n p e f r
c l c e s a p t e e u i
x d l s g l o i r t r t
x u i i e d s n t d e e
v s b e n e s e o e t x
i d r b i r i t t s u c
o e u e t s d o u a r o
b v m a r p e p m n v n
i i p t i a n t A c e m
i l r i c c d i n t l u
t s o s i h u m n i f n
f h c s s i m o u s r i
r o u i d n q s m s a c
a f r m e p u s d i u a
t e a e i e i e e q d t
e n v v m r i r t u e u
r q i i a p n m e i m s
l u t r r e s o m s f e
e i e g i t e n p c e s t
```

Figure F.4: A 14th century columnar transposition.

Figure F.4 is part of a theological manuscript from Bavaria and dated to the 14th century. It reads columnwise:

> *mcccxxvi [1326] obiit frater leupoldus de vilshofen qui hunc librum procuravit ecclesie beatissime virginis genitricis dei marie in alderspach in perpetuum possidendum qui in se continet optimos sermones per totum Annum de tempore et de sanctis si quiscum furetur vel fraudem fererit exconmuni-catus est.*[1]

More generally, the letters of the message may be arranged in some geometrical pattern which has to be read according to previously fixed rules (the key), as in Figure F.5. Can you discover the message? Wilkins (1641) describes several others, and concludes: *"All these kinds may be varied unto divers other more intricate transpositions, according as a mans fancy or occasion shall lead him.".*

---

[1] In 1326, Father Leopold of Vilshofen passed away, who procured this book for the church of the Blessed Virgin Mary, mother of God, for eternal keeping. It contains the best sermons for the whole Year about the clerical seasons and the saints. If someone steals it or removes it fraudulently, he is excommunicated.

```
W  m  r  p  i  t  a  h  h  s  c  t  e  i  n  p  k  e
h  a  t  h  f  o  n  o  i  h  k  f  t  o  e  n  i  l
a  n  o  e  r  r  o  c  g  t  t  t  h  m  n  v  r  l
e  a  u  o  m  h  t  e  i  n  l  e  n  e  t  t  e  s
```

Figure F.5: A transposition cipher by Wilkins.

Figure F.6: The Runic columnar transposition in Verne's *Voyage to the Centre of the Earth*.

Serious cryptographic use of columnar transpositions was made by the German military just before its final defeat in the Second World War. They introduced a columnar transposition system called *Rasterschlüssel 44*, from August 1944 to the end in May 1945. It uses a 24 x 25 grid like a grille (Section F.4), with 10 white cells per row into which the plaintext is written in the usual writing direction, and 15 black cells. The ciphertext is read out columnwise in an irregular sequence of columns. The grid and the column sequence are changed daily, the start cells for writing and reading vary with each message and indicators for these positions are included in the ciphertext. It was hard to use and error-prone, but also much more difficult to break than the ENIGMA by the cryptanalysts in the US and at Bletchley Park, who called it *"practically unbreakable"* and said *"it defeated our cryptographers."* Taking into account the time and effort required versus the purely tactical value, they eventually gave up its cryptanalysis.

Columnar transpositions have also made it into literature. In Jules Verne's classic *Voyage au centre de la terre*[2], the hero, a German professor

---

[2]*Voyage to the Centre of the Earth*

named Lidenbrock, has discovered by chance a piece of parchment with Runic writing on it (Figure F.6). He first transcribes it into our letters

```
m.rnlls    esreuel    seecJde
sgtssmf    unteief    niedrke
kt,samn    atrateS    Saodrrn
emtnaeI    nuaect     rrilSa
Atvaar     .nscrc     ieaabs
ccdrmi     eeutul     frantu
dt,iac     oseibo     KediiI
```

and then begins his guessed plaintext attack, assuming the presumed author's name Saknussem to appear in the cryptogram. Lo and behold, we see it indeed in the first letters of the first nine of the $7 \cdot 3 = 21$ blocks, starting with the S in the third line of the last column, and then reading the first letter of each block against the usual direction.

Thus decryption proceeds by starting with the I as the last letter in the fourth block of the first column. This is the last block of seven letters—Saknussem forgot about padding, if the starting point was meant as a secret key. The last letters are read from right to left and bottom to top, until the final s of the first block. Then reading resumes at the I, the very last letter of the cryptogram.

Particularly convenient is Lidenbrock's capital S, while Runic writing does not distinguish between small and capital letters. With this much help from the author (Verne, not Saknussem), the brilliant Lidenbrock cannot help but recover the plaintext:

> In Sneffels Ioculis craterem kem delibat
> umbra Scartaris Julii intra calendas
> descende, audas viator, et terrestre centrum
> attinges. Kod feci. Arne Saknussem[3]

These instructions by Saknussem send Lidenbrock, his young nephew and a tough Icelandic guide off on a phantastic trip towards the centre of the earth—one of the voyages announced by *Verne Holidays* but still not available for booking.

---

[3]Audacious traveller, descend into the crater of Sneffels Yokul which the shadow of Scartaris caresses during the first days of July, and you will reach the centre of the earth. Which I did. Arne Saknussem

F.3. Breaking a columnar transposition

When the frequency distribution of some ciphertext y is close to that of English, one may suspect that it was produced from some English plaintext x by a transposition. If it comes indeed from a $r \times c$ columnar transposition, this fact is easy to find out. Namely, a digram (= two adjacent letters) $x_i x_{i+1}$ in x is mapped to ciphertext letters y_j and y_{j+r} for some unknown j.

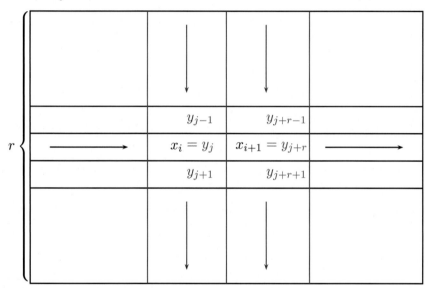

The first step is to prepare a list of digram frequencies $f_{d,\text{Eng}}$ (including contacts across words) for all digrams d. Table C.9 does this for the most frequent ones. Thus $f_{\text{th,Eng}} = 2.09\ \%$ means that the digram $d = \text{th}$ occurs about $0.0209 \cdot 335006 \approx 7002$ many times in the COCA corpus (Table A.19), since the text consists of 335007 letters and one fewer digram. The next step is to guess the number $r = 2, 3, \ldots$ of rows, and for each digram $d = d_1 d_2 \in A^2$, where A is the alphabet, to note how often it occurs with distance r:

$$t^*_{d,y} = \# \{j : y_j = d_1 \text{ and } y_{j+r} = d_2\}.$$

This is normalized as $t_{d,y} = t^*_{d,y}/(\ell - 1)$, when y has ℓ letters in total. Finally, one computes the statistical distance (15.85)

$$\Delta_{\text{digram}}(y, \text{Eng}) = \frac{1}{2} \sum_{d \in A^2} |t_{d,\text{Eng}} - t_{d,y}|$$

of the two digram distributions. This distance will be small at the value of r which is the actual number of rows, and also at its integer multiples. Some minor disturbances are created by digrams that are split onto two (consecutive) rows in the plaintext, and by possible divisions of the plaintext into blocks that were encrypted separately. But these vagaries do not seriously affect the method.

Generally speaking, the combination of substitution and transposition can increase security drastically. However, a columnar transposition composed with a simple substitution can still be solved by the method above. Namely, the nine most frequent letters etaionsrh in English account for about 13.8 % of all digrams (Table C.9). After guessing the value substituted for e, one uses the digram frequencies among the nine most frequent ciphertext letters to guess the substitutions for some of the letters. The number of possibilities for c and r is usually quite small, say at most 20 or 100 for each of them. This corresponds to a key space of 400 or 10000 elements, which is easy to search exhaustively by any computer at hand.

This cryptanalytic method can also be applied to grilles (see below), with appropriate modifications, in order to determine (vertically or horizontally) adjacent holes. And trigrams can be used for holes in one row or one column, separated by a single space.

F.4. Grilles

In a *grille cipher*, the plaintext letters are distributed in an $n \times n$ array of n rows and n columns, for some number n. The secret key consists of an $n \times n$ mask plus the reading direction. The mask has a certain number of holes in it, and one side is marked as "top". The writing of the ciphertext proceeds in four stages. In the first one, the mask is put on an empty array, the "top" at the top, and the holes are filled by consecutive letters of the plaintext in the agreed reading direction. In the second stage, the mask is rotated clockwise by a quarter turn, and again the holes are filled in, with the next plaintext letters. For stages three and four, we rotate each time by a quarter turn.

As in Section F.2, the indefatigable Jules Verne helps us with a literary example. In his 1885 novel *Mathias Sandorf*, he presents the following 6×6 array of ciphertext:

i	h	n	a	l	z
a	r	n	u	r	o
o	d	x	h	n	p
a	e	e	e	i	l
s	p	e	s	d	r
e	e	d	g	n	c

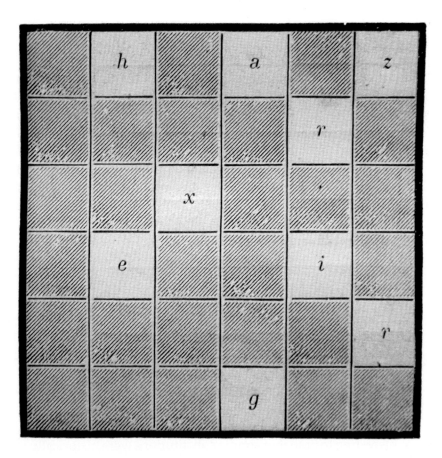

Figure F.7: The stolen grille from Verne's *Mathias Sandorf*. It is shown in the first position over the ciphertext on page 443.

It is the first part of a message intended to start a Hungarian revolution against Austrian dominance; two further parts follow. The ruthless enemies of our brave (and fictitious) hero of the revolution, Mathias Sandorf, have mercilessly killed the pigeon carrying the note. Next they cannot decipher the message and resort to one of the cruder methods of codebreaking: burglary and theft of the grille. Figure F.7 shows the first nine letters decrypted. The crude breakers need some more thinking before they discover that the text is written in reverse, and that these are actually the last nine letters (in reverse). The enterprising reader can recover the last 36 letters of the message. The complete plaintext, properly adorned with punctuation marks etc., reads:

```
Tout est prêt. Au premier signal que vous
nous enverrez de Trieste, tous se lèveront
```

```
en masse pour l'indépendance de la Hongrie.
Xrzah.⁴
```

Sandorf & friends are arrested, the revolution fails (and the Hungarians had to wait until 1918 for their independence), and the next 500 pages are filled with Sandorf's adventures and revenge. The story culminates in a glorious finale where a whole island is blown to smithereens—with all the bad guys on it.

The basic idea of the grille goes back to Girolamo Cardano (1501–1576). He proposed in 1550 to cut holes in a cardboard mask, write each word of the message through one of the holes, remove the mask, and fill in the remaining blanks with an arbitrary text. It is hard to do this without an awkward result.

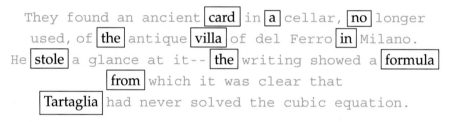

The holes in the mask are over the boxed words. When you put it on the text, you read the plaintext: `Cardano the villain stole the formula from Tartaglia`.

It is interesting to note that simple substitutions were historically first used for single letters and later for words (in which case they are called codebooks). With grilles, it was the other way around: Cardano wrote whole words in the holes, and the *Cardano grilles* use single letters, as in the *Sandorf* example above, plus the rotations which Cardano did not have.

Notes F.1. Plutarch's *Vitae parallelae* presents a sequence of biographies of pairs of people, one Greek and one Roman personality. The goal is a comparison of the two cultures. Lucius Cornelius Sulla Felix (138–78 BC) is Lysander's parallel.

The word skytale also occurs in other meanings related to "stick". An example is Aristophanes' comedy *The Birds*, in which the Greek women try to get their men to stop their interminable wars, by a total sex strike. In one scene, the skytale stands for a Spartan messenger's burgeoning erection.

The weak basis for the cryptographic claims about the skytale was first pointed out by Leopold (1900) and more recently by West (1988) and Kelly (1998). Leopold's work has

---

⁴Everything is ready. At the first signal that you send us from Trieste, everybody will rise in masses for the independence of Hungary. Xrzah. [The last five letters are dummies to fill the square.]

been largely ignored in the cryptographic literature. Lecoy de la Marche (1875) and Roy (1926) mention a skytale used in Burgundy in 1431.

F.2. Figure F.2 is from the title page of the manuscript Latin 1868 at the Bibliothèque Nationale de France, Paris. Its 192 leaves contain documents attributed to Hieronymus Stridonius (also: Hieronymus Sophronius Eusebius), a Christian saint who lived about 345-420. Figure F.3 is from a codex preserved at the Kantonsbibliothek Thurgau, Frauenfeld, Switzerland, which presents on 113 leaves the *Exordium Parvum*. The document is explained in Waddell (1999); see pages 294–297 and 451–452. See also Griesser (1952). Special thanks go to Dr. Barbara Schmid for her expert advice. The columnar text in Figures F.4 is in the Codex Clm 2598 at the Bayerische Staatsbibliothek, München.

Wilkins' 1641 transposition cipher in Figure F.5 is in his Chapter VI, *Secret writing with the common letters, by changing of their places*, page 68, and the quote is on page 69. The ciphertext has to be read in the following sequence $(1,1),(4,1),(1,18),(4,18),(2,1)$, $(3,1),(2,18),\ldots$. Wilkins gives the decryption as *"Wee shall make an irruption upon the enemie from the North at ten of the clock this night."* If you decrypt this yourself, you will find that even the designer of this simple system has made two encryption errors.

Rasterschlüssel 44 "defeated our cryptographers" according to Major Hugh Skillen (1995), as cited in Cowan (2004), page 133. Hinsley *et al.* (1984), Volume 3, Part 2, Appendix 15, calls it "practically unbreakable".

F.3. Barker (1961) gives a systematic treatment of columnar transposition cryptanalysis.

F.4. The reader is encouraged to copy the mask in Figure F.7 onto sturdy paper, cut out the nine holes as indicated, copy the ciphertext with appropriate magnification, and find the plaintext herself. The reading direction is the opposite of the usual one: from right to left, and from bottom to top. The whole decryption progress is detailed on pages 56–63 of Verne's novel.

The grille idea is from Cardano (1550), Book 17. Your car probably has a *Cardan axle* in its transmission gear, and a famous formula for solving cubic equations bears Cardano's name. This is in fact, a mathematical cloak-and-dagger story. Niccolò Fontana Tartaglia (c. 1499–1557) had found this formula, and reluctantly told it to Cardano in 1539 after obtaining a holy oath of secrecy. And then Cardano published it in his own name ... A clear tradeoff between honor and fame. Cardano had seen the formula in ealier unpublished work of Scipione del Ferro (1465–1526) and gave credit to him and Tartaglia. In the awkward grille, plaintext and ciphertext present opposite views of the ensuing controversy.

Exercises.

EXERCISE F.1. *We consider a columnar transposition with r rows and c columns on a plaintext x with ciphertext y. If x_{jk} and y_{jk} are the letters of x and y, respectively, in row j and column k, then $y_{jk} = x_{kj}$. In a cipher, we read both $x = x_0x_1 \cdots$ and $y = y_0y_1 \cdots$ row-wise from top to bottom. Given some $i < rc$, find the value $\tau(i)$ with $y_i = x_{\tau(i)}$. See Example A.3 (ii).*

EXERCISE F.2. *Let n be an even number, and consider an $n \times n$ grille for single letters.*

(i) *Each square of the $n \times n$ array should be visible exactly once through a hole during the four rotations. Show that there are exactly $n^2/4$ holes in the mask.*

(ii) *Show that there are exactly $4^{n^2/4+1}$ $n \times n$ grilles, given a fixed reading direction.*

(iii) Perform the same analysis for odd n. What do you do with the center field?

Don't never prophesy, — onless ye know.
JAMES RUSSELL LOWELL (1862)

Il est remarquable qu'une science qui a commencé
par la considération des jeux, se soit élevée
aux plus importans objets des connaissances humaines.[5]
PIERRE SIMON LAPLACE (1814)

Random number generation is too important to be left to chance.
ROBERT R. COVEYOU (1969)

The emotional laws of probability
are quite different from the mathematical ones.
CELIA MARGARET FREMLIN (1985)

Jedenfalls bin ich überzeugt, daß *der* nicht würfelt.[6]
ALBERT EINSTEIN (1926)

The legend that every cipher is breakable is of course absurd,
though still widespread among people who should know better.
JOHN EDENSOR LITTLEWOOD (1953)

Once world was run according to entrails of bird.
Extispicy. Now it is run according to Number,
and probability is placed ahead of knowledge and learning.
PHILIP KERR (1995)

The warden said, "Hey, buddy, don't you be no square.
If you can't find a partner use a wooden chair."
ELVIS PRESLEY (1957)

The cryptographer developed—on the spot—
a geometrical progress of numbers and corresponding letters
that would take the best cryps Spaulding knew a week to break.
ROBERT LUDLUM (1974)

---

[5]It is remarkable that a science [probability theory] which began with the consideration of games of chance, should have moved up among the most important objects of human knowledge.

[6]At any rate, I am convinced that *He* is not playing at dice.

Chapter 11

Random generation

Random numbers and random bit strings are essential in many areas of computer science, from sorting, routing in networks, and computer algebra to cryptography. Most computer systems provide a function like `rand` that delivers elements which look "random" in some sense. However, it is hard to come up with a practical and inexpensive way to generate truly random numbers. One can think of measuring radioactive decay, current machine clock time or disk usage, user input like key stroke timing or mouse movement, but these are either expensive or not very random. What else can you imagine?

The most popular type of random generators, based on linear congruential generation, is successfully used in many applications. But it is not good enough for cryptography. Yao (1982) expresses a wish: *"Wouldn't it be nice if there existed a pseudorandom number generator that is fit for all applications?"* So cryptographers had to invent their own notion, called (computational) pseudorandom generators, which are the main topic of this chapter. The resulting concepts may be viewed as *computational probability theory*.

Such a generator takes a small amount of true randomness as input and produces a large amount of pseudorandomness. The defining property is that these pseudorandom elements cannot be told apart from truly random ones by any efficient algorithm.

In this chapter, we first define and illustrate this notion of "distinguishing" between pseudorandom and truly random elements, then see that it is essentially equivalent to "predicting the next element", and finally discuss two specific generators, by Nisan & Wigderson (1994) and by Blum, Blum & Shub (1986).

© Springer-Verlag Berlin Heidelberg 2015
J. von zur Gathen, *CryptoSchool*, DOI 10.1007/978-3-662-48425-8_17

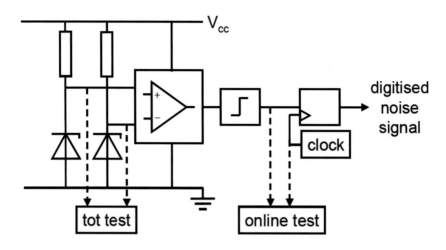

Figure 11.1: A physical random generator with noisy diodes.

11.1. True random generators

Randomness is a vital ingredient for cryptography, from the generation of random keys to the challenges in identification schemes. There are two types of methods for generating the required randomness. Both are inconvenient, expensive, and potentially insecure.

A *software-based generator* measures some process such as

○ the system clock,

○ key stroke or mouse movements,

○ system or network parameters,

○ the contents of certain registers,

○ user input.

All of these have their problems. A 1 GHz machine running uninterrupted for a whole year (good luck!) goes through $365 \cdot 24 \cdot 60 \cdot 60 \cdot 10^9$ or about $2^{54.8}$ cycles. So even if we took that as random, we would only get about 54 bits. In a more realistic situation, say a smartcard engaged in an identification protocol, we can at best expect a few usable bits, certainly not enough for any reasonable protocol. Key strokes and mouse movements can possibly be observed. Some versions of GnuPG require a new user to execute about 15 seconds of energetic mouse pushing. That's ok, but you would not be prepared to do this every time you withdraw money from an ATM. System parameters and register contents might be predicted or simulated.

Another method employs user-generated passwords, but again one can expect only a few "random" bits.

The second type of method are *hardware-based generators* which measure some physical process, such as

○ radioactive decay—but plutonium-endowed laptops face a problem with user acceptance,

○ sector access times in a sealed hard disk,

○ semi conductor thermal noise,

○ ring oscillators sampled at independent rates.

○ capacitor charge or noisy diodes.

To illustrate the kind of electrical engineering that goes into this, we give one example. Killmann & Schindler (2008) analyze a physical random bit generator whose basic description reads as follows.

> *The random source of the generator consists of two equal noisy diodes. For example, Zener diodes have a reverse avalanche effect (depending on diode type at 3–4 Volts or about 10 V) and produce more than 1mV noisy voltage at about 10 MHz. The Flicker Noise in Schottky diodes is associated with static current flow in both resistive and depletion regions (caused by traps due to crystal defects and contaminants, which randomly capture and release carriers). Both diodes provide symmetric input to an operational amplifier to amplify the difference of noise voltages. The output of the operational amplifier is provided to a Schmitt trigger, where the mean voltage of the amplifier meets the threshold of the Schmitt trigger. The output signal of the Schmitt trigger consists of zero and one signals of random length. This signal is latched to the digitized random signal with a clock, which should be at least 20 times slower than the output signal of the Schmitt trigger.*

A variant of this design is shown in Figure 11.1 and a commercial hardware implementation on a USB stick in Figure 11.2. The *tot test* checks for total failure of the process (say, only zeroes are output) and the *online test* tests the randomness quality of the generated bits at runtime. Standardized software packages are available for such tests, here appropriately implemented in hardware. The analysis mentioned includes a model of the physical process, yielding a parametrized model. Experimental estimates

Figure 11.2: A USB stick with the noisy diode generator.

of the parameters lead to bounds on the output entropy. This is determined to be close to maximal, and the *online test* results are consistent with this finding.

A physical generator will often apply some software postprocessing to harden its output. One task is to remove *correlation*, when the probability that a bit is 1 depends on the previous bits. The *bias* of a bit sequence is $p - 1/2$ if individual bits equal 1 with probability p. Then a bit equals 0 with probability $1 - p$. Von Neumann suggested how to remove a nonzero bias: we group the sequence into consecutive pairs, and take 10 to mean 1, 01 to mean 0, and discard 00 and 11.

The noisy diode generator does not require such post-processing and produces over 30 kbit/sec. Beginning in Section 11.2, we discuss *pseudo-random generators*. They produce from short truly random seeds large amounts of randomness, which are good enough for any efficient (cryptographical) algorithm. In practical applications, the physical generator might be used for obtaining such seeds.

In the following, we briefly describe the Linux pseudorandom generator, or *Linux generator* for short. It is part of the kernel in all Linux distributions. It collects data from certain system events in its primary pool of 512 bytes, together with an estimate of the entropy provided by each event. A *secondary entropy pool* of 128 bytes is filled from the primary one and system events. Both pools are regularly updated, using the hash function SHA-1 (Section 7.5). The device /dev/random extracts pseudorandom bits from the secondary pool. If n bits are extracted, then the entropy estimate is reduced by n. On such a call, if the total entropy estimate in both pools is below n, the system waits until more entropy is gathered; see Figure 11.3.

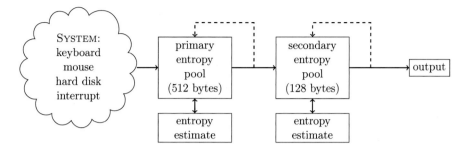

Figure 11.3: The `/dev/random` generator.

On booting, a script, not part of the kernel, writes 512 random bits from the previous session into the primary pool. During operation, the following operating system events contribute to the pools:

event	keyboard	mouse	hard drive	interrupt
type spec	8	12	3	4

The bottom row shows the number of bits for specifying the type of the event, say, 8 for one of $2^8 = 256$ possible key strokes. The times (in milliseconds since booting) of each type of event are recorded as 32-bit words $t_0, t_1, \ldots$. If such an event occurs four or more times, the following iterated discrete derivatives are calculated modulo 2^{32}:

$$\Delta_n^1 = t_n - t_{n-1},$$
$$\Delta_n^2 = \Delta_n^1 - \Delta_{n-1}^1,$$
$$\Delta_n^3 = \Delta_n^2 - \Delta_{n-1}^2,$$
$$m_n = \min\{\Delta_n^1, \Delta_n^2, \Delta_n^3\},$$
$$E_n = \lfloor \log_2((m_n)_{[19\,..\,30]}) \rfloor.$$

Here $(m_n)_{[19\,..\,30]}$ is the integer represented in binary by 12 low-order bits of m_n, excepting the least significant bit. E_n represents an estimate of the entropy. The two 32-bit words (type, E_n) are used to update the entropy pools via a feedback shift register which we do not describe here. Experimentally, it seems that about 2 bits of entropy are added per second on a largely idle computer. Randomness extraction is done via a process involving SHA-1, whose description we also forego.

Before we go on to present further generators, Table 11.4 already presents some speed measurements; see Burlakov *et al.* (2015) and Exercise 11.4 for details.

	byte entropy max = 8 [bits]	runtime [μs]	throughput [kB/s]
noisy diode generator			
no postprocessing	7.99963	16308400	31.39486
AES postprocessing	7.99963	36524300	14.01806
`/dev/random`			
in the field	7.99979	9.169×10^{10}	5.584×10^{-3}
in the lab	7.99948	2.671×10^{12}	1.917×10^{-4}
Littlewood	6.47244	15206550	33.66970
linear congruential	7.99969	2644039	193.64313
Blum-Blum-Shub	7.99962	17708350	28.91291
RSA			
$e = 2^{16} + 1$, 1400 bit/round	7.99966	267604	1913.27484
$e = 3$, 1 bit/round	7.99963	70103838	7.30345
Nisan-Wigderson	7.99961	2731227	187.46153

Table 11.4: Various generators producing 512 kB of output.

11.2. Pseudorandom generators

A *pseudorandom generator* is a deterministic algorithm $\mathcal{A}$ with (random) inputs from a small set X and outputs in a large set Y which are "indistinguishable" from uniformly random elements of Y. This notion is defined in the next section. Often we will have $X = \{0,1\}^k$ and $Y = \{0,1\}^n$ for some $k < n$, and then $\mathcal{A}$ is a pseudorandom bit generator.

Thus the idea of pseudorandom generators is to take a small amount of randomness (a random element of X) to produce a large amount of pseudorandomness (an element of Y). These new elements are not truly random, but behave as if they were for the intended application.

A detailed discussion of "classical" random generation is in Knuth (1998). Knuth presents a large array of *statistical tests* for pseudorandomness, and NIST (2010) provides a standardized battery of tests. It seems hard to describe a general strategy for employing these tests; one has to decide each time anew which tests are appropriate for the purpose at hand. In general, failing a test indicates a weakness in the generator, but passing one particular test—or all of them—is inconclusive.

A standard measure of randomness is the *entropy* of the source (Definition A.9). For n uniformly random bits, it equals n, and a value close to this indicates good randomness. But on the one hand it is computationally hard to measure the entropy, and on the other hand excellent pseudorandom generators (Sections 11.5, 11.6 and 11.8) have an entropy that is asymptotically zero.

In practical terms, one can measure in a (long but finite) sequence of generator output bits whether 0 and 1 occur roughly equally often (*equidistribution*). Or more generally, one can test whether certain patterns occur roughly as often as they would in truly random sequences. As an example, one can divide the sequence into successive bytes and study the resulting *byte distribution* on the 256 bytes and its entropy, the *byte entropy*. This is done by the software for the noisy diode generator of Section 11.1. Experiments with any generator in this chapter, excepting Littlewood, certify a byte entropy close to maximal (Table 11.4), also for the cryptographically weak linear congruential generator described below. More generally, one may test the equidistribution of the bit strings of any fixed length, but this can again not certify (pseudo)randomness.

In contrast to the underlying notion of *statistical randomness*, we will develop a theory of *computational pseudorandomness*. This is the right approach for cryptographical applications. We will see a "universal test", namely predicting the next pseudorandom element, and establish a strong connection with computational complexity, the theory that asks how "hard" it is to solve a given problem. Such a pseudorandom generator is good enough for any polynomial-time algorithm, since if it fails to behave like a truly random one, we have an efficient solution for the underlying problem, contradicting its assumed hardness. This is an analog of the axiomatic security in Chapter 9 for the present question. In a sense, randomness in many instances, for example in primality testing, only has to resist *nature*, but in cryptography it has to resist *malicious adversaries*—a much harder task.

What is a random element, say a random bit? Is 0 a random bit? Is 1? These nonsensical questions indicate that there is no reasonable way to talk about the randomness of an individual bit, or any finite bit string. One can define randomness for infinite strings. For our purposes, it is more useful to talk about "potentially infinite strings", namely machines that produce individual bits. Then one can have such a machine produce arbitrarily long strings of "random" elements. When X is a finite set, a (uniform) truly random generator for X would produce (without any input) a uniformly random element of X, so that each element of X has the same probability $1/\#X$ of occurring.

The most popular pseudorandom generators are the *linear congruential pseudorandom generators*, first proposed by Lehmer (1951). On input a security parameter n, we choose randomly an n-bit modulus $m \in \mathbb{N}$, also s, t and a *seed* x_0 in $\mathbb{Z}_m$, and define

$$(11.1) \qquad x_i = s x_{i-1} + t \text{ in } \mathbb{Z}_m$$

for $i \geq 1$. These are good enough for many purposes, for example in primality testing, computer algebra, and numerical integration, but not for cryptography. Suppose that ALICE and BOB are part of a cryptographic network that uses Schnorr's identification scheme; see Section 10.2 for details. Each time ALICE identifies herself to BOB, he sends her a random number c as part of the protocol. Now, suppose that BOB makes the mistake of taking the c's provided by his machine's `rand` command in C, which is based on a linear congruential generator. If EVE listens in to the traffic and observes several consecutive values of c, she can predict future values of c, as described below. Then the identification scheme is completely broken. The same would happen if a bank computer used such a generator to produce individual transaction numbers. After observing a few of them, an adversary is able to determine the next ones.

In the generator (11.1), we have

$$\begin{aligned} x_i &= sx_{i-1} + t \text{ in } \mathbb{Z}_m, \\ x_{i+1} &= sx_i + t \text{ in } \mathbb{Z}_m. \end{aligned}$$

In order to eliminate s and t, we subtract and find

$$x_i - x_{i+1} = s(x_{i-1} - x_i) \text{ in } \mathbb{Z}_m.$$

Similarly we get

$$x_{i+1} - x_{i+2} = s(x_i - x_{i+1}) \text{ in } \mathbb{Z}_m.$$

Multiplying by appropriate quantities, we obtain

$$\begin{aligned} (x_i - x_{i+1})^2 &= s(x_i - x_{i+1})(x_{i-1} - x_i) \\ &= (x_{i+1} - x_{i+2})(x_{i-1} - x_i) \text{ in } \mathbb{Z}_m. \end{aligned}$$

Thus from four consecutive values $x_{i-1}, x_i, x_{i+1}, x_{i+2}$ we get a multiple

$$m' = (x_i - x_{i+1})^2 - (x_{i+1} - x_{i+2})(x_{i-1}x_i)$$

of m. If the required gcds are 1, then we can also compute guesses s' and t' for s and t, respectively. We can then compute the next values $x_{i+3}, x_{i+4}, \ldots$ with these guesses and also observe the generator. Whenever a discrepancy occurs, we refine our guesses. One can show that after a polynomial number of steps one arrives at guesses which produce the same sequence as the original generator (although the actual values of s, t, and m may be different from the guessed ones).

Such a generator is useless for cryptographic purposes, since we can *predict* the next value after having seen enough previous ones. With an

n-bit modulus m, the entropy is only $4n$ bits at initialization and never increases afterwards. The cryptanalysis just described can be viewed as a test which this generator fails miserably—but it is not included in any standard test battery.

There are variations of these generators that compute internally $x_0, x_1,$ ... in $\mathbb{Z}_m = \{0, \dots, m-1\}$ as before, but publish only the middle half (or the top half) of the bits of x_i. These generators are also insecure; they fall prey to a short vector attack, see Section 13.5. One may also take just one bit, say the least significant bit of x_i. It is not known whether this yields pseudorandom bits.

In the *RSA generator*, we have an n-bit $N = pq$ and e as in the RSA notation of Figure 2.15, and a random seed $x_0 \in \mathbb{Z}_N^\times$. We define $x_1, x_2, \dots \in \mathbb{Z}_N^\times$ by $x_{i+1} = x_i^e$ and the generator output to consist of the least significant k bits of $x_0, x_1, \dots$. For some constant c depending on e and $k = c \cdot n$, this yields indeed a pseudorandom generator.

For the *Littlewood pseudorandom number generator*, we pick (small) integers $n < d$, which are publicly known, and an n-bit string x as (truly random) seed. We can also consider x as an integer in binary, and $2^{-n}x$ is the rational number with binary representation $0.x$.

Output is the sequence of the dth bits of the binary representation of $\log_2((x+i)2^{-n}) = \log(x+i) - n$ for $i = 0, 1, \dots$. Thus with $n = 10$, $d = 14$ and seed $x = 0110100111$, the first five pseudorandom bits are 11001, produced according to the entries (all in binary) of the following table.

i	$(x+i)2^{-n}$	$\log_2(x+i) - n$
0	0.0110100111	-1.010001101000011
1	0.0110101000	-1.010001011010011
2	0.0110101001	-1.010001001100100
3	0.0110101010	-1.010000111110101
4	0.0110101011	-1.010000110000110

Littlewood (1953), page 23, proposed this number generator, actually with $n = 5$ and $d = 7$ in its decimal version and for use in a key addition encryption scheme. He says that "it is sufficiently obvious that a *single* message cannot be unscrambled".

This may look quite attractive, but is flawed. Table 11.4 exhibits its statistical weakness. Wilson (1979) showed a first attack, and Stehlé (2004) breaks the original system and even apparently stronger variants. His approach relies on modern cryptanalytic techniques including lattice basis reduction and Coppersmith's root finding method (Section 13.9).

According to Kerckhoffs's Principle (A.1), a cryptographic system must be assumed to be known to an adversary; its security relies only on the secret keys. Thus it must be infeasible to guess the keys. Weak random generation can make systems insecure, and this has been noted and exploited in the past; see Notes 3.5.

A blatant example of weak random generation is the *dual-EC generator* in a standard introduced by the US NIST in 2012. An alleged backdoor allows to efficiently cryptanalyze SSL/TLS connections. Some details are in Exercise 11.3, where you break the generator yourself. The standard prescribes two points P and Q on an elliptic curve (Chapter 5) and performs computation with them and a user-chosen secret seed. The crux is that if $a = \operatorname{dlog}_P Q$ is known, so that $Q = aP$, it is easy to completely break the generator and predict all future outputs. A public debate arose about the suspicion that—under NSA's influence—the NIST had chosen P and a, and then prescribed $Q = aP$ in the standard. In fact, when John Kelsey from the NIST asked in 2004 where Q came from, the reply was that *"Q is (in essence) the public key for some random private key"*. This is precisely the standard concept of first generating a and then publishing $Q = aP$, as we also use it in Chapters 4 and 5. The reply adds that NSA prohibited the use of a randomly generated Q and forbade a public discussion of this issue.

In their elliptic curve standards (Section 5.6), NIST uses a generally accepted way of avoiding suspicion of a maliciously generated value: publish a seed and a way of generating the value from it by applying a secure hash function. This was never done for the Q under discussion.

11.3. Distinguishers

We now turn to methods which, if properly implemented and used in conjunction with a secure seed generator as in Section 11.1, avoid insecurities as described above. For starters, we want to formalize the notion that the elements generated by a pseudorandom generator should look "random". The idea is that no efficient algorithm should be able to distinguish between these elements and truly random ones. We assume the basic notions of probability theory from Section 15.16.

Suppose that we have a random variable X on $\mathbb{B}^n = \{0,1\}^n$, and a probabilistic algorithm $\mathcal{A}$ with n-bit inputs $x \in \mathbb{B}^n$ and one bit of output. This gives a random variable $\mathcal{A}(X)$ with values in $\mathbb{B} = \{0,1\}$ whose underlying distribution consists of X and the internal randomization in $\mathcal{A}$. For a bit $b \in \mathbb{B}$, we have

$$\operatorname{prob}\{b \xleftarrow{\boxdot} \mathcal{A}(X)\} = \sum_{x \in \mathbb{B}^n} \operatorname{prob}\{x \xleftarrow{\boxdot} X\} \cdot \operatorname{prob}\{b \xleftarrow{\boxdot} \mathcal{A}(x)\}.$$

An important special case is the uniform distribution U_n on $\mathbb{B}^n$. The *expected value* of $\mathcal{A}$ on X is, as in (15.83),

$$\mathcal{E}_{\mathcal{A}}(X) = \sum_{b \in \mathbb{B}} b \cdot \text{prob}\{b \xleftarrow{\boxdot} \mathcal{A}(X)\} = \text{prob}\{1 \xleftarrow{\boxdot} \mathcal{A}(X)\}.$$

When $\mathcal{A}$ is deterministic, then $\text{prob}\{1 \longleftarrow \mathcal{A}(x)\} = \mathcal{A}(x)$ is either 0 or 1.

EXAMPLE 11.2. Let $\mathcal{A}$ be the deterministic algorithm which outputs $\mathcal{A}((x_1, \ldots, x_6)) = x_3$ on any input $(x_1, \ldots, x_6) \in \mathbb{B}^6$. Then for the uniform random variable U_6 on $\mathbb{B}^6$ we have

$$\mathcal{E}_{\mathcal{A}}(U_6) = \text{prob}\{1 \xleftarrow{\boxdot} \mathcal{A}(U_6)\} = \text{prob}\{1 \xleftarrow{\boxdot} U_1\} = \frac{1}{2}.$$

The U_1 here is the third component of $U_6 = U_1 \times U_1 \times U_1 \times U_1 \times U_1 \times U_1 = U_1^6$. ◇

The pseudorandom generators that we define below cannot produce truly random values. But we want their values to be practically indistinguishable from random ones, namely ε-distinguishable only with tiny ε by any efficient $\mathcal{A}$; see Definitions 15.86 and 9.17.

EXAMPLE 11.3. Suppose that n is even and X takes only values with exactly $n/2$ ones: if $x \in \mathbb{B}^n$ and $\text{prob}\{x \xleftarrow{\boxdot} X\} > 0$, then $w(x) = \frac{n}{2}$. Here $w(x)$ is the Hamming weight of x, that is, the number of ones in x. Then the following deterministic algorithm $\mathcal{A}$ distinguishes between X and the uniform variable U_n on $\mathbb{B}^n$: on input x, return 1 if $w(x) = n/2$ else return 0. We have

$$\mathcal{E}_{\mathcal{A}}(X) = \text{prob}\{1 \xleftarrow{\boxdot} \mathcal{A}(X)\} = \text{prob}\{\frac{n}{2} \xleftarrow{\boxdot} w(X)\} = 1,$$

$$\mathcal{E}_{\mathcal{A}}(U_n) = \text{prob}\{1 \xleftarrow{\boxdot} \mathcal{A}(U_n)\} = \text{prob}\left\{\frac{n}{2} \xleftarrow{\boxdot} w(U_n)\right\}$$

$$= 2^{-n} \cdot \#\{x \in \mathbb{B}^n : w(x) = \frac{n}{2}\} = 2^{-n}\binom{n}{n/2}.$$

Stirling's formula (3.57) says that $n! \approx \sqrt{2\pi n}(n/e)^n$. Substituting this into the binomial coefficient and ignoring all minor terms, we find

$$\mathcal{E}_{\mathcal{A}}(U_n) \approx 2^{-n}\frac{\sqrt{2\pi n}(\frac{n}{e})^n}{\pi n(\frac{n}{2e})^n} = 2^{-n}\frac{2^n}{\sqrt{\pi n/2}} = \frac{1}{\sqrt{\pi n/2}}.$$

Thus

$$\text{adv}_{\mathcal{A}}(U_n, X) = |\mathcal{E}_{\mathcal{A}}(U_n) - \mathcal{E}_{\mathcal{A}}(X)| \approx 1 - \frac{1}{\sqrt{\pi n/2}} \geq \varepsilon$$

for any ε with, say $\frac{\sqrt{\pi}-1}{\sqrt{\pi}} > 0.43 \geq \varepsilon > 0$, as soon as $n \geq 2$. For $n = 100$, X and U_n are 0.9-distinguishable. ◇

DEFINITION 11.4. *A bit generator (or generator for short) is a function* $f\colon \mathbb{B}^k \longrightarrow \mathbb{B}^n$ *for some $k < n$. The corresponding random variable on $\mathbb{B}^n$ is $f(U_k)$.*

EXAMPLE 11.5. We consider the generator

$$f\colon \mathbb{B}^3 \longrightarrow \mathbb{B}^6$$

given by the following table

x	$f(x)$
000	001101
001	001011
010	011010
011	010110
100	101100
101	100101
110	110100
111	110010

Each image word in $f(\mathbb{B}^3)$ has Hamming weight 3. We can easily distinguish the random variable $X = f(U_3)$ from U_6 by the distinguisher $\mathcal{A}$ from the previous Example 11.3. Namely, on input $y \in \mathbb{B}^6$, $\mathcal{A}$ returns 1 if $w(y) = 3$ and 0 otherwise. Then

$$\mathcal{E}_\mathcal{A}(U_6) = 2^{-6} \cdot \binom{6}{3} = \frac{5}{16}, \quad \mathcal{E}_\mathcal{A}(X) = 1, \quad \mathrm{adv}_\mathcal{A}(U_6, X) = 1 - \frac{5}{16} = \frac{11}{16}.$$

Thus $\mathcal{A}$ is a $\frac{11}{16}$-distinguisher. In such a small example, one can find other distinguishing properties. The following illustrates a general construction that we will see a little later.

We can use the fourth bit of y to distinguish U_6 from $f(U_3)$, by comparing it to the value of 0 or 1 which occurs less often in the first three positions, called the *minority*. Thus this algorithm $\mathcal{B}$ returns for $y \in \mathbb{B}^6$

$$\mathcal{B}(y) = \begin{cases} 1 & \text{if } y_4 = \mathrm{minority}(y_1, y_2, y_3), \\ 0 & \text{otherwise.} \end{cases}$$

Since both values for y_4 are equally likely in U_6 and independent of y_1, y_2, y_3, we have $\mathcal{E}_\mathcal{B}(U_6) = 1/2$.

We now calculate $\mathcal{E}_{\mathcal{B}}(X) = \text{prob}\{1 \xleftarrow{\boxdot} \mathcal{B}(X)\}$. There are eight values of y which occur as values of X, each with probability $1/8$.

y	$\text{prob}\{1 \xleftarrow{\boxdot} \mathcal{B}(y)\}$
001101	1
001011	0
011010	1
010110	1
101100	0
100101	1
110100	0
110010	1

Therefore $\mathcal{E}_{\mathcal{B}}(X) = 5/8$, $\text{adv}_{\mathcal{B}}(U_6, X) = |\mathcal{E}_{\mathcal{B}}(U_6) - \mathcal{E}_{\mathcal{B}}(X)| = |\frac{1}{2} - \frac{5}{8}| = \frac{1}{8}$, and $\mathcal{B}$ is an $\frac{1}{8}$-distinguisher between U_6 and X. This is quite ok, but not as good as the distinguisher $\mathcal{A}$ from above. $\diamondsuit$

We now define pseudorandom generators. To this end, we consider a family $g = (g_k)_{k \in \mathbb{N}}$ of Boolean functions g_k with

$$g_k \colon \mathbb{B}^k \longrightarrow \mathbb{B}^{n(k)},$$

where $n(k) > k$ for all $k \in \mathbb{N}$. Thus each family member g_k is a generator from $\mathbb{B}^k$ to $\mathbb{B}^{n(k)}$. On input a uniformly random $x \in \mathbb{B}^k$, it produces a (much) longer output $y = g_k(x) \in \mathbb{B}^{n(k)}$ which should look "random". For any $k \in \mathbb{N}$, the random variable $X = g_k(U_k)$ assumes the value $y \in \mathbb{B}^{n(k)}$ with probability

$$\text{prob}\{y \xleftarrow{\boxdot} X\} = 2^{-k} \cdot \#\{x \in \mathbb{B}^k \colon g_k(x) = y\}.$$

At most 2^k many y's have positive probability. Since $k < n(k)$, only "very few" values y actually occur, and X is "very far" from the uniform random variable. But still it might be quite difficult to detect this difference. However, it is always possible to detect some difference. For example, we may choose some $x_0 \in \mathbb{B}^k$ and $y_0 = g_k(x_0)$, so that $\text{prob}\{y_0 \xleftarrow{\boxdot} X\} \geq 2^{-k}$, and take an algorithm $\mathcal{A}$ which on input $y \in \mathbb{B}^{n(k)}$ returns 1 if $y = y_0$ and 0 otherwise. Then

$$\mathcal{E}_{\mathcal{A}}(X) \geq 2^{-k} \gg 2^{-n(k)} = \mathcal{E}_{\mathcal{A}}(U_{n(k)}).$$

Thus $\mathcal{A}$ distinguishes somewhat between the two distributions, but its distinguishing power $2^{-k} - 2^{-n(k)} \approx 2^{-k}$ is exponentially small in k. We can't be bothered with such tiny (and unavoidable) differences, and allow any distinguishing power that is negligible, that is, smaller than any inverse polynomial (Section 15.6).

Now the generators we consider have to be efficient, but there must not exist efficient distinguishers. This leads to the following notion.

DEFINITION 11.6. *A family* $g = (g_k)_{k \in \mathbb{N}}$ *with* $g_k \colon \mathbb{B}^k \longrightarrow \mathbb{B}^{n(k)}$ *and* $n(k) > k$ *for all k is a* pseudorandom generator *if*

○ *it can be implemented in time polynomial in k,*

○ *for all probabilistic polynomial-time algorithms $\mathcal{A}$ using polynomially many samples, the advantage* $\mathrm{adv}_{\mathcal{A}}(U_{n(k)}, g_k(U_k))$ *(Definition 15.86) of $\mathcal{A}$ on $g_k(U_k)$ is a negligible function of k.*

Such a generator can be used in any efficient (polynomial-time) algorithm that requires truly random bits. Namely, if it was ever observed that the algorithm does not perform as predicted for truly random inputs, then the algorithm would distinguish between $U_{n(k)}$ and the pseudorandom generator; but this is ruled out. The efficiency requirement implies that $n(k)$ is polynomial in k. This illustrates the difference between abstract and computational views. Pseudorandom sequences are substantially different from truly random ones in the sense of information theory; for example, their entropy goes to zero for long outputs. But there is no practical (polynomial-time) way of detecting this difference. For our purposes, pseudorandomness is good enough, and it has the great advantage that we can produce it efficiently (under hardness assumptions). However, it requires a small amount of true randomness (from U_k), and without such help, no deterministic algorithm can produce (pseudo)randomness.

This is, quite appropriately, an "asymptotic" notion. It does not depend on the first hundred (or hundred million) g_k's, only on their eventual behavior. We have seen many cryptosystems, such as RSA, which can be implemented for arbitrary key lengths. However, there are also cryptosystems like AES which have fixed input lengths and are not part of an infinite family.

We now want to define a "finite" version of generators. It should be applicable to individual Boolean functions such as $g \colon \mathbb{B}^3 \longrightarrow \mathbb{B}^6$ from Example 11.5. In Definition 11.6 we did not specify the notion of "algorithm". The reader should think, as usual, of Turing machines or appropriate random access machines. For our finite version, Boolean circuits are appropriate. There is a technical issue with *uniformity* here; see Section 15.6.

DEFINITION 11.7. *Let $k < n$ and s be integers, $\varepsilon \geq 0$ real, and $f \colon \mathbb{B}^k \longrightarrow \mathbb{B}^n$ a generator. A probabilistic Boolean circuit $\mathcal{C}$ of size s (number of gates, including inputs) and with advantage* $\mathrm{adv}_{\mathcal{C}}(U_n, f(U_k)) \geq \varepsilon$ *is called an* (ε, s)-distinguisher *between U_n and $f(U_k)$. The circuit $\mathcal{C}$ has several inputs from $\mathbb{B}^n$, and we compare $\mathcal{C}$'s average output when either all inputs are uniform or all drawn from $f(U_k)$. The function f is called an (ε, s)-* resilient pseudorandom generator *if no such $\mathcal{C}$ exists.*

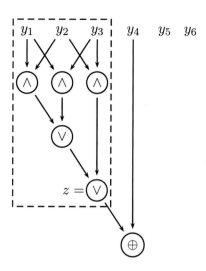

Figure 11.5: A distinguisher circuit using *and* gates $\wedge$ and *or* gates $\vee$.

If $\varepsilon' \geq \varepsilon$ and $s \geq s'$, then (ε, s)-resilience implies (ε', s')-resilience.

EXAMPLE 11.5 CONTINUED. We take $f \colon \mathbb{B}^3 \longrightarrow \mathbb{B}^6$ as above, and implement the two distinguishers as Boolean ciruits. We start with the second one, and first compute $z = (w(y_1, y_2, y_3) \geq 2)$ in the circuit of Figure 11.5 within dashed lines, and then output $z \oplus y_4$. This circuit has 12 gates and is therefore a $(\frac{1}{8}, 12)$-distinguisher between U_6 and $f(U_3)$. Thus f is not a $(\frac{1}{8}, 12)$-resilient pseudorandom generator.

To implement the first distinguisher "$w(y) = 3$" as a circuit, we first compute $u = \bigoplus_{1 \leq i \leq 6} y_i$. Thus $u = 1$ if and only if $w(y)$ is 1, 3, or 5. If we add the condition that

$$(w(h_1) \geq 2 \text{ and } w(h_2) \leq 1) \text{ or } (w(h_1) \leq 1 \text{ and } w(h_2) \geq 2),$$

where $h_1 = (y_1, y_2, y_3)$ and $h_2 = (y_4, y_5, y_6)$ are the two halves of y, then we compute precisely the Boolean function "$w(y) = 3$". We re-use the 5-gate circuit from Figure 11.5 twice in dashed lines and get the circuit in Figure 11.6. It has $2 \cdot 11 + 11 = 33$ gates, so that f is also not $(\frac{11}{16}, 33)$-resilient. $\diamondsuit$

11.4. Predictors

We consider probabilistic algorithms that try to predict the next value x_i of a sequence from the previous bits $x_1, \ldots, x_{i-1}$. A good predictor

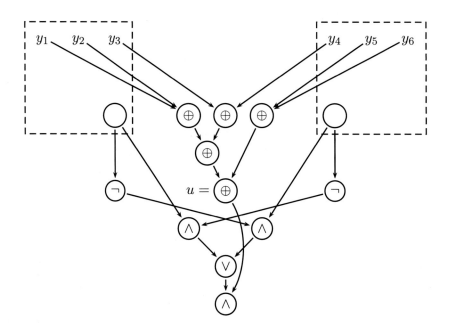

Figure 11.6: A circuit computing the bit "$w(y) = 3$".

can also be used as a distinguisher. The main result of this section is the converse: from any good distinguisher one can build a reasonably good predictor. The proof introduces an important tool: *hybrid* distributions which *interpolate* between two given distributions.

For two random variables X and Y on the same set B, we write

$$\text{prob}\{Y \xleftarrow{\boxdot} X\} = \sum_{b \in B} \text{prob}\{b \xleftarrow{\boxdot} Y\} \cdot \text{prob}\{b \xleftarrow{\boxdot} X\}$$

for the probability that both produce the same value. This generalizes the notion $b \xleftarrow{\boxdot} X$ in a natural way.

When X is a random variable on $\mathbb{B}^n$ and $i \leq n$, we want to consider the *ith successor bit* under X, namely the following one-bit random variable $X_i(y)$, for any $y \in \mathbb{B}^{i-1}$. Its value is 0 with the same probability as the one with which strings $(y, 0, z)$ occur under X, for any $z \in \mathbb{B}^{n-i}$, and 1 with the probability of $(y, 1, z)$ occurring under X. More precisely, for any $j \leq n$ and $w \in \mathbb{B}^j$, we let

$$
\begin{aligned}
\text{(11.8)} \qquad p(w, *) &= \text{prob}\{w \xleftarrow{\boxdot} (X_1, \ldots, X_j)\} \\
&= \text{prob}\{\{w\} \times U_{n-j} \xleftarrow{\boxdot} X\} \\
&= 2^{-n+j} \cdot \sum_{z \in \mathbb{B}^{n-j}} \text{prob}\{(w, z) \xleftarrow{\boxdot} X\}
\end{aligned}
$$

be the probability of w as an initial segment under X. Then for $b \in \mathbb{B}$, $i \geq 1$, and $y \in \mathbb{B}^{i-1}$, we set (arbitrarily) $\text{prob}\{b \xleftarrow{\boxdot} X_i(y)\} = 1/2$ if $p(y, *) = 0$, and otherwise

$$\text{prob}\{b \xleftarrow{\boxdot} X_i(y)\} = \frac{p((y, b), *)}{p(y, *)}.$$

DEFINITION 11.9. *Let $1 \leq i \leq n$ be integers.*

(i) *A predictor for the ith bit is a probabilistic algorithm with input from $\mathbb{B}^{i-1}$ and output in $\mathbb{B}$.*

(ii) *Let X be a random variable on $\mathbb{B}^n$, $(X_1, \ldots, X_{i-1})$ the corresponding truncated variable on $\mathbb{B}^{i-1}$, and $\mathcal{P}$ a predictor for the ith bit. Then the success rate $\sigma_\mathcal{P}(X)$ of $\mathcal{P}$ on X is*

$$\sigma_\mathcal{P}(X) = \sum_{y \in \mathbb{B}^{i-1}} p(y, *) \cdot \text{prob}\{\mathcal{P}(y) \xleftarrow{\boxdot} X_i(y)\}.$$

Its advantage (or prediction power) is $\text{adv}_\mathcal{P}(X) = |\sigma_\mathcal{P}(X) - 1/2|$. If $\text{adv}_\mathcal{P}(X) \geq \varepsilon$, then $\mathcal{P}$ is an ε-predictor for the ith bit of X.

(iii) *A family $(X_k)_{k \in \mathbb{N}}$ of random variables X_k on $\mathbb{B}^{n(k)}$ is (computationally) unpredictable if for any function i_k with $i_k \leq n(k)$, any probabilistic polynomial-time predictor for the i_kth bit of X_k has negligible advantage. Here, the predictor is an algorithm which takes as input k (encoded in unary) and $y \in \mathbb{B}^{i_k-1}$.*

Thus $0 \leq \sigma_\mathcal{P} \leq 1$. A simple and rather useless predictor is to output a uniformly random bit, independent of the input. It has success rate $1/2$ for any X and advantage 0.

If $\sigma_\mathcal{P}(X) \leq 1/2$, then flipping the output bit of $\mathcal{P}$ produces a predictor $\mathcal{P}'$ with $\sigma_{\mathcal{P}'}(X) = 1 - \sigma_\mathcal{P}(X) \geq 1/2$. For a "good" predictor $\mathcal{P}$, the goal is to make its advantage $\text{adv}_\mathcal{P}(X)$ as large as possible.

As in the previous section, we also have a finite version of this asymptotic notion. Now X is a random variable on $\mathbb{B}^n$, $1 \leq i \leq n$, and $\mathcal{P}$ is a probabilistic circuit of size s with $i - 1$ inputs and one output, and is called an (ε, s)-predictor (or ε-predictor) if $\text{adv}_\mathcal{P}(X) \geq \varepsilon$. We say that X is (ε, s)-unpredictable if no such i and $\mathcal{P}$ exist.

EXAMPLE 11.5 CONTINUED. We take $X = f(U_3)$ on $\mathbb{B}^6$. Since 0 and 1 occur equally often in each $f(x)$, we consider the "minority bit predictor" $\mathcal{M}_i$ for the ith bit. It predicts the bit that occurs less frequently in the history; if both occur equally often, it predicts 0 or 1, each with probability $1/2$.

This algorithm always predicts the sixth bit correctly, so that $\sigma_{\mathcal{M}_6}(X) = 1$. It can be implemented with 30 gates, and $\mathcal{M}_6$ is a $(1/2, 30)$-predictor. We now compute its quality as a predictor for the fourth bit:

$$\sigma_{\mathcal{M}_4}(X) = \sum_{y \in \mathbb{B}^3} p(y, *) \cdot \mathrm{prob}\{\mathcal{M}_4(y) \xleftarrow{\boxdot} X_4(y)\}.$$

We only have six $y \in \mathbb{B}^3$ with $p(y, *) > 0$.

y	$p(y, *)$	$X_4(y)$	$\mathcal{M}_4(y)$	$\mathrm{prob}\{\mathcal{M}_4(y) \xleftarrow{\boxdot} X_4(y)\}$
001	$1/4$	0, 1	1	$1/2$
011	$1/8$	0	0	1
010	$1/8$	1	1	1
101	$1/8$	1	0	0
100	$1/8$	1	1	1
110	$1/4$	0, 1	0	$1/2$

Therefore the success rate is

$$\frac{1}{4} \cdot \frac{1}{2} + \frac{1}{8} \cdot 1 + \frac{1}{8} \cdot 1 + \frac{1}{8} \cdot 0 + \frac{1}{8} \cdot 1 + \frac{1}{4} \cdot \frac{1}{2} = \frac{5}{8} > \frac{1}{2}.$$

The dashed box in Figure 11.5 computes the majority z of three bits in size 8, and therefore $\mathcal{M}_4 = z$ is a $(1/8, 8)$-predictor for the fourth bit under X. $\diamondsuit$

It is clear that a predictor can also serve as a distinguisher. Suppose that X is a random variable on $\mathbb{B}^n$, $1 \leq i \leq n$, and $\mathcal{P}$ is an ε-predictor for the ith bit under X. Then we consider the following method for obtaining an algorithm $\mathcal{A}$.

ALGORITHM 11.10. Distinguisher $\mathcal{A}$ from predictor $\mathcal{P}$.

Input: $y \in \mathbb{B}^n$, and i and $\mathcal{P}$ as above.
Output: 0 or 1.

1.   Compute $z \leftarrow \mathcal{P}(y_1, \ldots, y_{i-1})$.
2.   Return $z \oplus y_i \oplus 1$.

THEOREM 11.11. *If* $\mathcal{P}$ *is an* (ε, s)-*predictor for the* ith *bit under* X, *then* $\mathcal{A}$ *is an* $(\varepsilon, s + 2)$-*distinguisher between* U_n *and* X.

PROOF. The output of $\mathcal{A}$ is independent of the values of $y_{i+1}, \ldots, y_n$, and $\mathcal{A}$ has size $s + 2$. We have

$$
\begin{aligned}
\mathcal{E}_{\mathcal{A}}(X) &= \mathrm{prob}\{1 \xleftarrow{\boxdot} \mathcal{A}(X)\} \\
&= \mathrm{prob}\{X_i \xleftarrow{\boxdot} \mathcal{P}(X_1, \ldots, X_{i-1})\} \\
&= \sigma_{\mathcal{P}}(X) \geq \frac{1}{2} + \varepsilon.
\end{aligned}
$$

On the other hand, independently of what $\mathcal{P}$ computes, the probability that its output $\mathcal{P}(U_{i-1})$ equals a uniformly random bit from U_1 is $1/2$. Thus the advantage of $\mathcal{A}$ on X is

$$
\mathrm{adv}_{\mathcal{A}}(U_n, X) = |\mathcal{E}_{\mathcal{A}}(X) - \mathcal{E}_{\mathcal{A}}(U_n)| \geq \frac{1}{2} + \varepsilon - \frac{1}{2} = \varepsilon. \qquad \square
$$

It is quite surprising that also from any good distinguisher one can obtain a reasonably good predictor. This strong result uses the idea of *hybrid distributions*. Thus distinguishers and predictors are essentially equivalent. In other words, predicting the next bit is a "universal test" for pseudorandomness.

THEOREM 11.12. *Let* X *be a random variable on* $\mathbb{B}^n$, *and* $\mathcal{A}$ *an* (ε, s)-*distinguisher between* U_n *and* X. *Then there exists an* i *with* $1 \leq i \leq n$ *and an* $(\varepsilon/n, s + 2)$-*predictor for the* ith *bit under* X.

PROOF. For $0 \leq i \leq n$, we let $\pi_i \colon \mathbb{B}^n \longrightarrow \mathbb{B}^i$ be the projection onto the first i coordinates, and

$$
Y_i = \pi_i(X) \times U_{n-i}.
$$

Thus Y_i is the random variable on $\mathbb{B}^n$ where the first i bits are generated according to X, and the other $n - i$ according to the uniform distribution. These Y_i are *hybrid* variables, partly made up from X and partly from the uniform distribution, and they "interpolate" between the extremes $Y_0 = U_n$ and $Y_n = X$. For $1 \leq i \leq n$, let $e_i = \mathcal{E}_{\mathcal{A}}(Y_i)$. The distinguishing power of $\mathcal{A}$ is at least ε, so that $|e_0 - e_n| \geq \varepsilon$. By flipping the output bit of $\mathcal{A}$ if necessary, we may assume that $e_n - e_0 \geq \varepsilon$. Intuitively, this means that an output 0 of $\mathcal{A}$ indicates that the input is likely to come from U_n, and an output 1 that it comes from X. For our predictor, we choose $i \xleftarrow{\boxdot} \{1, \ldots, n\}$ at random and show that the expected advantage is at least ε/n. We have

$$
(11.13) \qquad \frac{1}{n} \sum_{1 \leq i \leq n} (e_i - e_{i-1}) = \frac{1}{n}(e_n - e_0) \geq \frac{\varepsilon}{n}.
$$

Thus for our random choice of i, the expected value of $e_i - e_{i-1}$ is at least ε/n. We now construct a predictor $\mathcal{P}$ for the ith bit under X.

ALGORITHM 11.14. Predictor $\mathcal{P}$.

Input: $y \in \mathbb{B}^{i-1}$.
Output: 0 or 1.

1. Choose $y_i, \ldots, y_n \xleftarrow{\boxed{\cdot}} \mathbb{B}$ uniformly and independently at random.
2. $y^* \longleftarrow (y, y_i, \ldots, y_n)$. [Thus $y^* \in \mathbb{B}^n$.]
3. $z \longleftarrow \mathcal{A}(y^*)$.
4. Return $y_i \oplus z \oplus 1$.

The intuition why this should work is as follows. If $\mathcal{A}$ outputs $z = 0$, then probably (y, y_i) comes from Y_{i-1}, since (probably) $e_{i-1} < e_i$, and if $z = 1$, then (y, y_i) is more likely to come from Y_i. Now Y_{i-1} and Y_i differ only in the ith place, where Y_i is derived from X, while Y_{i-1} has a uniformly random bit.

Thus we take $z = 1$ as an indication that y_i comes from X, and indeed output $y_i = y_i \oplus z \oplus 1$ as the prediction. On the other hand, $z = 0$ indicates that y_i is presumably from U_1, and that the opposite bit $y_i \oplus 1 = y_i \oplus z \oplus 1$ is a better prediction for the ith bit under X than y_i itself is.

The success rate of $\mathcal{P}$ on X is the expectation over the random choice of i that $\mathcal{P}$ returns the same bit as X_i would. Dropping the dependence on i, we have

$$
\begin{aligned}
&\sigma_{\mathcal{P}}(X) \\
&= \sum_{y \in \mathbb{B}^{i-1}} p(y, *) \cdot \mathrm{prob}\{\mathcal{P}(y) \xleftarrow{\boxed{\cdot}} X_i(y)\} \\
&= \sum_{y \in \mathbb{B}^{i-1}} p(y, *) \cdot \sum_{y_i \in \mathbb{B}} \mathrm{prob}\{y_i \oplus \mathcal{A}((y, y_i) \times U_{n-i}) \oplus 1 \xleftarrow{\boxed{\cdot}} X_i(y)\} \\
&= \sum_{y \in \mathbb{B}^{i-1}} p(y, *) \cdot \sum_{y_i \in \mathbb{B}} [\, \mathrm{prob}\{0 \xleftarrow{\boxed{\cdot}} \mathcal{A}((y, y_i) \times U_{n-i}), y_i \oplus 1 \xleftarrow{\boxed{\cdot}} X_i(y)\} \\
&\qquad\qquad + \mathrm{prob}\{1 \xleftarrow{\boxed{\cdot}} \mathcal{A}((y, y_i) \times U_{n-i}), y_i \xleftarrow{\boxed{\cdot}} X_i(y)\}\,] \\
&= \sum_{y \in \mathbb{B}^{i-1}} p(y, *) \cdot \sum_{y_i \in \mathbb{B}} [\, \mathrm{prob}\{0 \xleftarrow{\boxed{\cdot}} \mathcal{A}((y, y_i) \times U_{n-i})\} \\
&\qquad\qquad - \mathrm{prob}\{0 \xleftarrow{\boxed{\cdot}} \mathcal{A}((y, y_i) \times U_{n-i}), y_i \xleftarrow{\boxed{\cdot}} X_i(y)\} \\
&\qquad\qquad + \mathrm{prob}\{1 \xleftarrow{\boxed{\cdot}} \mathcal{A}((y, y_i) \times U_{n-i}), y_i \xleftarrow{\boxed{\cdot}} X_i(y)\}\,]
\end{aligned}
$$

$$= \sum_{y \in \mathbb{B}^{i-1}} p(y, *) \cdot \big[\operatorname{prob}\{0 \xleftarrow{\boxtimes} \mathcal{A}(\{y\} \times U_{n-i+1})\}$$

$$- \operatorname{prob}\{0 \xleftarrow{\boxtimes} \mathcal{A}((y, X_i(y))) \times U_{n-i}\} \cdot \frac{1}{2}$$

$$+ \operatorname{prob}\{1 \xleftarrow{\boxtimes} \mathcal{A}((y, X_i(y)) \times U_{n-i})\} \cdot \frac{1}{2} \big]$$

$$= \operatorname{prob}\{0 \xleftarrow{\boxtimes} \mathcal{A}(Y_{i-1})\} - \frac{1}{2}\operatorname{prob}\{0 \xleftarrow{\boxtimes} \mathcal{A}(Y_i)\}$$

$$+ \frac{1}{2}\operatorname{prob}\{1 \xleftarrow{\boxtimes} \mathcal{A}(Y_i)\}$$

$$= 1 - e_{i-1} - \frac{1 - e_i}{2} + \frac{e_i}{2} = \frac{1}{2} + e_i - e_{i-1} \geq \frac{1}{2} + \frac{\varepsilon}{n}.$$

Some explanations may be useful. In the second equation, we sum over the two possible values for y_i chosen in step 1 of $\mathcal{P}$. Now $\mathcal{P}(y) = y_i \oplus z \oplus 1$ and $X_i(y)$ take the same value in two cases:

$$z = 1 \quad \text{and} \quad y_i = X_i(y), \text{ or}$$
$$z = 0 \quad \text{and} \quad y_i \oplus 1 = X_i(y).$$

These two cases lead to the third equation. In the fourth equation, the first summand of the previous expression is split into the probability that 0 occurs as value of $\mathcal{A}((y, y_i) \times U_{n-1})$, without regard to $X_i(y)$, minus the probability that y_i occurs as value of $X_i(y)$—this is the complement to the condition $y_i \longleftarrow X_i(y)$. For the fifth equation, we use the fact that the event $y_i \xleftarrow{\boxtimes} U_1$ is independent of the other events, for both possible choices of y_i, and occurs with probability $1/2$. $\square$

COROLLARY 11.15. (i) *Suppose that each bit of the generator $f : \mathbb{B}^k \to \mathbb{B}^n$ is (ε, s)-unpredictable. Then $f(U_k)$ is $(\varepsilon, s+1)$-indistinguishable from U_n.*

(ii) *Suppose that the generator $g = (g_k)_{k \in \mathbb{N}}$ is such that random bits (that is, for a sequence $(i_k)_{k \in \mathbb{N}}$ with $i_k \xleftarrow{\boxtimes} \{1, \ldots, n(k)\}$, the i_kth bit of g_k) are computationally unpredictable. Then g is a pseudorandom generator.*

EXAMPLE 11.5 CONTINUED. We apply the hybrid construction to $X = f(U_3)$ on $\mathbb{B}^6$ and the first distinguisher $\mathcal{A}$ in Example 11.5, with $\mathcal{A}(y) = 1$ if and only if $w(y) = 3$. We have seen above that $\mathcal{E}_{\mathcal{A}}(X) = 1$ and $\mathcal{E}_{\mathcal{A}}(U_6) = 5/16$, and now have to calculate the expected value of $\mathcal{A}$ on the hybrid distributions $Y_i = \pi_i(X) \times U_{6-i}$. These distributions are depicted in Figure 11.7. At the back, we have $f(U_3) = Y_6$, a rugged landscape with eight peaks and valleys at level zero. The montains get eroded as we move forward, to Y_5, Y_4, and Y_3, until we arrive at $Y_2 = Y_1 = Y_0$, a uniformly flat seascape.

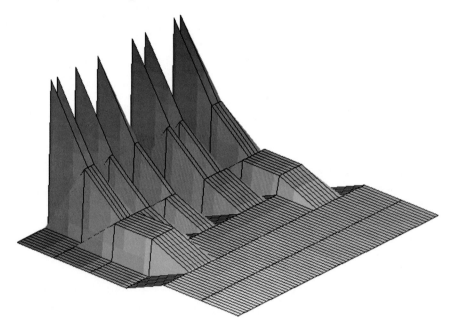

Figure 11.7: The seven hybrid distributions.

For $0 \le i \le 6$ and any $y \in f(\mathbb{B}^3)$, we denote by

$$c_i(y) = \binom{6-i}{3 - w(y_1, \ldots, y_i)}$$

the number of extensions $(z_{i+1}, \ldots, z_6)$ of $(y_1, \ldots, y_i)$ that lead to total Hamming weight 3, that is, with $w(y_1, \ldots, y_i, z_{i+1}, \ldots, z_6) = 3$. Then

$$\text{prob}\{y \xleftarrow{\boxtimes} \pi_i(X) \times U_{6-i}\} = \text{prob}\{(y_1, \ldots, y_i) \xleftarrow{\boxtimes} \pi_i(X)\} \cdot 2^{i-6}$$
$$= 2^{i-6} \cdot 2^{-3} \cdot \#\{x \in \mathbb{B}^3 : (f(x)_1, \ldots, f(x)_i) = (y_1, \ldots, y_i)\},$$

$$e_i = \mathcal{E}_{\mathcal{A}}(Y_i) = \text{prob}\{1 \xleftarrow{\boxtimes} \mathcal{A}(y_i)\} = \text{prob}\{3 \xleftarrow{\boxtimes} w(Y_i)\}$$
$$= \text{prob}\{3 \xleftarrow{\boxtimes} w((X_1, \ldots, X_i) \times U_{6-i})\}$$
$$= 2^{-3} \cdot 2^{-(6-i)} \cdot \#\{(x, y) \in \mathbb{B}^3 \times \mathbb{B}^{6-i} :$$
$$w(f(x)_1, \ldots, f(x)_i, y_{i+1}, \ldots, y_6) = 3\}$$
$$= 2^{i-9} \sum_{x \in \mathbb{B}^3} c_i(f(x)).$$

x	$f(x)$	c_0	c_1	c_2	c_3	c_4	c_5	c_6
000	001101	20	10	4	3	2	1	1
001	001011	20	10	4	3	1	1	1
010	011010	20	10	6	3	2	1	1
011	010110	20	10	6	3	2	1	1
100	101100	20	10	6	3	1	1	1
101	100101	20	10	6	3	2	1	1
110	110100	20	10	4	3	1	1	1
111	110010	20	10	4	3	2	1	1

i	0	1	2	3	4	5	6
e_i	$\frac{5}{16}$	$\frac{5}{16}$	$\frac{5}{16}$	$\frac{3}{8}$	$\frac{13}{32}$	$\frac{1}{2}$	1
$e_i - e_{i-1}$	0	0	$\frac{2}{32}$	$\frac{1}{32}$	$\frac{3}{32}$	$\frac{16}{32}$	

The tables give the values of the $c_i(f(x))$ and e_i.

We check that the sum of these differences equals $e_6 - e_0 = 11/16$. The largest of the differences is $e_6 - e_5 = 1/2$. Intuitively, this points to the minority bit predictor $\mathcal{M}_6$ for the last bit, from page 464, with success probability 1 and $1/2$ on $f(U_3)$ and U_6, respectively. But we want to trace the general construction. It yields the following predictor $\mathcal{P}$ for the sixth bit under $X = f(U_3)$. We have $e_6 - e_5 = 1/2 > 0$. On input $y \in \mathbb{B}^5$, $\mathcal{P}$ chooses $y_6 \in \mathbb{B}$ uniformly at random, calculates $z = \mathcal{A}(y, y_6)$, so that

$$z = \begin{cases} 1 & \text{if } w(y, y_6) = 3, \\ 0 & \text{otherwise,} \end{cases}$$

and outputs $y_6 \oplus z \oplus 1$. We claim that $\mathcal{P}(y) = \mathcal{M}_6(y)$ for any $y \in \pi_5(X)$. This follows from the following table of the values $(z, \mathcal{P}(y))$, where the second component in each the four entries indeed always equals $\mathcal{M}_6(y)$:

		$w(y)$	
		2	3
	0	$(1,1)$	$(0,0)$
y_6			
	1	$(0,1)$	$(1,0)$

In particular, we have $\sigma_{\mathcal{P}}(X) = \sigma_{\mathcal{M}_6}(X) = 1$, and $\mathcal{P}$ is also an $1/2$-predictor for X. However, $\mathcal{P}$ is not equal to $\mathcal{M}_6$, since on input $y = (0,0,0,0,0) \notin \pi_5(X)$, say, we have $\text{prob}\{1 \xleftarrow{\boxtimes} \mathcal{P}(y)\} = 1/2$ and $\text{prob}\{1 \xleftarrow{\boxtimes} \mathcal{M}_6(y)\} = 1$. $\diamond$

11.5. From short to long generators

If we have an (ε, s)-resilient pseudorandom generator $f \colon \mathbb{B}^k \to \mathbb{B}^n$ and $k < \ell \leq n$, then by composing with the projection $\pi \colon \mathbb{B}^n \to \mathbb{B}^\ell$ to the first ℓ bits we get a function $g = \pi \circ f \colon \mathbb{B}^k \longrightarrow \mathbb{B}^\ell$. It is also an (ε, s)-resilient pseudorandom generator, since any algorithm that distinguishes $g(U_k)$ from U_ℓ can also distinguish $f(U_k)$ from U_n, with the same size and quality, by ignoring the final $n - \ell$ bits.

Thus it is easy as pie to shorten generators. Can we also make them longer? This is less obvious, and this section is devoted to showing that it can indeed be achieved.

We take as our starting point a generator that is as short as possible, namely $f \colon \mathbb{B}^k \to \mathbb{B}^{k+1}$, and construct from it a generator $g \colon \mathbb{B}^k \to \mathbb{B}^{k+n}$ for any $n > 0$. To do this, we apply f iteratively to k-bit strings, save the first bit, and apply f again to the remaining k bits.

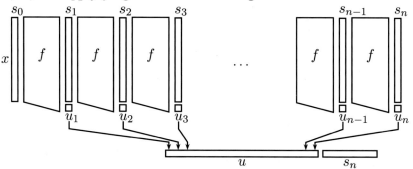

Figure 11.8: Long generator g.

So we have $f \colon \mathbb{B}^k \to \mathbb{B}^{k+1}$, $n > 0$, $s_0 = x \in \mathbb{B}^k$ and parse $f(s_{i-1})$ as $u_i s_i$ for $1 \leq i \leq n$, with $u_i \in \mathbb{B}$ and $s_i \in \mathbb{B}^k$. Then we take $g(x) = u_1 \cdots u_n s_n$. Figure 11.8 illustrates this construction, Algorithm 11.16 implements it and below we show that g is pseudorandom if f is.

ALGORITHM 11.16. Long generator g from short generator f.

Input: Positive integers k and n, $f \colon \mathbb{B}^k \to \mathbb{B}^{k+1}$, and a seed $x \in \mathbb{B}^k$.
Output: A string $g(x) \in \mathbb{B}^{k+n}$.

1. $s_0 \leftarrow x$.
2. For i from 1 to n do step 3.
3. $\quad u_i s_i \leftarrow f(s_{i-1})$, where $u_i \in \mathbb{B}$ and $s_i \in \mathbb{B}^k$.
4. Return $g(x) = u_1 \cdots u_n s_n$.

It is useful to have names for the functions computed in the algorithm, so that we can describe their actions on distributions. We denote by $\mathrm{id}_j \colon \mathbb{B}^j \to \mathbb{B}^j$ the identity function and define for $i \geq 1$ functions f_i that leave the first $i - 1$ bits unchanged and apply f to the last k bits (as in step 3), and their compositions (as in a partial loop):

$$f_i = \mathrm{id}_{i-1} \times f \colon \quad \begin{array}{ccc} \mathbb{B}^{k+i-1} & \longrightarrow & \mathbb{B}^{k+i}, \\ (x_1, \ldots, x_{k+i-1}) & \longmapsto & (x_1, \ldots, x_{i-1}, f(x_i, \ldots, x_{k+i-1})), \end{array}$$

$$g_i = f_i \circ \cdots \circ f_2 \circ f_1 \colon \quad \mathbb{B}^k \longrightarrow \mathbb{B}^{k+i}.$$

Thus $f = f_1 = g_1$ and $g = g_n$. We also set $g_0 = \mathrm{id}_k$.

The idea for proving that g is pseudorandom is to turn a distinguisher $\mathcal{A}$ between U_{k+n} and $g(U_k)$ into a distinguisher $\mathcal{B}$ between U_{k+1} and $f(U_k)$. We consider hybrid random variables $Y_0, Y_1, \ldots, Y_n$ which interpolate between $Y_n = g(U_k)$ and $Y_0 = U_n$. If $\mathcal{A}$ distinguishes well between Y_0 and Y_n, then it also distinguishes reasonably well between Y_i and Y_{i+1} for random i. But these adjacent distributions Y_i and Y_{i+1} are essentially like $f(U_k)$ and U_{k+1}, so that we can also distinguish between these two. By assumption, this can only be done with small advantage, so that also the advantage of the initial $\mathcal{A}$ is small.

ALGORITHM 11.17. From long to short distinguishers.

Input: $x \in \mathbb{B}^{k+1}$.
Output: 0 or 1.

1. Choose $i \xleftarrow{\boxdot} \{1, \ldots, n\}$.
2. Choose $y \xleftarrow{\boxdot} U_{n-i}$.
3. Execute $\mathcal{A}$ on input $(y, x_1, g_{i-1}(x_2, \ldots, x_{k+1})) \in \mathbb{B}^{k+n}$ and return its output.

LEMMA 11.18. *Suppose that f can be computed in size t and that $\mathcal{A}$ is a (ε, s')-distinguisher between U_{k+n} and $g(U_k)$. Then $\mathcal{B}$ given by Algorithm 11.17 is a $(\varepsilon/n, s' + n(t + 2))$-distinguisher between U_{k+1} and $f(U_k)$.*

PROOF. $\mathcal{A}$ is a circuit of size s' and we may assume that

$$\mathcal{E}_{\mathcal{A}}(g(U_k)) - \mathcal{E}_{\mathcal{A}}(U_{k+n}) \geq \varepsilon.$$

If the left hand quantity is at most $-\varepsilon$, then we flip the output bit of $\mathcal{A}$ to obtain the above inequality.

For $1 \leq i \leq n$, the right-hand part of the string in step 3 is given by the function $h_i = \mathrm{id}_1 \times g_{i-1} \colon \mathbb{B}^{k+1} \longrightarrow \mathbb{B}^{k+i}$, so that

$$h_i(y_1, y_2, \ldots, y_{k+1}) = (y_1, g_{i-1}(y_2, \ldots, y_{k+1}))$$

for all $(y_1, \ldots, y_{k+1}) \in \mathbb{B}^{k+1}$. Now consider an input $s_0 = x \in \mathbb{B}^k$ to Algorithm 11.16, $y = f(x) = (y_1, \ldots, y_{k+1}) \in \mathbb{B}^{k+1}$ and $s_1 = y' = (y_2, \ldots, y_{k+1}) \in \mathbb{B}^k$. The algorithm first produces $u_1 = y_1$ and then proceeds as it would with input y', with $i' = i-1$ running from 1 to $n-1$. Thus the value $g_i(x)$ after the ith round is the same as first applying f and then h_i. In other words, $h_i \circ f = g_i$ on $\mathbb{B}^k$ for $i \geq 1$ and it follows that

$$(11.19) \qquad h_i(f(U_k)) = g_i(U_k), \quad h_i(U_{k+1}) = U_1 \times g_{i-1}(U_k).$$

Here and in similar situations later, the uniform distributions like U_1 and U_k are taken independently. For any input $x \in \mathbb{B}^{k+1}$ to $\mathcal{B}$, we have

$$(11.20) \quad \mathrm{prob}\{1 \xleftarrow{\;\boxtimes\;} \mathcal{B}(x)\} = \frac{1}{n} \sum_{1 \leq i \leq n} \mathrm{prob}\{1 \xleftarrow{\;\boxtimes\;} \mathcal{A}(U_{n-i}, h_i(x))\}.$$

We consider for $0 \leq i \leq n$ the hybrid random variable

$$Y_i = U_{n-i} \times g_i(U_k)$$

with values in $\mathbb{B}^{k+n}$. Thus $Y_n = g_n(U_k)$ and $Y_0 = U_{k+n}$ are the two random variables between which $\mathcal{A}$ distinguishes. For any $i \leq n$ we have

$$\begin{aligned} Y_i &= U_{n-i} \times g_i(U_k) = U_{n-i} \times h_i(f(U_k)) \text{ if } i \geq 0, \\ Y_{i-1} &= U_{n-i+1} \times g_{i-1}(U_k) \\ &= U_{n-i} \times U_1 \times g_{i-1}(U_k) = U_{n-i} \times h_i(U_{k+1}) \text{ if } i \geq 1. \end{aligned}$$

Now let $e_i = \mathrm{prob}\{1 \xleftarrow{\;\boxtimes\;} \mathcal{A}(Y_i)\} = \mathcal{E}_{\mathcal{A}}(Y_i)$ for $0 \leq i \leq n$. The assumption about $\mathcal{A}$'s distinguishing power says that $e_n - e_0 \geq \varepsilon$. Then using (11.20) we have

$$\mathcal{E}_{\mathcal{B}}(f(U_k)) = \frac{1}{n} \sum_{1 \leq i \leq n} \mathcal{E}_{\mathcal{A}}(U_{n-i} \times h_i(f(U_k)))\}$$

$$= \frac{1}{n} \sum_{1 \leq i \leq n} \mathcal{E}_{\mathcal{A}}(Y_i) = \frac{1}{n} \sum_{1 \leq i \leq n} e_i,$$

$$\mathcal{E}_{\mathcal{B}}(U_{k+1}) = \frac{1}{n} \sum_{1 \leq i \leq n} \mathcal{E}_{\mathcal{A}}(U_{n-i} \times h_i(U_{k+1}))$$

$$= \frac{1}{n} \sum_{1 \leq i \leq n} \mathcal{E}_{\mathcal{A}}(Y_{i-1})\} = \frac{1}{n} \sum_{1 \leq i \leq n} e_{i-1},$$

$$\mathcal{E}_{\mathcal{B}}(f(U_k)) - \mathcal{E}_{\mathcal{B}}(U_{k+1}) = \frac{1}{n} \Big(\sum_{1 \leq i \leq n} e_i - \sum_{1 \leq i \leq n} e_{i-1} \Big)$$

$$= \frac{1}{n}(e_n - e_0) \geq \frac{\varepsilon}{n}.$$

Thus algorithm $\mathcal{B}$ has distinguishing power at least ε/n between $f(U_k)$ and U_{k+1}. We have to determine the size of $\mathcal{B}$. The random choices in steps 1 and 2 correspond to at most $2n$ further random input gates. For $h_i(x)$, we have to apply f exactly $i - 1 \leq n - 1$ times, using size at most nt. The execution of $\mathcal{A}$ takes another s' gates. The total comes to $s' + n(t + 2)$. $\square$

THEOREM 11.21. Let $f \colon \mathbb{B}^k \to \mathbb{B}^{k+1}$ be an (δ, s)-resilient generator that can be computed by a circuit of size t, let $n \geq 1$, and $g = g_n \colon \mathbb{B}^k \to \mathbb{B}^{k+n}$ as in Algorithm 11.16. Then g is an $(n\delta, s - n(t + 2))$-resilient generator, and can be computed with n applications of f.

PROOF.   Since f is (δ, s)-resilient, we have in Lemma 11.18 either $\varepsilon/n < \delta$ or $s' + n(t + 2) > s$, which is the claim. $\square$

It is straightforward to apply this construction to the asymptotic notion of pseudorandom generators, whose output cannot be distinguished by polynomial-size circuit families from the uniform distribution.

COROLLARY 11.22. Let $f = (f_k)_{k \in \mathbb{N}}$ be a pseudorandom generator with $f_k \colon \mathbb{B}^k \longrightarrow \mathbb{B}^{k+1}$, and $p \in \mathbb{Z}[t]$ a positive polynomial. Then the above construction yields a pseudorandom generator $g = (g_k)_{k \in \mathbb{N}}$ with $g_k \colon \mathbb{B}^k \longrightarrow \mathbb{B}^{k+p(k)}$.

Thus we have the pleasant result that from the smallest possible pseudorandom generators, which add only one pseudorandom bit, we can obtain pseudorandom generators with arbitrary polynomial expansion rate.

11.6. The Nisan-Wigderson generator

All known pseudorandom generators assume that some function is hard to compute, and then extend few random bits to many bits that look random to all efficient algorithms. The Nisan-Wigderson generator that we now describe starts from a fairly general assumption of this type, and produces a pseudorandom generator.

We first quantify when a Boolean function f is hard to approximate. Namely, a probabilistic Boolean circuit $\mathcal{A}$ can produce a random bit, which will equal the value of f with probability $1/2$. Now f is difficult if nothing essentially better is possible, with small circuits. More precisely, let $f\colon \mathbb{B}^n \longrightarrow \mathbb{B}$ be a Boolean function, $\varepsilon > 0$, and $s \in \mathbb{N}$. We say that f is (ε, s)-*hard* if for all algorithms (Boolean circuits) $\mathcal{A}$ with n inputs and size s, we have

$$|\operatorname{prob}\{f(x) = \mathcal{A}(x)\colon x \xleftarrow{\boxtimes} U_n\} - \frac{1}{2}| \leq \frac{\varepsilon}{2}.$$

If we have a hard function f, then the single bit $f(x)$, for random $x \in \mathbb{B}^n$, looks random to any efficient algorithm. We now show how to get many bits that look random by evaluating f at many different, nearly disjoint, subsets of bits of a larger input. The tool for achieving this comes from *design theory*, an area of combinatorics, and the theory of finite fields.

DEFINITION 11.23. *Let k, n, s, and t be integers. A (k, n, s, t)-design D is a sequence $D = (S_1, \ldots, S_n)$ of subsets of $\{1, \ldots, k\}$ such that for all $i, j \leq n$ we have*

(i) *$\#S_i = s$,*

(ii) *$\#(S_i \cap S_j) \leq t$ if $i \neq j$.*

EXAMPLE 11.24. We take $k = 9$, $n = 12$, $s = 3$, and $t = 1$, and arrange the nine elements of $\{1, \ldots, 9\}$ in a 3×3 square:

7	8	9
4	5	6
1	2	3

$\cdot$

The reason for doing this will be explained after Theorem 11.28. In each of the following four copies of the square, we have marked three subsets S_i: one with •, one with ■, and the third one with ♦.

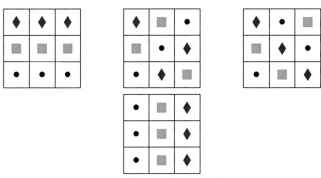

Thus $S_1 = \{1,2,3\}$, $S_2 = \{4,5,6\}$, $S_3 = \{7,8,9\}$, $S_4 = \{1,5,9\}$, $S_5 = \{3,4,8\}$, $S_6 = \{2,6,7\}$, $S_7 = \{1,6,8\}$, $S_8 = \{2,4,9\}$, $S_9 = \{3,5,7\}$, $S_{10} = \{1,4,7\}$, $S_{11} = \{2,5,8\}$, and $S_{12} = \{3,6,9\}$.

Now $D = \{S_1, \ldots, S_{12}\}$ is an $(9,12,3,1)$-design as one easily verifies. As an example, $S_1 \cap S_5 = \{3\}$ has only one element.   ◇

In design theory, one does not usually order the S_1, $\ldots$, S_n, but the above is more appropriate for our purposes. The general goal in design theory is to fix some of the four parameters and optimize the others, making n and s as large and k and t as small as possible.

If D is a (k,n,s,t)-design as above and $f\colon \mathbb{B}^s \longrightarrow \mathbb{B}$ a Boolean function, we obtain a Boolean function $f_D\colon \mathbb{B}^k \longrightarrow \mathbb{B}^n$ by evaluating f at the subsets of the bits of x given by S_1, $\ldots$, S_n. More specifically, if $x \in \mathbb{B}^k$, $1 \leq i \leq n$, and $S_i = \{v_1, \ldots, v_s\}$, with $1 \leq v_1 < v_2 < \cdots < v_s \leq k$, then the ith bit of $f_D(x)$ is $f(x_{v_1}, \ldots, x_{v_s})$.

EXAMPLE 11.24 CONTINUED. We consider the XOR function $f\colon \mathbb{B}^3 \longrightarrow \mathbb{B}$, so that $f(x_1, x_2, x_3) = x_1 \oplus x_2 \oplus x_3$ is the sum of x_1, x_2, and x_3 modulo 2. With the design from above, the value of $f_D\colon \mathbb{B}^9 \longrightarrow \mathbb{B}^{12}$ at $x = 011110001 \in \mathbb{B}^9$ is

$$f_D \left(\begin{array}{|c|c|c|} \hline 0 & 0 & 1 \\ \hline 1 & 1 & 0 \\ \hline 0 & 1 & 1 \\ \hline \end{array} \right) = 001001010100.$$

For example, the second of the twelve values is computed as $f_D(x)_2 = f(x_4, x_5, x_6) = f(110) = 1 \oplus 1 \oplus 0 = 0$.   ◇

We want to simplify the two parameters ε and s in our notion of (ε, s)-resilient pseudorandom generators. To this end, we—somewhat artificially—set $\varepsilon = n^{-1}$ and $s = n$. Thus we now consider pseudorandom generators $f\colon \mathbb{B}^k \longrightarrow \mathbb{B}^n$ for which there is no algorithm using time at most n and with

$$\mathrm{adv}_{\mathcal{A}}(U_n, f(U_k)) = |\mathcal{E}_{\mathcal{A}}(U_n) - \mathcal{E}_{\mathcal{A}}(f(U_k))| \geq n^{-1}.$$

This is seemingly more generous than the previous definition. One has to show that from a pseudorandom generator in the new sense one can construct one in the previous sense (with different values of k and n). Similarly, the *hardness* H_f of a Boolean function f is the maximal integer $H_f = h$ such that f is (h^{-1}, h)-hard.

THEOREM 11.25. *Let k, n, s be positive integers, $s \geq 2$, $t = \lfloor \log_s n \rfloor - 1$, $f\colon \mathbb{B}^s \longrightarrow \mathbb{B}$ with hardness at least $2n^2$, and D an (k, n, s, t)-design. Then $f_D\colon \mathbb{B}^k \longrightarrow \mathbb{B}^n$ is an (n^{-1}, n)-resilient pseudorandom generator.*

PROOF.   By Theorem 11.12, any ε-distinguisher between $X = f_D(U_k)$ and U_n can be transformed into a $\frac{\varepsilon}{n}$-predictor for some bit under X. So we now assume that we have a predictor $\mathcal{P}$ for the ith bit under X, for some $i \leq n$, with $\mathrm{adv}_{\mathcal{P}}(X) \geq \varepsilon \geq n^{-2}$, and derive a contradiction to our hardness assumption.

    To simplify the notation, we reorder the elements of $\{1, \ldots, k\}$ and assume that $S_i = \{1, \ldots, s\}$, so that the ith bit depends only on the first s components of the values of U_k. In order to separate out the dependence on the first s and the last $k - s$ bits, we write $U_k = U_s \times U_{k-s}$. As in (11.8), we let $p(y, *) = \mathrm{prob}\{y \xleftarrow{\boxdot} (X_1, \ldots, X_{i-1})\}$ be the probability that y occurs as an initial segment under X, for $y \in \mathbb{B}^{i-1}$. Then

$$
\begin{aligned}
1/2 + \varepsilon &\leq \sigma_{\mathcal{P}}(X) \\
&= \sum_{y \in \mathbb{B}^{i-1}} \mathrm{prob}\{y \xleftarrow{\boxdot} (X_1, \ldots, X_{i-1})\} \cdot \mathrm{prob}\{\mathcal{P}(y) \xleftarrow{\boxdot} X_i(y)\} \\
&= \sum_{\substack{x' \in \mathbb{B}^s, x'' \in \mathbb{B}^{k-s} \\ y = f_D(x', x'')_{1\ldots i-1} \in \mathbb{B}^{i-1}}} \mathrm{prob}\{x' \xleftarrow{\boxdot} U_s\} \cdot \mathrm{prob}\{x'' \xleftarrow{\boxdot} U_{k-s}\} \\
&\qquad\qquad\qquad\qquad\qquad \cdot \mathrm{prob}\{f(x') \xleftarrow{\boxdot} \mathcal{P}(y)\} \\
&= 2^{-(k-s)} \sum_{x'' \in \mathbb{B}^{k-s}} r(x''),
\end{aligned}
$$

where $f_D(x', x'')_{1\ldots i-1}$ stands for $(f_D(x', x'')_1, \ldots, f_D(x', x'')_{i-1}) \in \mathbb{B}^{i-1}$, and

$$r(x'') = 2^{-s} \sum_{\substack{x' \in \mathbb{B}^s \\ y = f_D(x', x'')_{1\ldots i-1}}} \mathrm{prob}\{f(x') \xleftarrow{\boxdot} \mathcal{P}(y)\}.$$

Thus the average of r over $\mathbb{B}^{k-s}$ is at least $1/2+\varepsilon$. Then there exists some value $z \in \mathbb{B}^{k-s}$ of x'' so that $r(z) \geq 1/2 + \varepsilon$; otherwise we would have

$$2^{k-s}(1/2 + \varepsilon) > 2^{k-s} \max_{x'' \in \mathbb{B}^{k-s}} r(x'') \geq \sum_{x'' \in \mathbb{B}^{k-s}} r(x'') \geq 2^{k-s}(1/2 + \varepsilon).$$

This is just an instance of the general fact that "not everybody can be below average". Now we fix such a z. Thus

$$(11.26) \qquad r(z) = 2^{-s} \sum_{\substack{x' \in \mathbb{B}^s \\ y = f_D(x',z)_{1...i-1}}} \mathrm{prob}\{f(x') \xleftarrow{\boxtimes} \mathcal{P}(y)\} \geq 1/2 + \varepsilon.$$

It may be hard to compute such a z, but it does not depend on any input and can be hard-wired into the circuit below, since this model of computation is nonuniform; see Section 15.6. We now have an algorithm for approximating f: on input x', compute y as above and use $\mathcal{P}(y)$ as an approximation for $f(x')$.

ALGORITHM 11.27. Circuit $\mathcal{A}$ that approximates f.

Input: $x' = (x_1, \ldots, x_s) \in \mathbb{B}^s$.
Output: 0 or 1.

1.   For $j = 1, \ldots, i-1$ do step 2.
2. $y_j \longleftarrow f_D(x', z)_j$, with $z \in \mathbb{B}^{k-s}$ satisfying (11.26).
3.   Return $\mathcal{P}(y_1 \ldots, y_{i-1})$.

We have to show that $\mathcal{A}$ approximates f well, and that it can be built with few gates. The latter seems implausible at first, since in step 2 we have to evaluate f at some point $w_i \in \mathbb{B}^s$, given by the bits of (x', z) in the positions contained in S_i. But isn't that hard? Yes, computing f at an arbitrary input is hard, but the whole set-up is designed so that these special evaluation problems become easy.

Let $1 \leq j < i$. Since $\#(S_i \cap S_j) \leq t = \lfloor \log_s n \rfloor - 1 \leq \lfloor \log_2 n \rfloor - 1$ and z is fixed, y_j depends on at most t bits. It is a general fact that any Boolean function on t bits (with one output) can be computed with size 2^{t+1}, say by writing it in disjunctive (or conjunctive) normal form. Thus y_j can be computed from x' with size $2^{t+1} \leq n$, and all of $y_1, \ldots, y_{i-1}$ can be computed with at most $n(i-1) \leq n^2$ operations.

What is the probability that $\mathcal{A}(x') = f(x')$, for $x' \longleftarrow U_s$? We are given our fixed z, and compute $y_1, \ldots, y_{i-1}$ correctly from x'. Thus

$\mathcal{A}(x') = \mathcal{P}(y_1, \ldots, y_{i-1})$, and

$$2^{-s} \sum_{x' \in \mathbb{B}^s} \text{prob}\{f(x') \xleftarrow{\;\boxdot\;} \mathcal{A}(x')\} = 2^{-s} \sum_{\substack{x' \in \mathbb{B}^s \\ y = f_D(x',z)_{1\ldots i-1}}} \text{prob}\{f(x') \xleftarrow{\;\boxdot\;} \mathcal{P}(y)\}$$

$$= r(z) \geq 1/2 + \varepsilon \geq 1/2 + n^{-2}.$$

This contradicts the assumption that $H_f \geq 2n^2$, and proves the claim. $\square$

11.7. Construction of good designs

As in many other fields of combinatorics, finite fields are the basis for an attractive solution. Let $\mathbb{F}_q$ be a finite field with q elements, so that q is a prime power, $t < q$ an integer,

$$\begin{aligned}
P &= \{f \in \mathbb{F}_q[x]\colon \deg f \leq t\}, \\
S_f &= \{(u, f(u))\colon u \in \mathbb{F}_q\} \subseteq L = \mathbb{F}_q^2 \text{ for } f \in P, \\
k &= \#L = q^2, \quad n = q^{t+1}.
\end{aligned}$$

THEOREM 11.28. *The collection of all these graphs S_f for $f \in P$ is a (k, n, q, t)-design.*

PROOF.   The only claim to verify is that $\#(S_f \cap S_g) \leq t$ for distinct f and $g \in P$. But $\#(S_f \cap S_g) \geq t + 1$ means that the two polynomials f and g of degree at most t have $t + 1$ values in common. Then $f - g$ is a polynomial of degree at most t with at least $t + 1$ roots, hence the zero polynomial, and we have $f = g$. $\square$

EXAMPLE 11.24 CONTINUED. We take $q = 3$ and $t = 1$, so that $k = 9 = q^2$, $n = 9 = q^{1+1}$, and $s = 3 = q$. The following picture shows this design.

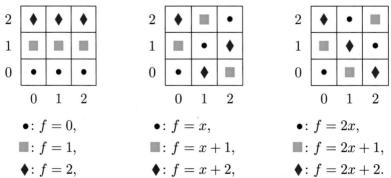

•: $f = 0$, 　　　•: $f = x$, 　　　•: $f = 2x$,

■: $f = 1$, 　　　■: $f = x + 1$, 　　　■: $f = 2x + 1$,

◆: $f = 2$, 　　　◆: $f = x + 2$, 　　　◆: $f = 2x + 2$.

Thus we find the first nine pieces of the design from Example 11.24.   ◇

In general, this construction does not provide the best possible design, but it is very simple and sufficient for our purposes.

$s = q$	t	$k = s^2$	$n = s^{t+1}$	$\binom{k}{s}$
3	1	9	9	28
3	2	9	27	84
4	2	16	64	1820

Figure 11.9: Some design parameters. Compare n to the number $\binom{k}{s}$ of all subsets of size s.

COROLLARY 11.29.

(i) *For any positive integers $s > t$, where s is a prime power, there exists an (s^2, s^{t+1}, s, t)-design.*

(ii) *For any positive integers k, n, t, and a prime power s with $k \geq s^2$ and $t \geq \log_s n - 1$ there exists a (k, n, s, t)-design.*

PROOF. In (i) we have recorded the above construction. For (ii), we use (i) and note that $n = s^{\log_s n} \leq s^{t+1}$. $\qquad\square$

COROLLARY 11.30. *Let n and s be positive integers, with s a prime power, and $f : \mathbb{B}^s \longrightarrow \mathbb{B}$ with hardness $H_f \geq 2n^2$. Then the Nisan-Wigderson generator is a pseudorandom generator from $\mathbb{B}^{s^2}$ to $\mathbb{B}^n$.*

In particular, if n is exponential in s, say $n = 2^{s/4}$, then we have a pseudorandom generator that turns short random strings into exponentially long pseudorandom ones. Even from the weaker requirement $n > s^2$, so that $H_f \geq 2(s^2 + 1)^2$ is needed, we obtain with Theorem 11.21 arbitrarily long pseudorandom generators. Further consequences of this result are discussed in the Notes.

11.8. The Blum-Blum-Shub generator

Blum, Blum & Shub (1986) propose the following pseudorandom *Blum-Blum-Shub* generator(BBS). We make substantial use of the notions and facts from Section 15.14. For an n-bit Blum integer N, an integer $\ell > 0$ as the bit-length of the desired output, and a seed $u = u_0^2 \in \square_N$ for $u_0 \xleftarrow{\;\boxdot\;} \mathbb{Z}_N^\times$, we define its output as

$$(11.31) \qquad \mathrm{BBS}_N^{(\ell)}(u) = (\mathrm{lsb}(u), \mathrm{lsb}(u^2), \mathrm{lsb}(u^4), \ldots, \mathrm{lsb}(u^{2^{\ell-1}})).$$

This means that we repeatedly square the input seed in $\mathbb{Z}_N$ and extract the least significant bit lsb before each squaring step. We claim that starting from a random square u, the resulting ℓ-bit sequence is indistinguishable from a uniformly random ℓ-bit sequence, provided that the factoring assumption Conjecture 3.33 holds for the distribution of N used in the system. For proving this claim, we take the following steps.

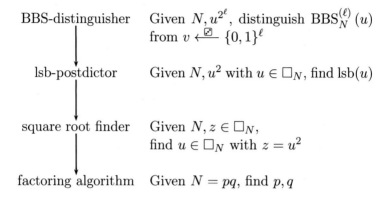

Figure 11.10: From a BBS-distinguisher to a factoring algorithm.

Blum *et al.* (1986) show that their BBS generatoris indistinguishable under the quadratic residuosity assumption, which says that it is hard to determine whether a given element of $\mathbb{Z}_N^\times$ is a square. Furthermore, Hofheinz *et al.* (2013) derive from a (hypothetical) distinguisher for their HKS encryption system (Section 9.10) a BBS-distinguisher that receives an additional input parameter $z \in \square_N$ whose value is u^{2^ℓ} if $\mathrm{BBS}^\ell(u)$ is to be tested and a random square if a random string chosen from U^ℓ is to be tested, where U is the uniform distribution on bits. According to Definition 9.17 (i), the distinguishing advantage of a BBS-distinguisher $\mathcal{B}$ is

(11.32)
$$\mathrm{adv}_{\mathrm{BBS},\mathcal{B}} = \left| \mathcal{E}_{\mathcal{B}}\left(N, u^{2^\ell}, \mathrm{BBS}^{(\ell)}(u) \colon u \xleftarrow{\;\boxempty\;} \square_N\right) - \mathcal{E}_{\mathcal{B}}\left(N, z \xleftarrow{\;\boxempty\;} \square_N, U^\ell\right)\right|.$$

Lsb-postdictor from BBS-distinguisher.   While a predictor guesses the next bit of a pseudorandom sequence (Definition 11.9 (i)), a *postdictor* guesses the previous bit. In this first reduction, we use a distinguisher for the BBS generator to construct an algorithm that given $z = u^2 \in \square_N$ with $u \in \square_N$ returns the least significant bit of u. Every square in $\square_N$ has four square roots, and exactly one of them is again in $\square_N$. The construction is based on hybrid distributions, as in the proof of Theorem 11.12 and Figure 11.7.

REDUCTION 11.33. Lsb-postdictor $\mathcal{L}$ from $\mathrm{BBS}^{(\ell)}$-distinguisher $\mathcal{B}$.
Input: $N \in \mathbb{Z}$, $z = u^2 \in \square_N$.

Output: A bit in $\{0,1\}$.

1. $k \xleftarrow{\boxdot} \{1,\ldots,\ell\}$.
2. $(v_1,\ldots,v_{k-1},b) \xleftarrow{\boxdot} \mathbb{B}^k$.
3. $v \leftarrow (v_1,\ldots,v_{k-1},b,\mathrm{lsb}(z),\mathrm{lsb}(z^2),\ldots,\mathrm{lsb}(z^{2^{\ell-k-1}})) \in \{0,1\}^\ell$.
4. $b^* \leftarrow \mathcal{B}(N,z^{2^{\ell-k}},v)$.
5.   Return $b \oplus b^* \oplus 1$.

LEMMA 11.34. *Let $\mathcal{B}$ be a $\mathrm{BBS}^{(\ell)}$-distinguisher with advantage ε. Then $\mathcal{L}$ as in Reduction 11.33 is an lsb-postdictor with advantage $\varepsilon/2\ell$.*

PROOF.   For $0 \le k \le \ell$, let $X^{(\ell-k)}$ be the random variable on $\mathbb{B}^{\ell-k}$ corresponding to the output distribution of $\mathrm{BBS}^{(\ell-k)}$ with input $u \xleftarrow{\boxdot} \square_N$. We define the random variable

$$Y_k = U^k \times X^{(\ell-k)}$$

on $\mathbb{B}^\ell$. Then Y_k is the *hybrid variable* where the first k bits come from the uniform distribution and the last $\ell - k$ bits are generated according to the output distribution of $\mathrm{BBS}^{(\ell)}$. We have $Y_0 = U^\ell$ and $Y_\ell = X^{(\ell)}$. Furthermore, for $k \le \ell$ let

$$e_k = \mathcal{E}_\mathcal{B}(N, z^{2^{\ell-k}}, Y_k)$$

be the expected value of the output of $\mathcal{B}$ on input a sample of a hybrid distribution Y_k. According to (11.32), the distinguishing advantage of $\mathcal{B}$ is

(11.35) $$\varepsilon = \mathrm{adv}_\mathcal{B} = |e_0 - e_\ell|.$$

If b as chosen in step 2 of Reduction 11.33 is in fact the least significant bit of $u = z^{1/2} \in \square_N$, then step 3 constructs a v from Y_{k-1}. That is, the first $k - 1$ bits of v are uniformly random, while the other bits correspond to the output distribution of $\mathrm{BBS}^{(\ell-k+1)}$. We have

(11.36) $$\begin{aligned} &\mathrm{prob}\{\mathcal{L}(N, u^2) = \mathrm{lsb}(u)\colon b = \mathrm{lsb}(u)\} \\ &= \mathrm{prob}\{b^* = 1\colon b = \mathrm{lsb}(u)\} \\ &= \sum_{1 \le k \le \ell} \frac{1}{\ell} \mathrm{prob}\{1 \xleftarrow{\boxdot} \mathcal{B}(N, u^{2^{\ell-k+1}}, Y_{k-1})\} \\ &= \frac{1}{\ell} \sum_{1 \le k \le \ell} \mathcal{E}_\mathcal{B}(N, u^{2^{\ell-k+1}}, Y_{k-1}) = \frac{1}{\ell} \sum_{1 \le k \le \ell} e_{k-1}. \end{aligned}$$

On the other hand, if $b \neq \text{lsb}(u)$, then the first k bits of v are uniformly random, and v will be from Y_k. Hence, we find

(11.37)

$$\text{prob}\{\mathcal{L}(N, u^2) = \text{lsb}(u) \colon b \neq \text{lsb}(u)\}$$
$$= \text{prob}\{b^* = 1 \colon b \neq \text{lsb}(u)\}$$
$$= \sum_{1 \leq k \leq \ell} \frac{1}{\ell}(1 - \text{prob}\{1 \xleftarrow{\;\square\;} \mathcal{B}(N, z^{2^{\ell-k}}, Y_k)\})$$
$$= \frac{1}{\ell} \sum_{1 \leq k \leq \ell} (1 - e_k).$$

The *success probability* $\sigma_{\mathcal{L}}$ of $\mathcal{L}$ is the probability that $\mathcal{L}$ guesses the least significant bit of u correctly. Then

(11.38)

$$\sigma_{\mathcal{L}} = \text{prob}\{\mathcal{L}(N, u^2) = \text{lsb}(u)\}$$
$$= \text{prob}\{b = \text{lsb}(u)\} \cdot \text{prob}\{\mathcal{L}(N, u^2) = \text{lsb}(u) \colon b = \text{lsb}(u)\}$$
$$\quad + \text{prob}\{b \neq \text{lsb}(u)\} \cdot \text{prob}\{\mathcal{L}(N, u^2) = \text{lsb}(u) \colon b \neq \text{lsb}(u)\}$$
$$= \frac{1}{2} \cdot \frac{1}{\ell} \sum_{1 \leq k \leq \ell} e_{k-1} + \frac{1}{2} \cdot \frac{1}{\ell} \sum_{1 \leq k \leq \ell} (1 - e_k)$$
$$= \frac{1}{2\ell} \sum_{1 \leq k \leq \ell} (e_{k-1} - e_k) + \frac{1}{2\ell} \sum_{1 \leq k \leq \ell} 1$$
$$= \frac{1}{2\ell}(e_0 - e_\ell) + \frac{1}{2}.$$

Here we use (11.36) and (11.37) as well as the fact that b is drawn uniformly at random, and therefore $\text{prob}\{b = \text{lsb}(u)\} = \text{prob}\{b \neq \text{lsb}(u)\} = 1/2$. Equations (11.35) and (11.38) finally give us the *advantage* ε' of $\mathcal{L}$, indicating how much better $\mathcal{L}$ can find $\text{lsb}(u)$ over simply guessing:

$$\varepsilon' = |\sigma_{\mathcal{L}} - \frac{1}{2}| = |\frac{1}{2\ell}(e_0 - e_\ell)| = \frac{\varepsilon}{2\ell}. \qquad \square$$

Square root finder from lsb-postdictor. Given x^2, we want to compute $x \in \square_N$, using a hypothetical lsb-postdictor. The simplest way would be to compute the bits in the binary representation one by one: $x_0 = \text{lsb}(x)$, $x_1 = \text{lsb}((x - x_0)/2)$, ... This does not work for two reasons. If $x_0 = 1$, we do not know how to compute $((x - x_0)/2)^2$ in $\mathbb{Z}_N$, which the lsb-postdictor requires. Furthermore, this postdictor only works with a (small) advantage for a uniformly random input, and on any fixed "worst-case" input it might return a uniformly random bit, not containing information about the lsb we seek.

The remedy is to run the postdictor on inputs of the form $(ax)^2 \in \mathbb{Z}_N$, where the reduction chooses $a \xleftarrow{\boxtimes} \mathbb{Z}_N$. The reduction can then compute $(ax)^2 = a^2 y$ and the postdictor is likely to return $\mathrm{lsb}(ax)$ with some (small) advantage. Once we have ax, $x = a^{-1} \cdot ax$ is easy to compute.

To explain the approach in more detail, we write $[z]_N$ for the residue in $\mathbb{Z}_N = \{0, \ldots, N-1\}$ of the rational number $z \in \mathbb{Q}$ modulo N. If $z = z_0/z_1$, with integers z_0 and z_1 and $\gcd(z_1, N) = 1$, then $[z]_N = z_0 \cdot z_1^{-1}$ in $\mathbb{Z}_N$. In the following, the gcd condition will be satisfied.

We first outline the idea of a binary division to approximate certain multiples of x with errors that get halved at each step, assuming that we have an exact lsb-postdictor $\mathcal{L}'$ that returns $\mathrm{lsb}(x)$ on any input $y = x^2$ for $x \in \square_N$. We have

$$(11.39) \qquad \left[\frac{x}{2}\right]_N = \frac{[x]_N + \mathrm{lsb}(x) \cdot N}{2}.$$

Let $a_0 \in \mathbb{Z}_N$, and for $t > 0$ define $a_t = [a_{t-1}/2]_N$. We define approximations $u_t N$ to $[a_t x]_N$ with $u_t \in \mathbb{Q}$ which get better and better with increasing t. Namely, we pick any $u_0 \in [0, 1)$ and set $u_t = (u_{t-1} + \mathrm{lsb}(a_t x))/2$. Then by (11.39), we have

(11.40)

$$\begin{aligned} [a_t x]_N - u_t N &= \left[\frac{a_{t-1} x}{2}\right]_N - \frac{N}{2}(u_{t-1} + \mathrm{lsb}(a_{t-1} x)) \\ &= \frac{[a_{t-1} x]_N + \mathrm{lsb}(a_{t-1} x) N}{2} - \frac{u_{t-1} N + \mathrm{lsb}(a_{t-1} x) N}{2} \\ &= \frac{[a_{t-1} x]_N - u_{t-1} N}{2}. \end{aligned}$$

Thus an approximation $u_{t-1} N$ to $[a_{t-1} x]_N$ gives in the next iteration an approximation $u_t N$ to $[a_t x]_N$ whose error is only half of the previous one. This allows to recover x after $\log_2 N$ stages.

However, the postdictor actually yields $\mathrm{lsb}(a_t x)$ only with small advantage ε. In order to boost it, we ask it for $\mathrm{lsb}(c_{t,i} x)$ for several known values of $c_{t,i}$ closely related to a_t. Then the average value returned indicates the correct value with better probability.

In most reductions, when we derive an algorithm $\mathcal{B}$ from some given algorithm $\mathcal{A}$ with success probability at least ε, we can bound the success probability of $\mathcal{B}$ as a function of ε without actually knowing ε. The next reduction is different. We assume that $\varepsilon \in \mathbb{Q}$ with denominator coprime to N and use random assignments of the form $u \xleftarrow{\boxtimes} \varepsilon \cdot [0 .. \varepsilon^{-1})$. This means that we choose uniformly an integer w with $0 \le w < \varepsilon^{-1}$ and set $u = \varepsilon \cdot w \in \mathbb{Q}$. If we do not know ε, we may try successive negative powers $2^{-1}, 2^{-2}, \ldots$ for ε.

The full-fledged square root finder $\mathcal{S}$ from an lsb-postdictor $\mathcal{L}$ with advantage at least $\varepsilon \in \mathbb{Q}$ looks as follows.

REDUCTION 11.41. Square root finder $\mathcal{S}$ from lsb-postdictor $\mathcal{L}$ with advantage at least $\varepsilon > 0$.

Input: An n-bit Blum integer N and $y \in \square_N$.
Output: $x \in \square_N$ with $x^2 = y$ or "failure".

1. $a_0, b \xleftarrow{\boxdot} \mathbb{Z}_N$.
2. $u_0 \xleftarrow{\boxdot} \frac{\varepsilon^3}{8} \cdot [0 .. \frac{8}{\varepsilon^3})$, $v \xleftarrow{\boxdot} \frac{\varepsilon}{8} \cdot [0 .. \frac{8}{\varepsilon})$.
3. $\alpha_0, \beta \xleftarrow{\boxdot} \mathbb{B}$.
4. For t from 1 to n do steps 5–10
5. $a_t \leftarrow [a_{t-1}/2]_N$, $u_t \leftarrow (u_{t-1} + \alpha_{t-1})/2$.
6. $A_t \leftarrow \{i \in \mathbb{Z} : |2i+1| \le 2\lceil n\varepsilon^{-2}\rceil\}$.
7.   For $i \in A_t$ do steps 8–9.
8. $c_{t,i} \leftarrow [(2i+1)a_t + b]_N, w_{t,i} \leftarrow \lfloor (2i+1)u_t + v \rfloor$.
9. $\alpha_{t,i} \leftarrow \mathcal{L}(N, c_{t,i}^2 y) + \beta + w_{t,i} \mod 2$.
10.   If $\sum_{i \in A_t} \alpha_{t,i} < \#A_t/2$ then $\alpha_t \leftarrow 0$ else $\alpha_t \leftarrow 1$.
11. $x \leftarrow [a_n^{-1} \lfloor u_n N + \frac{1}{2} \rfloor]_N$.
12. If $x^2 = y$ in $\mathbb{Z}_N$ then return x else return "failure".

LEMMA 11.42. *Let N be a Blum integer, $y \in \square_N$, and $\mathcal{L}$ a polynomial-time lsb-postdictor as above with advantage at least ε. Then the square root finder $\mathcal{S}$ from Reduction 11.41 on input N and y returns $x \in \square_N$ satisfying $x^2 = y$ in $\mathbb{Z}_N$ with success probability at least $2^{-9}\varepsilon^4$, and uses time polynomial in n and ε^{-1}.*

PROOF.   In the first three steps of Reduction 11.41, we choose two random elements a_0 and b of $\mathbb{Z}_N$ and then try to guess approximations u_0 and v as well as their least significant bits α_0 and β. We say the reduction guessed correctly, if

○ $|\frac{1}{N}[a_0 x]_N - u_0| \le \varepsilon^3/16$,

○ $|\frac{1}{N}[bx]_N - v| \le \varepsilon/16$,

○ $\alpha_0 = \mathrm{lsb}([a_0 x]_N)$,

○ $\beta = \mathrm{lsb}([bx]_N)$.

The corresponding probabilities are $\varepsilon^3/8$, $\varepsilon/8$, $1/2$, and $1/2$, respectively.

We now assume that this indeed happens. Furthermore, we assume $\varepsilon \le 1/2$ to be a rational number with denominator coprime to N. Then all quantities $[..]_N$ are well-defined, since only powers of 2 are introduced into denominators in the course of the reduction.

The loop starting in step 4 performs the previously mentioned binary division n times. In steps 6 to 10 we find the least significant bit α_t of $[a_t x]_N$ by majority decision. Although each $a_t x$ individually is uniformly random in $\mathbb{Z}_N$, we cannot rely on good luck n times in a row. Thus we do not query $\mathcal{L}$ for $\mathrm{lsb}([a_t x]_N)$ directly. Instead we compute $c_{t,i} = [(2i+1)a_t + b]_N$ for all integers i with $|2i+1| \leq m = 2\lceil n\varepsilon^{-2}\rceil$ from step 6. Since m is even, we have $\#A_t = m$. We define the integer $w'_{t,i}$ by

$$(11.43) \qquad \begin{aligned} [c_{t,i} x]_N &= [(2i+1)a_t x + bx]_N \\ &= (2i+1)[a_t x]_N + [bx]_N - w'_{t,i} N. \end{aligned}$$

Since any $[..]_N$ is an integer between 0 and $N-1$, we have $|w'_{t,i} N| \leq |2i+2|(N-1) < (m+1)N$ and hence $|w'_{t,i}| \leq m$. Furthermore,

$$(11.44) \qquad w'_{t,i} = \lfloor ((2i+1)[a_t x]_N + [bx]_N)/N \rfloor .$$

From (11.43) follows

$$\begin{aligned} \mathrm{lsb}([c_{t,i} x]_N) &= \mathrm{lsb}((2i+1)[a_t x]_N + [bx]_N - w'_{t,i} N) \\ &\equiv \mathrm{lsb}([a_t x]_N) + \mathrm{lsb}([bx]_N) + w'_{t,i} \bmod 2. \end{aligned}$$

Here we use that $2i+1$ and N are odd, hence $\mathrm{lsb}((2i+1)[a_t x]_N) = \mathrm{lsb}([a_t x]_N)$ and $-\mathrm{lsb}(w'_{t,i} N) \equiv \mathrm{lsb}(w'_{t,i}) \equiv w'_{t,i} \bmod 2$.

Given $\mathrm{lsb}([bx]_N)$ and $w_{t,i}$, a postdiction of $\mathrm{lsb}([c_{t,i} x]_N)$ as $\mathcal{L}(N, \mathrm{lsb}(c_{t,i}^2 y))$ with advantage ε allows us to postdict

$$(11.45) \qquad \mathrm{lsb}([a_t x]_N) \equiv \mathrm{lsb}([c_{t,i} x]_N) + \mathrm{lsb}([bx]_N) + w'_{t,i} \bmod 2$$

with the same advantage. Since $[a_t x]_N$ and $[bx]_N$ are unknown, we use their approximations u_t and v for computing an approximation $w_{t,i}$ to $w'_{t,i}$. This introduces some error which we analyze below.

Now we assume that the initial guesses of the algorithm are correct and for each $t \leq n$ we have $\alpha_t = \mathrm{lsb}(a_t x)$. Then by (11.40) the approximation error after the nth binary division is

$$|[a_n x]_N - u_n N| = 2^{-n}|[a_0 x]_N - u_0 N| = \frac{N}{2^n}\frac{1}{N}|[a_0 x]_N - u_0| \leq \frac{\varepsilon^3 N}{2^{n+4}} < \frac{1}{2}.$$

Thus we have $a_n x \equiv \lfloor u_n N + 1/2 \rfloor \bmod N$ and the output x is correct.

Next we investigate the probability that the least significant bit of $[a_t x]_N$ is correctly estimated in the tth iteration. From (11.44) and step 8, the error in approximating $w'_{t,i}$ by $w_{t,i}$ is

$$\Delta_{t,i} = \frac{1}{N}((2i+1)[a_t x]_N + [bx]_N) - ((2i+1)u_t + v).$$

Thus if $w_{t,i} \neq w'_{t,i}$, then $((2i+1)[a_t x]_N + [bx]_N)/N$ has distance at most $|\Delta_{t,i}|$ to some integer. If that happens, then, using (11.40), we have

$$\frac{1}{N}[c_{t,i}x]_N \leq |\Delta_{t,i}| = |\frac{1}{N}(2i+1)[a_t x]_N - (2i+1)u_t + \frac{1}{N}[bx]_N - v|$$

$$\leq |2^{-t}(2i+1)(\frac{1}{N}[a_0 x]_N - u_0)| + |\frac{1}{N}[bx]_N - v)|$$

$$\leq 2^{-t}|2i+1|\frac{\varepsilon^3}{16} + \frac{\varepsilon}{16} \leq \frac{\varepsilon}{8}.$$

We denote the event $[c_{t,i}x]_N/N \leq \varepsilon/8$ by "$\mathrm{Err}_{t,i}$". Thus $\mathrm{prob}\{w_{t,i} \neq w'_{t,i}\} \leq \mathrm{prob}\{\mathrm{Err}_{t,i}\}$. Since b is uniformly random in $\mathbb{Z}_N$, so is $c_{t,i}$, and $[c_{t,i}x]_N/N$ is uniformly random among the z/N with integral z satisfying $0 \leq z < N$. Hence

$$\mathrm{prob}\{\mathrm{Err}_{t,i}\} = \mathrm{prob}\{\frac{1}{N}[c_{t,i}x]_N \leq \frac{\varepsilon}{8}\} = \frac{\varepsilon}{8}.$$

The value for $\alpha_{t,i}$ computed in step 9, that is, the ith vote of the majority decision, is correct if $w_{t,i}$ as well as the result of $\mathcal{L}(N, c_{t,i}^2 x^2)$ are correct. To cover the error of the ith vote, we consider for fixed t and all $i \in A_t$ the random variable

$$X_i = \begin{cases} 1 & \text{if } \mathcal{L}(N, c_{t,i}^2 y) \neq \mathrm{lsb}(c_{t,i}x) \text{ or } \mathrm{Err}_{t,i}, \\ 0 & \text{otherwise.} \end{cases}$$

The majority decision for α_t in step 10 at stage t is correct unless

$$\frac{1}{m} \sum_{i \in A_t} X_i \geq \frac{1}{2}.$$

The expected value of X_i is

$$\mathcal{E}(X_i) = \mathrm{prob}\{\mathcal{L}(N, c_{t,i}^2 y) \neq \mathrm{lsb}(c_{t,i}x)\} + \mathrm{prob}\{\mathrm{Err}_{t,i}\}$$

$$\leq 1 - (\frac{1}{2} + \varepsilon) + \frac{\varepsilon}{8} = \frac{1}{2} - \frac{7\varepsilon}{8},$$

and for the variance we have $X_i^2 = X_i$ and

$$\mathrm{var}\, X_i = \mathcal{E}(X_i^2) - (\mathcal{E}(X_i))^2 \leq \mathcal{E}(X_i) \leq \frac{1}{2} - \frac{7\varepsilon}{8} < \frac{1}{2}.$$

With

$$\mu = \frac{1}{m} \sum_{i \in A_t} \mathcal{E}\left(X_i\right) \leq \frac{1}{2} - \frac{7\varepsilon}{8},$$

the majority decision is correct unless

$$\frac{1}{m} \sum_{i \in A_t} X_i - \mu \geq \frac{1}{2} - \left(\frac{1}{2} - \frac{7\varepsilon}{8}\right) = \frac{7\varepsilon}{8}.$$

For fixed t, the $c_{t,i}$ are pairwise independent, since a_t and b are independent,

$$(c_{t,i}, c_{t,j}) = \begin{pmatrix} 2i+1 & 1 \\ 2j+1 & 1 \end{pmatrix}\begin{pmatrix} a_t \\ b \end{pmatrix},$$

and the determinant of this matrix is $2(i-j) \neq 0$ in $\mathbb{Z}_N$ for integers $i \neq j$ with $|i|, |j| \leq m/2 < N/2$. Since $\mathrm{Err}_{t,i}$ depends on $c_{t,i}$, also the m random variables X_i are pairwise independent and by (15.80), we have $\mathrm{var}((\sum_i X_i)/m) = \sum_i \mathrm{var}(X_i)/m^2$. By Chebyshev's inequality (15.81), the probability for a wrong majority decision is at most

$$\mathrm{prob}\{|\frac{1}{m} \sum_{i \in A_t} X_i - \mu| \geq \frac{7\varepsilon}{8}\} \leq \left(\frac{7\varepsilon}{8}\right)^{-2} \mathrm{var}\left(\sum_{i \in A_t} X_i\right)$$

$$= \frac{64}{49m^2\varepsilon^2} \sum_{i \in A_t} \mathrm{var}\, X_i < \frac{32}{49m\varepsilon^2} \leq \frac{16}{49n} < \frac{1}{3n}.$$

Thus the probability for the single majority decision to fail at stage t is at most $1/3n$, and the probability that all majority decisions are correct is at least $1 - n \cdot 1/3n = 2/3$. Reduction 11.41 succeeds if the guesses for u_0, v in Step 2 and for α_0, β in Step 3 are successful, and all majority decisions are correct. Putting things together, its success probability σ is at least

$$\sigma \geq \frac{\varepsilon^3}{8} \cdot \frac{\varepsilon}{8} \cdot \frac{1}{2} \cdot \frac{1}{2} \cdot \frac{2}{3} > 2^{-9}\varepsilon^4$$

and the claim follows.   □

According to Theorem 15.94 (i), k independent repetitions increase the success probability σ of $\mathcal{S}$ to at least $1 - e^{-\sigma k}$.

Factoring algorithm from square root finder.   We now construct an algorithm $\mathcal{F}$ for factoring a Blum integer N from a square root finder $\mathcal{S}$.

REDUCTION 11.46. Factoring algorithm $\mathcal{F}$ from square root finder $\mathcal{S}$.

Input: Positive odd integer N.

Output: A proper factor p of N or "failure".

1. $v \xleftarrow{\boxdot} \mathbb{Z}_N^\times$.
2. If $\mathcal{S}(N, v^2)$ returns "failure" then return "failure" else $v^* \leftarrow \mathcal{S}(N, v^2)$.
3. If $v \in \pm v^*$ in $\mathbb{Z}_N$ then return "failure".
4. $p \leftarrow \gcd(v - v^*, N)$.
5. Return p.

LEMMA 11.47. *Let N be a Blum integer and $\mathcal{S}$ a square-root finder as above with success probability at least σ. Then $\mathcal{F}$ as in Reduction 11.46 returns a proper factor of N with probability $\sigma/2$.*

PROOF.   Lemma 3.8 (i) implies that if step 4 is reached, then a proper factor of N is returned. We have to estimate the probability for this to happen. The value v^* in step 2 is a square root of v^2 and is returned with probability at least σ.

As any square, v^2 has exactly four square roots, and v^* is among them, according to a distribution dictated by $\mathcal{S}$ and unknown to us. But v is uniformly distributed among the four values and, no matter which value v^* takes, v is in $\pm v^*$ with probability $1/2$, and the other two possibilities for v lead to success of $\mathcal{F}$, also with probability $1/2$. $\square$

We can now state the glorious final result.

THEOREM 11.48. *Under the factoring assumption for uniformly random Blum integers, the Blum-Blum-Shub generator is a pseudorandom generator.*

PROOF.   Lemmas 11.34, 11.42, and 11.47 imply that every BBS$^{(\ell)}$-distinguisher with advantage ε yields a factoring algorithm with success probability at least

$$\frac{1}{2} \cdot 2^{-10} \cdot \frac{1}{2\ell} \cdot \varepsilon^4 = \frac{\varepsilon^4}{2^{11}\ell}.$$

Assuming that factoring N is hard, all probabilistic polynomial-time factoring algorithms have negligible success probability. Hence there is no BBS$^{(\ell)}$-distinguisher with nonnegligible advantage. $\square$

Notes. I thank Daniel Loebenberger for suggestions on the presentation in this chapter.

11.1. The quote is from Killmann & Schindler (2008). The noisy diode generator is described in Bergmann (2014), including Figure 11.1, and marketed at about 12 Euro per USB stick in bulk. Frank Bergmann kindly provided a sample used for Table 11.4 and Figure 11.2. GnuPG is a successor to PGP. This free software was initially developed by Werner Koch, funded by the German government, and is available on many platforms.

The Linux device `/dev/random` is considered a reasonably secure generator (Killmann & Schindler 2011, Section 5.6.3). Besides it, Linux also includes `/dev/urandom` which has its own entropy pool and does not block when the entropy estimate is low. It is considered to be less safe than `/dev/random`. Gutterman et al. (2006) report about 1 bit/min of entropy on a running machine and several attacks on the Linux generator, none of which seems lethal. The two greatest practical dangers are lack of entropy at startup of embedded systems such as routers, and denial-of-service attacks in view of the small amount of randomness added.

Alexi et al. (1988) show that the least significant $\log n$ bits of the RSA generator yield pseudorandom bits under the standard assumption that factoring is hard; their method also plays a central role for the BBS generator in Section 11.8, where it is used only for the least significant bit. Steinfeld et al. (2006) show pseudorandomness for slightly fewer than $(1/2 - 1/e) \cdot n$ bits of output with public exponent e, under a less standard assumption. The experiments for Table 11.4 were done as a course project and appear in Burlakov et al. (2015). They were run under Linux Mint 17.1 Cinnamom 64-bit on a Lenovo ThinkPad with an Intel Core i7 at 2.3 GHz and 12 GB of RAM, using standard software packages and no attempt at optimization. An AES-based pseudorandom generator is not included in Table 11.4. Using AES-friendly hardware, it can be expected to have a highly competitive throughput.

11.2. Goldreich et al. (1999) show that under fairly standard complexity assumptions, it is computationally hard to approximate the entropy of a source by looking at samples from its distribution.

Besides considering (11.1), Lehmer (1951) also suggests to sample $x_0, x_1, \ldots$ at irregularly spaced intervals. The attack on linear congruential generators is due to Reeds (1977) and Boyar (1989). See also Stern (1987).

For a numerical approximation of an integral, say over a cube $[0, 1)^n \subseteq \mathbb{R}^n$, one may evaluate the integrand at many points $x = (x_1/m, \ldots, x_n/m) \in \mathbb{Q}^n \cap [0, 1)^n$, where the $x_i \in \mathbb{Z}_m$ are outputs of a random generator. Coveyou (1969) finds linear congruential generators to be most suitable and that lattice basis reduction would be useful to determine distributional properties of these x.

For the reply to Kelsey's question, see Green (2015).

In his contribution to the discussion on the revelations by Edward Snowden and the dual-EC generator, Michael Wertheimer (2015), then Director of Research at the NSA, does not state whether or not the NSA put a backdoor in the standard. He regrets that NSA supported it even after security researchers had discovered the potential backdoor; this may refer to Shumow & Ferguson (2007) or to earlier criticism voiced within the NIST committee. RSA Security, a market leader in security, installed in 2004 the generator as the default in its popular BSafe software, apparently receiving $10 million from the NSA in this deal. After the public revelations in 2013, the company warned against the use of its own product.

11.3. The notion of indistinguishability and the resulting concept of pseudorandom generators were introduced by Yao (1982) and Blum & Micali (1984).

11.4. The hybrid method originates from Goldwasser & Micali (1984). In Definition 11.9 (iii), the predictor actually also has to compute i_k from k, so that i_k depends "uniformly" (possibly randomly) on k; see Section 15.6. Usually, a "test" refers to a specific procedure. We use "universal test" in a different sense.

11.6. One can amplify the hardness of a function by XORing several copies. This is Yao's (1982) famous XOR lemma: let $f_1, \ldots, f_k \colon \mathbb{B}^n \longrightarrow \mathbb{B}$ all be (ε, s)-hard, $\delta > 0$, and $f \colon \mathbb{B}^{kn} \longrightarrow \mathbb{B}$ with

$$f(x_1, \ldots, x_k) = \bigoplus_{1 \leq i \leq k} f_i(x_i).$$

Then f is $(\varepsilon^k + \delta, \delta^2(1 - \varepsilon)^2 s)$-hard.

A thorough survey of design theory is in Beth *et al.* (1999).

11.7. Corollary 11.30 has the form: If there is a hard problem, then a pseudorandom generator exists. Most statements about the existence of pseudorandom generators have this shape. We have a substantial collection of problems that we think are hard, but unfortunately it is even harder to *prove* this. In fact, very few such results are known. On the other hand, almost all Boolean functions on s inputs require time at least $2^s/s$ to compute them exactly. This is easily proved by a counting argument; see Muller (1956) and Boppana & Sipser (1990), Theorem 2.4. Thus hard functions do exist; an unresolved difficulty is to find an explicit such function. But for our application we would have to solve a yet more difficult problem: to show that some functions are even hard to approximate.

One of the interesting consequences of Nisan and Wigderson's work is that this lamentable situation of relying on the hardness of functions is unavoidable: its converse also holds! If we can prove that something is a pseudorandom generator, then we have automatically proved some problem to be hard. A fundamental question in complexity theory is whether randomness is really useful for efficient algorithms; technically, whether $\mathsf{P} = \mathsf{BPP}$. The study of (the existence of) pseudorandom generators plays a basic role for this open question.

11.8. This section is based on the Master's Thesis of Zollmann (2013), which relies on the author's lecture notes.

The reduction from lsb-postdicting to BBS-distinguishing is from Yao (1982) and Blum & Micali (1984). Alexi, Chor, Goldreich & Schnorr (1988) show that an algorithm for finding the least significant bit of x given N and $x^2 \mod N$ implies an algorithm for finding x. A similar, yet simpler and more efficient reduction is given by Fischlin & Schnorr (2000). The text presents a slightly simplified version of the latter one, which comes at the cost of worse bounds for running time and success probability. The reduction from square root finder to factoring is from Rabin (1979).

Exercises.

EXERCISE 11.1 (Linear congruential generators).   *We consider linear congruential generators with $x_i = sx_{i-1} + t$ in $\mathbb{Z}_m$ (see (11.1)).*

   (i) *Compute the sequence of numbers resulting from*

      (a) *$m = 10$, $s = 3$, $t = 2$, $x_0 = 1$ and*

      (b) *$m = 10$, $s = 8$, $t = 7$, $x_0 = 1$.*

   What do you observe?

(ii) You observe the sequence of numbers

$$13, 223, 793, 483, 213, 623, 593, \ldots$$

generated by a linear congruential generator. Find matching values of m, s and t. Explain how you do this.

EXERCISE 11.2 (Linear feedback shift register). A linear feedback shift register is a linear Boolean function $f\colon \mathbb{B}^n \to \mathbb{B}$. It can be represented as $f(y_0, \ldots, y_{n-1}) = \bigoplus_{0 \le i < n} a_i y_i$ for some fixed $a_0, \ldots, a_{n-1} \in \mathbb{B}$. We obtain a mapping $\mathbb{B}^n \to \mathbb{B}^*$ by applying f recursively, starting with a seed $(x_0, \ldots, x_{n-1}) \in \mathbb{B}^n$ and

(11.49) $$x_i = f(x_{i-n}, \ldots, x_{i-1}) \text{ for } i \ge n.$$

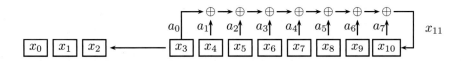

Figure 11.11: linear feedback shift register.

Figure 11.11 illustrates the fourth computation step, with $n = 8$. Such a structure is convenient to implement in hardware. Show that an LFSR that produces the same sequence of bits can be efficiently computed from $x_0, \ldots, x_{2n-1}$. Conclude that such an LFSR is not useful as a cryptographic pseudorandom generator.

[Hint: consider the smallest value of n for which (11.49) holds with some linear f for the given values. Set up a system of linear equations over $\mathbb{F}_2$ for the coefficients of this f, and show that it is nonsingular.]

EXERCISE 11.3 (NIST generator with backdoor). The dual-EC pseudorandom generator standardized in NIST (2012c) works as follows. The set-up consists of an n-bit prime number p, an elliptic curve E over $\mathbb{F}_p$ of prime order d, and finite points $P, Q \in E$. These data are fixed in the standard and publicly known. For any point $R = (u, v) \ne \mathcal{O}$ on E, we have the integer $x(R) = u \in \mathbb{Z}_p = \{0, \ldots, p-1\} \subseteq \mathbb{Z}$. Recall that given u, there are at most two values of $v \in \mathbb{Z}_p$ with $(u, v) \in E$. For an integer $t, \text{extract}(t) \in \mathbb{B}^{n-16}$ is the sequence of the $n - 16$ low order bits of t, possibly with leading zeroes.

Starting from a secret random integer seed s_0, the generator calculates for $i \ge 1$ values $s_i = x(s_{i-1}P)$ and $t_i = x(s_iQ)$. It then outputs $\text{extract}(t_i)$ as pseudorandom and uses s_i as the seed for the next iteration.

(i) Consider the three modifications where (P, Q) is substituted by $(P, -Q)$, $(-P, Q)$, and $(-P, -Q)$. How does the generator output change?

(ii) Suppose you know $a = \text{dlog}_P Q$, so that $Q = aP$ and let $a^{-1} \in \mathbb{Z}_d$ be the inverse of a. When does this inverse exist? Show that $x(a^{-1}s_iQ) = x(s_iP) = s_{i+1}$ and how to calculate this from generator outputs. Do this first for the modification of the generator where $\text{extract}'(t) \in \mathbb{B}^n$ is all of t. Then discuss how to adapt this to the actual generator, at the cost of some more computation. Conclude that knowledge of a breaks the generator.

(iii) In (ii), can you also compute s_i?

(iv) *Consider the variant of the generator where extract$'$ is used and $P, Q \in E$ are randomly generated by the user and then published. Can you reduce the discrete logarithm problem in E to some form of breaking this generator?*

EXERCISE 11.4 (Comparing generators). *You are to produce your own version of Table 11.4. On the webpage of this book you find a 512 kB binary file* `random_source.dat` *which is the output of a run of the noisy diode generator presented in Section 11.1.*

(i) *Compute the distribution of bytes, that is, count for every byte how often it occurs in the file.*

(ii) *Compute the byte entropy of the distribution. What is the theoretical value for uniformly random bits?*

Use the bits in the file sequentially to produce any random data that you need. In each case, state how many bits you use. For each of the six generators in (iii) through (viii), generate 512 kB of output, measure the time that this takes, and compute the byte entropy of your output. Do not include the set-up effort such as generating N in (vi) in your timings and start your timings with a "warm" generator, after outputting 10 kB which you discard. State which hardware and software environments you use and how you measure time. Before measurement, restart your machine and make sure no other processes run concurrently. Prepare a table with byte entropies, runtimes and two further columns. One of them contains the throughput in kB/sec, and the other one a normalized form of this. Namely, you take t_0 to be the throughput of the RSA generator with $e = 3$ and present for each other one its throughput divided by t_0.

(iii) *Analyze* `/dev/random` *or a similar source of random bytes on your machine. If the entropy pools fill very slowly, perform a second experiment with other processes running concurrently, maybe a computer game.*

(iv) *Analyze the linear congruential generator with $m = 2^{1279} - 1$ prime. Generate s, t and x_0 from the source of random bytes provided.*

(v) *Analyze the Littlewood generator with $n = 5$ and $d = 7$ in its decimal version.*

(vi) *Analyze the RSA generator for a 3000-bit RSA modulus N chosen as below, public exponent $e = 2^{16} + 1$ and a random seed, outputting 1400 bits per round. In addition, take $e = 3$ and produce one bit per round. To compute a prime factor p of the modulus, you generate a random 1500-bit number a from the file. Then you compute deterministically the next larger number $b \geq a$ such that $b = 3$ in $\mathbb{Z}_4$ and $\gcd(b, m) = 1$, where m is the product of all primes up to 1000. Then you test whether b is prime and return $p = b$ if it is and else repeat the choice of a. State the values of a and b that you used. Proceed similarly to find the second prime factor of N.*

(vii) *Analyze the Blum-Blum-Shub generator with the same N as in (vi).*

(viii) *Take the following function $f \colon \mathbb{B}^{11} \to \mathbb{B}$. On input $(x_0, \ldots, x_{10})$, it computes the sum y in $\mathbb{Z}_{2^6}$ of the two numbers with binary representations $(x_0, \ldots, x_5)$ and $(x_5, \ldots, x_{10})$, leading bits at left. The value of f is the XOR of the bits of y. (We do not claim that f is hard as stipulated for the generator.) Analyze the Nisan-Wigderson generator for a $(131^2, 131^3, 131, 2)$-design D constructed by the methods of Section 11.7. Use two random seeds for the generator $f_D \colon \mathbb{B}^{131^2} \to \mathbb{B}^{131^3}$ to produce the desired number of output bits.*

(ix) *Perform further experiments and interpret your results.*

EXERCISE 11.5 (Subset sum pseudorandom generator). *This generator by Rueppel & Massey (1985) uses the subset sum problem, which in full generality is* NP-complete *(Section 13.4). For security parameter n, it takes weights $w_0, \ldots, w_{n-1} \in \mathbb{Z}_m$ for some integer m and the outputs $x_0, x_1, \ldots$ in $\mathbb{B}$ of some LFSR with secret seed $x_0, \ldots, x_{n-1}$. Its ith output is $y_i = \sum_{0 \le j < n} x_{i+j} w_j \in \mathbb{Z}_m$, where we take $x_k \in \{0,1\} \subseteq \mathbb{Z}_m$.*

(i) *Consider the power series*

$$h_x = \sum_{i \ge 0} x_i t^i, \quad h_y = \sum_{i \ge 0} y_i t^i, \quad h_w = \sum_{0 \le i < n} w_{n-i-1} t^i$$

in t over $\mathbb{Z}_m$. Let r be the remainder of $h_x \cdot h_y$ on division by x^{n-1}. Show that $h_x \cdot h_w - r = x^{n-1} h_y$.

(ii) *Turn the last equation into a system of linear equations for the w_i, when sufficiently many y_i (n of them are enough) and the x_i are known.*

(iii) *There are 2^{2n} possibilities for the x_i. Show that by searching them exhaustively, you can compute the weights w_i from observing the generator's output at a cost of essentially 2^{2n}.*

(iv) *Large values of m, say with n bits, have been suggested. Conclude from the above that under this attack, such values do not provide more security then $m = 2$ or $m = 3$.*

For this attack and more, see von zur Gathen & Shparlinski (2004, 2008).

EXERCISE 11.6 (Clock randomness). *Suppose that on some machine, clock time is measured in nanoseconds, and that we take the current time, modulo 24 hours, to be a random value. How many random bits would this provide? How many, if we take the time modulo one hour? Modulo one minute?*

EXERCISE 11.7 (Resilience).

(i) *Can you show that $f \colon \mathbb{B}^3 \longrightarrow \mathbb{B}^6$ from Example 11.5 is not $(1/8, 11)$-resilient?*

(ii) *Determine some s so that f is not $(7/8, s)$-resilient.*

EXERCISE 11.8. *Let X_i be a random variable on a finite set A_i for $1 \le i \le k$, $X = X_1 \times \cdots \times X_k$ on $A = A_1 \times \cdots \times A_k$, let $1 \le j \le k$ and $\pi_j \colon A \longrightarrow A_j$ be the jth projection. Show that $X_j = \pi_j(X)$.*

EXERCISE 11.9 (Circuits).

(i) *How many Boolean functions $f \colon \mathbb{B}^n \to \mathbb{B}$ are there?*

(ii) *Prove that every Boolean function $f \colon \mathbb{B}^n \to \mathbb{B}$, can be computed with $2^{n+2} - 4$ gates, using only the binary gates $\wedge$ (AND), $\vee$ (OR) and the unary gate $\neg$ (NOT).*
   Hint: For $b_1, \ldots, b_n \in \mathbb{B}$ we have

$$f(b_1, \ldots, b_n) = \big(b_n \wedge f(b_1, \ldots, b_{n-1}, 1)\big) \vee \big((\neg b_n) \wedge f(b_1, \ldots, b_{n-1}, 0)\big).$$

(iii*) *Prove that every Boolean function $f \colon \mathbb{B}^n \to \mathbb{B}$, can be computed using $3 \cdot 2^{n-1} + n - 3$ gates. Hint: compute first the inverse of every input.*

EXERCISE 11.10 (Predictor and distinguisher). *We consider* $f : \mathbb{B}^3 \to \mathbb{B}^6$ *as in Example 11.5, and a circuit* $\mathcal{P}$ *with the following specification. It takes as input* $y \in \mathbb{B}^4$ *and returns a random bit if* $w(y) = 2$, *and otherwise* minority(y).

 (i) *Determine a Boolean circuit with which* $\mathcal{P}$ *can be implemented, and its size.*

 (ii) *Determine the prediction power of* $\mathcal{P}$ *as predictor for the fifth bit of* $f(U_3)$.

 (iii) *Describe the resulting distinguisher between* $f(U_3)$ *and* U_6, *and determine its distinguishing power.*

 (iv) *Compare to the predictors and distinguishers presented in the text.*

EXERCISE 11.11 (Probabilities). *Consider the following generator* $g \colon \mathbb{B}^3 \to \mathbb{B}^5$ *and let* $(X_1, \ldots, X_5) = g(U_3)$.

x	$g(x)$		x	$g(x)$
000	11100		100	00110
001	00101		101	11110
010	01011		110	01010
011	10101		111	01101

 (i) *Compute the distribution of the projection on the second to fourth bit, thus of* (X_2, X_3, X_4).

 (ii) *Compute a table of the probabilities* prob$\{b \xleftarrow{\;\boxdot\;} X_4(y)\}$ *for all possible initial sections* $y \in \mathbb{B}^3$ *and all* $b \in \mathbb{B}$.

EXERCISE 11.12 (Distinguishers). *We consider distinguishing circuits for* $f \colon \mathbb{B}^3 \to \mathbb{B}^6$ *from Example 11.5.*

 (i) *Find and depict a circuit for the minority distinguisher* $\mathcal{M}_6$ *with* $\mathcal{M}_6(y) =$ "$y_6 =$ minority $(x_1, x_2, x_3, x_4, x_5)$", *using at most 30 gates. Can you do with fewer gates?*

 (ii) *Can you find circuits for the distinguisher* $\mathcal{M}_4$ *with fewer than 12 gates? For the "$w(y) = 3$" distinguisher with fewer than 27 gates?*

EXERCISE 11.13 (Enhancing distinguishers). *Let* $\mathcal{A}$ *be a* ε-*distinguisher between a distribution* X *and the uniform distribution* U *on bit strings in* $\mathbb{B}^n$ *of length* n. *An algorithm* $\mathcal{B}$ *receives as input three bit strings* y_1, y_2, y_3 *and returns the majority of* $\mathcal{A}$*'s three answers. Is this* $\mathcal{B}$ *a better distinguisher between* X *and* U?

 (i) *Specify the definition. For which distributions does the new algorithm* $\mathcal{B}$ *give an exact distinction?*

 (ii) *Is the new algorithm better?*

   (a) *Prove that the following holds with* $q = \mathcal{E}_{\mathcal{A}}(X)$ *and* $r = \mathcal{E}_{\mathcal{A}}(U)$:
   $$\delta_{\mathcal{B}} = \Delta_{\mathcal{B}}(X^3, U^3) = \left| E_{\mathcal{B}}(X^3) - E_{\mathcal{B}}(U^3) \right| = \left| (3q^2 - 2q^3) - (3r^2 - 2r^3) \right|.$$

   (b) *Find the best lower bound for* $\delta_{\mathcal{B}}$ *that only depends on* ε. *Compare with* ε.

   (c) *Compare the old and the new distinction power by plotting*
   $$\left| \Delta_{\mathcal{B}}(X^3, U^3) \right| / \left| \Delta_{\mathcal{A}}(X, U) \right|.$$

EXERCISE 11.14 (Horoscopes). *You are to investigate whether horoscopes really pre-dict the future or if they act more like pseudorandom generators. There is nothing like "the one correct answer" for this task. Experiment on this question.*

Thus the generator is a horoscope H and the distinguisher is real life. Take a horoscope of last week (or last month or last year) and ask you friends to what extent the horoscope was true for them.

Several approaches to this task are possible. Here is just one suggestion.

Every horoscope consists of twelve parts $H_1, \ldots, H_{12}$, one for each zodiac sign S_i. For each S_i, prepare a sheet B_i that consists of H_i and a randomly chosen $H'_i \leftarrow$ ☑ $\{H_1, \ldots, H_{12}\}$. (Then $H_i = H'_i$ with probability $\frac{1}{12}$.) You roll a die to determine which one is the first on the sheet. Then you show the sheet B_i to as many of your friends as possible whose zodiac sign is S_i. Ask the friend to tell you to which extent (in percent) the two statements were true for her. In the end you compute the value of the distinguisher on both distributions H and H'.

What do you observe? Which further analysis could be interesting? Why? To which degree do your results fit with the concept of pseudorandom generators? Do you get a predictor?

EXERCISE 11.15 (Predictors). *Consider the following generator $g\colon \{0,1\}^3 \to \{0,1\}^6$:*

x	$g(x)$
000	100110
001	000011
010	110101
011	111111
100	011000
101	011000
110	001110
111	100101

(i) *Looking closely at the output it seems that the fourth bit is always 1 whenever the sum of the first three bits is odd, and almost never otherwise. Using this observation, construct a predictor both as a Boolean function and a circuit, and compute its size and prediction power.*

(ii) *Starting with the predictor from (i), construct a circuit that distinguishes $g(U_3)$ and U_6. How many additional gates are needed? What is the resulting distin-guishing power?*

(iii) *Construct a "perfect" 1/2-predictor for a bit of your choice, both as a Boolean function and as a circuit. Use as few gates as possible.*

EXERCISE 11.16 (Predictors). *Let $m = qs + 1$ with s odd and q even, and consider the linear congruential generator given by $x_i = sx_{i-1} + 1$ in $\mathbb{Z}_m = \{0, \ldots, m-1\}$. Let y_i be the least significant bit of x_i, let $j \geq 1$, and take a circuit $\mathcal{A}$ which returns $1 - z_{j-1}$ on input $z \in \mathbb{B}^{j-1}$. Show that $\mathcal{A}$ is an ε-predictor for y_j with*

$$\frac{1}{2} + \varepsilon = \frac{q(s+1)}{2m}.$$

Thus $\varepsilon = (q-1)/2m \approx 1/2s$, for large q.

EXERCISE 11.17 (Pseudorandom generator).
  Consider the function

$$f\colon \begin{array}{ccc} \mathbb{B}^k & \longrightarrow & \mathbb{B}^{k+1}, \\ (x_1,\ldots,x_k) & \longmapsto & (x_1,\ldots,x_k,\bigoplus_{i=1}^k x_i). \end{array}$$

 (i) Prove that for all $\varepsilon > 0$, f is an $(\varepsilon, k-1)$-resilient pseudorandom generator.

 (ii) Show that f is not $(\frac{1}{2}, k)$-resilient.

 (iii) Using (i) and Section 11.5, produce a value table of a pseudorandom generator $g\colon \mathbb{B}^3 \to \mathbb{B}^6$ that is $(\varepsilon, 2)$-resilient for all $\varepsilon > 0$.

 (iv) Describe a circuit for g.

EXERCISE 11.18 (Design from lines).   Let p be a prime number, $\mathbb{F}_p$ the field with p elements, and

  o $P = \mathbb{F}_p^2$,

  o $S_{a,b} = \{(x, ax + b)\colon x \in \mathbb{F}_p\} \subseteq P$, for $a, b \in \mathbb{F}_p$,

  o $D' = \{S_{a,b}\colon a, b \in \mathbb{F}_p\}$.

 (i) Arrange the elements of D' into a sequence D.

 (ii) Determine the uniquely determined values $k, n, s \in \mathbb{N}$ and the smallest possible value $t \in \mathbb{N}$ such that D is a (k, n, s, t)-design.

 (iii) Let $p = 3$ and consider the function

$$f\colon \begin{array}{ccc} \mathbb{B}^3 & \longrightarrow & \mathbb{B}, \\ x & \longmapsto & (x_1 \vee \neg x_2) \oplus (x_2 \wedge x_3) \end{array} \cdot$$

   Compute $f_D(x)$ for $x = 101010101$ and $x = 110010011$.

EXERCISE 11.19 (Nisan-Wigderson generator).   Let D be the design from Example 11.24 and let $f\colon \mathbb{B}^3 \to \mathbb{B}$ be the function with $f(001) = f(010) = f(100) = 1$ and $f(x) = 0$ otherwise.

 (i) Determine f_D for the arguments 010101000 and 000111000.

 (ii) Find some s such that f is not $(3/4, s)$-hard, and give a corresponding circuit.

 (iii) Find a positive real number $\varepsilon < 3/4$ and a natural number $s' < s$ such that f is not (ε, s')-hard, and give a corresponding circuit.

Let $\mathcal{P}$ be the predictor for the sixth bit with $\mathcal{P}(y_1,\ldots,y_5) =$ least significant bit of $y_1 + \cdots + y_5$.

 (iv) Prove that there are 344 matrices in $\mathbb{B}^{3\times 3}$ for which the number of lines (columns and rows) with ones only is even.

 (v) Prove that $\mathcal{P}$ is a 11/64-predictor for f_D. You may use the result of Exercise 11.19.

 (vi) Design an algorithm $\mathcal{A}$ which approximates f with

$$|\operatorname{prob}(\mathcal{A}(X) = f(X)) - \frac{1}{2}| \geq \frac{11}{64}.$$

EXERCISE 11.20 (Small Blum-Blum-Shub generator).

   (i) Determine $\square_7$, $\boxtimes_7$, $\square_{11}$, $\boxtimes_{11}$, $\square_{77}$, and $\boxtimes_{77}$.

   (ii) Draw a graph as in Figure 15.2 of the squaring map in $\mathbb{Z}_{77}$ for the arguments in $\square_{77} \cup \boxtimes_{77}$.

EXERCISE 11.21 (Large Blum-Blum-Shub generator).   Let p, q be the smallest prime numbers congruent to 3 modulo 4 and with $p \geq 2^9$ and $q \geq 2^{11}$, and let $N = p \cdot q$.

   (iii) Implement the Blum-Blum-Shub generator in $\mathbb{Z}_N$.

   (iv) Compute the first 50 bits $z_0, \ldots, z_{49}$ of the generator with seed 100 001.

Carry out a few statistical tests.

   (v) What is the mean value $\mathcal{E}(z_i)$ and the variance $\mathcal{E}(z_i^2) - \mathcal{E}(z_i)^2$ of the first 2^{13} bits. Do this for several random seeds. Compare with the theoretical values for a "truly" random sequence.

   (vi) How often does each bit in $\{0, 1\}$ occur among the first 2^{13} outputs?

   (vii) Now we combine two consecutive output bits into one value $z_i = 2y_{2i} + y_{i+1}$, for $i \geq 0$. How often does each of the four possible values occur among the first 2^{12} z_i's? What is the mean value and the variance of these z_i? Use several random seeds.

   (viii) Let $a, b \xleftarrow{\boxtimes} \{0, 1\}$ and $c = 2a + b$. Show that $\mathcal{E}(c) = 3/2$, using the linearity of $\mathcal{E}$. Using $a^2 = a$, show that $\mathcal{E}(c^2) = 7/2$ and the variance of c is $\mathcal{E}(c^2) - \mathcal{E}(c)^2$. Compare with your results from (v).

   (ix) Plot 1000 points (u, v), where the binary representations of u and v are 10 bits from the sequence produced in (vii). Do you see any regularities in the picture? Compare with a linear congruential generator with $x_i \leftarrow 313 x_{i-1}$ in $\mathbb{Z}_{2053}$ and output (z_{2i}, z_{2i-1}), where z_i is the integer represented by the lowest 10 bits of x_i.

   (x) Interpret your results.

EXERCISE 11.22 (Intervals). Prove equation (11.39).

EXERCISE 11.23 (Pseudorandom generators).   Use the method of Theorem 11.21 to give a detailed proof of the following claim.
      Let $p \in \mathbb{Z}[t]$ be a positive polynomial and $f = (f_k)_{k \in \mathbb{N}}$ a pseudorandom generator with $f_k \colon \mathbb{B}^k \longrightarrow \mathbb{B}^{k+1}$. Then the construction of Algorithm 11.16 yields a pseudorandom generator $g = (g_k)_{k \in \mathbb{N}}$ with $g_k \colon \mathbb{B}^k \longrightarrow \mathbb{B}^{k+p(k)}$.

Setting out to cryptanalyze a cryptogram, you begin first of all
by counting the symbols in it, and then count the number of
occurrences of each symbol and set down the totals individually.
[...] On cryptanalyzing a cipher, the most frequently occurring
letter is considered to be the letter *a*; [...] Then the first words
you try to work out in the message are the bigrams
—two-character words—through somehow trying to have
access to the most feasible combinations of their letters, until
you are sure you have discovered something correct in them.

IBN AD-DURAYHIM

Quant à Trithème, ça esté à la verité le premier qui a fait le
chemin aux autres, à tout le moins publiquemēt ; & ce en deux
grands & laborieux ouvrages, l'un imprimé, assauoir la
Polygraphie, & l'autre non, qui est la Steganographie, dõt le
precedãt n'est que comme un precurseur.[1]

BLAISE DE VIGENÈRE (1586)

The *Steganographia* purports to be, and perhaps really is to some
extent, about cryptography or ways of writing in cipher. It is
also, however, Cabalist angel magic. The first book is about
summoning district angels, or angels which rule over parts
of the earth; the second is about time angels who rule the hours
of the day and night; the third is about seven angels higher
than all these who rule the seven planets. Trithemius aims
at using this angelic network for the very practical purpose
of transmitting messages to people at a distance by telepathy;
he also seems to hope to gain from it knowledge
"of everything that is happening in the world".

FRANCES A. YATES (1964)

Si id lorum in manus hostium inciderat,
nihil quicquam coniectari ex eo scripto quibat.[2]

AULUS GELLIUS (c. 143)

---

[1]As to Trithemius, he is really the first who paved the way for others, and this in two great and laborious works, the one printed, namely the Polygraphia, and the other one not, which is the Steganographia, of which the previous one is just a predecessor.

[2]If this should fall into the hands of the enemies, they should not be able to draw anything from this writing.

Chapter G

Some cryptographers

ublishing results is a major goal of active researchers in most sciences. Not so in cryptography: some of the top work remains secret, at least for a while. We now describe some of the contributors to cryptography BDH (before Diffie & Hellman). Others have played a role in this text before, among them Caesar, Augustus, Vigenère, Alberti, Porta, Vernam, Shannon, Kasiski, Babbage, Friedman, Viète, Bacon, and Cardano, in order of appearance.

G.1. Arab cryptology

The Arab civilization was the first to produce systematic approaches to cryptology and codify them in books. Around the time that the long drawn-out demise of the Roman empire came to an end—with a whimper, not a bang—there arose from the Arab peninsula a new world power. مُحَمَّد (Mohammed, 570–632) was a businessman from Mecca. Starting in 610, he put together the Holy Qurān and the monotheistic religion of Islam, borrowing heavily from the Judaic and Christian writings. Muslim doctrine views this holy book as the words of God. He was forced to flee from Mecca to Medina in 622, the *Hijrah*, and enthusiastically welcomed in his new hometown. In spite of substantial initial resistance, mainly in Mecca, the new religion fired up enormous energies, and within a century its followers had conquered half of the "known"

© Springer-Verlag Berlin Heidelberg 2015
J. von zur Gathen, *CryptoSchool*, DOI 10.1007/978-3-662-48425-8_18

Figure G.1: Al-Kindī's classification of cryptosystems.

world: the Arab peninsula, Palestine, Egypt, North Africa, Spain, and Mesopotamia. One of his successors, Caliph أَلمَأمون (al-maʾmūn, reigned 813–833), built his capital Baghdad into a magnificent city. Besides the military and architectural successes, he also created an intellectual center, the *Bait al-Hikmah* (House of Wisdom). It housed the world's leading scientists. One of their important tasks was the acquisition, translation, and study of the Greek and Roman scientific texts. Without these successful efforts, the mathematical works of Pythagoras, Diophant, Euclid, and many others might well have gotten lost in the ensuing Dark Ages of Europe, where religious fundamentalism discouraged free thinking and independent research.

From the start, Arabic as the language of the Qurān was studied, and grammars and secretarial handbooks were produced. And we own the first systematic writings about cryptography to this period. أَلكِندِي (al-Kindī, 801–873), wrote the oldest surviving treatise on cryptography, including cryptanalytic techniques and the necessary statistical methods.

Al-Kindī was versed in many fields, including philosophy, medicine, astronomy, mathematics, linguistics, and music. He became the first director of the Bait al-Hikmah, and served under four caliphs. The first three supported his research, but the fourth one was unsympathetic to his ideas, had him removed from his office and publicly flogged. Al-Kindī's writings include about 290 titles in the various fields of his interests.

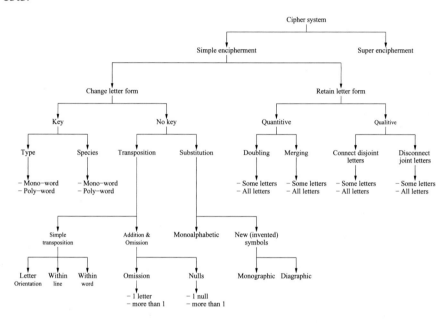

Figure G.2: English version of Figure G.1.

The oldest existing book on cryptography is his رِسَالَة فِي إِسْتِخْرَاج المُعَمَّى (Risālah fī ʾistiḫrāǧ al-muʿammā[1]); a manuscript was discovered in the 1980s and published by Mrayati *et al.* (1987). Al-Kindī gives a detailed classification of cryptographic systems. The original page is shown in Figure G.1, and an English translation in Figure G.2. This might be compared with Figure A.2.

His most advanced contribution is a clear description of frequency cryptanalysis of a simple substitution. He first explains how to set up a frequency table for his language from a sample text of 3667 letters.

Then he goes on to analyze a ciphertext by determining symbol frequencies in it and matching them with the sample frequencies. He is well aware of a qualitative version of Shannon's theory of the unicity distance (Section A.5), namely that the statistics can only be expected to work if the text is long enough. Al-Kindī writes:

---

[1]Manuscript on deciphering cryptographic messages

One way to solve an encrypted message, if we know its [original] language, is to find a [different clear] text of the same language long enough to fill one sheet or so and then we count [the occurrences of] each letter of it. We call the most frequently occurring letter the "first", the next most occurring the "second", the following most frequent the "third" and so on, until we finish all different letters in the plaintext [sample]. Then we look at the cryptogram we want to solve and we also classify its symbols. We find the most frequently occurring symbol and change it to the form of the "first" letter [of the plaintext sample], the next most common symbol is changed to the form of the "second" letter, and the following most common symbol is changed to the form of the "third" letter and so on, until we account for all symbols of the cryptogram we want to solve.

It could happen sometimes that the cryptogram is too short to have all different letters. The high and low [frequency] counts will not be correct, for high and low [frequency] counts are only correct in long enough messages to correspond to all places of frequent and rare occurrences so that if some letters are [too] few in one segment of the message, they will be [too] many in others. But if the cryptogram is too short, equivalence does not apply, letter ranks are not correct and [consequently] a second trick should be used to recover letters.

One such "trick" is the study of frequent and of impossible digrams and trigrams. Al-Kindī's work contains much more than mentioned here and stands out as a first not only in cryptanalysis, but also in statistics and computational linguistics.

There seem to be no surviving manuscripts actually documenting the use of cryptography in the Arab world around al-Kindī's time. He mainly writes about simple substitutions, and also mentions transpositions. One may wonder whether encrypted documents repose in the archives.

A great work like al-Kindī's does not come out of the blue. But few traces remain of his predecessors in the theory and practice of cryptography. The linguist الخَلِيل (al-Ḥalīl, 718–786) apparently wrote a book on cryptography; however, this is lost. He produced the oldest surviving dictionary. In it, he determined combinatorically the number of possible Arabic words and the number of permutations of a given word:

A two-letter word has two variations like qd, dq, and

sd, ḍṣ [= 2!]. *A three-letter word has* 6 *variations* [= 3!]; *the collection of the six variations is called a hexa-word, an example of which is:* ḍrb, ḍbr, brḍ, bḍr, rḍb, rbḍ. *A four-letter word has* 24 *variations which are* 6 *and the result is* 24 [= 4! = 4 × 3!] . . .

In the Western civilization, Gottfried Wilhelm Leibniz (1646–1716) was the first to study combinatorics systematically, except some earlier work by Ramón Llull (1235–1316).

Cryptography was an established topic in Arab science. إِبِن عَدلَان (ibn ʿAdlān, 1187–1268) wrote about frequency analysis of simple substitutions. He grouped the 28 letters of the Arabic alphabet into three categories: seven with high frequency, eleven occurring moderately often, and ten rare ones, similar to Table A.21. He indicated an explicit lower bound on the unicity distance (Section A.5): *The text to be solved should be [at least]* 90 *letters long approximately because letters will have circulated about three times.* Now 3 · 28 = 84 is close to 90, but it is not clear where the factor 3 comes from. Ibn Adlān gives a detailed description of the process of cryptanalysis of a simple substitution ciphertext, basically going through the steps of Poe's Legrand as in Section A.4.

إِبِن دُنَينِير (ibn Dunainīr, 1187–1229) used the fact that each Arabic letter also has a numerical value. One can encrypt a plaintext letter by two letters whose sum equals the plaintext, or whose sum equals twice the plaintext, and so on. For illustration, a clumsy Latin equivalent would be to write VIDI = 5, 1, 500, 1 as VX, IIIIV, CCCCC, IIIIV. إِبِن الدُرَيهِم (ibn al-Duraihim, 1312–1361) pushed cryptology even further. He wrote about eighty books on arithmetic, linguistics, theology, and cryptography. His مِفتَاح الكُنُوز فِي إِيضَاح المَرمُوز (miftāḥ al-kunūz fī iḍāḥ al-marmūz[2]) explains transposition ciphers and how to hide a secret as the third letter, say, in an innocuous text, as in secrecy through key ciphers with protection, with crypto as the plaintext. (This contrived example shows the difficulty of doing this in an elegant way.) One and a half centuries later, Trithemius proposed a similar method; see Section G.3.

G.2. Cicco Simonetta

An early cryptographic work dated 4 July 1474 is attributed to Cicco Simonetta, advisor and secretary to the first three Sforza dukes of Milan

---

[2]Treasured key for clarifying ciphers

(Section D.2). He fell into disgrace and was executed on 30 October 1480. His thirteen rules are the earliest substantial Western work on cryptanalysis that has survived. He explains how frequency analysis can distinguish between encrypted Latin and Italian texts, assumed to be written in a simple substitution and with spacing between words left intact. The main discriminant is that Italian words always end in a vowel, but in Latin many more terminal letters are possible. He then lists the most frequent monograms, digrams, and trigrams, and gives some more hints. As an example, the eighth of his thirteen rules says:

> Consydera litteram propositam, si in ea est aliqua dictio representata per unicam zifram, et conjectura quod illa zifra sit a, quia in litteris que sunt in latino raro contingunt dictiones unius littere tantum preter a prepositionem, ut supra dictum est.[3]

Finally he gives four pieces of advice to the cryptographer: write partly in Latin and partly in Italian, use dummies, employ two different alphabets, and a single letter to represent qu.

G.3. Johannes Trithemius

The most influential printed book on cryptography, Trithemius' *Polygraphia*, illustrates on its 300 leaves[5] the difficult transition from medieval magic to Renaissance rationality. Several cryptosystems are described, all original, and it provides an innovation—polyalphabetic substitutions—that was to play a major role in the cryptography of the next four hundred years. It also proposes the first systematic use of randomness in cryptography and became a standard reference in the cryptographic literature of the following centuries. This section first describes briefly his life and then his two cryptographic books, the *Polygraphia* and the *Steganographia*.

Johannes (Ioannes) Trithemius (1462–1516) was born as Johannes Cellers into a wine-growing family in Trittenheim on the Mosel River in Germany. He displayed high intelligence in school and at university, and entered the Benedictine monastery of Sponheim (near Bad Kreuznach) in early 1482. The circumstances—as described by him—were dramatic. With a companion, he had stayed overnight in the monastery, and set

---

[3]Check whether in the message to be deciphered there is a word represented by a single cipher symbol. Then conjecture that this symbol represents a, because Latin texts rarely contain one-letter words except the preposition a, as was said above.

[5]600 pages

Figure G.3: Portrait of Trithemius.

out in the morning. Heavy snowfall made them return, against his will, since he had a premonition that he might then stay. And so it happened. The next year, he was elected abbot—the monks must have been truly impressed by the youth's abilities.

These were times of religious upheavals, to culminate in the Reformation of Martin Luther and others. In many monasteries, discipline was lax and debauchery rampant, the monks had a great time, eating and drinking well and cajoling with whores. Some liked it, others did not—Trithemius was an important proponent of a reform movement whose goal was to bring back traditional morals to the monasteries. But his monks preferred the comfortable old ways, and during an absence of his they simply ousted him in 1505. He was not jobless for long: after a temporary position as advisor to Emperor Maximilian I, he became abbot of the St. Jacob monastery in Würzburg, in 1506.

Some sites of Trithemius' life are still around today: Trittenheim is a prosperous wine-growing village on the Mosel river, and the weary traveller can even rest in a guesthouse located in the house where Trithemius

Figure G.4: The title page of Trithemius' *Polygraphia* from 1518.

a Deus   a clemens
b Creator   b clementiffimus
c Conditor   c pius
o Opifex   o pijffimus
e Dominus   e magnus
f Dominator   f excelfus
g Confolator   g maximus
h Arbiter   h optimus
i Judex   i fapientiffimus
k Jlluminator   k inuifibilis
l Jlluftrator   l immortalis
m Rector   m eternus
n Rex   n fempiternus
o Jmperator   o gloriofus
p Gubernator   p fortiffimus
q Factor   q fanctiffimus
r Fabricator   r incôprehenfibilis
f Conferuator   f omnipotens
t Redemptor   t pacificus
v Auctor   v mifericors
x Princeps   x mifericordiffimus
y Paftor   y cunctipotens
z Moderator   z magnificus
zv Saluator   zv excellentiffimus
                           A

Figure G.5: The first of 192 pages of the *Ave Maria cipher* in Trithemius' *Polygraphia*.

was born. A grape was his escutcheon, and today *Trittenheimer Altärchen* is one of the better known German white wines—often a bit too sweet for this author's taste. His modern bronze statue adorns the bridge rampart. Of his first monastery, Sponheim, the mighty church from the 13th century still stands intact, dominating its valley in the Hunsrück hills. Its floor plan is a Greek cross with four arms of equal length. At the intersection of the four wings rises a massive octogonal tower. His main regret about Sponheim was the loss of the valuable library of over 2000 manuscripts and books that he had assembled. His second monastery, St. Jakob at Würzburg in the hills above the Main river, suffered severely in the Second World War, but is now attractively rebuilt. No trace of Trithemius' occupancy remains there. His grave in Würzburg (and most of the city) was destroyed in an RAF bombing raid on 16 March 1945, but his life-size sepulchral stone, carved by the famous Tilman Riemenschneider (c. 1460–1531) or his workshop, is prominently displayed today in the *Neumünsterkirche*.

Trithemius published extensively, mostly on religious and historical subjects. He had a liberal attitude towards historiography, and invented documents and their authors in some of his historical works. His view of the world was not always constrained by mere facts, and some of the sources, events, and documents in his historical works are—by now—known to be figments of his imagination. But the contemporary reader was suitably impressed by the abbot's erudition. One of Trithemius' invented writers, the *"Frankish historiographer Hunibaldus"*, even makes a spectral appearance in his main cryptographic work, the *Polygraphia*.

He also delved into the world of the occult, and is venerated by its current inhabitants as one of the great masters and writer of *"a main Renaissance manual of practical Cabala or angel-conjuring"*. Indeed, for fear of misuse, Trithemius hid his ideas behind smokescreens, and the style and contents of his books display obscure humbug that will look cabbalistic to some.

In Würzburg, he wrote his classic *Polygraphiae libri sex*[6], or *Polygraphia* for short. It was finished in 1508, but published only posthumously in 1518, the first printed book about cryptography. For this other work on cryptography, the *Steganographiae libri octo*[7], Trithemius finished three chapters in manuscript in 1499, before his *Polygraphia*. Much on its 180 pages reads like astrological and cabbalistic humbug (see Figure G.8) and generated a lot of controversy. Trithemius recounts his experience with malicious public peer reviews of an abstract of his *Steganographia*

---

[6]Six books on cryptography. Here "book" means "chapter".
[7]Eight books on steganography

that he communicated in a letter:

> *fuere nonnulli qui meas in dicta epistola pollicationes constanter asseuerarent friuolas, impossibiles, atque mendaces & a me inanis glorie causa impudēter excogitatas. Alij uero dicebant: Magna et miranda pollicetur iste abbas Tritemius: quae si potest, non aliter quam dæmonum ministerio potest: cum nature metas procul uideantur excedere. Si uero non potest: quis eum dubitabit esse mendacem, & ab omnibus uiris bonis & doctis merito refutandū? [...] epistolam [...] falsitate, mendacijs, iniurijs, & contumelijs multis plenam rescripsit: in qua nõ intellecte steganographiæ mētionem faciens, me prauis artibus deditum, magum & necromanticum, falso, mendaciter, & nimis iniuriose temeraria præsumptione proclamat.*[8]

Today, reviews tend to be less colorful. Such accusations posed a serious health threat in times of the Inquisition. Thus threatened with *publish and perish*, he prudently postponed publication, and it was not printed until 1606.

He was also afraid that it might fall into the wrong hands—an early anticipation of today's crypto policy debate. In rather modern terms, he points out the dual-use problem as with military hardware:

> *Bonum instrumentum est gladius [...]. Quod si ad læsionem cuiuscunque iniuriosè extenditur per abusum res bona in malum relaxatur. Vt ergo boni habeant quæ suis negotiis prudenter consulant, à nobis Polygraphia scribitur, & ne pravis delinquendi præstetur occasio, nucleus in cortice sub ænigmatibus variis occultatur.*[9]

---

[8]There were some who kept claiming that the proposals in this letter were frivolous, impossible, and mendacious, and that I had dreamed them up out of shameless self-aggrandizement. Others said: this abbot Trithemius promises weird and wonderful things. If he can do them, it is not possible but with the help of demons, for it would exceed the bounds of nature by far. But if he cannot do them, who will doubt that he is a charlatan who deserves to be rejected by all good and learned men. [...] he wrote a letter [...] full of falsehood, lies, injuries, and insults. In it he referred to the Steganographia which he had failed to understand, and in a wrong, lying, extremely libellous way, based on unjustified assumptions, he declared me to be an adept of vile arts, a wizard and a necromancer.

[9]The sword is a good instrument [...]. But if wielded to hurt someone unjustly, this abuse turns something good into something bad. Thus if good men have a means to perform their business prudently, as in the Polygraphia written by us, then also the depraved are offered a way of hiding a nut in a shell of various puzzles.

Trithemius' *Polygraphia* consists of six chapters, called *books* in the language of his times. The first two—which form the bulk of the text—comprise 384 tables each of letter-word pairs: Trithemius' *Ave Maria cipher*. Each table in Books 1 and 2 gives one Latin word for each of the 24 letters. The first two of these tables are shown in Figure G.5. Messages are encrypted letter by letter, using consecutive tables. The first four tables of Book 1 encrypt the word book as `Creator gloriosus illustrans terram`[10]. A longer message would result in a meaningless encantation which, however, might not sound too strange to an uncritical contemporary ear. Books 3 and 4 use phantasy words like `cadalan, cadelen, cadilin, ...` for the same purpose. In Book 4, each letter is encrypted by words that contain it as their second letter; thus `abrach bonefar comani akamech` now encrypts book.

The system is easy to use for all those who have a copy of the book, and thus affords little security. These tables of Trithemius already present fundamental progress in cryptography: polyalphabetic substitutions. Here, a letter is not encrypted by a single value, as in a simple substitution, or by choosing one encryption among a few, as in a homophonic substitution; see the top four lines of Tranchedini's nomenclator in Figure D.2 for the latter. Rather, his *Ave Maria cipher* consists of a sequence of simple substitutions, 384 of them in Book 1 and all with different words as encryptions of individual letters. One applies the first substitution to the first letter, the second one to the second letter, and so on. After exhausting the list, the 385th letter is encrypted with the first substitution again. Trithemius is aware of the drastic message expansion, but cheerfully turns it around into something pleasant: *"Nec te moueat lector quod multitudo uerborũ in aperto, sensum breuẽ cõtinet in occulto"*[11]. In many places, he suggests to the reader to set up her own encrypting tables.

Inventing polyalphabetic substitutions guarantees Trithemius a prominent place in cryptography's Hall of Fame. But in Book 5, the highlight of his work, he goes one step further and makes a crucial assist to the later Vigenère or key addition systems which—together with codebooks—was to dominate cryptography for the next four centuries. Namely, instead of 384 substitutions on 192 pages, he prints on a single page a 24 × 24 square of letters, the *tabula recta*[12] shown in Figure G.6.

---

[10] Glorious creator who illustrates the earth.

[11] It should not bother you, reader, that an open ciphertext of many words conveys a short plaintext secretly.

[12] forward table, encryption. The text at the bottom: In this canonical or forward table [for encryption] you find all the alphabets that can be obtained from our usual Latin alphabet by mutation or transposition. In total, the number of letters is twenty-four

Recta transpositionis tabula.

a	b	c	d	e	f	g	h	i	k	l	m	n	o	p	q	r	s	t	u	x	y	z	w
b	c	d	e	f	g	h	i	k	l	m	n	o	p	q	r	s	t	u	x	y	z	w	a
c	d	e	f	g	h	i	k	l	m	n	o	p	q	r	s	t	u	x	y	z	w	a	b
d	e	f	g	h	i	k	l	m	n	o	p	q	r	s	t	u	x	y	z	w	a	b	c
e	f	g	h	i	k	l	m	n	o	p	q	r	s	t	u	x	y	z	w	a	b	c	d
f	g	h	i	k	l	m	n	o	p	q	r	s	t	u	x	y	z	w	a	b	c	d	e
g	h	i	k	l	m	n	o	p	q	r	s	t	u	x	y	z	w	a	b	c	d	e	f
h	i	k	l	m	n	o	p	q	r	s	t	u	x	y	z	w	a	b	c	d	e	f	g
i	k	l	m	n	o	p	q	r	s	t	u	x	y	z	w	a	b	c	d	e	f	g	h
k	l	m	n	o	p	q	r	s	t	u	x	y	z	w	a	b	c	d	e	f	g	h	i
l	m	n	o	p	q	r	s	t	u	x	y	z	w	a	b	c	d	e	f	g	h	i	k
m	n	o	p	q	r	s	t	u	x	y	z	w	a	b	c	d	e	f	g	h	i	k	l
n	o	p	q	r	s	t	u	x	y	z	w	a	b	c	d	e	f	g	h	i	k	l	m
o	p	q	r	s	t	u	x	y	z	w	a	b	c	d	e	f	g	h	i	k	l	m	n
p	q	r	s	t	u	x	y	z	w	a	b	c	d	e	f	g	h	i	k	l	m	n	o
q	r	s	t	u	x	y	z	w	a	b	c	d	e	f	g	h	i	k	l	m	n	o	p
r	s	t	u	x	y	z	w	a	b	c	d	e	f	g	h	i	k	l	m	n	o	p	q
s	t	u	x	y	z	w	a	b	c	d	e	f	g	h	i	k	l	m	n	o	p	q	r
t	u	x	y	z	w	a	b	c	d	e	f	g	h	i	k	l	m	n	o	p	q	r	s
u	x	y	z	w	a	b	c	d	e	f	g	h	i	k	l	m	n	o	p	q	r	s	t
x	y	z	w	a	b	c	d	e	f	g	h	i	k	l	m	n	o	p	q	r	s	t	u
y	z	w	a	b	c	d	e	f	g	h	i	k	l	m	n	o	p	q	r	s	t	u	x
z	w	a	b	c	d	e	f	g	h	i	k	l	m	n	o	p	q	r	s	t	u	x	y
w	a	b	c	d	e	f	g	h	i	k	l	m	n	o	p	q	r	s	t	u	x	y	z

In hac tabula literarũ canonica siue recta tot ex uno & usuali nostro latinarum literarum ipsarum per mutationem seu transpositionẽ habes alphabeta, quot in ea per totum sunt monogrammata, uidelicet quater & uigesies quatuor & uiginti, quæ faciunt in numero D.lxxvi. ac per to tidẽ multiplicata, paulo efficiunt minus ꝗ quatuordecẽ milia.

o ij

Figure G.6: Trithemius' *tabula recta*.

The original manuscript, without a separate text called the *clavis* (key), only contained the lame hint that one can use a (single) different substitution with each correspondent:

> *Horum usus pro voto consistit in singulis, aut si malue-*
> *ris etiam in universis. Si tibi cum multis fuerit arcani com-*
> *mertium, alphabeta singula familiaribus tuis distribue singu-*
> *lis, tu verò memoriæ commenda quid vnicuique tradideris.*
> *Primo quidem alphabetum committatur primum, secundo*
> *secundum: tertio tertium, & ita consequenter fiat de reli-*
> *quis.*[13]

Thus the first correspondent would receive enc(book) = `cppl`. But Trithemius was only hiding the goodies. Warned by the violent attacks on his earlier manuscript of the *Steganographia*, he only put the well-known Caesar cipher into his writing.

His *clavis* explains a better usage. Instead of encrypting a message with just one alphabet, as in *"Horum usus"*, now the encryption changes with each character. Namely, one looks up each plaintext letter in the leftmost column (printed in red). Then for the first letter one takes the first column (printed in black), for the second the second column, and so on, as explained above. Thus book now is encrypted as `cqro`. His instructions in the *clavis* are very clear:

> *Est autem modus iste scribendi, ut in primo alphabeto*
> *nigro, capias occultæ sententiæ literam unam, de secundo*
> *aliam, de tertio tertiam, & sic consequêter usque ad finem.*[14]

Besides the *tabula recta* of Figure G.6, Trithemius also gives a *tabula aversa*[15] in which the alphabet is written backwards both in rows and columns. It equals the *tabula recta* mirrored around its center point, and actually its 24 columns give the same set of permutations as the *tabula recta*. In a further table designated *orchema* the order in the columns is different: first come the "odd" letters $c, e, g, \ldots, z$, then $a, w, b, d, \ldots,$ and

---

times twenty-four, or 576 in number. This, multiplied by the same, gives a little less than $14\,000$.

[13]Their use for encryption is by single columns, or if you want, also by all of them. If you have secret business with many people, distribute a single permutation to each. Be sure to commit to memory which one you gave to whom. The first permutation should be given to the first correspondent, the second to the second one, the third to the third, and so on with the remaining ones.

[14]There is also this mode of writing, where you take one letter of the plaintext from the first black alphabet, the second letter from the next one, the third from the third one, and so on until the end.

[15]backward table, decryption

similar variations of arithmetic sequences up and down the alphabet. The Greek ὄρχημα (orchema) means *dance*.

Trithemius' letters are waltzing around in the alphabet. We may read *orchema* as *random*. Trithemius translates it into Latin as *transilitio* meaning skip or jump around and he tells the reader repeatedly that she should construct and use her own permutations, following his examples. The *orchema* constitutes the first systematic use of randomness in cryptography. This can also be read into Alberti's disk (Section A.2), but only implicitly at best. Trithemius could not even have dreamed of the question of P vs. BPP, but he writes *"In orchemate nũerali magna latēt arcana, q̃ si quis proprie intellexerit, plura numerorũ mysteria penetrabit"*[16].

Trithemius is on the way to general polyalphabetic substitutions, consisting of a sequence of arbitrary simple substitutions, but he is lacking Vigenère's idea of a keyword-driven rotation among substitutions.

He suggests that the reader can make up further substitutions herself, and that there are as many substitutions as stars in the sky: this is quite a reasonable value, falling within the range of modern astronomical estimates. We can take this as further proof of Trithemius' excellent celestial connections.

Trithemius' table was taken up by Vigenère in 1587, and since then goes by the name of *Vigenère tableau*—so much for historical justice. Vigenère added the idea of using the columns not in their straight order, but in an irregular sequence driven by a key word (Chapters C and D). Trithemius' tables were a challenge to the typesetter—some alignments in Figure G.6 are rather imperfect, and other tables have columns running off. In the figure, the visually trained eye immediately notices the constant antidiagonals (from top right to bottom left) and the mathematically trained mind almost immediately recognizes it as the addition table modulo 24 (identifying his 24 letters a, b, c, . . . with the numbers 0, 1, 2, . . .). It actually took a couple of centuries to arrive at this "immediate" conclusion, as explained in Chapter B.

In summary, Book 5 of the *Polygraphia* made the following fundamental contributions to cryptography:

○ polyalphabetic substitutions with cyclic permutations (ordered alphabets), predated by Alberti,

○ the *tabula recta* as a mnemonic device, later called the *Vigenère tableau*,

---

[16]In random numbers lure great secrets, and if someone understands them properly, he will penetrate many mysteries of numbers.

○ polyalphabetic substitutions with random permutations; in particular, the first proposal for systematic use of randomness in cryptography.

The sixth and last book of the *Polygraphia* contains alphabets and number systems allegedly used in history, for example by Charlemagne. Some of these may well be fruits of Trithemius' imagination.

The *Polygraphia* went through many editions: 1518, 1550, 1564, 1571, 1600, 1613, 1621, 1637, 1671, 1676, 1721. Gabriel de Collange from Tours published an interesting French translation *Polygraphie* in 1561. He leaves out Trithemius' second book and thus has a total of five books. He adds translations of the *clavis* and an exposition by Adolph von Glauburg as Trithemius (1571) of "obscure" points in the *Polygraphia*. And then he contributes an invention of his own, an early piece of cryptographic hardware. He draws six diametrical strips in a circle and fills each of them with a shifted alphabet. A seventh movable strip is fastened to the center of the circle which can be turned about the center to align with any of the six fixed strips. Such rotating pieces of papercraft in books are called *volvelles*.

In order to encrypt a message, the user aligns the movable part with the first strip to read off the encryption (on the fixed strip) of the first letter (on the movable part). Then it is rotated counterclockwise by 30° for the second letter, and so on. Figure G.7 shows this beautifully crafted piece of paperwork. I tried turning one in the original book, and it worked perfectly well, more than four centuries later! (The paper is not soft enough to call this the first piece of cryptographic software.) With the volvelle as shown, `gabriel` is encrypted as `toqhayb`.

Trithemius' other work on cryptography, the *Steganographia*, was finished in 1499, before his *Polygraphia*. It got so much flak already from visitors to whom he gave the manuscript for perusal that it was published only in 1606, almost a century after his death. It was announced to have eight chapters (*books*), but only contains three. In his first brush with cryptography, Trithemius proposes the following system: you read only the letters in certain positions in certain words of the received message, say the third letter in each second word.

In our language, we have a homophonic substitution described by a key consisting of two integers r and s, where we take every sth letter in each rth word. This sentence is a steganographic sample relying on ugly deformations, natural nonsense, and ridiculous convolutions. Get it? As seen in this sample, it is hard to make up innocuous sentences. The ciphertexts are sometimes read in the forward direction, sometimes backwards.

Figure G.7: A volvelle in Collange's *Polygraphie*. The vertical strip labelled $a, b, c, \ldots, y, z, \&$ rotates about the center. The inscriptions read at the top *Third extended figure of the forward table* and *In this third planar figure, expanding [some columns of] the above-mentioned forward table & . . .* at the bottom.

Clavis.

Modus quoque 24. ficut præcedentes duo à fine incipit & unâ valente fecundum recipit alternatim otiofam, cumque verfus principium reverfum fuerit, denuò à fine fumitur exordium, & tunc eæ, quæ antea vacabant, occulto infèrviunt.

Conjuratio I.

Pyrichiel marfoys chameron nael peanos pury lames iameue famerufyn mearlo canorfon theory torfa

nealthis dilumeris maphroys carful meor thubra phorfotiel chrebonos aray pemalon layr toyfi uadiviel nemor rofeuafi cabti phroys amenada nachyr fabelronthis, poyl carepon vemy aaslotyn.

Clavis & Senfus.

mArFoYsNaElPuRyIaMeVeMeArLoThEoRyNeAlThIsMa PhRoYsMeOrPhOrSoTiElArAyLaYrVaDiViElRoSeVoSiPhRoYs NaChYrPoYlVeMy. *Senfus.* A fine primum alternatim primò, poftea aliud verfus principium.

Ioannes Trithemius Abbas Spanheimenfis Iacob Dracomio Præmonftratenfis Ordinis Canonico, S. & charitatem.

Eft Nobifcum Mathematicus Infignis Et Egregius Rhetor Tranquillisfimæ Vitæ Studio Feruens lugitur ; Habet Secum Codices Ferme Infinitos Exemplariumque Diverfòrum Inæftimabilem Thefaurum Rectæ Et Bonæ Vitæ Societatis Honeftæ Iucundus Benevolusque Doctor, In Refolvendis Dubiis Expertiffimis , Res Grandes Exponit Nomen Suum Eft Orphonius Regni Bofne Burgravius. Vale. 12. Calend. Aprilis.   ↓.

Tenor arcani latentis.

Brenger diß Briefs ift ein böfer Dib/ huet dich für eme.

Figure G.8: The 24th cipher in Book 1 of Trithemius' *Steganographia.*

Book 1 of the *Steganographia* gives 31 cryptosystems in which the system is composed (superenciphered) with itself. Figure G.8 shows five parts from his Example 24. At the top, we see his *Clavis* or *key: the 24th system begins, like the preceding ones, two words from the end and counts one and leaves one out. When arrived at the beginning, one restarts at the end, and then the words which were dummies before now show the plaintext.* This corresponds to using the key $(2, 2)$, starting twice at the end. Then comes the *Conjuratio I*, made up of phantasy words. Its decipherment yields the *Clavis & Sensus*, by the following rule. The *conjuratio* in all his examples is deciphered in the same way, namely by taking every second letter of every second word. The second words are given on lines 14 to 16 with the relevant letters printed in capitals. He usually identifies u and v, and i, j, and y. The *Sensus* is the same text as this second *clavis* but in standard Latin spelling, reading *"The first letter from the end alternatingly, then every other to the beginning."*. These are the same rules as the first *Clavis* at the beginning, only stated more concisely. The next item is the ciphertext, a fictitious letter from Trithemius to one Jacob Dracomius; sender and recipient are stated on the two lines beginning with *Ioannes* and ending in *charitatem*:

> Est Nobiscum Mathematicus Insignis Et
> Egregius Rhetor Tranquillissimae Vitae
> Studio Feruens Iugitur; Habet Secum Codices
> Ferme Infinitos Exemplariumque Diversorum
> Inaestimabilem Thesaurum Recte Et Bonae
> Vitae Societatis Honestae Iucundus
> Benevolusque Doctor, In Resolvendis Dubiis
> Expertissimis, Res Grandes Exponit Nomen
> Suum Est Orphonius Regni Bosne Burgravius.
> Vale.

The letter translates as:

> *We have with us [in our monastery] one famous mathematician and excellent speaker who combines many interests with a tranquil life. He carries with him old books [codices], a great expert in resolving doubtful cases. His name signals great things; he is Orphonius Burgravius. Sincerely. 12 April.* Ψ.

Following the rules given in the *Clavis*, the decryption is by reading the first letters (all in capitals) of each second word, starting at the end with Burgravius and going backwards. Having reached the

Clavis literarum & Tenor occultus.
Q R G. N A Q R. N E T. K D C. S R X A H Q D G S. K K X
d e r.   a l d e.   a p t.   v o n.   f e i l s d o r f. v v i
A. Q R B K D C. G R X C V N G I M O D G C. Q X R. C N
l. d e n. v o n. r e i n h a r t z b o r n. d i e. n a
H R.   N O H P V C X Q R C.
s e.   a b s c h n i d e n.
    Tenor. Der alde Apt von Feilsdorff/ will dem von Reinhartz-
born die Nase abschniden.

Figure G.9: Trithemius' example in Book 2, Chapter 12, of his *Steganographia.*

beginning, one starts at the end again with Bosme. The plaintext is spelled out in German at the end as the *Tenor arcani latentis* or meaning of the hidden secret: The bringer of this letter is a vile thief; beware of him.[17] This is a rather involved system, quite difficult for the sender to encrypt, with a substantial message expansion, easy to decrypt if you know the global key $(2, 2)$, but nothing diabolical about it. Reader beware: this is only the posthumous edition. Trithemius' original manuscript only contained two out of the five elements present in Figure G.8: the *Conjuratio I* and the ciphertext from *Ioannes* to *Aprilis*. The reader was expected to figure out the *Clavis & Sensus* by herself, from the general rule of taking every second letter of every second word in the *Conjuratio* to derive the decryption algorithm. This is a nontrivial task, and all printed editions contain the key. The first one appeared almost a hundred years after his death, due to Trithemius' troubles with the church authorities. These came from his cabbalistic-sounding encoding of the key, as in the *Conjuratio I*.

The second book adds to this principle the superencipherment of letters by a Caesar cipher. An example is in Figure G.9. The second, fourth and sixth lines are the intermediate decryptions obtained from the first letters of certain words in the ciphertext, much as in the previous example of Figure G.8; we do not show the details. The third, fifth, and seventh lines correspond to a Caesar shift of 12 in his 22-letter alphabet (lacking j, v, w, and y) and present the plaintext. It is repeated as the *Tenor* and translates as The old abbot of Feilsdorff wants to cut off the nose of the abbot of Reinhartzborn—monastic clemency in action.

---

[17]In modern German: Der (Über)Bringer dieses Briefes ist ein böser Dieb; hüte dich vor ihm.

The third book of the *Steganographia* enciphers a 25-letter alphabet by the 700 numbers between 26 and 725. Each letter has 28 possible encryptions, which are precisely those numbers $-x + 25 \cdot i$ in the range that are congruent to $-x$ modulo 25 for the letter numbered x. Here $2 \le i \le 29$. Thus a has number 0 and the 28 encryptions 50, 75, 100, ..., 725. Again, this is couched in cabbalistic and astrological terms, but this time so obscurely that it took almost 400 years to decipher the system.

The offset i as above, is associated to one of 28 celestial intelligences with names like Orifiel, Sadael, Pomiel, and Morifiel from orthodox Christian angelology associated in groups of four to each of the seven planets. Trithemius then gives precise instructions to perform some magical acts, like drawing images on paper, writing your friend's name on them, wrapping them in cloth and putting them in specified locations, like a side-entrance to your house, at astrologically defined times. And then, bang!, your friend receives the message. The strange incantations of the *Steganographia* called *conjurationes* couch cryptographic instructions in mysterious words. Just in case you are not up-to-date in this science: angels and other spirits, being bodiless, can transport messages immediately and secretly from anywhere to anywhere else. Celestial Federal Express. All this mumbo jumbo is stated drily as fact, dead pan.

The *Steganographia* was also printed again and again, and Daniel Defoe wrote in his 1726 *Essay Upon Literature*:

> it seems that Trithemius was the first that set down the Rules of it, which he hath perform'd, not only in his six Books of Polygraphy, *but more especially in his famous Treatise of* Steganography *which has made so much Noise in the World. Now though his Design was, in part, to reveal this useful Secret, yet was he not willing to make it indifferently intelligible to all Sorts of Persons; his End being only to instruct the Learned, and the Ministers of State; and, therefore, to deter the common People from reading his Books, he pretended to a Familiarity with evil Spirits, and make use of some strange Bastard Hebrew Names; such as* Pamersiel, Camuel, &c. (...)

To a modern mind it is hard to understand how the same person can write a substantial book about evangelical express message delivery and then also one of the foundational books in cryptography. Was Trithemius' thinking half shadowed by the Dark Ages and half illuminated by the upcoming Renaissance?

Alphabet Harmonique.

A	1	vt,re,mi,fa.	N	13	mi,fa,re,vt.
B	2	vt,re,fa,mi.	O	14	mi,fa,vt,re.
C	3	vt,mi,re,fa.	P	15	mi,re,fa,vt.
D	4	vt,mi,fa,re.	Q	16	mi,re,vt,fa.
E	5	vt,fa,re,mi.	R	17	mi,vt,re,fa.
F	6	vt,fa,mi,re.	S	18	mi,vt,fa,re.
G	7	re,vt,mi,fa.	T	19	fa,mi,re,vt.
H	8	re,vt,fa,mi.	V	20	fa,mi,vt,re.
I	9	re,mi,vt,fa.	X	21	fa,re,mi,vt.
K	10	re,mi,fa,vt.	Y	22	fa,re,vt,mi.
L	11	re,fa,mi,vt.	Z	23	fa,vt,re,mi.
M	12	re,fa,vt,mi.		24	fa,vt,mi,re.

Figure G.10: Mersenne's musical cipher.

G.4. Marin Mersenne

The Franciscan monk Marin Mersenne (1588–1648) corresponded with the leading mathematicians of his times. He studied the *Mersenne numbers* of the form $M_n = 2^n - 1$; see Section 3.2. His main published work, the massive *Harmonie universelle* from 1636, presents on its 1582 pages a large edifice of musical theory. He proposes to transmit secret messages by music: *"l'on peut discourir auec vn autre en ioüant de l'Orgue, de la Trompette, de la Viole, de la Fleute, du Luth & des autres instrumens, sans que nul puisse entendre le discours, que celuy qui sçait le secret."*[18]

---

[18]One can converse with another person by playing the organ, the trumpet, the viola, the flute, the lute, and other instruments, without anybody being able to understand the

As one possibility, he encodes the 23 letters of his alphabet by the $4! = 24$ permutations of the four notes `ut`, `re`, `mi`, and `fa` (Figure G.10).

> *Ces* 24 *changemens monstrent que l'on peut faire vingt-quatre chants differents avec quatre chordes d'vne Epinette, quatre tuyaux d'Orgue, ou autres quatre Sons, sans repeter deux fois vn mesme Son; la Quinte donne six vingt chants tous differents: [...] auec l'Octave l'on peut exprimer tous les characteres des Chinois, pourueu qu'ils ne surpassent pas le nombre de quarante mille trois cens vingt.*[19]

His method requires a pretty good ear and will usually produce a cacophony. It has little cryptographic content, but shows clear mastery of the technique of expressing one alphabet in terms of another, that is, of simple substitutions.

G.5. Athanasius Kircher and Kaspar Schott

Athanasius Kircher (1601–1680) was a German Jesuit who published extensively in many areas of science. His 1663 *Polygraphia* presents codebooks in Latin, Italian, French, Spanish, and German with a numerical code. Messages in any of the five languages could be encrypted with it, and then be decrypted in any of the five, thus providing both secrecy and translation. Kircher was actually more interested in the latter. A popular topic at his times was a "universal language" from and into which any language can be translated. This was also a goal of Wilkins. It was generally abandoned about a century later as infeasible, but has now resurfaced as the meta languages of machine translators.

Kircher's disciple Kaspar (Gaspar) Schott (1608–1666) also wrote prodigiously about scientific subjects, including an *Organum Mathematicum*. His four-volume *Magia Naturalis* includes cryptography, and his *Schola Steganographica* describes many cryptosystems including Kircher's *cistula*[20] where index cards provide alphabetic permutations. Such boxes are visible in the *Combinatio Tabellarum Steganographicarum* in the frontispiece. The *Schola*'s title inspired that of the present text.

---

conversation except the one who knows the secret.

[19]These 24 cadences show that one can produce twenty-four different melodies with four strings of a spinet, four pipes of an organ, or four other sounds without repeating twice the same sound; the quint (five sounds) gives 120 different melodies [...] with an octave one can express all characters of the Chinese, provided there are not more than forty thousand three hundred and twenty [=8!] of them.

[20]little box

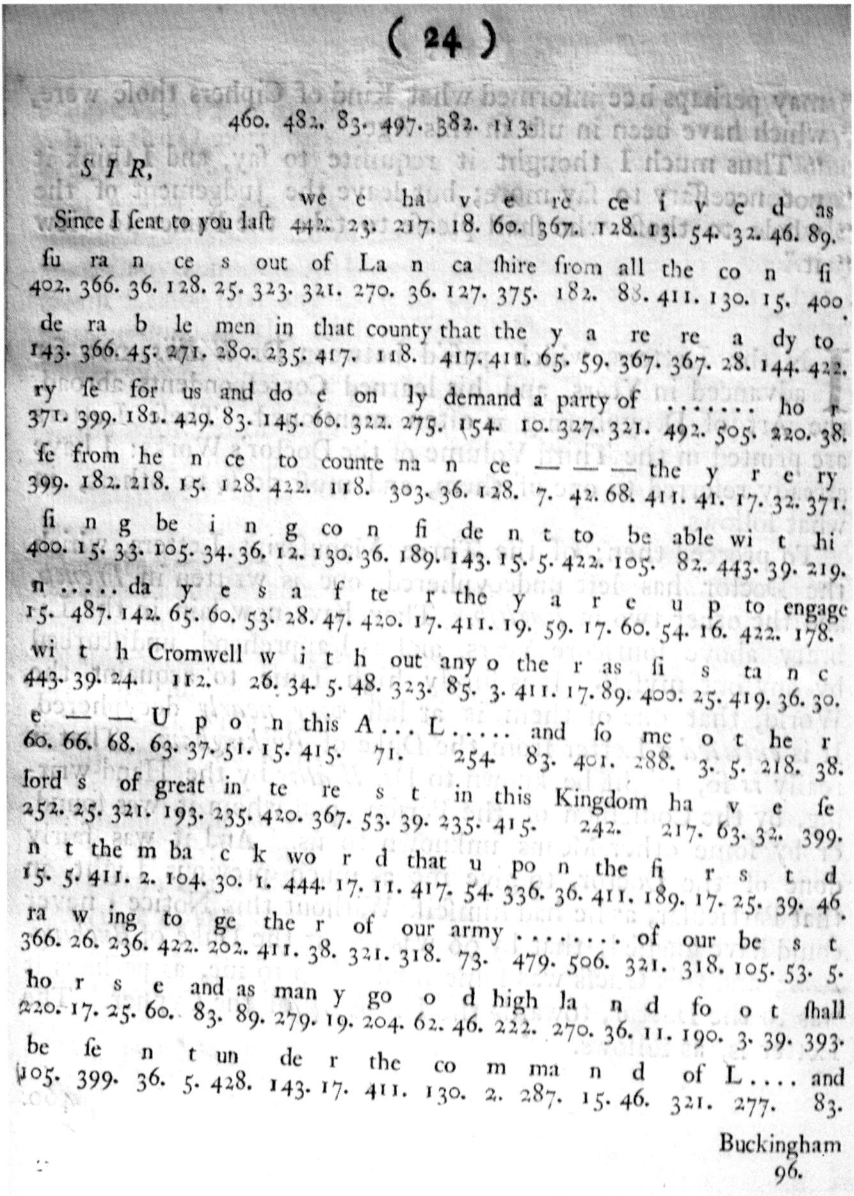

(24)

460. 482. 83. 497. 582. 113.

SIR,

                we c ha v e re ce i v e d as
Since I fent to you laſt 44². 23. 217. 18. 60. 367. 128. 13. 54. 32. 46. 89.

fu ra n ce s out of La n ca ſhire from all the co n fi
402. 366. 36. 128. 25. 323. 321. 270. 36. 127. 375. 182. 83. 411. 130. 15. 400.

de ra b le men in that county that the y a re re a dy to
143. 366. 45. 271. 280. 235. 417. 118. 417. 411. 65. 59. 367. 367. 28. 144. 422.

ry fe for us and do c on ly demand a party of . . . . ho r
371. 399. 181. 429. 83. 145. 60. 322. 275. 154. 10. 327. 321. 492. 505. 220. 38.

fe from he n ce to counte na n ce — — — the y r e ry
399. 182. 218. 15. 128. 422. 118. 303. 36. 128. 7. 42. 68. 411. 41. 17. 32. 371.

fi n g be i n g co n fi de n t to be able wi t hi
400. 15. 33. 105. 34. 36. 12. 130. 36. 189. 143. 15. 5. 422. 105. 82. 443. 39. 219.

n . . . . . da y e s a f te r the y a re u p to engage
15. 487. 142. 65. 60. 53. 28. 47. 420. 17. 411. 19. 59. 17. 60. 54. 16. 422. 178.

wi t h Cromwell w i t h out any o the r as fi s ta n c
443. 39. 24. 112. 26. 34. 5. 48. 323. 85. 3. 411. 17. 89. 400. 25. 419. 36. 30.

e — — U p o n this A . . . L . . . . . and fo me o the r
60. 66. 68. 63. 37. 51. 15. 415. 71. 254. 83. 401. 288. 3. 5. 218. 38.

lord s of great in te re s t in this Kingdom ha v e fe
252. 25. 321. 193. 235. 420. 367. 53. 39. 235. 415. 242. 217. 63. 32. 399.

n t the m ba c k wo r d that u po n the fi r s t d
15. 5. 411. 2. 164. 30. 1. 444. 17. 11. 417. 54. 336. 36. 411. 189. 17. 25. 39. 46.

ra w ing to ge the r of our army . . . . of our be s t
366. 26. 236. 422. 202. 411. 38. 321. 318. 73. 479. 506. 321. 318. 105. 53. 5.

ho r s e and as man y go o d high la n d fo o t ſhall
220. 17. 25. 60. 83. 89. 279. 19. 204. 62. 46. 222. 270. 36. 11. 190. 3. 39. 393.

be fe n t un de r the co m ma n d of L . . . . and
405. 399. 36. 5. 428. 143. 17. 411. 130. 2. 287. 15. 46. 321. 277. 83.

Buckingham
96.

Figure G.11: An encrypted letter, with the solution above the ciphertext.

K E Y.

1 k	19 y	37 p	55	82 able	134 duke	217 ha
2 m	20 —	38 r	56	83 and	140 desire	218 he
3 o	21	39 t	57	85 any	142 da	219 hi
4	22	40 w	58	88 all	143 de	220 ho
5 t	23 e	41 y	59 a	89 as	144 di	222 high
6	24 h	42 —	60 e	96 Buckingham	145 do	232 if
7 —	25 s	43	61 i	104 ba	154 demand	235 in
8 —	26 w	44	62 o	105 be	160 en	236 ing
9	27	45 b	63 u	107 bo	178 engage	240 the King
10 a	28 a	46 d	64 w	108 bu	181 for	242 kingdom
11 d	29 c	47 f	65 y	112 Cromwell	182 from	252 lord
12 g	30 c	48 h	66 —	113	189 fi	254 L......
13 i	31	49 k	67	114 Colonel	190 fo	258 London
14	32 e	50	68 —	118 county	193 great	270 la
15 n	33 g	51 o	69	127 ca	202 ge	271 le
16 p	34 i	52	70	128 ce	203 gi	272 li
17 r	35	53 s	71 A....	130 co	204 go	275 ly
18 u	36 n	54 u	73 army	131 cu	205 gu	277 L ....

279

Figure G.12: Part of the codebook used in Figure G.11.

G.6. John Wallis

John Wallis (1616–1703), Savilian Professor at Oxford and the most famous English mathematician before Isaac Newton, worked as a cryptanalyst for Oliver Cromwell and the Republic and also for King Charles II after the monarchy's restoration. Few people managed to be in office under both regimes. Some of the historical background is explained in Section H.3.

He wrote on 11 January 1697 to his colleague Gottfried Ludwig Mencke (1683–1744) in Leipzig, explaining his long-standing interest in cryptanalysis and how this is a matter of great labor and cleverness:

> Hanc ego rem primum aggressus eram, jam ante annos plus quinquaginta (quum simpliciores fuerint Ciphræ quam nunc dierum) & quasi ludendo, in facili Ciphra, (quæ prima fuit quam ego unquam videram,) quam duarum horarum spatio superavi. Postea vero subinde, ut ferebat occasio, difficiliores aggresso, res successit non male.[21]

---

[21] I first attacked this question over fifty years ago (how much simpler ciphers were then than in our days), and almost as a game I solved an easy cipher (which was the first that I saw) in two hours. And after this, when the occasion arose, I attacked more difficult ones, with good success.

Wallis includes two long letters that he deciphered, one of them a report from de Teil, the French ambassador in Poland, to his King, dated 8 July 1689.

Figure G.11 shows the first half of a letter from the Duke of Buckingham, and Figure G.12 part of its key, essentially a one-part codebook with homophones for some individual letters. The undated message deals with military preparations. It was apparently partly cryptanalyzed by Wallis, and Davys (1737), disciple and admirer of Wallis, presented the break as in Figure G.11. He adds several comments, including that $71\,254 = A \ldots L \ldots$ is probably Argyle Lauderdale, but that this is not certain.

Notes. An example of a major discovery kept under wraps is IBM's discovery of differential cryptanalysis, divulged only eighteen years later; see Section 6.2.

G.1. For a more complete story of Arab cryptography, see Al-Kadi (1992). Al-Kindī's *"One way to solve..."* is from Mrayati *et al.* 1987, page 216. Mohammed's name including some of his ancestors is محمَّد ابن عَبد اللَّه ابن عَبد المُطَّلِب ابن هَاشِم ابن عَبد مَنَاف ال-القُرَيشى (muhammad ibn ʿabd āl-lah ibn ʿabd āl-muttalib ibn hāšim ibn ʿabd manāf āl-quraīšy), and al-Kindī is أبُو يُوسُف يَعقُوب ابن إسحَاق ابن الصَبَاه ابن عِمرَان ابن إسمَاعِيل الكِندِى (ʾabū yūsuf yaʿqūb ibn ʾishāq ibn al-sabāh ibn ʿimrān ibn ʾismāʿīl al-kindī). Al-Halīl, also called al-Farāhīdī, is أبُو عَبد الرَّحمَان الخَلِيل ابن أحمَد ابن عُمَر ابن تمَّام الفَرَاهِيدِى (ʾabū ʿabd al-rahmān al-halīl ibn ʾahmad ibn ʿumar ibn tammām al-farāhīdī), ibn ʿAdlān is عَفِيف الدِين عَلِي ابن عَدلَان النحُوِى (ʿafīf al-dīn ʿalī ibn ʿadlān al-nhūī), and ibn Dunainīr is إبرَاهِيم ابن محمَّد ابن إبرَاهِيم ابن عَلِي ابن هِبَة اللَّه ابن يُوسُف ابن نَصِر ابن (ʾibrāhīm ibn muhammad ibn ʾibrāhīm ibn ʿalī ibn hibah āl-lah ibn أحمَد ابن دُنَينِير اللخمِي yūsuf ibn nasr ibn ʾahmad ibn dunainīr al-lahmī). Al-Duraihim is عَلِي ابن محمَّد ابن عَبد العَزِيز تَاج الدِين ابن الدُرَيهِم (ʿalī ibn muhammad ibn ʿabd al-ʿazīz tāğ al-dīn ibn al-duraihim).

Latin uses I, V, X, L, C, M for 1, 5, 10, 50, 100, 1 000, respectively. The larger values come before smaller ones, but a combination ab with $a < b$ stands for $b - a$. Thus VX = 10 − 5 = V.

G.2. Cicco Simonetta's rules are edited in Perret (1890).

G.3. The drawing of Trithemius is now in the Musée Condé, Château de Chantilly, just north of Paris. The inscription may have been added later. It reads:

> IOHANNES TRITEMI, von Tritenheim geboren A-o [Anno] 1460 Benediktiner ordens zu Spanheim, Apt geworden seines Alters im 29 Jahr unter bapst alexander dem 6 und Kays[er] Maximilian HB. [4]

The drawing is attributed to Hans Brosamer (1494–1554), who used the monogram HB on other paintings. But Figure G.3 might also be due to Hans Burkmair (1473–1531). Spamer (1894) says it was produced after an unfinished drawing by Hans Holbein

---

[4]Iohannes Trithemius, born in Trittenheim in the year 1460 [correct: 1462], abbot of the Benedictine order at Sponheim at age 29 under Pope Alexander VI. and Emperor Maximilian. HB.

(c. 1465-c. 1524). For the last name Cellers, see Volk (1950). In the *Pinax* at the beginning of his Polygraphia, Trithemius recalls his father Johann de Heidenberg. For the grape as Trithemius' coat-of-arms, see Arnold (1991), page 6: at his sepulchral stone in Würzburg, at the Sponheim altar (page 20), and on the *Polygraphia's* title page at bottom left. Trithemius' bronze statue is by W. Henning.

Trithemius describes at the end of Book 5 of the *Polygraphia* a (fictitious) number system of the Franks, based on Roman numerals and allegedly provided to him by the equally fictitious historian Hunibaldus.

The British historian Yates (1972), page 108, writes about Trithemius' *"angel-conjuring"*. In the view of Noel L. Brann (1999), *"in his angel-free* Polygraphia *(published 1518), Trithemius furnished his cryptic applications with theoretical buttressing derived from ancient hermetic, Pythagorean, and kabbalistic principles. In this way he effectively transformed his mystical theology into magical theology."* Trithemius' complaints about *"fuere nonnulli"* are in the *Polygraphia's* dedication to Emperor Maximilian. The *sword* quote is from a 1613 edition of the *Polygraphia*, edited by Adolph Glauburg, on page 37. The admonition not to bother about the message expansion is from the 1518 *clavis*, section *in explanationem*, seventh page. The *"Horum usus ..."* is in the introduction to Chapter 5 of Trithemius (1518). The *"Est autem"* is from the *clavis* to Book 5.

In the *clavis* to Book 5, Trithemius suggests that the readers *"tot poterunt alphabeta componi, quot stellæ numerantur in firmamento cæli"* (= can compose as many alphabets as there are stars in the firmament). There are $24! \approx 6.2 \times 10^{23}$ many substitutions in Trithemius' alphabet, and astronomers estimate the number of stars at around 10^{22} to 10^{24}; see http://www.esa.int/Our_Activities/Space_Science/Herschel/ How_many_stars_are_there_in_the_Universe.

Trithemius' alphabet in Figure G.6 omits our letters j and v, as usual at his time. The inclusion of k and w makes it more of a German than of a Latin alphabet. It is not clear what his remark on $24^3 (= 13\,824)$ refers to.

For *orchema*, see Frisk (1973), volume II, page 433. The word comes from the verb ὀρχεῖσθαι (orchesthai) for *to dance* or *to skip*; the word ορχηστρα (orchestra) has the same origin and means *dance floor* or *orchestra*. Trithemius' *transilitio* is in his *clavis* to Book 5. Trithemius' remark on the mysteries of random numbers is in his *clavis* to Book 5. Cryptanalysis of simple substitutions as pioneered by Arab scientists (Section G.1) and the choice of one among several possibilities in a homophonic substitution (say 3 or 7 for b in Figure D.2) predate Trithemius and may also be considered as using randomness, maybe even in a systematic way. Did Trithemius know about Alberti's treatise? He produced catalogs of his extensive library, but they are lost. Lehmann (1961) presents an extensive reconstruction of this collection. It does not contain Alberti's work nor any other books of cryptographic relevance except Trithemius' own.

Trithemius' various orchema substitutions are not uniformly random, but contain only a small amount of randomness. He takes every second or every fourth letter in turn, or breaks the alphabet into blocks and uses some in normal and some in reverse order. Translated from letters into numbers for transparency, his examples include the following four simple substitutions of the plaintext letters in the top row:

0	1	2	3	4	5	6	7	8	9	10	11	12	13	14	15	16	17	18	19	20	21	22	23
2	4	6	8	10	12	14	16	18	20	22	0	23	1	3	5	7	9	11	13	15	17	19	21	23
21	17	13	9	5	1	0	20	16	12	8	4	19	15	11	7	3	23	22	18	14	10	6	2
4	5	6	7	8	3	2	1	0	13	14	15	16	17	12	11	10	9	23	22	21	20	19
15	16	17	18	19	20	21	22	23	14	13	12	11	10	9	8	7	6	5	4	3	2	1	0

This is reminiscent of codebooks (Section D.1) which come in several flavors: ordered (one-part) or completely random (two-part) or somewhere in between, as in Figures D.3 and D.4 with internally ordered pieces shuffled at random. Trithemius' orchematic substitutions follow the latter principle.

For mobiles and volvelles in books, see Lindberg (1979) and Helfand (2002). The printed cipher disks of Alberti (1568) and Porta (1602) could, in principle, also be used to make a volvelle. Similarly, Schott (1665) prints on his pages 95 and 99 concentric disks with do-it-yourself instructions for cipher disks. The idea of volvelles may go back to the Mallorcan philosopher Ramón Llull (c. 1232–1316).

The eight books of the *Steganographia* are mentioned on page 39, and the fourth book on page 51 of Trithemius (1616); see also Fabricius (1858), volume III, page 440. With the key $r = s = 2$ and leaving out the first word, `This sentence` ... encodes `stegano`. In general, if $y_{k,j}$ denotes the jth letter of the kth word of the plaintext y, then the ith letter x_i of the ciphertext x is simply $y_{t+is,r}$. Book 3 of the *Steganographia* was deciphered in a tour de force by Ernst (1996) and Reeds (1998). Orifiel from angelology: see Borchardt (1972), page 45. Defoe's comment is on page 298 in Furbank's edition. The fourth letter `B` in line 3 of Figure G.9 is erroneous for the correct encryption `C` of the plaintext n.

G.4. Besides musical cryptography and musicology, Mersenne's *Harmonie Universelle* also discusses extensively combinatorics, mechanics, universal languages, and many other topics.

G.5. McCracken (1948) gives a description of the contents of Kircher's *Polygraphia*.

Der *Beweis* ist die Rechnung, deren Resultat
der zu beweisende Satz ist. Rechnen und Denken ist eins.[22]
NOVALIS (1798)

In math, absolute proof is the goal, and once something is proved,
it is proved forever, with no room for change.
SIMON SINGH (1997)

"Take nothing on its looks; take everything on evidence.
There's no better rule."
CHARLES DICKENS (1861)

The whole of Mathematics consists in the organization
of a series of aids to the imagination in the process of reasoning.
ALFRED NORTH WHITEHEAD (1898)

"What you say is rather profound, and probably erroneous,"
he said, with a laugh.
JOSEPH CONRAD (1899)

Those intervening *ideas*, which serve to shew the Agreement
of any two others, are called *proofs*; and where the Agreement
or Disagreement is by this means plainly and clearly perceived,
it is called *Demonstration*, it being *shewn* to the Understanding,
and the Mind made to see that it is so. A quickness in the mind
to find out these intermediate *Ideas*, (that shall discover
the Agreement or Disagreement of any other), and to apply
them right, is, I suppose, that which is called *Sagacity*.
JOHN LOCKE (1690)

Obviousness is always the enemy to correctness.
BERTRAND RUSSELL (1956)

Two roads diverged in a wood and I –
I took the one less traveled by,
And that has made all the difference.
ROBERT FROST (1916)

---

[22]The *proof* is the computation whose result is the theorem to be proven. Computing and thinking are one.

Chapter 12

Proof systems and zero knowledge

In an *interactive proof system*, a prover called PAULA knows some fact and tries to convince a verifier VICTOR of its truth. This might be some property of a secret key, say, that its least bit is 0. Such proof systems proceed in several rounds. In each round, typically PAULA sends some information to VICTOR, he comes up with a (randomly chosen) challenge, and PAULA replies to it. VICTOR studies this answer and is either happy or unhappy with it. If he is happy in each round, he accepts PAULA's claim to be true, and otherwise he rejects it.

The system is *complete* if VICTOR ends up being convinced of every true claim of PAULA's, and *sound* if PAULA can cheat him into being convinced of a false fact only with negligible probability.

Such a system has the *zero knowledge* property if VICTOR learns nothing more than the fact claimed by PAULA. In particular, we assume that VICTOR is a random polynomial-time computer, so that everything he can calculate is within BPP, and that he cannot verify the claimed fact by himself, because the problem is (presumably) not in BPP. Then he will not be able to convince anyone else of this fact.

How can we make this formal? The wonderful idea of Goldwasser, Micali & Rackoff, the inventors of this theory, is to stipulate a random polynomial-time simulation of the protocol's *transcript*, consisting of all messages interchanged. Then after executing the protocol with PAULA, everything that VICTOR can do is still in BPP, since he might as well simulate the protocol all by himself.

We discuss two variants, namely perfect and computational zero knowledge. The first one is too demanding, but it turns out that all problems in NP have a proof system with the second property.

© Springer-Verlag Berlin Heidelberg 2015
J. von zur Gathen, *CryptoSchool*, DOI 10.1007/978-3-662-48425-8_19

12.1. Interactive proof systems

Suppose that PAULA knows a secret, say the solution to a computationally hard problem. She wants to convince VICTOR that she knows it, but without revealing any other knowledge to VICTOR! For example, VICTOR should not be able to convince anyone else later that he knows this secret. So clearly PAULA cannot just tell her secret to VICTOR, but she has to find some roundabout way of convincing him without giving too much away.

How can we make this work?

As usual, everything that PAULA and VICTOR do is described by probabilistic algorithms. VICTOR is only allowed polynomial time, because the secrets that we will deal with are (presumably) hard to find under this constraint but may be easy if VICTOR is more powerful. We impose no restriction on PAULA's computing power. PAULA and VICTOR exchange messages according to some fixed *protocol*, and in between they do some private computations, using private random strings. In the end, VICTOR either accepts or rejects a run of a protocol, meaning that PAULA has convinced him of her claim or not.

We start with *graph isomorphism* as an example. We recall that an (undirected) graph $G = (V, E)$ with n vertices has $V = \{1, \ldots, n\}$ as its vertex set and a set $E \subseteq \binom{V}{2}$ of unordered pairs $\{i, j\} \subseteq V$ as edges. An *isomorphism* between two such graphs $G_0 = (V_0, E_0)$ and $G_1 = (V_1, E_1)$ is a bijection $\pi \colon V_0 \longrightarrow V_1$ so that $\pi(E_0) = E_1$, that is, $\{i, j\} \in E_0$ if and only if $\{\pi(i), \pi(j)\} \in E_1$, for all $i \neq j \in V_0$. We also write $\pi \colon G_0 \longrightarrow G_1$, $G_1 = \pi(G_0)$, and $G_0 \simeq G_1$. Thus $\pi \in \mathrm{Sym}_n$; see Section 15.2 for permutations. An isomorphism π of a graph G to itself, so that $\pi(G) = G$, is called an *automorphism*. These automorphisms form a group $\mathrm{Aut}(G) \subseteq \mathrm{Sym}_n$.

EXAMPLE 12.1. We let $n = 4$ and $\pi \in \mathrm{Sym}_4$ with $\pi(1) = 3$, $\pi(2) = 1$, $\pi(3) = 4$, $\pi(4) = 2$. Then π is an isomorphism between the two graphs

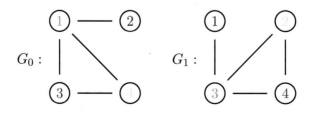

Corresponding vertices under π have the same color. We have $\mathrm{Aut}(G_0) = \{\mathrm{id}, \sigma\} \subseteq \mathrm{Sym}_4$, where σ interchanges 3 and 4 and leaves 1 and 2 fixed. ◇

In the *cycle representation*, we have $\pi = (1, 3, 4, 2)$ and $\sigma = (1)(2)(3, 4) = (3, 4)$ in Example 12.1. We encode $G = (V, E)$ by the binary representations of n and each entry of E, listed in alphabetical order. (We might also encode each number in unary.)

The following simple example exhibits many of the salient features of interactive proofs. PAULA wants to convince VICTOR that two given graphs are not isomorphic. For isomorphic graphs, she could simply show VICTOR an isomorphism to convince him. (We will see cleverer ways later.) But now it does not make sense to "show a nonisomorphism". It is not immediate how VICTOR could convince himself, even with help from PAULA, that the graphs are not isomorphic. The problem GRAPH NONISOMORPHISM of pairs of nonisomorphic graphs is in coNP, but otherwise its complexity status is unclear. It is conjectured to be neither coNP-complete nor to be in NP, even less in P (Section 15.6), and the conjecture would show that there is no polynomial-sized advice that is sufficient to prove nonisomorphism. In the example, cleverly chosen interactive advice from powerful PAULA is sufficient to convince VICTOR.

EXAMPLE 12.2. PAULA claims that the two graphs

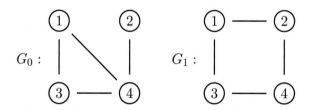

on $n = 4$ vertices are not isomorphic. In one round, VICTOR chooses randomly $\pi \xleftarrow{\boxtimes} \mathrm{Sym}_4$ and $i \xleftarrow{\boxtimes} \{0, 1\}$, say $\pi = (1, 2, 3)$ and $i = 0$, and computes $H = \pi(G_i) = \pi(G_0)$:

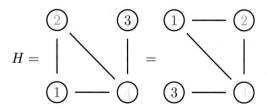

He sends this H to PAULA. If $H \simeq G_0$, then PAULA sets $j = 0$, else she sets $j = 1$. She passes j to VICTOR. Here we get $j = 0$. VICTOR checks whether $i = j$. It is true in this case.

These steps are repeated n times, with independent choices of π and i in each round. In the end VICTOR accepts if and only if all his checks were satisfactory.   ◊

We make the following observations.

- If $G_0 \not\simeq G_1$, then PAULA can answer so that VICTOR will always accept.

- If $G_0 \simeq G_1$, so that PAULA's claim is false, then the H she sees is isomorphic both to G_0 and to G_1. No matter how she chooses her j, it only has a chance of $1/2$ of matching VICTOR's randomly chosen i. Therefore VICTOR will accept in the end only with probability 2^{-n}. PAULA needs to get really lucky in order to fool VICTOR, as lucky as tossing n heads in a row with a fair coin.

- The communication pattern of the system consists of n repetitions of the messages exchanged. The sequence: first H, then j, is vital for the protocol's validity. But the various rounds could be intermingled. VICTOR could even send the sequence of all his H, and PAULA then replies with the corresponding sequence of her j.

- In the example VICTOR only uses polynomial time, whereas PAULA might need more than that for solving graph (non-)isomorphism problems.

We will always assume that VICTOR uses only polynomial time, whereas PAULA is "omnipotent": she can compute anything that can be computed at all. A proof system consists of a polynomial number of rounds, where typically the same type of messages is exchanged in each round.

DEFINITION 12.3. *An interactive proof system consists of two interacting algorithms PAULA and VICTOR. They get the same input but use separate memory and random strings. PAULA is omnipotent and VICTOR uses random polynomial time. When one of the two parties has finished its computations, it sends a message to its counterpart and switches to waiting mode. Then it is the other's turn. VICTOR decides at the end whether he accepts or rejects, and outputs the result "accept" or "reject". The system is an interactive proof system for a problem $X \subseteq \{0,1\}^*$ if it is:*

- *complete: VICTOR accepts every $x \in X$ with probability at least $2/3$,*

- *sound: for any algorithm that PAULA might use and for any $x \in \{0,1\}^* \setminus X$, VICTOR accepts x with probability at most $1/3$.*

The class IP consists of those languages that have such an interactive proof system.

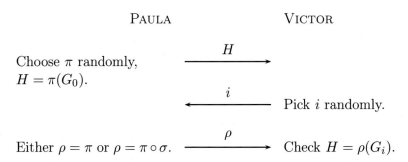

Figure 12.1: One round of the graph isomorphism protocol 12.5.

This definition of completeness refers to the situation where both PAULA and VICTOR follow the protocol, but for soundness, we allow PAULA to cheat in an arbitrary way. Some attempts at cheating may be easy to detect, for example, if she sends $j = 6$ in the above protocol; in this case, VICTOR will reject. So we may as well assume that even a cheating PAULA sends messages of the type (number, graph, ...) stipulated in the protocols.

The usual notion of a proof fits into this framework as follows. PAULA computes a proof of the fact of which she wants to convince VICTOR, and sends this proof to him. He then checks this proof. Under our definition, he is only allowed random polynomial time for this. Thus we have a single one-way communication. By a famous result of Shamir (1992), the class IP equals PSPACE, a "huge" complexity class encompassing NP and (presumably) much more.

In what follows, the "facts" to be proven are single combinatorial statements like "these two graphs are nonisomorphic". One could also ask about "proofs" and proof systems for general facts in mathematical theories, say the Chinese Remainder Theorem 15.44 in number theory, but we will not deal with such questions.

Next we consider an example where PAULA has some secret knowledge, namely the complement of GRAPH NONISOMORPHISM.

PROBLEM 12.4. GRAPH ISOMORPHISM.

Instance: Two graphs G_0, G_1 with the same set of vertices $\{1, \ldots, n\}$.
Question: Is $G_0 \simeq G_1$?

If PAULA has an isomorphism $\sigma \colon G_1 \longrightarrow G_0$, then she can easily convince VICTOR that $G_0 \simeq G_1$ by telling him σ. But can she achieve this without revealing any information about σ?

INTERACTIVE SYSTEM 12.5. Graph Isomorphism.

Input: Two graphs G_0 and G_1 with the same vertex set $\{1, \ldots, n\}$.
PAULA's knowledge: an isomorphism $\sigma \colon G_1 \longrightarrow G_0$.

1. Execute n rounds of steps 2 through 5.
2.   PAULA chooses a random permutation π of $\{1, \ldots, n\}$. She computes $H \longleftarrow \pi(G_0)$ and sends H to VICTOR.
3.   VICTOR generates a random $i \xleftarrow{\boxdot} \{0, 1\}$ and sends this to PAULA.
4.   PAULA computes

$$\rho \longleftarrow \begin{cases} \pi & \text{if } i = 0, \\ \pi \circ \sigma & \text{if } i = 1, \end{cases}$$

    and sends ρ to VICTOR. [Then $H = \rho(G_i)$.]
5.   VICTOR checks whether $H = \rho(G_i)$ holds.
6. VICTOR returns "accept" if all his checks were satisfactory, otherwise "reject".

The messages in one round are illustrated in Figure 12.1.

EXAMPLE 12.6. We illustrate the above proof system by the following graphs G_0 and G_1 with vertex set $\{1, 2, 3, 4\}$:

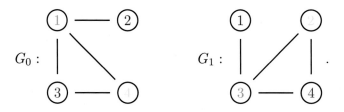

Then $\sigma = (1, 2, 4, 3)$ is an isomorphism from G_1 to G_0. In one round, PAULA chooses $\pi \xleftarrow{\boxdot} \mathrm{Sym}_4$, say $\pi = (1, 2, 3)$. She sends the following graph H to VICTOR:

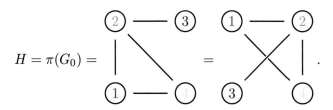

VICTOR generates $i = 1$ at random. PAULA computes $\rho = \pi \circ \sigma = (1, 3, 2, 4)$, since $i = 1$, and sends ρ to VICTOR. VICTOR checks that

$\rho(G_1) = H$:

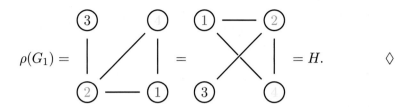

$$\rho(G_1) = \quad \Big| \diagup \Big| \quad = \quad \diagslash\!\!\!\!\diagdown \Big| \quad = H. \qquad \diamond$$

The graph isomorphism Interactive System 12.5 is complete: if $G_0 \simeq G_1$, then an isomorphism $\sigma\colon G_1 \longrightarrow G_0$ exists, and PAULA can compute it. Then $H = \rho(G_i)$ in each round, so VICTOR accepts in the end.

It is also sound: if $G_0 \not\simeq G_1$, then H is isomorphic to at most one of G_0 and G_1, say to G_j, and PAULA can send a satisfactory reply only if $i = j$, no matter how hard she tries to cheat. Since VICTOR follows the protocol, his choice of i is random and PAULA can fool him in one round only with probability at most $1/2$. (For example, she might pass him an isomorphic copy H of G_1 and succeeds if VICTOR chooses $i = 1$.) It follows that VICTOR accepts with probability at most $2^{-n} < 1/3$ for $n \geq 2$. The sequence here is crucial: first PAULA has to send H and is now committed to a single value of j if $G_0 \not\simeq G_1$, but still has both choices for j if $G_0 \simeq G_1$. Then VICTOR challenges her with his value i. Intuitively, VICTOR learns nothing about σ, since all he sees is the random permutation $\pi \circ \sigma$ (if $i = 1$) and he does not know π. The power of the zero knowledge idea in the next section is to make this intuition rigorous.

12.2. Zero knowledge

What is *knowledge*? Philosophers, psychologists, biologists, computer scientists, and many others have poured over this question. How can we formalize the notion of how much knowledge VICTOR gains by executing an interactive protocol with PAULA? It turns out that it is easier to say what "no knowledge" is rather than what "knowledge" is. (Well, we all know this from exams ;-).)

But what does VICTOR "know"? Being a random polynomial-time machine, he knows what such machines know, formalized as any language in BPP. We think of this as "trivial knowledge". Anything beyond this is nontrivial, and we want to express the idea that VICTOR gains no nontrivial knowledge. We assume that all that VICTOR learns is contained in the messages exchanged during the protocol. (For example, in step 5 of the graph isomorphism Interactive System 12.5, it might take PAULA a bit longer to compute ρ when $i = 1$ compared to when $i = 0$; it is

assumed that VICTOR does not observe this.) The genial idea of Goldwasser *et al.* (1989) is to say: well, if VICTOR could generate this message
traffic himself, in random polynomial time, then he has gained no nontrivial knowledge. We formalize this concept in the following as "zero
knowledge".

DEFINITION 12.7. *The transcript of an execution of an interactive protocol is the list of all inputs and messages exchanged.*

At the end of an execution of the graph isomorphism protocol, the
transcript T looks as follows:

$$T = \Big(\underbrace{(G_0, G_1)}_{\text{input}}, \underbrace{(H_1, i_1, \rho_1)}_{\substack{\text{messages} \\ \text{in round 1}}}, \ldots, \underbrace{(H_n, i_n, \rho_n)}_{\substack{\text{messages} \\ \text{in round } n}}, \underbrace{\text{``accept'' or ``reject''}}_{\substack{\text{VICTOR's} \\ \text{output}}}\Big).$$

Such a transcript is a random variable depending on the probabilistic
choices made by PAULA and VICTOR.

We formalize zero knowledge by saying that VICTOR might produce such transcripts by himself, or rather that a random polynomial-
timesimulator SAM might do this. Before the general definition, we look
at how this works for the graph isomorphism Interactive System 12.5.

ALGORITHM 12.8. SAM's simulator for Interactive System 12.5.

Input: Two isomorphic graphs G_0, G_1 with vertex set $\{1, \ldots, n\}$.
Output: A transcript T.

1. Initialize T with (G_0, G_1).
2. For n times do steps 3–5
3.   Choose $i \xleftarrow{\boxed{\text{\tiny ?}}} \{0, 1\}$ and $\rho \xleftarrow{\boxed{\text{\tiny ?}}} \text{Sym}_n$ uniformly at random.
4. $H \longleftarrow \rho(G_i)$.
5.   Append (H, i, ρ) to T.
6. Append "accept" to T, and output T.

We will show that there is no way to tell the difference between such
a simulated transcript and a real one. We denote by $T_{\text{PAULA,VICTOR}}$ a
transcript produced by PAULA and VICTOR, while following the protocol
and assuming $G_0 \simeq G_1$, and by T_{SAM} one as produced by SAM. We will
prove that for any two isomorphic graphs G_0 and G_1 and any conceivable
transcript T for these inputs, the probability that PAULA and VICTOR
generate T is equal to the probability of SAM producing it:

$$\text{prob}\{T_{\text{PAULA,VICTOR}}(G_1, G_2) = T\} = \text{prob}\{T_{\text{SAM}}(G_1, G_2) = T\}.$$

We also allow SAM to fail, but at most half of the time.

DEFINITION 12.9. *An interactive proof system* PAULA/VICTOR *for a problem* X *has* perfect zero knowledge for VICTOR *if there exists a random polynomial-time simulator* SAM *which on input* x *either produces a transcript* $T_{\text{SAM}}(x)$ *or "failure"; the latter with probability at most* $1/2$. *For any input* $x \in X$ *and any conceivable transcript* T *we require that*

$$\text{prob}\{T_{\text{PAULA,VICTOR}}(x) = T\} = \text{prob}\{T_{\text{SAM}}(x) = T : \text{not "failure"}\}.$$

We note that there is no requirement for those inputs x that are not in X. Intuitively, zero knowledge means that transcripts for $x \in X$ are easy to generate, without any intervention from PAULA.

THEOREM 12.10. *The graph isomorphism protocol 12.5 has perfect zero knowledge for* VICTOR.

PROOF. We consider an input $x = (G_0, G_1)$ consisting of two isomorphic graphs on n vertices, and let $\mathcal{L} = \{(H, i, \rho) : H = \rho(G_i), i \in \{0,1\}, \rho \in \text{Sym}_n\}$ be the set of all possible "lines" of the transcript, excepting the input and output line. We have $\#\mathcal{L} = 2 \cdot n!$, and a transcript T is of the form $T = ((G_0, G_1), L_1, \ldots, L_n, \text{"accept"})$ with $L_1, \ldots, L_n \in \mathcal{L}$.

Now in each round $j \leq n$, SAM chooses i and ρ uniformly at random to generate his line $L_{\text{SAM},j}(x)$. Therefore we get $\text{prob}\{L_{\text{SAM},j}(x) = L_j\} = 1/(2 \cdot n!)$ for any $L \in \mathcal{L}$, and SAM's transcript equals any given T with probability $(1/(2 \cdot n!))^n$. On the other hand, PAULA and VICTOR choose $\pi \xleftarrow{\text{☑}} \text{Sym}_n$ and $i \xleftarrow{\text{☑}} \{0,1\}$ uniformly at random in each round and compute $H = \pi(G_1)$ as well as

$$\rho = \begin{cases} \pi, & \text{if } i = 0, \\ \pi \circ \sigma, & \text{if } i = 1. \end{cases}$$

PAULA's isomorphism σ is fixed. Since $\pi \in \text{Sym}_n$ is uniformly random, so is $\pi \circ \sigma$. Therefore we get for any line $L \in \mathcal{L}$ that

$$\text{prob}\{\text{PAULA and VICTOR produce } L\} = \frac{1}{2 \cdot n!}.$$

Thus the individual lines of SAM and of PAULA/VICTOR are identically distributed. Since the different rounds are independent, we have

$$\text{prob}\{T_{\text{PAULA,VICTOR}}(x) = T\} = \left(\frac{1}{2 \cdot n!}\right)^n$$

for any transcript T. Thus SAM never fails and the two distributions on transcripts are identical. $\qquad\square$

It is important to note that although SAM (or VICTOR, who is also a random polynomial-time machine) can produce transcripts as well as PAULA and VICTOR can, he cannot play PAULA's part in an execution of the protocol. This is because of the sequence of messages. After giving away H, PAULA has to answer to any challenge i. SAM can do this for the value of i that he chose, but not for the other value, assuming that graph isomorphisms cannot be found in random polynomial time.

We recall that for completeness, we assumed PAULA and VICTOR to follow the protocol, while for soundness, we had to allow a cheater P for PAULA who may do anything that she wants. For zero knowledge, we also consider a cheater V instead of VICTOR, who may or may not follow the protocol. Can V gain any knowledge from the exchange, or can his transcripts also be simulated?

For example, V could always choose $i = 1$, and then for any transcript T we would have

$$\text{prob}\{T_{\text{PAULA, V}}(G_0, G_1) = T\} = \begin{cases} \left(\frac{1}{n!}\right)^n, & \text{if all } i \text{ in } T \text{ equal } 1, \\ 0, & \text{otherwise.} \end{cases}$$

This is quite different from $\text{prob}\{T_{\text{SAM}}(G_0, G_1) = T\}$. Can such a V learn anything from the execution of the protocol? We have the task to construct a simulator SAM whose transcripts look like those of PAULA and V. Since V is quite arbitrary, this is only possible if SAM has sufficient information about V. Namely, we assume that SAM can run V's algorithm step by step.

DEFINITION 12.11. *An interactive proof system* PAULA/VICTOR *has perfect zero knowledge if for every* V *instead of* VICTOR *there exists a simulator* SAM *who generates transcripts, and for each* $x \in X$ *and transcript* T *we have*

$$\text{prob}\{T_{\text{PAULA,V}}(x) = T\} = \text{prob}\{T_{\text{SAM}}(x) = T \colon \text{not ``failure''}\}.$$

SAM *is allowed to run* V *step by step and to "restart"* V *whenever required.*

If an interactive proof system has perfect zero knowledge, then it has perfect zero knowledge for VICTOR, since the definition includes the special case V = VICTOR. For the graph isomorphism system 12.5, the following simulator works.

ALGORITHM 12.12. Simulator for V instead of VICTOR.

Input: VICTOR's input.
Output: A transcript T.

1. Initialize T with (G_0, G_1).
2. Execute n rounds of steps 3 through 7.
3.   Repeat steps 4–5
4.   Generate random j and ρ, $H = \rho(G_j)$.
5.   Run V with first message H to generate i.
6.   Until $i = j$
7.   Append (H, i, ρ) to the transcript T.
8. Append "accept" to T, and output T.

THEOREM 12.13. *The interactive proof system 12.5 for graph isomorphism has perfect zero knowledge.*

PROOF. Let V play VICTOR's part in the protocol, possibly cheating. We consider the simulator SAM of Algorithm 12.12 which, of course, calls V as a subroutine.

SAM generates transcripts line by line. Each line is a triple $L = (H, j, \rho) \in \mathcal{L}$. It is *valid* if $H = \rho(G_j)$. For a new line, SAM generates a random valid line (H, j, ρ) (namely, random j and ρ, and $H = \rho(G_j)$), runs V on the previous part of the transcript plus H, and observes the challenge i that V produces. If $i = j$, SAM appends the line to the transcript. If $i \neq j$, then the line is discarded and V "reset" to the previous line, that is, the values of all of V's variables are set to the old state, after the previous line. The expected number of choices that SAM has to make for each line is 2, since $i = j$ is satisfied with probability $1/2$.

Now we prove that $\text{prob}_{\text{PAULA, V}}\{T\} = \text{prob}_{\text{SAM}}\{T\}$ for each transcript T, by induction along the protocol. Namely, for each partial transcript $T = ((G_0, G_1), L_1, \ldots, L_k)$ consisting of k lines, we show that the probabilities under (PAULA, V) and under SAM are identical.

For $k = 0$, the transcript contains only the input (G_0, G_1), and the probabilities are identical. Now we let $k \geq 1$ and fix a partial transcript $T = ((G_0, G_1), L_1, \ldots, L_{k-1})$. It is sufficient to show that a valid line $L = (H, i, \rho)$ is generated as L_k with the same probability by (PAULA, V) and by SAM.

By definition, i and ρ determine H, but H and i do not determine ρ. Namely, if $\tau \in \text{Aut}(G_0)$, then $\rho(G_0) = \rho \circ \tau(G_0)$. We denote by

$$A_H = \{\rho \in \text{Sym}_n \colon \rho(G_0) = H\}$$

the set of isomorphisms from G_0 to H.

For two graphs $H = \tau(G_0)$ and $H' = \tau'(G_0)$, the map $\rho \longmapsto \tau'\tau^{-1}\rho$ is a bijection from A_H to $A_{H'}$. In particular, all these sets have the same cardinality $\#A_H = \# \text{Aut}(G_0)$. For any $\rho, \rho' \in \text{Sym}_n$ we have

$$\rho(G_0) = \rho'(G_0) \iff \rho^{-1}\rho' \in \text{Aut}(G_0).$$

VICTOR chooses $i = 0$ with probability $1/2$, but V is not required to follow the protocol and may have his own strategy for choosing i. For any graph H isomorphic to G_0, let q_H be the probability that V chooses $i = 0$ after seeing the partial transcript T and H. Thus he chooses $i = 1$ with probability $1 - q_H$. A valid line $L = (H, 0, \rho)$ is chosen by (PAULA, V) with probability $q_H/n!$, and $(H, 1, \rho)$ with probability $(1 - q_H)/n!$.

	$i = 0$	$i = 1$
(PAULA, V)	$\dfrac{q_H}{n!}$	$\dfrac{1 - q_H}{n!}$
SAM	$\dfrac{q_H}{2 \cdot n!}$	$\dfrac{1 - q_H}{2 \cdot n!}$

Table 12.2: Probabilities for a valid line (H, i, ρ).

What is SAM's probability for choosing such a line? Since he chooses his $j \xleftarrow{\boxtimes} \{0, 1\}$ uniformly at random and independently of all other choices, we have $\text{prob}\{i = j\} = 1/2$, no matter which value V has for i. Therefore the probability that he does not append any line in $\ell - 1$ attempts is $2^{-(\ell-1)}$. For any particular line $L = (H, j, \rho)$, his probability of choosing it is $1/2 \cdot n!$.

The probability in one attempt that SAM appends a line $(H, 0, \rho)$ is $q_H/(2 \cdot n!)$, namely, $1/n!$ for the choice of ρ (and $H = \rho(G_0)$), and $q_H/2$ for the choice of $i = j = 0$. Similarly, the probability for $(H, 1, \rho)$ is $(1 - q_H)/(2 \cdot n!)$. These probabilities are assembled in Table 12.2. Therefore the probability that $(H, 0, \rho)$ gets written by SAM is

$$\sum_{\ell \geq 1} 2^{-\ell+1} \frac{q_H}{2 \cdot n!} = \frac{q_H}{n!},$$

the same as for (PAULA,V). Similarly, $(H, 1, \rho)$ gets written eventually with probability $(1 - q_H)/n!$ in both scenarios.

By induction, we assume that the partial transcript $T = ((G_0, G_1), L_1, \ldots, L_{k-1})$ is equally likely under both distributions, and we have now shown this for any $((G_0, G_1), L_1, \ldots, L_{k-1}, L_k)$. $\square$

Next we describe a perfect zero knowledge proof system for square-ness. We use the notation from Section 15.14. Thus $N = pq$ is a Blum integer, $\square_N$ and $\boxtimes_N$ in $\mathbb{Z}_N^\times$ are the subgroups of squares and antisquares, respectively, and $J = \square_N \cup \boxtimes_N \subset \mathbb{Z}_N^\times$ is the subgroup of elements a whose Jacobi symbol $\left(\frac{a}{N}\right)$ equals 1. The latter property can be tested in poly-nomial time. Then $\#\square_N = (p-1)(q-1)/4$, $\#J = 2\#\square_N$, and squaring provides two bijections $\square_N \to \square_N$ and $\boxtimes_N \to \square_N$. Input to the following protocol is a square $x \in \square_N$. PAULA wants to prove to VICTOR that she knows some y such that $y^2 = x$, without revealing y to VICTOR.

INTERACTIVE SYSTEM 12.14. Squareness.

Input: A Blum integer N and $x \in \mathbb{Z}_N^\times$.
PAULA's knowledge: $y \in J$ with $y^2 = x$.

1. Execute $n = \lceil \log_2 N \rceil$ rounds of steps 2 through 5.
2.   PAULA chooses $u \xleftarrow{\boxdot} J$. She computes $v = u^2$ and sends v to VICTOR.
3.   VICTOR chooses $i \xleftarrow{\boxdot} \{0,1\}$ and sends this to PAULA.
4.   PAULA computes $z = y^i u$ and sends z to VICTOR. [Then $z \in J$.]
5.   VICTOR checks whether $z^2 = x^i v$ holds.
6. VICTOR returns "accept" if all his checks were satisfactory, else "re-ject".

THEOREM 12.15. *The above is a perfect zero knowledge protocol for squareness modulo N.*

PROOF. We have to show three properties. Completeness is clear, since VICTOR's check will always be satisfactory if PAULA follows the protocol.

For soundness, we assume that x is not a square. In step 2, a cheating PAULA may send a v which is a square, or one which is not. In the first case, xv is not a square and for VICTOR's choice $i = 1$ in step 3, she cannot send a z in step 4 which produces VICTOR's ok in step 5. In the second case, v is not a square and no z produces VICTOR's ok for $i = 0$. For either choice of PAULA's v, VICTOR catches the cheat with probability $1/2$. The probability for this not to happen in n rounds is 2^{-n}. This bounds the success probability for PAULA to cheat VICTOR into believing (incorrectly) that x is a square.

For the perfect zero knowledge property, we have a V in VICTOR's place. The simulator SAM generates transcripts line by line, as in the proof of Theorem 12.13. Each line is a triple $L = (v, i, z) \in \square_N \times \{0,1\} \times J$. It is *valid* if $z^2 = x^i v$. For a new line, SAM takes $j \xleftarrow{\boxdot} \{0,1\}$, $z \xleftarrow{\boxdot} J$, and $v = z^2 x^{-j}$. He runs V on the previous part of the transcript plus v and observes the challenge i that V produces. If $i = j$, SAM appends (v, j, z) to the transcript, otherwise he discards the new line and resets V to the previous line.

Using the notation from the proof of Theorem 12.13, it is sufficient to show that a valid line (v, i, z) is generated after a partial transcript T by (PAULA, V) and by SAM with the same probability.

Now i and z determine $v = z^2 x^{-i}$, but i and v do not determine z, since vx^i has exactly two square roots in J. For any $v \in \square_N$, let q_v be the probability that V chooses $i = 0$ after seeing T and v. Thus he chooses $i = 1$ with probability $1 - q_v$. Any valid line $(v, 0, z)$ is chosen by (PAULA, V) with probability $q_v / \#J$, and any $(v, 1, z)$ with probability $(1 - q_v)/\#J$.

As to SAM's choices, the probability that $i = j = 0$ is $q_v/2$, and $(1 - q_v)/2$ for $i = j = 1$. He has $\#J$ choices for z, and his probability for some $(v, 0, z)$ is $q_v/2\#J$, and $(1 - q_v)/2\#J$ for some $(v, 1, z)$.

These probabilities are like those of Table 12.2, with $n!$ replaced by $\#J$. the rest of the proof of Theorem 12.13 also carries over and shows the perfect zero knowledge property of the Interactive System 12.14.   □

EXAMPLE 12.16. We illustrate the interactive system for the toy example $N = 21$ and refer to Figure 15.3. PAULA wants to prove that $x = 4 \in \mathbb{Z}_{21}^{\times}$ is a square, and has the square root $y = 5 \in \boxtimes_{21}$. In one round, she chooses $u \xleftarrow{\boxtimes} J$ with probability $1/\#J = 1/6$. Suppose her choice is $u = -4$. She sends $v = (-4)^2 = -5 \in \square_{21}$ to VICTOR. He chooses, say, $i = 1$ at random and transmits it. PAULA calculates $z = 5^1 \cdot (-4) = 1$ and sends it to VICTOR. In turn, he checks that $1^2 = 4^1 \cdot (-5)$ and accepts in this round. Some V instead of VICTOR may choose $i = 0$ with some probability, say, $q_v = q_{16} = 3/5$, after seeing a partial transcript T and v. Then prob$\{$V chooses $i = 1\} = 2/5$. The rest of this round, for $i = 1$, is described above. The probability of $(-5, 1, 1)$ occurring for (PAULA, V) is

$$\frac{1}{\#J} \cdot (1 - q_{16}) = \frac{1}{6} \cdot \frac{2}{5} = \frac{1}{15}.$$

The simulator SAM chooses (j, z) uniformly at random, each value, say $(1, 1)$ with probability $1/2\#J = 1/12$, and then sets $v = z^2 x^{-j}$, so that $v = 1^2 \cdot 4^{-1} = -5$ for that choice. SAM watches V's doings and appends his $(-5, 1, 1)$ to T if $i = 1$, which happens with probability $1 - q_{16} = 2/5$ in one attempt of SAM's. He discards his choice if $i \neq j$; this happens $\ell - 1$ times in a row with probability $2^{-\ell+1}$. Therefore SAM eventually appends $(-5, 1, 1)$ to T with probability

$$\frac{1}{12} \cdot \sum_{\ell \geq 1} 2^{-\ell+1} \cdot (1 - q_{16}) = \frac{1}{12} \cdot (1 + 1/2 + 1/4 + \cdots) \cdot \frac{2}{5} = \frac{1}{15},$$

the same as for (PAULA, V).   ◊

12.3. Bit commitment

In various protocols, we have seen the importance of the sequence in time in which computations are done. We now look for a cryptographic implementation of the following situation. PAULA computes a bit $b \in \{0,1\}$ and gives it to VICTOR. She does not want him to read it right away, but only at a later point, of her choosing. So she writes b on a piece of paper, puts it into a strongbox and locks it, and gives the strongbox to VICTOR. Later, at a time she chooses, she gives the key to VICTOR, who can then unlock the strongbox and read b. The point is that while he is in possession of the locked strongbox, PAULA cannot change b.

This translates into the following cryptographic problem. We need a function comm: $\{0,1\} \times X \to Y$, where X (the set of keys) and Y are publicly known finite sets, so that for all $(b,x) \in \{0,1\} \times X$, PAULA can send $y = \text{comm}(b,x)$ to VICTOR with the following properties.

- *Revealing.* VICTOR can determine b efficiently from (x,y). Thus ALICE can send x to VICTOR in order to reveal b. VICTOR checks whether indeed $x \in X$.

- *Perfect binding.* No message $x' \in X$ makes VICTOR determine the wrong bit $b \oplus 1$ from (x', y).

- *Efficiency.* All computations can be carried out in time polynomial in a security parameter n.

- *Computational concealing.* It is hard for VICTOR to determine b from y.

This task is called *perfect bit commitment*. In order to commit a longer word, PAULA can commit each bit independently. We present two examples.

RSA bit commitment. PAULA chooses $N = pq$ and e as in the RSA notation of Figure 2.15, $x_0 \xleftarrow{\boxtimes} \{1, \ldots, (N-1)/2\}$, and $x = 2x_0 + b$ in $\mathbb{Z}_N$ for $b \in \{0,1\} \subseteq \mathbb{Z}_N$, repeating the choice of x_0 in the unlikely case that $\gcd(x, N) \neq 1$. She computes $y = x^e \in \mathbb{Z}_N^\times$ and sends $\text{comm}(b,x) = (N,e,y)$ to VICTOR. To reveal b, she sends (p,q,b,x) to VICTOR. He then verifies that p and q are prime and $N = pq$, computes $\varphi(N)$, and checks that $e \in \mathbb{Z}_{\varphi(N)}^\times$, $x^e = y$, and $x = b$ modulo 2. If one of these does not hold, he rejects the commitment.

We verify the required properties.

○ *Revealing.* PAULA sends (p, q, b, x).

○ *Perfect binding.* VICTOR's checks guarantee that p and q are prime, $N = pq$, and $\gcd(e, \varphi(N)) = 1$. If $x' \in \mathbb{Z}_N^\times$ also satisfies $(x')^e = y$, then $(x'/x)^e = 1$, and this implies that $x'/x = 1$, so that $x' = x$. We use the fact that the map $\mathbb{Z}_N^\times \longrightarrow \mathbb{Z}_N^\times$ sending z to z^e is injective (Corollary 15.62).

○ *Efficiency.* Is clear.

○ *Computational concealing.* The least significant bit of x is a hard core bit for RSA, and determining it is as hard as breaking RSA; see Section 3.5. This qualifies as hard.

Squareness bit commitment. We can also use the presumed difficulty of deciding squareness in $\mathbb{Z}_N^\times$, where $N = pq$ is a Blum integer. We recall from the previous section the sets $\square_N$ and $\boxtimes_N$ in $\mathbb{Z}_N^\times$. The assumption is that, given $x \in \square_N \cup \boxtimes_N$, it is hard to tell whether $x \in \square_N$. PAULA chooses p and q and publishes $N = pq$ and some $z \in \boxtimes_N$, that is, z is a nonsquare modulo p and modulo q. In order to commit a bit b, PAULA chooses $x \xleftarrow{\boxtimes} \square_N \cup \boxtimes_N$ at random and sends $y = \text{comm}(b, x) = z^b x^2 \in \mathbb{Z}_N^\times$ to VICTOR. In order to reveal the bit, she sends b and x to VICTOR, who checks that indeed $y = z^b x^2$. If this fails to hold, he rejects the commitment.

Does this have the four required properties? We have $y \in \square_N \cup \boxtimes_N$ and

$$y \in \square_N \iff \exists u \in \mathbb{Z}_N^\times \quad u^2 = y = z^b x^2$$
$$\iff \exists u \in \mathbb{Z}_N^\times \quad z^b = (x^{-1}u)^2 \iff b = 0.$$

The last equivalence follows from the facts that $(x^{-1}u)^2 \in \square_N$, $1 = z^0 \in \square_N$, $z = z^1 \in \boxtimes_N$, and $\square_N \cap \boxtimes_N = \varnothing$.

○ *Revealing.* After receiving x, it is easy for VICTOR to tell whether $y = x^2$ or $y = zx^2$.

○ *Perfect binding.* After PAULA has sent $y = z^b x^2$, we have to check that $y = z^{1-b} v^2$ is impossible for any $v \in \mathbb{Z}_N^\times$ with which PAULA might try to cheat. But

$$y \in \square_N \iff b = 0 \iff 1 - b = 1 \iff \forall v \in \mathbb{Z}_N^\times \quad z^{1-b} v^2 \in \boxtimes_N,$$

-4 so that always $y \neq z^{1-b} v^2$.

○ *Efficiency.* Is clear.

○ *Computational concealing.* Determining b from y means determining whether $y \in \square_N$, which is hard by assumption.

EXAMPLE 12.17. -4 We take $N = 3 \cdot 7 = 21$, and $z = 5$ in $\mathbb{Z}_{21}^{\times}$. Then $z = 2$ in $\mathbb{Z}_3$ and $z = 5$ in $\mathbb{Z}_7$ is a nonsquare modulo 3 and modulo 7, so that $z \in \boxtimes_{21}$. To commit a bit b, PAULA transmits $y = z^b x^2$ for a random $x \in \square_{21} \cup \boxtimes_{21}$. If $x = -4$ is chosen, then $y = \mathrm{comm}(0, -4) = 5^0 \cdot (-4)^2 = -5 \in \square_{21}$, and $y = \mathrm{comm}(1, -4) = 5^1 \cdot (-4)^2 = -4 \in \boxtimes_{21}$; see Figure 15.4.

In order to encode a pair $b = b_1 b_2 \in \{0,1\}^2$ of bits, we work with pairs $a = (a_1, a_2) \in (\mathbb{Z}_{21}^{\times})^2$. Besides $N = 21$, also $(z_1, z_2) = (5, -1) \in (\boxtimes_{21})^2$ is public (see Figure 15.4). In order to commit $b = (1, 0)$, say, PAULA chooses some $x \in (\square_{21} \cup \boxtimes_{21})^2$, say $x = (-4, 4)$, and sends $y = \mathrm{comm}((1,0), -4, 4) = (5^1 \cdot (-4)^2, (-1)^0 \cdot 4^2) = (-4, -5)$. To reveal b, PAULA sends $((1,0), -4, 4)$, and VICTOR checks that $y_1 = -4 = 5^1 \cdot (-4)^2 = z_1^1 x_1^2$ and $y_2 = -5 = z_2^0 x_2^2$. More such commitments are in Example 12.19. ◇

We may relax the binding condition slightly.

Computational binding. It is hard for PAULA to compute y and x' which make VICTOR determine the wrong bit $1 - b$.

One can devise several *computational bit commitment* protocols that satisfy the first two plus this new requirement, for example, using hash functions.

12.4. Computational zero knowledge

It would be nice to have perfect zero knowledge proofs of arbitrary problems in NP. Assuming standard conjectures in computational complexity, this is not possible. But a natural relaxation of our requirements achieves this goal. Instead of asking SAM to produce a transcript distribution equal to that of (PAULA,V), we only require the two to be polynomially indistinguishable.

We start with an example, the 3-colorability of graphs. As usual, $G = (V, E)$ is an undirected graph with vertex set V and edge set $E \subseteq \binom{V}{2}$ consisting of unordered pairs of vertices. A *k-coloring* of G is a map $c \colon V \longrightarrow \{1, \ldots, k\}$ so that $c_i \neq c_j$ for all edges $\{i, j\} \in E$, where we write $c_i = c(i)$. That is, two vertices connected by an edge receive different colors. If such a k-coloring exists, then G is *k-colorable*. We set

$$k\text{-COLOR} = \{G \colon G \text{ is } k\text{-colorable}\}.$$

Then 2-COLOR $\in$ P, and k-COLOR is NP-complete for any $k \geq 3$.

We now exhibit a zero knowledge protocol for 3-COLOR. The three colors are encoded as 00, 01, and 10. We employ a two-bit commitment scheme with which PAULA can commit two bits $a = a_1 a_2 \in \{0,1\}^2$ by choosing a random $x \in X$ in a set X and sending $y = \text{comm}(a, x)$ to VICTOR. She can reveal a later by sending x. This commitment scheme is assumed to have the three properties discussed in Section 12.3.

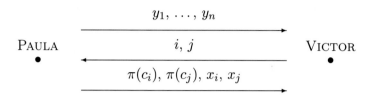

$$y_1, \ldots, y_n$$

PAULA ● i, j VICTOR ●

$$\pi(c_i), \pi(c_j), x_i, x_j$$

Figure 12.3: One round of the 3-COLOR proof system.

INTERACTIVE PROOF 12.18. 3-COLOR.

Input: A graph $G = (V, E)$ with vertex set $V = \{1, \ldots, n\}$ and $m = \#E$ edges.

PAULA's knowledge: a 3-coloring $c \colon V \longrightarrow \{00, 01, 10\} = C$ of G. We write $c_i \in C$ for the color of $i \in V$.

1. Execute mn rounds of steps 2 through 5.
2. PAULA chooses a random permutation $\pi \xleftarrow{\boxdot} \text{Sym}_3$ of C and n random keys $x_1, \ldots, x_n \xleftarrow{\boxdot} X$, and computes $y_i \leftarrow \text{comm}(\pi(c_i), x_i)$ for $1 \leq i \leq n$. She sends $(y_1, \ldots, y_n)$ to VICTOR.
3. VICTOR chooses a random edge $\{i, j\} \xleftarrow{\boxdot} E$ and sends it to PAULA.
4. PAULA sends $(\pi(c_i), \pi(c_j), x_i, x_j)$ to VICTOR.
5. VICTOR checks that $\pi(c_i)$ and $\pi(c_j)$ are the committed two-bit strings, that both are not 11, and that $\pi(c_i) \neq \pi(c_j)$.
6. VICTOR returns "accept" if all his checks are satisfactory, else "reject".

EXAMPLE 12.19. PAULA has the following 3-coloring of G:

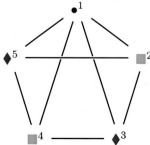

Thus $n = 5$, $m = 8$,

$$V = \{1, 2, 3, 4, 5\},$$
$$E = \{\{1, 2\}, \{1, 3\}, \{1, 4\}, \{1, 5\}, \{2, 3\}, \{2, 5\}, \{3, 4\}, \{4, 5\}\},$$

and c is given by

i	1	2	3	4	5
c_i	00	01	10	01	10

We use the two-bit commitment from Example 12.17, with public $z = (5, -1)$. In one round of the protocol, PAULA might choose the permutation $\pi(00) = 01$, $\pi(01) = 00$, $\pi(10) = 10$ of C, and random $x_1 = (1, -4), x_2 = (5, 4), x_3 = (-4, -1)$, $x_4 = (5, -5)$, $x_5 = (4, 5)$ in $X = (\square_{21} \cup \boxtimes_{21})^2$. See Figure 15.4. Since $\pi(c_1) = 01$, we have $y_1 = (z_1^0 \cdot x_{11}^2, z_2^1 \cdot x_{12}^2) = (1^2, (-1) \cdot (-4)^2) = (1, 5)$. PAULA computes in this way

$$(y_1, y_2, y_3, y_4, y_5) = ((1, 5), (4, -5), (-4, 1), (4, 4), (-4, 4))$$

and sends this to VICTOR. If he challenges her with the edge $\{i, j\} = \{2, 3\}$, PAULA returns $(00, 10, (5, 4), (-4, -1))$, and VICTOR's check is satisfactory. ◊

We do not know whether this system has perfect zero knowledge. But it satisfies the following weaker requirement.

DEFINITION 12.20. *A proof system has the* computational zero knowledge *property if for every* V *instead of* VICTOR *there exists a random polynomial-time simulator* SAM *such that the distributions on transcripts generated by* PAULA / V *and by* SAM *are polynomially indistinguishable.*

We recall that a polynomial distinguisher between two probability distributions X and Y is a random polynomial-time machine $\mathcal{A}$ with nonnegligible distinguishing quality between X and Y, that is, for some $k \geq 1$ and sufficiently large n, we have

$$\text{adv}_{\mathcal{A}}(X_n, Y_n) = |\mathcal{E}_{\mathcal{A}}(X_n) - \mathcal{E}_{\mathcal{A}}(Y_n)| \geq \frac{1}{n^k},$$

where X_n and Y_n are X and Y, respectively, restricted to inputs of length n. See Definition 9.17 and (15.83). Definition 12.20 stipulates that such an $\mathcal{A}$ should not exist.

THEOREM 12.21. *The 3-Color Interactive Proof 12.18 has the computational zero knowledge property.*

PROOF.   We first verify completeness and soundness of the proof system. Clearly, if $G \in$ 3-COLOR, then PAULA can answer all of VICTOR's challenges correctly, and VICTOR will end up being convinced that indeed $G \in$ 3-COLOR.

Now assume that $G \notin$ 3-COLOR, and that some cheating P instead of PAULA tries to convince VICTOR of the contrary. For every $c \colon V \longrightarrow \{00, 01, 10\}$, there is at least one edge $\{i, j\} \in E$ with $c_i = c_j$. In each round, VICTOR chooses such an edge with probability at least $1/m$. Then P will not be able to send any (c_i', c_j', x_i', x_j') which is consistent with her previously sent y_i and y_j and which VICTOR will check to be correct. Here we use the properties of the underlying commitment scheme. Since VICTOR's choices are independent (although P's need not be), the probability of being cheated in mn rounds is at most

$$\left(1 - \frac{1}{m}\right)^{mn} \leq e^{-n},$$

which is exponentially small. This proves soundness.

For computational zero knowledge, we construct an appropriate simulator SAM. We only do this for the case where an honest VICTOR follows the rules of the protocol. SAM builds the transcript $(G, L_1, \ldots, L_{mn})$ line by line. Any line $L = (y_1, \ldots, y_n, i, j, \pi(c_i), \pi(c_j), x_i, x_j)$ consists of the data transmitted, as in Figure 12.3.

For a new line L, SAM chooses an edge $\{i, j\} \xleftarrow{\;\boxtimes\;} E$ at random, two distinct random colors $d_i, d_j \xleftarrow{\;\boxtimes\;} \{00, 01, 10\}$, and $x_1, \ldots, x_n \xleftarrow{\;\boxtimes\;} X$. He computes for $1 \leq k \leq n$

$$z_k = \begin{cases} \operatorname{comm}(00, x_k), & \text{if } k \neq i, j, \\ \operatorname{comm}(d_i, x_i), & \text{if } k = i, \\ \operatorname{comm}(d_j, x_j), & \text{if } k = j, \end{cases}$$

and appends

$$L = (z_1, \ldots, z_n, i, j, d_i, d_j, x_i, x_j)$$

to be previous partial transcript.

Certainly SAM's transcripts are different from those generated in the protocol, since most $z_1, \ldots, z_n$ are commitments of 00, while PAULA's $y_1, \ldots, y_n$ will commit the three colors in roughly equal proportion.

The following transcript parts are chosen with the same distribution in the protocol and in SAM's simulation: $i, j, d_i, d_j, x_i, x_j, z_i, z_j$. The only difference is in the z_k with $k \neq i, j$. But any algorithm which distinguishes between SAM's z_k and PAULA's $y_k = \text{comm}(\pi(c_k), x_k)$ can be used to distinguish between $\text{comm}(00, x)$ and $\text{comm}(d, x')$ for random $x, x' \in X$ and $d \in \{01, 10\}$. This would contradict the concealing property of comm, according to which one cannot achieve such a distinction in random polynomial time.   □

One can also show that the computational zero knowledge property holds when VICTOR is replaced by a potentially cheating V.

We have seen only a few examples of zero knowledge proofs in this text. The following important result of Goldreich, Micali & Wigderson (1991) shows that the concept is widely applicable, in particular to the problems we usually encounter in cryptology.

THEOREM 12.22. *Any problem in* NP *has a computational zero knowledge proof system.*

Notes 12.1. See Köbler *et al.* (1993) for the complexity of graph isomorphism.

The acceptance probability $2/3$ in Definition 12.3 could be replaced by any constant c with $1/2 < c < 1$ for the completeness property, and similarly for soundness (see Exercise 12.2). In the literature one even finds the requirement $c = 1$, for example in Stinson (1995); it is not clear whether this yields the same notion. In general, one may allow VICTOR more computing power than just random polynomial time. We choose this restriction because all our examples satisfy it.

The catchy term "zero knowledge", coined by Goldwasser *et al.* (1989), is a felicitous choice. Adverbial use in "this protocol is zero knowledge" is linguistically a bit heavy-handed, but has become common.

12.2. In the proof of Theorem 12.13, $A_{\rho(G_0)}$ is the left coset $\rho \cdot \text{Aut}(G_0)$ of $\text{Aut}(G_0)$ in Sym_n. In general, $\text{Aut}(G_0)$ is not a normal subgroup. With G_0 and σ as in Example 12.1 and $\tau = (2, 4) \in \text{Sym}_4$, we have $\tau^{-1}\sigma\tau \notin \text{Aut}(G_0)$.

12.3. The RSA bit commitment was suggested by Birgitta Fricke, and the use of hash functions for computational bit commitments (Exercise 12.9) by Alfred Junklewitz and Dirk Held, all three students in a crypto course in 1999.

12.4. Fortnow (1987) and Aiello & Hastad (1987) show that NP-complete problems do not have perfect zero knowledge protocols unless the polynomial hierarchy collapses to the second level. The latter is considered unlikely. The famous 4-color theorem, conjectured in 1852, states that every planar graph G (which can be drawn in the plane without intersecting edges) is 4-colorable.It was proven by Appel & Haken (1977). For the complexity of graph coloring, see Garey & Johnson (1979), problem GT4. The zero knowledge proof system for 3-COLOR is from Goldreich *et al.* (1991). More information about interactive proofs and their subtleties can be found in Goldreich (2001).

Exercises.

EXERCISE 12.1 (NP $\subseteq$ IP). *Let X be a problem in NP. Using a commitment scheme, design an interactive proof system for X with computational zero knowledge. State precisely your assumptions on the commitment scheme.*

EXERCISE 12.2 (Accept/reject probabilities.). *Let $0 < \varepsilon < 1/2$.*

(i) *In Definition 12.3, replace the acceptance probability 2/3 for completeness by $1/2 + \varepsilon$. Show that this does not change IP.*

(ii) *Similarly, replace 1/3 for soundness by ε, and show that this does not change IP.*

EXERCISE 12.3 (Perfect zero knowledge). *Quisquater et al. (1989) explain zero knowledge to kids. Translate this story into a protocol in the style of the present section. Does it have perfect zero knowledge for VICTOR? For V? Either prove that it does or argue why this is not likely or not the case.*

EXERCISE 12.4 (RSA and zero knowledge). PAULA *has secret and public RSA keys (N, d) and (N, e), respectively. With the following perfect zero knowledge proof system, she can convince VICTOR that e is invertible modulo $\varphi(N)$.*

INTERACTIVE PROOF SYSTEM. INVERTIBILITY OF THE PUBLIC RSA EXPONENT.

Input: A number $N \in \mathbb{N}$ and an exponent $e \in \{0, \dots, N-2\}$.
PAULA's knowledge: $d \in \{0, \dots, N-2\}$ with $ed = 1$ in $\mathbb{Z}_{\phi(N)}$.

1. Repeat $\lceil \log_2 N \rceil$ times steps 2 through 6.
2.   VICTOR chooses $y \xleftarrow{\boxdot} \mathbb{Z}_N^\times$ and sends it to PAULA.
3.   PAULA chooses $x \xleftarrow{\boxdot} \mathbb{Z}_N^\times$ and sends it to VICTOR.
4.   VICTOR chooses $i \xleftarrow{\boxdot} \{0, 1\}$ and sends it to PAULA.
5.   PAULA computes $z \leftarrow (xy^i)^d$ in $\mathbb{Z}_N^\times$ and sends it to VICTOR.
6.   VICTOR verifies $z^e = xy^i$ in $\mathbb{Z}_N^\times$.
7. VICTOR accepts if and only if each verification was successful.

(i) *Show that this interactive proof system is complete.*

(ii) *Show that the system is sound. See Section 15.13 for some properties of exponentiation in finite groups.*

(iii) *A possible protocol line is some (y, x, i, z) with $x, y, z \in \mathbb{Z}_N^\times$, $i \in \{0, 1\}$, and $z^e = xy^i$ in $\mathbb{Z}_N^\times$. Determine the number of possible protocol lines and prove that for a given input (N, e), all lines occur with the same probability if PAULA and VICTOR follow the protocol.*

(iv) *Describe a simulator SAM which generates uniformly distributed possible protocol lines without knowing the inverse of e in $\mathbb{Z}_{\varphi(N)}$. Show that the proof system has perfect zero knowledge for VICTOR.*

(v) *Does the system have perfect zero knowledge?*

(vi) *Have you used anywhere the property that N is a product of two distinct primes, as in RSA? Does the system have the desired properties for any composite N? For any N?*

EXERCISE 12.5 (Nonsquares and zero knowledge). *We consider the following interactive proof system for the complement of the Squareness Interactive System 12.14.*

INTERACTIVE PROOF SYSTEM. NONSQUARES.

Input: An integer $N \in \mathbb{N}$ with $N = p \cdot q$ for some (unknown) prime numbers p and q
      and an $x \in \mathbb{Z}_N^\times$.
PAULA's information: p and q. [She can test in polynomial time whether a given
      element of $\mathbb{Z}_N^\times$ is a square; see Theorem 15.68 (iv) and (15.70).]

1. Repeat $\lceil \log_2 N \rceil$ times steps 2–4.
2.   VICTOR chooses $y \xleftarrow{\boxdot} \mathbb{Z}_N^\times$ and $i \xleftarrow{\boxdot} \{0, 1\}$ randomly, computes $z \leftarrow x^i y^2$
       in $\mathbb{Z}_N^\times$, and sends z to PAULA.
3.   PAULA tests whether z is a square in $\mathbb{Z}_N^\times$. If so, she sets $j \leftarrow 0$ else $j \leftarrow 1$. She
       sends j to VICTOR.
4.   VICTOR verifies whether $i = j$.
5. VICTOR accepts PAULA's proof if and only if his verification was true in each round.

   (i) *Show that this interactive proof system for $\mathbb{Z}_N^\times \setminus \square_N$ is complete and sound.*

  (ii) *Explain whether this protocol has perfect zero knowledge or not.*

 (iii) *What can you say about the following statement: if a problem has an interactive proof system, then so does its complement?*

EXERCISE 12.6 (RSA bit commitment).

   (i) *Check that the RSA bit commitment from Section 12.3 has the required properties. You may assume that in the RSA notation of Figure 2.15, it is hard to compute the least significant bit of x from N, e, and y.*

  (ii) *What goes wrong when ALICE sends only (N, y) in the first message, and (b, d, e, x) in the second message?*

 (iii) *The most expensive computation is to find p and q. Can ALICE use the same N several times?*

EXERCISE 12.7 (RSA bit commitment example). PAULA *sends the following RSA commitment to* VICTOR, *as in Exercise 12.6.*

$$N = 2699470789124172845359624063455394203394843863321 8809,$$

$$e = 14478952961650663963011492799422589920688060814917363,$$

$$y = 11597036552209975536364589730166726242713932167909764.$$

Later VICTOR *gets the key* $p = \text{nextprime}(2^{87})$ *and* $q = \text{nextprime}(3^{55})$ *and*

$$x = 89891409740203780725914344721835089405157340576788 42.$$

Which bit did PAULA *commit to?*

EXERCISE 12.8 (Discrete logarithm bit commitment). *We have a cyclic group $G = \langle g \rangle$ of order d, in which we assume discrete logarithms to be hard, and consider the following bit commitment scheme. Let $m = \lfloor d/2 \rfloor$ and $M = \{0, \ldots, m\} \subseteq \mathbb{Z}_d$. PAULA hides the bit 0 by sending some g^a with $a \in M$ and the bit 1 by g^a with $a \in \mathbb{Z}_d \setminus M$. More formally, we set $X = \{g^a : a \in M\}$, $Y = G$, and comm: $\{0, 1\} \times X \to Y$ with comm$(b, x) = x \cdot g^{mb}$.*

   (i) *Show that this scheme is perfectly binding.*

   (ii) *For the leading bit problem of the discrete logarithm, input is some $x \in G$, and the output is 0 if $\mathrm{dlog}_g x \in M$ and 1 otherwise. Reduce DL_G to the leading bit problem by binary search.*

  (iii) *Prove that the commitment scheme has the computational concealing property.*

EXERCISE 12.9 (Bit commitment via hash functions). *Let $n \in \mathbb{N}$ and $h \colon \{0, 1\}^{2n} \to \{0, 1\}^n$ be a hash function.*

  (a) *First bit commitment: in an element $(p, q, e, d, z) \in \{3, \ldots, 2^n\}^5 = X$, the first four entries p, q, e, and d are as in the RSA notation of Figure 2.15, $N = pq$, and $z < N/2$. We denote as z_2 in $\{0, 1\}^{2n}$ the binary representation of $(z^2$ in $\mathbb{Z}_N)$, padded with leading zeroes. To commit the bit $b \in \{0, 1\}$, we calculate*

$$y = \mathrm{comm}(b, (p, q, e, d, z)) = (N, (2z + b)^e \in \mathbb{Z}_N, h(z_2)).$$

  (b) *Second bit commitment: to commit the bit $b \in \{0, 1\}$ with a key $x \in \{0, 1\}^{2n-1} = X$, we compute*

$$y = \mathrm{comm}(b, x) = (x_1, \ldots, x_n, h(b, x)).$$

What is the cost of creating and verifying a commitment according to these schemes? Choose one of these commitment schemes and state the properties of the hash function (see Section 7.1) so that the commitment scheme

   (i) *conceals. [It is impossible to compute b.]*

   (ii) *Computationally conceals. [It is impossible to compute b in polynomial time.]*

  (iii) *Can be revealed.*

  (iv) *Is binding. [The bit b is uniquely determined by y.]*

   (v) *Is computationally binding. [Given y, it is hard to compute a key for $1 - b$.]*

  (vi) *Conceals. [It is impossible to compute b.]*

 (vii) *Computationally conceals. [It is impossible to compute b in polynomial time.]*

Justify your answers in detail. Some of the five properties may be unattainable.

EXERCISE 12.10. *The 3-COLOR proof system of Section 12.4 uses a bit commitment scheme that is computationally concealing and binding.*

   (i) *Show that this system does not provide perfect zero knowledge for VICTOR.*

A bit commitment scheme is perfectly concealing if the distribution of $\{\mathrm{comm}(b, x) : x \xleftarrow{\boxtimes} X\}$ is independent of $b \in \{0, 1\}$.

   (i) *Consider the 3-COLOR proof system with a perfectly concealing and perfectly binding bit commitment. Prove that this provides perfect zero knowledge for VICTOR.*

Au défaut de la force, il faut employer la ruse.[1]
CHRISTOPHER LAYER (1722)

It was the first time that I saw a man hanged,
drawn and quartered, which is a beasty business.
PHILIP KERR (2002)

The King hath note of all that they intend,
By interception which they dream not of.
[...] Why, how now, Gentlemen?
What see you in those papers that you lose
So much complexion?
WILLIAM SHAKESPEARE (1599)

"I was confident," Hyde, the friend of the Stuarts,
wrote in 1659, "that the devil himself
cannot decypher a letter that is well written."
JAMES WESTFALL THOMPSON AND SAUL K. PADOVER (1937)

Für den heutigen Bearbeiter von noch nicht entzifferten Akten
der vergangenen Zeit ist es leichter, wenn die
Sicherheitsempfehlungen damals nicht befolgt worden sind.[2]
ERICH HÜTTENHAIN (1974)

Ob euch schon vorgemeldte Stuck zum thail dunckel,
verborgen, oder vnuollkommen geduncken, so last euch doch
solliches nicht abschrecken, in Betrachtung des Spruchs Christi:
Daß das Perlin nit für die Saw geworffen soll werden.
Oder daß der ihenige so Gehaimnussen beschreibt
vn manigklichen offenbaret, gleich ist ainer Hennen,
die ain Ay legt vnnd frißt es wider.[3]
SAMUEL ZIMMERMAN (1579)

---

[1]Lacking force, you have to use deception.

[2]For someone working on as yet undecrypted documents from the past it is easier if the security instructions were not followed at the time.

[3]If the previous pieces appear to you partly dark, hidden, or imperfect, then do not be intimidated by this, in view of Christ's saying: do not cast pearls before swine. Or that he who discloses secrets and explains them to the multitude is like a hen that lays an egg and eats it right away.

Chapter H

People using cryptography

tars and starlets, luminaries and dignitaries from all scenes have tried their hand at cryptography, or somehow used it for their purposes. This chapter tells some anecdotes about nonprofessional users of cryptography. Several of them are well-known for their exploits in some other field.

H.1. Interception, Black Cabinets, and Martin Luther

Messages over longer distances were always prone to interception. We can distinguish four periods. From antiquity to the 18th century, the messengers were captured, their letters taken and read, and deciphered if written in a cipher that had been broken. In the second period, post offices, after humble beginnings in the 16th century, carried the bulk of the mail until late last century. Several governments installed *Black Cabinets* where mail was routinely opened, copied, and closed again so that the recipient would not even realize the interference. These Black Cabinets usually housed the country's leading cryptanalysts who got a lot of exercise from the encrypted letters, mainly diplomatic ciphers.

In the third period, from the mid-19th to the late 20th century, telegram cables and radio carried a lot of traffic. These were open for many (with cables) or all (with radio) to listen in, and confidential messages had to be encrypted. Two famous encryption failures, the Zimmermann telegram and the ENIGMA, are discussed in Chapters I and J.

The fourth period of communication started recently and relies on large networks, mainly the internet, still using wires or radio for node-to-node connections. Many of the systems are quite insecure and extensively monitored, and often users care little about their privacy. However, the development of modern cryptography as presented in this book has largely been driven by the need for security in those situations where it really matters. The reader will be well aware of the convenience and the pitfalls of internet banking and ATMs, and also about the massive interception, storage, and evaluation of data by secret services.

Going back to the second period, the Black Cabinets were particularly important in Austria, France, and England over the centuries. Only rarely did the public or the government object. One of two notable exceptions is *"Gentlemen do not read each other's mail"*, with which US Secretary of State Henry Lewis Stimson (1867–1950) closed the US Black Cabinet in 1926, which had been directed by Herbert Osborne Yardley (1869–1958).

The second exception is a ruckus created by an irascible and vociferous German monk in 1529. Indeed, Martin Luther (1483–1546) was in conflict with Duke George of Saxony (1471–1539), who was using in an anti-protestant campaign a letter sent by Luther to the reformer Wenzeslaus Linck (1483–1547) at Nürnberg, intercepted and opened at the Duke's chancellery. Luther published in 1529 a polemic exegesis of Psalm 7 about innocence persecuted and its enemy who fell into a hole that he had dug himself. He likens letter stealing to money stealing: *"Ja brieffe sind nicht güter? Lieber, Wie wenn es sich begebe, das mir odder dir an eim brieffe mehr denn an tausent gulden gelegen were? Solt nicht solcher brieff so werd vnd lieb sein als tausent gulden? Dieb ist ein dieb, er sey gelt dieb odder brieffe dieb ... Ey danck habt, lieber Herr."*[1] Today, people do not complain this loudly about their emails being read.

H.2. Hernán Cortés

Hernán Cortés (1485–1547), the Spanish *conquistador* of Aztec Mexico,

---

[1] Well, letters are not goods? My dear, suppose it happened that I or you valued some letter more than a thousand guilders. Should not this letter be worth and dear just as a thousand guilders? A thief is a thief, be he a money thief or a letter thief ... yeah, thank you, my Lord.

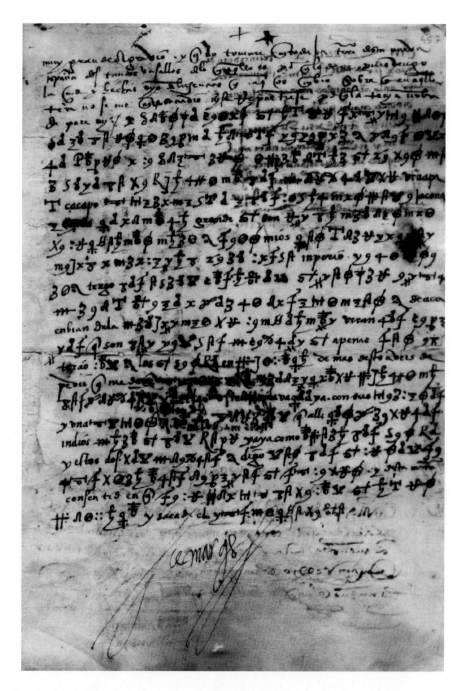

Figure H.1: The fourth page of Hernán Cortés' letter from 25 June 1532.

wrote to his representative Francisco Núñez in Spain on 25 June 1532 from Cuernavaca in Mexico. This letter deals with Cortés' financial arrangements—real estate deals involving whole provinces. Most of the text on its last four pages is encrypted, and was presumably decrypted at the time. However, the *Archivo de Indias* at Sevilla conserves only the original, which resisted several attempts at deciphering. It may well be the first documented cryptography in America.

The *Museo Nacional* in Mexico published in 1924 in the leading local newspapers a competition to decipher it—an early forerunner of today's crypto challenges. The prolific writer Francisco de Asís Monterde García Icazbalceta (1925) succeeded, and was presented the prize of 200 pesos in November 1925. The cipher had $1, 2,$ or 3 symbols (49 in total) for each of the 24 letters, plus *ll* and *que*. Monterde solved it by frequency analysis, aided by the separation of words. He explains how difficult the first inroads were, and how easy the rest: *"El resto fue sencillo: pude leer la carta con facilidad, reemplazando todos los signos por letras; pero había tenido que emplear varias horas de la noche, durante tres meses, para llegar a ese resultado."*[2]

Figure H.1 shows the last of Cortés' four pages. The first lines of the ciphertext, beginning on line 5, read:

> [...] *Y quando veais q ay se siente quán-*
> *to yo lo encarezco, aueis de dezir q pues as´*
> *lo hazen y me quiren agrauiar q me den po-*
> *r todo lo de Guaxaca los pueblos de* Vruapa
> *y* Cacapo *e Tiripitío y los Matalcingos e* Jacona
> *y* [word crossed out] *Coyuca la* grande *q son en la c [p] rouincia*
> *de Mechoacán c [p]ara q sean* mios *con yuredic[i]on*
> *ceuil y criminal mero misto* inperio *de la [ge]ne-*
> *ral* tengo *los otros vasollos q nonbré en el*
> *preuillejo y por la uisitación q de aca*
> enbian dela *prouincia de Mechoacán* veran *los vezi-*
> nos *q son [son] destos pveblos q* apenas *son el*
> tercio *más q* los *q tengo en* Guaxaca. [...]

> > El Marques.[3]

---

[2]The rest was straightforward: I read the letter with ease, replacing each symbol by a letter; but I had had to spend several hours each night, during three months, to arrive at that result.

[3]And when you [Núñez] see that they [the Spanish court] realize how much I desire it, then you have to say that they should do it that way, and if they want to please me, they should give me for the whole of Oaxaca the villages of Uruapa and Cacapo, Tirip-itío, the Matalcingas, Jacona and [word crossed out] greater Coyuca which are in the province of Michoacán. They should be mine with full powers in civil and criminal ju-risdiction. I have the other vassals that I listed in the privilege. The inspection team that they will send from over there to the province of Michoacán will see that the inhabitants

Only the words in slanted type are encrypted in the original.

H.3. Margaret Kennedy

The stories of this and the next section play in the England of religious upheavals and clashes within the kingdom and Scotland. King Henry VIII (1491–1547) had separated the Church of England from Rome, because the Pope would not grant him a divorce from Catherine of Aragon (1485–1536). For two centuries, the *papists* fought to get back into power. His daughter Queen Mary I of England (1516–1558) from the first marriage was wed to King Philip II of Spain (see Section D.3 for one of his ciphers) and briefly reintroduced catholicism. This required some resolute politics which earned her the name of *Bloody Mary*. Her half-sister Elizabeth I (1533–1603), Queen of England, daughter of Henry VIII and Anne Boleyn (1501–1536), reverted to the Church of England. Mary Stuart, Queen of Scots (1542–1587), aspired to the throne, was put under house arrest and finally sentenced to death; cryptanalysis of her ciphers played a major role in the justification of the verdict.

Mary Stuart's son James I (1566–1625), King of England, and his son Charles I (1600–1649) ruled next. They had a fair amount of trouble with Catholics, Scots, Guy Fawkes (1570–1606) and his pyrotechnic plans, and the Parliament, which under Oliver Cromwell's (1599–1658) leadership took the unfriendly step of beheading the King. The Republic following the regicide only lasted a decade, and Charles II (1630–1685) of the House of Stuart became King in 1660 and strengthened the Anglican Church. The religious and Scottish troubles continued, and Lady Kennedy's letter in Figures H.2 and H.3 is a witness from these times.

The next English King, James II (1633–1701), was a Catholic and promoted the *popish* cause substantially. He was chased out of the country in an unbloody revolution. Eventually England, since 1707 the United Kingdom including Scotland, ran out of Royal blood and the throne passed in 1714 to the Elector of Hanover, as George I (1660–1727).

This German King of England never became popular—he did not speak English (no TOEFL test was required for Royal jobs and he did not bother to learn the language), went home regularly to Hanover for many months of holidays, and—worst of all—kept ugly German mistresses. Neither this step-king nor his Prime Minister, Sir Robert Walpole (1676–1745), spoke Latin well—but it was their only common language, and thus state affairs were conducted in broken Latin for over a decade. The historian William Cobbett (1811) describes this:

of those villages are hardly a third more than those that I own in Oaxaca.

As the king could not readily speak English, nor sir Robert Walpole French, the minister was obliged to deliver his sentiments in Latin; and as neither could converse in that language with readiness and propriety, Walpole was frequently heard to say, that during the reign of the first George, he governed the kingdom by means of bad Latin.

FOR THE EARLE OF LAWDERDALL, SOLE SECRETARY OF STAT

FOR SCOTLAND.

Hamilton, 2 of Febru^{ry}·

M. D. L.

I cannot say much more now nor I did yesterday, tho I doe not so undervalue my friends being railed at as you seeme to doe ; yet if that were all, I should care the less, but really : $^{dis}_{132}$: $^{con}_{102}$: $^{ten}_{388}$: $^{n}_{60}$: $^{t}_{16}$: is more universall nor you can imagine, in so much as almost you shall speake to none of no quality : $^{in}_{214}$: $^{Scotland}_{383}$: but they appear more dissatisfied nor the : $^{Presbytry}_{330}$: $^{an}_{67}$: $^{s}_{30}$: who you know have most cause ; and they make no bones to say : 250 : $^{c}_{46}$: $^{h}_{11}$: 40 : $^{n}_{60}$: $^{g}_{22}$: $^{e}_{23}$: $^{from}_{162}$: 42 : $^{r}_{69}$: $^{so}_{357}$: $^{it}_{149}$: 212 : 103 : shall be most wellcome, and, as I wrote before : $^{R}_{352}$: takes all the paines he can to lay the obloquy off : $^{him}_{295}$: $^{self}_{380}$: $^{up}_{412}$: $^{on}_{296}$: $^{Lauderdale}_{259}$: and tho : $^{R}_{352}$: $^{do}_{124}$: $^{no}_{277}$: good to himselfe, he makes : $^{Lauderdale}_{259}$: very : $^{ha}_{189}$: $^{te}_{388}$: $^{d}_{27}$: so that I would very much : $^{doubt}_{134}$: $^{his}_{194}$: $^{sa}_{354}$: $^{f}_{29}$: $^{t}_{16}$: $^{y}_{h}$: 211 : $^{g}_{21.42}$: $^{Scotland}_{383}$: I shall say no more but wish : $^{us}_{410}$: a : $^{good}_{186}$: $^{me}_{273}$: $^{it}_{212}$: $^{in}_{215}$: but dare not hope for it : $^{my}_{264}$: $^{dear}_{133}$: 147 : adieu.

M. K.

Figure H.2: A letter from Lady Margaret Kennedy to John Lauderdale, dated 2 February 1667.

Besides being linguistically challenged, George I was plagued by a Scottish invasion in 1715 and the continuing aspirations of James III (1688–1766) —the Old Pretender or His Catholic Majesty, depending on the point of view—to regain the throne of England. His exile in Rome was a hotbed for conspiratory characters, and we will meet two of them in the next section: Christopher Layer and John Plunket.

Lady Margaret Kennedy (c. 1630–c. 1685), *"a lady of great repute for*

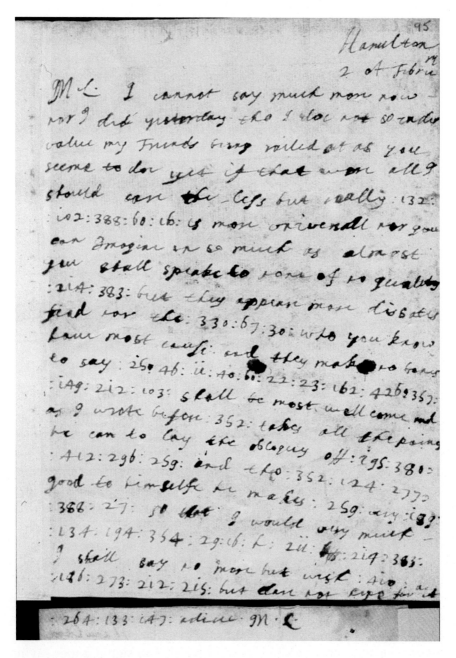

Figure H.3: The original of Lady Kennedy's letter. The last line is on the back of the page in the manuscript.

knowledge and religion," consorted with the high-flyers of her times. She was rich, daring, and well-educated, and carried on an affair with John Maitland, Duke of Lauderdale (1616–1682), for several years in the 1660s. Lauderdale was of Scottish nobility and for some time the second most powerful man in the country, after the king. One of his jobs was as Secretary of State for Scotland, and his residence in Highgate, London.

> *Lauderdaile had, of a long tyme, entertained with Ladie Margaret Kennedie, daughter to the Earle of Cassills, ane intimacie which had grown great enough to become suspicious, in a person who loved not, as some said, his own Ladie. This Ladie had never married, and was always reputit a wit, and the great patron of the Presbyterians, in which persuasion she was verie bigot; and the suspicion increased much, upon her living in the Abbey, in which no woman els lodged; nor did the commissioner [Lauderdale] blush to goe openlie to her chamber, in his night gowne: whereupon her friends haveing challenged her for that unusuall commerce, and having represented to her the open reprehensions and raileries of the people, received no other answer, than that her vertue was above suspision; as realie it was, she being a person whose religion exceeded as far her witt, as her parts exceeded others of her sex.*

She often sojourned at the place of her cousin, the Duchess of Hamilton (1631–1716), in Scotland. In correspondence between Lauderdale and his friend Sir Robert Moray (c. 1608–1673), the two ladies Kennedy and Hamilton are called "our wives"—an interesting concept of a *ménage à quatre*. An unflattering contemporary characterization of Lauderdale reads: *"In private life he was the type of all that was worst in Charles's court."*

Inspite (or because) of all this, Lauderdale had an active sex life, including a relation of many years with Elizabeth Murray (1626–1698), in parallel to the Margaret Kennedy affair. His wife was not amused. After her death in 1671, he married Elizabeth. Margaret Kennedy was not amused. *"Well stricken in years"*, she married Bishop Gilbert Burnet (1643–1715) in the same year, possibly out of spite. He was considerably younger and poorer than she was, and hurried to sign a paper saying that he did not claim any of her fortune.

Burnet was one of the leading Anglican clerics of his period, at times top advisor at the court of King Charles II and friend of Lauderdale, then falling from grace and going into exile on the continent. He was a member of the Royal Society, and among his friends were the mathematician (and cryptanalyst) John Wallis (Section G.6) and the chemist Robert Boyle (1627–1691). In 1675, under heavy fire from Lauderdale, he defended himself with some disclosures about unconstitutional activities of Lauderdale. It was assumed, quite unnecessarily, that this information was provided by his wife, Lauderdale's former flame.

Some of Bishop Burnet's ecclesiastical writings and political speeches in 1701 to 1703 were bitterly opposed by Bishop Atterbury, whom we will meet again as a defendant in Section H.4. During their affair, Lady Margaret provided Lauderdale with important information about the unrest in Scotland, where she resided, while he was at the court in London.

Lady Margaret and Lauderdale partially enciphered their correspondence using a numerical codebook. It is evidence that knowledge of such cryptographic systems was widely spread at the time, and people wanting to protect their little secrets used them. In her 1667 letter (Figures H.2 and H.3), she writes about the widespread discontent in Scotland, and at the end, wishes for a *"good meeting"* between the two of them. Its 54 code words show just seven repetitions—too few for an attack. But a cryptanalyst has all the plaintext parts and presumably further letters at his disposal.

H.4. Christopher Layer

The Anglican vs. Catholic background to our next cryptographic episode has been told at the beginning of Section H.3. We now present one of the rare cases of well-documented cryptanalytic action before the 20th century. In the following, we first tell the story, including the legal proceedings, and then discuss some cryptographic issues. Only at the end, an analysis of the evidence casts doubt on some conclusions of the Royal security agency.

The two main characters in this Jacobite conspiracy were Christopher Layer (1683–1723) and Francis Atterbury (1663–1732), Lord Bishop of Rochester. Layer was sentenced to death, partly given away by bad cryptography. By profession he was a lawyer, a good one, and grossly immoral, quarrelsome, and unscrupulous. With a small band of conspirators, their plan was to put together a gang of soldiers, seize the Tower, Mint, and Bank in London, and also the Royal family. He explained his plot to the *Pretender* James III, exiled in Rome and eager to become His Catholic Majesty, in the summer of 1721.

Two of Layer's lady friends betrayed him, and his papers were seized in a brothel run by Elizabeth Mason. They contained many encrypted messages. Some of the keys were also found, and other missives were deciphered. The step-King of England, George I, was not amused. Layer's trial before the court of King's bench was opened on 21 November 1722. After eighteen hours in total, he was pronounced guilty. Layer was ably defended by the lawyers Hungerford and Ketelbey, with legal shenanigans that sound like straight out of a John Grisham thriller; they even took a misspelling of Layer's first name as reason for a number of motions. To no avail; his sentence—not hampered by human rights considerations—read:

> You C. L. be led to the place from whence you came, and from thence you are to be drawn to the place of execution, and there you are to be hanged by the neck, but not till you are dead, but you are to be cut down alive, and your bowels to be taken out, and burnt before your face; your head is to be severed from your body, and your body to be divided into four quarters; and that your head and quarters be disposed of where his Majesty shall think fit.

On 17 May 1723, at Tyburn, this dreadful sentence was executed—a "beasty business". *Braveheart* aficionados can picture the gory scene. Layer faced his death courageously.

The plot was an affair of the highest importance. Two committees, of the House of Commons and of the House of Lords, investigated it, and William Pulteney, First Earl of Bath, published in 1722 an extensive report. Layer's co-conspirators John Plunket (or Plunkett) and George Kelly were not tried before a court of justice, but rather the House of Commons made this one of the rare cases where it exercised its judicial powers in addition to its legislative ones. Consequently, these Jacobites did not receive a court sentence, but rather a parliamentary *Act to inflict Pain and Penalties* on them. The beginning of the Act against Plunket is shown in Figure H.4; the hole in the middle is due to frequent folding. The text reads:

> *Whereas in the years 1721 and 1722. A Detestable and Horrid Conspiracy was formed and carried on by diverse Traitors for Invading Your Majesties Kingdoms with Foreign Forces, for Raising an Insurrection and Rebellion against your Majesty, for Seizing the Tower and City of London, and for laying violent hands upon your Majesty's Most Sacred Person and upon His Royall Highness the Prince of Wales*

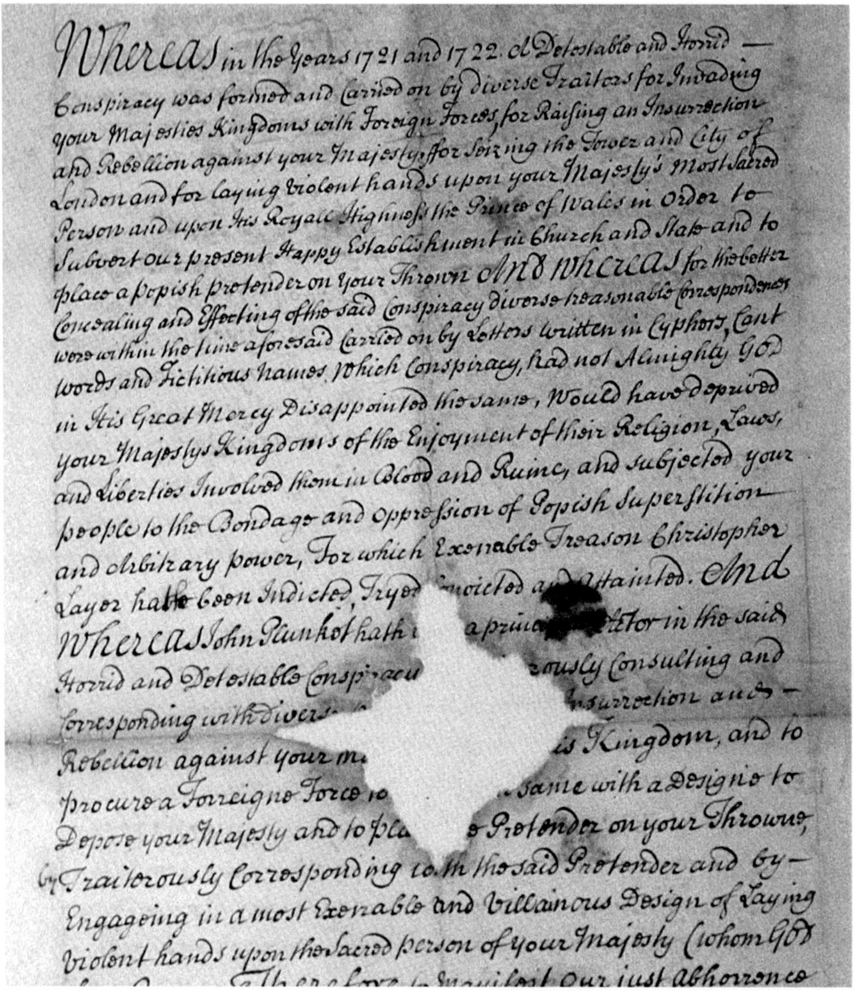

Figure H.4: Original Act of the House of Commons "to inflict pains and penalties on John Plunket".

in Order to Subvert our present Happy Establishment in Church and State and to place a Popish Pretender on your Thrown: And Whereas for the better Concealing and Effecting of the said Conspiracy diverse treasonable Correspondences were within the time aforesaid Carried on by Letters written in Cyphors, Cant words and Fictitious Names [...]. For which execrable Treason Christopher Layer hath been Indicted, Tried, Convicted and Attainted.

The Act against Plunket goes on to describe the penalties. It passed on third reading in the House of Lords on 29 April 1723 by 87 against 34 votes, was signed by the King, and became law.

Kelly, another commoner in the plot, and Plunket received a milder sentence than Layer, namely deportation from England. The conspiracy also involved some nobles, among them the Duke of Norfolk (1683–1732), the Duke of Ormonde (1665–1745), Lord North and Grey (1678–1734), the Earl of Orrery (1674–1731), and Bishop Atterbury.

Only the latter was tried, and that before the House of Lords, of which he was a member. He had refused to recognize the jurisdiction of the House of Commons. The evidence against him was scant, eventually boiling down to three enciphered signatures on "traitorous letters". He was not allowed to cross-examine either Kelly, who allegedly wrote those letters as dictated by Atterbury, or the decipherers. There was some circumstantial evidence related to the encryptions, but his lawyers forcefully pointed out that the evidence would not hold up in a court of law. However, this was politics, and the Lords condemned Atterbury (with an 83 to 43 vote on the third reading) to lose his job and leave the country. He went to France on 18 June 1723.

We now look at some details of cryptographic interest and will learn the following:

○ cryptography was considered a standard tool of conspirators,

○ the government maintained a "black cabinet" of able cryptanalysts, who regularly broke the conspirators' ciphers,

○ letters were routinely opened and copied at the post office, and passed to the crypto section,

○ several types of cryptography were used: mainly (mixed) codebooks, but also grilles and columnar substitutions,

○ the correctness of decipherments was established not by convincing argumentation, but by the raw power of majority vote.

The House of Commons Committee that investigated Layer's case explains in its 1722 *Report* how easy it was to decipher his encryptions:

> The Letter to Dumville, as well as the Copy of the late Duke of Ormond's Letter, was writ partly in Cypher, and among the Words out of Cypher several fictitious Names were made use of, which the Committee observe is the Case also in several others of the intercepted Letters referr'd to them.

C 52.

Cypher Numeral, Plünket's.

A,, B,, C,. D,, E,. F,, G,, H,, I,, K,, L,, M,, N,, O,, P,, Q, R,, S,, T,, U,, X,, Y,, Z,,
10,, 14,, 19,, 1,, 16,, 11,, 2,, 15,, 13,. 12,, 17,, 4,, 18,, 9,, 7,, 5,, 8,, 6,, 20,, 22,, 21., 23,, 3,,
46,, 24,, 45,, 25,, 26,, 44,, 43,, 27,, 42,, 28,, 41,. 29,, 40,, 30,, 39,, 31,, 38,, 32,, 37,, 33,, 36,, 34,, 35,,
Ba. be. bi. bo. bu.   Ca. ce. ci, co, cu.   Da. de. di. do. du.   Fa. fe. fi. fo. fu.   Ga. ge. gi. go. gu.   Ha. he. hi. ho. hu.
47. 48. 49. 50. 51.   52. 53. 54. 55. 56.   57. 58. 59. 60. 61.   62. 63. 64. 65. 66.   67. 68. 69. 70. 71.   72. 73. 74. 75. 76.
Ka. ke. ki. ko. ku.   La. le. li. lo. lu.   Ma. me. mi. mo. mu.   na. ne. ni. no. nu.   Pa. pe. pi. po. pu.
77. 78. 79. 80. 81.   82. 83. 84. 85. 86.   87. 88. 89. 90. 91.   92. 93. 94. 95. 96.   97. 98. 99. 100. 101.
Qua. que. qui. quo. quu.   Ra. re. ri. ro. ru.   Sa. se. si. fo. fu.   Ta. te. ti. to. tu.
102. 103. 104. 105. 106.   107. 108. 109. 110. 111.   112. 113. 114. 115. 116.   117. 118. 119. 120. 121.
Va. ve. vi. vo. vu.   Xa. xe. xi. xo. xu.
122. 123. 124. 125. 126.   127. 128. 129. 130. 131.

England	———— 133	*Indies*	———— 199
His *Britifh* Majefty.	——— 239	Peace	———— 227
His *Catholick* Majefty	——— 238	War	——— 191
The Emperor	—— 237	Treaty	———— 251
Miniftry	——— 215	Secret	———— 248
Minifter	———— 216	Secretary	——— 247
France	———— 183	Duke	——— 164
States of *Holland*	——— 176	*Anjou*•	——— 142
Spain	——— 179		

Figure H.5: One of the many small codebooks used by Layer and his co-conspirators.

> It was reasonable to expect, that in managing Correspon-
> dences of so hazardous a Nature, all sort of Art and Indus-
> try should be used, and all the Help of Cyphers and Jargon
> called in, to disguise the real Designs, and to conceal the true
> Names of the Persons concerned, in order to their avoiding
> the Danger of legal Conviction: But your Committee like-
> wise observe, that several of these Disguises are so gross and
> obvious, that they only serve to betray themselves; others
> of them are explained by the Skill of different Decypherers,
> agreeing in the same Explication, which Explication is again
> confirmed by Facts unknown to those Persons at the time of
> the Decyphering. Others are explained by Cyphers and Lists
> of fictitious Names, seized on the Conspirators themselves,
> as well as by comparing the several Parts of their Correspon-
> dence together; and others again by direct Informations upon
> Oath. And, as the Degrees of Evidence, in a Search of this Na-
> ture, must be various, the Committee have taken all the Care
> they can, to distinguish what appears to them fully proved,
> from what is supported by strong and probable Conjectures
> only.

Figure H.5 shows part of a codebook of the co-conspirator Plunket, originally in his handwriting. About another codebook, the Report says that

> among *Plunket's Papers was found a large Cypher of*

COPY *of a Key directed to* Mr. Burton.

A	B	C	D	E	F	G	H	i	J	K	L	M	N	O	P	Q	R	S	T	V	U	W	X	Y	Z
30	29	28	27	26	5	25	24	23	22	21	6	20	19	18	17	16	7	15	14	13	8	12	11	10	9
10	11	12	13	14	9	15	16	17	18	19	8	20	21	22	23	24	7	25	26	27	6	28	29	30	5

ba	be	bi	bo	bu	ca	ce	ci	co	cu	da	de	di	do	du	fa	fe	fi	fo	fu	ga		ge		gi		go		gu
31	32	33	34	35	31	32	33	34	35	36	37	38	39	40	36	37	38	39	40	41		42		43		44		45

ha	he	hi	ho	hu	Ja	Je	Ji	Jo	Ju	ka	ke	ki	ko	ku	la	le	li	lo	lu	ma	me	mi	mo	mu
41	42	43	44	45	46	47	48	49	50	46	47	48	49	50	51	52	53	54	55	51	52	53	54	55

na	ne	ni	no	nu	pa	pe	pi	po	pu	qua	que	qui	quo	qu'	ra	re	ri	ro	ru	sa	se	si	so	su
56	57	58	59	60	56	57	58	59	60	61	62	63	64	65	61	62	63	64	65	66	67	68	69	70

ta	te	ti	to	tu	va	ve	vi	vo	vu	za	ze	zi	zo	zu	cha	che	chi	cho	chu	pour.	acause.	car.	d'autant
66	67	68	69	70	71	72	73	74	75	71	72	73	74	75	76	77	78	79	80	76	77	78	79
																					Parceq		
																					80		

Accommodment - - - 100	Baron de - - - - - 114	Confeil - - - - - - 129
Anglois - - - - - - 101	Barriere - - - - - 115	Commerce - - - - 130
Angleterre - - - - 102	Bataille - - - - - 116	Compté de - - - 131
Allemagne - - - - 103	Bataillon - - - - 117	Courrier - - - - - 132
Allemans - - - - - 104	Baviere - - - - - 118	Confpiration - - - 133
Allié - - - - - - - 105	Brandenburg - - - 119	Cologne - - - - - 134
Alliance - - - - - 106	Baltique - - - - - 120	Cour de - - - - - 135
Ambaffadeur de - - 107	La Grand Bretagne 121	
Ami - - - - - - - 108		Danemarq - - - - 136
Archiduc - - - - - 109	Cabinet - - - - - 122	Declaration - - - - 137
Archiducheffe - - 110	Cardinal - - - - - 123	Defaite - - - - - 138
Armeé - - - - - - 111	Campagne - - - - 124	Doparriere de - - 139
Artillerie - - - - - 112	Cavallerie - - - - 125	Diffention - - - - 140
Autriche - - - - - 113	Czar - - - - - - - 126	Duc de - - - - - 141
	Congres - - - - - 127	Doge de - - - - - 142
	Convoy - - - - - 128	

Duché - - - - - - 143		
Defenfe - - - - - 144	Flotte - - - - - - - - - 158
Envoyé - - - - - 145	Florence - - - - - - - 159
Electeur de - - - - 146	France - - - - - - - - 160
Enemi - - - - - - 147	Francois - - - - - - - 161
Empereur - - - - 148	Frontiere - - - - - - - 162
Empire - - - - - 149	Grand - - - - - - - - 163
Efpagne - - - - - 150	Garantie - - - - - - - 164
Efpagnol - - - - 151	Garnifon - - - - - - - 165
Etats Gen.ᵃᵘˣ - - 152	Guerre - - - - - - - - 166
Efcadre - - - - - 153	General - - - - - - - - 167
Ecoffe - - - - - 154	Genes - - - - - - - - 168
Efcadron - - - - 155	Governeur - - - - - - 169
Expref - - - - - 156	Galere - - - - - - - - 170
Extraordinaire - - 157	Garand - - - - - - - - 171

Figure H.6: Initial section of a codebook used in the Layer conspiracy. At top are letters with two encodings each, then digrams consonant-vowel and five words, then an ordered codebook with a total of 189 French words, and 14 dummies.

> *Names with fictitious Names over-against them, sworn to be all in* Plunket's *own Hand-writing, which Cypher tallies with, and explains an original Letter of the 23d of July 1722, directed to Mr.* Digby *at Paris, and signed J.* Rogers, *which was stopt at the Post-Office, and is likewise sworn to be* Plunket's *Hand-writing.*

In his examination on 25 January 1723, Plunket lamely claims

> *that the Cipher No. 2. is not his Hand-Writing; that it belong'd to Hugh Thomas, and he found it among Thomas's Books. Being ask'd whether Hugh Thomas ever sign'd Plunket, he says he does not know that ever he did.*

A main difficulty in symmetric cryptography, namely that of communicating safely the secret keys, is illustrated by other codebooks such as the one in Figure H.6, of which the Report drily states:

> *Copies of these Cyphers having been taken at the Post-Office, it appears that one of them was mark'd, A Key and Cypher, with Mr.* Farmer *and* Jerry; *and another, D, O, and J; the first of which is probably a Cypher between the Pretender*

and Jernegan, and the latter between him and the late Duke of Ormonde.

During the trial of Bishop Atterbury, the cryptanalyst Willes testified as witness. The report describes how he refused to justify his claims:

Atterbury showed more than ordinary eagerness to grapple with him. Willes quietly asserted that he had properly decyphered the arrested letters, given to him for that purpose. 'Pray, sir,' said the bishop (who had failed to obtain the production of the key), 'will you explain to me your process of decyphering?' 'No, my lord,' was the reply, 'I will not. It would tend to the discovery of my art, and to instruct ill-designing men to contrive more difficult cyphers.'.

Edward Willes (1693–1773) was the second successor to John Wallis as head of the *Secret Department of the Secretaries of State*, or *Deciphering Branch*. His initial salary was £200 a year, which was raised to £500 after his successful work in the case against Layer, Atterbury, and others. While cryptographers like Trithemius and Viète faced the spectre of potential persecution by the Catholic church, the Anglican state church looked more benevolently upon such diabolical activities: Willes became Canon of Westminster in 1724, Bishop of St. David's in 1742, and Bishop of Bath and Wells in 1743. Before 1800, three of his sons and three of his grandsons had joined the office as cryptanalysts. Anthony Corbiere, born 1687, was another cryptanalyst working on the Layer and Atterbury ciphers. He worked as Decipherer from 1719 to 1743, with a substantial annual income of £800 in his later years.

A second report, written by a House of Lords committee in 1723, explains the cryptanalysts' successful work:

The Lords Committee observing that some Paragraphs of the Letters referr'd to them were writ originally in Cypher, thought it proper to call the Decypherers before them, in order to their being satisfied of the Truth of the Decyphering. The Account they receiv'd from those Persons, was, that they have long been versed in this Science, and are ready to produce Witnesses of undoubted Reputation, who have framed Letters in Cypher, on purpose to put them to a Tryal, and have constantly found their decyphering to agree with the Original Keys which had been concealed from them; it was likewise confirmed to the Committee, that Letters decyphered by one or other of them in England, had exactly agreed with the decyphering of the same Letters performed by Persons in foreign Parts, with whom they could have no Communication; and that in some instances after they had decyphered Letters for the Government, the Keys of those Cyphers had been seized, and upon comparing them, had agreed exactly with their decyphering.

With respect to the intercepted Letters in Question, they alledged that in the Cypher used by George Kelly, they find the Words ranged in an alphabetical Order, answering the progressive Order of the Figures by which they are expressed, so that the farther the initial Letter of any Word is removed from the Letter A, the higher the Number is by which such Word is denoted; [...]

As to the Truth of the Decyphering, they alledged that several Letters written in this Cipher had been decyphered by them separately, and being many Miles distant in the Country, and the other in Town, and yet their decyphering agreed; that Facts unknown to them and the Government at the time of their deciphering had been verified in every Circumstance by subsequent Discoveries, as particularly that of H—'s Ship coming in Ballast to fetch O— to England, which had been so decyphered by them two Months before the Government had the least notice of Halstead's having left England: That a Supplement to this Cypher having been found among Dennis Kelly's Papers the latter end of July, agreed with the Key they had formed of that Cypher the April before: That the decyphering of the Letters signed Jones, Illington, and 1378, being afterwards applied by them to others written in the same Cypher, did immediately make pertinent Sense, and such as had an evident Connexion and Coherence with the Parts of those Letters that were out of Cypher, tho' the Words in Cypher were repeated in different Paragraphs and differently combined. And they insist that these several Particulars duely weighed amount to a Demonstration of the Truth of their decyphering.

A major point of contention in both Houses was in how far the decipherments were correct. The decipherers stated in parliament that their work was correct, and gave the plausible arguments that some messages had been deciphered twice independently with identical results, and that for others the key had been obtained later and found to agree. But the defendants were not allowed to interrogate them closely.

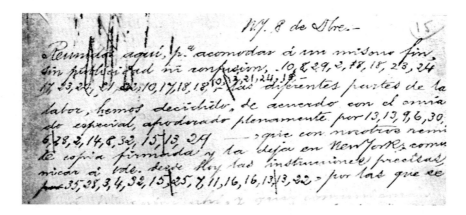

Figure H.7: The first lines of Martí's plan for the Cuban uprising

Besides the numerous occurrences of codebooks, we find two other kinds of cryptosystems: in one cipher the codeword *pattern* encrypts *cut-paper*, which presumably means a grille cipher. In another place, the conspirators correspond about *Paper writ Column-ways*, which presumably refers to a columnar transposition.

Bishop Atterbury was convicted on the evidence of three *"treason-able"* letters signed Jones, Illington, and 1378 that he allegedly dictated. It is disconcerting to see a lack of internal consistency in the decipherers' work. As an example, one encrypted letter talks about Jones and Illington as if they were distinct persons. Can Jones be Illington? The counsel for the defence pointed out problems with the deciphering (although not the inconsistencies just mentioned), but their requests to cross-examine the cryptanalysts got steam-rolled by the political majority. It would be interesting to calculate the unicity distance (Section D.6) for these codebooks, but the lack of a concordance between the many codes and letters in the report does not seem to allow this.

H.5. José Martí

José Martí, the Cuban national hero who led the uprising against Spanish colonial occupation in February 1895, used a Vigenère cipher with key word habana = $8, 1, 2, 1, 15, 1$ and his 28-letter alphabet a $= 1, b, \ldots,$ l, ll, m, n, ñ, o,.... Figure H.7 shows the first lines of his famous *Plan de Alzamiento para Cuba*[4], written on 8 December 1894 in his New York City exile with two fellow revolutionaries:

---

[4]Plan of uprising for Cuba

N.Y. 8 de Dbre. Reunidos aquí, para acomodar á un mismo
fin, sin publicidad ni confusión, b.g.ya con un término
cierto —las diferentes partes de la labor, hemos decidido,
de acuerdo con el enviado especial, apoderado plenamente
por el general gomez—, que con nosotros remite copia
firmada y la deja en New York, comunicar á Vds. desde
hoy las instrucciones precisas, y ya como finales, por
las que se deberán Vds. guiar ahí.[5]

Martí commits a cardinal sin in cryptography: mixing plaintext and ciphertext. But not even such a blunder could stop Cuban independence.
The plaintext was written first, leaving space for the ciphertext which
was then inserted by another hand. On line 3, too little, and on line 6,
too much space had been left within the plaintext.

Notes H.1. *The Reader of Other Gentlemen's Mail* is David Kahn's title for his colorful biography of Yardley, who wrote the highly controversial book *The American Black Chamber*
after his abrupt dismissal from US government services. In it, he talked freely about the
US breaks into Japanese codes, especially at a 1923 disarmament conference. In the eyes
of many of his colleagues, this stigmatized him as a traitor.

H.2. The modern province of Oaxaca is south-east of Mexico City, Michoacán south-
west, and Guanajuato north. The towns of Uruapan, Zacapu, Jacona, and Coyuca de
Catalán are all in today's province of Michoacán, and Irapuato in Guanajuato. Tiripetio
is just south-east of Morelia. Many thanks go to Florian Luca from the Universidad de
Morelia for help with the Morelian geography.

H.3. The story of how Elizabeth's secretary Walsingham trapped Mary Stuart with the
help of a double-agent, deciphered letters and forged messages encrypted in her code
is told in Kahn (1967), pages 121–124. The last of the Stuarts was Queen Anne (1665–
1714), second daughter of James I. The Act of Settlement of 1701 provided that Sophia,
Princess of Hanover and the granddaughter of James I, or her descendants, would wear
the crown. Upon the death of Queen Anne, who outlived her seventeen children, the
crown descended to the Elector of Hanover, who ascended the British throne as George I.
He was thus a great-grandson of James I. A step-king is to his step-people like a step-
father to his step-children. Walpole became Earl of Oxford in 1742. His peace politics
furthered colonial trade and the economy of the United Kingdom. The *bad latin* is from
Cobbett (1811), column 573.

61 letters of Lady Margaret Burnet were published in the handsome volume of Burnet
(1828), 41 of them partially encrypted. Her name is "Burnet" only on the title, elsewhere
she is "Kennedy". The description of her affair is on its pages ii–iii, and her letter in

---

[5]New York, 8 December. United here in order to concentrate the different aspects of
our enterprise towards our common goal, with neither publicity nor confusion *but with
a sure outcome*, we have decided, in agreement with the special envoy, fully empowered
by *General Gómez*—of which he gave us a signed copy to be left in New York—to communicate to you precise instructions *to be taken as final* according to which you shall act
...

Figure H.2 is on page 60. The original in Figure H.3 is kept at the Advocates' Library in Edinburgh, now part of The National Library of Scotland.

H.4. The Jacobites are named after James = Jacob = Iago, as are Santiago and San Diego. The description of Layer's character is not a lawyer joke, but from the *Dictionary of National Biography*.

Tyburn was a place of public execution until 1783. Today, Tyburn Way is near Marble Arch and Speakers' Corner at the north-east corner of Hyde Park in London. The macabre humor of the day called *London's grisly overseers* the heads displayed on high poles from where they could "see" all of London; see Kerr (2002), page 263. Layer's sentence is reproduced in Howell (1816), volume XVI, columns 320–321. The report from the House of Commons was published as a book, jointly with that of the House of Lords, in Pulteney (1722). The latter consists of the two reports plus numerous appendices, a total of about 480 pages, but without a continuous pagination. Our quotations are, in sequence, from page 6 (The letter to *Dumville* ...), page 27 (among Plunket's papers ...), C.73 on page 40 of Appendix C (Cipher No. 2 ...), page 70 (Copies of these Cyphers ...), and pages 9/10 of the Lords' Report. The exchange between Atterbury and Willes is from Doran (1877), pages 417–418. Figure H.5 is on page 24 of Appendix C.52, and Figure H.6 on page 18 of Appendix H.36.

Cipher No. 2 is C.73 on page 40 of Appendix C and Figure H.6 is H.36 on page 18 of Pulteney (1722). The Lords Committee observing: pages 9–10 of Lord's Report in Pulteney (1722). The *pattern=cut-paper* occurs on page 23 of Pulteney (1722), in cipher B.Y.1, and *Column-ways* on page 38. The letter D.23 in Pulteney (1722) from J. Hatfield to Howel and dated May 10, 1722, includes *"Jones promises to be a good Customer [...] Illington is gone to the Country."* To me, it seems plausible that Atterbury = Illington ≠ Jones, that Atterbury was heavily involved in the conspiracy, and that the proof offered by the King's counsel would not have held up in an impartial court of justice. Legal finagling.

Layer's plot is mentioned in Stanhope (1839), volume II, chapter XII, pages 34–56; Cobbett (1811), volume VIII, columns 44–353; Howell (1816), volume XVI, Nos. 463 and 464, columns 93–696; and on pages 170–171 of Kahn (1967). Kahn gives an excerpt from the *Journals of the House of Lords* which agrees with part of our quotation from the Report. The Deciphering Branch with Willes and Corbiere is described on pages 127–129 of Ellis (1958). The court proceedings generated so much attention that a 300-page German translation was published as Layer (1723).

The text of *An Act to inflict Pains and Penalties on John Plunkett* is printed in Howell (1816), column 468. Figure H.4 shows the original act (or a draft of it), which is now at The National Archives in London. Some plotters considered Atterbury to be the head of the conspiracy, together with Lord North and Grey. See Pulteney (1722), page 38.

Der Urquell aller Mathematik sind die ganzen Zahlen.[6]

HERMANN MINKOWSKI (1907)

Unsere Wissenschaft ist, im Gegensatze zu anderen, nicht auf eine einzelne Periode der menschlichen Geschichte gegründet, sondern sie hat die Entwickelung der Cultur auf allen ihren Stufen begleitet. Die Mathematik ist mit der griechischen Bildung ebenso verwachsen, wie mit den modernsten Aufgaben des Ingenieursbetriebes. Sie reicht nicht nur den vorwärtsschreitenden Naturwissenschaften die Hand, sondern sie participiert gleichzeitig an den abstracten Untersuchungen der Logiker und Philosophen.[7]

FELIX KLEIN (1900)

One who knows the enemy and knows himself will not be endangered in a hundred engagements.

SUN TZU (6th century BC)

Comme conclusion, nous inscrirons la phrase souvent répétée par un des chefs du Service du Chiffre aux armées: «Chiffrez bien, ou ne chiffrez pas. En transmettant du clair, vous ne donnez qu'un renseignement à l'ennemi et vous savez lequel; en chiffrant mal, vous lui permettez de lire toute votre correspondance et celle de vos amis.»[8]

COLONEL MARCEL GIVIERGE (1924)

---

[6]Integral numbers are the fountainhead of all mathematics.

[7]Our science is, in contrast with others, not founded on a single period of human history, but has accompanied the development of civilization through all its stages. Mathematics is as much interwoven with Greek culture as with the most modern problems in engineering. It not only lends a hand to the progressing natural sciences, but participates at the same time in the abstract investigations of logicians and philosophers.

[8]In conclusion, we will highlight the maxim often repeated by one of the heads of the Army Cipher Service: "Encrypt well or do not encrypt at all. In transmitting plaintext, you give only one piece of information to the enemy, and you know what it is; in encrypting badly, you permit him to read all your correspondence and that of your friends."

Chapter 13

Integral lattices

The methods we discuss in this chapter deal with computational aspects of the *geometry of numbers*, a mathematical theory initiated by Hermann Minkowski in the 1890s. This theory produces many results about Diophantine approximation, convex bodies, embeddings of algebraic number fields in $\mathbb{C}$, and the ellipsoid method for rational linear programming. In cryptology it serves three purposes.

Cryptanalysis: It was used to break cryptosystems. In the 1980's, the first generation of public key cryptosystems besides RSA, the *subset sum* system, was obliterated by this attack. For many types of new systems, one has to consider carefully potential attacks using this methodology.

Security reductions: If a system like the Diffie-Hellman key exchange or RSA encryption is secure, it is not clear that partial information like the leading bits of a Diffie–Hellman key or of a prime factor in an RSA modulus are also secure. Lattice technology provides proofs that this is indeed the case.

Cryptography: Since 1996, the method has been used to devise cryptosystems that have (provably under a hardness axiom) a desirable property that no previous system had: breaking an "average instance" is as difficult as breaking a "hardest instance". These notions are discussed in Section 15.7. Section 13.13 presents a cryptosystem which is secure if a certain lattice problem is hard in the worst case.

13.1. Lattices and short vectors

We start with basic concepts used throughout this chapter.

© Springer-Verlag Berlin Heidelberg 2015

J. von zur Gathen, *CryptoSchool*, DOI 10.1007/978-3-662-48425-8_21

DEFINITION 13.1. *Let $a_1, \ldots, a_\ell \in \mathbb{R}^n$ be linearly independent row vectors over $\mathbb{R}$. Then*

$$L = \sum_{1 \leq i \leq \ell} \mathbb{Z}a_i = \{ \sum_{1 \leq i \leq \ell} r_i a_i : r_1, \ldots, r_\ell \in \mathbb{Z} \}$$

is the lattice *(or $\mathbb{Z}$-module) generated by $a_1, \ldots, a_\ell$. These vectors form a* basis *of L. If $\ell = n$, we call L a* full-rank *lattice. A subset $K \subseteq L$ which is also a lattice is a* sublattice *of L.*

This notion is quite similar to that of the $\mathbb{R}$-vector space in $\mathbb{R}^n$ spanned by $a_1, \ldots, a_\ell$, for which we write $V = \mathrm{span}_\mathbb{R}(a_1, \ldots, a_\ell) = \mathrm{span}_\mathbb{R}(L)$. The advantage over the integers $\mathbb{Z}$ is that we can talk about integer solutions and interesting concepts like divisibility and modular reduction. The drawback is that we cannot divide by an arbitrary nonzero integer and therefore have no analog of Gauß' algorithm for solving systems of linear equations. As a substitute we have the basis reduction algorithm, which is somewhat trickier to use.

Let $x = (x_1, \ldots, x_n) \in \mathbb{R}^n$. The *norm* (or 2-norm, or Euclidean norm) of x is given by

$$(13.2) \qquad \|x\| = \|x\|_2 = \left(\sum_{1 \leq i \leq n} x_i^2 \right)^{1/2} = (x \star x)^{1/2} \in \mathbb{R}_{\geq 0},$$

where $x \star y = \sum_{1 \leq i \leq n} x_i y_i \in \mathbb{R}$ is the usual *inner product* of two vectors $x = (x_1, \ldots, x_n)$ and $y = (y_1, \ldots, y_n)$ in $\mathbb{R}^n$ (sometimes written as (x, y), or $\langle x, y \rangle$, or $x \cdot y^T$ in the literature). The vectors x and y are *orthogonal* if $x \star y = 0$. A square integer matrix U is *unimodular* if and only if $\det(U) \in \pm 1$.

THEOREM 13.3. *Let $A, B \in \mathbb{R}^{\ell \times n}$ both have linearly independent rows. Then the row vectors generate the same lattice if and only if there is a unimodular $\ell \times \ell$ matrix U with $A = UB$.*

PROOF.   We denote as L and M the lattices generated by the rows of A and B, respectively. Then $A = UB$ with a unimodular matrix U implies that $L \subseteq M$. Also, $B = U^{-1}A$ and hence $M \subseteq L$. We conclude that $L = M$.

On the other hand, assume that $L = M$. Then there exist matrices $V, W \in \mathbb{Z}^{\ell \times \ell}$ with $A = VB$ and $B = WA$. It follows that $(VW - I)A = 0$. Since the rows of A are linearly independent, we conclude that $VW = I$, $W = V^{-1}$, $\det V \cdot \det W = 1$, and $\det V = \det W \in \pm 1$, since both are integers. $\square$

We can transform any basis A without changing the lattice generated by applying the following elementary row operations:

1. Swap two rows of A.

2. Multiply a row by the constant -1.

3. Add an integer multiple of a row a_j to another one, in other words set $a_i \leftarrow a_i + ka_j$ with $i \neq j$ and $k \in \mathbb{Z}$.

Actually any unimodular matrix U can be obtained from the identity matrix by performing the above transformations repeatedly (Exercise 13.1).

DEFINITION 13.4. *Let L be the lattice generated by the rows of the matrix $A \in \mathbb{R}^{\ell \times n}$. The norm of L is $|L| = \det(AA^T)^{1/2} \in \mathbb{R}$.*

Theorem 13.3 implies that the norm is well defined, namely, it is independent of the choice of the generators of L. In the case $\ell = n$, the definition simplifies to $|L| = |\det(A)|$. Geometrically, $|L|$ is the volume of the parallelepiped spanned by $a_1, \ldots, a_\ell$ and often called $\mathrm{vol}(L)$. For $\ell = n$, Hadamard's inequality 15.79 says that $|L| \leq \|a_1\| \cdots \|a_n\|$.

EXAMPLE 13.5. We let $\ell = n = 2$, $a_1 = (12, 2)$, $a_2 = (13, 4)$ and $L = \mathbb{Z}a_1 + \mathbb{Z}a_2$. Figure 13.1 shows some lattice points of L near the origin of the plane $\mathbb{R}^2$. The norm of L is

$$|L| = \left| \det \begin{pmatrix} 12 & 2 \\ 13 & 4 \end{pmatrix} \right| = 22$$

and equals the area of the gray parallelogram in Figure 13.1. We have $22 \leq \|a_1\| \cdot \|a_2\| = 74\sqrt{5} \approx 165.469$. Another basis of L is $b_1 = (1, 2)$ and $b_2 = (11, 0) = 2a_1 - a_2$, and b_1 is a "shortest" vector in L. We have $22 \leq \|b_1\| \cdot \|b_2\| = 11\sqrt{5} \approx 24.597$. $\Diamond$

Exercises 13.3 and 13.4 present a different way of describing lattices.

DEFINITION 13.6. *Let $L \subset \mathbb{R}^n$ be an ℓ-dimensional lattice and $1 \leq i \leq \ell$. The ith successive minimum $\lambda_i(L)$ is the smallest real number so that there exist i linearly independent vectors in L, all of length at most $\lambda_i(L)$.*

In particular, $\lambda_1(L)$ is the length of a shortest nonzero vector in L. An element $x \in L$ is a *shortest* vector if $\|x\| = \lambda_1(L)$. We have $\lambda_1(L) \leq \lambda_2(L) \leq \cdots \leq \lambda_n(L)$. For $z \in \mathbb{R}^n$ and $r > 0$, we consider the n-dimensional ball

(13.7) $$\mathcal{B}_n(z, r) = \{x \in \mathbb{R}^n : \|x - z\| \leq r\}$$

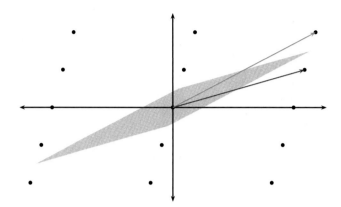

Figure 13.1: The lattice in $\mathbb{R}^2$ generated by $(12, 2)$ (red) and $(13, 4)$ (blue).

of radius r centered at z. Two famous theorems of Minkowski say that

$$\lambda_1(L)^n \le \lambda_1(L) \cdots \lambda_n(L) \le \frac{2^n |L|}{\text{vol}(\mathcal{B}_n(0, 1))}.$$

and then estimates for $\text{vol}(\mathcal{B}_n(0, 1))$ (see (13.111)) imply that

(13.8) $$\lambda_1(L)^n \le \lambda_1(L) \cdots \lambda_n(L) \le n^n |L|.$$

A natural question is to compute a shortest vector in a given lattice. As reported in Section 13.12, this problem is NP-hard under randomized reductions, and there is no hope for efficient algorithms. But for a plethora of applications, it is sufficient to compute a "reasonably short" vector, a problem solved in polynomial time by the famous lattice basis reduction algorithm of Lenstra, Lenstra & Lovász (1982), sometimes called the *LLL*- or L^3-algorithm. Their "short vector" is guaranteed to be off by not more than a specified factor, which depends on the dimension but not the lattice itself.

13.2. Lenstra, Lenstra & Lovász' lattice basis reduction

The result of this and the next section is embodied in Corollary 13.21 and used throughout this chapter. At first reading, the two sections might be skipped, and referred back to when necessary.

We assume familiarity with the notions and facts from Section 15.15, and first exhibit a connection between the Gram-Schmidt orthogonal basis (GSO) $(b_1^*, \ldots, b_\ell^*)$ and the successive minima of the lattice.

THEOREM 13.9. *Let $L \subseteq \mathbb{R}^n$ be a lattice with basis $b_1, \ldots, b_\ell \in \mathbb{R}^n$, Gram-Schmidt orthogonal basis $b_1^*, \ldots, b_\ell^* \in \mathbb{R}^n$, and successive minima $\lambda_1(L), \ldots, \lambda_\ell(L)$. Then for $1 \leq i \leq \ell$ we have*

$$\min\{\|b_i^*\|, \|b_{i+1}^*\|, \ldots, \|b_\ell^*\|\} \leq \lambda_i(L).$$

PROOF. According to Definition 13.6, there are linearly independent vectors $x_1, \ldots, x_\ell \in L$ with $\|x_i\| = \lambda_i(L)$ for $1 \leq i \leq \ell$. We write each vector in terms of the given basis as $x_i = \sum_{1 \leq j \leq \ell} r_{i,j} b_j$ with $r_{i,j} \in \mathbb{Z}$ for $1 \leq i, j \leq \ell$.

For $i \leq \ell$, let $c_i = \max\{1 \leq j \leq \ell : r_{i,j} \neq 0\}$. Since not all linearly independent vectors $x_1, \ldots, x_i$ lie in the $(i-1)$-dimensional vector space spanned by $b_1, \ldots, b_{i-1}$, there is some $k \leq i$ such that $c_k \geq i$. For such a k, we obtain $\lambda_i(L) \geq \lambda_k(L) = \|x_k\|$ and we can thus write

$$x_k = \sum_{1 \leq j \leq c_k} r_{k,j} b_j = \sum_{1 \leq j \leq c_k} r_{k,j}^* b_j^*$$

with $r_{k,j}^* \in \mathbb{Z}$ for $1 \leq j < c_k$ and $r_{k,c_k}^* \in \mathbb{Z} \setminus \{0\}$. Additionally, it follows from the triangle inequality that for all those j we have $\|x_k\| \geq |r_{k,j}^*| \cdot \|b_j^*\|$. This implies that $\lambda_i(L) \geq \lambda_k(L) = \|x_k\| \geq |r_{k,c_k}^*| \cdot \|b_{c_k}^*\| \geq \|b_{c_k}^*\|$ with $c_k \geq i$, and the theorem is proven. $\square$

COROLLARY 13.10. *Let $L \subseteq \mathbb{R}^n$ be a lattice with basis $(b_1, \ldots, b_\ell)$ and Gram-Schmidt orthogonal basis $(b_1^*, \ldots, b_\ell^*)$. Then for any nonzero $x \in L$ we have*

$$\min\{\|b_1^*\|, \ldots, \|b_\ell^*\|\} \leq \|x\|. \qquad \square$$

Our goal is to compute a short vector in the lattice L generated by $b_1, \ldots, b_\ell$. If the Gram-Schmidt orthogonal basis is a basis of L, then the lemma says that one of the b_i^* is a shortest vector. But usually the b_i^* are not even in L, as in Example 15.77. This problem motivates the following definition of a reduced basis as an "almost orthogonal" basis of L.

DEFINITION 13.11. *Let $b_1, \ldots, b_\ell \in \mathbb{R}^n$ be linearly independent and $b_1^*, \ldots, b_\ell^*$ the corresponding Gram-Schmidt orthogonal basis. Then $b_1, \ldots, b_\ell$ is reduced if and only if $\|b_i^*\|^2 \leq 2\|b_{i+1}^*\|^2$ for $1 \leq i < \ell$.*

The basis in Example 15.77 is reduced. Corollary 13.10 provides a lower bound on the length of nonzero vectors in L, in terms of the GSO. With a reduced basis, we get a similar, though somewhat weaker, bound in terms of the original basis.

THEOREM 13.12. *Let* $b_1, \ldots, b_\ell \in \mathbb{R}^n$ *be a reduced basis of the lattice* L. *Then* $\|b_1\| \leq 2^{(\ell-1)/2} \cdot \lambda_1(L)$.

PROOF.   By Definition 13.11 we have

$$\|b_1\|^2 = \|b_1^*\|^2 \leq 2\|b_2^*\|^2 \leq 2^2\|b_3^*\|^2 \leq \cdots \leq 2^{\ell-1}\|b_\ell^*\|^2.$$

For nonzero $x \in L$, Corollary 13.10 implies that

$$2^{-(\ell-1)/2}\|b_1\| \leq \min\{\|b_1^*\|, \ldots, \|b_\ell^*\|\} \leq \|x\|,$$

so that $\|b_1\| \leq 2^{(\ell-1)/2} \cdot \lambda_1(L)$. $\square$

At first sight, the approximation factor $2^{(\ell-1)/2}$ might look exponential. But it has only a linear number $(\ell-1)/2$ of bits and is small enough for a plethora of applications. A more general bound on each $\|b_i\|$ is shown in Theorem 13.43.

We now present an algorithm that computes a reduced basis of a lattice $L \subseteq \mathbb{Z}^n$ from an arbitrary basis. One can use this to find a reduced basis of a lattice in $\mathbb{Q}^n$, by multiplying with a common denominator of the given basis vectors. For $\mu \in \mathbb{R}$, we write $\lceil \mu \rfloor = \lfloor \mu + 1/2 \rfloor$ for the integer nearest to μ, with $\lceil a + 1/2 \rfloor = a + 1$ for $a \in \mathbb{Z}$.

ALGORITHM 13.13. Basis reduction.

Input: Linearly independent row vectors $a_1, \ldots, a_\ell \in \mathbb{Z}^n$.
Output: A reduced basis $b_1, \ldots, b_\ell$ of the lattice $L = \sum_{1 \leq i \leq \ell} a_i \mathbb{Z} \subseteq \mathbb{Z}^n$.

1. For $i = 1, \ldots, \ell$ do steps $b_i \leftarrow a_i$.
2. Compute the GSO $B^* \in \mathbb{Q}^{\ell \times n}$, $M \in \mathbb{Q}^{\ell \times \ell}$, as in (15.73) and (15.75).
3. $i \leftarrow 2$.
4. While $i \leq \ell$ do steps 5–10
5.   For $j = i - 1, i - 2, \ldots, 1$ do step 6–6
6. $b_i \leftarrow b_i - \lceil \mu_{ij} \rfloor b_j$,   update the GSO,   [replacement step]
7.   If $i > 1$ and $\|b_{i-1}^*\|^2 > 2\|b_i^*\|^2$ then
8.   exchange b_{i-1} and b_i and update the GSO,   [exchange step]
9. $i \leftarrow i - 1$.
10.   Else $i \leftarrow i + 1$.
11. Return $b_1, \ldots, b_\ell$.

EXAMPLE 13.5 CONTINUED. *Table 13.2 traces the algorithm on the lattice of Example 13.5, and Figure 13.3 depicts the vectors* b_i *in the computation. We start with* $b_1 = a_1 = (12, 2)$ *(red) and* $b_2 = a_2 = (13, 4)$ *(blue). In the second row of Table 13.2,* b_2 *is replaced by* $x = b_2 - \lceil 41/37 \rfloor b_1 =$

step	$\begin{pmatrix} b_1 \\ b_2 \end{pmatrix}$	M	$\begin{pmatrix} b_1^* \\ b_2^* \end{pmatrix}$	action
4	$\begin{pmatrix} 12 & 2 \\ 13 & 4 \end{pmatrix}$	$\begin{pmatrix} 1 & 0 \\ \frac{41}{37} & 1 \end{pmatrix}$	$\begin{pmatrix} 12 & 2 \\ -\frac{11}{37} & \frac{66}{37} \end{pmatrix}$	row 2 ← row 2 − row 1
5	$\begin{pmatrix} 12 & 2 \\ 1 & 2 \end{pmatrix}$	$\begin{pmatrix} 1 & 0 \\ \frac{4}{37} & 1 \end{pmatrix}$	$\begin{pmatrix} 12 & 2 \\ -\frac{11}{37} & \frac{66}{37} \end{pmatrix}$	exchange rows 1 and 2
4	$\begin{pmatrix} 1 & 2 \\ 12 & 2 \end{pmatrix}$	$\begin{pmatrix} 1 & 0 \\ \frac{16}{5} & 1 \end{pmatrix}$	$\begin{pmatrix} 1 & 2 \\ \frac{44}{5} & -\frac{22}{5} \end{pmatrix}$	row 2 ← row 2 − 3 · row 1
6	$\begin{pmatrix} 1 & 2 \\ 9 & -4 \end{pmatrix}$	$\begin{pmatrix} 1 & 0 \\ \frac{1}{5} & 1 \end{pmatrix}$	$\begin{pmatrix} 1 & 2 \\ \frac{44}{5} & -\frac{22}{5} \end{pmatrix}$	

Table 13.2: Trace of the basis reduction Algorithm 13.13 on the lattice of Example 13.5.

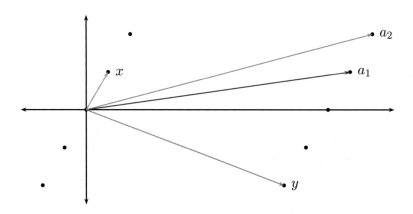

Figure 13.3: The vectors computed by the basis reduction Algorithm 13.13 for the lattice of Example 13.5.

$(1, 2)$ (green). Then $||b_1^*||^2 = 148 > 242/37 = 2 \cdot ||b_2^*||^2 \approx 6.541$, and b_1 and b_2 are exchanged in the third row. In the last row, $y = b_2 - \lceil 16/5 \rfloor b_1 = a_1 - 3x = (9, -4)$ (brown) is computed, and the algorithm returns the reduced basis $x = (1, 2)$ and $y = (9, -4)$. We can see clearly in Figure 13.3 that the final $b_1 = x$ (the green vector) is much shorter than the two input vectors a_1, a_2, and that the computed basis x, y (the green and the brown vectors) is nearly orthogonal.

In the above example, the final b_1 is actually a *shortest* vector. This seems to happen quite often, but Theorem 13.12 only guarantees that the norm of the first vector in the computed basis is larger by a factor of at most $2^{(\ell-1)/2}$ than the norm of a shortest vector, where ℓ is the dimension of the lattice.

The following bound will be used for Theorem 13.129.

LEMMA 13.14. *Let $L \subseteq \mathbb{R}^n$ be an n-dimensional lattice with reduced basis $b_1, \ldots, b_n$ and corresponding GSO $b_1^*, \ldots, b_n^*$. Then*

$$\lambda_n(L) \leq \sqrt{n} \max\{\|b_1^*\|, \ldots, \|b_n^*\|\}.$$

PROOF.   Let M be the maximum as above. According to step 6 of Algorithm 13.13, for any $j < i \leq n$ the projection of b_i onto b_j^* has length at most $\|b_j^*\|/2 \leq M/2$, and $\|b_i\| = \|b_i^*\| \leq M$. Thus

$$\|b_i\|^2 \leq \sum_{1 \leq j < i} \|b_j^*\|^2/4 + \|b_i^*\|^2 \leq ((n-1)/4 + 1)M^2 < nM^2.$$

Since $b_1, \ldots, b_n$ is a basis of L, we have

$$\lambda_n(L) \leq \max\{\|b_1\|, \ldots, \|b_n\|\} \leq \sqrt{n}M. \qquad \square$$

13.3. Cost estimate for basis reduction

THEOREM 13.15. *Algorithm 13.13 correctly computes a reduced basis of $L \subseteq \mathbb{Z}^n$ and runs in polynomial time. It uses $O(n^4 m)$ arithmetic operations on integers whose bit length is $O(nm)$, if the norm of each given generator for L has bit length at most m.*

The idea of the estimate on the number of arithmetic operations is as follows. Each execution of steps 6 or 7–10 has polynomial cost, and it is sufficient to bound the number of passes through step 7–10 with an exchange. In fact, at first glance it is not obvious that the algorithm terminates at all, since the decrease in step 9 and increase in step 10 of i might continue forever. The crucial point then is to exhibit a value D with the following properties: it is always a positive integer, reasonably small in the beginning, and does not change in the algorithm except that at each exchange step it decreases (at least) by a factor of 3/4. Therefore only few exchange steps can happen.

To structure the somewhat lengthy proof, we first investigate in the following two lemmas how the GSO of $(b_1, \ldots, b_\ell)$ changes in steps 6 and 7–10.

LEMMA 13.16. (i) *We consider one execution of step* 6 *for* i, j *with* $1 \le j < i \le \ell$. *Let* $B, B^* \in \mathbb{Q}^{\ell \times n}, M \in \mathbb{Q}^{\ell \times \ell}$ *and* $C, C^* \in \mathbb{Q}^{\ell \times n}, N \in \mathbb{Q}^{\ell \times \ell}$ *be the matrices of the* b_k, b_k^*, μ_{kh} *before and after the replacement, respectively, and* $E = (e_{kh}) \in \mathbb{Z}^{\ell \times \ell}$ *the matrix which has* $e_{kk} = 1$ *for all* k, $e_{ij} = -\lceil \mu_{ij} \rfloor$, *and* $e_{kh} = 0$ *otherwise. Then*

$$C = EB, \ C^* = B^*, \ and \ N = EM.$$

(ii) *The following invariant holds before each execution of step* 6:

$$|\mu_{ih}| \le \frac{1}{2} \ for \ j < h < i.$$

(iii) *The Gram-Schmidt orthogonal basis* $b_1^*, \ldots, b_\ell^*$ *does not change in step* 6, *and after the loop in steps* 5–6 *we have* $|\mu_{ih}| \le 1/2$ *for* $1 \le h < i$.

PROOF. (i) Let $\lambda = \lceil \mu_{ij} \rfloor$. The equality $C = EB$ is just another way of saying that b_i is replaced by $b_i - \lambda b_j$ and all other b_k remain unchanged. Since $j < i$, for any $k \le \ell$ the space spanned by $b_1, \ldots, b_k$ remains the same, and hence the orthogonal vectors $b_1^*, \ldots, b_\ell^*$ are not changed, which means that $B^* = C^*$. Now the third claim follows from the equations analogous to (15.76),

$$EMB^* = EB = C = NC^* = NB^*,$$

and the fact that B^* has rank ℓ.

(ii) The invariant is trivial at the beginning of the loop, and we suppose that it holds before step 6. Now the multiplication by E subtracts λ times row j from row i of M, which does not affect the μ_{ih} with $j < h$. Finally, μ_{ij} is replaced by $\mu_{ij} - \lambda \mu_{ij}$, which has absolute value at most $1/2$, by the choice of λ, and the invariant holds again before the next execution of step 6.

(iii) is immediate from (i) and (ii). □

The effect of one replacement step 6 is illustrated in Figure 13.4 for $\ell = 8$, $i = 6$, and $j = 3$. Each · represents a number which is absolutely less that $1/2$; it is not modified in the current or any later replacement in the loop 5–6. Each ∘ represents a number that is changed in step 6, and the • is μ_{63}, which is reduced to not more than $1/2$ in absolute value by it. (This is also the position of the lonely nonzero off-diagonal entry of E.) The values of ∗ are arbitrary and not changed by the current or any later replacement in the loop 5–6.

$$\begin{pmatrix} 1 & 0 & \cdots & \cdots & \cdots & \cdots & \cdots & 0 \\ \cdot & 1 & \ddots & & & & & \vdots \\ \cdot & \cdot & 1 & \ddots & & & & \vdots \\ \cdot & \cdot & \cdot & 1 & \ddots & & & \vdots \\ \cdot & \cdot & \cdot & \cdot & 1 & \ddots & & \vdots \\ \circ & \circ & \bullet & \cdot & \cdot & 1 & \ddots & \vdots \\ * & * & * & * & * & * & 1 & 0 \\ * & * & * & * & * & * & * & 1 \end{pmatrix}$$

Figure 13.4: The effect of one replacement step on the μ_{ij}.

LEMMA 13.17. *Suppose that b_{i-1} and b_i are exchanged in step 8, and denote by $c_1, \ldots, c_\ell$ and $c_1^*, \ldots, c_\ell^*$ the values of the vectors and their Gram-Schmidt orthogonal basis after the exchange, respectively. Then*

(i) $c_k^* = b_k^*$ *for $k \in \{1, \ldots, \ell\} \setminus \{i-1, i\}$,*

(ii) $\|c_{i-1}^*\|^2 < \frac{3}{4}\|b_{i-1}^*\|^2$,

(iii) $\|c_i^*\| \le \|b_{i-1}^*\|$.

PROOF.   (i) For such k, we have $b_k = c_k$ and $\sum_{1 \le h < k} \mathbb{R}b_h = \sum_{1 \le h < k} \mathbb{R}c_h$, and so Theorem 15.78 (ii) implies that $c_k^* = b_k^*$.

(ii) The vector c_{i-1}^* is the component of b_i orthogonal to $U = \sum_{1 \le h < i-1} \mathbb{R}b_h$, that is, its projection to the orthogonal space $U^\perp$. Since $b_i = b_i^* + \sum_{1 \le h \le i-1} \mu_{i,h} b_h^*$, we have $c_{i-1}^* = b_i^* + \mu_{i,i-1} b_{i-1}^*$ and

$$\|c_{i-1}^*\|^2 = \|b_i^*\|^2 + \mu_{i,i-1}^2 \|b_{i-1}^*\|^2 < \frac{1}{2}\|b_{i-1}^*\|^2 + \frac{1}{4}\|b_{i-1}^*\|^2 = \frac{3}{4}\|b_{i-1}^*\|^2,$$

by the condition for the exchange, the orthogonality of b_i^* and b_{i-1}^*, and the fact that $|\mu_{i,i-1}| \le 1/2$, by the previous lemma.

(iii) We let $u = \sum_{1 \le h < i-1} \mu_{i-1,h} b_h^*$. Then the vector c_i^* is the component of $b_{i-1} = b_{i-1}^* + u$ orthogonal to $U + b_i \mathbb{R}$. Now Theorem 15.78 implies that $u \in U \subseteq U + b_i \mathbb{R}$. Thus c_i^* is the component of b_{i-1}^* orthogonal to $U + b_i \mathbb{R}$, and hence $\|c_i^*\| \le \|b_{i-1}^*\|$.   □

LEMMA 13.18. *At the beginning of each iteration of the loop in steps 4–10, the following invariants hold:*

$$|\mu_{kh}| \le \frac{1}{2} \text{ for } 1 \le h < k < i, \quad \|b_{k-1}^*\|^2 \le 2\|b_k^*\|^2 \text{ for } 1 \le k < i.$$

PROOF.   The claim is trivial at the beginning of the algorithm. So we
assume that the invariants hold at the beginning of steps 5–6 and prove
that they hold again at the end of steps 7–10. Lemma 13.16 implies that
the first invariant also holds for $k = i$ immediately before steps 7–10, and
since an exchange does not affect the μ_{kh} for $k < i - 1$, the first invariant
holds after steps 7–10 in any case. Again by Lemma 13.16, the b_k^* do not
change in steps 5–6, and the second invariant is still valid immediately
before steps 7–10. Now an exchange in steps 7–10 does not affect the b_k^*
for $k \notin \{i - 1, i\}$, by Lemma 13.17, and the second invariant holds again
after steps 7–10 in any case as well.   □

In particular, Lemma 13.18 implies that the basis $b_1, \ldots, b_\ell$ is reduced
upon termination of the algorithm, and it remains to bound the number
of iterations of the loop in steps 4–10. At any stage in the algorithm and
for $1 \leq k \leq \ell$, we consider the matrix

$$B_k = \begin{pmatrix} b_1 \\ \vdots \\ b_k \end{pmatrix} \in \mathbb{Z}^{k \times n}$$

comprising the first k vectors, their Gram matrix $B_k \cdot B_k^T = (b_j \star b_h)_{1 \leq j, h \leq k} \in \mathbb{Z}^{k \times k}$, and the *Gram determinant* $d_k = \det(B_k \cdot B_k^T) \in \mathbb{Z}$.
For convenience, we let $d_0 = 1$.

LEMMA 13.19. *For* $1 \leq k \leq \ell$, *we have* $d_k = \prod_{1 \leq h \leq k} \|b_h^*\|^2 > 0$.

PROOF.   Let $1 \leq k \leq \ell$, let M_k be the upper left $k \times k$ submatrix of the
transition matrix M, and

$$B_k^* = \begin{pmatrix} b_1^* \\ \vdots \\ b_k^* \end{pmatrix} \in \mathbb{R}^{k \times n}.$$

Then $\det M_k = 1$, $B_k^* \cdot (B_k^*)^T \in \mathbb{R}^{k \times k}$ is a diagonal matrix with diagonal
entries $\|b_1^*\|^2, \ldots, \|b_k^*\|^2$, $B_k = M_k B_k^*$, and

$$\begin{aligned} d_k = \det(B_k B_k^T) &= \det(M_k B_k^* (B_k^*)^T (M_k)^T) \\ &= \det(M_k) \cdot \det(B_k^* (B_k^*)^T) \cdot \det(M_k^T) = \prod_{1 \leq h \leq k} \|b_h^*\|^2. \end{aligned}$$

□

LEMMA 13.20.   (i) In steps 5–6, none of the d_k changes.

(ii) If b_{i-1} and b_i are exchanged in step 7–10 and d_k^* denotes the new value of d_k, for any k, then $d_k^* = d_k$ for $k \neq i-1$ and $d_{i-1}^* \leq \frac{3}{4} d_{i-1}$.

PROOF.   (i) follows from Lemmas 13.16 and 13.19.

(ii) For $k \neq i-1$ the effect of the exchange on the matrix B_k is a multiplication by a $k \times k$ permutation matrix, whose determinant is 1 or -1. Thus $d_k^* = d_k$. Lemma 13.19 says that $d_{i-1} = \prod_{1 \leq h < i} \|b_h^*\|^2$, and Lemma 13.17 yields $d_{i-1}^* \leq \frac{3}{4} d_{i-1}$.   □

So now we have found our desired loop variant $D = \prod_{1 \leq k < \ell} d_k$, and can bound the number of arithmetic operations. We take the m-bit number $R = \max\{\|a_1\|, \ldots, \|a_\ell\|\}$ and note that $\ell \leq n$. Steps 1–3 take $O(n^3)$ operations in $\mathbb{Z}$. With the notation as in Lemma 13.16, one execution of step 6 amounts to computing the matrix products EB and EM, at a cost of $O(n)$ operations. Thus the number of operations in $\mathbb{Z}$ used in the loop in steps 5–6 is $O(\ell n)$. If an exchange happens in steps 7–10, then only b_{i-1}^*, b_i^*, and rows and columns $i-1$ and i of the transition matrix M change, and they can be updated using $O(n)$ operations, which is dominated by the cost for the loop 5–6. Write $R = 2^m$. We always have $1 \leq D \in \mathbb{Z}$, and its initial value D_0, at the start of the algorithm, satisfies

$$D_0 = \|a_1^*\|^{\ell-1} \|a_2^*\|^{\ell-2} \cdots \|a_{\ell-1}^*\| \leq \|a_1\|^{\ell-1} \|a_2\|^{\ell-2} \cdots \|a_{\ell-1}\| \leq R^{\ell(\ell-1)/2},$$

since a_i^* is a projection of a_i for all i. By Lemma 13.20, D does not change in steps 5–6, and decreases at least by a factor of $3/4$ if an exchange happens in steps 7–10, so that the number of such exchange steps is bounded by $\log_{4/3} D_0 \in O(\ell^2 m)$. At any stage in the algorithm, let $e \in \mathbb{N}$ denote the number of exchange steps performed so far and e^* the number of times where the else-branch in steps 7–10 has been taken. Since i is decreased by one in an exchange step and increased by one otherwise, the number $i + e - e^*$ is constant throughout the loop of steps 4–10. Initially, it equals 2, and hence $\ell + 1 + e - e^* = 2$ at termination, where $e \in O(\ell^2 m)$ by the above. Thus the total number of iterations of the loop in steps 4–10 is $e + e^* = 2e + \ell - 1 \in O(\ell^2 m)$, and we get a total of $O(\ell^3 nm) \subseteq O(n^4 m)$ operations in $\mathbb{Z}$, as claimed in Theorem 13.15.

Table 13.5 traces Algorithm 13.13 and the values of the Gram determinants d_k and of their product D on a three-dimensional lattice.

The intermediate results have bit length at most $O(nm)$, as claimed in Theorem 13.27. A proof can be found in von zur Gathen & Gerhard (2013), Section 16.3.

step	$\begin{pmatrix} b_1 \\ b_2 \\ b_3 \end{pmatrix}$	$\begin{pmatrix} \mu_{21} & \\ \mu_{31} & \mu_{32} \end{pmatrix}$	$\begin{pmatrix} \|b_1^*\|^2 \\ \|b_2^*\|^2 \\ \|b_3^*\|^2 \end{pmatrix}$	$\begin{matrix} d_1, d_2 \\ D \end{matrix}$	action
4	$\begin{pmatrix} 1 & 1 & 1 \\ -1 & 0 & 2 \\ 3 & 5 & 6 \end{pmatrix}$	$\begin{pmatrix} \frac{1}{3} & \\ \frac{14}{3} & \frac{13}{14} \end{pmatrix}$	$\begin{pmatrix} 3 \\ \frac{14}{3} \\ \frac{9}{14} \end{pmatrix}$	$3, 14$ 42	$\mathrm{rep}(3,2)$
4	$\begin{pmatrix} 1 & 1 & 1 \\ -1 & 0 & 2 \\ 4 & 5 & 4 \end{pmatrix}$	$\begin{pmatrix} \frac{1}{3} & \\ \frac{13}{3} & \frac{-1}{14} \end{pmatrix}$	$\begin{pmatrix} 3 \\ \frac{14}{3} \\ \frac{9}{14} \end{pmatrix}$	$3, 14$ 42	$\mathrm{rep}(3,1)$
5	$\begin{pmatrix} 1 & 1 & 1 \\ -1 & 0 & 2 \\ 0 & 1 & 0 \end{pmatrix}$	$\begin{pmatrix} \frac{1}{3} & \\ \frac{1}{3} & \frac{-1}{14} \end{pmatrix}$	$\begin{pmatrix} 3 \\ \frac{14}{3} \\ \frac{9}{14} \end{pmatrix}$	$3, 14$ 42	$\mathrm{ex}(3,2)$
5	$\begin{pmatrix} 1 & 1 & 1 \\ 0 & 1 & 0 \\ -1 & 0 & 2 \end{pmatrix}$	$\begin{pmatrix} \frac{1}{3} & \\ \frac{1}{3} & \frac{-1}{2} \end{pmatrix}$	$\begin{pmatrix} 3 \\ \frac{2}{3} \\ \frac{9}{2} \end{pmatrix}$	$3, 2$ 6	$\mathrm{ex}(2,1)$
4	$\begin{pmatrix} 0 & 1 & 0 \\ 1 & 1 & 1 \\ -1 & 0 & 2 \end{pmatrix}$	$\begin{pmatrix} 1 & \\ 0 & \frac{1}{2} \end{pmatrix}$	$\begin{pmatrix} 1 \\ 2 \\ \frac{9}{2} \end{pmatrix}$	$1, 2$ 2	$\mathrm{rep}(2,1)$
6	$\begin{pmatrix} 0 & 1 & 0 \\ 1 & 0 & 1 \\ -1 & 0 & 2 \end{pmatrix}$	$\begin{pmatrix} 0 & \\ 0 & \frac{1}{2} \end{pmatrix}$	$\begin{pmatrix} 1 \\ 2 \\ \frac{9}{2} \end{pmatrix}$	$1, 2$ 2	

Table 13.5: Trace of the basis reduction Algorithm 13.13 on the lattice $L = \mathbb{Z}(1,1,1) + \mathbb{Z}(-1,0,2) + \mathbb{Z}(3,5,6)$. We have $d_1 = \|b_1^*\|^2$, $d_2 = \|b_1^*\|^2\|b_2^*\|^2$, $D = d_1 d_2$, and $(\det L)^2 = d_3 = \|b_1^*\|^2\|b_2^*\|^2\|b_3^*\|^2 = 9$ throughout. Only the relevant values of the μ_{ij} and the squares of the norms of the b_i^* are given, and we have abbreviated a replacement $b_i \leftarrow b_i - \lceil \mu_{ij} \rceil b_j$ by $\mathrm{rep}(i,j)$ and an exchange of b_i and b_{i-1} by $\mathrm{ex}(i, i-1)$ in the "action" column.

COROLLARY 13.21. *Given linearly independent vectors $a_1, \ldots, a_\ell \in \mathbb{Z}^n$ whose norm has bit length at most m, the basis reduction algorithm 13.13 computes a reduced basis $b_1, \ldots, b_\ell$ of $L = \sum_{1 \le i \le \ell} \mathbb{Z}a_i$. Furthermore, b_1 is a "short" nonzero vector in L with*

$$\|b_1\| \le 2^{(\ell-1)/2}\lambda_1(L) = 2^{(\ell-1)/2}\min\{\|y\| : 0 \ne y \in L\}.$$

The algorithm uses $\mathrm{O}(n^6 m^3)$ *bit operations.*

PROOF.   The claim follows immediately from Theorem 13.15 and the remark before this corollary, noting that one arithmetic operation in $\mathbb{Z}$ (addition, multiplication, division with remainder, or gcd) on integers of length k can be performed with $\mathrm{O}(k^2)$ bit operations.   □

Suppose that each entry of the a_i has at most k bits. Then $m \leq (\log_2 n)/2 + k$, the input size N is at most $N \approx n^2 k$, and the algorithm's running time becomes $\mathrm{O}^\sim(n^2 N^2)$. In our applications, each a_i will typically be "sparse", with length only about $n + k$ rather than nk.

13.4. Breaking subset sum cryptosystems

We now have established the basic tools for this chapter, lattices and basis reduction, and come to the cryptologic applications. Two successful cryptanalyses are presented for starters, namely of the subset sum cryptosystem and of truncated linear congruential generators.

The *subset sum problem* (or *knapsack problem*) asks to decide the following question.

(13.22)   Given $t_0, t_1, \ldots, t_n \in \mathbb{N}$, are there $x_1, \ldots, x_n \in \{0, 1\}$ with $t_0 = \sum_{1 \leq i \leq n} t_i x_i$?

EXAMPLE 13.23. The input $(1215, 366, 385, 392, 401, 422, 437)$ means that we ask whether there exist $x_1, \ldots, x_6 \in \{0, 1\}$ such that $366 x_1 + 385 x_2 + 392 x_3 + 401 x_4 + 422 x_5 + 437 x_6 = 1215$.   ◇

This problem is NP-complete, as is a slight generalization of it, called the knapsack problem. If the decision question can be answered efficiently, one can also compute a solution, for example, by setting $x_n = 0$ if $(t_0, \ldots, t_{n-1})$ also has a solution, and $x_n = 1$ otherwise. After Diffie & Hellman (1976) invented public key cryptography, Merkle & Hellman (1978) proposed a public key encryption scheme based on the subset sum problem. The computations in this system were much less voluminous than for other systems such as RSA, and its higher throughput seemed to promise a bright future. Its connection to an NP-complete problem seemed to bode well for its security. Several other such systems were proposed, based on versions of the knapsack problem. But the roof fell in when Shamir (1984) broke the Merkle & Hellman system, and subsequently proposed improved schemes have suffered the same fate. Basis reduction has played a major role in some of these cryptanalyses.

The encryption scheme works as follows. BOB publishes his public key $t_1, \ldots, t_n$ for such a *subset sum cryptosystem* (or knapsack cryptosystem). When ALICE wants to send him n bits $x_1, \ldots, x_n$ secretly, she encodes them as $t_0 = \sum_{1 \le i \le n} t_i x_i$ and sends t_0. Deciding a general problem $(t_0, \ldots, t_n)$ is NP-complete and hence infeasible, but the idea now is to use a special type of problem for which decoding is easy with some secret additional knowledge, but hopefully hard without the secret. These special subset sum problems start with "superincreasing" $s_1 \ll s_2 \ll \cdots \ll s_n$ as the summands. A trivial example is $s_i = 2^{i-1}$, where the solution $(x_1, \ldots, x_n)$ is just the binary representation of s_0. More generally, it is sufficient to have $s_i > \sum_{1 \le j < i} s_j$ for all i; the solution is then unique and easy to calculate. The "easiness" is then hidden by multiplying the s_i with a random number r modulo another random number $m > \sum_{1 \le i \le n} s_i$ with $\gcd(r, m) = 1$ to obtain the public t_i for $i \ge 0$:

$$(13.24) \qquad\qquad t_i = r s_i \text{ in } \mathbb{Z}_m.$$

BOB's secret key $\mathsf{sk} = (r, m)$ allows him to multiply t_0 by r^{-1} modulo m and then solve an easy subset sum problem. At first sight, the t_i look like a general subset sum problem, but the cryptanalysts' work then showed that this hiding does not work. Of course, the breaking of these schemes does *not* mean that large instances of an NP-complete problem can be solved routinely; the "superincreasing" subset sum problem is just too special.

EXAMPLE 13.25. BOB takes $m = 1009$ and $r = 621$, his secret $s_1, \ldots, s_6$ as follows, and publishes $t_1, \ldots, t_6$.

i	s_i	t_i
1	2	233
2	3	854
3	7	311
4	15	234
5	31	80
6	60	936

If ALICE wants to send the bit string $x = 010110$ to BOB, she encrypts this as $t_0 = t_2 + t_4 + t_5 = 1168 = 159$ in $\mathbb{Z}_{1009}$. BOB computes $s_0 = r^{-1} t_0 = 13 \cdot 159 = 49$ in $\mathbb{Z}_{1009}$, and solves the easy subset sum problem $49 = 3 + 15 + 31$, from which he recovers x. $\Diamond$

We start by connecting subset sum problems to short vector problems. For Example 13.23, we consider the lattice $L \subseteq \mathbb{Z}^7$ generated by the rows of the matrix

$$
\begin{pmatrix}
1215 & 0 & 0 & 0 & 0 & 0 & 0 \\
-366 & 1 & 0 & 0 & 0 & 0 & 0 \\
-385 & 0 & 1 & 0 & 0 & 0 & 0 \\
-392 & 0 & 0 & 1 & 0 & 0 & 0 \\
-401 & 0 & 0 & 0 & 1 & 0 & 0 \\
-422 & 0 & 0 & 0 & 0 & 1 & 0 \\
-437 & 0 & 0 & 0 & 0 & 0 & 1
\end{pmatrix}
\in \mathbb{Z}^{7\times 7}.
$$

Algorithm 13.13 then computes the short vector $y = (0,0,0,1,1,1,0) \in L$, and indeed $1215 = 366 \cdot 0 + 385 \cdot 0 + 392 \cdot 1 + 401 \cdot 1 + 422 \cdot 1 + 437 \cdot 0$. Let a_i be the ith row vector, for $0 \leq i \leq 6$, so that $a_6 = (-437,0,0,0,0,0,1)$ as an example, $x_0 = 1$, and $x = (0,0,1,1,1,0) \in \{0,1\}^6$ the solution vector. Then $y = \sum_{0 \leq i \leq 6} a_i x_i$. Thus basis reduction solves this particular subset sum problem.

It is clear how to generalize this approach, and indeed that works reasonably well in reasonable situations. In order to obtain a proven result, we proceed as above, except that we multiply the first column by some large number. This has the effect of making the short vector unique in most cases, as specified below.

ALGORITHM 13.26.   Short vectors for subset sums.

Input: Positive integers $t_0, t_1, \ldots, t_n$.
Output: $(x_1, \ldots, x_n) \in \mathbb{Z}^n$ or "failure".

1. Let $M = \lceil 2^{n/2} n^{1/2} \rceil + 1$.
2. If $t_0 < \sum_{1 \leq i \leq n} t_i / 2$ then $t_0 \leftarrow \sum_{1 \leq i \leq n} t_i - t_0$.
3. For $0 \leq i \leq n$, let $a_i \in \mathbb{Z}^{n+1}$ be the ith row of the matrix

$$
\begin{pmatrix}
t_0 M & 0 & 0 & \cdots & 0 \\
-t_1 M & 1 & 0 & \cdots & 0 \\
-t_2 M & 0 & 1 & \cdots & 0 \\
\vdots & \vdots & \vdots & \ddots & \vdots \\
-t_n M & 0 & 0 & \cdots & 1
\end{pmatrix}
\in \mathbb{Z}^{(n+1)\times(n+1)}.
$$

4. Let $L \subseteq \mathbb{Z}^{n+1}$ be the lattice generated by $a_0, \ldots, a_n$. Run the basis reduction Algorithm 13.13 on this basis to receive a short nonzero vector $y = (y_0, \ldots, y_n) \in L$.
5. If $y_0 = 0$ and there is some $\delta \in \pm 1$ with $\delta y \in \{0,1\}^{n+1}$, then

$$
x \leftarrow \begin{cases}
(1 - \delta y_1, \ldots, 1 - \delta y_n) & \text{if the condition in step 2 is} \\
 & \text{satisfied for the input } t, \\
(\delta y_1, \ldots, \delta y_n) & \text{otherwise.}
\end{cases}
$$

else return "failure".
6. Return x.

Now let $(x_1, \ldots, x_n) \in \{0, 1\}^n$ be a solution of (13.22). Then

$$z = a_0 + \sum_{1 \le i \le n} x_i b_i = (0, x_1, \ldots, x_n) \in L$$

is a vector with norm $\|z\| \le \sqrt{n}$, which is small compared to the $t_i M$. The idea of the algorithm is to compute a reduced basis of L and to hope that the resulting short vector is (essentially) z.

In the remainder of this section, our goal is to show that if $t_1, \ldots, t_n$ are randomly chosen (as keys should be) within some large enough bound C, then Algorithm 13.26 returns the correct answer with high probability. We consider the following set of solvable subset sum problems:

$$E = \{(t_0, \ldots, t_n) \in \mathbb{Z}^{n+1} \colon \exists x \in \{0, 1\}^n \; t_0 = \sum_{1 \le i \le n} x_i t_i > 0$$

$$\text{and } 1 \le t_i \le C \text{ for } 1 \le i \le n\}.$$

THEOREM 13.27. *Let $\varepsilon > 0$, $n \ge 4$, let $C \ge \varepsilon^{-1} 2^{n(n + \log_2 n + 5)/2}$ be an integer, and consider inputs $t = (t_0, t_1, \ldots, t_n) \in E$ to Algorithm 13.26, where $(t_1, \ldots, t_n)$ is chosen uniformly at random in $T = \{1, \ldots, C\}^n$. Then the algorithm correctly returns a solution x to the subset sum problem t with probability at least $1 - \varepsilon$.*

PROOF. We set

$$E^* = \{t \in E \colon t_0 \ge \sum_{1 \le i \le n} t_i / 2\}.$$

Step 2 converts an input $t \in E \setminus E^*$ into one in E^* and conserves solvability, since

$$\sum_{1 \le i \le n} t_i (1 - x_i) = \sum_{1 \le i \le n} t_i - t_0.$$

For $t \in E$, there might be several solutions x. We fix one. Then

(13.28) $$z = (0, x_1, \ldots, x_n) = a_0 + \sum_{1 \le i \le n} x_i a_i \in L$$

with the vectors a_i from step 3, and $0 < \|z\| \leq n^{1/2}$, so that $\pm z$ are two "short" vectors in L. By Corollary 13.21, the output y of step 4 satisfies $\|y\| \leq 2^{n/2} \lambda_1(L) \leq 2^{n/2} n^{1/2} < M$. We consider the sublattice

$$K = \{v = (v_0, \ldots, v_n) \in L \colon v_0 = 0\}$$

of L. Thus K corresponds to the general solutions $t_0 = \sum_{1 \leq i \leq n} v_i t_i$ with all $v_i \in \mathbb{Z}$. For any $v = \sum_{0 \leq i \leq n} r_i a_i \in L$ with $r_0, \ldots, r_n \in \mathbb{Z}$, we have

$$v_0 = \sum_{0 \leq i \leq n} r_i a_{i,0} = M \cdot \left(r_0 t_0 - \sum_{1 \leq i \leq n} r_i t_i \right).$$

If $v_0 \neq 0$, then $\|v\| \geq |v_0| \geq M$. In particular, if $y_0 \neq 0$, then $M > \|y\| \geq |y_0| \geq M$. From this contradiction we conclude that $y_0 = 0$ and $y \in K$.

We now write $\mathbb{Z}u = \{ru \colon r \in \mathbb{Z}\}$ for the integer line through $u \in \mathbb{Z}^{n+1}$, and x_t, L_t, and K_t to indicate the dependence of the solution x and of the lattices L and K on $t \in E^*$. Furthermore, we extend x_t to $(0, x_t) = (0, (x_t)_1, \ldots, (x_t)_n) \in \mathbb{Z}^{n+1}$, and set

$$F = \{(v, t) \in \mathbb{Z}^{n+1} \times E^* \colon v \in K_t \setminus \mathbb{Z}(0, x_t), \|v\| \leq M, (t_1, \ldots, t_n) \in T\}.$$

We let $\pi_2 \colon F \longrightarrow E^*$ be the second projection. If $t \in E^*$ and $\pi_2^{-1}(t) = \varnothing$, then the only elements v of K_t with $\|v\| \leq M$ are the integer multiples of $(0, x_t)$, and y is one of them. Thus $y = \delta \cdot (0, x_t)$ for some nonzero $\delta \in \mathbb{Z}$. Let B be the basis of L returned by basis reduction, which includes y. Since $0 \neq (0, x_t) \in K_t \subseteq L_t$, $(0, x_t)$ has a unique representation as a linear combination over $\mathbb{Z}$ of the elements of B. Over $\mathbb{Q}$, we also have a unique such representation, and $(0, x_t) = \delta^{-1} y$. It follows that $\delta^{-1} \in \mathbb{Z}$ and $\delta \in \pm 1$. Furthermore, $x_i \in \{0, 1\}$ for $1 \leq i \leq n$, the conditions in step 5 are satisfied, and the solution x for the input t is returned in step 6.

It remains to show that $\pi_2^{-1}(t)$ is empty most of the time. We achieve this by proving that F is small; then also $\pi_2(F)$ is small.

For $(v, t) \in F$, we can write $v \in L$ as

$$v = (0, v_1, \ldots, v_n) = ua_0 + \sum_{1 \leq i \leq n} v_i a_i = \left(ut_0 M - \sum_{1 \leq i \leq n} v_i(t_i M, v_1, \ldots, v_n) \right)$$

$$= \left(ut_0 - \sum_{1 \leq i \leq n} v_i t_i, v_1, \ldots, v_n \right)$$

with some $u, v_1, \ldots, v_n \in \mathbb{Z}$. The factor M in the first coordinate may be removed because it is multiplied by 0. Then

$$|ut_0| = |\sum_{1 \le i \le n} v_i t_i| \le \sum_{1 \le i \le n} t_i \|v\|,$$

$$|u| \le (\sum_{1 \le i \le n} t_i)/t_0 \cdot \|v\| \le 2M,$$

using the condition in E^*. Thus for the shortened vector $v^* = (v_1, \ldots, v_n) \in \mathbb{Z}^n$ we have

$$\|v^*\| = \|v\| \le M, \quad |u| \le 2M, \quad v^* \notin \mathbb{Z}x_t,$$

$$(13.29) \qquad\qquad ut_0 = \sum_{1 \le i \le n} v_i t_i.$$

Since $v^* \notin \mathbb{Z}x_t$, we have $v \ne ux_t$ and can take some j with $1 \le j \le n$ and $v_j \ne u \cdot (x_t)_j$. Combining (13.28) and (13.29), we find

$$0 = \sum_{1 \le i \le n} v_i t_i - ut_0 - u\left(\sum_{1 \le i \le n} t_i(x_t)_i - t_0\right) = \sum_{1 \le i \le n} t_i \cdot (v_i - u(x_t)_i).$$

The coefficient of t_j in this linear equation is nonzero, so that t_j is uniquely determined. Hence for all $v \in \mathbb{Z}^{n+1}$ and taking the $2^n - 1$ possible values of x into account, we have

$$\#\{t \colon (v, t) \in F\} \le (2^n - 1)C^{n-1},$$

$$\#F \le \sum_{\substack{v^* \in \mathbb{Z}^n \\ \|v\| \le M}} 2^n C^{n-1} \le (2M+1)^n 2^n C^{n-1}$$

$$\le (2(2^{n/2}n^{1/2} + 2) + 1)^n 2^n C^{n-1}$$

$$\le 2^{n(n + \log_2 n + 5)/2} C^{-1} \cdot C^n \le \varepsilon C^n.$$

The penultimate inequality follows from the fact that

$$n \ge 4 \Longrightarrow k = \frac{n + \log_2 n + 2}{2} \ge 4$$

$$\Longrightarrow 5 \cdot 2^{-k} \le \frac{5}{16} < 0.414 < \sqrt{2} - 1$$

$$\Longrightarrow 2^k + 5 < 2^{k+1/2}.$$

We let p be the probability that Algorithm 13.26 returns "failure" on input $(t_0, \ldots, t_n) \in E$, where $(t_1, \ldots, t_n) \xleftarrow{\square} \{1, \ldots, C\}^n$. Then

$$p \le \frac{\#\pi_2(F)}{C^n} \le \frac{\#F}{C^n} \le \varepsilon. \qquad\qquad \square$$

The proof also shows that for $t \in E^* \setminus \pi_2(F)$, the two solution vectors $\pm(0, x_t)$ are the shortest vectors in L_t, and besides their integral multiples there is no vector in L_t of length at most $2^{(n-1)/2} n^{1/2}$.

Furthermore, we can take some $\gamma > 0$ and replace the condition on C by

$$(13.30) \qquad\qquad C \geq 2^{(1/2 + \gamma) n^2}$$

for large enough n.

EXAMPLE 13.31. For $n = 6$ and $\varepsilon = 1/10$, we can take $C = 36238786559$. We ran 100 examples with $(t_1, \ldots, t_6) \xleftarrow{\boxtimes} T = \{1, \ldots, C\}^6$ and $x \xleftarrow{\boxtimes} \{0, 1\}^6 \setminus \{(0, \ldots, 0)\}$, and the algorithm returned x in all cases. $\diamond$

The *density* $\delta(t)$ of a subset sum problem $t = (t_0, \ldots, t_n)$ in (13.22) is

$$\delta(t) = \frac{n}{\log_2(\max_{1 \leq i \leq n} t_i)},$$

assuming that $t_i \geq 2$ for some i. A subset sum cryptosystem encrypts n bits $x_1, \ldots, x_n$ into the single number $t_0 = \sum_{1 \leq i \leq n} x_i t_i$, whose bit length is on average about $\max_{1 \leq i \leq n} \{\log_2 t_i\}$. Thus $\delta(t)$ is roughly the information rate

$$\frac{\text{length of plaintext}}{\text{length of ciphertext}}.$$

When we take ϵ to be a constant, we can interpret Theorem 13.27 and (13.30) as saying that Algorithm 13.26 solves almost all subset sum problems t with

$$\delta(t) \leq \frac{2}{n}.$$

In practice, the algorithm performs much better, and seems to solve most subset sum problems with

$$(13.32) \qquad\qquad \delta(t) < 0.9408.$$

The superincreasing nature of $s_1 \ll s_2 \ll \cdots$ requires s_n to be large, and then also m and the t_i are large. In Merkle & Hellman's original proposal, they suggest $n = 100$ and m of 200 bits, with density 0.5.

In the above subset sum attack, we have not assumed any property that makes the problem easy for the owner of a secret, except that the density be low. Such problems are called *low-density subset sums*. Of course, the attack does not work well on general subset sums, since the problem is NP-complete.

EXAMPLE 13.33. The three examples in the text have the following densities.

	n	$\max\{\log_2 t_i\}$	$\delta(t)$
Example 13.23	6	$\log_2 437 \approx 8.771$	0.684
Example 13.25, t_i	6	$\log_2 60 \approx 5.907$	1.016
Example 13.25, s_i	6	$\log_2 936 \approx 9.870$	0.608

$\diamond$

The transmission rate is never more than 1. In the second line, the secret 111111 is transmitted as the 7-bit number 118, and the transmission rate is 6/7.

13.5. Truncated linear congruential pseudorandom generators

A further cryptanalytic use of basis reduction is to break certain pseudorandom number generators. The *linear congruential generator* produces

$$(13.34) \qquad x_i = sx_{i-1} + t \text{ in } \mathbb{Z}_m$$

for $i \geq 1$ from a modulus m and s, t, and a random seed x_0, all in $\mathbb{Z}_m$. This popular generator is cryptographically insecure. Namely, after observing just four consecutive outputs, one can (essentially) determine s, t, and m; see Section 11.2 for details. The generator is too nice to give up on it just because it is broken. We change it somewhat and hope that the modification will be secure. Namely, instead of outputting all of x_i, we only use part of it, say the top half of its bits. More generally, we take an integer approximation parameter $\alpha \geq 1$ and for $i \geq 0$ output an α-approximation y_i to x_i with

$$(13.35) \qquad |x_i - y_i| \leq \alpha.$$

There are many such y_i, and we need a deterministic way of determining one of them. A natural choice is

$$(13.36) \qquad y_i = \left\lfloor \frac{x_i}{\alpha} \right\rfloor \cdot \alpha;$$

see Section 15.3. Then (13.35) is satisfied, and

$$(13.37) \qquad x_i = \left\lfloor \frac{x_i}{\alpha} \right\rfloor \alpha + z_i = y_i + z_i$$

with $0 \leq z_i < \alpha$ is the usual division with remainder of x_i by α. As an example, suppose that m is a k-bit number, we take some $\ell < k$ and $\alpha = 2^\ell$. Then $y_i = \lfloor x_i \cdot 2^{-\ell} \rfloor \cdot 2^\ell$. The trailing ℓ bits of y_i are zero, and we might as well output $\lfloor x_i \cdot 2^{-\ell} \rfloor$ instead of y_i. Due to carries, this truncation is not robust under addition or subtraction, as explained below. Thus we cannot take this intuitive version of approximation by the leading bits, and rather work with (13.35). An adversary sees s, t, m, and $y_0, \ldots, y_{n-1}$. There are two attack goals: *prediction* produces the next value y_n, and *key recovery* yields x_0 (from which all other values follow as well).

We use the symmetric system of representatives modulo m

$$R_m = \{-\lfloor m/2 \rfloor, \ldots, \lfloor (m-1)/2 \rfloor\}$$

from Section 15.12, where $u \text{ srem } m \in R_m$ is the representative of $u \in \mathbb{Z}$ and $u = (u \text{ srem } m)$ in $\mathbb{Z}_m$. For an approximation parameter α and $u \in \mathbb{R}$, the α-*vicinity* of u is the set of integers whose distance from u is at most α:

(13.38) $V_\alpha(u) = \{v \in \mathbb{Z} : |u - v| \leq \alpha\}.$

If u and $v \in \mathbb{Z}$ are positive k-bit integers and their first $k - \ell$ bits agree, then $|u - v| < 2^{\ell+1}$ and $v \in V_{2^{\ell+1}}(u)$. But due to carries, the converse may be false. As an example, we take $k = 6$, $0 \leq \ell \leq 4$, $47 = (101111)_2 \in V_1(48) \subseteq V_{2^\ell}(48)$, and $48 = (110000)_2 \in V_1(47) \subseteq V_{2^\ell}(47)$. But the two (or more) leading bits of the 6-bit integers 47 and 48 do not agree.

In order to understand the vicinity $V_\alpha(u)$ of some $u \in \mathbb{Z}$, we suppose that both have at most k bits and $\alpha = 2^\ell$ for some integer $\ell < k$. We ignore the effect of long blocks of 1's or of 0's as discussed above. Then $v \in V_\alpha(u)$ means that $|u - v|$ is an ℓ-bit number. In Figure 13.6, the boxes are supposed to contain the bit representations of u and v, both positive and their leading bits to the left. Thus the high-order $k - \ell$ bits of u and v are equal, and the lower ℓ bits are unrelated. We may think of v as providing $k - \ell$ bits of information about u.

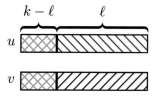

Figure 13.6: Two k-bit integers in the $2^{k-\ell}$-vicinity of each other.

We first show that key recovery from $y_0, \ldots, y_{n-1}$ is usually possible when $t = 0$ in (13.34), which we now assume. At the end of this section, we reduce the general case to this one. The unknown integers $x_0, \ldots, x_{n-1}$ satisfy

(13.39)
$$x_i = s x_{i-1} \text{ in } \mathbb{Z}_m,$$
$$x_i = s^i x_0 \text{ in } \mathbb{Z}_m \text{ for } 0 \le i < n.$$

We consider the lattice $L = L_{s,m}$ spanned by the rows $a_0, \ldots, a_{n-1} \in \mathbb{Z}^n$ of the following $n \times n$ matrix:

(13.40)
$$A = \begin{pmatrix} m & 0 & 0 & \cdots & 0 \\ -s & 1 & 0 & \cdots & 0 \\ -s^2 & 0 & 1 & \cdots & 0 \\ \vdots & \vdots & \vdots & \ddots & \vdots \\ -s^{n-1} & 0 & 0 & \cdots & 1 \end{pmatrix}.$$

The matrix has nonzero entries only on its diagonal and in the first column. L does not change if we replace an entry $-s^i$ by $-\sigma_i$ with $\sigma_i = s^i$ in $\mathbb{Z}_m$. As above, we write

(13.41)
$$z_i = x_i - y_i \text{ with } |z_i| \le \alpha$$

for each i. The z_i are unknown, and our task is to find them. Equation (13.39) implies that

$$z_i = x_i - y_i = s^i(y_0 + z_0) - y_i$$
$$= s^i z_0 + (s^{i-1} y_0 - y_i) \text{ in } \mathbb{Z}_m.$$

This is a set of linear congruences, but in contrast to the homogeneous congruences (13.39), they are inhomogeneous with (known) constants

(13.42)
$$c_i = s^i y_0 - y_i.$$

The attack on the generator proceeds in three steps:

o If the vector $z = (z_0, \ldots, z_{n-1}) \in L$ is "small", then z can be computed with a reduced basis of L by standard Gaussian elimination over $\mathbb{Q}$.

o Show with tools from lattice technology that $\|z\|$ is "small" for "almost all" generators.

o Extend the attack to inhomogeneous generators.

The lattice basis reduction Algorithm 13.13 works on n linearly independent vectors in $\mathbb{Z}^n$, and the first element b_1 of the reduced basis that it produces satisfies $\|b_1\| \le 2^{(n-1)/2} \lambda_1(L)$. We now need a generalization which gives a bound on each $\|b_i\|$ in terms of the successive minima $\lambda_i(L)$ from Definition 13.6.

THEOREM 13.43. *Let $L \subseteq \mathbb{R}^n$ be the lattice generated by its reduced basis $(b_1, \ldots, b_\ell) \in \mathbb{R}^{\ell \times n}$. Then for all $i \leq \ell$ we have*

$$\|b_i\| \leq 2^{(\ell-1)/2} \cdot \lambda_i(L) \leq 2^{(\ell-1)/2} \lambda_\ell(L).$$

PROOF.   For $i = 1$, this is Theorem 13.12. We now assume that $i \geq 2$. By Definition 13.11, we have $\|b_i^*\|^2 \leq 2^{j-i}\|b_j^*\|^2$ for $i \leq j$. By Theorem 13.9 there is some $k \geq i$ with $\|b_k^*\| \leq \lambda_i(L)$. From the Gram-Schmidt relation (15.73) and the properties of the Gram-Schmidt coefficients $\mu_{i,j}$ in Lemma 13.18, we obtain

$$\|b_i\|^2 = \|b_i^* + \sum_{1 \leq j < i} \mu_{i,j} b_j^*\|^2 \leq \|b_i^*\|^2 + \frac{1}{4}\sum_{1 \leq j < i} \|b_j^*\|^2$$
$$\leq \|b_k^*\|^2 (2^{k-i} + \frac{1}{4}\sum_{1 \leq j < i} 2^{k-j})$$
$$= \|b_k^*\|^2 (2^{k-i} + 2^{-2+k-i+1} \sum_{0 \leq h \leq i-2} 2^h)$$
$$= \|b_k^*\|^2 (2^{k-i} + 2^{k-i-1}(2^{i-1} - 1))$$
$$< \|b_k^*\|^2 (2^{k-1} + 2^{k-2}) \leq \lambda_i(L)^2 \cdot 2^{k-1},$$

since $i \geq 2$. In the sum, we have replaced j by $h = i - 1 - j$. The inequalities imply the claim. $\square$

We consider the matrix A in (13.40), $c \in \mathbb{Z}^n$, and the lattice L spanned by the rows of A.

LEMMA 13.44. *There is at most one $z \in \mathbb{Z}^n$ with $Az = c$ in $\mathbb{Z}_m^n$ and*

$$(13.45) \qquad\qquad \|z\| \leq \frac{m}{\lambda_n(L) \cdot (2^{(n+1)/2} + 1)}.$$

Given A, c, and m, one can determine in polynomial time whether such a z exists, and if so, compute it.

PROOF.   We apply the basis reduction algorithm 13.13 to the lattice L generated by the rows of A. It returns a reduced basis $B = (b_1, \ldots, b_n)$. By Theorem 13.3, $U = BA^{-1} \in \mathbb{Z}^{n \times n}$ is unimodular with $B = UA$. Since $\det U \in \pm 1$, U is also invertible over $\mathbb{Z}_m$. For $z \in \mathbb{Z}^n$, we have

$$(13.46) \qquad Az = c \text{ in } \mathbb{Z}_m^n \iff Bz = UAz = Uc \text{ in } \mathbb{Z}_m^n.$$

We take R_m and srem as in (15.12). Setting $c_i' = ((Uc)_i \text{ srem } m) \in R_m$, (13.46) holds with $c' = (c_1', \ldots, c_n')$ replacing Uc. Furthermore, for

$i \leq n$ and $z \in \mathbb{Z}^n$ satisfying (13.45), Theorem 13.43 and the Cauchy-Schwarz inequality of Theorem 15.72 yield

$$|b_i \star z| = |\sum_{1 \leq j \leq n} b_{ij} z_j| \leq \|b_i\| \cdot \|z\|$$

$$\leq 2^{(n-1)/2} \lambda_n(L) \cdot \frac{m}{\lambda_n(L)(2^{(n-1)/2+1} + 1)} = \frac{m}{2 + 2^{-(n-1)/2}} < \frac{m}{2},$$

$$b_i \star z = c_i' \text{ in } \mathbb{Z}_m \iff b_i \star z = c_i' \text{ in } \mathbb{Z}.$$

Since $b_1, \ldots, b_n$ are linearly independent over $\mathbb{Q}$, $z \in \mathbb{Q}^n$ is uniquely determined by the latter n conditions and can be computed by Gaussian elimination over $\mathbb{Q}$. It is a solution if and only if it has integer coordinates. $\qquad\square$

Lemma 13.44 with c as in (13.42) and (13.41) imply that if

$$(13.47) \qquad\qquad \alpha \leq \frac{m}{\lambda_n(L) \cdot (2^{(n+1)/2} + 1)},$$

then the approximated generator with $t = 0$ can be broken. In (13.47), we have to analyze $\lambda_n(L)$. More specifically, we show an upper bound on $\lambda_n(L)$ for almost all $s \in \mathbb{Z}_m$.

To this end, we need a new tool, namely the *dual lattice* L^* of a lattice $L \subseteq \mathbb{R}^n$, which is defined as

$$L^* = \{v \in \mathbb{R}^n : x \star v \in \mathbb{Z} \text{ for all } x \in L\}.$$

We recall the transpose A^T of a matrix A.

LEMMA 13.48. *If $A = (a_1, \ldots, a_n) \in \mathbb{R}^{n \times n}$ is nonsingular and L the lattice generated by the rows of A, then $B = (A^T)^{-1} \in \mathbb{R}^{n \times n}$ is a basis of the dual lattice L^*.*

PROOF. We let M be the lattice generated by the rows of $B = (b_1, \ldots, b_n)$ and take any $x \in L$. Then there exist $r_1, \ldots, r_n \in \mathbb{Z}$ such that $x = \sum_{1 \leq i \leq n} r_i a_i$. Then for all $j \leq n$

$$x \star b_j = \sum_{1 \leq i \leq n} r_i (a_i \star b_j) = r_j \in \mathbb{Z}.$$

Hence $b_1, \ldots, b_n \in L^*$ and $M \subseteq L^*$.

For the reverse containment, we take any $v \in L^*$. Since $\det B = (\det A)^{-1}$ is nonzero, M spans over $\mathbb{R}$ the whole $\mathbb{R}^n$. Thus there are $r_1, \ldots, r_n \in \mathbb{R}$ with $v = \sum_{1 \leq i \leq n} r_i b_i$. For all $j \leq n$ we have

$$\mathbb{Z} \ni v \star a_j = \sum_{1 \leq i \leq n} r_i (b_i \star a_j) = r_j.$$

It follows that $v \in M$, $M \subseteq L^*$, and $M = L^*$.   □

The lemma implies that

(13.49) $$(L^*)^* = L \text{ and } |L^*| = |L|^{-1};$$

see Exercises 13.15 and 13.16.

We now consider for $i \leq n$ the unit vectors $e_i = (0, \ldots, 0, 1, 0, \ldots, 0) \in \mathbb{Z}^n$, with 1 in position i, and

$$u = \frac{1}{m}(1, s, s^2, \ldots, s^{n-1}) \in \mathbb{Q}^n.$$

LEMMA 13.50. *The dual L^* of $L_{s,m}$ in (13.40) is generated by $u, e_2, \ldots, e_n$.*

PROOF.   The inverse of A is the matrix

$$A^{-1} = \begin{pmatrix} \frac{1}{m} & 0 & \cdots & 0 \\ \frac{s}{m} & 1 & \cdots & 0 \\ \vdots & 0 & \ddots & 0 \\ \frac{s^{n-1}}{m} & 0 & \cdots & 1 \end{pmatrix}.$$

By transposing the matrix we obtain by Lemma 13.48 a basis of L^*.   □

We use the following fact without proof.

THEOREM 13.51. *If L is an n-dimensional lattice, then $1 \leq \lambda_1(L^*) \cdot \lambda_n(L) \leq n$.*

Thus a lower bound on $\lambda_1(L^*)$ yields an upper bound on $\lambda_n(L)$. We next derive such a lower bound for most $s \in \mathbb{Z}_m$. For notational simplicity, we study the lattice $M = mL^*$ generated by the rows of

$$\begin{pmatrix} 1 & s & s^2 & \cdots & s^{n-1} \\ 0 & m & 0 & \cdots & 0 \\ 0 & 0 & m & \cdots & 0 \\ \vdots & \vdots & \vdots & & \vdots \\ 0 & 0 & 0 & \cdots & m \end{pmatrix}.$$

For any nonzero $v \in M$ whose first coordinate vanishes, we have $\|v\| \geq m$, so that v is "long". For any nonzero $t \in \mathbb{Z}$, the shortest vector in M with first coordinate t is

$$v = (t, st \text{ srem } m, s^2 t \text{ srem } m, \ldots, s^{n-1} t \text{ srem } m).$$

We want to show that for most s, all such vectors are "long". We may assume that $t \neq 0$ in $\mathbb{Z}_m$, since otherwise $\|v\| \geq m$, and also $t \in R_m$. Thus we consider, for a positive bound $C < m$, the set

$$E_C = \{s \in \mathbb{Z} \colon \begin{array}{l} |s^i t \text{ srem } m| < C \text{ for } 0 \leq i < n \text{ and} \\ \text{some } t \in \mathbb{Z} \text{ with } \gcd(t, m) = 1 \end{array} \}$$

of exceptional multipliers s. We will later assume m to be prime, so that the gcd condition corresponds to t srem $m \neq 0$. We have $\lambda_1(M) \geq C$ for all $s \in \mathbb{Z}_m \setminus E_C$.

LEMMA 13.52. Let $n \geq 2$ and $s \in E_C$. Then there exist $d_0, \ldots, d_{n-1} \in \mathbb{Z}$, not all divisible by m, with

$$(13.53) \qquad \sum_{0 \leq i < n} d_i s^i = 0 \text{ in } \mathbb{Z}_m,$$

$$|d_i| < (nC)^{1/(n-1)} + 2 \text{ for all } i < n.$$

PROOF. We let t be as in the definition of E_C, set $D = \lceil (nC)^{1/(n-1)}/2 \rceil$, and consider the set of all linear combinations

$$c = \sum_{0 \leq i < n} u_i t s^{i-1},$$

with $u_i \in \mathbb{Z}$ and $|u_i| \leq D$ for all i. Since $nC \leq (2D)^{n-1}$, we have for all such c

$$|c \text{ srem } m| \leq nDC \leq (2D)^{n-1} \cdot D = 2^{n-1} D^n,$$

and there are at most $(2D)^n + 1$ values of c srem m. The number of possible u is

$$(2D + 1)^n \geq 2^n D^n + n2^{n-1} D^{n-1} + 1 > (2D)^n + 1.$$

By the pigeon hole principle, there are two different vectors u with the same value of c srem m. We take their difference and divide it by t to obtain $d_0, \ldots, d_{n-1}$ satisfying (13.53) and

$$|d_i| \leq 2D < 2((nC)^{1/(n-1)}/2 + 1). \qquad \square$$

So far, the modulus m has been arbitrary. Now we restrict m to be a prime number. Then $\mathbb{Z}_m$ is a field, and a polynomial of degree i has at most i roots. (This is not generally true for composite m; the polynomial $t^2 - 1$ has four roots 1, 3, 5, and 7 in $\mathbb{Z}_8$.)

THEOREM 13.54. *Let m be a k-bit prime, $n \geq 9$, $\varepsilon > 0$, $2^{5n} \leq m^{1-\varepsilon}$,*

$$\ell \leq (1 - \varepsilon)(1 - \frac{1}{n})(k - 1) - 3n,$$

and $\alpha = 2^{\ell}$. Given s and m and α-approximations $y_0, \ldots, y_{n-1}$ to the output of the generator (13.34) with $t = 0$, the generator can be broken in polynomial time for all but at most $m^{1-\varepsilon}$ values $s \in \mathbb{Z}_m$.

PROOF.   For any $d_0, \ldots, d_{n-1} \in \mathbb{Z}$ with at least one d_i not divisible by m, the polynomial $\sum_{0 \leq i < n} d_i x^{i-1}$ has at most $n-1$ roots in $\mathbb{Z}_m$. It follows from Lemma 13.52 that

$$\#E_C \leq (n - 1)(2((nC)^{1/(n-1)} + 2) + 1)^n = (n - 1)(2(nC)^{1/(n-1)} + 5)^n.$$

We set

$$C = \frac{1}{n} \left(\frac{m^{1-\varepsilon}}{n \cdot 3^n} \right)^{(n-1)/n}.$$

Then $C < m$, and

$$n \cdot 15^n \leq 32^n \leq m^{1-\varepsilon},$$

$$5 \leq \frac{m^{(1-\varepsilon)/n}}{3 \cdot n^{1/n}} = (nC)^{1/(n-1)},$$

$$\#E_C < n \cdot 3^n (nC)^{n/(n-1)} = m^{1-\varepsilon}.$$

For all $s \in \mathbb{Z}_m \setminus E_C$, any nonzero vector in $M = mL^*$ has length at least C. It follows that $\lambda_1(L^*) \geq C/m$ and from Theorem 13.51 that

$$\lambda_n(L) \leq \frac{nm}{C}$$

for these s. Now (13.47) implies that the α-approximated generator with multiplier s is broken if

$$(13.55) \qquad \alpha \leq \frac{mC}{nm(2^{(n+1)/2} + 1)} = \frac{m^{(1-\varepsilon)(1-1/n)}}{n^{3-1/n}3^{n-1}(2^{(n+1)/2} + 1)}.$$

The denominator depends only on n and can be bounded as

$$n^{3-1/n}3^{n-1}(2^{(n+1)/2} + 1) \leq 8^n$$

for $n \geq 9$. Since $m > 2^{k-1}$, it is therefore sufficient to have

$$\alpha \leq m^{(1-\varepsilon)(1-1/n)}8^{-n},$$

$$\ell = \log_2 \alpha \leq (1 - \varepsilon)(1 - \frac{1}{n})(k - 1) - 3n. \qquad \square$$

This result is almost optimal in the following sense. We think of k as being large and of ε as small. Then the upper bound on $\ell = \log_2 \alpha$ is roughly $(1 - 1/n)k$, so that the approximations y_i only have about k/n bits of information about x_i. We use n approximations, with a total of about k bits, to reconstruct the k-bit seed x_1. Fewer than k bits do not contain enough information to determine a general k-bit number.

The number $m^{1-\varepsilon}$ might look like being close to m. But for $\varepsilon = 1/10$ and a modulus m of cryptographic size, say with 1000 bits, it has 900 bits. The probability for a random $s \in \mathbb{Z}_m$ to be in the exceptional set is only 2^{-100}.

We have broken the generator when $t = 0$, and now reduce the general case of (13.34) with arbitrary t to this one. Let $x'_i = x_{i+1} - x_i$ for $i \geq 0$. Then

$$x'_{i+1} = x_{i+2} - x_{i+1} = (sx_{i+1} + t) - (sx_i + t) = s(x_{i+1} - x_i) = sx'_i \text{ in } \mathbb{Z}_m,$$

so that the sequence $x'_1, x'_2, \ldots$ satisfies (13.34) with $t = 0$. Their approximations can be recovered from the original ones, as described below, with a loss of two bits.

We have to cope with the following issue. In the standard formulation (13.34), we take $\{0, 1, \ldots, m - 1\}$ as representatives of $\mathbb{Z}_m$, and these integers are approximated in the generator. Thus instead of x'_i, we have to use

$$(13.56) \qquad x^*_i = \begin{cases} x'_i = x_{i+1} - x_i & \text{if } x_{i+1} - x_i \geq 0, \\ x'_i + m = x_{i+1} - x_i + m & \text{otherwise.} \end{cases}$$

Then $x^*_0, x^*_1, \ldots$ satisfy (13.34) with $t = 0$. From approximations y_i to x_i, as observed for the attack, we have to determine approximations to the x^*_i.

According to the case distinction in (13.56), we set

$$(13.57) \qquad y^*_i = \begin{cases} y_{i+1} - y_i & \text{if } x_{i+1} - x_i \geq 0, \\ y_{i+1} - y_i + m & \text{otherwise.} \end{cases}$$

In the first case, we have

$$|x^*_i - y^*_i| = |x_{i+1} - y_{i+1} - (x_i - y_i)|$$
$$\leq |x_{i+1} - y_{i+1}| + |x_i - y_i| \leq 2\alpha.$$

In the second case, we find similarly

$$|x^*_i - y^*_i| = |x_{i+1} - y_{i+1} - (x_i - y_i) + m - m|$$
$$\leq |x_{i+1} - y_{i+1}| + |x_i - y_i| \leq 2\alpha.$$

In our attack, we are only given the y_i and do not know the sign of $x_{i+1} - x_i$. But we can (almost) deduce it. Namely, if y_i and y_{i+1} differ by at least 2α, say $y_i \geq y_{i+1} + 2\alpha$, then $x_i \geq y_i - \alpha \geq y_{i+1} + \alpha \geq x_{i+1}$ and we have the sign. If $|y_i - y_{i+1}| < 2\alpha$, we do not know this sign and pursue both possibilities. Hopefully the y_i are sufficiently random so that this undesirable branching occurs only rarely; see Exercise 13.17.

Thus we finally take

$$
y_i' = \begin{cases} y_{i+1} - y_i & \text{if } y_{i+1} \geq y_i + 2\alpha, \\ y_{i+1} - y_i + m & \text{if } y_{i+1} \leq y_i - 2\alpha, \\ \text{both } y_{i+1} - y_i \text{ and } y_{i+1} - y_i + m & \text{if } |y_{i+1} - y_i| < 2\alpha, \end{cases}
$$

and call the algorithm for Theorem 13.54 with s, m, $t = 0$, and 2α for α and the 2α-approximations $y_0, \ldots, y_{n-1}$. It either returns "failure" or some v so that $x_i' = s^i v$ in $\mathbb{Z}_m$ for $i \geq 0$, and $x_{i+1} = x_i + s^i v$. Suppose that y_{n-1} and z_{n-1} satisfy (13.36) and (13.37), and set $y = y_{n-1} + \lfloor s^n v / \alpha \rfloor \cdot \alpha$. Then

$$
x_n - y = x_{n-1} - y_{n-1} + s^n v - \lfloor s^n v / \alpha \rfloor \cdot \alpha,
$$
$$
0 \leq x_n - y < 2\alpha,
$$

and y_n, defined by (13.36) for $i = n$, equals either y or $y - \alpha$. Thus we can predict the next value of the approximate generator, namely the approximation (13.36) of the original generator (13.34), in the sense that it equals one of these two values. Furthermore, y equals the 2α-approximation given by (13.36) with 2α instead of α.

13.6. Close vectors

Given a lattice $L \subseteq \mathbb{R}^n$ and $u \in \mathbb{R}^n$, the *distance* of u to L is

$$
\text{dist}(u, L) = \min\{\|u - x\| : x \in L\}.
$$

An element $x \in L$ is a *closest vector* to u if $\|u - x\| = \text{dist}(u, L)$. The *closest vector problem* (CVP) is to compute such an x, given u and a basis of L. In the approximate *close vector problem* α-CVP, we are also given some $\alpha \geq 1$ and have to compute $x \in L$ with $\|u - x\| \leq \alpha \cdot \text{dist}(u, L)$. The following heuristic method reduces 2α-CVP to the similarly defined problem α-SVP. The claim is unproven, but the algorithm seems to work in practice, at least in high dimensions.

We write $u = (u_1, \ldots, u_n)$ and place the $n \times n$ matrix A representing a basis of L into the top right part of the following matrix:

$$(13.58) \qquad A' = \begin{pmatrix} 0 & & & \\ \vdots & & A & \\ 0 & & & \\ 1 & u_1 & \cdots & u_n \end{pmatrix}.$$

Using an algorithm for α-SVP, we compute a nonzero $y = (y_0, y_1, \ldots, y_n)$ in the lattice $M \subseteq \mathbb{R}^{n+1}$ generated by the rows of A', with $\|y\| \leq \alpha \cdot \lambda_1(M)$. If y_0 divides $y_1, \ldots, y_n$, we return $u - y_0^{-1} \cdot (y_1, \ldots, y_n)$, and otherwise "failure".

For $z \in L$, we have $(1, u - z) \in M$. We let z be a vector in L closest to u, and assume that all $y \in M$ with $\|y\| \leq \alpha \cdot \lambda_1(M)$ are integer multiples of $(1, u - z)$. Then in particular, $\pm(1, u - z)$ are the only two shortest vectors in M. The short vector y then satisfies

$$y = y_0 \cdot (1, u - x) = (y_0, y_{n+1}u - y_{n+1}x),$$
$$x = u - y_0^{-1}(y_1, \ldots, y_n),$$
$$\|u - x\| = |y_0^{-1}| \cdot \|(y_1, \ldots, y_n)\| < \|y\| \leq \alpha\lambda_1(M)$$
$$\leq \alpha\|(1, u - z)\| \leq 2\alpha\|u - z\| = 2\alpha \cdot \mathrm{dist}(u, L).$$

and the algorithm returns correctly.

The assumption does not always hold, but sufficiently often to make this *embedding method* a useful heuristic tool.

The following algorithm solves $2^{n/2}$-CVP inductively. It takes the $(n-1)$-dimensional sublattice $M \subset L$ generated by the first $n-1$ basis vectors of L, the $\mathbb{R}$-subspace generated by M, and $y \in L$ minimizing the distance between $\mathbb{Z} + y$ and u. Then it recursively applies this step, and computes a vector $x \in L$ which is close to u.

ALGORITHM 13.59. Nearest hyperplane.

Input: A reduced basis $B = (b_1, \ldots, b_\ell)$ of an ℓ-dimensional lattice L in $\mathbb{R}^n$, and $u \in \mathrm{span}_{\mathbb{R}}(L) \subseteq \mathbb{R}^n$.
Output: $x \in L$ with $\|u - x\| \leq 2^{\ell/2} \mathrm{dist}(u, L)$.

1. Compute the GSO $(b_1^*, \ldots, b_\ell^*)$ of $(b_1, \ldots, b_\ell)$; see Section 15.15.
2. $c \leftarrow u \star b_\ell^* / (b_\ell^* \star b_\ell^*)$.
3. $c' \leftarrow \lceil c \rfloor$,
   $v \leftarrow u - (c - c')b_\ell^*$,
   $y \leftarrow c'b_\ell$.
4. If $\ell = 1$, then return $x = y$. Else let M be the lattice generated by $b_1, \ldots, b_{\ell-1}$. Call the algorithm recursively to return $z \in M$ close to $v - y$.

5. Return $x = y + z$.

THEOREM 13.60. *The output x of the nearest hyperplane algorithm 13.59 satisfies $\|u - x\| < 2^{\ell/2} \operatorname{dist}(u, L)$. It runs in polynomial time.*

PROOF.   We show the first claim and simultaneously

(13.61) $$\|u - x\| \leq \frac{(2^\ell - 1)^{1/2}}{2} \|b_\ell^*\|$$

by induction on ℓ. For $\ell = 1$, we have $b_1^* = b_1$, $y \in L$ is a vector closest to u, and $x = y$. Furthermore, $u \in \operatorname{span}_{\mathbb{R}}(L) = \mathbb{R} \cdot b_1 = \mathbb{R} \cdot b_1^*$, say $u = c^* b_1^*$ with $c^* \in \mathbb{R}$. Then $c = c^*$, $u = c b_1^*$, and

$$\|u - x\| = \|(c - c')b_1^*\| \leq \|b_1^*/2\| = \frac{1}{2}\|b_1^*\|.$$

Now we assume $\ell \geq 2$. We have

(13.62) $$u - v = (c - c') \cdot b_\ell^* \in b_\ell^* \cdot \mathbb{R},$$

(13.63) $$\|u - v\| \leq \frac{\|b_\ell^*\|}{2}.$$

We let $Z = \operatorname{span}_{\mathbb{R}}(M)$ be the $\mathbb{R}$-vector space spanned by $b_1, \ldots, b_{\ell-1}$. By the definition of the GSO, b_ℓ^* is the projection of b_ℓ to $Z^\perp$, and b_ℓ^* is orthogonal to Z. We will repeatedly use decompositions $t = t_0 + t_1 b_\ell^*$ of various $t \in \mathbb{R}^n$, with $t_0 \in Z$ and $t_1 \in \mathbb{R}$. Then $t_0 \star (t_1 b_\ell^*) = 0$ and $\|t\|^2 = \|t_0\|^2 + \|t_1 b_\ell^*\|^2$. Now

(13.64) $$v - y = u - (c - c')b_\ell^* - c'b_\ell = (u - cb_\ell^*) + c'(b_\ell^* - b_\ell).$$

Since $u - cb_\ell^*$, $b_\ell^* - b_\ell \in Z$, we have

(13.65) $$v - y \in Z.$$

In particular, this input condition is satisfied in the recursive call. By Definition 13.11, we have

(13.66) $$\|b_{\ell-1}^*\|^2 \leq 2\|b_\ell^*\|^2.$$

Using (13.61) inductively and (13.62), (13.66), and $z \in Z$, we find

(13.67) $$\begin{aligned}\|u - x\|^2 &= \|u - v + v - y - z\|^2 = \|u - v\|^2 + \|v - y - z\|^2 \\ &\leq \frac{\|b_\ell^*\|^2}{4} + \frac{2^{\ell-1} - 1}{4} \|b_{\ell-1}^*\|^2 \leq \left(\frac{1}{4} + \frac{2^{\ell-1} - 1}{2}\right) \|b_\ell^*\|^2 \\ &= \frac{2^\ell - 1}{4} \|b_\ell^*\|^2,\end{aligned}$$

which proves (13.61). This implies that

$$(13.68) \qquad \qquad \|u - x\| < 2^{\ell/2-1} \|b_\ell^*\|.$$

Let $w \in L$ be a point in L closest to u. We now consider two cases. In the first one, we have $w \in Z + y$, so that $w - y \in Z$. We claim that $w - y \in L \cap Z = M$ is a point in M closest to $v - y$. So suppose that $m \in M$ satisfies

$$(13.69) \qquad \quad \|(v - y) - m\| \leq \|(v - y) - (w - y)\| = \|v - w\|.$$

Since $v - y \in Z$ by (13.64) and $v - w = (v - y) - (w - y) \in Z$, both vectors in (13.69) are in Z. Using (13.62), we find

$$
\begin{aligned}
(13.70) \qquad \|u - y - m\|^2 &= \|v - y - m + u - v\|^2 \\
&= \|v - y - m\|^2 + \|u - v\|^2 \leq \|v - w\|^2 + \|u - v\|^2 \\
&= \|v - w + u - v\|^2 = \|u - w\|^2.
\end{aligned}
$$

Since w is a closest point in L to u and $y + m \in L$, we have equality in (13.70), and hence in (13.69), as claimed. We also have

$$\|v - w\|^2 \leq \|v - w\|^2 + \|u - v\|^2 = \|v + u - v - w\|^2 = \|u - w\|^2.$$

By induction it follows that

$$
\begin{aligned}
(13.71) \qquad \|v - x\| = \|(v - y) - z\| &\leq 2^{(\ell-1)/2} \|(v - y) - (w - y)\| \\
&= 2^{(\ell-1)/2} \|v - w\| \leq 2^{(\ell-1)/2} \|u - w\|.
\end{aligned}
$$

Furthermore, we have $u - v = (c - c')b_\ell^*$ and (13.70) implies that $\|u - v\| \leq \|u - w\|$. Also, $v - x = (v - y) - z \in Z$ by (13.64). Now it follows with (13.71) that

$$
\begin{aligned}
\|u - x\|^2 = \|u - v\|^2 + \|v - x\|^2 &\leq (1 + 2^{\ell-1}) \|u - w\|^2 \\
&< 2^\ell \|u - w\|^2 = 2^\ell \operatorname{dist}(u, L)^2,
\end{aligned}
$$

as claimed.

In the second case, we have $w \notin Z + y$. We first claim that $\|w - y\| \geq \|b_\ell^*\|$. Setting $t = w - y$, we have $t \in L$ and $t \notin Z$. We consider the two representations

$$(13.72) \qquad \qquad t = \sum_{1 \leq i \leq \ell} r_i b_i = \sum_{1 \leq i \leq \ell} s_i b_i^*,$$

with uniquely determined $r_i \in \mathbb{Z}$ and $s_i \in \mathbb{R}$ for all $i \leq \ell$. Since $b_1^*, \ldots, b_{\ell-1}^* \in Z$ and $b_\ell^* \in Z^\perp$, we have $\|t\| \geq |s_\ell| \cdot \|b_\ell^*\|$. Furthermore, from (15.74) we have $b_\ell - b_\ell^* \in Z$. We set $r = t - r_\ell b_\ell^*$ and $s = t - s_\ell b_\ell^*$. Then $r, s \in Z$, and also $(r_\ell - s_\ell) b_\ell^* = s - r \in Z$. Since $b_\ell^* \in Z^\perp$, this shows that $r_\ell - s_\ell = 0$ and $s_\ell = r_\ell \in Z$. Since $t \notin Z$, we have $s_\ell \neq 0$, hence $|s_\ell| \geq 1$ and $\|t\| \geq \|b_\ell^*\|$, as claimed. Next, we have

$$w - v = (w - y) - (v - y) = s + s_\ell b_\ell^* - v + y$$

and $s - v + y \in Z$, by (13.65). Thus also $\|w - v\| \geq \|b_\ell^*\|$. Now (13.62) and the triangle inequality imply that

$$\|u - w\| + \frac{\|b_\ell^*\|}{2} \geq \|w - u\| + \|u - v\| \geq \|w - u + u - v\| \geq \|b_\ell^*\|,$$
$$\|u - w\| \geq \frac{\|b_\ell^*\|}{2}.$$

Finally, (13.68) shows that

$$\|u - x\| \leq 2^{\ell/2-1}\|b_\ell^*\| \leq 2^{\ell/2}\|u - w\| = 2^{\ell/2}\,\mathrm{dist}(u, L). \qquad \square$$

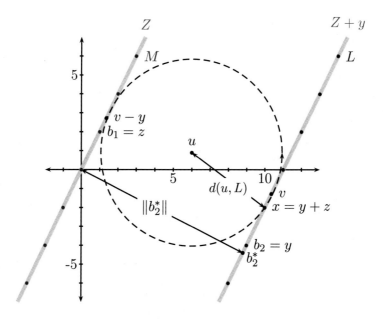

Figure 13.7: Trace of Algorithm 13.59 on the reduced basis $(b_1, b_2) = ((1,2),(9,-4))$ and the target vector $u = (6, 0.9)$.

EXAMPLE 13.73. We take the reduced basis $(b_1, b_2) = ((1,2),(9,-4))$ of the lattice L in Example 13.5 and Table 13.2 with $\ell = n = 2$, and

$u = (6, 0.9)$. In Figure 13.7, one sees four candidates in L that look close to u. Which one is the closest? We have

$$b_1^* = b_1,$$
$$b_2^* = (44/5, -22/5),$$
$$u = \frac{39}{25}b_1^* + \frac{111}{220}b_2^*,$$
$$c' = \left\lceil \frac{111}{220} \right\rceil = 1,$$
$$v = u - (\frac{111}{220} - 1) \cdot b_2^* = (6, 0.9) + \frac{109}{220} \cdot (\frac{44}{5}, \frac{-22}{5})$$
$$= (10.36, -1.28),$$
$$y = (9, -4),$$
$$v - y = (10.36, -1.28) - (9, -4) = (1.36, 2.72).$$

Thus $v - y = 1.36 \cdot (1, 2) \in Z = (1, 2) \cdot \mathbb{R} = b_1 \cdot \mathbb{R}$. The vector $z \in M = (1, 2) \cdot \mathbb{Z}$ closest to $v - y$ is $z = (1, 2)$. This is returned by the recursive call, and $x = y + z = (10, -2)$ is the algorithm's output. We have $\|u - x\| = \|(-4, 2.9)\| = \sqrt{24.41} \approx 4.94$.

Figure 13.7 depicts the trace of the algorithm. The circle around u with radius $\text{dist}(u, L)$ crosses the lattice (and the line $Z + y$) in the point x. Three further lattice points are just outside of the circle, while the point v is inside of it.   ◇

In the example, the algorithm actually finds the closest point to u. The general guarantee allows for an approximation factor of 2. It is quite typical of algorithms in this area to perform better in practice than guaranteed by the generally valid theorems.

13.7. The hidden number problem

In this section, we use a close vector algorithm to solve the following *hidden number problem*. We have a prime p, and want to find an unknown integer s, given some high-order bits of st_i in $\mathbb{Z}_p$ for various random t_i. This will be applied to the Diffie-Hellman problem in the next section.

We are given $t_1, \ldots, t_n \in \mathbb{Z}_p^\times$, a positive integer α, and some $u_i \in V_\alpha(st_i \text{ srem } p)$ for each $i \leq n$, and want to compute $s \in R_p$ (see (15.12)). We consider the lattice L spanned by the rows $a_0, \ldots, a_n$ of the following

$(n+1) \times (n+1)$ matrix:

$$
(13.74) \qquad A = \begin{pmatrix} 1/\alpha & t_1 & t_2 & \cdots & t_n \\ 0 & p & 0 & \cdots & 0 \\ 0 & 0 & p & & 0 \\ \vdots & \vdots & \vdots & \ddots & \vdots \\ 0 & 0 & 0 & \cdots & p \end{pmatrix}
$$

Then

$$
y = (s/\alpha, st_1 \text{ srem } p, \ldots, st_n \text{ srem } p) \in L.
$$

It would be sufficient to compute y, since its first coordinate discloses s, but we do not know how to do this. However, the known vector

$$
u = (0, u_1, \ldots, u_n) \in \mathbb{Z}^{n+1}
$$

is close to y:

$$
(13.75) \qquad \qquad \|u - y\| \leq \sqrt{n+1} \cdot p/\alpha.
$$

We now show that a close vector algorithm will actually find s, for most inputs $t_1, \ldots, t_n$.

ALGORITHM 13.76. Finding a hidden number.

Input: A prime p, positive integers α and n, and $t = (t_1, \ldots, t_n) \in (\mathbb{Z}_p^\times)^n$
      and $u = (u_1, \ldots, u_n) \in \mathbb{Z}^n$, so that there exists an (unknown)
      $s \in \mathbb{Z}_p$ with $u_i \in V_\alpha(t_i s \text{ srem } p)$ for all $i \leq n$.
Output: $s^* \in R_p$ with

$$
(13.77) \qquad u_i \in V_\alpha(t_i s^* \text{ srem } p) \text{ for all } i \leq n,
$$

    or "failure".

1. Run the basis reduction Algorithm 13.13 on the basis A in (13.74) and return a reduced basis B.
2. Let L be the lattice generated by B. Call the nearest hyperplane algorithm 13.59 to return some $x = (x_0, \ldots, x_n) \in L$ which is $2^{(n+1)/2}$-close to u.
3. $s^* \leftarrow x_0 \cdot \alpha$.
4. If (13.77) holds, then return s^* else return "failure".

THEOREM 13.78. *Let $p \geq 2^{36}$ be prime, $\lambda = (\log_2 p)^{1/2}$, $\varepsilon = \lambda^{-1}$, $\alpha \geq 2^{5\lambda}$, $n = \lfloor \lambda/2 \rfloor$, and assume $s \in \mathbb{Z}_p$ as specified. There exists a set $E \subset (\mathbb{Z}_p^\times)^n$ with $\#E \leq p^{n(1-\varepsilon)}$ such that for all inputs with $t \in (\mathbb{Z}_p^\times)^n \setminus E$, Algorithm 13.76 correctly computes $s^* = s$. The algorithm runs in polynomial time.*

PROOF. With $\beta = 2^{(n+1)/2}$, $\gamma = \beta\sqrt{n+1} \cdot p/\alpha$, and using (13.75), step 2 returns $x \in L$ with

$$\|u - x\| \leq \beta \cdot \|u - y\| \leq \gamma,$$

(13.79)
$$x = \sum_{0 \leq i \leq n} r_i a_i = (r_0/\alpha, r_0 t_1 + r_1 p, \ldots, r_0 t_n + r_n p),$$

where $r_0 = s^*$. The coefficients $r_0, \ldots, r_n \in \mathbb{Z}$ can be read off x, but we only need this for r_0. The close vector algorithm is deterministic, so that $t = (t_1, \ldots, t_n)$ fixes a unique $r_0(t) = x_0 \alpha = s^*$. If $s^* = s$, we have s, as desired. Our goal now is to show that this is usually the case.

There are some choices of $t = (t_1, \ldots, t_n)$ for which the algorithm does not work. For example, if all t_i are equal (which is improbable but possible) and all u_i are equal, then there is not enough information to determine s. At best, we can hope for an average-case result, which works for most t. We call

$$E = \{t \in (\mathbb{Z}_p^\times)^n \colon r_0(t) \neq s\}$$

the set of unlucky values. The algorithm works for all choices of t outside of E, and it is sufficient to show that E is small.

We take some $t \in E$ and some corresponding $u_1, \ldots, u_n$ with $u_i \in V_\alpha(st_i \text{ srem } p)$. Then (13.38) says that there exist $c_1, \ldots, c_n \in \mathbb{Z}$ so that

$$|u_i - (t_i s + c_i p)| \leq p/\alpha,$$

and (13.79) implies that

$$|u_i - x_i| \leq \gamma$$

for all $i \geq 1$. Let $d_i = u_i - (t_i s + c_i p)$ and $e_i = (r_0 - s)t_i$. Then

$$e_i = t_i(r_0 - s) = t_i r_0 - t_i s = x_i - r_i p - (u_i - d_i - c_i p)$$
$$= x_i - u_i + d_i + (-r_i + c_i)p,$$
$$|e_i \text{ srem } p| = |(x_i - u_i + d_i) \text{ srem } p| \leq |x_i - u_i + d_i|$$
$$\leq |x_i - u_i| + |d_i| \leq \gamma + p/\alpha < 2\gamma.$$

For any nonzero $k \in \mathbb{Z}_p$, multiplication by k is a permutation of $\mathbb{Z}_p^\times$, and for any $i \leq n$

$$\#\{t_i \in \mathbb{Z}_p^\times \colon |kt_i \text{ srem } p| \leq 2\gamma\} = 4\gamma.$$

Now $k = r_0(t) - s \neq 0$ in $\mathbb{Z}_p$, and allowing for an arbitrary nonzero value of k, we have

$$\#E \leq (p-1)(4\gamma)^n < p^{n+1}(8 \cdot 2^{(n+1)/2}\sqrt{n+1} \cdot \alpha^{-1})^n.$$

Some calculations now show that this quantity is at most $p^{n(1-\epsilon)}$. First the bound on p implies that $\lambda \geq 6$. Hence

$$\frac{2}{\lambda - 2} \leq \frac{3}{\lambda},$$
$$12 + 2\log_2(\lambda + 2) \leq 3\lambda,$$

with equalities for $\lambda = 6$. This implies the following, using $\mu = \log_2 \alpha$.

$$\epsilon\lambda^2 + 3 + \frac{1}{2} + \frac{1}{2}\log_2(\frac{\lambda}{2} + 1) \leq \lambda + 3 + \frac{1}{2} + \frac{1}{2}(\log_2(\lambda + 2) - 1)$$

$$= \lambda + \frac{1}{4}(12 + 2\log_2(\lambda + 2)) \leq \lambda + \frac{1}{4} \cdot 3\lambda = \frac{11}{8}\lambda,$$

$$(\frac{1}{n} + \epsilon)\lambda^2 + 3 + \frac{n+1}{2} + \frac{1}{2}\log_2(n+1)$$

$$\leq (\frac{1}{\frac{\lambda}{2} - 1} + \epsilon)\lambda^2 + 3 + \frac{\lambda + 2}{4} + \frac{1}{2}\log_2(\frac{\lambda}{2} + 1)$$

$$\leq \frac{2}{\lambda - 2}\lambda^2 + \frac{\lambda}{4} + \frac{11}{8}\lambda < \frac{3}{\lambda} \cdot \lambda^2 + 2\lambda = 5\lambda \leq \mu,$$

$$(13.80) \quad (1 + \epsilon n)\lambda^2 + 3n + \frac{n(n+1)}{2} + \frac{n}{2}\log_2(n+1) \leq \mu n,$$

$$p^{n+1} = p^{1+\epsilon n} \cdot p^{n(1-\epsilon)},$$

$$\#E \leq p^{n+1}(8 \cdot 2^{(n+1)/2}(n+1)^{1/2} \cdot \alpha^{-1})^n \leq p^{n(1-\epsilon)}.$$

The polynomial runtime of the algorithm follows from Corollary 13.21 and Theorem 13.60. □

If $t \notin E$, then the algorithm correctly returns $s^* = s$. But if $t \in E$, then s^* might be different from s but still satisfy the condition (13.77). The algorithm does not distinguish between these two cases, since the input might as well correspond to s^*.

How strong (or weak) is Theorem 13.78? We made assumptions involving constants like $1/2$ or 5, but these choices are somewhat arbitrary. We now discuss the situation, ignoring constants. The number of bits in p (and s) is about $\log_2 p = \lambda^2$. Each u_i provides about $\mu = \log_2 \alpha \approx \lambda$ bits of information about s, and we have $n \approx \lambda$ such values, with a total of $\approx \lambda^2$ bits. Thus the information sufficient for the algorithm's success is also necessary, up to constant factors. In this sense, the result is optimal.

Can we trade a smaller μ for a larger n, or vice versa? The answer is no. Each summand in (13.80) has to be at most μn, in particular (disregarding constants)

$$\lambda^2 = \log_2 p, \quad \varepsilon n \log_2 p, \quad n^2 \le \mu n.$$

If we want to stay close to the information-theoretic barrier, we need $\mu n = \log_2 p$. Then we require $\varepsilon \le n^{-1}$ and $n \le \mu$. The latter comes from the uncertainty factor of about n bits for close vectors; this must be less than the precision with which u_i approximates st_i srem p. Thus $\mu \ge \lambda$ is required. Increasing μ will require larger n and reduce ε, which is undesired.

In the next section, we apply the algorithm with a uniformly random choice $t \xleftarrow{\boxdot} (\mathbb{Z}_p^\times)^n$. We have the following consequence.

COROLLARY 13.81. Let $p \ge 2^{22}$ be prime, $\lambda = (\log_2 p)^{1/2}$, $\alpha = \lceil 2^{5\lambda} \rceil$, and $n = \lfloor \lambda/2 \rfloor$. If $t \xleftarrow{\boxdot} (\mathbb{Z}_p^\times)^n$ is chosen randomly, then the success probability that $s^* = s$ in Algorithm 13.76 is at least

$$1 - p^{-1/4} > 1 - 1/40 > \frac{1}{2}.$$

PROOF. We set $\varepsilon = \lambda^{-1}$, so that $\lambda/2 - 1 \le n$, $p^{1-\varepsilon} \le p - 1$, and

$$2^\lambda \left(\frac{p}{p-1}\right)^{\lambda/2-1} < \left(4\left(1 + \frac{1}{p-1}\right)\right)^{\lambda/2} < 5^{\lambda/2}.$$

By Theorem 13.78, the failure probability of the algorithm is at most

$$\frac{p^{n(1-\varepsilon)}}{(p-1)^n} = \left(\frac{p^{1-\varepsilon}}{p} \cdot \frac{p}{p-1}\right)^n \le \left(p^{-\lambda^{-1}} \cdot \frac{p}{p-1}\right)^{\lambda/2-1}$$

$$= p^{-1/2+\lambda^{-1}} \left(\frac{p}{p-1}\right)^{\lambda/2-1} = p^{-1/2} \cdot 2^\lambda \left(\frac{p}{p-1}\right)^{\lambda/2-1}$$

$$\le p^{-1/2} \cdot 5^{\lambda/2} = 2^{-\frac{1}{2}(\log_2 p - (\log_2 5)/2 \cdot \sqrt{\log_2 p})}$$

$$< 2^{-(\log_2 p)/4} = p^{-1/4} < 1/40 < 1/2. \qquad \square$$

13.8. Security of Diffie-Hellman

We recall the Diffie-Hellman Problem 2.21 (DH_G): we have a cyclic group $G = \langle g \rangle$ with generator g, are given $A = g^a$ and $B = g^b$ (but do not know the exponents a and b), and have to compute g^{ab}. Then (g^a, g^b, g^{ab}) are a DH triple; see Definition 2.25.

DH_G is assumed to be hard in general. This assumption provides the security of the Diffie-Hellman key exchange 2.20. For $G = \mathbb{Z}_p^\times$, we need a prime p of about 3000 bits at current security requirements (Table 5.12). In many applications, only a small part of the common key is used, say the leading 256 bits as a shared AES key. We proceed to show that the leading $5 \cdot \sqrt{3000} \approx 274$ bits are secure.

We use the symmetric residues R_p of $\mathbb{Z}_p$ and srem p as in (15.12), and show that finding α-approximations of several Diffie-Hellman solutions for $\alpha \approx 2^\mu$ with $\mu = 5(\log_2 p)^{1/2}$ is as hard as finding all of g^{ab}. So we assume a "leading bits" algorithm which computes an α-approximation $u \in V_\alpha(g^{ab} \text{ srem } p)$ of g^{ab} from p, g, g^a, and g^b, and design an algorithm which takes the same type of inputs, and outputs g^{ab}.

ALGORITHM 13.82. Reduction from DH to leading bits of DH.

Input: A prime p, a generator g of $\mathbb{Z}_p^\times$, and $A, B \in \mathbb{Z}_p^\times$.
Output: Some $W \in \mathbb{Z}_p^\times$, likely to solve the DH problem for A, B, or "failure".

1. $\lambda \leftarrow (\log_2 p)^{1/2}$,
   $\alpha \leftarrow s\lceil 2^{5\lambda} \rceil$,
   $n \leftarrow \lfloor \lambda/2 \rfloor$.
2. $r \xleftarrow{\boxtimes} \mathbb{Z}_{p-1}$,
   $C \leftarrow Ag^r$.
3. For $1 \le i \le n$ do steps 4 and 5.
4. $\qquad d_i \xleftarrow{\boxtimes} \mathbb{Z}_{p-1}$,
   $\qquad D_i \leftarrow Bg^{d_i}$,
   $\qquad t_i \leftarrow C^{d_i}$.
5. $\qquad$ Call a leading bits algorithm for C and D_i to return
   $\qquad u_i \in V_\alpha(y_i \text{ srem } p)$, where (C, D_i, y_i) is a DH triple.
6. Call the hidden number Algorithm 13.76 with input $t = (t_1, \dots, t_n)$
   and $u = (u_1, \dots, u_n)$ to return $s \in \mathbb{Z}_p^\times$ or "failure". In the latter case
   return "failure".
7. Return $W = sB^{-r} \in \mathbb{Z}_p^\times$.

THEOREM 13.83. Let $p \ge 2^{36}$ be a k-bit prime and $G = \mathbb{Z}_p^\times$. The output W of the algorithm solves the DH$_G$ problem for A, B with probability at least $1/(4 \log_2 k)$. It uses polynomial time plus at most $\sqrt{k}/2$ calls to a leading bits algorithm for DH$_G$.

PROOF. The number of generators of $\mathbb{Z}_p^\times$ is $\varphi(p-1) \ge p/2\ln\ln p$; see Remark 4.15 (i). Since $A \ne 0$, $r \in \mathbb{Z}_{p-1}$ is uniformly random and g generates $\mathbb{Z}_p^\times$, $C = Ag^r$ is a uniformly random element of $\mathbb{Z}_p^\times$. Therefore it is a generator of $\mathbb{Z}_{p-1}^\times$ with probability $\varphi(p-1)/p \ge 1/2\ln\ln p$. We now assume C to be a generator. Then $t_1, \ldots, t_n$ are independent uniformly random elements of $\mathbb{Z}_p^\times$. We write $a = \mathrm{dlog}_g A$, $b = \mathrm{dlog}_g B$, and $E = g^{ab} \in \mathbb{Z}_p^\times$, so that (A, B, E) is a DH triple, and want to show that $W = E$ with high probability.

Since $C = g^{a+r}$ and $D_i = g^{b+d_i}$, $(C, D_i, g^{(a+r)(b+d_i)})$ is a DH triple. Let $s' = g^{(a+r)b} = EB^r$. Then $y_i = g^{(a+r)(b+d_i)} = s'C^{d_i} = s't_i$, and in step 6 we have $u_i \in V_i(s't_i \mathrm{\,srem\,} p)$. The conditions of Corollary 13.81 are satisfied, and in step 6 we have $s = s'$ with probability at least $1/2$, and then also $W = EB^r \cdot B^{-r} = E$. Thus the success probability of the reduction is at least

$$\frac{1}{2 \ln\ln p} \cdot \frac{1}{2} \geq \frac{1}{4 \log_2 k}. \qquad \square$$

The following axiomatic security statement is the upshot of this and the two previous sections. It says that if DH in $\mathbb{Z}_p^\times$ is secure against efficient attacks, then so are the leading $5\sqrt{\log_2 p}$ bits of DH.

COROLLARY 13.84. *Let $p \geq 2^{36}$ be a k-bit prime, $G = \mathbb{Z}_p^\times$, and $\alpha = \lceil 2^{5\sqrt{k}} \rceil$. If DH_G is secure against polynomial-time attacks with success probability at least $1 - 1/(4 \log_2 k)$, then DH_G is also secure against polynomial-time α-approximations.*

If $p = 2\ell + 1$ is a safe k-bit prime, with ℓ either prime or product of two odd primes, each with at least $k/4$ bits (see Section 3.4), then the claim is also true with the weaker assumption of security against success probability at least $1 - 2p^{-1/4}$. For $\ell = qr$, this follows from Theorem 13.83 and

$$\varphi(p - 1) = (q - 1)(r - 1) = p - (q + r - 1) \geq p(1 - 2p^{-1/4}).$$

We do not know how to check efficiently whether (A, B, W) is indeed a DH triple; this is the Decisional Diffie-Hellman Problem 2.23. But we can run the reduction repeatedly and take a majority vote to increase the success probability.

While the failure probability of Algorithm 13.76 involves the input randomization, Theorem 13.83 applies to any input of Algorithm 13.82. In our usual notion of attack, the leading bits algorithm $\mathcal{A}$ for DH_G is only required to have some nonnegligible success probability σ. The reduction only works correctly if all n calls to $\mathcal{A}$ in step 5 return a correct answer. Again, repeating the calls to $\mathcal{A}$ a polynomial number of times and taking the majority vote on the outcomes increases the success probability to exponentially close to 1.

13.9. The Coppersmith method

This section uses lattice basis reduction to find "small" integer roots of integer polynomials. The idea is due to Coppersmith (1997) and has been extended by May (2003) and others. The method will be applied to the cryptanalysis of RSA under special circumstances.

In order to recover an RSA plaintext from public and transmitted data, one has to compute x with $x^e = y$ in $\mathbb{Z}_N$, given only N, e, and y. In other words, one has to find a root modulo N of the polynomial $f = t^e - y \in \mathbb{Z}_N[t]$, where t is a variable. Polynomials of small degree can be factored efficiently modulo a prime number or over the integers, and the roots correspond to linear factors. But no efficient algorithm is known for factoring polynomials modulo composite numbers. In fact, factoring the modulus is efficiently reducible to factoring polynomials, even a completely innocuous-looking one like $f = t$; see Exercise 3.16.

Indeed, we are quite familiar with this: finding roots different from ± 1 of $f = t^2 - 1$ modulo composite numbers lies at the heart of the pseudoprimality test (Algorithm 3.7) and the random squares factoring method (Algorithm 3.48). The idea now is to reduce root-finding for $t^e - y$ modulo N to root-finding over the integers. Of course, this only works efficiently in special cases, namely when e is small. In fact, we use this method to find roots modulo an unknown factor of N, and linear polynomials are good enough for this application.

For $u = (u_0, \ldots, u_{n-1}) \in \mathbb{R}^n$, we have the 1-norm of u

$$(13.85) \qquad \|u\|_1 = \sum_{1 \leq i < n} |u_i|.$$

A famous result relating it to the 2-norm $\|\cdot\|$ from (13.2) is the Cauchy-Schwarz inequality in Theorem 15.72 (ii):

$$(13.86) \qquad \|u\|_1 \leq n^{1/2} \cdot \|u\|.$$

In the following, we identify a polynomial

$$(13.87) \qquad g = \sum_{0 \leq i < n} g_i t^i \in \mathbb{Z}[t]$$

with its coefficient vector $(g_0, \ldots, g_{n-1}) \in \mathbb{Z}^n$. Thus

$$\|g\|_1 = \sum_{0 \leq i < n} |g_i|,$$

$$\|g\| = \Big(\sum_{0 \leq i < n} g_i^2 \Big)^{1/2}.$$

For any integers w and M we have

(13.88) $$w = 0 \text{ in } \mathbb{Z}_M \text{ and } |w| < M \Longrightarrow w = 0.$$

The following lemma says that small roots modulo M of a small polynomial are actually roots in $\mathbb{Z}$. Such roots can then efficiently be found.

LEMMA 13.89. Let $g \in \mathbb{Z}[t]$ have degree less than n, let c and M be positive integers with

$$n^{1/2} \cdot \|g(c \cdot t)\| < M,$$

and let $r \in \mathbb{Z}$ satisfy $|r| \le c$ and $g(r) = 0$ in $\mathbb{Z}_M$. Then $g(r) = 0$ in $\mathbb{Z}$.

PROOF. We write g as in (13.87) and apply the Cauchy-Schwarz inequality (13.86) to $g(c \cdot t) = \sum_{0 \le i < n} g_i c^i t^i$. This yields

$$|g(r)| = \Big| \sum_{0 \le i < n} g_i r^i \Big| \le \sum_{0 \le i < n} |g_i r^i| \le \sum_{0 \le i < n} |g_i c^i|$$
$$= \|g(c \cdot t)\|_1 \le n^{1/2} \|g(c \cdot t)\| < M.$$

Since also $g(r) = 0$ in $\mathbb{Z}_M$, we have $g(r) = 0$ by (13.88). □

The lemma suggests a roundabout way of hunting down the small roots r modulo N of $f \in \mathbb{Z}[t]$: design a small polynomial g whose modular roots include those of f, say $g = fh$ for some polynomial h, and which has small norm. Then we can find the integer roots of g, which include the small modular ones by the lemma. If we allow h to have degree up to k, then this amounts to choosing for h an integral linear combination of $f, ft, ft^2, \ldots, ft^k$. Unfortunately, there usually is no g as desired. But the following relaxation leads to a result.

We choose some positive integer k. If $0 \le j \le k$ and $f(r) = 0$ in $\mathbb{Z}_N$, then $N^{k-j} f(r)^j = 0$ in $\mathbb{Z}_{N^k}$, and we take for g a linear combination of these $N^{k-j} f^j$. So we let

$$h_{ij} = N^{k-j} f^j t^i \in \mathbb{Z}[t]$$

for integers i and j with $0 \le i < e = \deg f$ and $0 \le j \le k$. Then

$$h_{ij}(r) = 0 \text{ in } \mathbb{Z}_{N^k}$$

for all i, j. What we have gained is the much larger bound N^k instead of just N on $\|g(c \cdot t)\|$ that we can allow for M in Lemma 13.89. We then use basis reduction to compute an integral linear combination g of the $h_{ij}(c \cdot t)$.

The approach as sketched works for an arbitrary polynomial f. For our application, linear polynomials are sufficient, and we only present this case. Our goal is to factor N. We know that f has a small root r modulo an unknown divisor m of N. Thus we have an integer N and a monic linear polynomial $f = t + w \in \mathbb{Z}[t]$ and want to find its small roots modulo m. The main algorithmic ingredient is the following procedure that builds a polynomial g with r as a root modulo m^k.

ALGORITHM 13.90. Small polynomial with high-order roots.

Input: A monic linear polynomial $f \in \mathbb{Z}[t]$, positive integers N, c, and k,
        and real μ with $0 < \mu \leq 1$.
Output: $g \in \mathbb{Z}[t]$.

1. $n \leftarrow \lceil k/\mu \rceil$.
2. $h_i \leftarrow \begin{cases} N^{k-i} f^i & \text{for } 0 \leq i \leq k, \\ t^{i-k} f^k & \text{for } k < i < n. \end{cases}$
3. Form the $n \times n$ matrix A whose rows are the coefficient vectors of $h_0(ct), \ldots, h_{n-1}(ct)$.
4. Apply the basis reduction algorithm 13.13 to the rows of A, with output $B = UA$ and $U \in \mathbb{Z}^{n \times n}$ unimodular. Let $(u_0, \ldots, u_{n-1}) \in \mathbb{Z}^n$ be the top row of U.
5. Return $g = \sum_{0 \leq i < n} u_i h_i$.

EXAMPLE 13.91. We trace the algorithm on the inputs $f = t + 53$, $N = 2183$, $c = 6$, $k = 4$, and $\mu = 1/2$. In step 1, we have $n = 8$. The polynomials h_i are

$$h_0 = 22709885409121,$$
$$h_1 = 10403062487\,t + 551362311811,$$
$$h_2 = 4765489\,t^2 + 505141834\,t + 13386258601,$$
$$h_3 = 2183\,t^3 + 347097\,t^2 + 18396141\,t + 324998491,$$
$$h_4 = t^4 + 212\,t^3 + 16854\,t^2 + 595508\,t + 7890481,$$
$$h_5 = t^5 + 212\,t^4 + 16854\,t^3 + 595508\,t^2 + 7890481\,t,$$
$$h_6 = t^6 + 212\,t^5 + 16854\,t^4 + 595508\,t^3 + 7890481\,t^2,$$
$$h_7 = t^7 + 212\,t^6 + 16854\,t^5 + 595508\,t^4 + 7890481\,t^3.$$

The 8×8 matrix A has as its rows the coefficients at $t^0, t^1, t^2, \ldots, t^7$ of

$h_i(6t)$ and looks as follows:

22709885409121	0	0	0	0	0	0	0
551362311811	62418374922	0	0	0	0	0	0
13386258601	3030851004	171557604	0	0	0	0	0
324998491	110376846	12495492	471528	0	0	0	0
7890481	3573048	606744	45792	1296	0	0	0
0	47342886	21438288	3640464	274752	7776	0	0
0	0	284057316	128629728	21842784	1648512	46656	0
0	0	0	1704343896	771778368	131056704	9891072	279936

Step 4 returns B and U, and the first rows of B and U are

$$b_0 = (-2163672, -4246020, 3044412, 315792,$$
$$-970704, 1127520, 2612736, 279936),$$
$$u = (0, 2, -500, 52065, -1435989, 16363, -156, 1),$$

respectively. The top left entry of A has fourteen decimal digits, while all entries of B have at most seven digits, as we see in b_0. Furthermore

$$\|b_0\| = 540 \cdot \sqrt{141277958} \approx 6418461.853.$$

The algorithm's output then is

$$g = t^7 + 56t^6 + 145t^5 - 749t^4 + 1462t^3 + 84567t^2 - 707670t - 2163672$$
$$= (t - 6) \cdot (t + 53) \cdot (t^2 + 13t + 63) \cdot (t^3 - 4t^2 + 29t + 108). \qquad \Diamond$$

LEMMA 13.92. *The output of Algorithm 13.90 satisfies* $\deg g < n$ *and*

$$\|g(ct)\| \leq 2^{(n-1)/4} N^{k(k+1)/2\ell} c^{(n-1)/2}.$$

For any $r \in \mathbb{Z}$ *and a divisor* m *of* N, *we have*

$$f(r) = 0 \text{ in } \mathbb{Z}_m \implies g(r) = 0 \text{ in } \mathbb{Z}_{m^k}.$$

The algorithm uses time polynomial in $\log(N \cdot \|f\|)$ *and* k/μ.

PROOF.   Since $\deg h_i = i$ for all $i < n$, A is a lower triangular matrix. Its determinant is the product of the diagonal entries:

$$\det A = \prod_{0 \leq i \leq k} c^i N^{k-i} \prod_{k < i < n} c^i = N^{k(k+1)/2} c^{n(n-1)/2}.$$

Let $a_0, \ldots, a_{n-1}$ be the rows of A, and let b_0 be the first vector in the reduced basis B. Then $b_0 = \sum_{0 \leq i < n} u_i a_i$, and the polynomial whose coefficients form b_0 is

$$\sum_{0 \leq i < n} u_i h_i(ct) = g(ct).$$

Since $c \geq 1$, we have $\|g(t)\| \leq \|g(ct)\|$. Using Definition 13.4, (13.8), and Theorem 13.12 we have

$$\|g\| \leq \|g(ct)\| = \|b_0\| \leq 2^{(n-1)/4}(\det A)^{1/n} = 2^{(n-1)/4}N^{k(k+1)/2\ell}c^{(n-1)/2},$$

as claimed. For the second assertion, we have

$$m^{k-i} \mid N^{k-i},$$
$$m^i \mid f(r)^i$$

for $0 \leq i \leq k$, and hence

$$m^k \mid h_i(r)$$

for $0 \leq i < n$. This implies that m^k divides $\sum_{0 \leq i < n} u_i h_i(r) = g(r)$. $\qquad \square$

THEOREM 13.93. *Let f, N, c, k, and μ be an input for Algorithm 13.90, g the output, and*

(13.94) $$\delta \geq \frac{1}{2} \log_N((\frac{k}{\mu} + 1)2^{k/2\mu}),$$

(13.95) $$c \leq N^{\mu^2 - \mu(\mu+2\delta)/k},$$

and $m \geq N^\mu$ be a divisor of N. Then the set R of all integer roots of g has at most $\lceil k/\mu \rceil$ elements and contains all $r \in \mathbb{Z}$ with $f(r) = 0$ in $\mathbb{Z}_m$ and $|r| \leq c$. R can be computed in polynomial time.

PROOF. Let $r \in \mathbb{Z}$ with $f(r) = 0$ in $\mathbb{Z}_m$ and $|r| \leq c$. Lemma 13.92 implies that $g(r) = 0$ in $\mathbb{Z}_{m^k}$. We claim that $\|g(ct)\| < m^k/(n-1)^{1/2}$. Since $\deg g < n$, we then have $g(r) = 0$ in $\mathbb{Z}$ by Lemma 13.89.

Since $k/\mu \leq n < k/\mu + 1$, we have

$$n^{1/2}2^{(n-1)/4} < (\frac{k}{\mu} + 1)^{1/2}2^{k/4\mu}N^\delta,$$

$$\delta + \frac{k(k+1)}{2\ell} + (\mu^2 - \frac{\mu(\mu+2\delta)}{k}) \cdot \frac{n-1}{2}$$

$$\leq \delta + \frac{k(k+1)}{2k/\mu} + (\mu^2 - \frac{\mu(\mu+2\delta)}{k}) \cdot \frac{k}{2\mu}$$

$$= \delta + \frac{k\mu + \mu}{2} + \frac{k\mu}{2} - \frac{\mu+2\delta}{2} = k\mu.$$

Lemma 13.92 and our assumptions imply that

$$n^{1/2}\|g(ct)\| \leq n^{1/2}2^{(n-1)/4}N^{k(k+1)/2\ell}c^{(n-1)/2}$$

$$< N^\delta \cdot N^{k(k+1)/2\ell}N^{(\mu^2-\mu(\mu+2\delta)/k)\cdot(n-1)/2} \leq N^{k\mu} \leq m^k. \quad \square$$

When we apply this theorem, we choose k large enough so that the summand $\mu(\mu + 2\delta)/k$ in the bound on c becomes quite small compared to μ^2.

13.10. Security of leading bits of an RSA prime

In Section 3.8, we briefly mention side channel attacks. They are typically run against implementations of cryptosystems like RSA on small devices, say smartcards. Such attacks perform physical measurements at certain components of the device, and can discover information about the secret key stored on it. In particular, they may reveal certain bits of secret data, like p or d in RSA.

Suppose that an adversary discovers in this way the most significant half of the bits of p. At first sight, it is not clear how to use this. But this section shows how to factor N completely and efficiently with this partial information.

This kind of result has two interpretations. On the one hand, it says that already moderately successful side channel attacks can be lethal. On the other hand, if we assume RSA to be secure, then just the top half of p is also secure. In Section 3.9, we have seen an even stronger result. Namely, just a single bit, the least significant one, is secure.

THEOREM 13.96. Let $p < q$ be primes, $N = pq \geq 2653$, and $v \in \mathbb{Z}$ with $|q - v| \leq N^{1/4}/2$. Given N and v, one can compute q in polynomial time.

PROOF. Let $f = t + v \in \mathbb{Z}[t]$, $r = q - v$, and $\alpha = \log_2 N$. Then $f(r) = 0$ in $\mathbb{Z}_q$ and $|r| \leq N^{1/4}/2$. Since $p < q$, we have $q > N^\mu$ for $\mu = 1/2$. We set

$$k = \lceil \alpha \rceil,$$
$$\delta = \frac{1}{2} + \frac{\log_2(4\alpha + 6)}{2\alpha},$$
$$c^* = N^{\mu^2 - \mu(\mu + 2\delta)/k} = N^{1/4 - (1/4 + \delta)/k},$$
$$c = \lfloor c^* \rfloor.$$

(13.94) is satisfied, since $\log_N x = (\log_2 x)/\alpha$ for all positive real x and

$$\frac{1}{2} \log_N((\frac{k}{\mu} + 1)2^{k/2\mu}) = \frac{1}{2\alpha} \log_2((2k + 1) \cdot 2^k)$$
$$\leq \frac{1}{2\alpha}(k + \log_2(2(\alpha + 1) + 1))$$
$$\leq \frac{1}{2\alpha}(\alpha + 1 + \log_2(2\alpha + 3)) = \delta.$$

To check that $|r| \leq c$, we have $\alpha \geq 11.373$ and

$$\log_2(4\alpha + 6) \leq \frac{\alpha}{2},$$

$$\delta \leq \frac{1}{2} + \frac{\alpha/2}{2\alpha} = \frac{3}{4},$$

$$\log_2 |r| \leq \log_2 \frac{N^{1/4}}{2} = \frac{\alpha}{4} - 1 = \alpha(\frac{1}{4} - \frac{1}{\alpha})$$

$$\leq \alpha(\frac{1}{4} - \frac{1}{k}) \leq \alpha(\frac{1}{4} - \frac{1/4 + \delta}{k}) = \log_2 c^*,$$

$$|r| \leq c^*,$$

$$|r| \leq c.$$

We now call Algorithm 13.90 with inputs f, N, c, k, and μ as above and output g. Lemma 13.92 says that $\deg g \leq \lceil k/\mu \rceil - 1 = 2k - 1$, and the coefficient size of g is polynomially bounded. Thus we can compute the set R of integer roots of g in polynomial time. Furthermore, $\#R \leq 2k \leq 2\alpha + 2$. The conditions of Theorem 13.93 are satisfied, so that $r = q - v \in R$ and $\gcd(N, r + v) = q$. We can perform this gcd computation for all $r^* \in R$ in polynomial time and thus find q.   □

EXAMPLE 13.97. We take $N = 2183 = 37 \cdot 59$ from Example 3.47. Then $N^{1/4}/2 < 3.42$ and $\alpha = \log_2 N \approx 11.09$. Thus $k = \lceil \alpha \rceil = 12$, $\mu = 1/2$,

$$\delta = \frac{1}{2} + \frac{\log_2(4\alpha + 6)}{2\alpha} \approx 0.75,$$

$$c^* = N^{1/4 - (1/4 + \delta)/k} \approx 3.59,$$

$$c = 3.$$

We are also given $v = 56$ with the guarantee that $|q - v| \leq N^{1/4}/2 < 3.42$. Then $|q - v| \leq c$. We call Algorithm 13.90 with inputs $f = t + 56$, N, c, k, and $\mu = 1/2$. In step 1, we have $\alpha = \lceil k/\mu \rceil = 24$, and form a 24×24 matrix A. The output is a polynomial $g \in \mathbb{Z}[x]$ of degree 23 which factors over $\mathbb{Z}$ as $g = (x - 3) \cdot (x - 56) \cdot h$, where $h \in \mathbb{Z}[x]$ is irreducible. Thus $R = \{3\}$ and $\gcd(56 + 3, N) = 59$. We have found the factor $q = 59$ of N, and then $p = N/59 = 37$.

Although the condition $N \geq 2653$ of Theorem 13.96 is not satisfied, the algorithm produces the desired result. No surprise! Basis reduction often works better than guaranteed.

We look at some experiments that delineate by how much the algorithm works better than guaranteed. The top six lines in Table 13.8 take $k = 12$, as required, and then deliver q even if the bound c on $|q - v|$ is as large as 8. The "roots" column presents all linear factors of g, with multiplicities.

For $k = 11$, the method works up to $c = 7$; only two results are shown. For $k \in \{9, 10\}$, $c = 7$ is also solved correctly, and for $4 \leq k \leq 8$, the method works for $3 \leq c \leq 6$. Two of the results are shown for $k = 5$, and all of them at $k = 4$. Details of the calculations for the very last line are presented in Example 13.91.   ◇

k	v	c	roots
12	56	3	$(t-3)^3$
	55	4	$(t-4)^3$
	54	5	$(t-5)^2$
	53	6	$(t-6)(t+53)^2$
	52	7	$(t-7)(t+52)^{10}$
	51	8	$(t-8)(t+51)^{10}$
11	56	3	$(t-3)^3$
	52	7	$(t-7)(t+52)^9$
5	56	3	$t-3$
	53	6	$(t-3)(t+53)^3$
4	56	3	$(t-3)(t+56)$
	55	4	$t-4$
	54	5	$t-5$
	53	6	$(t-6)(t+53)$

Table 13.8: Some experiments for factoring 2183.

COROLLARY 13.98. *Let $p < q$ be primes and $N = pq \geq 2653$. If N is hard to factor, then it is hard to find an approximation to q to within $N^{1/4}/2$.*

This method has many further applications, to recover the plaintext, factors of N, or the secret exponent d, under the assumption that one of these quantities is small (half of the typical bit length) or that some information about them (typically half the bits) is known. We present one more example in the next section and refer the reader to May (2010) for a substantial list of further results and a bibliography.

13.11. Security of leading bits of a secret CRT-RSA exponent

Section 3.7 discusses the Chinese Remainder version CRT-RSA of RSA, where the $n/2$-bit prime factors p and q of N are kept as part of the secret key. The exponents d and e are reduced modulo $p - 1$ and $q - 1$, respectively, to obtain d_p, d_q, e_p, and e_q. Then the RSA exponentiation can be performed with only one eighth of the cost of the standard method.

As explained in Section 3.7, this RSA variant falls prey to fault attacks. The side channel attacks mentioned in the previous section can also be used. How many bits of the $n/2$-bit d_q are sufficient? We show that slightly more than the top half are enough, provided that the public exponent e is small.

LEMMA 13.99. Let p, q, $N = pq$ be as in the RSA notation 2.15, k and v positive integers with $k \neq 0$ in $\mathbb{Z}_p$ and

$$(13.100) \qquad\qquad |kq - v| \leq N^{1/4}.$$

Given N and v, we can compute q in polynomial time.

THEOREM 13.101. In the RSA notation p, q, N, e, d of Figure 2.15, assume that $N^{1/4} < p < q$ and $1 < e \leq N^\alpha$ for some α with $0 < \alpha \leq 1/4$, and let $v \in \mathbb{Z}$ be an approximation of $d_q \in \mathbb{Z}_{q-1}$ with

$$|d_q - v| \leq N^{1/4-\alpha}.$$

Given N and v, one can factor N in polynomial time.

PROOF.   We set $c = \lfloor N^{1/4} \rfloor$. Then $e \leq c$. Since $ed_q = 1$ in $\mathbb{Z}_{q-1}$, there exists a (unique) positive integer k with

$$ed_q - 1 = k(q - 1).$$

Then

$$k = \frac{ed_q - 1}{q - 1} < e\frac{d_q}{q - 1} < e \leq c \leq N^{1/4}.$$

Hence p does not divide k. We now take $w = ev - 1$ as an approximation of kq. Indeed,

$$|kq - w| = |ed_q - 1 + k - (ev - 1)| = |e(d_q - v) + k|$$
$$\leq N^\alpha \cdot N^{1/4-\alpha} + c - 1 = N^{1/4} + c - 1,$$
$$|kq - w| \leq c + c - 1 = 2c - 1.$$

It follows that at least one of the three values $w^* \in \{w, w + c, w - c\}$ satisfies

$$|kq - w^*| \leq c \leq N^{1/4}.$$

We perform the calculations of Lemma 13.99 with each of them and receive y_0, y_1, y_2, at least one of which equals kq. Then q is among the three values $\gcd(y_i, N)$ for $i \in \{0, 1, 2\}$. $\qquad\square$

COROLLARY 13.102. *As in Theorem 13.101, we take the RSA notation* p, q, N, d, e *and* $0 < \alpha \leq 1/4$ *with* $N^{1/4} < p < q$ *and* $1 < e \leq N^\alpha$, *and assume that N is hard to factor. Then it is hard to find an approximation to d_q to within $N^{1/4-\alpha}$.*

EXAMPLE 13.103. Some years ago, parts of the German online banking system used a 1024-bit RSA modulus N, between 2^{1023} and 2^{1024}, and a fixed public exponent $e = 2^{16} + 1 = 65\,337 = 2^{1024/64} + 1$. For each such N, we have $2^{16} + 1 \leq N^\alpha$ with $\alpha < 0.016$. We can apply the corollary and conclude that it is hard to approximate d_q to within $N^{1/4-\alpha} > 2^{239.7}$.

If d is sufficiently random, then d_q is likely to have about 512 bits. If N is hard to factor, then it is also hard to find the leading $512 - 239 = 273$ bits of d_q. $\qquad\Diamond$

13.12. The complexity of short vector problems

In Section 15.7 we discuss the issue of worst-case vs. average-case complexity. In a spectacular breakthrough, Ajtai showed in 1996 the equivalence of average and worst case for a certain question concerning short vectors—the first result of this type.

Cryptographic systems were built, based on this and similar ideas, which are secure if certain lattice problems are hard in the worst case. In Section 13.13, we explain some of the ingredients of one such system. Its security is based on a short vector problem which is not known to be in P nor to be NP-hard. It has been around since the early 1980s, and nobody has yet found a polynomial-time solution. Until further notice, we may assume that such a solution does not exist.

The main point is that the lattice technology presents cryptographers a new tool with desirable properties that no previous system possesses. This area is the object of intense research at the time of writing, and we can expect further progress.

Until now in this chapter, we have discussed "positive" aspects of lattices, namely how to solve computational problems and how to use such results. We now turn to "negative" aspects, namely questions for which we have no efficient solution.

Cyptography has a wonderful way of turning the positive/negative aspects upside down. Inventors and users of a cryptosystem broken by a lattice computation may not share the "positive" view. But "negative" results mean hard problems, which may serve as the security anchor of cryptosystems. The remainder of this chapter is devoted to this aspect.

Concepts of hard problems fall into the purview of *complexity theory* (or *computational complexity*), briefly outlined in Section 15.6, whose notes point to textbooks where some of the complexity classes below are defined.

α	$\log \alpha$	class
$2^{n \log^2 \log n / \log n}$	$n \log^2 \log n / \log n$	P
$2^{n \log\log n / \log n}$	$n \log\log n / \log n$	BPP
$\vdots$	$\vdots$	$\vdots$
$\sqrt{n}$	$\frac{1}{2} \log n$	NP $\cap$ co-NP not NP-hard
$\sqrt{\dfrac{n}{\log n}}$	$\frac{1}{2}(\log n - \log\log n)$	NP $\cap$ co-AM not NP-hard
$\vdots$	$\vdots$	$\vdots$
$n^{1/\log\log n}$	$\log n / \log\log n$	hard
$n^{1/(\log n)^{\varepsilon}}$	$(\log n)^{1-\varepsilon}$	NP-hard
constant	0	NP-hard

Table 13.9: Complexity of α-approximations to SVP.

For a real $\alpha \geq 1$, we consider the problem α-SVP of computing an α-approximation to a shortest vector. The input is a basis of a lattice $L \subseteq \mathbb{R}^n$, and the output is a short vector $v \in L \setminus \{0\}$ with $\|v\| \leq \alpha \cdot \lambda_1(L)$. These are *function problems* and not members of the usual complexity classes, which consist of *decision problems*. The "class" entries of Table 13.9 must be taken with this proviso. The *gap* approach of the next section turns these function problems into decision problems which are then bona fide members of standard complexity classes.

We have the complexity results of Table 13.9 for α-SVP. For ease of comparison, we give both α and $\log \alpha$ as parameters, and leave out constant factors where appropriate. We have co-NP $\subseteq$ co-AM, and the claims "not NP-hard" hold under standard complexity assumptions.

Some of the reductions used are probabilistic, and the ε means "for any $\varepsilon > 0$". The basis reduction we use in this text gives $\alpha = 2^{(n-1)/2}$ (Corollary 13.21), which in the notation of Table 13.9 would be presented as $\alpha = 2^n$ and $\log_2 \alpha = n$. The first two entries of the table are slightly smaller than this.

The upshot is that we can efficiently α-approximate SVP with α

slightly smaller than the 2^n from Corollary 13.21, but not for $\alpha = \sqrt{n}$. The gap between these two values is substantial.

All the above are stated as computational problems, whose answer is a vector with certain properties. But the complexity classes consist of decision problems, whose answer is a single bit. A "promise problem" X consists of two disjoint subsets X_{yes} and X_{no} of $\{0, 1\}^*$. An algorithm solving X (probably) answers "yes" or "no", respectively, on inputs from $X_{\text{yes}} \cup X_{\text{no}}$. On other inputs, it may return any answer. The SVP problems are brought into this format by the GapSVP$_\alpha$ problem of Definition 13.105.

13.13. Lattice cryptography: the Regev cryptosystem

In this section, we present various problems related to lattices, including an encryption scheme by Regev (2009), and reductions between them:

(13.104)   GapSVP $\leq$ BDD $\leq$ LWE $\leq$ distinguish encryptions.

All the problems and reductions come with certain parameters which are not shown here, but will be defined as we go along, from left to right in (13.104). Our hardness axiom is that GapSVP cannot be solved efficiently. The chain (13.104) of reductions shows that under this assumption the cryptosystem is secure. For one of the reductions, namely BDD $\leq$ LWE, we do not present the details.

We perform exact computations on real inputs and numbers like $\sqrt{n}$. This is neither possible nor required. In practice, inputs are presented exactly, via rational numbers or even in binary, and all operations can be carried out in polynomial time with sufficient precision so that all errors introduced are negligible. Making this explicit would only obscure the real issues.

In this section, all lattices $L \subset \mathbb{R}^n$ are assumed to have dimension n. We start by defining the two problems GapSVP and BDD. The following relaxation of the question "is $\lambda_1(L) \leq d$?" does not insist on the correct answer if $\lambda_1(L) > d$, but only when $\lambda_1(L)$ is substantially bigger than d, by a factor of $\alpha(n)$. It does not ask to compute a shortest vector, but only whether its length satisfies those bounds.

DEFINITION 13.105. *For a function* $\alpha \colon \mathbb{N} \longrightarrow \mathbb{R}$ *with* $\alpha(n) \geq 1$ *for all* n, *we define the* α-*gap shortest vector problem GapSVP$_\alpha$ as follows. Input is a basis* A *of an* n-*dimensional lattice* L *and a positive real number* d. *The answer is*

$$\begin{cases} yes & \text{if } \lambda_1(L) \leq d, \\ no & \text{if } \lambda_1(L) \geq \alpha(n) \cdot d. \end{cases}$$

When $d < \lambda_1(L) < \alpha(n) \cdot d$, any answer is permitted.

Using a subroutine for the promise problem $GapSVP_\alpha$, we can efficiently approximate $\lambda_1(L)$ to within a factor $2 \cdot \alpha(n)$ by binary search on d. $GapSVP_\alpha$ is in NP for any α and in P for $\alpha = 2^{\varepsilon n}$ for any $\varepsilon > 0$.

The *bounded distance decoding* (BDD) problem comes from coding theory, where one asks for the decoding of a received message y, given a bound on the distance from y to the code.

DEFINITION 13.106. *For a function* $\gamma \colon \mathbb{N} \longrightarrow \mathbb{R}_{>0}$, *the* γ-*bounded distance decoding problem* BDD_γ *takes as input a basis of a lattice* L *in* $\mathbb{R}^n$ *and some* $t \in \mathbb{R}^n$ *with* $\mathrm{dist}(t, L) < \gamma(n)\lambda_1(L)$, *and asks to find a vector in* L *that is closest to* t. *If* $\mathrm{dist}(t, L) \geq \gamma(n)\lambda_1(L)$, *any lattice vector solves the problem.*

When α increases, $GapSVP_\alpha$ becomes easier, and when γ decreases, BDD_γ becomes easier, that is,

$$(13.107) \qquad\qquad BDD_\gamma \leq BDD_{\gamma'}$$

if $\gamma \leq \gamma'$. Thus the behavior of the two problems with respect to their parameters is inverse to each other.

We recall some basic notions of real probability theory. In contrast to the finite probability spaces of Section 15.16, we do not assign probabilities to individual numbers. Rather we have a "nice" density function $\rho \colon \mathbb{R}^n \to \mathbb{R}$ with $\int_{\mathbb{R}^n} \rho(x)\, dx = 1$ and define the probability of a *measurable* subset $A \subseteq \mathbb{R}^n$ as $\int_A \rho(x)\, dx$. These (Lebesgue-)measurable sets include all products of intervals, and complements and countable unions of measurable sets. The "niceness" of ρ is defined so that all required integrals are well-defined.

The *normal distribution* $N(0, v^2)$ on $\mathbb{R}$ with mean 0 and variance v^2 is given by the density function $\exp(-(x/v)^2/2)/\sqrt{2\pi}v$. Adding two samples from two normal distributions with variances v_1^2 and v_2^2, respectively, gives a sample from a normal distribution with mean 0 and variance $v_1^2 + v_2^2$. For our purposes, we scale the normal density as

$$(13.108) \qquad\qquad v = r/\sqrt{2\pi}$$

and extend it to dimension $n \geq 1$ to obtain, for a positive integer n and positive real r, the *Gaussian density function* $\delta_r^{(n)}$

$$\delta_r^{(n)} \colon \mathbb{R}^n \longrightarrow \mathbb{R},$$
$$x \longmapsto r^{-n} \cdot e^{-\pi(\|x\|/r)^2}.$$

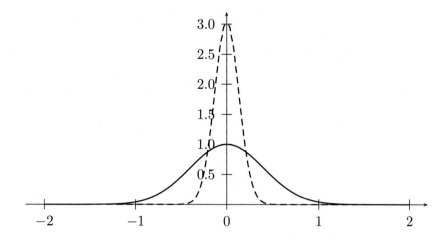

Figure 13.10: Plot of the functions $\delta_r^{(1)}$ on $\mathbb{R}$ for $r = 1$ (solid line) and $r = 1/3$ (dashed line).

The total volume of $\mathbb{R}^n$ under $\delta_r^{(n)}$ is

$$\int_{\mathbb{R}^n} \delta_r^{(n)}(x) \, dx = 1;$$

see Exercise 3.22. Thus we can define the continuous *Gaussian distribution* $D_r^{(n)}$ on $\mathbb{R}^n$ by its density function $\delta_r^{(n)}$, and $D_r^{(n)}(A) = \int_A \delta_r^{(n)}(x) \, dx$ for a measurable set $A \subseteq \mathbb{R}^n$ is the probability that some $x \in A$ is chosen if $x \xleftarrow{\boxtimes} D_r^{(n)}$. The chance for $x \xleftarrow{\boxtimes} D_r^{(n)}$ to be large decreases exponentially:

(13.109) $$\text{prob}\{\|x\| > tr\} = D_r^{(n)}(\mathbb{R}^n \smallsetminus \mathcal{B}_n(0, tr)) \leq e^{-t^2}$$

for positive real t.

In several algorithms, we need to generate uniform samples from a ball (13.7) or a Gaussian distribution. This can be done efficiently, given samples from the uniform distribution U on the real unit interval $[0 \mathinner{.\,.} 1)$.

PROPOSITION 13.110. *Given access to samples from U, one can efficiently generate samples from*

(i) *a Gaussian distribution* $G_r^{(n)}$,

(ii) *the uniform distribution on a ball* $\mathcal{B}_n(z, r)$.

PROOF. For (i), we take two independent samples $z_1, z_2 \xleftarrow{\boxtimes} U$, set $u_i = 2z_i - 1$ for $i \in \{1, 2\}$, $s = u_1^2 + u_2^2$ and repeat the selection if $s \geq 1$;

this happens with probability $1 - \pi/4 \approx 0.215$. Otherwise we set $t = \sqrt{(-2\ln s)/s}$ and return $x_i = u_i t$ for $i \in \{1, 2\}$. We claim that x_1 and x_2 are two independent samples from the normal distribution $N(0, 1)$. Once we know this, we can obtain a sample from $G_r^{(1)}$ as $y = rx_1/\sqrt{2\pi}$ by (13.108), and a sample $(y_1, \ldots, y_n)$ of $G_r^{(n)}$ via n independent samples $y_1, \ldots, y_n$ of $G_r^{(1)}$.

Since $s < 1$, $(u_1, u_2) \in \mathbb{R}^2$ is a uniformly distributed point in the unit disk $\mathcal{B}_2(0, 1)$. Using polar coordinates, we write $(u_1, u_2) = (w \cos\varphi, w \sin\varphi)$ with $s = w^2$ and $0 \le \varphi < 2\pi$. Then $(x_1, x_2) = (w' \cos\varphi, w' \sin\varphi)$ with $w' = \sqrt{-2\ln s}$. Furthermore, w and φ are independent, hence so are w' and φ, and φ is uniformly distributed in $[0 \mathinner{..} 2\pi)$. We have for any $y \in [-1, \mathinner{..} 1)$

$$\text{prob}\{w' \le y\} = \text{prob}\{-2\ln s \le y^2\} = \text{prob}\{s \ge e^{-y^2/2}\}$$
$$= 1 - e^{-y^2/2},$$

since s is uniform in $[0 \mathinner{..} 1)$. The uniformity of φ now implies that x_1 and x_2 are independent and normally distributed.

For (ii), we consider $\mathcal{B}_n(0, 1)$, take independent normal samples $y_1, \ldots, y_n \xleftarrow{\boxtimes} N(0, 1)$ and $u \xleftarrow{\boxtimes} U$, set $y = (y_1, \ldots, y_n) \in \mathbb{R}^n$ and $s = \|y\|$, and return $x = y \cdot u^{1/n}/s \in \mathbb{R}^n$. This works because the distribution of y has a density that depends only on $\|y\|$, and thus $1/s \cdot y$ is uniformly distributed on the sphere $\{z \in \mathbb{R}^n \colon \|z\| = 1\}$, and x in $\mathcal{B}_n(0, 1)$. Then also $r^n x + z$ is uniform in $\mathcal{B}_n(z, r)$. $\qquad\square$

The next lemma says that two balls close to each other have a nonnegligible intersection. The volume of an n-dimensional ball $\mathcal{B}_n(z, r)$ is

$$(13.111) \qquad \text{vol}_n(\mathcal{B}_n(z, r)) = \frac{\pi^{n/2} r^n}{\Gamma((n/2) + 1)},$$

where the gamma function Γ satisfies $\Gamma(x) = (x - 1)\Gamma(x - 1)$ for positive real $x \ne 1$, $\Gamma(1) = 1$, and $\Gamma(1/2) = \sqrt{\pi}$. Thus for a positive integer m, $\Gamma(m) = (m - 1)!$, and thus Γ interpolates the factorial function. For large real x, we have $\Gamma(x + 1) \approx \sqrt{2\pi x}(x/e)^x$, and thus, using (15.93)

$$\frac{\Gamma(x + 1/2)}{\Gamma(x)} \approx \frac{\left(\frac{2x+1}{2e}\right)^{(2x+1)/2}}{\left(\frac{x}{e}\right)^x} \approx \left(1 + \frac{1}{2x}\right)^x \cdot \sqrt{x} \cdot e^{-1/2} \approx \sqrt{x}.$$

In fact, the quotient approaches $\sqrt{x}$ rapidly from below, and for real $x \ge 2$ we have

$$(13.112) \qquad \frac{\Gamma(x + 1/2)}{\Gamma(x)} > 0.93\sqrt{x}.$$

LEMMA 13.113. *Let $n \geq 8$, $z \in \mathbb{R}^n$, $r > 0$, $\delta = \|z\|/r \leq \sqrt{(\log_2 n)/n}$, and B_0 and B_1 be the two balls of radius r centered at 0 and z, respectively. Then*

$$\frac{\mathrm{vol}_n(B_0 \cap B_1)}{\mathrm{vol}_n(B_0)} \geq \frac{1}{n^{3/2}}.$$

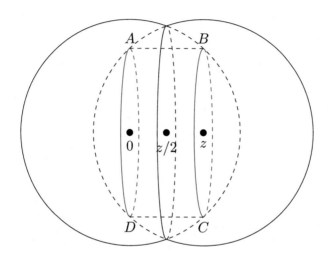

Figure 13.11: Two shifted balls containing a cylinder.

PROOF. Let $q = \mathrm{vol}_n(B_0 \cap B_1)/\mathrm{vol}_n(B_0)$. The volumes scale with r^n, and we may assume that $r = 1$. We consider the ball $B_2 = \mathcal{B}_n(z/2, 1 - \delta/2)$. For any $y \in B_2$ we have

$$\|y\| \leq \|y - z/2\| + \|z/2\| \leq 1 - \delta/2 + \delta/2 = 1,$$

so that $y \in B_0$. Similarly, $y \in B_1$, so that B_2 is contained in $B_0 \cap B_1$ and

$$q \geq \frac{\mathrm{vol}_n(B_2)}{\mathrm{vol}_n(B_0)} = (1 - \frac{\delta}{2})^n.$$

If $\delta \leq (\log_2 n)/n$, then with (15.92) we have

$$q \geq (1 - \frac{\log_2 n}{2n})^n \geq 2^{-2 \cdot (\log_2 n)/2} = \frac{1}{n}.$$

For $\delta > (\log_2 n)/n$, the bound from (15.92) becomes too small. Instead, we consider the following cylinder C contained in $B_0 \cap B_1$. We may assume that $z = (\delta, 0, \ldots, 0)$. Then the base of C is an $(n-1)$-dimensional ball B^* of radius $\sqrt{1 - \delta^2}$ containing the origin and perpendicular to the

x_1-axis, and C's height is δ. Figure 13.11 shows 3-dimensional balls at left and right. The intersection of their borders (two 3-dimensional spheres) is a circle (= 2-dimensional sphere) with center $z/2$ and indicated by the ellipse in the center. Imagine two soap bubbles stuck together at this vertical circle. The two smaller circles with center "0"= $(0,0,0)$ and z represent the base and the top, respectively, of the cylinder C (in blue).

Any point y in C is of the form $y = (y_1, y')$ with $0 \leq y_1 \leq \delta$ and $y' \in B^*$, so that indeed $\|y\|^2 \leq \delta^2 + (1 - \delta^2) = 1$, and similarly $\|y - z\|^2 \leq 1$. Thus $C \subseteq B_0 \cap B_1$ and using (15.92) and (13.112), we have

$$q \geq \frac{\mathrm{vol}_n(C)}{\mathrm{vol}_n(B_0)} \geq \frac{\delta \cdot \mathrm{vol}_{n-1}(B^*)}{\mathrm{vol}_n(B_0)} = \frac{\delta \cdot \pi^{(n-1)/2}(\sqrt{1 - \delta^2})^{n-1}\Gamma(\frac{n+2}{2})}{\Gamma(\frac{n+1}{2})\pi^{n/2}}$$

$$> \frac{0.93\,\delta(1 - \delta^2)^{(n-1)/2}\sqrt{n/2}}{\sqrt{\pi}} \geq \frac{0.93 \cdot \sqrt{n}\log_2 n}{\sqrt{2\pi} \cdot n} \cdot (1 - \frac{\log_2 n}{n})^{(n-1)/2}$$

$$> \frac{0.93\log_2 n}{\sqrt{2\pi n}} \cdot 2^{-\log_2 n} > \frac{\log_2 n}{3n^{3/2}} \geq \frac{1}{n^{3/2}}.$$

The two bounds together imply the claim.   □

We consider the (nonuniform) distributions X_0 and X_1 on $B_0 \cup B_1$, where X_0 is uniform on B_0 and zero on $B_1 \setminus B_0$; similarly for X_1. The statistical distance (or total variation distance) in (15.85) becomes the following:

$$(13.114)\quad \Delta(X_0, X_1) = \sup_{A \subseteq B_0 \cup B_1} \{|\operatorname{prob}\{A \xleftarrow{\boxtimes} X_0\} - \operatorname{prob}\{A \xleftarrow{\boxtimes} X_1\}|\},$$

so that $0 \leq \Delta(X_0, X_1) \leq 1$. The bound of Lemma 13.113 provides an upper bound for the statistical distance between X_0 and X_1.

COROLLARY 13.115. *Let n, z, r, δ, B_0, and B_1 be as in Lemma 13.113. Then*

$$\Delta(X_0, X_1) \leq 1 - \frac{1}{n^2}.$$

PROOF.   For any $A \subseteq B_0 \cup B_1$ and $i \in \{0, 1\}$, we have

$$X_i(A) = \operatorname{prob}\{A \xleftarrow{\boxtimes} X_i\} = \frac{\mathrm{vol}_n(A \cap B_i)}{\mathrm{vol}_n(B_i)} = \frac{\mathrm{vol}_n(A \cap B_i)}{\mathrm{vol}_n(B_0)},$$

$$|X_0(A) - X_1(A)| = |X_0(A \setminus (B_0 \cap B_1)) - X_1(A \setminus (B_0 \cap B_1))|.$$

Thus the supremum in (13.114) is attained at some $A \subseteq (B_0 \cup B_1) \setminus (B_0 \cap B_1)$, namely at one of the hollowed-out balls on the sides in Figure 13.11, say the left-hand one $D = B_0 \setminus B_1$. Then $X_1(D) = 0$,

$\text{vol}_n(D) = \text{vol}_n(B_0) - \frac{1}{2} \text{vol}_n(B_0 \cap B_1)$, and

$$\Delta(X_0, X_1) = \frac{\text{vol}_n(D)}{\text{vol}_n(B_0)} \le 1 - \frac{1}{2n^{3/2}} < 1 - \frac{1}{n^2}. \qquad \square$$

We recall the symmetric system of representatives R_m for an integer $m \ge 2$ and $\text{srem} : \mathbb{Z} \to R_m$ from (15.12). If $u \in \mathbb{Z}$ and $u \, \text{squo} \, m = \lceil u/m \rfloor$, then

(13.116) $$u = (u \, \text{squo} \, m) \cdot m + (u \, \text{srem} \, m)$$

is the division of u by m with symmetric remainder $u \, \text{srem} \, m$, satisfying $u \, \text{squo} \, m = \lceil u/m \rfloor \in \mathbb{Z}$ and $-m/2 \le u \, \text{srem} \, m < m/2$. We now generalize this to "symmetric division of a vector by a lattice basis".

For $v = (v_1, \ldots, v_n) \in \mathbb{R}^n$, we write $\lceil v \rfloor = (\lceil v_1 \rfloor, \ldots, \lceil v_n \rfloor) \in \mathbb{Z}^n$ for the rounded vector. Given a basis A for an n-dimensional lattice $L \subseteq \mathbb{R}^n$, we set

$$v \, \text{squo} \, A = \lceil vA^{-1} \rfloor,$$
$$v \, \text{srem} \, A = v - (v \, \text{squo} \, A) \cdot A.$$

Then

$$v \, \text{squo} \, A \in \mathbb{Z}^n,$$
$$(v \, \text{squo} \, A) \cdot A \in L,$$
$$v = (v \, \text{squo} \, A) \cdot A + (v \, \text{srem} \, A),$$

in analogy with (13.116). We think of $(v \, \text{squo} \, A) \cdot A$ as being a vector in L that is somewhat close to v, and $v \, \text{srem} \, A$ is its difference with v. Squo and srem depend on A, not just on L. If $v \in L$, say $v = w \cdot A$ with $w \in \mathbb{Z}^n$, then

$$v \, \text{squo} \, A = \lceil vA^{-1} \rfloor = \lceil wA \cdot A^{-1} \rfloor = \lceil w \rfloor = w,$$
$$v \, \text{srem} \, A = 0.$$

With this v and arbitrary $u \in \mathbb{R}^n$, we have

$$(u + v) \, \text{squo} \, A = \lceil (u+v)A^{-1} \rfloor = \lceil uA^{-1} + vA^{-1} \rfloor$$
$$= \lceil uA^{-1} + w \rfloor = \lceil uA^{-1} \rfloor + w$$

(13.117) $$= (u \, \text{squo} \, A) + (v \, \text{squo} \, A),$$
$$(u + v) \, \text{srem} \, A = u + v - (u \, \text{squo} \, A + v \, \text{squo} \, A) \cdot A$$
$$= u \, \text{srem} \, A + v \, \text{srem} \, A = u \, \text{srem} \, A.$$

Thus addition of a lattice vector does not change the value of srem A. After these preparations, we can present the first reduction in (13.104). For simplicity, we fix n and write α for $\alpha(n)$.

REDUCTION 13.118. From GapSVP to BDD.

Input: An instance (B, d) of GapSVP_α with $B \in \mathbb{R}^{n \times n}$ and $\alpha > 2\sqrt{n/\log n}$.
Output: "yes" or "no".

1. Do $2n^3$ times steps 2 through 5.
2. $s \xleftarrow{\boxtimes} \mathcal{B}_n(0, d\sqrt{n/\log n})$.
3. $t \leftarrow s$ srem B.
4.   Call an algorithm for BDD_γ with input (B, t), parameter $\gamma = \alpha^{-1}\sqrt{n/\log n}$, and output v.
5.   If $v \neq t - s$ then return "yes".
6. Return "no".

THEOREM 13.119. *Let $\alpha\colon \mathbb{N} \to \mathbb{R}$ with $\alpha(n) > 2\sqrt{n/\log n}$ for all n. Then Reduction 13.118 is a probabilistic polynomial-time reduction from GapSVP_α to BDD_γ with $\gamma(n) = \alpha(n)^{-1}\sqrt{n/\log n}$, correctly answering with overwhelming probability at least $1 - 2^{-n}$.*

PROOF.   The ball sampling in step 2 can be done efficiently by Proposition 13.110 (ii) and the claimed time bound follows.

To verify correctness of the reduction, we first take a "no" instance (B, d) of GapSVP_α, so that $\lambda_1(L) > \alpha d$, where B generates the lattice L and we drop the argument n of α and d.

We abbreviate $d' = d\sqrt{n/\log n}$. Since $t - s \in L$,

$$\operatorname{dist}(t, L) \leq \|s\| \leq d' \leq \frac{\lambda_1(L)\sqrt{n/\log n}}{\alpha}.$$

Thus (B, t) is a valid instance of BDD_γ. Since $\lambda_1(L) \geq \alpha d > 2d'$, two distinct vectors in L with distance at most d' from t would lead to a nonzero vector in L of length less than $\lambda_1(L)$. Thus $t - s$ is the unique vector in L closest to t, and $v = t - s$ is returned by all calls in step 4. Reduction 13.118 therefore reaches step 6 and returns "no", as desired.

Now we come to a "yes" instance (B, d) of GapSVP_α. The hypothetical BDD algorithm $\mathcal{A}$ knows (B, t), γ, and the method by which t is generated, but does not know s. The following argument is purely information-theoretic. It does not matter what $\mathcal{A}$ does, nor whether its input conditions are satisfied by the input.

We pick a shortest nonzero vector z in L, so that $\|z\| = \lambda_1(L) \leq d$. Besides the choices of s and t in steps 2 and 3, we consider another experiment: choose $s_0 \xleftarrow{\boxtimes} \mathcal{B}_n(0, d')$, $s' = z + s_0$, $t' = s'$ srem B, and call $\mathcal{A}$ with input t' (and B, which does not vary) to receive v'. Now s'

is a sample from the uniform distribution on $\mathcal{B}_n(z, d')$, whose statistical distance to that on $\mathcal{B}_n(0, d')$ is at most $1 - 1/n^2$ by Corollary 13.115, since $\|z\|/d' \leq \sqrt{(\log n)/n}$. We apply the same functions srem B and $\mathcal{A}$ to s and s', and this does not increase the statistical distance of distributions. Thus

$$\text{prob}\{\mathcal{A}(t) = t - s\} - \text{prob}\{\mathcal{A}(t') = t' - s'\} \leq 1 - \frac{1}{n^2}.$$

Furthermore,

$$\text{prob}\{\mathcal{A}(t) = t' - s'\} \leq 1 - \text{prob}\{\mathcal{A}(t') = t' - s_0\}$$
$$= 1 - \text{prob}\{\mathcal{A}(t) = t - s\},$$

since (s_0, t') are distributed exactly like (s, t). It follows that

$$\text{prob}\{\mathcal{A}(t) = t - s\} \leq 1 - \frac{1}{2n^2}.$$

Thus the probability to return "yes" in all $2n^3$ executions of step 5 is at most $(1 - 1/2n^2)^{2n^3} \leq e^{-n} < 2^{-n}$. $\square$

For our next problem LWE, we first translate the Gaussian distribution to the unit interval and then discretize it modulo some integer m. From the Gaussian distribution $D_r^{(1)}$ on $\mathbb{R}$ and its density $\delta_r^{(1)}$, we obtain a density function γ_r and a probability measure G_r^* on the shifted real unit interval $J = [-1/2 .. 1/2)$ by considering all integer translates $A + j$ of a measurable set $A \subseteq J$:

$$\gamma_r(x) = \sum_{j \in \mathbb{Z}} \delta_r^{(1)}(x + j),$$

(13.120)

$$G_r^*(A) = \int_A \gamma_r(x)\, dx = \sum_{j \in \mathbb{Z}} \int_{A+j} \delta_r^{(1)}(x)\, dx = \sum_{j \in \mathbb{Z}} D_r^{(1)}(A + j).$$

Due to (13.109), the infinite sums are well-defined. In Figure 13.12, the areas under the bell curve of Figure 13.10 are added to show the density on J. For $r = 1$, the area in Figure 13.10 above $[-1/2 .. 1/2)$ forms the main contribution, those above $[-3/2 .. -1/2)$ and $[1/2 .. 3/2)$ small pieces, and the others are too tiny to be visible. The density looks somewhat close to the uniform one, while for $r = 1/3$, only the central piece from Figure 13.10 contributes substantially and $\gamma_{1/3}$ is quite different from uniform.

We disretize this distribution modulo an integer m, assumed odd for simplicity, as follows. We take the symmetric representation $R_m =$

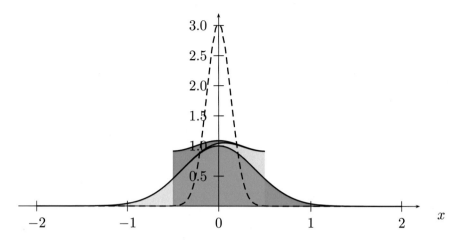

Figure 13.12: Plot of the functions γ_r (see (13.120)) on $\mathbb{T} = [-1/2 .. 1/2)$ for $r = 1$ (solid lines) and $r = 1/3$ (dashed line). The pieces outside of this interval are for illustration only.

$\{-\lfloor m/2 \rfloor, \ldots, \lfloor m/2 \rfloor\}$ of $\mathbb{Z}_m$ (see (15.12)), and for $i \in R_m$, take $c_i = (2i - 1)/2$ and the interval $I_i = [c_i .. c_{i+1})$. These I_i form a partition of the real interval $[-m/2 .. m/2)$ into m subintervals. Dividing these subintervals by m gives $I_i/m = \{a/m \colon a \in I_i\}$; these form a partition of $[-1/2 .. 1/2)$. We now take $G_{r,m}$ to be the Gaussian measure of I_i/m:

(13.121) $$G_{r,m}(i) = G_r^*(I_i/m) \text{ for } i \in \mathbb{Z}_m.$$

Then $\sum_{i \in \mathbb{Z}_m} G_{r,m}(i) = 1$. Furthermore, for an n-dimensional lattice L and real $r > 0$, we define the discrete Gaussian probability distribution $G_{r,L}$ on L with parameter r as

$$G_{r,L}(x) = \frac{\delta_r^{(n)}(x)}{\delta_r^{(n)}(L)}$$

for $x \in L$, where $\delta_r^{(n)}(L) = \sum_{x \in L} \delta_r^{(n)}(x)$ is finite and positive. $G_{r,L}$ can be efficiently sampled for large enough r.

LEMMA 13.122. *There is a probabilistic polynomial-time algorithm that on input a basis of an n-dimensional lattice L with GSO $a_1^*, \ldots, a_n^*$ and $r > \max\{\|a_i^*\| \colon 1 \leq i \leq n\} \cdot \sqrt{\ln(2n+4)/\pi}$, produces samples from $G_{r,L}$.*

We need a further lattice invariant, defined via the dual lattice L^* of L. The total Gaussian probability $\delta_r^{(n)}(L) = \sum_{x \in L} \delta_r^{(n)}(x) = 1$ has a large summand for $x = 0$, but the other summands become small when

r gets large. In particular, $\delta_r^{(n)}(L \smallsetminus \{0\}) > \delta_{r'}^{(n)}(L \smallsetminus \{0\})$ if $r > r'$, see Figure 13.12. In the following, we replace r by $1/t$, so that $\delta_{1/t}^{(n)}(L \smallsetminus \{0\}) > \delta_{1/t'}^{(n)}(L \smallsetminus \{0\})$ if $t' > t$. The next parameter is the smallest value where $\delta_{1/t}^{(n)}(L^*)$, excepting the value at $x = 0$, becomes less than a threshold ε.

DEFINITION 13.123. *Let* $L \subseteq \mathbb{R}^n$ *be an* n-*dimensional lattice and* $\varepsilon > 0$. *The smoothing parameter* $\eta_\varepsilon(L)$ *is the smallest* t *so that*

$$\delta_{1/t}^{(n)}(L^* \setminus \{0\}) = \sum_{x \in L^* \setminus \{0\}} \delta_{1/t}^{(n)}(x) \le \varepsilon.$$

For small ε, this means that the Gaussian measure $\delta_{1/t}^{(n)}$ on L^* is concentrated at 0; see Figure 13.12. One can show that, intuitively, taking a "random" point in L^* and adding Gaussian noise from $\delta_{1/t}^{(n)}$ to it with $t \ge \eta_\varepsilon(L)$ yields a distribution whose statistical distance to the "unifom distribution on $\mathbb{R}^n$" is at most $\varepsilon/2$. The smoothing parameter enjoys the following properties.

LEMMA 13.124. *Let* $L \subseteq \mathbb{R}^n$ *be an* n-*dimensional lattice and* $\varepsilon > 0$.

(i) *If* $t \ge \eta_\varepsilon(L)$, *then* $\delta_{1/t}^{(n)}(L^* \setminus \{0\}) \le \varepsilon$.

(ii) $\eta_{2^{-n}}(L) \le \sqrt{n}/\lambda_1(L^*)$.

(iii) *For any function* f *with* $f(n) \in \omega(\sqrt{\log n})$ *there exists a negligible function* ε *so that* $\eta_{\varepsilon(n)}(\mathbb{Z}^n) \le f(n)$.

(iv) *If* $0 < \varepsilon < 1$, $t \ge \eta_\varepsilon(L)$ *and* $z \in \mathbb{R}^n$, *then*

$$\frac{1 - \varepsilon}{1 + \varepsilon} \le \frac{\delta_{1/t}^{(n)}(L + z)}{\delta_{1/t}^{(n)}(L)} \le 1.$$

Using the distribution $G_{r,m}$ on $\mathbb{Z}_m = R_m$, we are now ready to describe the distribution that plays a central role in this business. We rename $G_{r,m}$ as $G_{\beta,m}$, with β playing the role of r; the reason for this will become apparent later. We have positive integers n and odd m, real $\beta > 0$, and $s \in \mathbb{Z}_m^n$. We obtain a probability distribution $W_{\beta,m,s}$ on $\mathbb{Z}_m^n \times \mathbb{Z}_m$ by choosing $a \xleftarrow{\boxtimes} \mathbb{Z}_m^n$, $e \xleftarrow{\boxtimes} G_{\beta,m}$, and returning the sample $(a, a \star s + e) \in \mathbb{Z}_m^n \times \mathbb{Z}_m$.

We next define the problem *Learning With Errors* (LWE). Namely, we are given positive integers m and n, several pairs (a, b) with uniformly and independently chosen $a \xleftarrow{\;\text{\tiny$\boxtimes$}\;} \mathbb{Z}_m^n$ and $b \in \mathbb{Z}_m$, and want to find a single $s \in \mathbb{Z}_m^n$ under the guarantee that the errors

$$(13.125) \qquad\qquad v = b - a \star s \in \mathbb{Z}_m$$

for the given pairs follow a Gaussian distribution (*LWE search problem*). The *LWE distinguishing problem* is to distinguish such (a, b) from uniformly random $(a, b) \xleftarrow{\;\text{\tiny$\boxtimes$}\;} \mathbb{Z}_m^n \times \mathbb{Z}_m$.

In the hidden number problem of Section 13.8, we have to determine an unknown integer u from good approximations to ut in $\mathbb{Z}_p$, for various randomly chosen t. LWE is a multidimensional generalization of this, except that the errors are not guaranteed to be small, but are likely to be small, by following a Gaussian distribution with mean 0. One might also call LWE "the hidden vector problem.

DEFINITION 13.126. *Let* $\beta, m \colon \mathbb{N} \longrightarrow \mathbb{R}$ *with* $\beta(n) > 0$ *and odd integral* $m(n) \geq 2$ *for all* n. *The* learning with errors search problem $LWE_{\beta(n),m(n)}$ *is to determine* $s \in \mathbb{Z}_{m(n)}^n$, *given access to any number, polynomial in* n, *of samples* $A = (a, b) \in \mathbb{Z}_{m(n)}^n \times \mathbb{Z}_{m(n)}$ *according to* $W_{\beta(n),m(n),s}$. *In the* learning with errors distinguishing problem *one can request many samples* $A \in \mathbb{Z}_{m(n)}^n \times \mathbb{Z}_{m(n)}$. *They are either all uniformly random or all follow* $W_{\beta(n),m(n),s}$. *The distinguisher should return 0 in the first and 1 in the second case. As usual, searching and distinguishing only have to be correct with a certain probability. The distinguishing problem comes in two flavors: in* worst-case distinguishing, *a correct answer is expected for any* $s \in \mathbb{Z}_{m(n)}^n$. *In* average-case distinguishing, *a correct answer is expected for randomly chosen* $s \xleftarrow{\;\text{\tiny$\boxtimes$}\;} \mathbb{Z}_{m(n)}^n$.

Regev (2009) presents a reduction which is the core of a reduction from BDD to LWE presented below. For a lattice $L \subseteq \mathbb{R}^n$ and $\delta > 0$, he considers the problem $\text{CVP}_{\delta,L}$ which on input $t \in \mathbb{R}^n$ with $\text{dist}(t, L) \leq \delta$ asks to return a vector in L that is closest to t. If $\text{dist}(t, L) > \delta$, then one may return any vector in L. Thus $\text{BDD}_\gamma(L, *)$ and $\text{CVP}_{\gamma\lambda_1(L),L}$ have essentially the same I/0 specification. We leave out the details of the following reduction from Regev (2009).

THEOREM 13.127. *Let* $\beta, \varepsilon, m \colon \mathbb{N} \to \mathbb{R}$ *with* ε *negligible,* $0 < \beta(n) < 1$, *and* $m(n) \geq 2$ *an integer for all* n. *For any* n-*dimensional lattice* L *and* $r \geq \sqrt{2}m(n) \cdot \eta_\varepsilon(L^*)$, *one can solve* $\text{CVP}_{\beta(n)m(n)/(\sqrt{2}r),L}$ *efficiently using an oracle for* $LWE_{\beta(n),m(n)}$ *and polynomially many samples from* G_{r,L^*}.

REDUCTION 13.128. $BDD_\gamma \leq LWE_{\beta,m}$ with $m \geq (2^{n-1}n\log n)^{1/2}$ and $\beta = 4\sqrt{n}\gamma$.

Input: (B,t) with $B \in \mathbb{R}^{n \times n}$ generating the lattice $L \subseteq \mathbb{R}^n$ and $t \in \mathbb{R}^n$ satisfying $\mathrm{dist}(t, L) \leq \gamma\lambda_1(L)$.

Output: v in L closest to t.

1. Compute a reduced basis $D \in \mathbb{R}^{n \times n}$ of the dual lattice L^* and its GSO $d_1, \ldots, d_n$.
2. $r_0 \leftarrow \max\{\|d_1\|, \ldots, \|d_n\|\} \cdot \sqrt{\log n}$,
   $m_0 \leftarrow (2^{n-1}n\log n)^{1/2}$,
   pick any integer m with $m_0 \leq m < 2n^2 m_0$,
   $\beta \leftarrow 4\sqrt{n}\gamma$,
   $N \leftarrow n/2 + 4\log n$,
   $\varepsilon \leftarrow 2^{-n}$.
3. For i from 0 to N do steps 4 and 5.
4. $\quad r_i \leftarrow 2^i r_0$.
5. $\quad$ Call the reduction of Theorem 13.127 with input t to return a solution $v_i \in L$ of $CVP_{\beta m/(\sqrt{2}r_i), L}(t)$ using an oracle for $LWE_{\beta,m}$.
6. Let $j \in \{0, \ldots, N\}$ be such that $\mathrm{dist}(t, v_j)$ is minimal among all $\mathrm{dist}(t, v_i)$ for $0 \leq i \leq N$.
7. Return $v = v_j$.

THEOREM 13.129. For $0 < \gamma < 1/(4\sqrt{n})$, Reduction 13.128 works correctly as specified and uses polynomial time.

PROOF. The claim about the running time is clear, using Lemma 13.48 to compute D. We have to verify that the application of Theorem 13.127 in step 5 yields a closest vector v_i for at least one $i \leq N$. We denote the maximal length in the GSO for L^* in step 2 as

$$M = \max\{\|d_1\|, \ldots, \|d_n\|\}.$$

By Theorem 13.9, we have $\|d_n\| \leq \lambda_n(L^*)$, and from Definition 13.11 that $\|d_i\| \leq 2^{(n-i)/2}\|d_n\| < 2^{n/2}\lambda_n(L^*)$ for all $i \leq n$. Thus

$$(13.130) \qquad M \leq 2^{n/2}\lambda_n(L^*).$$

Furthermore, we set

$$R = \frac{m\sqrt{2n}}{\lambda_1(L)}$$

and claim that

$$(13.131) \qquad r_0 = M\sqrt{\log n} \leq R \leq r_N = 2^N r_0.$$

The lower bound on R follows from (13.130) and Theorem 13.51. For the upper bound, Theorem 13.51 and Lemma 13.14 imply that

$$R = \frac{m\sqrt{2n}}{\lambda_1(L)} \leq \lambda_n(L^*)m\sqrt{2n} \leq mnM\sqrt{2} \leq 2^{n/2}n^4\sqrt{\log n}M \leq 2^N r_0$$

for $n \geq 3$. Now (13.131) implies that there exists a unique $j \leq N$ with $R \leq r_j < 2R$. We write $r = r_j$ and verify the conditions for Theorem 13.127 in the iteration for $i = j$. It will follow that v_j in step 6 is indeed a vector in L closest to t, and the reduction's output is correct.

Since $r < 2R$ and t satisfies the input condition of BDD_γ, we have

$$\mathrm{dist}(t, L) \leq \gamma\lambda_1(L) = \frac{\beta m}{\sqrt{2} \cdot 2R} < \frac{\beta m}{\sqrt{2}\,r},$$

and t is a legitimate instance of $\mathrm{CVP}_{\beta m/\sqrt{2}\,r,L}$.

Using Lemma 13.124 (ii), we have

$$r \geq R = \frac{\sqrt{2n}\,m}{\lambda_1(L)} \geq \sqrt{2}\,m\eta_\varepsilon(L^*),$$

so that the lower bound on r holds as required in Theorem 13.127.

Since $\log n > \ln(2n + 4)/\pi$ for $n \geq 2$, Theorem 13.51 and (13.130) imply that

$$r \geq R \geq \frac{2^{n/2}\sqrt{n/2}\sqrt{\log n}\sqrt{2n}}{\lambda_1(L)}$$

$$\geq 2^{n/2}\lambda_n(L^*)\sqrt{\log n} > M \cdot \sqrt{\frac{\ln(2n + 4)}{\pi}}.$$

Thus by Lemma 13.122, G_{r,L^*} can be sampled efficiently. It follows that $v_j \in L$ is a closest vector to t. $\square$

For simplicity, we fixed n in the reduction. Actually, it shows that $\mathrm{BDD}_\gamma \leq \mathrm{LWE}_{\beta,m}$ for any $\gamma, \beta, m \colon \mathbb{N} \to \mathbb{R}$ with $0 < \gamma(n) < 1/(4\sqrt{n})$, $\beta(n) = 4\sqrt{n}\gamma(n)$, and $m(n)$ as in step 2.

The next two lemmas present reductions first from the worst-case search version of LWE to a worst-case distinguishing problem, and then from the latter to its average-case version. This is useful for our cryptographic applications.

We describe the idea of the first reduction only for the special case where m is a small prime number. For the general case, see Brakerski et al. (2013).

LEMMA 13.132. *Let $\beta, p \colon \mathbb{N} \longrightarrow \mathbb{R}$ with $0 < \beta(n) < 1$ and $p(n)$ a prime number satisfying $n \leq p(n) \leq n^2$ for all n. Then there is a probabilistic polynomial-time reduction from solving the search problem $\mathrm{LWE}_{\beta(n), p(n)}$ with overwhelming probability to distinguishing between $W_{\beta(n), p(n), s}$ and $U_{\mathbb{Z}^n_{p(n)} \times \mathbb{Z}_{p(n)}}$ for unknown $s \in \mathbb{Z}^n_{p(n)}$ with overwhelming advantage.*

PROOF. We suppress the argument n of β and p, and let $1 \leq i \leq n$. We show how to compute the ith coordinate s_i of $s \in \mathbb{Z}^n_p$. We take the ith unit vector $v = (0, \ldots, 0, 1, 0, \ldots, 0) \in \mathbb{Z}_p$ with a 1 in position i, and do the following for all $k \in \mathbb{Z}_p$.

We choose several $r \xleftarrow{\boxtimes} \mathbb{Z}_p$, set

$$(13.133) \qquad (a', b') = (a + rv, b + rk),$$

run the assumed distinguisher on these samples, and return $s_i = k$ if the majority of its answers equals 1. An output of 1 means that (a', b') is likely to come from $W_{\beta, p, s}$. We have

$$(13.134) \quad b' - a' \star s = b + rk - (a + rv) \star s = b - a \star s + r \cdot (k - s_i).$$

If (a, b) is uniformly random, then so is (a', b'), so that the answer 0 is more likely than 1. Now assume that $(a, b) \xleftarrow{\boxtimes} W_{\beta, p, s}$. If $k = s_i$, then (a', b') equals (a, b) and is distributed according to $W_{\beta, p, s}$. If $k \neq s_i$, then (a', b') is uniformly random. Thus the distinguisher's answers are likely to be 1 if and only if $k = s_i$. $\qquad \square$

The following lemma reduces worst-case distinguishing with large advantage to average-case distinguishing with small advantage.

LEMMA 13.135. *Let $\beta, m \colon \mathbb{N} \longrightarrow \mathbb{R}$ with $0 < \beta(n) < 1$ and $m(n)$ an integer for all n, and $U_n = U_{\mathbb{Z}^n_{m(n)} \times \mathbb{Z}_{m(n)}}$. There is a probabilistic polynomial-time reduction from distinguishing between $W_{\beta(n), m(n), s}$ and U_n for an arbitrary unknown $s \in \mathbb{Z}^n_{m(n)}$ with overwhelming advantage to distinguishing between $W_{\beta(n), m(n), t}$ and U_n for uniformly random $t \xleftarrow{\boxtimes} \mathbb{Z}^n_{m(n)}$ with nonnegligible advantage.*

PROOF. We split the task into two reductions. Suppose that the algorithm $\mathcal{A}$ distinguishes between $W_{\beta, m, t}$ for $t \xleftarrow{\boxtimes} \mathbb{Z}^n_m$ and U_n with nonnegligible advantage ε, dropping the argument n of m and β. Thus the random variable Y_1 representing the result of $\mathcal{A}$ on inputs from $W_{\beta, m, t}$ as above has expected value $\mathcal{E}(Y_1) = (1 + \varepsilon)/2$, and Y_0 on uniform inputs has $\mathcal{E}(Y_0) = (1 - \varepsilon)/2$. We run $k = 2\lceil 2n\varepsilon^{-2} \rceil + 1$ independent executions of $\mathcal{A}$ and take a majority vote.

Suppose all inputs come from $W_{\beta,m,t}$. Then the experiment corresponds to the random variable $Z_1 = kY_1 = Y_1 + \cdots + Y_1$ (k summands) and $\mathcal{E}(Z_1) = k(1+\varepsilon)/2$. The vote is correct if $Z_1 > (k-1)/2 = \mathcal{E}(Z_1) - (k\varepsilon + 1)/2$. By (15.82), the probability for the event $|Z_1 - \mathcal{E}(Z_1)| > (k\varepsilon + 1)/2$ is at most

$$2e^{-2((k\varepsilon+1)/2)^2/k} \leq 2e^{-k\varepsilon^2/2} \leq 2e^{-2n} < 2^{-n}.$$

An analogous argument works in the case of uniform inputs. Thus the majority decision distinguishes between $W_{\beta,m,t}$ and U_n with overwhelming advantage.

Now we have to distinguish between $W_{\beta,m,s}$ and U_n for a fixed (but unknown) $s \in \mathbb{Z}_m^n$. We replace each sample $(a,b) \in \mathbb{Z}_m^n \times \mathbb{Z}_m$ by $(a,b_u) = (a, b + a \star u)$ for independent $u \xleftarrow{\boxdot} \mathbb{Z}_m^n$. Then $t = s + u \in \mathbb{Z}_m^n$ is uniformly random. If $(a,b) \xleftarrow{\boxdot} W_{\beta,m,s}$, then $(a,b_u) \xleftarrow{\boxdot} W_{\beta,m,s+u}$. If $(a,b) \xleftarrow{\boxdot} U_n$, then also $(a,b_u) \xleftarrow{\boxdot} U_n$. The distinguisher from the previous paragraph, between $W_{\beta,m,t}$ and U_n with overwhelming advantage and applied to the samples (a,b_u), provides a distinguisher between $W_{\beta,m,s}$ and U_n with overwhelming advantage. $\square$

The following public-key encryption scheme is due to Regev (2009). It uses the discrete Gaussian distribution $G_{\beta,m}$ from (13.121).

SCHEME 13.136. Regev cryptosystem.

Set-up.

Input: integer n.
Output: integers k and m.

1. Choose (pairwise distinct) random prime numbers $p_1, p_2, \ldots$ with $n^2/2 \leq p_i \leq 2n^2$ for all i until $m = \prod_i p_i$ satisfies $m_0 \leq m < 2n^2 m_0$, with $m_0 = (2^{n-1} n \log n)^{1/2}$. All integer arithmetic is performed in the symmetric representation R_m of $\mathbb{Z}_m$.
2. Choose μ with $0 < \mu < 1$, e.g., $\mu = 1/10$.
3. $k \leftarrow \lceil (1+\mu)(n+1) \log m \rceil$.
4. $\beta \leftarrow 1/\sqrt{n} \log^2 n$.

Keygen.

5. $s \xleftarrow{\boxdot} \mathbb{Z}_m^n$.
6. $\mathsf{sk} \leftarrow s$.
7. $a_1, \ldots, a_k \xleftarrow{\boxdot} \mathbb{Z}_m^n$.
8. $e_1, \ldots, e_k \xleftarrow{\boxdot} G_{\beta,m}$.

9. For $i = 1, \ldots, k$ do
10. $b_i \leftarrow a_i \star s + e_i$ in $\mathbb{Z}_m$.
11. $\mathsf{pk} \leftarrow ((a_1, b_1), \ldots, (a_k, b_k))$ in $(\mathbb{Z}_m^n \times \mathbb{Z}_m)^k$.

Encryption enc.

Input: message $c \in \{0, 1\}$.
Output: $\mathsf{enc}_{\mathsf{pk}}(c) \in \mathbb{Z}_p^n \times \mathbb{Z}_p$.

12. $S \xleftarrow{\boxdot} 2^{\{1, \ldots, k\}}$. [$S$ is a uniformly random subset of $\{1, \ldots, k\}$.]
13. $a \leftarrow \sum_{i \in S} a_i \in \mathbb{Z}_m^n$.
14. $b \leftarrow \sum_{i \in S} b_i \in \mathbb{Z}_m$.
15. Return $\mathsf{enc}_{\mathsf{pk}}(c) \leftarrow (\sum_{i \in S} a_i, c \cdot \lfloor m/2 \rfloor + b)$.

Decryption dec.

Input: $(a, b) \in \mathbb{Z}_m^n \times \mathbb{Z}_m$.
Output: $\mathsf{dec}_{\mathsf{sk}}(a, b) \in \{0, 1\}$.

16. $t \leftarrow b - a \star s \in \mathbb{Z}_m$.
17. If $|t| < m/4$ then $\mathsf{dec}_{\mathsf{sk}}(a, b) \leftarrow 0$ else $\mathsf{dec}_{\mathsf{sk}}(a, b) \leftarrow 1$.
18. Return $\mathsf{dec}_{\mathsf{sk}}(a, b)$.

A first task is to show that the procedures work correctly with overwhelming probability, that is, $\mathsf{dec}_{\mathsf{sk}}(\mathsf{enc}_{\mathsf{pk}}(c)) = c$ for any bit c is highly likely.

To this end, we consider for nonnegative $\ell \leq k$ the distribution $\ell G_{\beta, m}$ on $\mathbb{Z}_m$ which is obtained by summing ℓ independent copies of $G_{\beta, m}$. Thus if $G_{\beta, m}(u)$ is the probability for $u \in \mathbb{Z}_m$ to occur under $G_{\beta, m}$, and similarly for $(\ell G_{\beta, m})(v)$, then

$$(\ell G_{\beta, m})(v) = \sum_{\substack{u_1, \ldots, u_\ell \in \mathbb{Z}_m \\ u_1 + \cdots + u_\ell = v}} G_{\beta, m}(u_1) \cdots G_{\beta, m}(u_\ell).$$

In particular, $(0G_{\beta, m})(0) = 1$.

LEMMA 13.137. For $\ell \leq k$, we have

$$\mathrm{prob}\{|e| < m/4 : e \xleftarrow{\boxdot} \ell G_{\beta, m}\} > 1 - \delta(n)$$

for the negligible function δ with $\delta(n) = n^{-(\log_2 n)/1600}$.

PROOF. We abbreviate $J = [-1/2 .. 1/2)$. A sample e from $\ell G_{\beta, m}$ is given by independently choosing $d_1, \ldots, d_\ell \xleftarrow{\boxdot} G_\beta^*$ from the continuous Gaussian distribution G_β^* on J from (13.120), and setting $e =$

$\sum_{1 \leq i \leq \ell} \lfloor m d_i \rceil$ in R_m. The distance of e to $|\sum_{1 \leq i \leq \ell} m d_i|$ in R_m is at most $\ell/2 \leq k/2 < m/32$. Furthermore, $\sum_{1 \leq i \leq \ell} d_i$ in J (calculated modulo 1) is distributed like $G^*_{\sqrt{\ell} \cdot \beta}$, since the variance of the Gaussians behaves additively, and

$$(13.138) \qquad \sqrt{\ell} \beta \leq \frac{\sqrt{(1+\mu)(n+1) \log_2(2n^2)}}{\sqrt{n} \log_2^2 n} \leq \frac{3}{\log_2 n}$$

for $n \geq 4$. From (13.109), we have

$$\text{prob}\{|\sum_{1 \leq i \leq \ell} d_i \bmod 1| \geq 1/16\} \leq e^{-(\log_2 n/48)^2} < n^{-(\log_2 n)/1600} = \delta(n),$$

which is negligible in n. Thus $\text{prob}\{|\sum \lfloor m d_i \rceil \text{ in } R_m| < m/16\} \geq 1 - \delta(n)$ and the claim follows.   □

THEOREM 13.139. *For any $c \in \{0,1\}$, we have*

$$\text{dec}_{\text{sk}}(\text{enc}_{\text{pk}}(c)) = c$$

with overwhelming probability.

PROOF.   For $c = 0$, step 14 calculates

$$(13.140) \qquad b = \sum_{i \in S} b_i = \sum_{i \in S} (a_i \star s + e_i) = a \star s + \sum_{i \in S} e_i.$$

By Lemma 13.137, $\sum_{i \in S} e_i$ is likely to be small, and step 17 works correctly. Similarly, also $\text{enc}_{\text{pk}}(1)$ is decrypted correctly.   □

We now obtain from a distinguisher $\mathcal{A}$ between encryptions of 0 and of 1 with nonnegligible probability a distinguisher $\mathcal{B}$ between $W_{\beta,m,s}$ and the uniform distribution. This is the last link in the chain (13.104) and we then conclude that no such $\mathcal{A}$ exists under our axiom that GapSVP is hard. In other words, the cryptosystem is proven to be secure under the axiom.

REDUCTION 13.141. Distinguisher $\mathcal{B}$ between $W_{\beta,m,s}$ (with unknown s) and uniform from ε-distinguisher $\mathcal{A}$ between encryptions of 0 and of 1.

Input: Integers n, k, and m as in the Set-up of Scheme 13.136, and
$$((a_1, b_1), \ldots, (a_k, b_k)) \in (\mathbb{Z}_m^n \times \mathbb{Z}_m)^k.$$
Output: 0 or 1.

1. $t \leftarrow \lceil 2^{11} n k^2 \varepsilon^{-2} \rceil$.

2. For $1 \leq i \leq k$ and $1 \leq j \leq t$ do steps 3 through 5.
3. $A_i \leftarrow (a_i, b_i)$.
4.   Call $\mathcal{A}$ to receive $u_{i,j} = \mathcal{A}((A_i, (\text{enc}_{A_i}(0)_j)))$, where the t encryptions of 0 with $\text{pk} = A_i$ are executed independently for all j.
5.   Choose $(a, b)_{i,j} \xleftarrow{\boxdot} \mathbb{Z}_m^n \times \mathbb{Z}_m$ independently and call $\mathcal{A}$ to receive $v_{i,j} = \mathcal{A}((A_i, (a, b)_{i,j}))$.
6. Compute the averages

$$u \leftarrow \frac{1}{kt} \sum_{\substack{1 \leq i \leq k \\ 1 \leq j \leq t}} u_{i,j},$$

$$v \leftarrow \frac{1}{kt} \sum_{\substack{1 \leq i \leq k \\ 1 \leq j \leq t}} v_{i,j}.$$

7. If $|u - v| \geq \varepsilon/16$ then return 1 else return 0.

THEOREM 13.142. *Suppose that $\mathcal{A}$ distinguishes encryptions of 0 from encryptions of 1 in polynomial time and with nonnegligible advantage at least $\varepsilon(n)$. Then $\mathcal{B}$ from Reduction 13.141 works in polynomial time and for a nonnegligible fraction at least $\varepsilon(n)/4$ of all s has an overwhelming advantage in distinguishing $W_{\beta,m,s}$ from uniform.*

PROOF.   We first modify $\mathcal{A}$ slightly to obtain a simplifying additional property. We let

$$E_0 = \mathcal{E}_{\mathcal{A}}(\text{pk}, \text{enc}_{\text{pk}}(0))$$

be the probability that $\mathcal{A}$ returns 1 on input a public key pk and an encryption of 0. The probability space consists of the random choices for secret and public keys, in the encryption, and within $\mathcal{A}$. Similarly, we have

$$E_1 = \mathcal{E}_{\mathcal{A}}(\text{pk}, \text{enc}_{\text{pk}}(1)).$$

By assumption, we have

$$|E_0 - E_1| \geq \varepsilon,$$

dropping the argument of $\varepsilon(n)$. We may also consider inputs $(\text{pk}, (a, b))$ to $\mathcal{A}$ where pk is properly distributed and $(a, b) \xleftarrow{\boxdot} \mathbb{Z}_m^n \times \mathbb{Z}_m$. We write

$$E_{\text{unif}} = \mathcal{E}_{\mathcal{A}}(\text{pk}, U_{\mathbb{Z}_m^n \times \mathbb{Z}_m}).$$

We modify $\mathcal{A}$ into another algorithm $\mathcal{A}'$. If $|E_0 - E_{\mathrm{unif}}| \geq \varepsilon/2$, then $\mathcal{A}'$ is $\mathcal{A}$. Otherwise, we have $|E_1 - E_{\mathrm{unif}}| \geq \varepsilon/2$, and on input $(\mathsf{pk}, (a, b))$, $\mathcal{A}'$ calls $\mathcal{A}$ on $(\mathsf{pk}, (a, (m-1)/2+b))$. This transformations maps encryptions of 0 to encryptions of 1, preserving distributions, and uniform to uniform. Using E' for $\mathcal{A}'$ like E for $\mathcal{A}$, we have $|E_0' - E_{\mathrm{unif}}'| \geq \varepsilon/2$. Now we rename $\mathcal{A}'$ as $\mathcal{A}$ and have the additional property that

$$(13.143) \qquad\qquad |E_0 - E_{\mathrm{unif}}| \geq \varepsilon/2.$$

For any $s \in \mathbb{Z}_m^n$, the distribution of public keys pk belonging to $\mathsf{sk} = s$ follows $W_{\beta,m,s}$. We define $E_0(s)$ like the expectation E_0 of $\mathcal{A}$, but under the condition that $\mathsf{pk} \xleftarrow{\;\boxtimes\;} W_{\beta,m,s}$, and similarly $E_{\mathrm{unif}}(s)$. Thus $\mathcal{E}(E_0(s)\colon s \xleftarrow{\;\boxtimes\;} \mathbb{Z}_m^n) = E_0$. We take

$$Y = \{s \in \mathbb{Z}_m^n \colon |E_0(s) - E_{\mathrm{unif}}(s)| \geq \varepsilon/4\}.$$

By (13.143), we have

$$(13.144) \qquad |\mathcal{E}(E_0(s)\colon s \xleftarrow{\;\boxtimes\;} \mathbb{Z}_m^n) - \mathcal{E}(E_{\mathrm{unif}}(s)\colon s \xleftarrow{\;\boxtimes\;} \mathbb{Z}_m^n)| \geq \varepsilon/2,$$

and thus $\#Y \geq m^n \cdot \varepsilon/4$ by an averaging argument (15.88). Therefore Y contains a nonnegligible fraction of all s and it is sufficient to show that $\mathcal{B}$ distinguishes well for $s \in Y$.

First we consider the case where $\mathcal{B}$'s input samples $A_i = (a_i, b_i)$ come from $W_{\beta,m,s}$. Since u and v are the averages of $E_0(A_i)$ and $E_{\mathrm{unif}}(A_i)$, respectively, over these samples and $|E_0(s) - E_{\mathrm{unif}}(s)| \geq \varepsilon/4$, an averaging argument as above shows that

$$|E_0(A_i) - E_{\mathrm{unif}}(A_i)| \geq \varepsilon/8$$

occurs with probability at least $\varepsilon/8$ among these samples A_i.

We claim that if such an A_i is chosen, then $\mathcal{B}'s$ estimates u and v for $E_0(A_i)$ and $E_{\mathrm{unif}}(A_i)$, respectively, are likely to have a small error. To see this, fix some $i \leq k$ and let $u_i = t^{-1} \sum_{1 \leq j \leq t} u_{i,j}$. According to (15.91), the probability that $u_i \leq E_0(A_i) - \varepsilon/64k$ is at most

$$e^{-2(tE_0(A_i)-t(E_0(A_i)-\varepsilon/64k))^2/t} = e^{-2^{-11}t\varepsilon^2 k^{-2}} \leq e^{-n}.$$

Using a similar argument for $u_i \geq E_0(A_i) - \varepsilon/64k$ and for the corresponding v_i, it follows that

$$|E_0(A_i) - u_i|, |E_{\mathrm{unif}}(A_i) - v_i| \leq \frac{\varepsilon}{64k},$$

$$|E_0(s) - u|, |E_{\mathrm{unif}}(s) - v| \leq \frac{\varepsilon}{64},$$

$$\frac{\varepsilon}{16} \leq u - v,$$

with overwhelming probability. Therefore $\mathcal{B}$ accepts and returns 1.

Now let $\mathcal{B}$'s input samples A_i be uniformly random. Lemma 13.145 below, applied to $H = \mathbb{Z}_m^n \times \mathbb{Z}_m$, with m^{n+1} elements, shows that with probability exponentially close to 1, the distribution of $\mathrm{enc}_A(0)$ is exponentially close to $U_{\mathbb{Z}_m^n \times \mathbb{Z}_m}$. Thus there exists some $c > 0$ so that

$$|E_0(A_i) - E_{\mathrm{unif}}(A_i)| > 2^{-cn}$$

occurs only with exponentially small probability. For all other A_i, $\mathcal{B}'s$ estimates of $E_0(A)$ and $E_{\mathrm{unif}}(A)$ differ by at most $\varepsilon/32 + 2^{-cn} < \varepsilon/16$, and $\mathcal{B}$ returns 0. $\square$

We have used the following special case of the *leftover hash lemma* from Impagliazzo & Zuckermann (1989), which is also shown in Regev (2009), Claim 5.3, and in Exercise 13.23 below.

LEMMA 13.145. *Let H be a finite additive (commutative, Abelian) group with d elements, $\ell \geq 1$, $g = (g_1, \ldots, g_\ell) \in H^\ell, S \subseteq \{1, \ldots, \ell\}$ and $h_{g,S} = \sum_{i \in S} g_i$. Choosing S uniformly at random gives a random variable h_g with values in H, and we consider the statistical distance $\Delta_g = \Delta(h_g, U_H)$. Then*

$$\mathcal{E}(\Delta_g \colon g \xleftarrow{\boxtimes} H^\ell) \leq (d/2^\ell)^{1/2}.$$

In particular,

$$\mathrm{prob}\{\Delta_g \geq (d/2^\ell)^{1/4} \colon g \xleftarrow{\boxtimes} H^\ell\} \leq (d/2^\ell)^{1/4}.$$

COROLLARY 13.146. *If GapSVP_α is hard for $\alpha(n) = 4(n \log n)^{3/2}$, then the Regev cryptosystem 13.136 is indistinguishable under chosen plaintext attacks (IND-CPA).*

PROOF. This statement is the ultimate goal of the long sequence of reductions in this section, and we now walk through them backwards.

We start with a hypothetical distinguisher $\mathcal{A}$ between encryptions of 0 and of 1. Theorem 13.142 says that for a nonnegligible fraction of $s \in \mathbb{Z}_m^n$, one can efficiently distinguish between $W_{\beta,m,s}$ and uniform. Lemma 13.135 augments the fraction to overwhelming, and by Lemma 13.132 one can then also solve the search version of $\mathrm{LWE}_{m,\beta}$. This in turn implies an efficient solution to BDD_γ for $\gamma = \beta/(4\sqrt{n}) = (4n \log^2 n)^{-1}$ by Theorem 13.129, and of GapSVP_α with $\alpha = \sqrt{n/\log n}/\gamma = 4(n \log n)^{3/2}$, contradicting the hypothesis. Thus no such hypothetical distinguisher exists. $\square$

We have presented somewhat simplified versions of some reductions. The intrepid reader will find improved results with better key sizes and computing times in the literature, as well as different constructions and open questions. An important issue is to reduce the value of α in the hardness assumption about GapSVP_α.

Notes. Minkowski (1910) describes the *geometry of numbers* that he invented.

13.1. For (13.8), see Nguyen (2010), Theorem 5.

13.2. Using the (equivalent) language of quadratic forms, Lagrange (1773), pages 698–700, describes basis reduction in two dimensions. This is also in Gauß (1801), article 171. The Gram-Schmidt orthogonalization procedure is from Schmidt (1907), §3, who states that Gram (1883) has given essentially the same formulas. Hadamard (1893) proved Theorem 15.79. The geometrical idea is that the volume $|\det A|$ of the polytope spanned by $a_1, \ldots, a_n \in \mathbb{R}^n$ is maximal when these vectors are mutually orthogonal; in this case it is equal to $\|a_1\| \cdots \|a_n\|$.

The more general notion of a *δ-reduced basis* is defined by requiring that $\|b_i^*\| \leq 1/(\delta - 1/4) \cdot \|b_{i+1}^*\|^2$ in Definition 13.11. For $1/4 < \delta \leq 1$, the basis reduction algorithm, suitably modified, also works for this concept.

13.3. Using fast integer arithmetic, with $\mathsf{M}(k) \in \mathrm{O}(k \log k \log \log k)$ bit operations to multiply k-bit integers, the cost of Algorithm 13.13 comes to $\mathrm{O}((n^4 m)\,\mathsf{M}(nm) \log(nm))$ or $\mathrm{O}^{\sim}(n^5 m^2)$ bit operations. For a sparse matrix, with $m \approx k + \log_2 n$ and input size $N = n^2 + kn$, the cost is $\mathrm{O}^{\sim}(nN^2)$.

Many improvements to the algorithm have been found. The current record is due to Storjohann (1996), who achieves $\mathrm{O}^{\sim}(n^{3.381} m^2)$ operations with a modular approach and fast matrix multiplication. Nguyen & Stehlé (2009) present a floating-point method that uses $\mathrm{O}(n^6 m(n+m))$ bit operations for n-dimensional lattices in $\mathbb{R}^n$ with each basis vector of m-bit Euclidean length, and $\mathrm{O}^{\sim}(n^5 m(n + m))$ with fast arithmetic.

The collection of Nguyen & Vallée (2010), arising from a conference celebrating the 25th anniversary of the lattice basis reduction algorithm, presents the history, theory, and applications of this ground-breaking method. It can serve as an excellent companion to this chapter.

13.4. The Merkle & Hellman subset sum encryption scheme was first attacked by Shamir in 1982, see also Shamir (1984), and Adleman (1983). Our presentation is based on Lagarias & Odlyzko (1985) and Frieze (1986). Theorem 13.27 is from Frieze (1986), which is based on Lagarias & Odlyzko (1985). Among the subset sum cryptosystems, the one by Chor & Rivest (1988) survived for some time, but was broken by Vaudenay (1998).

The bound of Lagarias & Odlyzko (1985) is $\delta(t) < 0.645$ in (13.32). Coster *et al.* (1992) improved it to $\delta(t) < 0.9408$.

13.5. The linear congruential generator is due to Derrick Lehmer (1951); the truncated generator was proposed by Knuth, and the lattice attack is due to Frieze, Håstad, Kannan, Lagarias & Shamir (1988). They prove versions of Lemma 13.44 for arbitrary linear congruences and of Theorem 13.54 for squarefree moduli, and mention the difficulty when m is not squarefree. Theorem 13.51 is from Banaszczyk (1993); an upper bound of n^2 instead of n is in Lagarias *et al.* (1990). For Theorem 13.54, we assume m to be a prime. A similar result holds for squarefree m. The *table-makers problem* is the issue that two real numbers may be close to each other but differ in many of their binary or decimal digits, due to carries.

13.6. The nearest hyperplane algorithm is from Babai (1986). Kannan (1987) presents a survey of various approximate CVP computations. The approximation factor $2^{\ell/2}$ in Theorem 13.60 can be improved to 2^s with $s \in O(\ell \log^2 \log \ell / \log \ell)$. See Nguyen & Stern (2000), Section 2.4.

13.7. The hide-and-seek paper of Shparlinski (2002) presents rather general versions of the hidden number problem, many other applications, and a different algorithm, based on close vector algorithms with smaller approximation factors (see Notes to Section 13.6.)

13.9. May (2010) presents more details on Coppersmith's method. The first polynomial-time algorithm for factoring polynomials in $\mathbb{Q}[t]$ is in Lenstra et al. (1982). Their title shows that they invented basis reduction expressly for this purpose. All integral roots of a polynomial in $\mathbb{Z}[t]$ with degree n and m-bit coefficients can be found with $O^\sim(n^2 m)$ operations; see von zur Gathen & Gerhard (2013), Theorem 15.21. For root finding in finite fields, see Section 14.5 of the latter text. Lemma 13.89 is based on Howgrave-Graham (1997), Section 2; see also May (2010), Theorem 2.

13.10. Odlyzko (1990) notes that basis reduction *"turns out to be fast in practice, and usually finds a reduced basis in which the first vector is much shorter than is guaranteed [by Theorem 13.12]. (In low dimensions, it has been observed empirically that this algorithm usually finds the shortest nonzero vector in a lattice.)"*.

13.11. Lemma 13.99 is from Howgrave-Graham (1997). The results of this section are due to Blömer & May (2003). May (2005) uses the Coppersmith method to obtain a deterministic reduction from factoring N to determining d, while Section 3.5 provides a probabilistic reduction. Other attacks that work with similar tools include Boneh et al. (1998), where $e < N^{1/2}$ and some leading (or trailing) bits of d are known; the number of bits required depends on the size of e. Blömer & May (2003) extend this to $e \leq N^{0.725}$, using basis reduction heuristics.

13.12. This section assumes more about complexity classes than we can provide in this text. The "standard" complexity assumptions include $\mathsf{P} \neq \mathsf{NP} \neq \mathsf{co\text{-}NP}$. For detailed information and references, corresponding results for the closest vector problem, and further questions, see the surveys of Micciancio & Goldwasser (2002), Khot (2010), Regev (2010), and Peikert (2014).

The first entry in Table 13.9, with $\log \alpha \in O(n \log^2 \log n / \log n)$, is from Schnorr (1987), and the second entry from Ajtai et al. (2001).

The bottom part of Table 13.9 describes hardness results. Van Emde Boas showed already in 1980 that SVP is NP-hard if one takes the max-norm $\|v\|_\infty = \max\{|v_1|, \ldots, |v_n|\}$. Then Ajtai proved in 1997 that SVP ($= 1$-SVP) in the usual Euclidean norm is indeed NP-hard, under randomized reductions, and Khot (2005) that GapSVP_α is NP-hard for any constant α. NP-hardness also holds for $1 + n^{-\varepsilon}$ for any $\varepsilon > 0$, and even for $\log \alpha = (\log n)^{1/2-\varepsilon}$ with $\varepsilon > 0$. Khot (2005) and Haviv & Regev (2012) show that for any $\varepsilon > 0$, SVP_α with $\log \alpha = (\log n)^{1-\varepsilon}$ is hard unless all problems in NP have quasi-polynomial time randomized algorithms, and that $\log \alpha = \log n / \log\log n$ yields a hard problem unless NP is in random subexponential time. See also Micciancio (2012).

The middle part of Table 13.9 displays values of α for which GapSVP_α is neither known to be easy (in BPP) nor, presumably, NP-hard. Integer facorization (Chapter 3), discrete logarithms (Chapter 4) and graph isomorphism (Chapter 12) share these properties. These problems are still generally assumed to be "hard". Regev's cryptosystem Scheme 13.136 even assumes hardness somewhat "higher up", around $\alpha = n^{3/2}$.

Aharonov & Regev (2005) show $\text{GapSVP}_{\sqrt{n/\log n}} \in \text{NP} \cap \text{co-NP}$, and Goldreich & Goldwasser (2000) that $\text{GapSVP}_{\sqrt{n/\log n}} \in \text{NP} \cap \text{co-AM}$, and thus the problem is unlikely to be NP-hard, since otherwise the polynomial hierarchy would collapse. This works for a special type of reductions (smart Cook reductions); see also Lagarias *et al.* (1990).

A ground-breaking result of Ajtai's is the construction of a class $\mathcal{C}$ of lattices so that solving SVP for a random lattice in $\mathcal{C}$ in random polynomial time means that this can also be done in the worst case.

In a promise problem, we only expect a correct answer if the "promise" that the input is in $X_{\text{yes}} \cup X_{\text{no}}$ is kept.

13.13. I am grateful to Oded Regev for patiently helping me to get Reduction 13.128 straight. Many thanks also go to Chris Peikert for discussions on the material of this section.

Ajtai & Dwork (1996), Goldreich, Goldwasser & Halevi (1997), Ajtai (2008), Regev (2009), Peikert (2009), and others built cryptographic systems based on Ajtai's and related ideas. The decision versions GapSVP and GapCVP of the more standard SVP and CVP problems were formulated in the 1998 conference version of Goldreich & Goldwasser (2000). The cryptosystem of this section is due to Regev (2009), published in 2005, where he introduced the LWE problem. It continues a stream of results, among them Micciancio & Regev (2007) and Gentry, Peikert & Vaikuntanathan (2008).

General multidimensional normal distributions are given by a vector of means and a covariance matrix. The general one-dimensional Gaussian distribution D with mean m and variance r^2 is defined via $\text{prob}\{(x - m)/r \leq y : x \leftarrow D\} = r^{-1}\int_{-\infty}^{y} \exp(-\pi(t/r)^2)\,dt$ for all $y \in \mathbb{R}$. The "error function" $\text{erf}(x) = 2^{-1/2}\int_{-x}^{x} e^{-t^2}\,dt$ is closely related to the normal and Gaussian distributions. This integral is, in a specific sense, not expressible by elementary functions like rational functions, exp, and log. Gauß (1809) introduced the distribution named after him. He had used it in 1801, together with his least squares method, to eliminate measurement errors in his astronomical calculations and to learn the trajectory of the asteroid (dwarf planet) Ceres. This made him famous also outside of the mathematicians' world and garnered him a university professorship. He was already "learning with errors" two centuries ago.

The sampling methods of Proposition 13.110 and their history are described in Knuth (1998), 3.4.1, and Devroye (1986), Section V.4; the method for normal samples essentially goes back to Box & Muller (1958). For (i), we take uniform samples in the square $[-1..1)$ and reject until $s < 1$, that is, one of them lies inside the unit ball. This is an instance of the general rejection method. It does not yield an efficient method for sampling a higher-dimensional unit ball, because its volume is an exponentially decreasing fraction of the hypercube $[-1..1)^n$ (see (13.111)) and exponentially many hypercube samples are needed to get one point inside the ball.

For the approximations in (13.112), see (15.92) and compare with Stirling's formula (3.57). The cylinder construction in Lemma 13.113 is Lemma 3.6 in Goldreich & Goldwasser (2000). They also show an upper bound of $2\,\text{vol}_{n-1}(\mathcal{B}(0, \sqrt{1 - \varepsilon^2/4}))/\text{vol}_n(B_0)$ for the quotient q in Lemma 13.113. The Reduction 13.118 $\text{GapSVP} \leq \text{BDD}$ follows their approach in showing $\text{GapSVP} \in \text{co-AM}$. For our application in the proof of Theorem 13.119, we do not have a good lower bound on δ and need the inscribed ball for small δ. In Corollary 13.115, we replace the denominator $2n^{3/2}$ by n^2 only for convenience.

Theorem 13.119 is from Peikert (2009), as reformulated by Lyubashevsky & Micciancio

(2009). The BDD problem was formulated in Liu *et al.* (2006), who show that BDD_γ is NP-hard for $\gamma > 1/\sqrt{2}$. The statistical distance of distributions does not increase under (randomized) mappings: see Micciancio & Goldwasser (2002), Proposition 8.10.

Under the continuous Gaussian distribution $D_r^{(n)}$, the countable set L has probability zero for any lattice L. But $\delta_r^{(n)}(L)$ is finite and positive. The algorithm of Lemma 13.122 is from Brakerski *et al.* (2013), Lemma 2.3; see also Gentry *et al.* (2008).

There is no "unifom distribution on $\mathbb{R}^n$". But for a matrix $B \in \mathbb{R}^{n \times n}$ with row vectors $b_1, \ldots, b_n$, one can consider the *fundamental parallelepiped* $P = \{\sum_i r_i b_i \colon -1/2 \le r_i < 1/2$ for all $i\}$ of the lattice L generated by B, the uniform distribution U_P on P, and the n-dimensional analog $G_r^{(n)*}$ of G_r^* in (13.120). Micciancio & Regev (2007) show that for $\varepsilon > 0$ and $t \ge \eta_\varepsilon(L)$, the statistical distance between U_P and $G_{1/t}^{(n)*}$ on P is at most $\varepsilon/2$.

Micciancio & Regev (2007) introduced the smoothing parameter η and showed Lemma 13.124. (ii) is their Lemma 3.2, (iii) is a special case of their Lemma 3.3, and (iv) is implicit in their Lemma 4.4. Other cryptographic constructions have been proposed, for example, hash functions whose collision resistance is based on the hardness of lattice problems (Regev 2004, see Peikert 2014). This also includes one-way functions, trapdoor functions, identification schemes, and signatures.

We take Theorem 13.127 to imply the existence of an algorithm as called in step 5 of Reduction 13.128. It takes as input a basis of a lattice L and $t \in \mathbb{R}^n$ and numbers β, m, r. As output it produces some $v \in L$. It is allowed to use an oracle for $LWE_{\beta,m}$ and a sampler for G_{r,L^*} if the condition in Lemma 13.122 is satisfied. If in addition the lower bound on r in Theorem 13.127 and the CVP input condition are satisfied, then v solves $CVP(t)$.

In Definition 13.126, $m(n)$ is odd only for convenience, for example in (13.121).

Some exercises in this chapter are courtesy of Alex May.

Exercises.

EXERCISE 13.1 (Transforming bases).   Let $B \in \mathbb{R}^{n \times m}$ be a basis of the lattice L. *Express each of the following matrix operations on B as a right multiplication by a unimodular matrix U, that is, an integer matrix with $\det U \in \pm 1$:*

 (i) *Swap the order of the columns of B.*

 (ii) *Multiply a column by -1.*

 (iii) *Add an integer multiple of a column to another column, that is, set $b_i \leftarrow b_i + ab_j$ where $i \ne j$ and $a \in \mathbb{Z}$.*

 (iv) *Show that any unimodular matrix can be expressed as a sequence of these three elementary integer column transformations.*

EXERCISE 13.2. *Prove Theorem 15.78.*

EXERCISE 13.3 (Discrete sets).   *A subset $S \subseteq \mathbb{R}^n$ is called discrete if for any $x \in S$ there is a real number $r > 0$ such that $S \cap \mathcal{B}_n(x, r) = \{x\}$. Show that $\mathbb{Z}^n$ and $A = \{1/n \colon n \in \mathbb{N}_{\ge 1}\} \subseteq \mathbb{R}$ are discrete, but $A \cup \{0\}$ is not.*

EXERCISE 13.4 (Discrete groups). *Let L be a subgroup of the additive group $\mathbb{R}^n$.*

(i) *Show that L is discrete if and only if for some $r > 0$ we have $L \cap \mathcal{B}_n(0, r) = \{0\}$.*

Let $B \in \mathbb{R}^{n \times m}$ be the matrix with rows $b_1, \ldots, b_m \in \mathbb{R}^n$ and write L for the set of all integral linear combinations of the b_i's.

(ii) *Show that for $b_1 = 1$ and $b_2 = \sqrt{2} \in \mathbb{R}^1$, L is not discrete. You may want to use the rational approximations obtained from the continued fraction $(1, 2, 2, 2, \ldots)$ for $\sqrt{2}$; see Notes 15.9 and Hensley (2006), Section 1.6.*

(iii) *Show that L is discrete if*

(a) *$b_1, \ldots, b_m \in \mathbb{Q}^n$, or*

(b) *$b_1, \ldots, b_m$ are linearly independent. Hint: Use your result from (i) and consider the region $\{\sum_{1 \le i \le m} x_i b_i : |x_i| < 1\}$.*

(iv) *Let $L \subseteq \mathbb{Z}^n$ be a discrete subgroup and $\ell = \dim \operatorname{span}_{\mathbb{R}}(L)$. Thus there exist linearly independent vectors $a_1 \ldots, a_\ell \in L$. Construct inductively linearly independent vectors $b_1, \ldots, b_m \in L$ so that $L = \sum_{1 \le i \le m} \mathbb{Z} b_i$.*

(v) *Show that a subset $L \subseteq \mathbb{R}^n$ is a lattice if and only if it is a discrete additive subgroup of $\mathbb{R}^n$.*

EXERCISE 13.5 (Lattices and the gcd). *Let a, b, c be positive integers and consider the lattice $L = a\mathbb{Z} + b\mathbb{Z}$ spanned by the vectors (a) and (b) in $\mathbb{R}^1$.*

(i) *Show that $L = \gcd(a, b)\mathbb{Z}$. Hint: Extended Euclidean Algorithm.*

(ii) *Conclude that a shortest nonzero vector in L has length $\gcd(a, b)$.*

EXERCISE 13.6 (Gcd revisited). *Let a, b, c be positive integers and consider the lattice $L \subseteq \mathbb{Z}^3$ spanned by the row vectors of*

$$B = \begin{pmatrix} 1 & 0 & ca \\ 0 & 1 & cb \end{pmatrix} \in \mathbb{Z}^{2 \times 3}.$$

(i) *Do some experiments with the lattice L. Select, say, 100 pairs (a, b) randomly, where a and b are at most $C = 100$ and check for which (large) values of c the basis reduction algorithm always yields a basis of the form*

$$B = \begin{pmatrix} x_1 & x_2 & 0 \\ s & t & \pm c \gcd(a, b) \end{pmatrix},$$

of L, with $sa + tb \in \pm \gcd(a, b)$.

(ii) *Try also the values $C = 500$, $C = 1000$ and $C = 5000$. Hand in a table of values of c for which your experiment succeeded.*

You are to prove that for $c > C = (2(a^2 + b^2))^{1/2}$, the above basis reduction always computes the correct solution.

(iii) *Show that every vector $v \in L$ is of the form $(v_1, v_2, c(v_1 a + v_2 b))$.*

(iv) *Take any such vector with $v_1 a + v_2 b \neq 0$. Show that then $\|v\|^2 \ge c^2$.*

(v) *Now consider a reduced basis $\bar{B} = (\bar{b}_1, \bar{b}_2)^T$. Then $\|\bar{b}_1\| \le \sqrt{2}\lambda_1(L)$ and $\|\bar{b}_1\| \le \sqrt{2}\|v\|$ for any nonzero vector $v \in L$. Consider the vector $(-b, a, 0)$ and show that $\|\bar{b}_1\| \le C$.*

(vi) *Conclude that for $c > C$ the vector $\bar{b}_1$ is of the form $(x_1, x_2, 0)$.*

Thus the reduced basis has the form $\bar{B} = \begin{pmatrix} x_1 & x_2 & 0 \\ s & t & cg \end{pmatrix} \in \mathbb{Z}^{2 \times 3}$ with $g \in \mathbb{Z}$.

Furthermore, there is a unimodular transformation $U \in \mathbb{Z}^{2 \times 2}$ with $\bar{B} = UB$, $U = \begin{pmatrix} x_1 & x_2 \\ s & t \end{pmatrix}$, $d = \det U = x_1 t - x_2 s \in \pm 1$, and inverse $U^{-1} = \begin{pmatrix} t & -x_2 \\ -s & x_1 \end{pmatrix}$.

(vii) *Argue that $U \cdot (ca, cb)^T = (0, cg)^T$ and conclude that $g \in \pm \gcd(a, b)$.*

(viii) *Compare your result to your experiments from (i) and (ii).*

EXERCISE 13.7 (Gram-Schmidt orthogonalization). *In the Gram-Schmidt orthogonal basis of Section 15.15, we construct, given a basis $B \in \mathbb{R}^{n \times m}$ of the $\mathbb{R}$-vector space $V = \operatorname{span}(B)$, an orthogonal basis B^* by defining $b_1^* = b_1$, $b_i^* = b_i - \sum_{j<i} \mu_{i,j} b_j^*$ with $\mu_{i,j} = \langle b_i, b_j^* \rangle / \langle b_j^*, b_j^* \rangle$.*

(i) *Show that for $i_1 \neq i_2$ the vectors $b_{i_1}^*$ and $b_{i_2}^*$ are orthogonal.*

(ii) *Show that for $i < j$ the vectors b_i and b_j^* are orthogonal.*

(iii) *Consider the vector space $V = \operatorname{span}(B)$, spanned by the basis*

$$B = \begin{pmatrix} 2 & 1 & 2 \\ 0 & 2 & 1 \\ 0 & 0 & 2 \end{pmatrix}.$$

Compute an orthogonal basis of V.

(iv) *Is your orthogonal basis of V also a basis of the lattice generated by the rows of B? Justify your answer.*

(v) *Define the orthogonal projection operator of $\mathbb{R}^n$ to $\operatorname{span}(b_i^*, \ldots, b_n^*)$ as*

$$\pi_i(x) = \sum_{i \leq j \leq n} \frac{\langle x, b_j^* \rangle}{\langle b_j^*, b_j^* \rangle} b_j^*.$$

Show that $b_i^* = \pi_i(b_i)$.

(vi) *Design a method which returns an orthonormal basis, that is, an orthogonal basis B^* with $\|b_i^*\| = 1$ for all i.*

EXERCISE 13.8 (Orthogonal sublattices). *You are to show that, although not every lattice has an orthogonal basis, every integer lattice has an orthogonal sublattice. More specifically, show that for any nonsingular $B \in \mathbb{Z}^{n \times n}$ generating a lattice L and $d = |\det(B)|$, we have $d\mathbb{Z}^n \subseteq L$. Consider a vector $v = dy \in d\mathbb{Z}^n$. State Cramer's rule for solving a system of linear equations and use it to show that $v \in L$.*

EXERCISE 13.9 (Short vectors for subset sums). *We fix $B = 100\,000$ (in decimal), choose $a_1, \ldots, a_n \in \{1, \ldots, B\}$ and $x_1, \ldots, x_n \in \{0, 1\}$ uniformly at random, with $(x_1, \ldots, x_n) \neq (0, \ldots, 0)$. For several choices of n, run Algorithm 13.26 ten times (or 100 times) and record the success rate r with which it produces x.*

(i) *For which n does Theorem 13.27 say that the algorithm will usually be successful? Same for (13.32).*

(ii) *Choose several n to test this prediction, calling a success if $r \geq 80\%$.*

(iii) *Explain your findings.*

EXERCISE 13.10 (Subset sum density).   *Suppose that $s_1, \ldots, s_n$ is superincreasing. Determine upper and lower bounds, if any, on $\delta(s_1, \ldots, s_n)$.*

EXERCISE 13.11 (The basis reduction algorithm).   *In this exercise you will do several experiments with the lattice basis reduction algorithm.*

   (i) *Implement the basis reduction algorithm in a programming language of your choice. Hand in the source code. Hint: Try to work bottom up. Implement the vector arithmetic first, afterwards scalar products and the $\mu_{i,j}$. Build from that the GSO, which in turn is used by the size-reduction and the exchange-step. Once you have all this, start writing the basis reduction algorithm. It is helpful to employ a computer algebra system for this task.*

   Consider the lattice L spanned by the basis $B = \begin{pmatrix} 2 & 1 & 5 & 8 \\ 7 & 2 & 5 & 5 \\ 2 & 3 & 1 & 1 \\ 5 & 8 & 9 & 9 \end{pmatrix}$.

   (ii) *Minkowski's theorem states that for any lattice we have $\lambda(L) \leq \sqrt{n} \det(L)^{1/n}$. Determine both values in the example.*

   (iii) *What is the length of the shortest vector in the output of the basis reduction algorithm? Compare to (ii) and the general upper bound.*

   (iv) *Determine the integer $D = \prod_{1 \leq i \leq 4} \|b_i\|^2$.*

   (v) *Which upper bound on the number of iterations does the running time analysis of the basis reduction algorithm yield?*

   (vi) *What is the value of D upon finding a reduced basis?*

   (vii) *Give an upper bound on the number of iterations based on the initial and final values of D.*

   (viii) *How many iterations are actually executed?*

EXERCISE 13.12 (Subset sum cryptosystem).   *We consider the following subset sum cryptosystem. The pairs AA, AB, $\ldots$, AZ, BA, BB, $\ldots$, BZ, $\ldots$, ZA, ZB, $\ldots$, ZZ of letters are identified with the 10-bit binary representations of the numbers $0, \ldots, 26^2 - 1 = 675$. For example, the pair SO corresponds to the bit string $x_9 x_8 \cdots x_0 = 0111100010$. Longer messages are broken into two-letter blocks and each block is treated separately. You should use a computer algebra system for this exercise.*

CRYPTOSYSTEM.

   1. *The secret key is $s_0, \ldots, s_9, m, r \in \mathbb{N}$ with $s_{i+1} \geq 2s_i$ for $0 \leq i \leq 8$, $m > \sum_{0 \leq i \leq 9} s_i$, and $\gcd(r, m) = 1$.*
   2. *The public key is $t_i = r s_i$ in $\mathbb{Z}_m$ for $0 \leq i \leq 9$.*
   3. *A bit string $x = x_9 x_8 \cdots x_0$ is encrypted as $t = \sum_{0 \leq i \leq 9} x_i t_i \in \mathbb{N}$.*
   4. *To decrypt a ciphertext t, you compute $u \in \mathbb{N}$ such that $u = r^{-1} t$ in $\mathbb{Z}_m$. Then $u = \sum_{0 \leq i \leq 9} x_i s_i \in \mathbb{N}$, and you can reconstruct $x_9, x_8, \ldots, x_0$ from t.*

   (i) *Write procedures for encryption and decryption, and check them with the key $s_0 = 1$ and $s_{i+1} = 2s_i + 1$ for $0 \leq i \leq 8$, $m = 9973$, and $r = 2001$, on the message COGITOERGOSUM.*

(ii) Prove that $u = \sum_{0 \le i < 9} x_i s_i$ actually holds in step 4.

(iii) Now an eavesdropper knows the public key

i	9	8	7	6	5	4	3	2	1	0
t_i	5340	11669	4835	1418	9708	13853	5927	1964	9981	3991

and has intercepted a ciphertext consisting of the blocks

$$29513, 21863, 33739, 41433, 25942, 31988, 37702, 7345, 57937$$

in decimal representation. Try basis reduction to find the original message. This need not work for all blocks.

EXERCISE 13.13 (Subset sum cryptosystem). We have a subset sum cryptosystem with $m = 437$ and $r = 204$. Bob's private key is $s = (2, 6, 10, 26, 68, 161)$.

(i) Compute Bob's public key t.

Now Alice wants to send the string $x = (0, 1, 0, 1, 1, 0)$.

(ii) Encrypt x with Bob's public key obtaining y.

(iii) Describe in detail how Bob will decrypt the encrypted message y and do the decryption.

EXERCISE 13.14 (Breaking truncated linear congruential generators). We consider the truncated homogenous linear congruential generator with $x_i = s x_{i-1}$ in $\mathbb{Z}_m$. We are given $m = 1009$, $\ell = \lceil \log_2(m)/2 \rceil = 5$, $\alpha = 2^\ell$, and $s = 25$. The sequence $y = (y_1, y_2, \ldots)$ with $y_i = \lfloor x_i/\alpha \rfloor$ is

$$0, 10, 21, 25, 30, 8, 13, 13, 24, 14, 7, 6, 15, 28, 10, 3, 17, 25, 0, 15, 12, \ldots$$

Your task is to break this generator completely. To do so, you will recover the sequence z_i with $x_i = y_i \alpha + z_i$.

(i) Write down the matrix

$$A = \begin{pmatrix} m & 0 & 0 & 0 & 0 & 0 \\ -s & 1 & 0 & 0 & 0 & 0 \\ -s^2 & 0 & 1 & 0 & 0 & 0 \\ -s^3 & 0 & 0 & 1 & 0 & 0 \\ -s^4 & 0 & 0 & 0 & 1 & 0 \\ -s^5 & 0 & 0 & 0 & 0 & 1 \end{pmatrix}.$$

(ii) Compute the sequence $c_i = (s^{i-1} y_1 - y_i) \cdot \alpha$ in $\mathbb{Z}$ for $1 \le i \le 6$.

(iii) Using lattice basis reduction compute a reduced basis V and a unimodular transformation U such that $V = UA$.

(iv) Compute Uc and take the symmetric system of representatives modulo m of your result.

(v) Now solve $Vz = Uc$ using Gaussian elimination, obtaining the z_i.

(vi) Finish by writing down the sequence x_i.

(vii) Compute the next 10 values of the above sequence of y's.

(viii) Argue that you have broken the generator.

(ix) Explain in detail why you had to use basis reduction at all.

(x) Play around with your algorithm. Try different values of m, ℓ and s, and report on the successes and failures of your algorithm.

EXERCISE 13.15 (Dual of full-rank lattices).   Let $B \in \mathbb{R}^{n \times n}$ be nonsingular and L the full-rank lattice generated by its rows. Consider its dual $L^* = \{v \in \mathbb{R}^n : \forall u \in L \quad uv \in \mathbb{Z}\}$. Lemma 13.48 shows that the rows of $(B^T)^{-1}$ form a basis of L^*.

(i) Show that $(L^*)^* = L$.

(ii) Prove that $|L^*| = |L|^{-1}$.

(iii) Show that $\lambda_1(L)\lambda_1(L^*) \leq n$. Hint: Use $\lambda_1(L) \leq \sqrt{n} \det |L|^{1/n}$.

(iv) Some examples of dual lattices:

    (a) Compute the dual of the lattice $2\mathbb{Z}^n$.

    (b) Compute the dual of the lattice spanned by the basis

$$\begin{pmatrix} 1 & 2 \\ 3 & 4 \end{pmatrix}.$$

EXERCISE 13.16 (Dual lattices).   Let $\ell \leq n$, let L be a lattice generated by the basis $B \in \mathbb{R}^{\ell \times n}$, and let L^* be its dual.

(i) Prove that $D = (BB^T)^{-1}B$ is a basis of L^*. Hint: We have $\operatorname{span}_\mathbb{R}(B) = \operatorname{span}_\mathbb{R}(D)$ and $DB^T = I$, where I is the identity matrix.

(ii) Show that $(L^*)^* = L$.

(iii) Prove that $|L^*| = |L|^{-1}$.

(iv) Compute a basis of L^*, where L is the lattice generated by the basis

$$B = \begin{pmatrix} 1 & 2 & 3 \\ 4 & 5 & 6 \end{pmatrix}.$$

EXERCISE 13.17 (Branching in generator attack).   At the end of Section 13.5, a difficulty arises when $|y_i - y_{i+1}| < 2\alpha$. You are to approximate the probability with which this happens for random values. So we have a k-bit random x_i, $\alpha = 2^\ell$, and $y_i = \lfloor x_i/\alpha \rfloor$.

(i) Check that $|y_i - y_{i+1}| < 2\alpha$ is roughly equivalent to $|x_i - x_{i+1}| < 2\alpha^2$.

(ii) Prove that the probability of the latter event is about $2^{2\ell+1-k}$ for a random k-bit x_{i+1}.

(iii) Take $k = 1000$, $n = 100$, and $\varepsilon = 0.1$. What is the largest ℓ satisfying the condition in Theorem 13.54? What is the percentage of exceptional s?

(iv) Pretending that x_i and x_{i+1} are uniformly random, what is the probability for the event in (i) to happen?

(v) Give an upper bound on the probability that there exists some $i \leq n$ for which (i) happens.

EXERCISE 13.18 (The travelling salesman problem). *In the travelling salesman problem (TSP) you are given some cities and (integer) distances between them, and have to find the shortest route covering all cities or decide whether there exits a tour that is no longer than a given value. The latter decision problem is NP-complete. In the Euclidean Travelling Salesman Problem, the cities are given by integer coordinates in the plane, and the distance is the Euclidean distance. This problem is NP-hard but not known to be in NP. The difficulty is in deciding whether one tour is shorter than another one. This amounts to telling whether one sum of square roots (of integers) is smaller than another such sum. It is conceivable that two such sums might differ only by an exponentially small amount, and a conjecture states that this is not possible. More precisely we let $\delta(n, C)$ be the smallest distance between two such sums with at most n terms $\sqrt{a}$, with $1 \le a \le C$.*

(i) *Give a basis for a lattice L so that two close sums as above yield a short vector in L. Hint: Construct a lattice similar to the one used for the subset sum cryptosystem by using squarefree integers only.*

(ii) *For several values of n and C compute short vectors in L and see whether they yield two sums with small difference. Use $n = 5, 10, 15, 20$ and $C = 100, 200, 500$. What can you conclude about $\delta(n, C)$?*

EXERCISE 13.19 (An inequality of norms). *Let $f \in \mathbb{Z}[t] = \sum_{0 \le i \le n} f_i t^i$ be a polynomial of degree n and let $\sigma(f) = \#\{i \le n: f_i \ne 0\}$ be the sparsity of f. Show that $\|f\|_1 \le \sqrt{\sigma(f)}\|f\|_2$. Hint: Use the Cauchy-Schwarz inequality $g \star h \le \|g\| \cdot \|h\|$ (Theorem 15.72), where g and h are the coefficient vectors of two polynomials.*

EXERCISE 13.20 (GapSVP$_\alpha$). *Definition 13.105 describes the GapSVP$_\alpha$ problem.*

(i) *Give an algorithm that approximates $\lambda_1(L)$ by binary search on d using a subroutine for GapSVP$_\alpha$.*

(ii) *How well does your algorithm approximate $\lambda_1(L)$?*

EXERCISE 13.21 (2-dimensional balls). *Take $z \in \mathbb{R}^2$, consider the two balls $B_0 = \mathcal{B}_2(0, 1)$ and $B_1 = \mathcal{B}_2(z, 1)$, the uniform distributions U_0 and U_1 on them, and $d = \Delta(U_1, U_2)$.*

(i) *Draw a picture of the two balls for $z = (\sqrt{2}, 1)$. Where in the picture do you find the statistical distance $\Delta(X, Y)$?*

(ii) *Compute $\Delta(X, Y)$. Hint: You need a bit of basic calculus here. Parametrize the balls by appropriate functions in one variable and compute some areas.*

(iii) *What do you observe when you vary the radius and the distance? Perform experiments.*

(iv) *Compare your results to Lemma 13.113.*

EXERCISE 13.22 (Intersection of two balls). *You are to turn the approximate inequality for the cylinder in Lemma 13.113 into an explicit one. Insert upper and lower bounds on $\Gamma(x)$, say $\Gamma(x) = \sqrt{2\pi/x}(x/e)^{x+\varepsilon}$ with $0 < \varepsilon < 1/12x$ from Artin (1964), Chapter 3 (see also Graham et al. 1994, Section 9.6), into the proof of Lemma 13.113. Does this lead to a better result? Can you improve the last inequality in (13.8)?*

EXERCISE 13.23 (Left over hash lemma). *You are to prove Lemma 13.145, following Regev (2009). For any $g = (g_1, \ldots, g_\ell) \in H^\ell$ and $h \in H$, let*

$$q_g(h) = 2^{-\ell} \cdot \#\{x_1, \ldots, x_\ell \in \{0, 1\}\colon \sum_{i \in s} x_i g_i = h\}$$

be the probability for h to occur as a random subset sum of the g_i's. Consider q_g as a vector with d components $q_g(h)$ for $h \in H$, and show the following.

(i) *The square of its norm satisfies*

$$\|q_g\|^2 = \sum_{h \in H} q_g(h)^2 = \mathrm{prob}\{\sum_{1 \leq i \leq \ell} x_i g_i = \sum_{1 \leq i \leq \ell} x_i' g_i \colon x, x' \xleftarrow{\boxdot} \{0, 1\}^\ell\}$$

$$\leq 2^{-\ell} + \mathrm{prob}\{\sum_{1 \leq i \leq \ell} x_i g_i = \sum_{1 \leq i \leq \ell} x_i' g_i \colon x, x' \xleftarrow{\boxdot} \{0, 1\}^\ell, x \neq x'\}.$$

(ii) *For $x \neq x' \in \{0, 1\}^\ell$, $\mathrm{prob}\{\sum_{1 \leq i \leq \ell} x_i g_i = \sum_{1 \leq i \leq \ell} x_i' g_i \colon g \xleftarrow{\boxdot} H^\ell\} = d^{-1}$.*

(iii) *$\mathcal{E}(\|q_g\|^2 \colon g \xleftarrow{\boxdot} H^\ell) \leq 2^{-\ell} + d^{-1}$.*

(iv)

$$\mathcal{E}(\Delta(q_g, U_H) \colon g \xleftarrow{\boxdot} H^\ell)$$
$$\leq \mathcal{E}(d^{1/2}(\sum_{h \in H}(q_g(h) - d^{-1})^2)^{1/2} \colon g \xleftarrow{\boxdot} H^\ell)$$
$$= d^{1/2} \cdot \left(\mathcal{E}(\sum_{h \in H} q_g(h)^2 \colon g \xleftarrow{\boxdot} H^\ell) - d^{-1}\right) \leq d^{1/2} 2^{-\ell/2}.$$

Hint: $\|\cdot\|_1 \leq \sqrt{n}\, \|\cdot\|_2$.

Wilson's reaction to the Zimmermann telegram alone
justifies the conclusion that it was one of the maladroit
as well as montrous blunders in modern diplomatic history.
ARTHUR S. LINK (1965)

Speculation was uncontrolled as to how it [the Zimmermann
telegram] had been intercepted: it was rumoured that the
messenger had been caught by American guardsmen on the
Mexican border; that a copy had been taken from von Bernstorff
at Halifax; that it was in a mysterious box seized by the British
on the ship which Bernstorff sailed on.
COLONEL EDWARD MANDELL HOUSE (1926)

In fact, the Germans had throughout the war as efficient
a cryptographic service as any belligerent can desire,
and the greatest triumphs of their own intelligence work
were due to the efficiency of their cryptanalytical staff.
ELLIS MARK ZACHARIAS (1946)

If Tyranny and Oppression come to this land,
it will be in the guise of fighting a foreign enemy.
JAMES MADISON (1787)

Marina said with conviction: "You're all mad." Mitchell said:
"That's what a lunatic asylum inmate says about
the outside world—and he may well be right."
ALISTAIR MACLEAN (1977)

Chapter I

The Zimmermann telegram

here is no single event that decided the outcome of the First World War 1914–1918. But the entry of the USA as a belligerent—after long hesitation—was a decisive factor in the success of the *Entente*, originally led by France and Great Britain. And the (in)famous telegram discussed in this paper played a role in changing the anti-war attitude in large parts of the US population and giving President Thomas Woodrow Wilson the popular and political majority for entry into the war on the side of the *Entente*, clenching its victory. The cryptanalytic solution has been called *"the greatest intelligence coup of all time."*

The telegram is an instructive display of German failures and British successes, both in cryptology and in diplomacy.

I.1. Capturing the *Magdeburg*'s codebooks

We start with a tale from the early stages of the British cryptographic bureau, concerning a marvellous gift they received and which got them started on their breaks into the German cipher systems. The story begins less than a month after the German military had embarked on the adventure that would eventually lead to their downfall, by attacking Belgium

© Springer-Verlag Berlin Heidelberg 2015
J. von zur Gathen, *CryptoSchool*, DOI 10.1007/978-3-662-48425-8_22

Figure I.1: Eleven codewords from the *Signalbuch*.

and France.

In the middle of the night of 26 August 1914, the German light cruiser *SMS Magdeburg* was sailing in a Baltic Sea flotilla intending to wreak havoc on the Russian ships in the Gulf of Finland. She followed the leading ship, the light cruiser *Augsburg*, who tried to sneak south around a suspected Russian mine field. But she lost visual contact in a dense fog, and just as she was turning around from a southerly to an easterly course, she ran aground in shallow waters off the northern tip of the Estonian island of Osmussaar (Swedish: Odensholm), at 12.37 am. After desperate attempts to get her off, also with the help of the torpedo boat V-26, her captain Richard Habenicht ordered her to be blown up, around 9.00 am. By mistake, the fuses were lit too early, and the men had less then five minutes to abandon ship.

The *Magdeburg* had four copies of the codebook *Signalbuch der Kaiserlichen Marine* on board; see Figure I.1 for an exerpt.[1] One was burned in time. One was jettisoned overboard. Radioman Second Class Neuhaus jumped overboard with the third one and was not seen again. And the fourth one—was forgotten.

By then, Russian ships had arrived. Lieutenant Galibin of the tor-

---

[1] *Signalbuch der Kaiserlichen Marine* = signal book of the Imperial Navy. The plaintext words are: insult, to scold; dishonorable, disgrace; umbrella, to protect (against); umbrella (folding) anchor; battle; to offer a battle; to accept a battle; to evade a battle; battle begins; during the battle; after the battle.

pedo boat *Lejtenant Burakov* boarded the *Magdeburg* and found the code-book in captain Habenicht's cabin. Later, Russian divers also recovered the two other codebooks from the clear waters with a depth of less than ten meters.

The Russian military command immediately recognized the impor-tance of their bounty, and offered it to England, the major naval power of the *Entente*. After a trip on board the *H.M.S. Theseus* from Alexan-drovsk (now: Polyarny) to Hull in England, the Russian ambassador Count Alexander Konstantinovich Benckendorff (1849–1917) handed the *Signalbuch* to Winston Churchill (1874–1965), First Lord of the Admiralty, on 13 October 1914. The British cryptographers then put this gift to good use.

The German military command never recognized the importance of their loss. The commanding admiral downplayed the possibility of the codebooks having been recovered. An investigation by Prinz Heinrich von Preußen (1862–1929), the German emperor's younger brother, came to the opposite conclusion, but was ignored. The very Lieutenant Gal-ibin, retriever of the captain's codebook, was captured in August 1915 and told about his feat. He was ignored. On several occasions, British naval forces happened to be right there where a German fleet was to steam through. Such circumstantial evidence was ignored as well.

Fleet Paymaster Charles Rotter had worked on German naval sig-nals since he joined the British Naval Intelligence Division in 1911. This division was substantially reinforced shortly after the war's outbreak. The main player was James Alfred Ewing (1855–1935), an engineer by profession, among whose achievements are the design of seismic instru-ments, the discovery of hysteresis in magnetic materials, and studies of the structure of metals. After teaching in Tokyo, Dundee and Cambridge UK, he was Director General of Naval Education at the Royal Naval Col-lege in Dartmouth from 1902 on. He came to cryptography by accident, when on 4 August 1914, just after the start of World War I, his friend Admiral Sir Henry Oliver (1865–1965), Director of the Intelligence Divi-sion of the Admiralty War Staff, showed him some intercepted German cipher telegrams. Ewing said he would look at them, and the Admiral interpreted this quite liberally. Soon after, intercepted cipher messages were pouring into Ewing's office, often over two thousand per day. He acquired a large room numbered "40" in the Admirality's Old Build-ing, and even after a move into new quarters this cryptographic office was called "Room 40"—a name that does not give away much. Af-ter some startup difficulties, they routinely broke German military and diplomatic ciphers.

The *Signalbuch* that arrived at Room 40 contained between its heavy lead covers hundreds of pages with three-column entries, as shown in Figure I.1. Thus *Schlacht* (battle) would be encoded as QPJ (usually) or 66164 (less often). But this did not break the intercepts except some items of lesser importance like weather reports.

The clue arrived in the form of the *Handelsschiffsverkehrsbuch*[2] seized from a commercial vessel in Australian waters. This also contained a (different) list of codewords, and in addition a superencipherment by which each individual letter of a codeword was changed into another letter, via a simple substitution. Charles Rotter in Room 40 had the flash of insight that the same might be applied to the *Signalbuch* codewords. But the usual frequency cryptanalysis is hard on codewords, for lack of redundancy. But then the Germans helped out by sending a sequence of messages whose consecutive serial numbers they encoded. That was enough to reveal the superencipherment. Alastair Denniston, a scholar of German in Room 40, commented coolly: "Their folly was greater than our stupidity.".

From then on, Room 40 read most of the German naval signals. However, a participant like Lieutenant Filson Young, on board the battle cruiser *HMS Lion* from November 1914 to May 1915, bitterly complained about the Admiralty's inefficiency in using this valuable material, only a small portion of which actually reached the Grand Fleet.

I.2. The telegram

The most spectacular coup of Room 40 gave US President Thomas Woodrow Wilson the popular and political majority for entry into the war on the side of the Entente, thus ensuring their victory. Hoping to break the stalemate of the bloody trench battles in Northern France and Belgium, the German military wanted in January 1917 to force Great Britain into submission by cutting her lifelines to North America with all-out submarine attacks. A major concern was that this might lead the USA into the war, while an isolationist attitude had hitherto kept them out of it.

The Germans tried to create a diversion by dragging the Mexicans into the fray. Arthur Zimmermann (1864–1940) had become Secretary of State for Foreign Affairs on 22 November 1916, the first in this position not from nobility: *"He had not a von to his name, and his training and early career had been in the consular service, realm of the non-vons.".* He sent a top secret message to the German minister Heinrich Julius Ferdinand von Eckardt (1861–1944) in Mexico, via the German ambassador

---

[2]merchant navy codebook

Graf Johann Heinrich Andreas Hermann Albrecht von Bernstorff (1862–1939) in Washington. He offered, if war with the USA broke out, money to the Mexican President Venustiano Carranza Garza (1859–1920) and consent for Mexico to regain the states of Texas, New Mexico, and Arizona, which had been conquered by the USA in the war of 1848. The idea of the Mexican army beating US military might was as ridiculous then as it would be today. The telegram was encrypted and sent from Berlin to von Bernstorff in Washington, who forwarded it to Mexico.

The telegram was eventually deciphered by Room 40, then directed by Captain (later Admiral Sir) William Reginald "Blinker" Hall (1870–1943), and passed to the US ambassador in London, Walter Hines Page (1855–1918). President Thomas Woodrow Wilson (1856–1924) gave it to the US Press for publication on 1 March 1917, and the ensuing public outcry led the US Congress to declare war against Germany on 6 April 1917.

In this section, we present the wording of the telegram and a related message to von Bernstorff. The next section deals with questions of transmission and cryptography, and Section I.4 with the political fallout.

Figures I.2 and I.3 show the original handwritten draft, from the archives of the German Foreign Office, of the notorious *Zimmermann telegram*. Its text, beginning on line 7 of the right hand column, reads:

> *Ganz geheim. Selbst entziffern.*
>
> *[Wir beabsichtigen, am 1. Februar uneingeschränkten U-Boot Krieg zu beginnen. Es wird versucht werden, Amerika trotzdem neutral zu halten.*
>
> *Für den Fall, daß dies nicht gelingen sollte, schlagen wir Mexico auf folgender Grundlage Bündnis vor: Gemeinsame Kriegführung. Gemeinsamer Friedensschluß. Reichlich finanzielle Unterstützung und Einverständnis unsererseits, daß Mexico in Texas, Neu-Mexico, Arizona früher verlorenes Gebiet zurückerobert. Regelung im einzelnen Euer Hochwohlgeboren überlassen.*
>
> *Euer Hochwohlgeboren wollen Vorstehendes Präsidenten streng geheim eröffnen sobald Kriegsausbruch mit Vereinigten Staaten feststeht und Anregung hinzufügen, Japan von sich aus zu sofortigem Beitritt einzuladen und gleichzeitig zwischen uns und Japan zu vermitteln.*
>
> *Bitte Präsidenten darauf hinweisen, daß rücksichtslose Anwendung unserer U-Boote jetzt Aussicht bietet, England in wenigen Monaten zum Frieden zu zwingen.]*

This translates into English as:

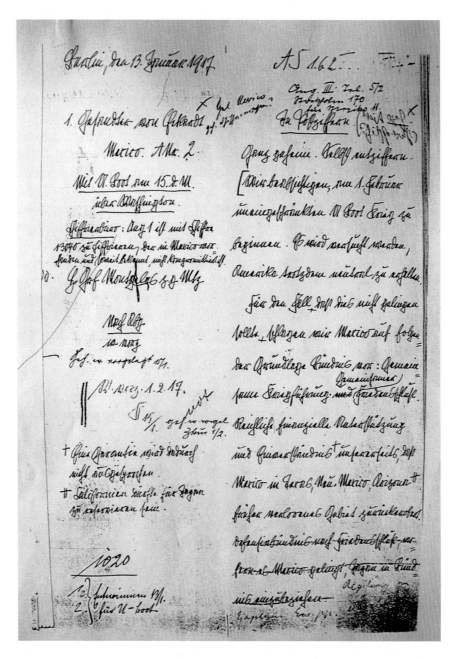

Figure I.2: The first page of a draft of the Zimmermann telegram, as prepared at the German Foreign Office.

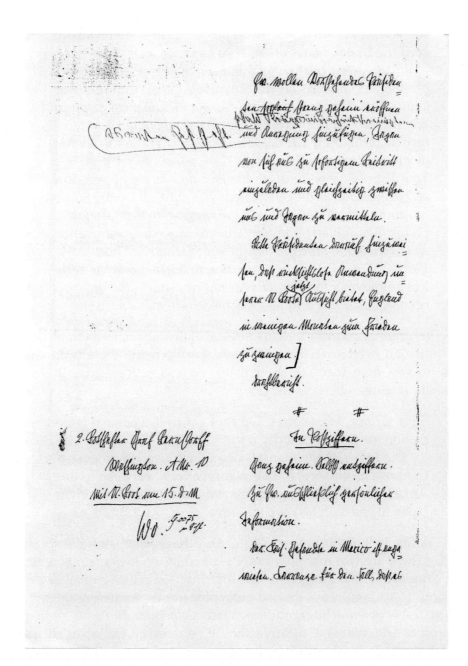

Figure I.3: The second and final part of the Zimmermann telegram, and the first part of the separate message to von Bernstorff.

> Most secret. Decipher yourself.
> [We intend to begin on the first of February unrestricted submarine warfare. We shall endeavour in spite of this to keep the United States of America neutral.
> In the event of this not succeeding, we make Mexico a proposal of alliance on the following basis: Conduct war jointly. Conclude peace jointly. Substantial financial support and consent on our part for Mexico to reconquer lost territory in Texas, New Mexico, and Arizona. The settlement in detail is left to your Excellency.
> Your Excellency will inform the President of the above most secretly as soon as the outbreak of war with the United States of America is certain, and add the suggestion that he should, on his own initiative, invite Japan to immediate adherence, and at the same time mediate between Japan and ourselves.
> Please call the President's attention to the fact that the ruthless employment of our submarines now offers the prospect of compelling England in a few months to make peace.]

The original record contains several notes about encryption and transmission, which we discuss below. Furthermore, there is another note to von Bernstorff which explains the instructions given to von Eckardt. It reads:

> In cipher. Most secret. Decrypt yourself. Personal information for your Excellency only.
> The Imperial envoy in Mexico is instructed to propose to Carranza an alliance, in case war breaks out between us and America, and to suggest to him at the same time to invite Japan to enter, on his own initiative. [3]

The first half of this note is in Figure I.3, the second half is not shown here. There are two marginal notes expanding on the contents which were not sent with the telegram. The first, inserted at the German *Einverständnis* (= consent), says that no guarantee (for reconquering the three US states) is given. The second one reads *Californien dürfte für Japan zu reservieren sein*, that is, California should be reserved for Japan. It had also been taken by the USA in the 1848 war, and its mention indicates

---

[3] The original text: In Postziffern. Ganz geheim. Selbst entziffern. Zu Euer Hochwohlgeboren ausschließlich persönlicher Information.

Der Kais. Gesandte in Mexico ist angewiesen, Carranza für den Fall, daß es zwischen uns und Amerika zum Kriege kommt, ein Bündnis anzutragen und ihm gleichzeitig nahezulegen, Japan von sich aus zum Beitritt einzuladen.

a discussion at the German Foreign Office about whether they should throw in California as a bonus—it would not increase their cost.

California does not appear in the decryption of the telegram. But somewhat mysteriously, Millis (1935) mentions California in this context. Friedman & Mendelsohn note this and ask: *"Is it possible that the Germans were reserving California as bait for Japan?"* Good guess!

The Zimmermann telegram has always played a major role in the American historiography of the First World War, and a very minor one in the German view. The basic difference is that on one side it is regarded as an evil and immoral plot, and on the other side as a legitimate if stupid diplomatic enterprise in times of war. Inexact translations of the central phrase have contributed to this rift; the noncommittal *"Einverständnis, daß Mexico ... zurückerobert"* = *consent for Mexico to reconquer ...* has usually become the exhortation of an *understanding* (or even *undertaking*) that Mexico is to reconquer ...

I.3. Transmission and cryptanalysis

We will notice two points of general cryptographic interest:

- ○ the difficulty of communicating secret keys, resolved only by Diffie & Hellman (Section 2.6),

- ○ the cryptographic sin of sending encryptions of the same message in two cryptosystems.

The German Foreign Office used two codebooks in the transmission, one called 0075 (or 7500) for Berlin-Washington and the other one 13040 for Washington-Mexico. The 11000-term codebook 0075 had been brought to the USA in November 1916 by the commercial U-boat *Deutschland*. It was a two-part codebook (ciphertext numbers assigned randomly to plaintext words; see Section D.6) and had not been sent to Mexico.

Codebook 13040 consisted of about 11000 terms, to which 3-, 4-, or 5-digit encryptions were assigned. The encrypted telegram is shown in Figure I.4. The codebook had 100 terms per page, numbered from 00 to 99 in their alphabetical order. Four pages were printed on one sheet, and these sheets could be rearranged to vary the code; the encoding of a term consisted of the page number plus its number on the page. The shorter codewords served for numbers, dates, common phrases, and grammatical inflections. Common terms like *Komma* or *Stop* were sprinkled on each page. Some pages were given two numbers, so that frequencies of

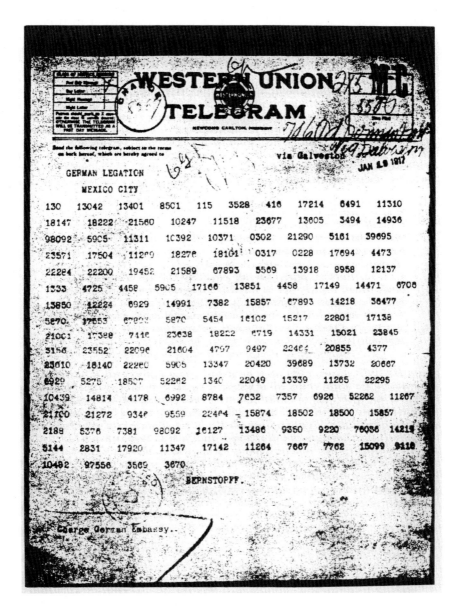

Figure I.4: The Zimmermann telegram in code 13040, as sent from Washington to Mexico.

words on that page could be halved. Our taxonomy (Section D.6) classifies it as mixed codebook with homophones. It seems that no original copy has survived. Room 40 had broken 13040 already in early 1915, while 0075 was still secure in early 1917. Figure I.5 shows a page from the well-used notebook in which the British cryptanalysts entered plaintexts that they had found. It has $6 \times 4 = 24$ boxes labelled with red numbers from 385 to 408 and each providing space for ten pencilled entries numbered 0 through 9. Each of these can contain a two-digit label, say from 90 to 99 for the second box in the bottom row, labelled 396 and enlarged in Figure I.6. At its bottom are the two entries 89 *Vereinigten Staaten* and 95 *V.S. v. Amerika*. And indeed we find 39695 = *US of America* on the fifth and 39689 = *United States* on the thirteenth line of the telegram in Figure I.4.

We can see a partial alphabetic order even in the 153 words of the Zimmermann telegram:

14814	einladen	22049	sich
14936	eingeschränkten	22096	Sie
14991	Einverständnis	22200	stop
15021	einzeln	22260	sobald
15099	Empfang	22284	sollte
		22295	sofortiger

Unteutonic alphabetical levity seems to have flipped 14814/14936 and 22284/22295; frequent words like *stop* often occur out of order. The German original in Figures I.2 and I.3 gives clear instructions: send the message to von Eckardt from Berlin in 13040, and the one for von Bernstorff in 0075. This is in perfect agreement with the availability of the codes in the two embassies. In fact, we can even follow the process leading to this decision: at top right in line 5, the scribe has noted *In Postziffern* (= in transmission cipher), and someone else has noted in parentheses *Mit geh. Chiffre vers.* (= to be sent with secret cipher), in the centre, crossreferenced to this note, someone has penned the question: *Hat Mexico geh. Chiffre vorliegen?*[4], and this interchange leads to the clear instruction at left to send the missive in 13040:

> *Chiffrierbüro: Ang. 1 ist mit Chiffre 13040 zu chiffrieren, der in Mexico vorhanden und, soweit bekannt, nicht kompromittiert ist.*[5]

---

[4]Is the secret cipher available in Mexico?
[5]To the cipher bureau: Document 1 is to be encrypted with code 13040, which is available in Mexico and, as far as is known, not compromised.

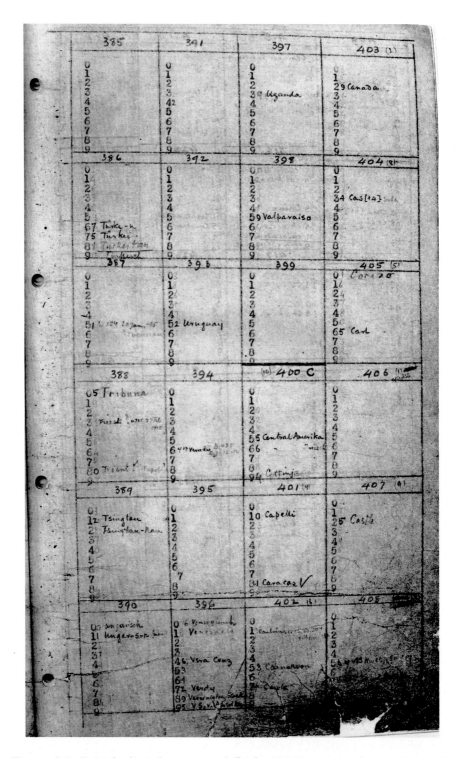

Figure I.5: British decipherment of Code 13040, entries beginning with 385 through 408.

Similarly in agreement with the availability of the codes is the note *0075* to the left of the message to von Bernstorff. The British *HMS Telconia* had cut on 5 August 1914 the two transatlantic cables off the German coast, linking Germany to America. This severely limited the German options for communication. The need to use two codebooks for one message illustrates the difficulty of establishing shared secret keys over an open line.

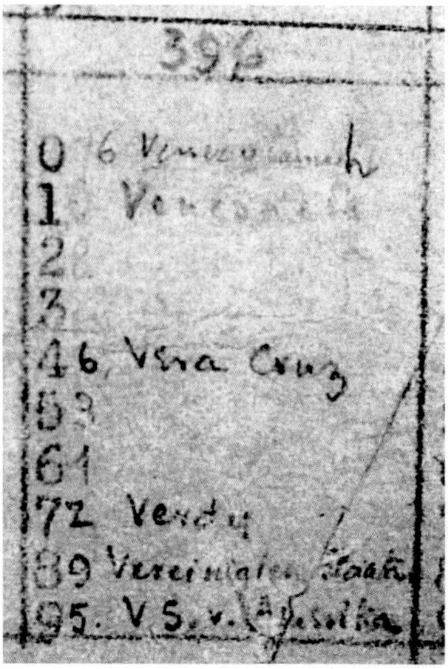

Partly due to deliberate misinformation by Hall, the transmission to Washington remained shrouded in mystery for a long time. Among the conjectures were Swedish diplomatic cables, German radio stations, and even German submarines. In fact, the German government had persuaded the US ambassador James Watson Gerard (1867–1951) in Berlin to allow the use of US diplomatic cables for instructions in *peace negotiations*. And indeed the encrypted telegram went this way to Washington, inspite of its less than peaceful intention.

A central source about the British cryptanalytic effort against the Zimmermann telegram is a note composed by Nigel de Grey (1886–1951) on 31 October 1945. He was the main codebreaker in Room 40 dealing with the telegram and wrote:

Figure I.6: The 396 box in Figure I.5.

> *The version of the telegram upon which we worked was the version in 13040, which reached us from the Cable office in transit [...] we had been at work some time on 13040. Only one person worked on it for many months then two and later three. It was a long code, our experience of book building was at its beginnings and there were many gaps unfilled. [...] We could at once read enough for Knox to see that the telegram was important. Together he and I worked solidly all the morning upon it. [...] Work [...] was slow and laborious.*

Now de Grey obfuscated the issue—as befits an able cryptographer—by writing in the same note that *the version that went through Bernstorff's office was in 7500 so far as I recollect*. There are two interpretations of this remark: either the telegram was sent from Berlin to Washington both in 13040 and in 0075 (a capital crime in cryptography), possibly over different channels, or de Grey's recollection failed him and only the second message to von Bernstorff was sent in 0075. De Grey also explains a cloak-and dagger action undertaken at Hall's suggestion:

> *Although we had the 13040 version and knew von Eckardt had no 7500 book, without disclosing our drop copy source, we could not produce it. Nor could we prove that the telegram had actually been delivered in Mexico to the German Legation and had not been faked in London. The only thing therefore was to steal a copy in Mexico City in the form delivered to the German Legation. We had two chances (a) the cable copy (b) the copy sent from Washington by Bernstorff which we banked on being also in 13040. Hence the delay till the end of February. How we succeeded in stealing the copy I never knew but money goes a long way in Mexico and steal it we did.*

Berlin knew that the Zimmermann telegram would go from Washington to Mexico in code 13040. Good practice would have forbidden to send it in code 0075 from Berlin to Washington. A further consideration is that the telegram had been transmitted in code, and its plaintext published. A professional cipher bureau would have considered the possibility that the encrypted version was also known to the enemy cryptanalysts and inferred that the code was then insecure. However, the German Foreign Office considered code 0075 secure still in February 1918.

The US cryptographers of the Signal Security Agency (MI-8) reviewed in 1945 the German codes of World War I and concluded:

> *in spite of [some] defects the German codes were distinctly better than those of other governments which MI-8 studied during the war [... They] were much better, it must be admitted, than the corresponding systems in use by the United States Army at the beginning of the war.*

I.4. The drama unfolds

The salient dates in the history of the Zimmermann telegram are given in Table I.7. At an Imperial war conference, the "ruthless" employment of total U-boat warfare was decided on 9 January, and the foreign minister Zimmermann signed the message on 13 January.

9 January	Imperial U-boat decision
13 January	Zimmermann signs message
16 January	telegram(s) from Berlin to Washington in 13040 (and 0075?)
17 January	partial decrypt of 13040 message at Room 40
19 January	telegram from Washington to Mexico in 13040
31 January	Germany declares unrestricted U-boat warfare
3 February	Wilson breaks diplomatic relations with Germany
10 February	Room 40 receives 13040 message from Mexico
22 February	Hall gives complete decrypt to Page
24 February	Wilson receives the telegram
1 March	story published in US newspapers
3 March	Zimmermann admits responsibility by a press communiqué
6 April	US congress declares war on Germany

Table I.7: The Zimmermann chronology in early 1917.

The two British cryptographers, Dillwyn Knox (1884–1943) and Nigel de Grey, dealing with the telegram worked feverishly on their task, but progress was slow. The first partial decrypt was handed to Admiral Hall, the head of Room 40, around 10.30 a.m. on 17 January. Right away, it was clear to everybody that the telegram was a bombshell that could serve to draw the US into the war—on the *Entente* side, of course. Three problems had to be addressed:

○ how to prove authenticity of the telegram,

○ how to prove correctness of the decryption,

○ how to safeguard the secret of Room 40.

Admiral Hall had a brilliant idea. He charged the British envoy Edward Thurstan in Mexico City with obtaining copies of all recent telegrams to the local German embassy, who did so through a Mexican telegraph office clerk. He may have paid for it, or "stole it he did", as de Grey says—in any case, Hall had the Zimmermann telegram as received in Mexico City in his hands on 10 February. The clever move paid off handsomely.

Now it was time for a series of subtle diplomatic moves. How to hand this god-sent message to the US government without raising suspicion about its authenticity? There was a sense of urgency. At the German announcement of unrestricted U-boat attacks, President Wilson had broken off diplomatic relations and sent ambassador von Bernstorff packing. But he kept stalling with the declaration of war that the *Entente* hoped for.

Finally, on 22 February 1917, Hall gave the telegram and its decipherment, completed on 19 February, to Walter Hines Page (1855–1918), the US ambassador in London, with the British Foreign Secretary Arthur James Balfour (1848–1930) being present. President Wilson had the message on 24 February. The US State Department found at the Washington office of Western Union the encrypted Zimmermann telegram that had travelled over its own lines. Indignation ran high in the White House at this abuse of American generosity. On 28 February, they obtained from Western Union a copy of the Washington to Mexico message, shown in Figure I.4. US Secretary of State Robert Lansing gave the story to E. M. Hood of Associated Press, and it hit the newspaper headlines on 1 March. A wave of patriotism swept through the nation, as even the South-Westerners and Westerners realized that the war was not as far away as they had thought. But some skeptics still thought this might all be a British ruse. On 1 March, Lansing cabled to Page in London *"the original message which we secured from the telegraph office in Washington"*, and de Grey deciphered this 13040 version at the Admiralty under the eyes of Edward Bell (1882–1924), a secretary at the American embassy. Actually, this almost ended in desaster. De Grey had brought an incomplete version of the codebook, and had to extemporize many codewords—which he knew by heart and, luckily for him, Bell did not ask to check in the codebook. Conjurer's magic in cryptography. It was more than enough to convince Wilson, but might not have been enough for a suspicious outsider. However, Zimmermann obliged again and came to rescue. An official German press communiqué appeared on 3 March 1917 in the papers. It stated that the German envoy in Mexico had been instructed to offer, in case of a US declaration of war against Germany, an alliance to Mexico. The communiqué also speculated how the Americans might have obtained the telegram, and proposed that this was most likely by treason on US territory.

President Wilson had won his election on 7 November 1916 with the slogan "He kept us out of war". Germany's declaration of unrestricted U-boat warfare changed his mind, but not yet that of the population. Zimmermann achieved this with his telegram. Even *"the German-Americans, of whom he had such fond hopes, retreated across their hyphen to take their stand, somewhat sullenly, on the American side."*

The secret of Room 40 was well guarded. Wild speculations abounded of how the message had been given away by treason or stolen in Mexico, or a messenger intercepted on the Rio Grande frontier. Nobody suspected the Berlin-Washington transmission, or deciphering of a code.

The rest is history: the massive deployment of American troops and arms, effective in early 1918 after almost a year of armament, helped to push the weakened German military over, enfeebled by a starved economy and disillusioned population.

Notes. Most of the material here on the Zimmermann telegram is from Kahn (1967) and von zur Gathen (2007). The *"greatest coup"* is on the first page of Kahn (1999). In Tuchman (1958), page 156, it is the more modest *"greatest coup of the war."*

I.1. The *Magdeburg*'s story is based on the vivid and detailed account in Kahn (1991), chapter 2. A copy of the *Signalbuch* is at the *Militärgeschichtliches Archiv* in Freiburg, Germany. Young (1922) describes his war experiences, which included signals intelligence.

Rotter had worked on radio intercepts at shorter wavelengths, up to 1 000 m, while Ewing's office mainly dealt with long-distance transmissions at larger wavelengths, up to 15 000 m (Hiley 1987).

I.2. The literature about the Zimmermann telegram is substantial. Among the first works were the (auto-)biographies of Bernstorff (1920), Hendrick (1922), and House (1926). Next came the cryptographic analysis of Friedman & Mendelsohn (1938), the political circumstances in Tuchman (1958), and the comprehensive treatment in Kahn (1967), Chapter 9. Further contributions were Kahn's publication of memoranda by Bell and de Grey, and Nassua (1992) who studied the reaction of the German press in the USA, and also the debates in the Reichstag committee. Freeman (2006) elucidates many details of cryptographic interest, and Boghardt (2012) gives a detailed analysis of this affair, including the newspaper reports in the US and Mexico. The German word for the Zimmermann telegram is *Mexiko-Depesche* (= Mexico dispatch). Zimmermann *"had not a von:"* Tuchman (1958), page 108.

Hall's involvement in the Zimmermann decode was not made public until 1956, when James' book appeared. Hall wrote in 1932 an account of his work in Room 40, but the British Admirality did not permit its publication. Black Chambers have a long tradition of being family affairs . . . and Hall's father had been the first Director of the Intelligence Division of the British Admiralty.

In Strother (1918), page 153, the reader is enticed by the remark that *"the story of the Zimmermann note cannot yet be told."* But James Alfred Ewing (1855–1935), the founder of Room 40, gave a lecture on *Some Special War Work* on 13 December 1927 at the University of Edinburgh, which *"disturbed the serenity of Admiralty circles"* so much that they prohibited publication of even newspaper articles about it. Ewing was Principal of the University of Edinburgh from 1917 to 1929. His 1927 lecture is described in Ewing (1939), pages 245–246; see Jones (1979).

The author is grateful for the kind permission of the *Politisches Archiv des Auswärtigen Amtes* to reprint Figures I.2 and I.3 from their archive, and to Dr. Maria Keipert for her help with obtaining copies from the archive. The Zimmermann telegram is reprinted in Deutsche Nationalversammlung (1920), II. Beilagen, Teil VIII, No. 235, pages 337–338. This contains, in addition to the text as given here, instructions to forward it to Mexico, and minor variations like "erhalten" for "halten", and in punctuation.

For completeness, we transcribe some other parts of the telegram. In the left margin of Figure I.2, we find below the signature of Graf Montgelas: *Nach Abg.[ang] w[ieder] vorl[egen] geh.[orsamst] w[ieder] vorgelegt 15/1. W.[ieder] vorl[age] 1.2.17. [initial of von Kemnitz] 15/1. geh.[orsamst] w.[ieder] vorgel.[egt] Z[entral] bur[eau] 1/2..* That is, an instruction to present the document again after the sending of the message, answered as *most obediently presented again on 15 January*. The same day, von Kemnitz asks to see it again on 1 February, and the Central Office has most obediently executed this order. The bottom note in the margin of Figure I.3 reads: *2. Botschafter Graf Bernstorff Washington. ANr. 10 Mit U-Boot vom 15. d. M. W.[ieder]v.[orlage] Lg[?] 0075 Schl.[?].* I cannot read the words before and after the 0075.

The *bait* is from Friedman & Mendelsohn (1938), page 46. The *undertaking* is in James (1956), page 141. It is *understanding* in Hendrick (1922), Volume III, Chapter XII, page 333; FRUS (1931), page 147; Friedman & Mendelsohn (1938), page 1; Ewing (1939), page 205; Tuchman (1958), pages 146 and 202; and Kahn (1967), page 292.

I.3. A statement that Zimmermann made in the Reichstag and the official record of Deutsche Nationalversammlung (1920), II. Beilagen, Teil VIII, footnote to No. 235, page 337, also confirm transmission via the US State Department.

There are no copies of codes 13040 or 0075 in the *Politisches Archiv* and no copy of the telegrams is in the US State Department archives (Friedman & Mendelsohn (1938), page 25). The British decipherment of 13040 in Figure I.5 was presumably filled over time and carries no date. Documents in the *Politisches Archiv* prove that the telegram arrived in Washington via US diplomatic channels, as shown in von zur Gathen (2007).

De Grey's 1945 note is published in Kahn (1999). The *laborious* quote by de Grey is from Kahn (1999), pages 153–154. James (1956) states on page 136 that the Reverend William Montgomery had been working with de Grey on the cryptanalysis, and this has been taken up in the literature. Alfred Dillwyn (Dilly) Knox, a classical scholar, had been in Room 40 since the beginning of 1915, and worked on German codes again at Bletchley Park, one war later. His biography by Fitzgerald (1977) contradicts (on page 144) de Grey's memory: *"For the Zimmermann solution Dilly felt professional admiration, but also some professional jealousy. 'Can't we buy something from the post office?' became his plaintive murmur in all kinds of situations, even quite inappropriate ones, as for instance when things were left behind at a picnic. He had no such good fortune, in 1917, with his own assignment. He had been detailed to work exclusively on the special flag-code."* Jones (1979), page 87, quotes from a 1927 talk given by Ewing: *"The Zimmermann message was sent in a difficult cipher of which one of my staff, Mr Nigel de Grey, had made a special study. Though the first version was imperfect, the general sense was clear enough to make him instantly appreciate its immense political importance. Later he succeeded in making out the full text."* Thus we have James, Fitzgerald, and Ewing vs. de Grey; it seems that the eyewitness account is more reliable than the words of the biographers.

The *steal it we did* is from de Grey (1917–1945), see also Kahn (1999), page 155; not everybody has the grandeur to pay for what they steal. The "drop copy" refers to a telegram intercepted in London. Presumably (a) means the copy of (b) kept at the Mexican telegraph office. The sending date of 16 January is in the telegram, printed in Deutsche Nationalversammlung (1920), II. Beilagen, Teil II, Nr. 57, pages 45–47; Hendrick (1922), volume III, page 336; Friedman & Mendelsohn (1938), pages 30–31. The forwarding date of 19 January is visible in Figure I.4, and stated in von Bernstorff (1920), pages 358 and 379–380. The assessment of codes is on page 53 of Signal Security Agency (1945).

I.4. Hendrick (1922), Volume III, Chapter XII, page 343, and Lansing (1935), page 227, say that the cipher message had been obtained at the Washington cable office. According to Kahn (1967), page 293, it came from the State Department files. Lansing, pages 226–232, further describes the events around the telegram's receipt in Washington.

Hendrick (1922), Volume III, Chapter XII, page 345, contains Page's report from 2 March:

"Bell [. . .] deciphered it from the German code which is in the Admiralty's posses-sion." The telegram in Figure I.4 has ten repeated codewords: 4458 (*gemeinsame*), 5870 (comma), 5904 (*Krieg*, three times), 6929 (*und*), 15857 (*dass*), 18222 (stop), 22464 (Präsi-dent), 52262 (Japan), 67893 (Mexico), 98092 (*U-Boot*). One could design other plaintexts with the same pattern of repetitions. Without a fairly large body of messages and deci-phered codewords (which the British had but de Grey forgot), Bell would have had no more reason to reject such false solutions than de Grey's correct one. The dates of initial and complete cryptanalysis are in James (1956), pages 136 and 141.

The press communiqué *Eine amtliche deutsche Mitteilung* entitled *Ein eventueller deutscher Bündnisvorschlag an Mexiko,* was dated 2 March, distributed by the official news agency *Wolffs Telegraphisches Büro,* and printed in the morning editions of most German news-papers on 3 March 1917. The *"hyphen-retreat"* is in Tuchman (1958), page 186.

Not all computational devices have analogues in the word
pictures that scientists use to describe reality, but Feynman's did.
JAMES GLEICK (1992)

Now, my own suspicion is that the universe is not only queerer
than we suppose, but queerer than we *can* suppose.
JOHN BURDON SANDERSON HALDANE (1928)

The electron is infinite, capricious, and free,
and does not at all share our love for algorithms.
YURI IVANOVITCH MANIN (1977)

"I'm really not into computers, Jay. I don't know much.
I do know the key to the code was the product of
two long prime numbers, each about a hundred digits, right?"
"Yes, that's correct. It's called the RSA cryptosystem."
"Right, for Rivest, Shamir, and Adleman from MIT.
That much I know. I also understand that even using a
sophisticated computer to decipher the code it would take
forever," she recalled. "Something like three point eight billion
years for a two-hundred-digit key, right?"
"[...] but there has to be a bug in that logic somewhere," he
said, loosening his dark green silk tie. "Vee, it's much warmer
than I thought. Would you mind if I removed my jacket?"
"Of course not. You're so formal," she remarked.
KATHERINE COFFARO (1984)

A proposition to reduce arithmetic to the dominion of mechanism
—to substitute an automaton for a compositor—
to throw the powers of thought into wheel-work
could not fail to awaken the attention of the world.
DIONYSIUS LARDNER (1834)

Chapter 14

Quantum computation

We have seen a large diversity of topics in this text, but they all have one aspect in common: they are somehow related to things you can do on your computer at home. We now leave this homely world and discuss a new type of computation, namely quantum computing. It has two central features:

- it seems to be more powerful than classical computation,

- it is not clear whether it will ever work in practice.

Its main interest for cryptography is the fact that if large-scale quantum computers can be built, then we will have to reconsider many of our systems. Peter Shor has invented quantum algorithms that factor integers and compute discrete logarithms in polynomial time. A practical realization of these methods—which is not yet on the horizon—would pull out the rug from under RSA and group based cryptography. Quantum computing dates back to suggestions of Manin (1980) and Feynman (1982).

14.1. Qubits

Classical computation can be modelled by Boolean circuits. These take bits as input, have gates that operate on them, and produce output bits.

We describe quantum computation in a similar way: we have *qubits* (quantum bits), *quantum gates* that operate on them, and output bits obtained by a *measurement*.

We start by considering the complex vector space $\mathbb{C}^k$ for any $k \geq 1$. The *conjugate* of a complex number $\alpha = \beta + i\gamma$, with $\beta, \gamma \in \mathbb{R}$ and $i^2 = -1$, is $\overline{\alpha} = \beta - i\gamma$, and we have $\alpha\overline{\alpha} = \beta^2 + \gamma^2 \geq 0$. The *inner product* on $\mathbb{C}^k$ maps two vectors $a = (a_0, \ldots, a_{k-1})$ and $b = (b_0, \ldots, b_{k-1})$ to the

© Springer-Verlag Berlin Heidelberg 2015
J. von zur Gathen, *CryptoSchool*, DOI 10.1007/978-3-662-48425-8_23

complex number

$$\langle a|b \rangle = \sum_{0 \leq i < k} \overline{a_i} b_i,$$

using Dirac's bra-ket notation $\langle \cdot | \cdot \rangle$. The inner product is sesquilinear, so that

$$\langle a|b \rangle = \overline{\langle b|a \rangle},$$
$$\langle \alpha a + \alpha' a'|b \rangle = \overline{\alpha}\langle a|b \rangle + \overline{\alpha'}\langle a'|b \rangle,$$
$$\langle a|\alpha b + \alpha' b' \rangle = \alpha\langle a|b \rangle + \alpha'\langle a|b' \rangle,$$

for numbers $\alpha, \alpha' \in \mathbb{C}$ and vectors $a, a', b, b' \in \mathbb{C}^k$. The *length* (or 2-*norm*) of $a \in \mathbb{C}^k$ is

$$\|a\| = \langle a|a \rangle^{1/2} = \left(\sum_{0 \leq i < k} \overline{a_i} a_i \right)^{1/2}.$$

This is a nonnegative real number; we always take square roots of positive real numbers to be positive. The *unit sphere*

$$\mathcal{S}_k = \{a \in \mathbb{C}^k \colon \|a\| = 1\}$$

consists of the *unit vectors*, with length 1. Two vectors a and b are *orthogonal* if $\langle a|b \rangle = 0$. An *orthonormal basis* of $\mathbb{C}^k$ consists of k vectors $u_0, \dots, u_{k-1}$ in $\mathbb{C}^k$ so that for all i, j

$$\langle u_i|u_j \rangle = \begin{cases} 1 & \text{if } i = j, \\ 0 & \text{otherwise.} \end{cases}$$

EXAMPLE 14.1.   (i) An orthonormal basis of $\mathbb{C}^2$ is $u_0 = (1, 0)$, $u_1 = (0, 1)$.

(ii) Another orthonormal basis of $\mathbb{C}^2$ is

$$u_0 = \frac{1}{\sqrt{2}}(1, 1), \quad u_1 = \frac{1}{\sqrt{2}}(1, -1).$$

As a four-dimensional real space, $\mathbb{C}^2$ is hard to draw in a real paper plane, but luckily these two vectors are in the real part, and shown in Figure 14.1. This will be our standard basis in the following.

(iii) Yet another orthonormal basis of $\mathbb{C}^2$ is

$$u_0 = \frac{1}{2}(1 + i, \sqrt{2}), u_1 = \frac{1}{2}(1 - i, i\sqrt{2}).$$

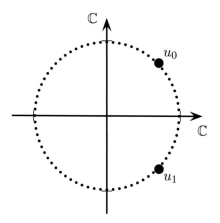

Figure 14.1: The basis (ii) of $\mathbb{C}^2$.

(iv) An orthonormal basis for $\mathbb{C}^4$ is

$$v_0 = \frac{1}{2}(1,1,1,1), \quad v_1 = \frac{1}{2}(1,-1,1,-1),$$

$$v_2 = \frac{1}{2}(1,1,-1,-1), \quad v_3 = \frac{1}{2}(1,-1,-1,1).$$

$\Diamond$

In order to define qubits, we now take an arbitrary orthonormal basis of $\mathbb{C}^2$; there are uncountably many such bases. We designate its two elements by $|0\rangle$ and $|1\rangle$. The $|0\rangle$ has nothing to do with the zero vector $(0,0) \in \mathbb{C}^2$, but rather this should remind of the two bits 0 and 1 in classical computation.

An arbitrary vector a in $\mathbb{C}^2$ can be written as $a = a_0 \cdot |0\rangle + a_1 \cdot |1\rangle$, with $a_0, a_1 \in \mathbb{C}$. The square of its length is

$$\begin{aligned}
\langle a, a \rangle &= \langle a_0|0\rangle + a_1|1\rangle, a_0|0\rangle + a_1|1\rangle \rangle \\
&= \overline{a_0}a_0\langle 0|0\rangle + \overline{a_0}a_1\langle 0|1\rangle + a_0\overline{a_1}\langle 1|0\rangle + \overline{a_1}a_1\langle 1|1\rangle \\
&= |a_0|^2 + |a_1|^2,
\end{aligned}$$

since $|0\rangle$ and $|1\rangle$ are orthonormal. Thus a is a unit vector if and only if $|a_0|^2 + |a_1|^2 = 1$. Now a qubit (quantum bit) is just a unit vector in $\mathbb{C}^2$.

DEFINITION 14.2. (i) Let $|0\rangle$, $|1\rangle$ be an orthononormal basis of $\mathbb{C}^2$. A 1-qubit system, or qubit, is a unit vector

$$a = a_0|0\rangle + a_1|1\rangle \in \mathcal{S}_2 \subset \mathbb{C}^2,$$

with

$$a_0, a_1 \in \mathbb{C} \text{ and } |a_0|^2 + |a_1|^2 = 1.$$

(ii) The measurement μ takes as input a qubit a as above and produces as output a random variable $\mu(a)$ on $\{0, 1\} = \mathbb{B}$ with

$$\text{prob}\{0 \longleftarrow \mu(a)\} = |a_0|^2, \quad \text{prob}\{1 \longleftarrow \mu(a)\} = |a_1|^2.$$

The qubit $a_0|0\rangle + a_1|1\rangle$ is called a superposition of the two base states $|0\rangle$ and $|1\rangle$.

Since $|a_0|^2$ and $|a_1|^2$ are nonnegative real numbers that sum to 1, $\mu(a)$ is indeed a random variable on $\mathbb{B}$. In contrast to random variables or Boolean gates, a qubit is "read-once": it can only be measured once. The huge state space of a qubit is then reduced to (a probability distribution on) just 0 or 1.

EXAMPLE 14.1 CONTINUED. (ii) We take

$$a = \frac{\sqrt{3}}{2}|0\rangle + \frac{1}{2}|1\rangle = \frac{1}{2\sqrt{2}}(\sqrt{3}+1, \sqrt{3}-1) \in \mathcal{S}_2.$$

Then

$$\text{prob}\{0 \longleftarrow \mu(a)\} = \frac{3}{4}, \quad \text{prob}\{1 \longleftarrow \mu(a)\} = \frac{1}{4}.$$

Note that this also depends on the basis $|0\rangle, |1\rangle$, not just on a.   ◇

The qubit cat $= \frac{1}{\sqrt{2}}|0\rangle + \frac{1}{\sqrt{2}}|1\rangle$ is called *Schrödinger's cat*. If we write $|0\rangle = |\text{dead}\rangle$ and $|1\rangle = |\text{alive}\rangle$, then this cat is both dead and alive, each with probability $1/2$. This is just one of the many quantum phenomena that are inconsistent with our macroscopic experience. (Schrödinger: *"Every cat I have seen was either dead or alive, but never both."*)

In order to define systems of several qubits, we consider the *tensor product $a \otimes b$* of two vectors $a = (a_1, \dots, a_k) \in \mathbb{C}^k$ and $b = (b_1, \dots, b_\ell) \in \mathbb{C}^\ell$, for some $k, \ell \geq 1$. It is defined as the vector

(14.3)
$$\begin{aligned} a \otimes b = (&a_0 b_0, a_0 b_1, \dots, a_0 b_{\ell-1}, a_1 b_0, \dots, a_1 b_{\ell-1}, \\ & \dots, a_{k-1} b_0, \dots, a_{k-1} b_{\ell-1}) \in \mathbb{C}^{k\ell}, \end{aligned}$$

obtained by writing the rows of the $k \times \ell$ matrix

$$\begin{pmatrix} a_0 b_0 & \cdots & a_0 b_{\ell-1} \\ a_1 b_0 & \cdots & a_1 b_{\ell-1} \\ \vdots & \vdots & \vdots \\ a_{k-1} b_0 & \cdots & a_{k-1} b_{\ell-1} \end{pmatrix}$$

one after the other. This matrix is called the *Kronecker product of a and b*. The tensor product (14.3) is a *simple tensor* and the linear space spanned by them is the *tensor product* of the two spaces: $\mathbb{C}^{k\ell} = \mathbb{C}^k \otimes \mathbb{C}^\ell$. A basis is given by the tensor products of bases for each component.

EXAMPLE 14.1 CONTINUED. With $k = \ell = 2$, the tensor product of the basis u_0, u_1 from Example 14.1 (ii) with itself is the basis (iii) of $\mathbb{C}^4 = \mathbb{C}^2 \otimes \mathbb{C}^2$.

$$
\begin{aligned}
u_0 \otimes u_0 &= \frac{1}{\sqrt{2}}(1,1) \otimes \frac{1}{\sqrt{2}}(1,1) = \frac{1}{2}(1,1,1,1) = v_0, \\
u_0 \otimes u_1 &= \frac{1}{\sqrt{2}}(1,1) \otimes \frac{1}{\sqrt{2}}(1,-1) = \frac{1}{2}(1,-1,1,-1) = v_1, \\
u_1 \otimes u_0 &= \frac{1}{\sqrt{2}}(1,-1) \otimes \frac{1}{\sqrt{2}}(1,1) = \frac{1}{2}(1,1,-1,-1) = v_2, \\
u_1 \otimes u_1 &= \frac{1}{\sqrt{2}}(1,-1) \otimes \frac{1}{\sqrt{2}}(1,-1) = \frac{1}{2}(1,-1,-1,1) = v_3. \qquad \diamond
\end{aligned}
$$

The tensor product is bilinear:

$$(\alpha a + \alpha' a') \otimes b = \alpha(a \otimes b) + \alpha'(a' \otimes b)$$

for $\alpha, \alpha' \in \mathbb{C}, a, a' \in \mathbb{C}^k$ and $b \in \mathbb{C}^\ell$, and similarly for the other factor. Furthermore, it turns the bra-ket into the product of complex numbers:

$$\langle a \otimes b | a' \otimes b' \rangle = \langle a | a' \rangle \cdot \langle b | b' \rangle$$

for $a, a' \in \mathbb{C}^k$ and $b, b' \in \mathbb{C}^\ell$. In particular,

$$\| a \otimes b \|^2 = \| a \|^2 \cdot \| b \|^2,$$

and hence the tensor product of two unit vectors is a unit vector. Furthermore, the tensor product of two orthonormal bases is an orthonormal basis; we have seen an example above.

When we iterate this construction, starting with $\mathbb{C}^2$, we obtain

$$\mathbb{C}^{2^n} = \mathbb{C}^2 \otimes \cdots \otimes \mathbb{C}^2 = (\cdots ((\mathbb{C}^2 \otimes \mathbb{C}^2) \otimes \mathbb{C}^2) \otimes \cdots \otimes \mathbb{C}^2).$$

If we have an orthonormal basis u_0, u_1 for $\mathbb{C}^2$, then the 2^n tensor products

$$u_{x_0} \otimes u_{x_1} \otimes \cdots \otimes u_{x_{n-1}} = (\cdots ((u_{x_0} \otimes u_{x_1}) \otimes u_{x_2})) \cdots \otimes u_{x_{n-1}}$$

form an orthonormal basis for $\mathbb{C}^{2^n}$, where $x_0, \ldots, x_{n-1}$ runs through all n-bit vectors. This is conveniently treated by the ket notation, where we abbreviate

$$|x\rangle = u_{x_0} \otimes \cdots \otimes u_{x_{n-1}}$$

for an n-bit vector $x = (x_0, \ldots, x_{n-1}) \in \mathbb{B}^n = \{0, 1\}^n$. This extends our previous notation $|0\rangle = u_0$, $|1\rangle = u_1$ for $n = 1$.

EXAMPLE 14.1 CONTINUED. (i) From the basis $|0\rangle = \frac{1}{\sqrt{2}}(1,1)$ and $|1\rangle = \frac{1}{\sqrt{2}}(1,-1)$, we get the basis

$$|00\rangle, |01\rangle, |10\rangle, |11\rangle$$

of $\mathbb{C}^4 = \mathbb{C}^2 \otimes \mathbb{C}^2$, as seen above.

(ii) Going one step further, we find the following basis for $\mathbb{C}^8 = \mathbb{C}^{2^3} = \mathbb{C}^2 \otimes \mathbb{C}^2 \otimes \mathbb{C}^2$:

$$
\begin{array}{llll}
|0\rangle \otimes |0\rangle \otimes |0\rangle & = |00\rangle \otimes |0\rangle & = |000\rangle \\
|1\rangle \otimes |0\rangle \otimes |0\rangle & = |10\rangle \otimes |0\rangle & = |100\rangle \\
|0\rangle \otimes |1\rangle \otimes |0\rangle & = |01\rangle \otimes |0\rangle & = |010\rangle \\
|1\rangle \otimes |1\rangle \otimes |0\rangle & = |11\rangle \otimes |0\rangle & = |110\rangle \\
|0\rangle \otimes |0\rangle \otimes |1\rangle & = |00\rangle \otimes |1\rangle & = |001\rangle \\
|1\rangle \otimes |0\rangle \otimes |1\rangle & = |10\rangle \otimes |1\rangle & = |101\rangle \\
|0\rangle \otimes |1\rangle \otimes |1\rangle & = |01\rangle \otimes |1\rangle & = |011\rangle \\
|1\rangle \otimes |1\rangle \otimes |1\rangle & = |11\rangle \otimes |1\rangle & = |111\rangle
\end{array}
$$

We calculate the coordinates of the sixth entry:

$$
\begin{aligned}
|101\rangle & = |10\rangle \otimes |1\rangle = \frac{1}{2}(1,1,-1,-1) \otimes \frac{1}{\sqrt{2}}(1,-1) \\
& = \frac{1}{2\sqrt{2}}(1,-1,1,-1,-1,1,-1,1) \in \mathbb{C}^8. \qquad \diamondsuit
\end{aligned}
$$

The 2^n basis vectors $|x\rangle$, for $x \in \{0,1\}^n = \mathbb{B}^n$, are called the *base states*.

DEFINITION 14.4. (i) *For $n \geq 1$, an n-qubit system (or n qubits) is a unit vector*

$$a = \sum_{x \in \mathbb{B}^n} a_x |x\rangle \in \mathbb{C}^{2^n},$$

with all $a_x \in \mathbb{C}$ and $\sum_{x \in \mathbb{B}^n} |a_x|^2 = 1$. The number a_x is the amplitude of a at the base state $|x\rangle$.

(ii) *The measurement μ of an n-qubit system $a = \sum_{x \in \mathbb{B}^n} a_x |x\rangle$ is a random variable $\mu(a)$ on $\mathbb{B}^n$, where for each $x \in \mathbb{B}^n$ we have*

$$\mathrm{prob}\{x \longleftarrow \mu(a)\} = |a_x|^2.$$

EXAMPLE 14.5. We take two qubits

$$a = \frac{\sqrt{3}}{2}|0\rangle + \frac{1}{2}|1\rangle, \quad b = \frac{i}{\sqrt{2}}|0\rangle - \frac{1}{\sqrt{2}}|1\rangle.$$

Then we have the 2-qubit

$$c = a \otimes b = \frac{1}{2\sqrt{2}} \cdot (\sqrt{3}i|00\rangle - \sqrt{3}i|01\rangle + |10\rangle - |11\rangle).$$

Measurement of c gives two bits:

x	prob$\{x \longleftarrow \mu(c)\}$
$\|00\rangle$	$3/8$
$\|01\rangle$	$3/8$
$\|10\rangle$	$1/8$
$\|11\rangle$	$1/8$

$\Diamond$

When a and b are qubits, then we can think of $a \otimes b$ as the 2-qubit whose first qubit is in state a, and whose second in state b. But for general 2-qubits such a decomposition is not possible. For example, the 2-qubit $c = \frac{1}{\sqrt{2}}(|00\rangle + |11\rangle)$ has no decomposition of the form

$$(14.6) \qquad \begin{aligned} c &= (a_0|0\rangle + a_1|1\rangle) \otimes (b_0|0\rangle + b_1|1\rangle) \\ &= a_0b_0|00\rangle + a_0b_1|01\rangle + a_1b_0|10\rangle + a_1b_1|11\rangle, \end{aligned}$$

since $a_0b_1 = 0$ and $a_0b_0 = a_1b_1 = 1/\sqrt{2}$ have no solution. The physicists call this *entanglement*: the first and second qubits are hopelessly entangled in c; they cannot be separated.

A bit is an element of $\{0, 1\}$, and n bits simply are n such things. That is not so in the quantum world. A qubit is an element of $\mathbb{C}^2$ (of length 1), and so n such things correspond to elements of $\mathbb{C}^2 \times \cdots \times \mathbb{C}^2 = \mathbb{C}^{2n}$. But an n-qubit system, or n qubits, corresponds to an element of $\mathbb{C}^{2^n}$! This is a much larger space, and its size is one of the sources of the power of quantum computation. (When we speak of n qubits, we always mean the latter.) Measuring an element of this huge space collapses it to just one of 2^n values.

14.2. Quantum circuits

We now have an idea of what quantum computers should work on, namely qubits. The next question is how they should work on them. In other words, what kind of gates should quantum circuits (or quantum computers) have?

These gates should transform one orthonormal basis into another, and thus unitary matrices are the right tool. A $k \times k$ matrix $U \in \mathbb{C}^{k \times k}$ is *unitary* if and only if $U\overline{U}^T = \mathrm{id}_k$, where $\overline{U}^T$ is the transpose of the conjugate of U and id_k the $k \times k$ unit matrix. A matrix is unitary if and only if its rows form an orthonormal basis. The condition says that $U^{-1} = \overline{U}^T$, and also $\overline{U}^T U = \mathrm{id}_k$.

EXAMPLE 14.7. (i) $H = \frac{1}{\sqrt{2}} \begin{pmatrix} 1 & 1 \\ 1 & -1 \end{pmatrix} \in \mathbb{C}^{2 \times 2}$

is unitary. It is called the 2×2 *Hadamard matrix*, and its rows are the basis vectors from Example 14.1 (ii).

(ii) $H_4 = \frac{1}{2} \begin{pmatrix} 1 & 1 & 1 & 1 \\ 1 & -1 & 1 & -1 \\ 1 & 1 & -1 & -1 \\ 1 & -1 & -1 & 1 \end{pmatrix}$ and $V_i = \frac{1}{2} \begin{pmatrix} 1 & 1 & 1 & 1 \\ 1 & i & -1 & -i \\ 1 & -1 & 1 & -1 \\ 1 & -i & -1 & i \end{pmatrix}$

are unitary 4×4 matrices, where $i^2 = -1$. H_4 is the 4×4 Hadamard matrix, whose rows are the basis vectors from Example 14.1 (iv), and V_i is the matrix of the 4×4 discrete Fourier transform (see Section 14.3). $\diamond$

THEOREM 14.8. *Unitary matrices preserve the length of vectors, that is, if $U \in \mathbb{C}^{k \times k}$ is unitary and $a \subset \mathbb{C}^k$, then $\|Ua\| = \|a\|$.*

PROOF.   Let $b \in \mathbb{C}^k$. Then

$$\langle b|Ua \rangle = \sum_{0 \le i < k} \overline{b_i}\,(Ua)_i = \sum_{0 \le i < k} \overline{b_i} \sum_{0 \le j < k} U_{ij} a_j$$

$$= \sum_{0 \le j < k} \Big(\sum_{0 \le i < k} \overline{b_i} U_{ij} \Big) a_j = \sum_{0 \le j < k} \Big(\sum_{0 \le i < k} (U^T)_{ji} \overline{b_i} \Big) a_j = \langle \overline{U^T} b|a \rangle.$$

Applying this with $b = Ua$, we have

$$\|Ua\|^2 = \langle Ua|Ua \rangle = \langle \overline{U^T} Ua|a \rangle = \langle a|a \rangle = \|a\|^2.$$

Since lengths are nonnegative, the claim follows. $\square$

In particular, a unitary matrix transforms a unit vector into a unit vector. Thus when we apply a unitary $2^n \times 2^n$ matrix to n qubits, we obtain n qubits again. Just as we number the coordinates of n qubits by the n-bit words $x \in \mathbb{B}^n$, we do this also with the rows and columns of unitary $2^n \times 2^n$ matrices.

If $U = (U_{xy})_{x,y\in\mathbb{B}^n} \in \mathbb{C}^{2^n \times 2^n}$ and $a = \sum_{z\in\mathbb{B}^n} a_z|z\rangle$, then the usual formula for the product of a matrix and a vector gives

$$U \cdot a = \sum_{x,y\in\mathbb{B}^n} U_{xy} a_y |x\rangle.$$

In particular, if $a = |y\rangle$ is a base state, then

$$(14.9) \qquad U \cdot a = \sum_{x\in\mathbb{B}^n} U_{xy} |x\rangle$$

is the column with index y of U.

EXAMPLE 14.10. We let $n = 1$ and take the Hadamard matrix

$$H = \frac{1}{\sqrt{2}} \begin{pmatrix} 1 & 1 \\ 1 & -1 \end{pmatrix} \begin{matrix} 0 \\ 1 \end{matrix}.$$

The small numbers are the 1-bit row and column numbers. Then

$$H \cdot |0\rangle = \frac{1}{\sqrt{2}} \begin{pmatrix} 1 & 1 \\ 1 & -1 \end{pmatrix} \begin{pmatrix} 1 \\ 0 \end{pmatrix} = \frac{1}{\sqrt{2}} \begin{pmatrix} 1 \\ 1 \end{pmatrix} = \frac{1}{\sqrt{2}} (|0\rangle + |1\rangle),$$

$$H \cdot |1\rangle = \frac{1}{\sqrt{2}} \begin{pmatrix} 1 & 1 \\ 1 & -1 \end{pmatrix} \begin{pmatrix} 0 \\ 1 \end{pmatrix} = \frac{1}{\sqrt{2}} \begin{pmatrix} 1 \\ -1 \end{pmatrix} = \frac{1}{\sqrt{2}} (|0\rangle - |1\rangle).$$

These values also follow from (14.9). For an arbitrary qubit $a = a_0|0\rangle + a_1|1\rangle$, we have

$$H \cdot a = \frac{1}{\sqrt{2}} ((a_0 + a_1)|0\rangle + (a_0 - a_1)|1\rangle),$$

$$H \cdot \left(\frac{1}{\sqrt{2}} (|0\rangle + |1\rangle) \right) = |0\rangle,$$

$$H \cdot \left(\frac{\sqrt{3}}{2}|0\rangle + \frac{1}{2}|1\rangle \right) = \frac{1}{2\sqrt{2}} \left((\sqrt{3}+1)|0\rangle + (\sqrt{3}-1)|1\rangle \right). \qquad \Diamond$$

The last but one line illustrates the phenomenon of *interference*. The input qubit cat $= \frac{1}{\sqrt{2}}(|0\rangle + |1\rangle)$ allows both measurement results $|0\rangle$ and $|1\rangle$, each with probability $1/2$. But $H \cdot$ cat equals $|0\rangle$; the possibility of measuring $|1\rangle$ has vanished by the application of H.

When we have a permutation $\pi: \mathbb{B}^n \longrightarrow \mathbb{B}^n$, the associated *permutation matrix* is the $2^n \times 2^n$ matrix M_π with

$$(M_\pi)_{xy} = \begin{cases} 1 & \text{if } y = \pi(x), \\ 0 & \text{otherwise,} \end{cases}$$

for any $x, y \in \mathbb{B}^n$. M_π is unitary:

$$
\begin{aligned}
(M_\pi \overline{M}_\pi^T)_{xz} &= \sum_{y \in \mathbb{B}^n} (M_\pi)_{xy} (\overline{M}_\pi^T)_{yz} = \sum_{y \in \mathbb{B}^n} (M_\pi)_{xy} (M_\pi)_{zy} \\
&= \sum_{\substack{y \in \mathbb{B}^n \\ y = \pi(x) \\ y = \pi(z)}} 1 \cdot 1 = \begin{cases} 1 & \text{if } x = z, \\ 0 & \text{otherwise.} \end{cases}
\end{aligned}
$$

For a base state z we have

$$
M_\pi |z\rangle = \sum_{w \in \mathbb{B}^n} (M_\pi)_{wz} |w\rangle
$$

and thus $\langle M_\pi | z \rangle_w = (M_\pi)_{wz}$ for a base state w. In particular, $M_\pi \cdot |\pi^{-1}(x)\rangle = x$ and M_π implements π^{-1} on the base states. It follows that $M_\pi^{-1} = M_\pi^T$ implements π on the base states.

For example, let π be the cyclic shift of the first three elements of $\mathbb{B}^2$:

$$
00 \overset{\pi}{\longmapsto} 01 \overset{\pi}{\longmapsto} 10 \overset{\pi}{\longmapsto} 00, \quad 11 \overset{\pi}{\longmapsto} 11.
$$

Then

$$
M_\pi = \begin{matrix} & \begin{matrix} 00 & 01 & 10 & 11 \end{matrix} & \\ \left(\begin{matrix} 0 & 1 & 0 & 0 \\ 0 & 0 & 1 & 0 \\ 1 & 0 & 0 & 0 \\ 0 & 0 & 0 & 1 \end{matrix}\right) & \begin{matrix} 00 \\ 01 \\ 10 \\ 11 \end{matrix} \end{matrix},
$$

$$
M_\pi \cdot |10\rangle = M_\pi \cdot \begin{pmatrix} 0 \\ 0 \\ 1 \\ 0 \end{pmatrix} = \begin{pmatrix} 0 \\ 1 \\ 0 \\ 0 \end{pmatrix} = |01\rangle.
$$

We will want to simulate arbitrary Boolean computations by quantum computations. But unitary matrices are invertible, and so quantum computations are *reversible*: given the computation and its output, we can reconstruct the input simply by reversing the computation. A Boolean function like the AND with $(x_1, x_2) \longmapsto x_1 \text{ AND } x_2$ is not reversible. This problem is remedied by the following construction.

DEFINITION 14.11. *Let $f \colon \mathbb{B}^m \longrightarrow \mathbb{B}^n$ be a Boolean function . Then*

$$
\pi_f \colon \mathbb{B}^{m+n} \longrightarrow \mathbb{B}^{m+n}
$$
$$
(x, y) \longmapsto (x, y \oplus f(x))
$$

is the reversible version of f. It is a permutation of $\mathbb{B}^{m+n}$, since $\pi_f(x, y)$ determines (x, y) uniquely. The unitary map

$$U_f = M_{\pi_f} : \mathbb{C}^{2^{m+n}} \longrightarrow \mathbb{C}^{2^{m+n}}$$

is the unitary version of f.

EXAMPLE 14.12.   (i) The reversible version of AND is

$$\pi_{\text{AND}} : \mathbb{B}^3 \longrightarrow \mathbb{B}^3$$
$$(x_1, x_2, y) \longmapsto (x_1, x_2, y \oplus (x_1 \text{ AND } x_2)),$$

and its unitary version is given by the 8×8 matrix

$$T = U_{\text{AND}} = \begin{pmatrix} 1 & 0 & 0 & 0 & 0 & 0 & 0 & 0 \\ 0 & 1 & 0 & 0 & 0 & 0 & 0 & 0 \\ 0 & 0 & 1 & 0 & 0 & 0 & 0 & 0 \\ 0 & 0 & 0 & 1 & 0 & 0 & 0 & 0 \\ 0 & 0 & 0 & 0 & 1 & 0 & 0 & 0 \\ 0 & 0 & 0 & 0 & 0 & 1 & 0 & 0 \\ 0 & 0 & 0 & 0 & 0 & 0 & 0 & 1 \\ 0 & 0 & 0 & 0 & 0 & 0 & 1 & 0 \end{pmatrix}.$$

This is the *Toffoli map*. We have, for example,

$$T \cdot \left(\frac{\sqrt{3}}{2} |010\rangle + \frac{1}{2} |111\rangle \right) = \frac{\sqrt{3}}{2} |010\rangle + \frac{1}{2} |110\rangle.$$

(ii) The identity function $x \longmapsto x$ has

$$\pi_{\text{id}} : \mathbb{B}^2 \longrightarrow \mathbb{B}^2$$
$$(x, y) \longmapsto (x, x \oplus y)$$

as reversible version, and the 4×4 matrix for U_{id} is

$$\text{cnot} = U_{\text{id}} = \begin{pmatrix} 1 & 0 & 0 & 0 \\ 0 & 1 & 0 & 0 \\ 0 & 0 & 0 & 1 \\ 0 & 0 & 1 & 0 \end{pmatrix}.$$

In quantum computation this is called the *cnot* or *controlled not*, because x controls whether y is negated or not: when $x = 0$, then the second component is y, and when $x = 1$, then it is (not y).

The 2-qubit $a = \frac{1}{\sqrt{2}}(|00\rangle + |10\rangle) = \frac{1}{\sqrt{2}}(|0\rangle + |1\rangle) \otimes |0\rangle$ is simple, but

$$\text{cnot } \cdot a = \frac{1}{\sqrt{2}}(|00\rangle + |11\rangle)$$

is not, as we have seen in (14.6).

(iii) We take the primitive 2^n-th root of unity $\omega = \exp(2\pi i/2^n) \in \mathbb{C}$, for which $\omega^{2^n} = 1$ and $\omega^k \neq 1$ for all k with $1 \leq k < 2^n$. In the group language of Section 15.12, ω has order 2^n. Then

$$\begin{pmatrix} 1 & 0 \\ 0 & \omega \end{pmatrix}$$

is a unitary matrix. It maps the qubit $|x\rangle$ to $\omega^x |x\rangle$. We will use the "controlled version" of this "phase shift", namely the matrix

$$S_n = \begin{pmatrix} 1 & 0 & 0 & 0 \\ 0 & 1 & 0 & 0 \\ 0 & 0 & 1 & 0 \\ 0 & 0 & 0 & \omega \end{pmatrix}$$

operating on two qubits. It maps $|x_0 x_1\rangle$ to $\omega^{x_0 x_1}|x_0 x_1\rangle$.   ◇

Boolean circuits are built by performing basic Boolean operations on 1 or 2 bits, like NOT, AND, OR, and $\oplus$. We build quantum circuits in the same way, by applying 2×2, 4×4, or 8×8 unitary transforms on 1, 2, or 3 "bits" of n qubits. As our basic operations we take $B = B_1 \cup B_2 \cup B_3$ with $B_1 = \{H\}$, $B_2 = \{\text{cnot}\} \cup \{S_n \colon n \in \mathbb{N}\}$, and $B_3 = \{T\}$. Thus the matrices in B_i act on i qubits.

When we have a unitary 2×2 matrix U, n qubits a and some $i \leq n$, then U acts on a by transforming the ith qubit according to U and leaving the others unchanged. We denote the result by $(U; i) \cdot a$. If $i = 1$, we also write $(U \otimes \text{id}_{n-1}) \cdot a$. Similarly, for a 4×4 matrix U and $i < j \leq n$, we get

(14.13) $(U; i, j) \cdot a$

by acting on the ith and jth bit by U, and doing nothing on the other bits. Again, for $i = 1$ and $j = 2$, we may write $(U \otimes \text{id}_{n-2}) \cdot a$. Finally, we also have $(U; i, j, k)$ for $U \in B_3$ and $1 \leq i < j < k \leq n$.

EXAMPLE 14.14. We take two qubits $a = \sum_{x_1, x_2 \in \mathbb{B}} a_{x_1, x_2} |x_1 x_2\rangle$ and apply $H \otimes \mathrm{id}_1$ to them:

$$
\begin{aligned}
(H \otimes \mathrm{id}_1) \cdot a &= \frac{1}{\sqrt{2}} \Big(\sum_{x_2 \in \mathbb{B}} (a_{0x_2} + a_{1x_2}) |0x_2\rangle \\
&\qquad + \sum_{x_2 \in \mathbb{B}} (a_{0x_2} - a_{1x_2}) |1x_2\rangle \Big) \\
&= \frac{1}{\sqrt{2}} ((a_{00} + a_{10})|00\rangle + (a_{01} + a_{11})|01\rangle \\
&\qquad + (a_{00} - a_{10})|10\rangle + (a_{01} - a_{11})|11\rangle), \\
(H \otimes \mathrm{id}_1) \cdot |11\rangle &= \frac{1}{\sqrt{2}} (|01\rangle - |11\rangle) = (H \cdot |1\rangle) \otimes |1\rangle, \\
(H \otimes \mathrm{id}_1) \cdot (b \otimes c) &= (H \cdot b) \otimes c.
\end{aligned}
$$

In the last line, we take $b, c \in \mathcal{S}_2$.   ◇

Another view is as follows. For two matrices $U \in \mathbb{C}^{k \times k}$ and $V \in \mathbb{C}^{\ell \times \ell}$, the tensor product (or Kronecker product) $U \otimes V$ is a $k\ell \times k\ell$ matrix obtained by replacing an entry at position (i, j) in U, where $1 \leq i, j \leq k$, by $U_{ij} \cdot V \in \mathbb{C}^{\ell \times \ell}$. For example,

$$
\begin{pmatrix} 1 & 2 \\ 3 & 4 \end{pmatrix} \otimes \begin{pmatrix} 5 & 6 \\ 7 & 8 \end{pmatrix} = \begin{pmatrix} 1 \cdot \begin{pmatrix} 5 & 6 \\ 7 & 8 \end{pmatrix} & 2 \cdot \begin{pmatrix} 5 & 6 \\ 7 & 8 \end{pmatrix} \\ 3 \cdot \begin{pmatrix} 5 & 6 \\ 7 & 8 \end{pmatrix} & 4 \cdot \begin{pmatrix} 5 & 6 \\ 7 & 8 \end{pmatrix} \end{pmatrix}
$$

$$
= \begin{pmatrix} 5 & 6 & 10 & 12 \\ 7 & 8 & 14 & 16 \\ 15 & 18 & 20 & 24 \\ 21 & 24 & 28 & 32 \end{pmatrix},
$$

$$
\begin{aligned}
H \otimes \mathrm{id}_1 &= \frac{1}{\sqrt{2}} \begin{pmatrix} 1 & 1 \\ 1 & -1 \end{pmatrix} \otimes \begin{pmatrix} 1 & 0 \\ 0 & 1 \end{pmatrix} \\
&= \frac{1}{\sqrt{2}} \begin{pmatrix} 1 & 0 & 1 & 0 \\ 0 & 1 & 0 & 1 \\ 1 & 0 & -1 & 0 \\ 0 & 1 & 0 & -1 \end{pmatrix}.
\end{aligned}
$$

The last equation in the previous example can be understood in this way.

If $b = c = |1\rangle$, then

$$(H \otimes \mathrm{id}_1) \cdot (|1\rangle \otimes |1\rangle) = \frac{1}{\sqrt{2}} \begin{pmatrix} 1 & 0 & 1 & 0 \\ 0 & 1 & 0 & 1 \\ 1 & 0 & -1 & 0 \\ 0 & 1 & 0 & -1 \end{pmatrix} \begin{pmatrix} 0 \\ 0 \\ 0 \\ 1 \end{pmatrix} = \frac{1}{\sqrt{2}} \begin{pmatrix} 0 \\ 1 \\ 0 \\ -1 \end{pmatrix}$$

$$= \frac{1}{\sqrt{2}}(|01\rangle - |11\rangle) = \frac{1}{\sqrt{2}}(|0\rangle - |1\rangle) \otimes |1\rangle$$

$$= (H \cdot |1\rangle) \otimes |1\rangle = (H \otimes \mathrm{id}_1) \cdot |11\rangle.$$

DEFINITION 14.15. *A quantum circuit $\mathcal{C}$ is a directed acyclic graph with some number n of input gates. Each gate is labelled with (U, L), where $U \in B_i$ for some $i \in \{1, 2, 3\}$ and L is a list of i distinct numbers between 1 and n. Each gate takes n qubits as input and produces n qubits as output, except for one designated "output gate" which has no output. When we feed an n-bit (classical) vector into $\mathcal{C}$, then $\mathcal{C}$ computes the random variable on n bits given by the measurement of the output gate.*

14.3. The quantum Fourier transform

In quantum computation, the *Discrete Fourier Transform* (DFT) is an important tool. We have an integer q, the *primitive qth root of unity* $\omega = \exp(2\pi i/q) \in \mathbb{C}$, and the linear map between q-dimensional vector spaces:

$$\mathrm{DFT}_\omega : \quad \begin{array}{ccc} \mathbb{C}^q & \longrightarrow & \mathbb{C}^q, \\ (a_0, \ldots, a_{q-1}) & \longmapsto & \frac{1}{\sqrt{q}}(a(\omega^0), \ldots, a(\omega^{q-1})), \end{array}$$

where $a = \sum_{0 \le i < q} a_i t^i \in \mathbb{C}[t]$. The matrix V_ω of this linear map is the Vandermonde matrix for $1, \omega, \ldots, \omega^{q-1}$:

$$V_\omega = \frac{1}{\sqrt{q}} \cdot \begin{pmatrix} 1 & 1 & 1 & \cdots & 1 \\ 1 & \omega & \omega^2 & \cdots & \omega^{q-1} \\ 1 & \omega^2 & \omega^4 & \cdots & \omega^{2(q-1)} \\ \vdots & \vdots & \vdots & \ddots & \vdots \\ 1 & \omega^{q-1} & \omega^{2(q-1)} & \cdots & \omega^{(q-1)^2} \end{pmatrix}$$

$$= \frac{1}{\sqrt{q}} \cdot (\omega^{xy})_{0 \le x, y < q} \in \mathbb{C}^{q \times q}.$$

EXAMPLE 14.16.   (i) For $q = 2$, we have $\omega = \exp(2\pi i/2) = -1$, and

$$V_{-1} = \frac{1}{\sqrt{2}} \begin{pmatrix} 1 & 1 \\ 1 & -1 \end{pmatrix} = H.$$

H is an old friend, namely the 2×2 Hadamard matrix.

(ii) For $q = 4$, we have $\omega = i$ and

$$V_i = \frac{1}{2} \begin{pmatrix} 1 & 1 & 1 & 1 \\ 1 & i & -1 & -i \\ 1 & -1 & 1 & -1 \\ 1 & -i & -1 & i \end{pmatrix},$$

as in Example 14.7 (ii).   ◊

The following lemma will not be used in what follows, and the reader may well skip its (easy) proof. But it provides an important intuition: the exponential sum $\sum_{0 \le j < q} \omega^{kj}$ is either huge, namely when $k = 0$, or tiny, namely when $k \ne 0$.

LEMMA 14.17.   Let $0 \le k < q$ and $\omega = \exp(2\pi i/q)$. Then

(i) $\sum_{0 \le j < q} \omega^{kj} = \begin{cases} q & \text{if } k = 0, \\ 0 & \text{otherwise.} \end{cases}$

(ii) V_ω is unitary.

PROOF.   (i) The claim is clear for $k = 0$, so that we may take $k \ge 1$. We have $x^q - 1 = (x - 1) \cdot \sum_{0 \le j < q} x^j$ in $\mathbb{C}[x]$. Now $\omega^{kq} - 1 = 0$ and $\omega^k - 1 \ne 0$, so that (i) follows by substituting ω^k for x.

(ii) We use $\overline{\omega} = \exp(-2\pi i/q) = \omega^{-1}$. We let $0 \le j, \ell < q$, and

$$\begin{aligned} u &= (V_\omega \cdot \overline{V_\omega}^T)_{j\ell} = \sum_{0 \le k < q} (V_\omega)_{jk} \cdot (\overline{V_\omega}^T)_{k\ell} \\ &= \sum_{0 \le k < q} q^{-1/2} \omega^{jk} \cdot q^{-1/2} \overline{\omega}^{\ell k} = q^{-1} \sum_{0 \le k < q} \omega^{(j-\ell)k}. \end{aligned}$$

If $j = \ell$, then $u = 1$, and if $j \ne \ell$, then $u = 0$, by (i). Thus indeed $V_\omega \cdot \overline{V_\omega}^T = \mathrm{id}_q$.   □

We fix the notation $q = 2^n$ and $\omega = \exp(2\pi i/q)$. The quantum version QFT_ω of DFT_ω maps the base state $|x\rangle$, for $x \in \mathbb{B}^n$, to

$$\mathrm{QFT}_\omega \cdot |x\rangle = q^{-1/2} \sum_{y \in \mathbb{B}^n} \omega^{xy} |y\rangle,$$

where x and y in the exponent are the numbers with binary representations x and y, respectively. In particular, $\text{QFT}_\omega \cdot |0^n\rangle = q^{-1/2} \sum_{y \in \mathbb{B}^n} |y\rangle$. The measurement of these n qubits gives the uniform random variable on $\mathbb{B}^n$, since

$$\text{prob}(y \longleftarrow \mu(\text{QFT}_\omega \cdot |0^n\rangle)) = q^{-1}$$

for all $y \in \mathbb{B}^n$. Thus we have randomness built into "deterministic" quantum circuits.

We now describe an efficient quantum circuit that computes the QFT when q is a power of 2. Our first step is a recursive description. Our usual notation is that $x = \sum_{0 \le j < n} x_j 2^j$ is the integer value corresponding to the bit vector $x = (x_0, \ldots, x_{n-1}) \in \mathbb{B}^n$; thus 011 has the value 6. It is convenient to write the target vector y in reverse order: if $y = (y_0 \ldots, y_{n-1}) \in \mathbb{B}^n$, then its integer value is $\bar{y} = \sum_{0 \le j < n} y_j 2^{n-1-j}$, so that $\overline{011} = 3$.

We split any vector $z = (z_0, \ldots, z_{n-1}) \in \mathbb{B}^n$ into its first $n-1$ and its last component: $z = (z_*, z_{n-1})$, with $z_* \in \mathbb{B}^{n-1}$. For the corresponding integer values, we have $z = z_* + 2^{n-1} z_{n-1}$, and $\bar{z} = 2\overline{z_*} + z_{n-1}$. We have

$$(14.18) \qquad \frac{1}{2^{(n-1)/2}} \sum_{y_* \in \mathbb{B}^{n-1}} \omega^{2x_* \overline{y_*}} |y_*\rangle \otimes \frac{1}{\sqrt{2}}(|0\rangle + \omega^x |1\rangle)$$

$$= \frac{1}{2^{n/2}} \sum_{y_* \subset \mathbb{B}^{n-1}} (\omega^{2x_* \overline{y_*}} |y_* 0\rangle + \omega^{2x_* \overline{y_*} + x} |y_* 1\rangle),$$

and claim that this equals

$$\text{QFT}_\omega \cdot |x\rangle = \frac{1}{2^{n/2}} \sum_{y \in \mathbb{B}^n} \omega^{x\bar{y}} |y\rangle.$$

So let $y = (y_*, y_{n-1}) \in \mathbb{B}^n$. If $y_{n-1} = 0$, then

$$2x_* \overline{y_*} = (x - 2^{n-1} x_{n-1}) \cdot 2\overline{y_*} \equiv 2x\overline{y_*} = x\bar{y} \bmod 2^n,$$

and therefore $\omega^{2x_* \overline{y_*}} = \omega^{x\bar{y}}$. Similarly, if $y_{n-1} = 1$, then

$$x\bar{y} = x(2\overline{y_*} + 1) = (x_* + 2^{n-1} x_{n-1}) \cdot 2\overline{y_*} + x \equiv 2x_* \overline{y_*} + x \bmod 2^n.$$

This shows the claim. The first factor in the tensor product is simply $\text{QFT}_\eta(x_*)$, where $\eta = \omega^2 = \exp(2\pi i/2^{n-1})$ is a primitive 2^{n-1}th root of unity.

We have to take care of the second factor in (14.18), namely the map

$$|x\rangle \longmapsto x_* \otimes \frac{1}{\sqrt{2}}(|0\rangle + \omega^x |1\rangle).$$

For $0 \le j < n$, we recall from Example 14.12 (iii) the unitary map on two qubits with

$$S_{n-j} \colon |z_0 z_1\rangle \longmapsto \omega^{2^j z_0 z_1} |z_0 z_1\rangle = \begin{cases} \omega^{2^j} |11\rangle & \text{if } z_0 z_1 = 11, \\ |z_0 z_1\rangle & \text{otherwise,} \end{cases}$$

for $(z_0, z_1) \in \mathbb{B}^2$, where $\omega^{2^j} = \exp(2\pi i / 2^{n-j})$. We apply first the Hadamard matrix to the last qubit, and then S_{n-j} to the jth and the last qubit, for $0 \le j \le n - 2$. In the notation of (14.13), we consider

$$(14.19) \quad (S_2; n-2, n-1) \cdots (S_{n-1}; 1, n-1) \cdot (S_n; 0, n-1) \cdot (H; n-1) \cdot |x\rangle.$$

We claim that the result after j steps, that is, after applying $(S_{n-j+1}; j - 1, n-1)$, equals $x_* \otimes \frac{1}{\sqrt{2}}(|0\rangle + \omega^{u_j} |1\rangle)$, where $u_j = \sum_{0 \le k < j} x_k 2^k$ corresponds to the first j bits of x.

For an inductive proof, we check that this is true for $j = 0$, since $u_0 = 0$. For the induction step, we have

$$(S_{n-j+1}; j - 1, n - 1) \cdot (x_* \otimes \frac{1}{\sqrt{2}}(|0\rangle + \omega^{u_{j-1}} |1\rangle))$$

$$= x_* \otimes \frac{1}{\sqrt{2}}(|0\rangle + \omega^{2^{j-1} x_{j-1}} \cdot \omega^{u_{j-1}} |1\rangle)$$

$$= x_* \otimes \frac{1}{\sqrt{2}}(|0\rangle + \omega^{u_j} |1\rangle).$$

THEOREM 14.20. *Let n be an integer and $\omega = \exp(2\pi i / 2^n)$. Then the quantum Fourier transform QFT_ω can be implemented on a quantum circuit with n qubits and $O(n^2)$ gates.*

PROOF. We have seen how to implement QFT_ω with one Hadamard gate, $n - 1$ gates S_k (with $2 \le k \le n$), and a recursive computation of QFT_{ω^2}. Thus we require a total of $(1 + n - 1) + (1 + (n-1) - 1) + \cdots = \sum_{1 \le \ell \le n} \ell = n(n+1)/2$ gates. $\square$

The circuit is illustrated in Figure 14.2 for $n = 4$.

The QFT of arbitrary order can be calculated efficiently and exactly for arbitrary q, not just powers of 2. This requires additional tools beyond the scope of this text.

Lemma 14.17 (i) says that the terms in the exponential sum either all point in the same direction, namely when $k = 0$, or otherwise they cancel each other; this is known as "interference". See also Example 14.10. A central tool in our quantum algorithms is an approximate version of

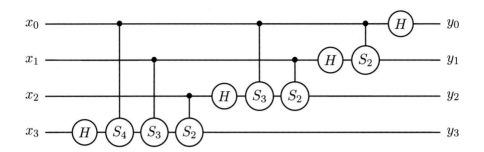

Figure 14.2: The quantum circuit for $\mathrm{QFT}_{\exp(2\pi i/16)}$.

this: when k is a small real number, then the sum is large. Namely, when all summands are contained in a half-circle, then we get a substantial contribution in the "middle direction", and the sum is substantially large. This is made precise in the following "half-circle lemma".

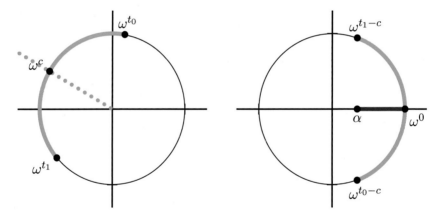

Figure 14.3: Estimating a partial exponential sum.

LEMMA 14.21 (Half-circle lemma).   Let $t_0 < t_1$ and q be real numbers, with q positive, $\omega = \exp(2\pi i/q)$, and $T \subset [t_0 \ldots t_1]$ a finite subset of the interval $[t_0 \ldots t_1]$ with m elements.

(i) If $2(t_1 - t_0) < q$ and $\alpha = \cos(\pi(t_1 - t_0)/q)$, then $\alpha > 0$ and

(14.22) $$\left| \sum_{t \in T} \omega^t \right| \geq m\alpha.$$

(ii) If $2(t_1 - t_0) = q$, $T' \subseteq T \cap [t_0 + s \ldots t_1 - s]$ has m' elements for some s with $0 < s \leq q/4$, and $\alpha' = \cos(\pi/2 \cdot (1 - 4s/q))$, then $\alpha' > 0$ and

$$\left| \sum_{t \in T} \omega^t \right| \geq m'\alpha'.$$

PROOF. (i) As illustrated in Figure 14.3, we "rotate" the sum so that its middle direction $c = (t_0 + t_1)/2$ becomes the x-axis:

$$\left| \sum_{t \in T} \omega^t \right| = |\omega^c| \cdot \left| \sum_{t \in T} \omega^{t-c} \right| \geq 1 \cdot \left| \Re(\sum_{t \in T} \omega^{t-c}) \right|$$

$$= \left| \sum_{t \in T} \Re(\omega^{t-c}) \right| = \left| \sum_{t \in T} \cos(2\pi/q \cdot (t-c)) \right|.$$

Here we use $\omega^x = e^{2\pi i x/q} = \cos(2\pi x/q) + i \sin(2\pi x/q)$ for a real number x, and $|z| \geq |\Re(z)|$ for any complex number z, where $\Re$ denotes the real part, and $\Re(z_1 + z_2) = \Re(z_1) + \Re(z_2)$.

Now for each argument of the cosines we have

$$-\frac{t_1 - t_0}{2} = t_0 - c \leq t - c \leq t_1 - c = \frac{t_1 - t_0}{2},$$

$$\left| \frac{2\pi}{q} \cdot (t - c) \right| \leq \frac{2\pi}{q} \cdot \frac{t_1 - t_0}{2} < \frac{\pi}{2}.$$

Hence $\alpha > 0$, each summand is at least α, and the claim follows.

For (ii), we have $|t - c| \leq q/4$ and $|2\pi/q \cdot (t - c)| \leq \pi/2$ for all $t \in T$, and $|t - c| \leq q/4 - s$ and $|2\pi/q \cdot (t - c)| \leq \pi/2 \cdot (1 - 4s/q) < \pi/2$ for all $t \in T'$. Thus $\cos(2\pi/q \cdot (t - c))$ is nonnegative for all $t \in T$, and at least α' for all $t \in T'$. $\square$

In Figure 14.3, the sum in (14.22) takes place in the green arc at the left. At the right, the rotated sum gives the same absolute value. The condition that $2(t_1 - t_0) < q$ guarantees that the right arc lies completely in the right half plane, and that all cosines are at least α. The red segment is the range of possible values of the cosines.

14.4. Polynomial-time integer factorization

In this and the next section, we will present in detail the two quantum algorithms of paramount relevance for cryptography: Shor's integer factorization and discrete logarithm computations. However, we omit the details of implementing basic arithmetic on a quantum computer. This includes the fundamental tasks from addition to modular exponentiation of numbers, and the Fourier transform see below. In principle, this must be done also for classical computation, but the rare reader who has actually designed Turing machines or Boolean circuits for these tasks is likely to remember this as fairly uninspiring but tiring drudgery. Thus we will use without proof the following result.

THEOREM 14.23. *Modular exponentiation* $(x, e, N, 0^n) \longmapsto (x, e, N, x^e)$, *where the last quantity is the bit string representing* x^e *in* $\mathbb{Z}_N$, *can be computed on a quantum circuit with* $4n + 3$ *qubits.*

This is a special case of the fact that any problem for which a probabilistic polynomial-time (classical) algorithm exists also has a polynomial-time quantum computation. The corresponding complexity classes are denoted by BPP (for *bounded-error probabilistic polynomial time*) and QBP (for *quantum bounded-error polynomial time*). The following containment follows easily.

THEOREM 14.24. BPP ⊆ QBP.

The task in this section is to factor an integer. This is deemed to be a difficult problem, and provides security in the RSA cryptosystem. The best known algorithms are based on the number field sieve and use $2^{O((n \log^2 n)^{1/3})}$ bit operations for n-bit integers. A 768-bit RSA number (230 decimal digits) has been factored with considerable effort. We have discussed this at length in Chapter 3. Quantum computers can factor integers in polynomial time. If they could ever be built to the required size, that would be bad news for RSA. As a small consolation, they would also ring the demise of group-based cryptography.

We describe a quantum circuit that computes a small multiple ℓ of the order $\text{ord}(x)$ of an element x of a finite group G, provided that $\text{ord}(x)$ is odd. When $G = \mathbb{Z}_N^\times$, Algorithm 3.53 is designed so that from ℓ we can obtain a proper factor of N.

The idea of the quantum algorithms below is to generate a distribution on states $|x\rangle$ with coefficients as in (14.22). There will be a notion of "helpful states" so that

○ given a helpful state, it is easy to solve the problem at hand,

○ each helpful state is reasonably likely to occur,

○ there are many helpful states.

The quantitative versions of the last two items have to be such that it is very likely that some helpful state occurs. In other words, each helpful state allows us to solve the problem, and there is enough help around.

We assume that the elements of G can be represented by n-bit strings, so that $\#G \leq 2^n$. However, the group order $\#G$ is not known to us. We identify an element x of G with n qubits $|x\rangle$, where $x \in \mathbb{B}^n$. Furthermore, the group multiplication is supposed to be efficiently computable. Then so is the exponentiation function

$$G \times \{0, 1, \ldots, 2^n - 1\} \longrightarrow G,$$
$$(x, \ell) \longmapsto x^\ell$$

see Section 15.12, and also its unitary version U_G can be computed by a polynomial-size quantum circuit. To have a more explicit bound, we assume that its size is $O(n^3)$. This is the case when G is $\mathbb{F}_r^\times$ or an elliptic curve over $\mathbb{F}_r$, for an n-bit prime power r. We use a primitive qth root of unity in $\mathbb{C}$ for some power q of 2, and have to invert q modulo $\mathrm{ord}(x)$. This is only possible if $\mathrm{ord}(x)$ is odd.

ALGORITHM 14.25. Quantum circuit for element order in a group.

Input: An integer n and an element $x \neq 1$ with odd order in a finite group G with fewer than 2^n elements, so that $x \in \mathbb{B}^n$. The circuit works on $4n$ qubits, beginning in state $|x, 0^{2n}, 0^n\rangle$.

Output: Either a multiple less than 2^{2n} of the order of x, or "failure".

1. Set $q = 2^{2n+1}$ and $\omega = \exp(2\pi i/q)$.
2. Apply $\mathrm{id}_n \otimes \mathrm{QFT}_\omega \otimes \mathrm{id}_n$.
3. Apply U_G', mapping $|x, \ell, 0^n\rangle$ to $|x, \ell, x^\ell\rangle$. Here U_G' acts like U_G on the first n and last n qubits, and like the identity on the middle $2n$ qubits. [Thus U_G' is $U_G \otimes \mathrm{id}_{2n}$ up to a change of coordinates, or $(U_G;$ $0, \ldots, n-1, 3n, \ldots, 4n-1)$ in a slight generalization of the notation on page 692.]
4. Apply $\mathrm{id}_n \otimes \mathrm{QFT}_\omega \otimes \mathrm{id}_n$.
5. Measure the current state a: $|x, u, v\rangle \longleftarrow \mu(a)$.

The next steps can be performed classically.

6. Compute the rational approximations s_1/k_1, $s_2/k_2, \ldots$ to u/q with the Extended Euclidean Algorithm 15.19 (EEA).
7. Compute $x^{k_1}, x^{k_2}, \ldots$. If one of them, say x^{k_i}, equals 1, return k_i as a multiple of the order of x. Else return "failure".

We shall prove the following result.

THEOREM 14.26. The size of the circuit 14.32 is $O(n^3)$. With probability $\Omega(1/\log\log k)$, it returns a multiple less than 2^{2n} of the order k of x. If $k \geq 31$, then the probability is at least $1/30 \ln\ln k$.

PROOF. The first statement is clear. For the second one, we have to analyze the output of the algorithm. Its base states are of the form $|x, \ell, v\rangle$.

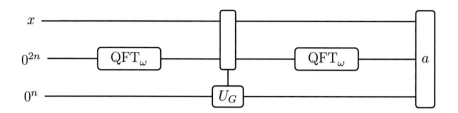

Figure 14.4: Quantum circuit for the order of x in G.

Since x never changes, we omit it in the following. The state after step 3 is

$$U_G \cdot \mathrm{QFT}_\omega \otimes \mathrm{id}_n \cdot |0^{2n}, 0^n\rangle = U_G \cdot \left(q^{-1/2} \sum_{0 \le \ell < q} |\ell, 0^n\rangle\right) = q^{-1/2} \sum_{0 \le \ell < q} |\ell, x^\ell\rangle.$$

Step 4 applies QFT_ω to the first $2n$ qubits; that is, it maps $|\ell\rangle$ to $q^{-1/2} \sum_{0 \le u < q} \omega^{u\ell} |u\rangle$. Thus the state a after step 4 is

$$a = q^{-1} \sum_{0 \le \ell < q} \sum_{0 \le u < q} \omega^{u\ell} |u, x^\ell\rangle.$$

We have to determine the probability $p = \mathrm{prob}(|u, v\rangle \longleftarrow \mu(a))$ with which a specific value $|u, v\rangle$ turns up, with $0 \le u < q$ and $v = x^{\ell_0}$ for some ℓ_0. Since $x^k = 1$, we only have to consider the values $0 \le \ell_0 < k$.

$$p = \left|q^{-1} \sum_{0 \le \ell < q\, x^\ell = v} \omega^{u\ell}\right|^2 = q^{-2} \left|\sum_{0 \le j < (q - \ell_0)/k} \omega^{u(jk + \ell_0)}\right|^2$$

$$\text{(14.27)} \qquad = q^{-2} \left|\omega^{u\ell_0} \sum_{0 \le j < j_0} \omega^{juk}\right|^2 = q^{-2} \left|\sum_{0 \le j < j_0} \omega^{q \cdot j(uk - b_u)/q} \cdot \omega^{jb_u}\right|^2$$

$$= q^{-2} \left|\sum_{0 \le j < j_0} \omega^{jb_u}\right|^2,$$

where $j_0 = (q - \ell_0)/k \le q/k$ and $b_u = uk \text{ srem } q$ is the symmetric remainder of uk modulo q, that is, the unique integer b_u with

$$\text{(14.28)} \qquad uk = b_u \text{ in } \mathbb{Z}_q, \quad -q/2 \le b_u < q/2;$$

see (15.12). For the fourth equation, we use $|\omega^{u\ell_0}| = 1$, and $\omega^q = 1$ for the last one.

We now consider the integer $\lambda_u = (uk - b_u)/q$, and call $|u, v\rangle$ *helpful* if

$$|b_u| \le k/2 \text{ and } \gcd(\lambda_u, k) = 1.$$

Our goal is to show the three required properties listed above: each helpful state solves the problem and is reasonably likely to occur, and there are many of them.

Let $|u, v\rangle$ be helpful. Then

(14.29)
$$0 \leq \lambda_u \leq ((q-1)k + k/2)/q < k, \quad |uk - \lambda_u q| = |b_u| \leq k/2,$$
$$\left| \frac{u}{q} - \frac{\lambda_u}{k} \right| \leq \frac{1}{2q} = \frac{1}{4 \cdot (2^n)^2} < \frac{1}{4k^2}.$$

Thus λ_u/k is a rational approximation with small denominator to u/q, and since $0 \leq \lambda_u < k$, and $\gcd(\lambda_u, k) = 1$, λ_u and k can be found from u and q via the EEA in steps 6 and 7; see Theorem 15.39.

Thus if a helpful state is measured, then one of the integers k_i in step 6 equals $\text{ord}(x)$. But in general we do not know how to check this property efficiently. The test in step 7 returns k_i if $x^{k_i} = 1$, that is, if k_i is an integer multiple of $\text{ord}(x)$; but it might be larger than $\text{ord}(x)$. This shows the first claimed property.

We can only claim this property for an eventual output of step 7. The point is slightly subtle: we only show for k that it is likely to occur as output, but we do not rule out other values. Even worse, they do not seem to occur in practice, but our proof does not rule out the possibility that some other multiple of k is more likely than the true value of k. The same applies to the gcd of k_i's from many executions of the algorithm.

The next step is to show that any helpful state is likely to occur. For an extreme case, imagine that $b_u = 0$. Then the probability is

$$q^{-2} j_0^2 = \left(\frac{q - \ell_0}{qk} \right)^2 \approx \frac{1}{k^2};$$

this counts as large in the present context.

In the general case, we claim that if $|u, v\rangle$ is helpful, then $p \geq 1/10k^2$. We distinguish four cases for b_u, namely according to which of the four intervals $-k/2 \ldots -k/4 \ldots 0 \ldots k/4 \ldots k/2$ it lies in. So first let $0 \leq b_u < k/4$. Then we take $t_0 = 0$ and $t_1 = q/4$ in the half-circle lemma 14.21(i). Each of the exponents jb_u in (14.27) lies in the interval $[t_0 .. t_1]$, $2(t_1 - t_0) = q/2 < q$, and $\alpha = \cos(\pi/4) = 1/\sqrt{2}$. We have

$$m = \#T = \#\{j \colon 0 \leq j < j_0\} = \lceil j_0 \rceil = \lceil \frac{q - \ell_0}{k} \rceil \geq \frac{q-k}{k} \geq \frac{q}{2k},$$

since $q \geq k^2 \geq 2k$. By Lemma 14.21 (i),

$$p \geq \frac{1}{q^2}(m\alpha)^2 \geq \frac{1}{q^2} \cdot \left(\frac{q}{2k} \right)^2 \cdot \frac{1}{2} = \frac{1}{8k^2} > \frac{1}{10k^2}.$$

As a second case, we take $k/4 \leq b_u \leq k/2$. We set $t_0 = 0$ and $t_1 = q/2$, so that $2(t_1 - t_0) = q$. For each exponent jb_u in (14.27) we have

$$t_0 = 0 \leq jb_u < \frac{q - \ell_0}{k} \cdot \frac{k}{2} \leq t_1.$$

We want to apply the half-circle lemma 14.21(ii) with $s = q/8$, and set

$$J' = \{j\colon \frac{q}{8b_u} \leq j \leq \frac{3q}{8b_u}\}, \quad \alpha' = \cos(\frac{\pi}{4}) = \frac{1}{\sqrt{2}}.$$

Then

$$m' = \#J' = \left\lfloor \frac{3q}{8b_u} \right\rfloor - \left\lceil \frac{q}{8b_u} \right\rceil + 1 \geq \frac{3q}{8b_u} - 1 - \left(\frac{q}{8b_u} + 1 \right) + 1$$

$$= \frac{q}{4b_u} - 1 \geq \frac{q}{4 \cdot \frac{k}{2}} - 1 = \frac{q}{2k} - 1 \geq \frac{q}{\sqrt{5}k};$$

the last inequality holds when $k \geq 19$, since then $q \geq k^2 \geq 19k$. For $j \in J'$ we have

$$t_0 + s = \frac{q}{8} \leq jb_u \leq \frac{3q}{8} = t_1 - s,$$

and from the lemma we conclude that

$$p \geq \frac{1}{q^2}(m'\alpha')^2 \geq \frac{1}{q^2} \cdot \left(\frac{q}{\sqrt{5}k} \right)^2 \cdot \frac{1}{2} = \frac{1}{10k^2}.$$

The remaining two cases, with negative b_u, are symmetrical to the previous ones. Thus $p \geq 1/10k^2$ for any helpful $|u, v\rangle$, which is the second claim.

How many helpful states are there? There are exactly k integers b_u with $|b_u| \leq k/2$, since k is odd. For each b_u, there is exactly one integer u with $u = b_u k^{-1}$ in $\mathbb{Z}_q$ and $0 \leq u < q$, since $\gcd(q, k) = 1$. Furthermore,

$$\lambda_u = \frac{uk - b_u}{q} = \frac{-b_u}{q} \text{ in } \mathbb{Z}_k.$$

As b_u runs from $-\lfloor k/2 \rfloor$ to $\lfloor k/2 \rfloor$, λ_u runs through the k values from 0 to $k - 1$. Exactly $\varphi(k)$ of these values satisfy $\gcd(\lambda_u, k) = 1$. We have $\varphi(k) \geq k/3 \ln\ln k$ for $k \geq 31$; see Notes 3.3. This is the number of possible values of u in helpful states, the number of possible values for v is k, and so the total number of helpful states $|u, v\rangle$ is at least $k^2/3 \ln\ln k$.

In conclusion, the probability of measuring a helpful state is at least

$$\frac{1}{10k^2} \cdot \frac{k^2}{3 \ln\ln k} = \frac{1}{30 \ln\ln k}. \qquad \square$$

In other words, the expected number of experiments until we find the order of x is $O(\log\log k)$, which is in $O(\log\log \#G)$.

EXAMPLE 14.30. We now trace a complete execution of the factoring algorithm and factor $N = 21$. (You can do this in your head, but that is besides the point.) We have the group $G = \mathbb{Z}_{21}^{\times}$ with $\#G \leq 21 < 2^5$, so that we take $n = 5$. We choose $x = 10$ at random in Algorithm 3.53. Then its step 4 computes $y = 10^{2^5} = 16$ in $\mathbb{Z}_{21}$.

The quantum circuit 14.25, called with input $x = y$ (now renamed), measures one of the spikes in Figure 14.5, say $\alpha \cdot |683, 4\rangle$ with $\alpha = 1/(683^2 + 4^2)^{1/2}$, and then computes 683 and the EEA of 1024 and 683:

$$1024 = 1 \cdot 683 + 341, \qquad 1024 - 1 \cdot 683 = 341,$$
$$683 = 2 \cdot 341 + 1, \qquad -2 \cdot 1024 + 3 \cdot 683 = 1,$$
$$341 = 341 \cdot 1.$$

The large quotient 341 would be unusual for random inputs, but is not a surprise here. The EEA yields the two approximations

$$1/1, 2/3$$

to $683/1024$, and step 7 calculates $x^1 = 16 \neq 1$ in $\mathbb{Z}_{21}$, $x^3 = 16^3 = 1$ in $\mathbb{Z}_{21}$, and the quantum circuit returns $\ell = 3$. Now Algorithm 3.53 takes up the ball again in its step 7, with $m = 3$ and

$$z_0 = 10^3 = 13 \text{ in } \mathbb{Z}_{21},$$
$$i = 1,$$
$$z = 13^2 = 1 \text{ in } \mathbb{Z}_{21},$$
$$r = \gcd(13 - 1, 21) = 3,$$

and finally returns the proper divisor 3 of 21.

The data flow is as follows:

$$x$$
$$y = x^{2^n}$$

$$\ell = \text{multiple of } \operatorname{ord}(y)$$

$$m = \operatorname{odd}(\ell)$$
$$u = x^{2^i \cdot m} = 1$$
$$z = x^{2^{i-1} \cdot m}$$
$$r = \gcd(z - 1, N)$$

Algorithm 3.53   Circuit 14.25   ◇

EXAMPLE 14.31. In Example 3.54, the quantum circuit 14.25 is called
with input $n = 5$ and $x = 16 \in G = \mathbb{Z}_{21}^{\times}$ (called y in the example) encoded
as $10000 \in \mathbb{B}^5$ and of odd order 3. Although its effect in the example may
look rather small, it is in fact the core of the method—namely the only
step that we do not know to do efficiently with classical computation. We
now illustrate its behavior in the example, using $q = 2^{10} = 1024$ (instead
of $q = 2048$, as chosen in the algorithm; see the Notes) and not a quantum
computer, but a classical simulation. The amplitude p in (14.27) is shown
in Figure 14.5 for $0 \le u < q$ and $j_0 = \lfloor q/k \rfloor$. Almost all values of u have

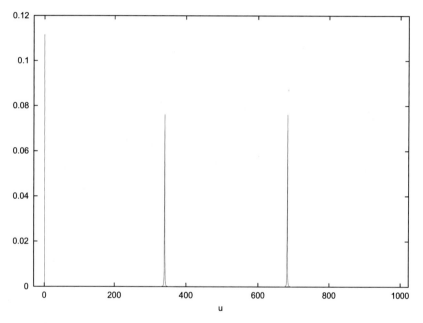

Figure 14.5: The amplitude p in (14.27), in percent.

tiny probability, except for the sharp spikes occurring at $\lambda \cdot q/k = \lambda \cdot 1024/3$
for the three integer values $0 \le \lambda < k$, that is, $\lambda = 0, 1$, or 2. The values
of u with largest probability (given in percent) are:

u	p
0	11.09
340	0.47
341	7.59
342	1.90
682	1.90
683	7.59
684	0.47

For each u, we have three possible values of v, namely $v = x^0, x^1, x^2$. The seven probabilities above, multiplied by 3, sum to over 93%, so that one of those u is quite likely to turn up.

Thus the measurement will most likely return a value of u near some $\lambda \cdot 1024/3$. In the sample run reported above, u was actually chosen at random according to this distribution, and indeed $u = 683$ is near $2 \cdot 1024/3 = 682\frac{2}{3}$. A choice near $1024/3$ would also be helpful, but not u near $0 \cdot 1024/3 = 0$; then $\lambda_u = 0$, $\gcd(\lambda_u, k) = k \neq 1$, and $|u, v\rangle$ is not helpful.

For $u = 683$, we have

$$
\begin{aligned}
b_u &= (uk \text{ srem } q) = (683 \cdot 3 \text{ srem } 1024) = 1, \\
|b_u| &= 1 \leq 3/2, \quad \gcd(\lambda_u, k) = \gcd(2, 3) = 1, \\
\left| \frac{u}{q} - \frac{\lambda_u}{k} \right| &= \left| \frac{682}{1024} - \frac{2}{3} \right| = \frac{1}{3 \cdot 1024} < \frac{1}{2 \cdot 3^2} = \frac{1}{2 \cdot k^2}.
\end{aligned}
$$

For the general estimate of $p \approx 7.59$ % in the proof, we are in the second case $k/4 = 3/4 \leq b_u = 1 \leq 3/2 = k/2$, and find

$$
m' = 257 \geq 152.6 \approx \frac{1024}{3\sqrt{5}},
$$

even though $k < 19$, and

$$
p \approx 0.076 \geq 0.031 \approx \frac{257^2}{1024^2 \cdot 2} \geq 0.011 \approx \frac{1}{10k^2}. \qquad \Diamond
$$

After previous quantum factorizations of 15 and 21, the currently largest number factored in this way is 143 (Xu *et al.* 2011). In contrast to the circuit-based model described in this chapter, they use *adiabatic quantum computation*, driven by a continuously varying Hamiltonian. This is implemented on a 4-qubit liquid-crystal NMR quantum computer. The qubits represent the four unknown bits in the binary representation of the factorization $N = p \cdot q = (1p_2 p_2 1) \cdot (1q_2 q_1 1)$. By hand, they find the equations $p_1 + q_1 = p_2 + q_2 = p_1 q_2 + p_2 q_1 = 1$. This leads to the required Hamiltonian. They run their adiabatic system to minimize $f(x, y) = (N - x \cdot y)^2$, and a measurement at the minimal value $f(x, y) = 0$ yields the factors 11 and 13.

14.5. Discrete logarithms

Quantum circuits are also powerful enough to solve discrete logarithm problems in polynomial time. We will describe an efficient generic "quantum group circuit" for the problem.

So we have a cyclic group $G = \langle g \rangle$ with d elements and $x \in G$, and want to determine $\operatorname{dlog}_g x$ as in Chapter 4. We do not want to factor d, but we can write it as $d = 2^e d'$ with integers e and d', and d' odd.

By the Chinese Remainder computation in Section 4.6, we can compute discrete logs separately in groups of orders 2^e (easy by the 2-adic method in Section 4.7) and d'. The latter is the core of the problem, and by replacing d with d', we may now assume that d is odd. We assume that the elements of G are coded as n-bit strings, with $2d \leq q = 2^n < 4d$, which is one bit more than necessary, and that we have efficient quantum circuits for multiplication and exponentiation in G. In particular, we have a quantum circuit U_G of size $\mathrm{O}(n^3)$ which computes the unitary version of $(g, x, \ell, j) \longmapsto g^\ell x^{-j}$. Circuit 14.32 below returns either $\operatorname{dlog}_g x$ or a proper factor of d. In the latter case, we compute via factor refinement (Exercise 15.11) the corresponding factorization of d into coprime factors and use the Chinese distribution of Section 4.6, solving the easier task of determining all the "small" discrete logarithms via the present circuit plus recursion if further factors of the group order turn up. Alternatively, we might assume d to be prime, according to the recommendation of Figure 4.6.

Our circuit works on $5n$ qubits, in three stages. The first $2n$ qubits equal $|g, x\rangle$ and are not changed in the algorithm. We leave these two arguments out in the following description. Thus we start with the all-zero input $|0^n 0^n 0^n\rangle$ in the last $3n$ positions, and first generate the uniform superposition

$$\frac{1}{d} \sum_{0 \leq \ell, j < d} |\ell, j, 0^n\rangle.$$

This can be achieved with the Quantum Fourier Transform QFT_η for $\eta = \exp(2\pi i/d)$. Then we apply the assumed group circuit to obtain

$$\frac{1}{d} \sum_{0 \leq \ell, j < d} |\ell, j, g^\ell x^{-j}\rangle.$$

Finally, we use $\mathrm{QFT}_\omega \otimes \mathrm{QFT}_\omega$ to map a basic $2n$ qubit $|\ell, j\rangle$ to

$$\frac{1}{q} \sum_{0 \leq u, v < q} \omega^{u\ell + vj} |u, v\rangle.$$

Thus the final state is

$$a = \frac{1}{dq} \sum_{0 \leq \ell, j < d} \sum_{0 \leq u, v < q} \omega^{u\ell + vj} |u, v, g^\ell x^{-j}\rangle,$$

and the output measurement is $\mu(a)$.

ALGORITHM 14.32. Quantum circuit for discrete logarithm.

Input: Two elements g and x of a finite group G with odd order $d = \#G$, where x is in the subgroup generated by g.

Output: Either $\mathrm{dlog}_g x$, or a proper divisor of d, or "failure".

1. Compute the integer n with $2d < q = 2^n < 4d$, $\omega = \exp(2\pi i/q)$, and $\eta = \exp(2\pi i/d)$. The quantum computationin steps 2–5 works on $5n$ qubits and the input $|g, x, 0^n, 0^n, 0^n\rangle$. It never changes the first $2n$ qubits, and only the action on the last $3n$ qubits is indicated in the following.
2. Apply $\mathrm{QFT}_\eta \otimes \mathrm{QFT}_\eta \otimes \mathrm{id}_n$.
3. Apply the circuit for G: $|\ell, j, 0^n\rangle \longmapsto |\ell, j, g^\ell x^{-j}\rangle$.
4. Apply $\mathrm{QFT}_\omega \otimes \mathrm{QFT}_\omega \otimes \mathrm{id}_n$ to obtain state a.
5. Measurement: $|u, v, y\rangle \longleftarrow \mu(a)$. [$a$ and its measurement have $5n$ qubits.]

The following steps are classical.

6. Compute $e \longleftarrow (ud - (ud \text{ srem } q))/q$. [The symmetric remainder srem is as in (15.12); thus e is an integer.]
7. Compute $e' = \gcd(e, d)$.
8. If e' is a proper divisor of d, then return e' and stop. If $e' = d$, then return "failure" and stop. [Now $e' = 1$.]
9. Compute $f = \lfloor vd/q \rceil$, the nearest integer to vd/q, and return $k = -fe^{-1}$ in $\mathbb{Z}_d$. [e is invertible in $\mathbb{Z}_d$, since $e' = 1$.]

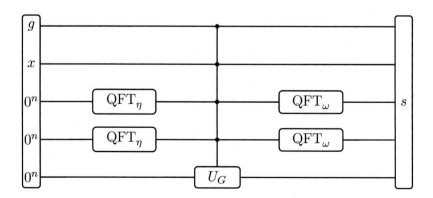

Figure 14.6: The quantum circuit for discrete logarithms.

Figure 14.6 shows the circuit.

THEOREM 14.33. *The circuit has size $O(n^3)$. If $n \geq 5$, then with proba-*

bility at least $1/420$, it does not fail and works correctly as specified.

PROOF.   The size of the QFT circuits is $O(n^2)$ by Theorem 14.20, and that of U_G is $O(n^3)$. We want to compute $k = \mathrm{dlog}_g\, x$. We consider a potential output state $|u, v, y\rangle$, with $0 \le u,\, v < q$, $y = g^m$ and $0 \le m < d$, and let

$$
\begin{aligned}
\gamma &= \gamma_{uv} = ud\,\mathrm{srem}\,q, \\
c &= c_{uv} = v + keq/d, \\
\lambda &= \lambda_{uv} = \lfloor c/q \rceil, \\
c' &= c'_{uv} = c\,\mathrm{srem}\,q = c - \lambda q,
\end{aligned}
$$

so that $-1/2 < c/q - \lambda \le 1/2$, and where e is as in step 6. Note that c and c' are, in general, not integers and that these quantities depend on u, v, and the input data, but not on y. The probability of some state $|u, v, g^m\rangle$ occurring under $\mu(a)$ is

$$
p = \left| \frac{1}{dq} \sum_{\substack{0 \le \ell, j < d \\ \ell - jk = m\,\mathrm{in}\,\mathbb{Z}_d}} \omega^{u\ell + vj} \right|^2.
$$

The equation in $\mathbb{Z}_d$ can be used to eliminate ℓ:

$$
\ell = jk + m - d \cdot \lfloor \frac{jk \mid m}{d} \rfloor,
$$

since $0 \le \ell < d$. Furthermore, we take the integer $b = e \cdot \lfloor (jk + m)/d \rfloor$, and the rational number

$$
\begin{aligned}
c''_j &= u\ell + vj - (um + jc) - bq \\
&= (ujk + um - ud\lfloor \frac{jk + m}{d} \rfloor) + vj - (um + ujk + vj - \frac{jk\gamma}{d}) \\
&\quad - \lfloor \frac{jk + m}{d} \rfloor (\gamma - ud) \\
&= (\frac{jk}{d} - \lfloor \frac{jk + m}{d} \rfloor) \cdot \gamma.
\end{aligned}
$$

Then we have

$$
p = \frac{1}{(dq)^2} \left| \sum_{0 \le j < d} \omega^{um + jc + c''_j + bq} \right|^2
$$

$$
(14.34) \qquad = \frac{1}{(dq)^2} |\omega^{um}|^2 \cdot \left| \sum_{0 \le j < d} \omega^{j(c' + \lambda q) + c''_j} \right|^2 = \frac{1}{(dq)^2} \left| \sum_{0 \le j < d} \omega^{jc' + c''_j} \right|^2,
$$

where we use $\omega^q = 1$. If $\gamma \geq 0$, then

$$-\gamma \leq \left(\frac{jk}{d} - \left\lfloor\frac{jk+d-1}{d}\right\rfloor\right) \cdot \gamma \leq c_j'' \leq \left(\frac{jk}{d} - \left\lfloor\frac{jk}{d}\right\rfloor\right) \cdot \gamma \leq \gamma,$$

and also when $\gamma \leq 0$, we have $|c_j''| \leq |\gamma|$.

Our goal now is twofold. Firstly, we want the sum in (14.34) to behave similar to the geometric sum $\sum_{0 \leq j < d} \omega^{jc'}$. For this, it is sufficient to have γ small. Secondly, we want the geometric sum to turn out large. As usual, it is sufficient to have c' small. This leads to the following notion of a *helpful state* $|u, v, y\rangle$:

(14.35) $$|c'| \leq 1/2 \quad \text{and} \quad |\gamma| \leq q/12.$$

We show quantitative versions of the three usual properties:

○ If a helpful state is measured, the output is either k or a proper divisor of d.

○ For any helpful state, the probability of measuring it is at least $1/15q^2$.

○ There are at least $q^2/28$ helpful states.

Suppose first that we have observed a helpful state $|u, v, y\rangle$. If $0 < u \leq q - 2$, then d does not divide e, because

$$e = \frac{ud - \gamma}{q} \leq \frac{(q-2)d + q/12}{q} = d - \frac{2d}{q} + \frac{1}{12} < d - \frac{2}{q} \cdot \frac{q}{4} + \frac{1}{12} < d.$$

Thus failure does not occur in step 8. The remaining value $u = q - 1$ is not part of a helpful state, since then

$$\gamma = (q-1)d \text{ srem } q = -d \text{ srem } q = -d \leq -q/4,$$

and the second condition in (14.35) is violated. We have $|c'| \leq 1/2$, and by the definition of f, $|vd/q - f| \leq 1/2$. Thus

$$|-ke + \lambda d - f| = \left|\frac{(v-c)d}{q} + \lambda d - f\right| = \left|\frac{-c'd}{q} + \frac{vd}{q} - f\right|$$

$$\leq \left|\frac{c'd}{q}\right| + \left|\frac{vd}{q} - f\right| \leq \frac{d}{2q} + \frac{1}{2} \leq \frac{1}{4} + \frac{1}{2} = \frac{3}{4}.$$

The left hand side is an integer, and hence zero. It follows that $ke = -f + \lambda d$ and $ke = -f$ in $\mathbb{Z}_d$. Since d does not divide e, the circuit does not fail, and the output—either e' or k—follows the specification. (Note that

$k = -fe^{-1}$ in $\mathbb{Z}_d$ makes sense and is correct if $e' = 1$, but not otherwise.) This shows the first claimed property.

Now we verify that p is large for a helpful state. We first suppose that $c' \geq 0$. Then we can apply the half-circle lemma 14.21(i) with

$$t_0 = -\frac{q}{12}, \quad t_1 = \frac{q}{3}.$$

We have $2(t_1 - t_0) = 5q/6 < q$, and $\alpha = \cos(5\pi/12) \approx 0.25882 > 0$. For each exponent in the sum we have

$$t_0 = -\frac{q}{12} \leq jc' + c''_j \leq \frac{d}{2} + \frac{q}{12} \leq \frac{q}{4} + \frac{q}{12} = t_1.$$

From the lemma we get

$$p \geq \frac{1}{(dq)^2}(d\alpha)^2 = \frac{\alpha^2}{q^2} \geq \frac{1}{15q^2}.$$

If $c' \leq 0$, then we use $t_0 = -q/3$ and $t_1 = q/12$, and find in the same way that $p \geq 1/15q^2$. This is the second property.

Finally, we want to count the number of helpful states $|u, v, y\rangle$, that is, those satisfying (14.35). Since $\gcd(d, q) = 1$, multiplication by d is a bijection of $\mathbb{Z}_q$. Therefore there are $2\lfloor q/12 \rfloor + 1 \geq q/6 - 1/3$ many nonzero u satisfying the second condition in (14.35). Now let $0 < u < q$, and $\rho = c_{u0}$ srem 1 be the fractional part of c_{u0}, with $1/2 \leq \rho < 1/2$. As v ranges from 0 to $q - 1$, $c'_{uv} = v + \rho + (c_{u0} - \rho) - \lambda_{uv}q$ runs through $\rho + i$ for the integers i with $-q/2 \leq i < q/2$. For exactly one value of v we have $|c'_{uv}| \leq 1/2$; see Exercise 14.6. (The values $\pm 1/2$ do not occur, since the reduced denominator of ρ is a divisor of d and hence odd.) There are d possible values for y, and we find a total of at least

$$d \cdot \left(\frac{q}{6} - \frac{1}{3}\right) \geq \frac{q}{4} \cdot \frac{q}{7} = \frac{q^2}{28}$$

helpful states, for $q \geq 14$.

Thus the probability of observing some helpful state is at least

$$\frac{q^2}{28} \cdot \frac{1}{15q^2} = \frac{1}{420}. \qquad \square$$

14.6. Outlook on quantum computation

Two new paradigms of computation started to emerge in the 1980s: quantum computation and molecular computation. Both offer enormous potential, but experimental realizations on a large scale still have to come.

In the history of quantum computers, a first concern was that quantum circuits are reversible, but Boolean circuits are not. This concern was removed by the invention of reversible versions for the classical Boolean operations (Bennett 1973, Toffoli 1980, 1981, Fredkin & Toffoli 1982). Then it was shown that quantum systems can simulate classical reversible Turing machines (Benioff 1980, 1982a, 1982b). Feynman (1982) asked about the reverse direction, and gave reasons why it might be exponentially expensive to simulate quantum computers classically. On the other hand, based on work of Deutsch (1985), Bernstein & Vazirani (1993) and Yao (1993) constructed universal quantum computers that can simulate in polynomial time any other quantum computer. Our definition of quantum circuits is geared towards the two instances presented. In fact, Barenco et al. (1995) have shown that cnot plus some simple 1-qubit gates are sufficient: quantum circuits with these gates can approximate arbitrary ones arbitrarily well.

It is not hard to simulate quantum computers with classical machines, if one is willing to accept an exponential slowdown. Thus in terms of what they can compute without restriction on time and space, classical and quantum computation are equally powerful. This reinforces belief in the *Church-Turing thesis* saying that everything that can somehow be computed by an algorithm can actually be computed by the specific and rigorously well-defined model of deterministic Turing machines.

But when we take efficiency into account, things may be different. On the one hand, quantum computers can simulate classical ones efficiently. In the shorthand of complexity theory, this leads to the inclusion $\mathsf{BPP} \subseteq \mathsf{QBP}$. On the other hand, large classes of quantum computations can be simulated classically (Valiant 2001). But possibly (and hopefully, for cryptography), not all. Namely, we have seen Shor's (1994, 1999) polynomial-time quantum algorithms for integer factorization and for discrete logarithms. An efficient classical simulation of this would be the death knell for much of cryptography. It was Shor's astounding result that attracted much attention to the power of quantum computing.

Previous results by Deutsch & Jozsa (1992), Grover (1996), Simon (1994), and others had already shown that quantum computing is, in a specific sense, more powerful than classical computing. To actually prove an exponential gap remains an open problem.

An important obstacle for implementations of quantum computers are the inherent imprecision and the interaction with the environment, leading to decoherence. In order to deal with this problem, algorithms for quantum error correction and fault-tolerant quantum gates have been invented (Calderbank et al. 1997, Calderbank & Shor 1996, Gottesman 1996, Shor 1996, Steane 1996).

There are also several suggestions to use quantum devices for cryptographic purposes; see Brassard (1994) and Brassard & Crépeau (1996) for overviews. Photons to be used for quantum key generation have been experimentally transmitted over a distance of almost 1 km (Buttler *et al.* 1998). However, a quantum bit commitment scheme was claimed to be unconditionally secure (Brassard *et al.* 1993), but turned out to be insecure (Lo & Chau 1997, 1998, Mayers 1997). This has generated some skepticism about security claims in quantum cryptography, and we do not treat the subject here.

Kalai (2011) puts forth arguments why quantum computers as required here may not be feasible in principle. He points out that nature seems to follow the laws of quantum physics, but that the behavior required of our quantum circuits is so substantially different (decoherence, error correlation) from nature that it may not be physically implementable.

Quantum computation is an exciting research area with interesting results. But the advent of scalable quantum computing is said to be in fifteen years from now—and has been since the 1980s. It remains a matter of opinion whether quantum computers or classical methods provide a bigger threat to number-theoretic cryptography.

Notes 14.1. Nielsen & Chuang (2004) is a comprehensive textbook on quantum computation. The read-once property of qubits is often expressed by saying that after a measurement the qubit is irrevocably lost. Mercy on its soul.

In the language of algebraic geometry, the mapping from $\mathbb{C}^{2n}$ to $\mathbb{C}^{2^n}$ taking $(a_0, \ldots, a_{n-1})$ to $a_0 \otimes \cdots \otimes a_{n-1}$ is called the Segre embedding. Its image, the Veronese variety, is not contained in any hyperplane, so that any point in $\mathbb{C}^{2^n}$ is a linear combination of images. Properly speaking, these notions have to be taken in the projective sense.

14.2. Our matrices like H, M_π, or V_i are defined in the given basis $|x\rangle_{x \in \mathbb{B}^n}$ of $\mathbb{C}^{2^n}$. If we take the same matrix in a different basis, then the effect will, in general, be different. For example, if we let

$$H' = \frac{1}{\sqrt{2}} \begin{pmatrix} 1 & 1 \\ 1 & -1 \end{pmatrix}$$

in the basis $((1,0), (0,1))$ and take the basis

$$|0\rangle = \frac{1}{2}(1 + i, \sqrt{2}), \quad |1\rangle = \frac{1}{2}(1 - i, i\sqrt{2})$$

from Example 14.1 (iii), then

$$H' \cdot |0\rangle = \frac{1}{2\sqrt{2}}(1 + \sqrt{2} + i, 1 - \sqrt{2} + i) = \frac{1}{2}(|0\rangle + (1 + i\sqrt{2})|1\rangle),$$

$$H' \cdot |1\rangle = \frac{1}{2\sqrt{2}}(1 + i(\sqrt{2} - 1), 1 - i(\sqrt{2} + 1)) = \frac{1}{2}(-|0\rangle + (1 - i\sqrt{2})|1\rangle).$$

14.3. The quantum computation for the Fourier transform with q a power of 2 is in Coppersmith (1994b), Ekert & Jozsa (1996) (an algorithm by Deutsch), and Cleve *et al.* (1998). For general q, it is in Mosca & Zalka (2003).

14.4. Theorem 14.23 is in Vedral *et al.* (1996), and Theorem 14.24 in Watrous (2009), Section IV.3.

In the argument after (14.29), instead of relying on Theorem 15.39, we could also use the stronger result mentioned in the Notes to Section 15.9. This would allow the denominator $2k^2$ in (14.29) and $q = 2^{2n}$.

14.6. Public-key cryptography currently relies on number theoretic problems like integer factorization and discrete logarithms. More progress on them can be expected such as the discrete logarithm advances mentioned in Section 4.8. It is therefore an important quest to design cryptosystems whose security relies on the presumed hardness of different types of problems. Section 13.13 (lattices) and Exercise 9.6 (error correcting codes) present examples. Scalable quantum computing would also bring down number-theoretic cryptography, but "post-quantum cryptography" is—to the quantum-skeptic—a misnomer for this quest.

Exercises.

EXERCISE 14.1 (Orthonormal tensor product). *Take* $(1, 0)$ *and* $(0, 1)$ *as orthonormal basis of* $\mathbb{C}$. *Compute the tensor product of this basis; this is an orthonormal basis of* $\mathbb{C}^4 = \mathbb{C}^2 \otimes \mathbb{C}^2$.

EXERCISE 14.2 (Orthonormal is unitary). *Let* $v = (v_1, \ldots, v_k)$ *be a vector space basis of* $\mathbb{C}^k$, *and* U *the matrix with rows* $v_1, \ldots, v_k$. *Show that* v *is orthonormal if and only if* U *is unitary.*

EXERCISE 14.3 (Tensor product bra-ket).

(i) *Let* $a \in \mathbb{C}^k$ *and* $b \in \mathbb{C}^\ell$ *be two unit vectors with exactly one coordinate 1 and all others 0. Show that* $a \otimes b$ *also has this property.*

(ii) *Let* $a, a' \in \mathbb{C}^k$ *and* $b, b' \in \mathbb{C}^\ell$ *be four vectors with coordinates as in (i). Show that*

(14.36) $$\langle a \otimes b, a' \otimes b' \rangle = \langle a, a' \rangle \cdot \langle b, b' \rangle$$

(iii) *Use the sesquilinearity of the inner product to conclude that (14.36) holds for all* a, a', b, b'.

EXERCISE 14.4 (Unitary group). *Show that for any* $k \in \mathbb{N}$ *the unitary* $k \times k$ *matrices over* $\mathbb{C}$ *form a group.*

EXERCISE 14.5 (Unitary matrices). *Let* $U \in \mathbb{C}^{k \times k}$ *be an invertible matrix. Show that the following are equivalent.*

(i) *For every orthonormal basis* v *of* $\mathbb{C}^k$, Uv *is an orthonormal basis.*

(ii) *There exists an orthonormal basis* v *of* $\mathbb{C}^k$ *for which* Uv *is an orthonormal basis.*

(iii) U *is unitary.*

EXERCISE 14.6 (Helpful states).   You are to verify the claim that for every a there exists exactly one b satisfying the second condition in (14.35) in the analysis of the discrete logarithm circuit. So take $0 \le a < q$, and for $0 \le b < q$ the quantities c_{ab}, λ_{ab}, and c'_{ab} as defined in the text. Also, let $b^* = (\lambda_{a0} + 1/2)q - c_{a0}$, and

$$\delta_b = \begin{cases} 1 & \text{if } b \ge b^*, \\ 0 & \text{otherwise.} \end{cases}$$

(i) Show $\lambda_{ab} = \lambda_{a0} + \delta_b$.

(ii) Show $c'_{ab} = c_{a0} + \lambda_{a0}q + b - \delta_b q$.

(iii) Show that $c'_{ab_1} - c'_{ab_2}$ is an integer not divisible by q, for all $b_1 \ne b_2$ in the range.

(iv) Show that $|c'_{ab}| \le q/2$.

(v) Conclude that the claim stated at the beginning is true.

EXERCISE 14.7 (Measuring helpful states).   You are to check in a different way that for a helpful state $|r, v\rangle$ in the integer factoring quantum algorithm, its probability p of being measured is reasonably large. In the notation of the proof of Theorem 14.26, we let $t_1 = \lfloor q/k \rfloor$. Then the summation range is $0 \le t < u$ with $u = t_1 + 1$ for $0 \le j_0 \le (q \text{ rem } k)$, and $u = t_1$ for $(q \text{ rem } k) < j_0 < k$. Remember that $e^{ix} = \cos x + i \sin x$ for real x.

(i) Prove that the real part of $\omega^{-ub_r/2} \sum_{0 \le t < u} \omega^{tb_r}$ is

$$\frac{\sin(\pi u b_r/q) \sin(2\pi b_r/q)}{1 - \cos(2\pi b_r/q)}$$

(ii) Show that with the first approximations $\sin x \approx x$ and $\cos x \approx 1 - x^2/2$, this evaluates to $u/2$.

(iii) Use a computer algebra system to compute the first 10 terms of the series expansion. Set $b_r = k/2$ and $q = ak$ and check that the expansion is of the form $ca + O(a^{-1})$ with $0 < c \le 1/2$.

(iv) What can you conclude about the constants β so that $p \ge \beta/k^2$?

EXERCISE 14.8 (Commuting).

(i) Show that in the circuit for QFT, any $(S_{n-j+1}; j - 1, n - 1)$ and $(S_{n-k+1}; k - 1, n - 1)$ as in (14.19) commute, for $0 \le j, k \le n - 2$.

(ii) Do $(H; n - 1)$ and $(S_n; 0, n - 1)$ commute?

The telegraphic communications of the Allies were monitored,
recorded, and deciphered, primarily by the *B-Dienst*,
the code and cipher bureau of the Naval High Command
that broke every one of the American diplomatic
and several naval codes and ciphers, and was reading
our encrypted signals most of the time during World War II.
LADISLAS FARAGO (1971)

The German Enigma was adequate in its day
to resist cryptanalysis had it been used properly.
LOUIS KRUH AND CIPHER A. DEAVOURS (1983)

... I was to go down to the deciphering headquarters at Bletchley
Park in Buckinghamshire. On no account was I to go in uniform,
it was far too secret. Having arrived at Bletchley Junction I
asked the taxi-driver to take me to Bletchley Park. 'Oh, the cloak
and dagger centre,' he replied — so much for secrecy! There I
met a number of people of all backgrounds; many were
university professors, some verging on the eccentric — indeed it
seemed, possibly falsely, that in many cases the more eccentric
you were, the better cryptographer you would be.
EWEN EDWARD SAMUEL MONTAGU (1977)

Quant aux cryptologues polonais, à eux seuls tout le mérite
et toute la gloire d'avoir mené à bien, techniquement,
cette aventure incroyable, grâce à leur science et leur ténacité,
inégalées dans aucun pays du Monde![1]
GUSTAVE BERTRAND (1973)

Türing Machine. In 1936 Dr. Turing wrote a paper on the design
and the limitations of computing machines. For this reason they
are sometimes known by his name. The umlaut is an unearned
and undesirable addition, due, presumably, to an impression
that anything so incomprehensible must be Teutonic.
BERTRAND VIVIAN BOWDEN (1953)

---

[1] As to the Polish cryptologists, to only them all the merit and all the glory for having completed, technically, this incredible adventure, thanks to their science and their tenacity, unequalled in any country of the world!

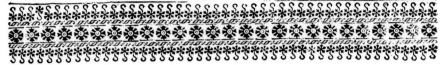

Chapter J

ENIGMA, Turing, and COLOSSUS

ENIGMA, Turing, COLOSSUS: what memorable names! How they shine compared to bland technocratic acronyms like RSA, DSA, or AES!

The ENIGMA was the cryptographic workhorse of the German military in World War II. It was originally broken by Polish mathematicians, who then handed their methods to French and British cryptographers. The latter eventually built up a large organization, whose most famous member was Alan Turing and whose cryptanalytic successes helped to shorten the war considerably. The team also designed COLOSSUS, the world's first electronic (valve) computer, for use in the cryptanalysis of a different German cipher machine.

J.1. ENIGMA

In the 1920s, a new type of cryptographic hardware emerged: electromechanical devices. As is often the case in the history of ideas, the time was ripe and the possibilities of such a cryptosystem were realized by four men in four countries around the same time. Apparently the US American Edward Hugh Hebern (1869–1952) was the first to have the idea, in 1917, but made a US Patent application only in 1924. The German Arthur Scherbius (1878–1929) applied for a patent on 23 February 1918,

© Springer-Verlag Berlin Heidelberg 2015
J. von zur Gathen, *CryptoSchool*, DOI 10.1007/978-3-662-48425-8_24

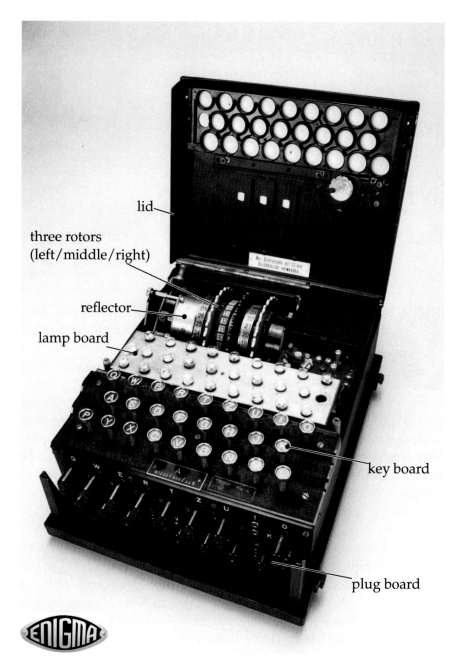

Figure J.1: An ENIGMA machine with open lid.

the Dutch Hugo Alexander Koch (1870–1928) on 7 October 1919, and the Swede Arvid Gerhard Damm (1869–1927) three days later.

Their common idea was to use an apparatus as described below. Hebern used just one rotor that turned by one position at each key stroke, thus implementing a Vigenère cipher with a 26-letter key. Scherbius took three rotors, and this machine became the ENIGMA. It was initially sold to the same clientele that was using commercial codebooks (Section D.5). The German military adopted it as a major cryptographic tool, beginning in 1926. Eventually the ENIGMA was used by various other German government agencies as well, including the post office, the railroad system, and the police. It went through several stages of development, some of which increased security and others decreased it, unwittingly. Our description in the following applies to one specific model. The number of ENIGMA machines built is estimated at around 100 000. Like Henry Ford's *Tin Lizzy*, it could be had in any color, provided the color was black.

The main parts of an ENIGMA are shown in Figure J.1:

- *plug board* (German: *Steckerbrett*),

- *key board*,

- *lamp board*,

- *rotors* (German: *Walzen*).

When you close the lid in Figure J.1, each of the 26 bulbs (the one at left in the middle row is a spare) is under a circular glass window in the lid with a letter printed on it, the grooved aluminum *rings* attached to the rotors show through the slits, and the letter at the top of a rotor through one of the three small square windows. To the left of the left rotor, the *reflector* (German: Umkehrwalze) is fixed.

Figure J.2 illustrates the principle on the six-letter alphabet QWERTZ. The arrangement of this fictitious machine corresponds to that in Figure J.1 except that we leave out the plug board, which was also absent from ENIGMA's early versions. When you press the Q key, contact is made with the battery at left. Current flows (in red) to the top contact of the entry (right) rotor. Here it is one out of five available rotors and named III. On all our rotors, contacts are labelled QWERTZ from top to bottom; in reality, the rotors are circular. Thus the internal wiring of III implements the permutation $N = $ (ERQ) (TWZ). Current exits at E which is in direct contact with the E on the middle rotor named I, whose permutation is $M = $ (EQT)(RZW). It exits I at Q and enters the left rotor IV,

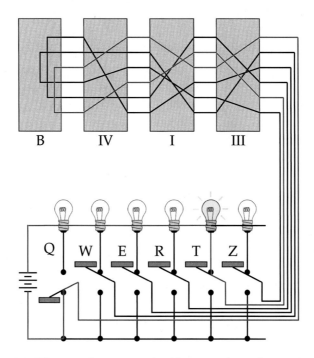

Figure J.2: Electrical wiring of a (fictitious) six-letter ENIGMA.

permuting as $L = (\texttt{ERZQ})(\texttt{T})(\texttt{W})$. It exits at $\texttt{E}$, enters the reflector B, permuting as $R = (\texttt{EZ})(\texttt{QT})(\texttt{RW})$, and then passes back through $\texttt{Z}$, $\texttt{R}$, and $\texttt{W}$ to
exit rotor III at $\texttt{T}$. Now current passes through the $\texttt{T}$ lamp which lights
up and closes the circuit with the battery. Then $\texttt{T}$ is the encryption of
$\texttt{Q}$ at this state of the machine. When the key $\texttt{R}$ is pressed next, the left
rotor III has rotated by one position (cyclically downwards in Figure J.2,
so that $\texttt{T–R}$ and $\texttt{Z–Q}$ are connected in III) and—assuming that rotor I has
not moved—the lamp $\texttt{Q}$ lights up.

The entry rotor turns by one letter (one 26th of a full turn) at each key
stroke. Each rotor has a notch in one of 26 positions. When it is in the
middle or at the left and the rotor to its right moves to the notched position, then the rotor turns by one letter. The *ring position* (German: *Ringstellung*) determines in which of 26 possible positions the ring is fixed
to the rotor. This is only used in determining the machine's initial state
before encryption.

The plug board consists of 26 connectors, some of which are connected in pairs. In the early days of the war, up to five pairs were connected, later exactly ten pairs. If $\texttt{E}$ is not plugged ("unsteckered"), then
current flows from the $\texttt{E}$ key on the key board to the $\texttt{E}$ connector on the
entry rotor. But if $\texttt{E}$ is plugged, say to $\texttt{Q}$, then current goes to the $\texttt{Q}$ plug

on the entry rotor. Then it transits the rotor assembly to exit at T, as above. If T is plugged to Z, this causes the Z lamp to light up, and the electrical circuit closes.

Two operators are required: FRITZ reads out the plaintext aloud (ALICE was not enlisted). EMIL types it into the ENIGMA, which he has set up with the current keys, and reads the ciphertext letter by letter back to FRITZ, who taps it in Morse code into his radio transmission unit. The recipients have to set up their ENIGMA in the same way, type in the ciphertext, and the plaintext lights up, letter by letter, to be copied down.

The setting used for encryption also serves for decryption, for the following reason. The encryption process can be viewed as a composition

$$(J.1) \qquad \pi = P^{-1} \circ N^{-1} \circ M^{-1} \circ L^{-1} \circ R \circ L \circ M \circ N \circ P$$

of the plug board permutation P, the three rotor permutations N, M, and L, and the reflector permutation R in the permutation group $\mathrm{Sym}_{\mathbb{A}}$ over the 26-letter alphabet $\mathbb{A} = \{A, B, C, \ldots, Y, Z\}$; see Section 15.2 for permutations. Now if E is sent to Q on the plug board, that is, $P(E) = Q$, then also $P(Q) = E$. That means that applying P twice does not change anything: $P \circ P$ is the identity, P is an involution, and $P^{-1} = P$. The same holds for R. When we take the composition $\pi \circ \pi$, adjacent terms cancel one after the other, and we also find $\pi \circ \pi$ to be the identity. Thus for any state, the ENIGMA permutation $\pi \in \mathrm{Sym}_{\mathbb{A}}$ is an involution. Since $R(x) \neq x$ for all letters $x \in \mathbb{A}$, we also have $\pi(x) \neq x$.

The original rules for setting up the ENIGMA contained a serious security flaw, which allowed Polish cryptanalysts to discover the rotor wirings; see Section J.3.

During much of its wartime use, a daily secret key was distributed on paper to all participants in a network of ENIGMAs. It consists of the rotor choice and order, the ring setting, and the plug board connections. The operator chooses two (supposedly random) 3-letter keys, the starting position, say HMD, and the secret session key, say YOS. He uses the grooves on the rings to move the rotors into position HMD, visible through the small windows. Then he encrypts YOS to obtain the public session key, say XMZ. He starts a transmission by sending HMD and XMZ, sets his machine to YOS, starts encrypting, and sends the resulting ciphertext. The recipient initializes his ENIGMA with the same daily secret key and starting position HMD, decrypts XMZ to obtain YOS, sets his machine to YOS, and decrypts the received ciphertext.

According to Kerckhoffs' Principle (A.1) and the early commercial availability, the ENIGMA system must be assumed to be known to the enemy. Security only relies on the secret key. It consists of three parts:

Figure J.3: Two ENIGMA rotors.

○ choice and sequence of rotors,

○ setting of rotors,

○ plug board connections.

Initially, a further secret ingredient was the internal wiring of the rotors. It would be unwise to rely on this for security, because then a single stolen or captured machine would jeopardize the whole system. The reflector was fixed. The other rotors came in a wooden box. Initially, there were three to choose from, which allows six possible permutations. A later version had five to choose from, giving $5 \cdot 4 \cdot 3 = 60$ possibilities. The three rotors of the German Navy ENIGMA could be chosen from a set of eight. The rotor setting was first changed monthly, later daily, and from mid-1942 on every eight hours. In 1943, a machine with four rotors was introduced. Each rotor could be set in one out of 26 positions at the beginning of a transmission. The entry rotor advanced by one position every time a key was pressed. Like an odometer, it advances after 26 turns the middle rotor by one position and this then moves the leftmost rotor by one after 26 turns. With 60 choices for a sequence of three out of five rotors, 26^3 positions for them, and $26!/(2^{10} \cdot 6! \cdot 10!) \approx 2^{47}$ choices for ten plug board connections, the size of the key space is

$$(J.2) \qquad 158\,962\,555\,217\,826\,360\,000 \approx 1.6 \cdot 10^{20} \approx 2^{67}.$$

At the time, 67-bit security against exhaustive key search was more than enough. The ENIGMA's downfall were insecure operating modes and, above all, faulty behavior of operators.

The second most common mistake of cryptosystem designers is to take a large key space as a guarantee of security. (The most common mistake is to take the designer's failure to break his own system as proof that everybody else will fail as well.)

J.2. Bletchley Park

No single event can be pinpointed that brought about Allied victory in the Second World War, but the British cryptanalysts at Bletchley Park played a vital role in many battles whose outcome eventually saved the world from brutal Nazi domination. It has been conjectured that *Ultra* shortened the war by two years, saving millions of lives on both sides.

The cryptanalytic success against the ENIGMA was started by a team of Polish cryptographers, including the mathematician Marian Rejewski. They had completely solved the then standard machine in 1939. Section J.3 describes in full detail their cryptanalysis of the ENIGMA rotors, which was completed in 1932. In the fall of 1938, they invented the *bomba*, an electromagnetic device simulating many ENIGMA runs on a fixed plaintext. On 25 July 1939, about a month before Hitler's blitzkrieg attack on Poland and while most people were still happy with the seeming success of appeasement politics at München, they were wise enough to share their secrets and machinery with French and British cryptographers. At the start of the war, they moved to France, and when the Germans also occupied that country, some fled to England. There, they were not allowed to participate in the British cryptanalytic effort.

A vital ingredient to the initial Polish ENIGMA break was a classical espionage coup by the French Secret Service. Hans-Thilo Schmidt, working in the *Chi-Stelle* of the *Reichswehrministerium* (cipher bureau of the Reich's Defense Ministry) offered his services in October 1932. Directed by Colonel Gustave Bertrand and under the codename Asché, he divulged many secrets. Among them were complete ENIGMA key schedules for certain periods, as discussed in Section J.3 below. The French secret agent Lemoine, captured and interrogated by the Germans, betrayed Asché, who was arrested at home in Fürstenwalde and executed in July 1943.

Figure J.4: The main building, the *manor*, at Bletchley Park, used by the administration, of course. The cryptanalysts worked in wooden huts. Umbrella and shorts illustrate the versatile weather of a Buckinghamshire summer day.

Alan Turing (1912–1954), a famous British mathematician and computer scientist, had proposed in 1937 a precise mathematical model of computers—the *Turing machine*—invented the idea that programs could be stored as data (namely, for his *universal Turing machine*), and proved that deceptively simple questions cannot be solved by any algorithm.

The British Foreign Office set up a team of cryptographers at Bletchley Park on 4 September 1939, one day after Hitler attacked Poland. A little later, Turing joined the team and became their intellectual leader. One of their main tasks became the breaking of the ENIGMA-encrypted communication between the German Navy headquarters at Kiel and the submarines in the North Atlantic. These inflicted crippling losses on Allied convoys from North America to Europe. Bletchley Park deciphered ENIGMA messages regularly until the German Navy introduced a fourth rotor for the ENIGMA in February 1942. After a long struggle, the British cryptanalysts resumed their successful reading in December 1942. The resulting stream of deciphered data ran under the name of *Ultra*. The unfortunate U-boat captain who had just radioed his position to headquarters did not know that the US *Catalina* bombers dropping depth charges all around him were secretly directed by the brain of a mathematical ge-

Figure J.5: COLOSSUS rebuilt at the Bletchley Park Museum.

nius.

At the highest level of secrecy, German military messages were enciphered on different systems, the *Siemens Geheimschreiber* and the *Lorenz SZ* (Schlüsselzusatz). The latter also used rotors, namely, twelve of them. But the principle differed from the ENIGMA's: the rotors generated a pseudorandom bit string (see Chapter 11), and each letter of the message was encoded by five bits, according to the standard *Baudot code*. These two bit streams were then added bitwise (XORed), just as one does in a one-time pad (Section 9.4). By a brilliant stroke of cryptanalytic genius, the mathematician Bill Tutte discovered this principle. He was unwittingly aided by a German radioman who transmitted essentially the same message twice with the same key. And then Bletchley Park, in collaboration with British Post Office engineers, set out for one of their main achievements: the world's first computer. This COLOSSUS had about 1500 valves. Its input was fed on rapidly moving paper tape, at right in Figure J.5 showing the replica now on display in the Bletchley Park museum.

J.3. Rotor cryptanalysis

There is a substantial amount of literature on the work at Bletchley Park and its effect on the war. Rather than add to this, we only present a beautiful piece of cryptanalysis in detail, actually pre-Bletchley. The reader with *horror mathesis* may want to skip this section.

The Polish cryptanalytic success against the ENIGMA was the basis for all subsequent work, and quite possibly the major effort at Bletchley Park would not even have been started without these previous results. Marian Rejewski conceived the basic mathematical ideas required for this cryptanalysis, aided by other Polish mathematicians. In 1932, he reconstructed the secret interior wiring of the ENIGMA rotors, and then they could read the German messages. Later they designed an electromechanical device they called *bomba* to aid in ENIGMA cryptanalysis. We present in detail the discovery of the rotor wirings, a clever piece of applied mathematics. It is sufficiently simple to be presented here, and sufficiently complicated to given an idea of the ingenuity required. The success of the approach is based on the following:

- interception of many encrypted messages,

- systematic cryptographic mistakes by the Germans,

- a French espionage success,

- Polish mathematical ingenuity.

The ENIGMA instructions, valid until 15 September 1938, provided a daily setting for the three rotors and the plug board. Then the operator had to choose a three-letter session key, say XIX, type it twice: XIX XIX, and read off the result: AJUOEZ. Then he set the three rotors to the corresponding positions X, I, and X, typed the message, and finally sent as ciphertext AJUOEZ followed by the encryption of the message. On a given day, all operators in a given net started in the same position, so that the permutation $A\colon \mathbb{A} \to \mathbb{A}$ corresponding to the first key stroke was identical for all of them. In our example we have $A(\mathrm{X}) = \mathrm{A}$. And also the next five permutations $B, C, D, E, F\colon \mathbb{A} \to \mathbb{A}$ are identical for everyone.

Table J.6 shows a list of six-letter beginnings of intercepts from a single day. We now deduce one ENIGMA rotor wiring from these intercepts.

Suppose that we see the encryption DZVUFA in some intercepted message, as recorded as number 8 among the 42 messages numbered $0, \ldots, 41$ in Table J.6. Then D and U are encodings of the same (unknown) letter x by the two permutations A and D, respectively. Here x is the first letter of the session key. In other words, $A(x) = \mathrm{D}$ and $D(x) = \mathrm{U}$ for some

	0	1	2	3	4	5
0	AJUOEZ	AOCORQ	AZUOFZ	BPTNXY	CLZHTK	CQJHPL
6	DDOUKH	DKIUGU	DZVUFA	EAAXCG	ECLXSB	FTXQLF
12	GGRIBW	GMYIYS	HYJLJL	ITMJLV	IWAJDG	JAGZCI
18	JERZVW	JZTZFY	KANMCE	KSAMZG	LLMDTV	MVUFHZ
24	NXSPOX	ONDBUC	PDGKKI	PICKWQ	QUGYNI	RSIAZU
30	SCPCSD	SCQCST	SRPCQD	TVLRHB	UCYGSS	VHFTMO
36	WSHEZN	XCSSSX	XFWSAJ	YLZVTK	ZDBWKM	ZLKWTR

Table J.6: 42 intercepts, sorted alphabetically, of the six-letter beginnings of messages with identical daily key.

unknown letter x. But then also $A(\text{D}) = x$, and $DA(\text{D}) = \text{U}$, where we write DA for $D \circ A$ from now on. Thus any intercepted message DZVUFA tells us that

$$DA(\text{D}) = \text{U}, \quad EB(\text{Z}) = \text{F}, \quad FC(\text{V}) = \text{A}.$$

We have thus obtained one value each of the three compositions DA, EB, and FC. If we have sufficiently many of these single values, we have the three permutations DA, EB, and FC completely in our hands. This is called the *characteristic* of the given day. Its determination from intercepts was the first step in the Polish solution of the rotor wirings.

As an example, we determine the first cycle of DA as follows:

x	A	O	B	N	P	K	M	F	Q	Y	V	T	R
$DA(x)$	O	B	N	P	K	M	F	Q	Y	V	T	R	A
k	0	25	3	24	26	20	23	11	28	39	35	33	29

An entry corresponds to the message number k; thus the second entry $DA(\text{O}) = \text{B}$ is derived from message number 25. One finds the second cycle in the same way, as well as the cycle presentations of the other two permutations:

$$
\begin{aligned}
& DA = (\text{AOBNPKMFQYVTR})\,(\text{CHLDUGIJZWEXS}), \\
(\text{J.3}) \quad & EB = (\text{BIWDKG})\,(\text{EVHMYJ})\,(\text{ACSZF})\,(\text{ORQPX})\,(\text{LT})\,(\text{NU}), \\
& FC = (\text{AGIUZKRWJLBMV})\,(\text{CQTYSXFOHNEPD}).
\end{aligned}
$$

The cycle lengths are $(13, 13)$ for DA and FC, and $(6, 6, 5, 5, 2, 2)$ for EB. A general theorem of Rejewski says that in any such product of involutions, each cycle length appears an even number of times; this certainly happens for our three permutations.

How can we get the individual permutations, like A, from the characteristic? If the system is properly used, then there is no easy way of doing so. But—fortunately for the Polish mathematicians—the German operators did not follow the rule of choosing session keys at random, but had a small set of preferred keys: three repeated letters, like JJJ, or three letters adjacent on the key board, usually from the outside towards the inside, like SDF, or three-letter female names like EVA. In a sufficiently large set of intercepts, there would then be repetitions: two operators had chosen the same key. The cryptanalyst takes one of those repetitions and assumes it to be one of the "preferred" keys. We will now see how to compute the involutions $A, \ldots, F$ from this assumption, and also that we get a way of checking its correctness.

This approach illustrates an important tool of the Polish and (later) British cryptanalysis: the *known-plaintext attack*, called a *crib* in those days. Here the 3-letter session key was not chosen at random, but with heavy preference on particular keys. At later stages of the war, there were stereotyped beginnings or endings of messages, such as salutations or signatures, or texts such as weather messages or ship positions. In fact, at some point the British laid mines near the German-occupied French coast just in order to intercept ENIGMA-encrypted messages from German minesweepers. Their text could be guessed, and then decipherment of these (uninteresting) messages yielded the (highly interesting) daily ENIGMA keys.

So suppose that some of the intercepts in Table J.6 have been intercepted several times, say PICKWQ. The cryptanalyst now makes the assumption that it corresponds to one of the "popular" keys, say that the session key generating them is JJJ. Thus $A(\text{J}) = \text{P}$. From this simple assumption about just a single value, the whole permutations unravel by magic. (Instead of assuming this, one might also try all twenty-five possibilities for $A(\text{P})$.) Namely, we also have $A(\text{P}) = \text{J}$, by the involutory property of A, and hence $D(\text{P}) = DA^2(\text{P}) = DA(\text{J}) = \text{Z}$. We present the start of the unravelling in the following diagram.

$$
\begin{array}{c c c}
 & A & D \\
1 & \text{J} \rightarrow \text{P} & \\
2 & \text{P} \rightarrow \text{J} & \\
3 & \text{J} \longrightarrow & \text{Z} \\
4 & & \text{P} \rightarrow \text{Z} \\
5 & & \text{Z} \rightarrow \text{P} \\
\end{array}
$$

$$
\begin{array}{c c c}
 & A & D \\
6 & \text{N} \longrightarrow & \text{P} \\
7 & \text{N} \rightarrow \text{Z} & \\
8 & \text{Z} \rightarrow \text{N} & \\
9 & \text{Z} \longrightarrow & \text{W} \\
10 & & \text{N} \rightarrow \text{W}
\end{array}
$$

Lines 1 through 4 have been explained. Lines 5 and 8 follow from 4 and 7, respectively, because A and D are involutions. Lines 3, 6, and 9 refer to DA and are part of the characteristic, with lines 3 and 9 occurring in the second cycle of DA, and line 6 in the first one; see (J.3). Lines 7 and 10 are new; line 7 is deduced as above:

$$
A(\text{N}) = D^2 A(\text{N}) = D(DA(\text{N})) = D(\text{P}) = \text{Z}.
$$

This process can now be repeated, and concludes the second step of the cryptanalysis. The first four permutations are determined as:

$A = $ (AX)(BW)(CT)(DQ)(EO)(FU)(GM)(HV)(IK)(JP)(LY)(NZ)(RS)

$B = $ (AQ)(BE)(CR)(DM)(FP)(GV)(HK)(IJ)(LN)(OS)(TU)(WY)(XZ)

$C = $ (AH)(BP)(CJ)(DL)(EM)(FI)(GO)(KY)(NV)(QW)(RT)(SZ)(UX)

$D = $ (AS)(BE)(CR)(DY)(FG)(HT)(IM)(JK)(LV)(NW)(OX)(PZ)(QU)

As an example, we can verify the first entry of DA in (J.3):

$$
DA(\text{A}) = D(A(\text{A})) = D(\text{X}) = \text{O}.
$$

We now know the six permutations $A, \ldots, F$, and next see how we can determine the wiring of the entry rotor. In the notation (J.1) we have

(J.4) $$ A = P^{-1}N^{-1}M^{-1}L^{-1}RLMNP. $$

The rotor N turns by 1 after every key stroke, the rotor M after 26 strokes, and L after 26^2 strokes—as in an odometer. The positions where this happens are set by the daily key, in a random fashion. For the cryptanalysis, we assume that the first five key strokes do not provoke a movement of M, and therefore not of L, either. This happens for 21 out of the 26 possibilities, which is good enough for our purposes. We can therefore abbreviate the fixed permutation

(J.5) $$ Q = M^{-1}L^{-1}RLM, $$

and (J.4) simplifies to

(J.6) $$ A = P^{-1}N^{-1}QNP. $$

Except for the plug board, this corresponds to Figure J.2. After typing one letter, the entry rotor advances by 1. The new permutation B is easily described using the cyclic shift

$$S = (\text{ABCDEFGHIJKLMNOPQRSTUVWXYZ}).$$

Namely, after P comes S, then the old rotor N, then the downshift S^{-1}, and the similarly on the way back. That is,

(J.7) $$B = P^{-1}S^{-1}N^{-1}SQS^{-1}NSP.$$

In the same way, we have

(J.8)
$$C = P^{-1}S^{-2}N^{-1}S^2QS^{-2}NS^2P,$$
$$D = P^{-1}S^{-3}N^{-1}S^3QS^{-3}NS^3P.$$

There are two more equations, for E and F, which we do not need at the moment. We know the left-hand sides of the four equations (J.6) through (J.8), but it is not clear how to determine P, N, and Q efficiently from them.

Next comes a further non-mathematical tool in cryptanalysis, besides the cribs. Namely, old-fashioned espionage. The German spy Asché, mentioned in Section J.2, provided secret material to Colonel Bertrand of the French Secret Service. Among it were the daily ENIGMA keys for some period in 1932, including the plug board connections P. Enough intercepted messages were available from those days, and now P and S can be brought to the left-hand side of the equations for $A, B, C,$ and D:

(J.9)
$$
\begin{aligned}
U &= PAP^{-1} &= N^{-1}QN, \\
V &= SPBP^{-1}S^{-1} &= N^{-1}SQS^{-1}N, \\
W &= S^2PCP^{-1}S^{-2} &= N^{-1}S^2QS^{-2}N, \\
X &= S^3PDP^{-1}S^{-3} &= N^{-1}S^3QS^{-3}N.
\end{aligned}
$$

The cycle structure of an involution consists of some transpositions, like (ab), and fixed points, like (c); see Section 15.2. If six connecting cables are used for the plug board, then P is the product of six transpositions and $26 - 2 \cdot 6 = 14$ fixed points. The reflector permutation R is a product of 13 (disjoint) transpositions. Equations (J.4) through (J.8) imply that A, B, C, D and Q are conjugates of R, and hence also products of 13 transpositions.

Pretending that we knew Q, the four equations (J.9) are of the form

(J.10) $$\rho = N^{-1}\sigma N,$$

where ρ and σ are known and N is unknown. Our solution will be by enumerating all possibilities for N. How many are there? To start with the worst case, if $\rho = \sigma = \mathrm{id}$, then (J.10) collapses to nothing and all 26! permutations are possible for N. If ρ and σ are products of 13 disjoint transpositions, there are $2^{13} \cdot 13! = 51\,011\,754\,393\,600 \approx 5 \cdot 10^{13}$ possibilities for N (see Exercise J.4), far too many for our purposes. However, a simple trick both eliminates the unknown Q and cuts down substantially the number of possible N.

Namely, we multiply the equations in (J.9) together in sequence:

$$
\begin{aligned}
UV &= N^{-1}\ QSQS^{-1}N, \\
VW &= N^{-1}S\ QSQS^{-2}N, \\
WX &= N^{-1}S^2QSQS^{-3}N.
\end{aligned}
$$

We eliminate the unknown QSQ by plugging in UV and VW:

$$
\begin{aligned}
VW &= N^{-1}SNUVN^{-1}S^{-1}N &= (N^{-1}SN)(UV)(N^{-1}SN)^{-1}, \\
WX &= N^{-1}SNVWN^{-1}S^{-1}N &= (N^{-1}SN)(VW)(N^{-1}SN)^{-1}.
\end{aligned}
$$

Thus $VW = \tau UV \tau^{-1}$ is a conjugate of UV by $\tau = N^{-1}SN$, and we determine all $2^{13} = 8192$ possible conjugations τ. For each τ, we also check whether $WX = \tau VW \tau^{-1}$, and keep only those which pass the test. There are two further equations—which we did not write down—that can be used as tests.

Only a few values, often a single one, of τ will survive those tests. We solve $\tau = N^{-1}SN$ for N. Since S is a single cycle of length 26, τ is also of this form and there are then exactly 26 solutions. There will be a few, often just one, values of N that are a product of 13 disjoint transpositions. This is then the wiring of the right-hand ENIGMA rotor!

The German procedures for the ENIGMA included a random placement of the three (later five and eight) rotors into the three (later four) positions. Thus each rotor occurred reasonably often as the rightmost one, and they could all be broken by Rejewski's method.

In the course of the war, several ENIGMA machines were captured by the British (and by the Russians at Stalingrad), and their rotor wiring was no secret anymore. But Rejewski's early break into the rotors was an important link in the chain leading to Ultra.

Notes J.1. The rotor movement differs slightly from that of an odometer because of *double-stepping*: when the middle and left rotor move forward simultaneously, the middle rotor also advances at the next key stroke.

There are $n!/(2^s \cdot (n-2s)! \cdot s!)$ ways of choosing s steckered pairs out of n letters. This number times $s!$ is the multinomial coefficient $\binom{n}{2 \cdots 2}$ which counts ordered sequences

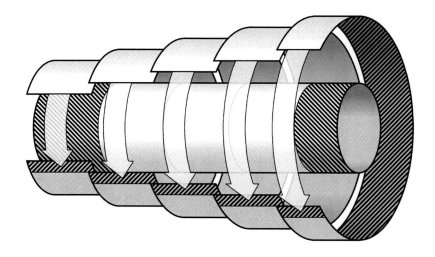

Figure J.7: Assembly of a paper ENIGMA.

of s mutually disjoint pairs. Using 12 plugged pairs decreases the key space by about 30%. A complete calculation, taking into account all possibilities for the inner wiring of the rotors and leading to $2 \cdot 10^{145}$ possible keys, is in Miller (2001). Although the rotor positions were determined by the ring positions and the secret session keys, with 26^6 possibilities, only the 26^3 possible positions at the start of encryption contribute to the key space.

J.2. Turing's work is related to ideas of the Austrian logician Kurt Gödel (1906–1978), namely his incompleteness result for axioms of number theory, and his *gödelization*, a numbering of all possible statements. Turing's famous *Halting Problem* takes as input a string which represents a program in any reasonable programming language, and as output you want to know whether it does not go on working forever (with all variables initially set to zero, say). Turing's undecidability result about this problem is devastating. It says that there exists no algorithmic method that can answer this question correctly. None at all! Not because programmers are stupid, but because it is inherently impossible! It somewhat resembles Heisenberg's *uncertainty principle*, which says that some reasonably posed problems in physics have no solution. After the war, he devised the *Turing test* of artificial intelligence: can you tell whether you are interacting with a human or a machine? If you cannot, then you are interacting with artificial intelligence. Half a century later, this remains an unfulfilled hope (or despair, depending on your outlook). Our distinguishers, for example between pseudorandom and truly random generators in Chapter 11, apply the same principle in a different setting.

The Polish bomba resembled a birthday icecream cake which goes by that name in Polish. It seems that German cryptanalysts developed an electro-mechanical machine for breaking the US M-209 (Figure A.17) along similar principles. See TICOM (1948), and

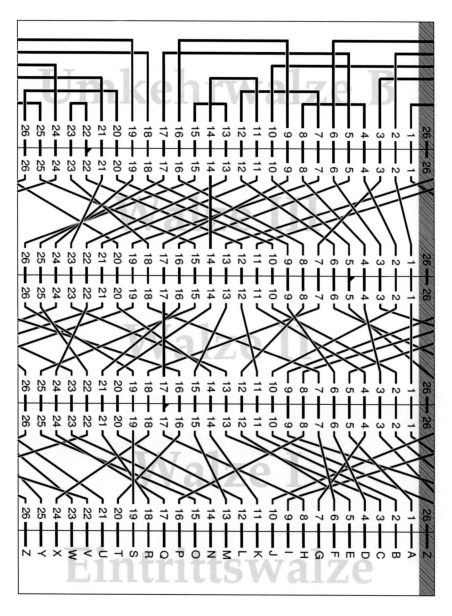

Figure J.8: Five rotors of the paper ENIGMA.

Schmeh (2004) for a presentation of one of the main inventors.

Schmidt offers services: Bertrand (1973), Chapter II *Une histoire merveilleuse*, page 23. Schmidt arrested: Chapter IX *Paris, Quelques nouvelles des anciens «amis»!*, page 252. The story of Bletchley Park remained a widely unknown secret until Winterbotham (1974). The earlier books by Kozaczuk (1967) in Polish and by Bertrand (1973) in French were not accessible to English-only readers. The official British history by Hinsley *et al.* (1984) largely neglected to mention the Polish contribution; see Rejewski (1982b)'s corrections.

The family of Hugh Sebag-Montefiore—who wrote a book about the ENIGMA—owned the property at Bletchley Park before it was sold to the government in the late 1930s.

After his work in Bletchley Park, William (Bill) Thomas Tutte (1917–2002) became a well-known researcher in graph theory and on matroids, inventing, among other things, the *Tutte graph polynomials*. He worked at the University of Waterloo in Canada. At Bletchley Park, the Lorenz SZ and its traffic were called *Tunny*, and the Siemens Geheimschreiber *Sturgeon*. The 1943 COLOSSUS model was replaced on 1 June 1944, just before D-day, by the 2500-valve COLOSSUS MARK 1. Its main purpose was to decipher radio traffic between the Berlin headquarters and German armies in Greece, North Africa, and Russia.

J.3. This section is largely based on Rejewski (1980, 1981, 1982a). Jerzy Rózycki and Henryk Zygalski worked with Rejewski on the ENIGMA.

In the intercepts of Table J.6, the rotor sequence is I, III, II (so that the wiring of rotor II is eventually cryptanalyzed), the basic setting (*Grundstellung*) is ALV (= 01 12 22), and the ring setting (*Ringstellung*) 16 11 13. There are no plug board connections, and the settings are not known to the cryptanalyst.

Exercises.

EXERCISE J.1 (Paper ENIGMA). *You are to build a paper model of the* ENIGMA *machine, using Jérémie Detrey's design in Figure J.7 and Figure J.8.*

   (i) *Find a stiff tube with a diameter of 34 mm. This could be a plumbing tube from your local hardware store, or even the interior tube of a kitchen paper roll. If your roll has a different size, you need to scale Figure J.8 appropriately.*

  (ii) *Copy Figure J.8 onto fairly stiff yet flexible paper. Scale it so that the width of the white (unshaded) area exactly equals the circumference of your tube.*

 (iii) *Cut your copy into five horizontal strips, labelled as* Umkehrwalze, Walze III, Walze II, Walze I, Eintrittswalze *(reflector, rotors III, II, I, entry rotor). Glue the shaded areas in Figure J.7 together as indicated. Fix the two outer strips to the roll; they must be aligned with each other. You should start with the* Umkehrwalze. *Then add* Walze III, Walze II, Walze I *making sure that they stay movable. Finally, add the* Eintrittswalze *aligned with the* Umkehrwalze.

EXERCISE J.2 (Number of intercepts). *The goal of this exercise is to see how many intercepts are necessary to determine the characteristic as the first step in Section J.3. Are the 42 intercepts of Table J.6 sufficient, if random message keys are chosen by the* ENIGMA *operators?*

(i) Prove the following formulas for any integer t and real number a with $|a| < 1$:

$$\sum_{1 \le k < t} k a^k = \frac{a - a^t(t - (t-1)a)}{(a-1)^2},$$

$$\sum_{k \ge 1} k a^k = \frac{a}{(a-1)^2},$$

$$\sum_{k \ge t} k a^k = \frac{a^t(t - (t-1)a)}{(a-1)^2}.$$

(ii) Suppose you toss a fair coin and stop as soon as both heads and tails have come up. Show that the expected number of tosses equals 3.

(iii) Suppose you choose uniformly at random one letter after another out of t letters (with replacement), and stop as soon as each of the t letters has been chosen at least once. Show that the expected number of choices is

$$t^{2-t} \sum_{0 \le j < t} (-1)^j \binom{t-1}{j} \frac{tj + 2t - j - 1}{(j+1)^2} (t - j - 1)^{t-1}.$$

This evaluates to about 100.2 for $t = 26$. [Hint: you should look up the Stirling numbers of the second kind in some book on combinatorics.]

(iv) In Rejewski's attack for reconstructing the ENIGMA rotor wiring, the German operators chose three letters as the message key (out of an alphabet of 26 letters). The Polish attack could be started when each letter had occurred somewhere in the first position, somewhere in the second position, and somewhere in the third position. What is the expected number of intercepted messages needed until this happens? Assume that the three letters were chosen uniformly at random in each case—in actual fact, the operators did not do this, which helped the cryptanalysts in another step of their work.

EXERCISE J.3 (Conjugates).   *Show that the relation of being conjugate is an equivalence relation, that is, it is reflexive, symmetric, and transitive. If we write $\sim$ for it, then you must prove that $\rho \sim \rho$, $\rho \sim \sigma \Rightarrow \sigma \sim \rho$, and $(\rho \sim \sigma$ and $\sigma \sim \tau \Rightarrow \rho \sim \tau)$ hold for all ρ, σ, τ.*

EXERCISE J.4 (Number of conjugates).   *Let ρ and σ be conjugate permutations, both with ℓ_i cycles of length i for $i \ge 1$, so that $\sum_{i \ge 1} i \ell_i$ equals the alphabet size.*

(i) Prove that there are exactly $\prod_{i \ge 1} i^{\ell_i} \ell_i!$ many τ with $\rho = \tau \sigma \tau^{-1}$.

(ii) We take some i with $\ell_i \ge 1$ and break one of the cycles of length i into two cycles of positive lengths j and k. Prove that the number of possible τ's for this new cycle structure equals

(J.11)
$$\frac{jk(\ell_j + 1)(\ell_k + 1)}{i \ell_i}$$

times the old number if $j \ne k$, and $i(\ell_j + 1)(\ell_j + 2)/4\ell_i$ times if $j = k$. Is the factor in (J.11) always smaller than 1, or always greater than 1?

EXERCISE J.5 (Rotor wiring). *You are to investigate the unravelling of the rotor wiring in general. In the example in the text, we have the product permutation DA and two values $A(\mathtt{J})$ and $D(\mathtt{J})$.*

(i) *Continue the unravelling of A and D for some more steps, so as to determine six more values each of A and D. Compare with the values on page 730.*

(ii) *Continuing to assume that message key $\mathtt{JJJ}$ gives $\mathtt{PICKWQ}$, determine six values each of B and E. Compare with the values on page 730.*

(iii) *Quite generally, let $\sigma, \tau \in \mathrm{Sym}_{\mathbb{A}}$ be involutions on an alphabet $\mathbb{A}$, and $\sigma(a_0) = b_0$. Suppose that a_0 and b_0 are in different cycles of the cycle presentation*

$$\tau\sigma = (a_0\, a_1 \ldots a_{m-2}\, a_{m-1})(b_0\, b_1 \ldots b_{n-2}\, b_{n-1}) \cdots$$

of $\tau\sigma$. Prove that $n = m$ and $\sigma(a_i) = \tau(a_{i+1}) = b_{m-i}$ for $0 \le i < m$, with index arithmetic modulo m. In other words, σ maps the first cycle in sequence to the reverse of the second cycle, and τ does the same with an offset of 1.

(iv) *Analyze similarly the case where $\sigma(a_0) = a_j$ for some j with $0 \le j < m$. Conclude that then at least one of the two involutions has a fixed point, that is, there exists some x with $\sigma(x) = x$ or $\tau(x) = x$.*

(v) *Conclude that the cycles of length at least 3 in the cycle representation occur in pairs of equal length, when σ and τ have no fixed points.*

EXERCISE J.6 (Rotor cryptanalysis). *You are to trace the cryptanalysis of the ENIGMA rotor, as described in the text.*

(i) *Verify the determination of DA, EB, and FC from the values in Table J.6.*

(ii) *Verify the determination of $A, \ldots, F$ under the assumption that the message key for message $\mathtt{PICKWQ}$ is $\mathtt{JJJ}$.*

EXERCISE J.7 (Optimal rotor sizes). *Design a reasonably efficient algorithm to solve the maximal Chinese modulus problem, in the following sense: you are given a bound B and a number n, and have to produce pairwise coprime integers $a_1, \ldots, a_n$ with $1 \le a_i \le B$ for all i, so that the product $a_1 \cdots a_n$ is maximal.*

(i) *Consider the weighted undirected graph $G = (V, E)$ with vertex set $V = \{1, 2, \ldots, B\}$, edge set $\{\{i, j\}: i, j \in V, i \ne j, \gcd(i, j) > 1\}$, and vertex weight $w(i) = \log_2(i)$ for $i \in V$. Show that the task is equivalent to finding a set of n independent vertices in G with maximum weight.*

(ii) *Finding a maximum independent set in an undirected graph is NP-complete (Garey & Johnson (1979), Section 3.1.3). Is the problem in (i) also NP-complete, say with integer weights?*

(iii) *The German cipher machine Lorenz SZ 40 (see Section J.2) used twelve rotors with 23, 26, 29, 31, 37, 41, 43, 47, 51, 53, 59, and 61 notches. By which factor can you improve the resulting period (= product of the rotor sizes) of about $1.6 \cdot 10^{19}$ without increasing the maximal number of notches? The rotor sizes must be pairwise coprime.*

(iv) *Same for the Siemens Geheimschreiber ("Sturgeon") with ten rotors of lengths $47, 53, 59, 61, 64, 65, 67, 69, 71$, and 73.*

(v) *Same for the Hagelin C-36 (= US Army M-209, Figure A.17) with six rotors of lengths $17, 19, 21, 23, 25,$ and 26.*

Obgenandter Plato zu einer zeit gefragt ward, wodurch
ein Mensch andere Thier vbertreff? geantwort hat,
Daß er rechnen kan vnd verstandt der zale hab.[1]

ADAM RISEN (1574)

Wherein the desire an hope of gain, maketh many willying
to sustaine some trauell. For aide of whom, I did sette forth
the first parte of *Arithmetike*. But if thei knewe how farre
this seconde parte, dooeth excell the firste parte,
thei would not accoumpte any tyme loste,
that were emploied in it. Yea thei would not thinke
any tyme well bestowed, till thei had gotten soche habilitie by it,
that it might be their aide in al other studies.

ROBERT RECORDE (1557)

Dixit Alchoarizmi: [...] Hec sunt igitur universa,
que necessaria sunt hominibus ex divisione et multiplicatione
in integro numero et in ceteris, que secuntur.
His peractis incipiemus narrare multiplicationem fractionum
et divisionem earum sive radices, si deus voluerit.[2]

أَلْخَوَارِزمِي (al-Ḥwārizmī, c. 830)

Smautf aurait aimé être ce bon élève récompensé. Son regret de
n'avoir pas fait d'études s'est avec les années transformé en une
passion maladive pour les quatre opérations [arithmétiques].[3]

GEORGES PEREC (1987)

---

[1]The aforementioned Plato was asked at one time in which way a human is superior to other animals. He answered, by the fact that he can calculate and has an understanding of numbers.

[2]Thus spoke Al-Khwarizmi: [...] So this is everything that is necessary for men concerning the division and multiplication with an integer, and the other things that are connected with it. Having completed this, we now begin to discuss the multiplication of fractions and their division, and the extraction of roots, if God so wills.

[3]Smautf would have liked to be this good student, well rewarded. His regret at having had no schooling has turned over the years into an unhealthy passion for addition, subtraction, multiplication, and division.

Chapter 15

The computer algebra toolbox

The meeting place of algebra and computer science is called *computer algebra*. We refer to von zur Gathen & Gerhard (2013) for a comprehensive introduction to it.

The tools in this chapter are fundamental not only for two core systems of cryptography, AES and RSA, but also for several others throughout this text. They fall into three categories:

- algorithms used to implement cryptographic procedures,

- facts from algebra used to show correctness of systems,

- algorithmic questions related to the security of some systems.

This chapter actually only deals with the first two aspects. The first sections deal with the following subjects:

- addition and multiplication (Section 15.1),

- division with remainder (Section 15.3),

- the Extended Euclidean Algorithm (Section 15.4),

- modular inverses (Section 15.5).

These are useful to understand AES. For the discussion of RSA, some more material is needed:

- Chinese Remainder Algorithm (Section 15.11),

- efficient exponentiation (Section 15.12),

- cost of arithmetic (Section 15.6).

Fundamental properties used in group-based cryptography are:

© Springer-Verlag Berlin Heidelberg 2015
J. von zur Gathen, *CryptoSchool*, DOI 10.1007/978-3-662-48425-8_25

○ theorems of Fermat, Euler, and Lagrange (Section 15.13).

Three further sections deal with the background to more specialized questions:

○ uniqueness of rational approximation (Section 15.9), for the Wiener attack on RSA (Section 3.6),

○ polynomial interpolation (Section 15.10), introducing Chinese remaindering, and used in secret sharing and many other places,

○ squares and the Jacobi symbol (Section 15.14), for the Blum-Blum-Shub pseudorandom generator (Section 11.8) and the Hofheinz-Kiltz-Shoup cryptosystem (Section 9.9).

Section 15.6 briefly discusses complexity theory, and the final two sections provide material on linear algebra (Section 15.15) and finite probability spaces (Section 15.16); the latter is used throughout the text.

A concept that permeates this chapter is the similarity

$$\text{integers} \longleftrightarrow \text{polynomials over a finite field.}$$

This is particularly strong if we consider

$$\text{binary representation of integers} \longleftrightarrow \mathbb{Z}_2[x].$$

This similarity carries through algorithmically for

○ modular arithmetic,

○ gcd, modular inverses,

but *not* for factorization which is easy for polynomials but presumably difficult for integers.

Much of the material of this chapter is taught in first and second year of computer science or mathematics at most universities. If you have the necessary background, you may well leave out the corresponding sections. If you are familiar with all these methods and results, you can skip the whole chapter. Congratulations!

For the experimental exercises in this text, a computer algebra system is useful, such as the open-source Sage at http://sagemath.org/ (see Nguyen 2009 and McAndrew 2011) or the educational software CrypTool at http://www.cryptool.org/en/.

15.1. Addition and multiplication

An integer a is represented in some integer base (or radix) $r \geq 2$ as follows:

$$a = \pm \sum_{0 \leq i < n} a_i r^i,$$

for some integer n and some integers a_i, where $0 \leq a_i < r$. Two representations are widely used: decimal ($r = 10$) and binary ($r = 2$). If we have $a = 155$ in decimal, then $a = 10011011$ in binary. This integer is a 3-digit and an 8-bit number.

In AES (Chapter 2), we also use the *hexadecimal* representation in base $r = 16$, where $\mathsf{A}, \mathsf{B}, \ldots, \mathsf{F}$ stand for the digits $10, 11, \ldots, 15$, and the decimal 155 is represented as 9B.

To add two positive numbers, we add them bit by bit (or digit by digit), starting at the low-order bit (or digit). At each step, a carry is produced, either 1 or 0. It is added together with the two input bits (or digits) at the next step. To multiply two such numbers $a = \sum_{0 \leq i < n} a_i r^i$ and $b = \sum_{0 \leq j < m} b_j r^j$, we can directly implement the formula $a \cdot b = \sum_{0 \leq i,j < n+m} a_i \cdot b_j \cdot r^{i+j}$. This is the *classical method* that you learnt—in its decimal version—in primary school. The example (2.3) reads

$$a = 155, \quad b = 205, \quad a + b = 360 \quad \text{in decimal},$$
$$(15.1) \quad a = 10011011, \quad b = 11001101, \quad a + b = 101101000 \quad \text{in binary}.$$

Hopefully, the result is the same for both representations. The multiplication example becomes:

$$
\begin{aligned}
a \cdot b \quad &= \quad 31775 & \text{in decimal}, \\
&= \quad 111110000011111 & \text{in binary}.
\end{aligned}
$$

The context will usually make it clear whether the string 11 means *eleven* or *three*.

The situation is even more confusing in AES, where an expression can be interpreted in three different environments. To avoid confusion, we often associate a "type" to each such expression by saying "in R", where R is the domain under consideration.

An important domain is the binary field $\mathbb{Z}_2 = \{0, 1\}$ of two elements. Addition and multiplication are as usual, except that the addition satisfies $1 + 1 = 0$. If we think of Boolean values $0 = \textit{false}$ and $1 = \textit{true}$, then addition is the *exclusive or* (XOR) and multiplication the *and*. In some chapters, we write $\mathbb{B}$ for $\mathbb{Z}_2$. Several general properties of such modular rings are explained in the next section.

A bit string like a in (15.1) can also be interpreted as a sequence of eight elements of $\mathbb{Z}_2$, that is, as an element of $\mathbb{Z}_2^8$. Under this interpretation, addition and multiplication are performed bitwise:

$$a + b = 01010110, \; a \cdot b = 10001001 \text{ in } \mathbb{Z}_2^8.$$

A third interpretation of strings of integers or bits is to consider them as polynomials over the set $\mathbb{Z}$ of integers. Under this interpretation we have

$$a = x^7 + x^4 + x^3 + x + 1, \quad b = x^7 + x^6 + x^3 + x^2 + 1,$$

$$a + b = 2x^7 + x^6 + x^4 + 2x^3 + x^2 + x + 2,$$

$$a \cdot b = x^{14} + x^{13} + x^{11} + 3x^{10} + 2x^9 + x^8 + 4x^7 + 3x^6 + x^5 + 2x^4 + 3x^3 + x^2 + x + 1.$$

Yet another view is to consider a and b as elements of $\mathbb{Z}_2[x]$, that is, as polynomials in the indeterminate x over $\mathbb{Z}_2$. Then their sum and product in this new interpretation are

$$a + b = x^6 + x^4 + x^2 + x,$$

$$a \cdot b = x^{14} + x^{13} + x^{11} + x^{10} + x^8 + x^6 + x^5 + x^3 + x^2 + x + 1 \text{ in } \mathbb{Z}_2[x].$$

It may be confusing to see four different interpretations of bit strings, with "multiplication" having a different meaning (and outcome) each time. However, the richness of structure is one of the things that makes them attractive for cryptography. Indeed, in AES the bit strings of data are interpreted in various domains. Thus we have to handle computations in several environments: the solutions will be similar, and it is convenient to have a common notion that encompasses all scenarios. The reader may usually think of R as being either the ring of integers or of polynomials over a field.

DEFINITION 15.2. *A ring is a nonempty set R with two operations (addition and multiplication) $+$ and $\cdot$ mapping $R \times R$ to R and satisfying the following for all $a, b, c \in R$.*

(i) *Both operations are associative:*

$$(a + b) + c = a + (b + c), \quad (a \cdot b) \cdot c = a \cdot (b \cdot c),$$

(ii) *there is an additive inverse $-a$:*

$$a + (-a) = (-a) + a = 0,$$

(iii) *both operations have an identity element 0 and 1, respectively:*

$$a + 0 = 0 + a = a, \quad a \cdot 1 = 1 \cdot a = a,$$

(iv) *addition is commutative: $a + b = b + a$,*

(v) *distributivity holds:*

$$(a + b) \cdot c = a \cdot c + b \cdot c,$$
$$a \cdot (b + c) = a \cdot b + a \cdot c.$$

(vi) *If also multiplication is commutative, so that*

$$a \cdot b = b \cdot a,$$

then R is a commutative ring. In this text, rings are usually commutative.

The examples of central interest are the familiar ring $\mathbb{Z}$ of integers, the two-element ring $\mathbb{Z}_2 = \{0, 1\}$, the RSA ring $\mathbb{Z}_N$ of integers modulo N, and the ring of polynomials $S[x]$, where S is itself a commutative ring. A polynomial $f \in S[x]$ is of the form

$$f = f_n x^n + f_{n-1} x^{n-1} + \cdots + f_0,$$

where $n \geq 0$ is an integer, $f_n, f_{n-1}, \ldots, f_0 \in S$, and x is an indeterminate. Rings like $\mathbb{Z}_2[x]$ play a major role in the AES system. Considering just $+$ on R, R is a group (Definition 15.51).

Both $\mathbb{Z}$ and $S[x]$ are infinite rings. In cryptography, however, we will want to use finite rings. The reduction from infinite to finite is provided by the modular arithmetic of the next section. Finally, we set up some basic terminology which we use throughout the text.

DEFINITION 15.3. *Let R be a commutative ring.*

(i) *For $a, b \in R$, a divides b if $b = ac$ for some $c \in R$.*

(ii) *An element $u \in R$ is a unit (or invertible) if there exists an element $v \in R$ such that $u \cdot v = 1$. We denote by $R^\times \subseteq R$ the set of units in R. It is a group using the multiplication in R.*

(iii) *Two elements a, $b \in R$ are associates if there exists a unit $u \in R$ such that $a = u \cdot b$. We then write $a \sim b$.*

(iv) *A nonzero element $u \in R$ is a zero divisor if there exists a nonzero element $v \in R$ such that $uv = 0$. Note that 0 is not a zero divisor.*

(v) *If $R \neq \{0\}$, it is an integral domain if it has no zero divisors. It is a field if every nonzero element is a unit.*

(vi) *For a positive integer n, the sum $n \cdot 1 = 1 + 1 + \cdots + 1$ of n terms is an element of R, often written as $n \in R$. If these are all nonzero, we define the characteristic $\operatorname{char}(R)$ of R as $\operatorname{char}(R) = 0$. Otherwise, $\operatorname{char}(R)$ is the smallest positive n for which $n \cdot 1 = 0$.*

For $R = \mathbb{Z}$, we have $R^{\times} = \{1, -1\}$, $a \sim b$ if and only if $b \in \{a, -a\}$, and no zero divisor, so that $\mathbb{Z}$ is an integral domain. For an element a of any ring R, we abbreviate

$$\pm a = \{a, -a\} \subseteq R.$$

The rings $\mathbb{Z}$, $\mathbb{Q}$, and $\mathbb{R}$ have characteristic 0, and $\operatorname{char}(\mathbb{Z}_p) = p$. For any field F, $\operatorname{char}(F)$ is either 0 or a prime number. For two elements a and b in a ring of prime characteristic p, we have

$$(a + b)^p = a^p + b^p,$$

since in the expansion of the left hand side, all binomial coefficients, except the first and last ones, are divisible by p and hence 0. It follows that for any positive integer i, we have

(15.4) $$(a + b)^{p^i} = a^{p^i} + b^{p^i}.$$

For a finite field $\mathbb{F}_q$ with q elements there is a positive integer n with $q = p^n$, where $p = \operatorname{char}(\mathbb{F}_q)$ is prime. Conversely, for every such q there is a field with q elements.

15.2. Homomorphisms and permutations

An arbitrary mapping $f \colon A \longrightarrow B$ between two sets A and B is

- *injective* (or *one-to-one*, or an *injection*) if $f(a) \neq f(b)$ for all $a, b \in A$ with $a \neq b$,

- *surjective* (or *onto*, or a *surjection*) if for all $b \in B$ there is some $a \in A$ with $f(a) = b$,

- *bijective* (or a *bijection*) if it is injective and surjective.

When A and B are finite with the same number of elements, then being one of injective or surjective implies the other property, and also bijectivity. A bijection from A to A is a *permutation*.

If A and B are equipped with some algebraic structure, then f is a *homomorphism* if it respects this structure. Say, we have additive groups A and B, then $f(0) = 0$ and $f(a + b) = f(a) + f(b)$ for all $a, b \in A$

is required. This leads to notions like a group, ring, or vector space homomorphism. It is an *isomorphism* if it is also bijective.

Some terminology relating to permutations of an alphabet (finite set) A is useful for cryptanalysis. Two such permutations ρ and σ can be composed, so that if $\rho(a) = b$ and $\sigma(b) = c$, then $(\sigma \circ \rho)(a) = \sigma(\rho(a)) = \sigma(b) = c$, for $a, b, c, \in A$. This provides the structure of a group (Definition 15.51) on the set of permutations, called the symmetric group Sym_A of A. If A has s elements, then Sym_A consists of $s! = 1 \cdot 2 \cdot 3 \cdots (s-1) \cdot s$ elements. $\mathrm{Sym}_{\{1,\dots,n\}}$ is also denoted as Sym_n.

We often write $\sigma\rho$ for $\sigma \circ \rho$. The inverse ρ^{-1} of ρ is again a permutation, with $\rho^{-1}(b) = a$ if and only if $\rho(a) = b$. If ρ and ρ^{-1} happen to coincide, so that $\rho(a) = \rho^{-1}(a)$ for all letters $a \in A$, then $\rho^2(a) = \rho\rho(a) = \rho\rho^{-1}(a) = a$ and hence $\rho^2 = \mathrm{id}$ is the identical permutation, which maps each letter to itself. Then ρ is called an *involution*.

We have $(\sigma\rho)^{-1} = \rho^{-1}\sigma^{-1}$, since

$$\rho^{-1}\sigma^{-1}(\sigma\rho) = \rho^{-1}\sigma^{-1}\sigma\rho = \rho^{-1}\rho = \mathrm{id};$$

we have used the associativity, and the uniqueness of an inverse. In other words, the inverse of a product is the product of the inverses, but in the inverse order. There are two useful data structures to represent a permutation σ. Taking the (fictitious) ENIGMA rotor III from Section J.1, the first is a *table of values* of σ:

(15.5)

a	Q	W	E	R	T	Z
$\sigma(a)$	E	Z	R	Q	W	T

The second one is the *cycle presentation*. It is obtained by taking the first letter, which is E in the example, then $\sigma(\mathrm{E})$, then $\sigma^2(\mathrm{E})$, and so on until we come back to E: $\sigma^i(\mathrm{E}) = \mathrm{E}$. These values form the first cycle $(\mathrm{E}\,\sigma(\mathrm{E})\,\sigma^2(\mathrm{E}))$ with $\sigma^3(\mathrm{E}) = \mathrm{E}$, which is (ERQ) in (15.5). Then the first letter not occurring here is taken, say T, and the cycle generated by T is formed. This is continued until all elements are exhausted.

We usually write the cycles in order of decreasing length, and order those of the same length alphabetically by the smallest element occurring in them. Thus σ as in (15.5) has the cycle presentation

(15.6)   (ERQ) (TWZ)

The *cycle structure* of σ is the sequence of cycle lengths in this presentation, $(3,3)$ in the example. The cycle $\sigma = (\mathrm{ABCD} \cdots \mathrm{XYZ})$ is the cyclic shift (or Augustus cipher) on $A = \{\mathrm{A, B}, \cdots, \mathrm{Z}\}$, and the Caesar system is $\mathrm{enc}_{\mathrm{Caesar}} = \sigma^3 = \sigma \circ \sigma \circ \sigma$.

A cycle (a) of length 1 is a *fixed point* of σ. It is usually not written in the cyclic presentation. A cycle (ab) of length 2 is called a *transposition* and has the special property that $(ab)^2 = \mathrm{id}$. More generally, the permutations σ with cycle structure $(2, \ldots, 2, 1, \ldots, 1)$, so that only transpositions and fixed points occur, are precisely those with the property that $\sigma^2 = \mathrm{id}$, that is, the involutions σ.

If ρ, σ, and $\tau \in \mathrm{Sym}_A$ satisfy $\rho = \tau \sigma \tau^{-1}$, then ρ is a *conjugate* of σ. Suppose that $\sigma(a) = b$. Then

$$\rho(\tau(a)) = \rho\tau(a) = \tau\sigma\tau^{-1}\tau(a) = \tau\sigma(a) = \tau(b).$$

Thus if $(x_1, x_2, \ldots, x_k)$ is a cycle of σ, so that $\sigma(x_i) = x_{i+1}$ for all i, including $\sigma(x_k) = x_1$, then $(\tau(x_1), \tau(x_2), \ldots, \tau(x_k))$ is a cycle of ρ. In particular, ρ and σ have the same cycle structure, and a conjugate of an involution is again an involution.

15.3. Division with remainder

In the ring $\mathbb{Z}$ of integers, we can in general not divide one element by another, but we can perform division with remainder: given integers a and b with $b \neq 0$ we can find integers q, r such that

$$(15.7) \qquad\qquad a = qb + r \text{ and } 0 \leq r < |b|.$$

These quantities are the *remainder* $a \operatorname{rem} b = r$ and the *quotient* $a \operatorname{quo} b = q$, where $q = \lfloor a/b \rfloor$ if $b > 0$. Division of 126 by 35 yields $126 = 3 \cdot 35 + 21$, so that $21 = 126 \operatorname{rem} 35$ and $3 = 126 \operatorname{quo} 35$.

As a further example, we have

$$
\begin{array}{r}
9797 : 123 = 79 \\
-8610 \\
\hline
1187 \\
-1107 \\
\hline
80
\end{array}
$$

Thus $9797 = 79 \cdot 123 + 80$, $9797 \operatorname{quo} 123 = \lfloor 9797/123 \rfloor = 79$, and $9797 \operatorname{rem} 123 = 80$. With the usual addition and multiplication, followed by taking the remainder modulo b, we obtain the ring $\mathbb{Z}_b$ of integers modulo b, which we identify with $\mathbb{Z}_b = \{0, 1, \ldots, b-1\}$. In the example, we have $97 \cdot 101 = 80$ in $\mathbb{Z}_{123}$. If (15.7) holds, we write $a = r$ in $\mathbb{Z}_b$.

The same method works for polynomials, where we take $|b| = \deg b$ for a nonzero polynomial b, and $|0| = -\infty$. Generally, we are given $a, b \in R[x]$, with R a ring and b nonzero, and want to find $q, r \in R[x]$ so that

$$(15.8) \qquad\qquad a = qb + r \text{ and } \deg r < \deg b.$$

A first problem is that such q and r do not always exist: it is impossible to divide x^2 by $2x+1$ with remainder in $\mathbb{Z}[x]$! Namely, the leading coefficient of any $q \cdot (2x + 1)$ is even, but that of x^2 is 1. However, if the leading coefficient $\operatorname{lc}(b)$ of b is invertible in R, then it has an inverse $v \in R$ with $\operatorname{lc}(b)v = 1$. For $R = \mathbb{Z}$, that still only allows 1 or -1 as leading coefficient, but when R is a field, division with remainder by any nonzero polynomial is possible.

Choosing $a = 3x^4 + 2x^3 + x + 5$ and the *monic* (leading coefficient 1) $b = x^2 + 2x + 3$ in $\mathbb{Z}[x]$, we have

$$
(15.9) \qquad
\begin{array}{l}
(3x^4+2x^3 \qquad\quad +x+5) : (x^2 + 2x + 3) = 3x^2 - 4x - 1. \\
\underline{-3x^4-6x^3-9x^2} \\
\qquad -4x^3-9x^2 \quad +x\ +5 \\
\qquad \underline{+4x^3+8x^2+12x} \\
\qquad\qquad -x^2+13x\ +5 \\
\qquad\qquad \underline{+x^2\ +2x\ +3} \\
\qquad\qquad\qquad 15x\ +8
\end{array}
$$

Thus $a \operatorname{quo} b = 3x^2 - 4x - 1$ and $a \operatorname{rem} b = 15x + 8$. As noted above, division by b is possible since b is monic. We now work in the ring $R[x]/(b)$ of polynomials modulo b, namely $\mathbb{Z}[x]/(x^2 + 2x + 3)$. The same approach works for ground rings other than $\mathbb{Z}$. The case $\mathbb{Z}_2[x]$ is extensively used in AES, among others, with $m = x^8 + x^4 + x^3 + x + 1$ (see Section 2.2). Then for a, b as in (2.6) and (15.1), we have

$$a \cdot b = x^{14} + x^{13} + x^{11} + x^{10} + x^8 + x^6 + x^5 + x^3 + x^2 + x + 1$$
$$= (x^6 + x^5 + x^3) \cdot m + x^4 + x^3 + x^2 + x + 1 \text{ in } \mathbb{Z}_2[x],$$

$$a \cdot b = x^4 + x^3 + x^2 + x + 1 \text{ in } \mathbb{Z}_2[x]/(m).$$

For integers $a, b \in \mathbb{Z}$, the following three properties are equivalent:

$$
\begin{array}{l}
a = b \text{ in } \mathbb{Z}_N, \\
(15.10) \qquad N \text{ divides } a - b, \\
a \operatorname{rem} N = b \operatorname{rem} N \text{ in } \mathbb{Z}.
\end{array}
$$

The mapping $\mathbb{Z} \to \mathbb{Z}_N$ with $a \mapsto$ "a in $\mathbb{Z}_N$" preserves the ring operations, namely $(a \text{ in } \mathbb{Z}_N) + (b \text{ in } \mathbb{Z}_N) = (a + b) \text{ in } \mathbb{Z}_N$, and similarly for $\cdot$. It is therefore a ring homomorphism; see Section 15.2. As an illustration, we have in the Caesar cipher (see Section A.3) $\operatorname{enc}_{caesar}(x) = x + 3$ in $\mathbb{Z}_{26} = \{0, \ldots, 25\}$ and $\operatorname{dec}_{caesar}(y) = y - 3$ in $\mathbb{Z}_{26}$. In particular, $\operatorname{enc}_{caesar}(24) = 1$ and $\operatorname{dec}_{caesar}(1) = 24$.

Again, the same works for polynomials over a field F. Let $m \in F[x]$ be a nonzero polynomial. The elements of the residue class ring $F[x]/(m)$

are the polynomials in $F[x]$ of degree less than $\deg m$. The sum of two elements is the sum in $F[x]$, while the product is the product in $F[x]$ reduced modulo m. We have the following equivalences for any $a, b \in F[x]$:

$$a = b \text{ in } F[x]/(m),$$
(15.11) $\qquad m \text{ divides } a - b,$
$$a \text{ rem } m = b \text{ rem } m \text{ in } F[x].$$

We usually identify $\mathbb{Z}_N$ with $\{0, 1, \ldots, N{-}1\}$ as above. But it is sometimes convenient to use the *symmetric system of representatives* modulo N

(15.12) $\qquad R_N = \{-\lfloor N/2 \rfloor, \ldots, \lfloor (N - 1)/2 \rfloor\}$

and to denote as $u \text{ srem } N \in R_N$ the representative of $u \in \mathbb{Z}$.

15.4. The Extended Euclidean Algorithm

The Euclidean Algorithm for the two integers 126 and 35 works as follows:

(15.13)
$$
\begin{aligned}
126 &= 3 \cdot 35 + 21, \\
35 &= 1 \cdot 21 + 14, \\
21 &= 1 \cdot 14 + 7, \\
14 &= 2 \cdot 7,
\end{aligned}
$$

and the last nonzero remainder, namely 7, is the greatest common divisor of 126 and 35. One of the most important applications is for exact arithmetic on rational numbers, where one wants to simplify $35/126$ to $5/18$ in order to keep the numbers small.

This algorithm can also be adapted to work for polynomials. An important application is the computation of modular inverses, discussed in Section 15.5. The integer version plays a central role in the RSA system (Section 2.5) and group cryptography (Chapter 4), while the polynomial version provides the SubBytes operation in AES (Section 2.2). It is useful to treat both scenarios in a common framework.

DEFINITION 15.14. *An integral domain R together with a function $d: R \to \mathbb{N} \cup \{-\infty\}$ is a Euclidean domain if for all $a, b \in R$ with $b \neq 0$, we can divide a by b with remainder, so that*

(15.15) $\qquad$ *there exist $q, r \in R$ such that $a = qb + r$ and $d(r) < d(b)$.*

Thus $q = a \text{ quo } b$ is the quotient and $r = a \text{ rem } b$ the remainder, although q and r need not be unique. Such a d is called a Euclidean function on R.

EXAMPLE 15.16. (i) $R = \mathbb{Z}$ and $d(a) = |a| \in \mathbb{N}$. Here the quotient and the remainder can be made unique by the additional requirement that $r \geq 0$.

(ii) $R = F[x]$, where F is a field, and $d(a) = \deg a$. We define the degree of the zero polynomial to be $-\infty$. It is easy to show uniqueness of the quotient and the remainder in this case.

(iii) $R = \mathbb{Z}[i] = \{a + ib : a, b \in \mathbb{Z}\}$, the ring of Gaussian integers, with $i = \sqrt{-1}$, and $d(a + ib) = a^2 + b^2$.

(iv) R a field, and $d(a) = 1$ if $a \neq 0$ and $d(0) = 0$.   ◊

The value $d(b)$ is never $-\infty$ except possibly when $b = 0$.

DEFINITION 15.17. *Let R be a ring and $a, b, c \in R$. Then c is a* greatest common divisor *(or* gcd*) of a and b if*

(i) $c \mid a$ *and* $c \mid b$,

(ii) *if $d \mid a$ and $d \mid b$, then $d \mid c$, for all $d \in R$.*

Similarly, c is called a least common multiple *(or* lcm*) of a and b if*

(i) $a \mid c$ *and* $b \mid c$,

(ii) *if $a \mid d$ and $b \mid d$, then $c \mid d$, for all $d \in R$.*

For example, 3 is a gcd of 12 and 15, and 60 is an lcm of 12 and 15 in $\mathbb{Z}$. In general, neither the gcd nor the lcm are unique, but all gcds of a and b are precisely the associates of one of them, and similarly for the lcms. The only units in $\mathbb{Z}$ are 1 and -1, and 3 and -3 are all gcds of 12 and 15 in $\mathbb{Z}$. For $R = \mathbb{Z}$, we may define $\gcd(a, b)$ as the unique nonnegative greatest common divisor and $\mathrm{lcm}(a, b)$ as the unique nonnegative least common multiple of a and b. As an example, for negative $a \in \mathbb{Z}$ we then have $\gcd(a, a) = \gcd(a, 0) = -a$. We say that two integers a, b are *coprime* (or *relatively prime*) if their gcd is a unit. In this case, we set $\gcd(a, b) = 1$.

LEMMA 15.18. *The gcd in $\mathbb{Z}$ has the following properties, for all $a, b, c \in \mathbb{Z}$.*

(i) $\gcd(a, b) = |a| \iff a \mid b$,

(ii) $\gcd(a, a) = \gcd(a, 0) = |a|$ *and* $\gcd(a, 1) = 1$,

(iii) $\gcd(a, b) = \gcd(b, a)$ *(commutativity)*,

(iv) $\gcd(a, \gcd(b, c)) = \gcd(\gcd(a, b), c)$ *(associativity),*

(v) $\gcd(c \cdot a, c \cdot b) = |c| \cdot \gcd(a, b)$ *(distributivity),*

(vi) $|a| = |b| \implies \gcd(a, c) = \gcd(b, c).$

For a proof, see Exercise 15.7.

The following algorithm computes not only the gcd but also a representation of it as a linear combination of the inputs. It generalizes the representation

$$7 = 21 - 1 \cdot 14 = 21 - (35 - 1 \cdot 21) = 2 \cdot (126 - 3 \cdot 35) - 35 = 2 \cdot 126 - 7 \cdot 35,$$

which is obtained by reading the lines of (15.13) from the bottom up. This important method is called the Extended Euclidean Algorithm and works in any Euclidean domain.

ALGORITHM 15.19. Extended Euclidean Algorithm.

Input: $a, b \in R$, where R is a Euclidean domain.
Output: $\ell \in \mathbb{N}$, $r_i, s_i, t_i \in R$ for $0 \le i \le \ell + 1$, and $q_i \in R$ for $1 \le i \le \ell$,
      as computed below.

1. $r_0 \leftarrow a, \quad s_0 \leftarrow 1, \quad t_0 \leftarrow 0,$
   $r_1 \leftarrow b, \quad s_1 \leftarrow 0, \quad t_1 \leftarrow 1.$
2. $i \leftarrow 1.$
3. While $r_i \ne 0$ do step 4
4. $q_i \leftarrow r_{i-1}$ quo $r_i,$
   $r_{i+1} \leftarrow r_{i-1} - q_i r_i,$
   $s_{i+1} \leftarrow s_{i-1} - q_i s_i,$
   $t_{i+1} \leftarrow t_{i-1} - q_i t_i,$
   $i \leftarrow i + 1.$
5. $\ell \leftarrow i - 1.$
6. If r_ℓ is a unit, replace (r_ℓ, s_ℓ, t_ℓ) by $(1, s_\ell/r_\ell, t_\ell/r_\ell).$
7. Return ℓ, r_i, s_i, t_i for $0 \le i \le \ell + 1$, and q_i for $1 \le i \le \ell.$

The algorithm terminates because the $d(r_i)$ are strictly decreasing nonnegative integers for $1 \le i \le \ell$, where d is the Euclidean function on R. The elements r_i for $0 \le i \le \ell + 1$ are the remainders and the q_i for $1 \le i \le \ell$ are the quotients in the Euclidean Algorithm. The elements r_i, s_i, and t_i form the ith row, for $0 \le i \le \ell + 1$. The central property is that $s_i a + t_i b = r_i$ for all i; in particular, $s_\ell a + t_\ell b = r_\ell$ is a gcd of a and b (see Lemma 15.21 below).

EXAMPLE 15.20. (i) As in (15.13), we consider $R = \mathbb{Z}$, $a = 126$, and $b = 35$. The following table illustrates the computation.

i	q_i	r_i	s_i	t_i
0		126	1	0
1	3	35	0	1
2	1	21	1	-3
3	1	14	-1	4
4	2	7	2	-7
5		0	-5	18

We can read from row 4 that $\gcd(126, 35) = 7 = 2 \cdot 126 + (-7) \cdot 35$.

(ii) We take $R = \mathbb{Z}_2[x]$, $a = m = x^8 + x^4 + x^3 + x + 1$, and $b = x^7 + x^4 + x^3 + x + 1$, as in the AES example of Section 2.2. Then the computation of the Extended Euclidean Algorithm goes as follows. Row $i+1$ is obtained from the two preceding ones by first computing the quotient $q_i = r_{i-1} \operatorname{quo} r_i$ and then for each of the three remaining columns by subtracting the quotient times the entry in row i of that column from the entry in row $i - 1$.

i	q_i	r_i	s_i	t_i
0		$x^8 + x^4 + x^3 + x + 1$	1	0
1	x	$x^7 + x^4 + x^3 + x + 1$	0	1
2	$x^2 + 1$	$x^5 + x^3 + x^2 + 1$	1	x
3	$x^4 + x^2 + x$	x	$x^2 + 1$	$x^3 + x + 1$
4	x	1	$x^6 + x^3 + x^2$ $+x + 1$	$x^7 + x^3$
5		0	$x^7 + x^4 + x^3$ $+x + 1$	$x^8 + x^4 + x^3$ $+x + 1$

We have $\ell = 4$, and from row 4, we find that a gcd of a and b is

$$1 = (x^6 + x^3 + x^2 + x + 1) \cdot a + (x^7 + x^3) \cdot b \text{ in } \mathbb{Z}_2[x]. \qquad \diamond$$

For a global view of the algorithm, it is convenient to consider the matrices

$$R_0 = \begin{pmatrix} s_0 & t_0 \\ s_1 & t_1 \end{pmatrix}, \quad Q_i = \begin{pmatrix} 0 & 1 \\ 1 & -q_i \end{pmatrix} \text{ for } 1 \le i \le \ell$$

in $R^{2\times2}$, and $R_i = Q_i \cdots Q_1 R_0$ for $0 \le i \le \ell$. The following lemma collects some invariants of the Extended Euclidean Algorithm. Here (r_ℓ, s_ℓ, t_ℓ) refer to the quantities just before step 6 of Algorithm 15.19.

LEMMA 15.21. *For $0 \le i \le \ell$, we have*

(i) $R_i \cdot \begin{pmatrix} a \\ b \end{pmatrix} = \begin{pmatrix} r_i \\ r_{i+1} \end{pmatrix}$,

(ii) $R_i = \begin{pmatrix} s_i & t_i \\ s_{i+1} & t_{i+1} \end{pmatrix}$,

(iii) $\gcd(a,b) \sim \gcd(r_i, r_{i+1}) \sim r_\ell$,

(iv) $s_i a + t_i b = r_i$ *(this also holds for $i = \ell + 1$)*,

(v) $s_i t_{i+1} - t_i s_{i+1} = (-1)^i$,

(vi) $\gcd(r_i, t_i) \sim \gcd(a, t_i)$,

(vii) $a = (-1)^i(t_{i+1}r_i - t_i r_{i+1})$ *and* $b = (-1)^{i+1}(s_{i+1}r_i - s_i r_{i+1})$.

PROOF.   For (i) and (ii) we proceed by induction on i. The case $i = 0$ is clear from step 1 of the algorithm, and we may assume $i \ge 1$. Then

$$Q_i \begin{pmatrix} r_{i-1} \\ r_i \end{pmatrix} = \begin{pmatrix} 0 & 1 \\ 1 & -q_i \end{pmatrix}\begin{pmatrix} r_{i-1} \\ r_i \end{pmatrix} = \begin{pmatrix} r_i \\ r_{i-1} - q_i r_i \end{pmatrix} = \begin{pmatrix} r_i \\ r_{i+1} \end{pmatrix},$$

and (i) follows from $R_i = Q_i R_{i-1}$ and the induction hypothesis. Similarly, (ii) follows from

$$Q_i \begin{pmatrix} s_{i-1} & t_{i-1} \\ s_i & t_i \end{pmatrix} = \begin{pmatrix} s_i & t_i \\ s_{i+1} & t_{i+1} \end{pmatrix}$$

and the induction hypothesis.

For (iii), let $i \in \{0, \ldots, \ell\}$. We conclude from (i) that

$$\begin{pmatrix} r_\ell \\ 0 \end{pmatrix} = Q_\ell \cdots Q_{i+1} R_i \begin{pmatrix} a \\ b \end{pmatrix} = Q_\ell \cdots Q_{i+1} \begin{pmatrix} r_i \\ r_{i+1} \end{pmatrix}.$$

Comparing the first entry on both sides, we see that r_ℓ is a linear combination of r_i and r_{i+1}, and hence any common divisor of r_i and r_{i+1} divides r_ℓ. On the other hand, $\det Q_i = -1$ and the matrix Q_i is invertible over R, with inverse

$$Q_i^{-1} = \begin{pmatrix} q_i & 1 \\ 1 & 0 \end{pmatrix},$$

and hence

$$\begin{pmatrix} r_i \\ r_{i+1} \end{pmatrix} = Q_{i+1}^{-1} \cdots Q_\ell^{-1} \begin{pmatrix} r_\ell \\ 0 \end{pmatrix}.$$

Thus both r_i and r_{i+1} are divisible by r_ℓ, and $r_\ell \sim \gcd(r_i, r_{i+1})$. In particular, this is true for $i = 0$, so that $\gcd(a, b) \sim \gcd(r_0, r_1) \sim r_\ell$.

The claim (iv) follows immediately from (i) and (ii), and (v) follows from (ii) by taking determinants:

$$s_i t_{i+1} - t_i s_{i+1} = \det \begin{pmatrix} s_i & t_i \\ s_{i+1} & t_{i+1} \end{pmatrix} = \det R_i$$

$$= \det Q_i \cdots \det Q_1 \cdot \det \begin{pmatrix} s_0 & t_0 \\ s_1 & t_1 \end{pmatrix} = (-1)^i.$$

In particular, this implies that $\gcd(s_i, t_i) \sim 1$ and that R_i is invertible. Now let $p \in R$ be a divisor of t_i. If p divides a, then it also divides $s_i a + t_i b = r_i$. On the other hand, if p divides r_i, then it divides $s_i a = r_i - t_i b$ and hence also a, since s_i and t_i are coprime. This proves (vi). For (vii), we multiply both sides of (i) by R_i^{-1} and obtain .

$$\begin{pmatrix} r_0 \\ r_1 \end{pmatrix} = R_i^{-1} \begin{pmatrix} r_i \\ r_{i+1} \end{pmatrix} = (-1)^i \begin{pmatrix} t_{i+1} & -t_i \\ -s_{i+1} & s_i \end{pmatrix} \begin{pmatrix} r_i \\ r_{i+1} \end{pmatrix},$$

using (ii) and (v), and the claim follows by writing this out as a system of linear equations. $\qquad\square$

COROLLARY 15.22. *Any two elements a and b of a Euclidean domain R have a greatest common divisor $c \in R$, and it is expressible as a linear combination $c = sa + tb$ with $s, t \in R$. The Extended Euclidean Algorithm computes c, s, and t.*

For the quadratic cost of this algorithm, see Theorem 15.33.

15.5. Modular inverses

The question of whether an element of a ring has a multiplicative inverse plays an important role in many cryptosystems and their analysis. We discuss it in this section, and recall the notions from Definition 15.3.

Units and zerodivisors have similar properties in some sense: the first can be multiplied to 1, the second to 0. Now 0 is neither a zerodivisor nor a unit, and 1 is a unit and never a zerodivisor (Exercise 15.4). Both types of elements have their use in cryptography. As examples, the exponents in RSA are units in $\mathbb{Z}_{\varphi(N)}$, and a zerodivisor in $\mathbb{Z}_N$ provides a factorization of N.

EXAMPLE 15.23. (i) $\mathbb{Z}^{\times} = \{1, -1\} = \pm 1$, $\mathbb{Q}^{\times} = \mathbb{Q} \setminus \{0\}$, $\mathbb{R}^{\times} = \mathbb{R} \setminus \{0\}$, $\mathbb{C}^{\times} = \mathbb{C} \setminus \{0\}$.

(ii) $(R[x])^{\times} = R^{\times}$, when R is an integral domain.

(iii) $\mathbb{Z}_{12}^{\times} \supseteq \{1, 5, 7, 11\}$, since $a^2 = 1$ in $\mathbb{Z}_{12}$ for $a = 1, 5, 7, 11$. We will see below (after Lemma 15.64) that actually equality holds. ◇

How can we test whether an element is a unit? And if so, compute its inverse? Do we have to try all elements of the ring as potential inverses, as prescribed in Definition 15.3. In RSA, we have to compute inverses modulo $\varphi(N) = (p-1)(q-1)$, a 3000-bit number. Trying all possibilities, roughly 2^{3000}, would be hopeless. Fortunately, we have essentially solved the problem: the Extended Euclidean Algorithm furnishes the complete answer.

ALGORITHM 15.24. Modular inverse in a Euclidean domain R.

Input: Elements $a, b \in R$, with $b \neq 0$.
Output: Either "a is not invertible" or an inverse s of a in $R/(b)$.

1. Call Algorithm 15.19 to obtain a representation $c = sy + tb$ of $c = \gcd(a, b)$.
2. If $c = 1$ then return s.
3. Else return "a is not invertible".

THEOREM 15.25. *Algorithm 15.24 works correctly as specified and in time* $O(n^2)$ *for n-bit inputs, in the notation of Theorem 15.33.*

PROOF. If $c = 1$, then $1 = sa + tb = sa$ in $R/(b)$, and s is indeed the inverse of a. Now suppose that $c \neq 1$ and that $ax = 1$ in $R/(b)$ for some $x \in R$. We have to show that this is impossible, so that a is not invertible. By the general version in R of (15.10) and (15.11), b divides $ax - 1$, so that $yb = ax - 1$ for some $y \in R$. Since c divides a and b, it also divides $ax - yb = 1$. This is impossible, since c is not a unit. □

COROLLARY 15.26. (i) *For an integer $N \geq 2$, the unit group $\mathbb{Z}_N^{\times}$ of invertible elements in $\mathbb{Z}_N$ is the set of all elements a coprime to N:*

$$\mathbb{Z}_N^{\times} = \{a \in \mathbb{Z}_N : \gcd(a, N) = 1\}.$$

(ii) *For a nonzero polynomial $m \in F[x]$ over a field F, the unit group $(F[x]/(m))^{\times}$ is the set of all polynomials coprime to m:*

$$(F[x]/(m))^{\times} = \{a \in F[x] : \gcd(a, m) = 1\}. \quad □$$

Of course, 0 cannot be invertible. Thus the "best" case would be if all elements of $\mathbb{Z}_N$ but 0 were invertible. By Corollary 15.26 this happens if and only if no integer a with $0 < a < N$ has a nontrivial common divisor with N. Equivalently, N has no proper divisor. A nonconstant polynomial $m \in F[x]$ over a field F has no proper divisor if and only if it is irreducible. In the language of Definition 15.3 (v), we can restate this as follows.

COROLLARY 15.27. (i) *For an integer $N \geq 2$, the ring $\mathbb{Z}_N$ is a field if and only if N is prime.*

(ii) *For a nonconstant polynomial $m \in F[x]$ over a field F, $F[x]/(m)$ is a field if and only if m is irreducible.* □

REMARK 15.28. *There is a subtlety here whose disregard sometimes causes confusion. We have defined the gcd for integers a and b, that is, elements of $\mathbb{Z}$. Thus a is not an element of $\mathbb{Z}_b$, but $a \bmod b$ is. There is no useful notion of a gcd in $\mathbb{Z}_b$. For example, when b is prime, then all nonzero elements of $\mathbb{Z}_b$ are units and the gcd of any two of them is associated to 1.*

15.6. Cost of algorithms and complexity of problems

Table 15.1 lists some things that we have to do, numbered by the steps in the RSA Cryptosystem 2.16. The third column describes the cost of

step	task	time	ref
1	generating primes	$O(n^4)$	Section 3.2
2	multiplying integers	$O(n^2)$	Sections 15.1 and 15.3
3	random integers		
	coprimality	$O(n^2)$	Section 15.5
4	modular inverse	$O(n^2)$	Section 15.5
7, 9	modular exponentiation	$O(n^3)$	Section 15.12
	correctness of RSA		Theorem 3.1

Table 15.1: Tools for the RSA cryptosystem.

these procedures for n-bit inputs. We now discuss this in more detail.

A basic tool is the language of *asymptotic time*. In contrast to absolutely fixed systems like AES, in RSA and many other cryptosystems there is a security parameter, say n (or several such parameters) which may be adjusted according to security requirements (see Table 5.12 for current suggestions). We have to evaluate both the cost of executing legitimately and of attacking the system in terms of n. Even for a basic task like multiplying integers it is not quite clear how to do this in the best way. Ideally, you would like to know how long such an operation takes on your machine. But if you have figured that out and then buy a new computer, do you have to start your figuring anew?

The methodology for obtaining machine-independent results is to ignore the constant factors depending on processor speed and operating system. This is made precise by the big-Oh notation "O".

DEFINITION 15.29.   (i) *For two functions* $f, g\colon \mathbb{N} \longrightarrow \mathbb{R}$, *we say that* $f \in O(g)$ *if there exist* c *and* N *so that*

$$|f(n)| \leq c \cdot |g(n)| \text{ for all } n \geq N.$$

Thus $O(g)$ *consists of those functions which grow in absolute value at most like a constant multiple of* f.

(ii) *A function* $f\colon \mathbb{N} \to \mathbb{R}_{\geq 0}$ *is* polynomially bounded *if there exists some* k *so that* $f \in O(n^{\bar{k}})$, *where* n *is the input of* f. *We then write* $f \in \operatorname{poly}(n)$.

(iii) *A probabilistic algorithm is* efficient *if its expected runtime, depending on the input length* n *and measured in basic operations (say, bit operations or, for some algebraic problems, operations in the ground domain), is polynomially bounded in* n. *We also call it a* polynomial-time algorithm.

(iv) *A computational problem* X *is* easy *if there exists a probabilistic polynomial-time algorithm that solves it; otherwise* X *is* hard.

(v) *The complexity class* BPP *(bounded error probabilistic polynomial time) consists of those decision problems* $X \subseteq \{0, 1\}^*$ *for which there exists a polynomial-time probabilistic algorithm* $\mathcal{A}$ *with output in* $\{0, 1\}$ *so that for all* $x \in \{0, 1\}^*$ *we have*

$$x \in X \Longrightarrow \operatorname{prob}\{\mathcal{A}(x) = 1\} \geq 2/3,$$
$$x \notin X \Longrightarrow \operatorname{prob}\{\mathcal{A}(x) = 1\} \leq 1/3.$$

The problems $X \in$ BPP *with a deterministic such* $\mathcal{A}$ *form the class* P *of (deterministic) polynomial time. (Thus* P $\subseteq$ BPP.)

Polynomial runtime is a central concept in the analysis of algorithms. If you double your computing power, then the size of problems solvable in time $O(n^k)$ increases noticeably, from n to $2^{1/k}n$, while for exponential time like 2^n, it only increases from n to $n + 1$. The "polynomial" always refers to polynomial in the input size. For this to mean $\text{poly}(n)$, we may provide the *unary* representation $1^n = 1 \ldots 1$ (n ones) of n as part of the input.

The concept "inverse" to poly is used to describe probabilities.

DEFINITION 15.30. *Let $\varepsilon \colon \mathbb{N} \to \mathbb{R}_{\geq 0}$ be a function. Then ε is nonnegligible if there exists some $k \geq 1$ so that $\varepsilon(n) \geq n^{-k}$ for $n \geq 2$. On the other hand, ε is negligible if it is not nonnegligible, that is, $\varepsilon(n) < n^{-k}$ for any k and large enough n.*

Thus ε is nonnegligible if and only if $\varepsilon^{-1} \in \text{poly}(n)$. For example, ε with $\varepsilon(n) = n^{-\log n}$ is negligible, but not exponentially small like 2^{-n}. At the other end of the spectrum, some ε is *overwhelming* if it is exponentially close to 1, so that $\varepsilon(n) \geq 1 - 2^{-cn}$ for some constant $c > 0$. Some related standard notation is used, but infrequently in this text. We also say that $f \in o(g)$ if for all $c > 0$ there is some N so that $|f(n)| \leq c \cdot |g(n)|$ for all $n \geq N$, and $f \in O^{\sim}(g)$ (pronounced "soft-oh") if $f \in g \cdot (\log n)^{O(1)}$. Furthermore, $f \in \Omega(g)$ if there exist $c > 0$ and N so that $f(n) \geq c \cdot |g(n)|$ for all $n \geq N$, and $f \in \omega(g)$ if for all $c > 0$ there exists an N so that $f(n) \geq c \cdot |g(n)|$ for all $n \geq N$.

It is usually clear what an "operation" in an algorithm $\mathcal{A}$ is, say a Boolean computation on two bits, or a step of a Turing machine $\mathcal{A}$. The *cost* $(c(\mathcal{A}))(x)$ is the number of operations used by $\mathcal{A}$ on input x. For an integer n, the (worst-case) cost $(\text{cost}(\mathcal{A}))(n)$ is the maximal $(c(\mathcal{A}))(x)$ over all inputs x of length n. For a "problem" X, the meaning of "$\mathcal{A}$ solves X" is also often clear. Then the *complexity* $\text{comp}(X) \colon \mathbb{N} \to \mathbb{R}$ of X is the minimal cost of all algorithms that solve X, so that $\text{comp}(X)(n) = \min\{(\text{cost}(\mathcal{A}))(n) \colon \mathcal{A} \text{ solves } X\}$. The runtime of any algorithm that solves X is an *upper bound* on $\text{comp}(X)$. But *lower bounds* on the complexity of some problem, implying that any algorithm that solves the problem takes at least that much time (or memory, communication, or some other cost measure), are much harder to come by. This is the purview of *complexity theory* (or *computational complexity*); see also Section 9.1.

Boolean circuits have (one-bit) input gates, and NOT, AND, OR, and XOR gates. The time that such a circuit takes is the number of gates in it including the input gates, usually called the *size* of the circuit. Then "algorithm" may also be taken to mean "family of Boolean circuits". In analogy to P, we can consider the class P_{circ} of all decision problems which can be decided by a family $(\mathcal{C}_n)_{n \in \mathbb{N}}$ of Boolean circuits $\mathcal{C}_n$, where $\mathcal{C}_n$ has n inputs and its size is polynomial in n. Circuits are a nonuniform model of computation, where C_n may be constructed in a manner totally different from that for C_{n-1}. Then $\mathsf{P} \subset \mathsf{P}_{\mathrm{circ}}$, but the two classes are not identical, because a Turing machine has only "one" behavior for all input sizes. This can be mended by stipulating that the circuits C_n have to be "uniformly constructed" in dependence on n. With the appropriate technical definitions, we have $\mathsf{P}_{\mathrm{circ}}(\text{uniform}) = \mathsf{P}$. Alternatively, we can allow Turing machines a special *advice tape*; this gives the complexity class $\mathsf{P}/\,\mathrm{poly}$, which equals $\mathsf{P}_{\mathrm{circ}}$.

The standard approach is to encode the objects under consideration in some fixed and easy way as binary strings. A problem X is a set of such strings. For such a decision problem, we are given an arbitrary string and have to decide whether it is in X.

In cryptography, probabilistic algorithms are ubiquitous, say in key generation. We take such an $\mathcal{A}$ outputting either 0 or 1. Then $\mathcal{E}_{\mathcal{A}}(x) = \mathrm{prob}\{\mathcal{A}(x) = 1\}$ is the probability of $\mathcal{A}$ returning 1 on input x, taken over the internal randomization of $\mathcal{A}$ (see Section 15.16). Similarly, one defines the expected runtime $c(\mathcal{A})(x)$ of $\mathcal{A}$ on input x, and $\mathrm{cost}(\mathcal{A})(n)$ as the maximal $c(\mathcal{A})(x)$ over all inputs x of length n. The relevant complexity class BPP (bounded-error probabilistic polynomial time) consists of those decision problems X for which a probabilistic (expected) polynomial-time algorithm $\mathcal{A}$ exists with

$$
\mathcal{E}_{\mathcal{A}}(x) \begin{cases} \geq 2/3 & \text{if } x \in X, \\ \leq 1/3 & \text{if } x \notin X. \end{cases}
$$

We also consider the expectation $\mathcal{E}_{\mathcal{A}}(X)$ of $\mathcal{E}_{\mathcal{A}}(x)$ as x is chosen from some distribution X (see Section 15.16). The required randomness can be provided to a Turing machine on a special infinite tape of random bits, and to a Boolean circuit by gates which deliver random bits. For simplicity, we do not count these gates for the size.

15.7. Worst case vs. average case

It is often necessary to distinguish between an algorithm's success in the worst case and on average. For factoring integers, the former means that N is given as an input, and the algorithm has to solve the problem for any such particular instance. This is the standard computer science notion for an algorithm to solve a problem. You may think of an adversary who submits the "toughest" N to the algorithm in order to test whether it can crack this with the claimed probability. The usual complexity classes are described in this fashion, with various cost measures and types of algorithms. A weaker notion is for an algorithm to solve a problem on average. Here the inputs are given by a distribution. For factoring, this might be an RSA key generator for N. Now an algorithm's success expectation is over its internal randomization and also over the input distribution. Its task is potentially easier: if chance throws a particularly "tough" N at it, it can just lean back and do nothing. As long as this happens rarely enough, its success probability will not be influenced by much. We can run a worst-case algorithm also on random inputs, obtaining the same estimates on the success probability and the running time. Thus the average case is never harder than the worst case:

$$\text{average case cost} \leq \text{worst case cost.}$$

Sometimes also the converse is essentially true; Algorithm 7.11 provides an example.

Thus a worst-case upper bound automatically gives one for the average case, and the latter might be easier to come by. For lower bounds, the situation is reversed. Proving a lower bound on the average is much stronger than proving it in the worst case. Not surprisingly, few such bounds are known. There are several approaches to building a complexity theory for the average case; the issues are tricky.

Given our civilization's current inability to prove large lower bounds, a natural question is whether at least one can show that the average case is about as hard as the worst case, without knowing absolutely how hard they are. Proving this for NP-complete problems, say, has been an open problem for a quarter of a century.

This problem is particularly relevant for cryptography. Namely, a cryptosystem that is secure for some special "hard" subset of the keys but vulnerable for an average key is quite useless.

Some reductions of Chapter 9 illustrate this. As an example, in the reduction 9.8 from the weak RSA problem to forging Gennaro-Halevi-Rabin (GHR) signatures, the same RSA modulus N is used on both sides. It is randomly generated for GHR, and in order to conclude the security of GHR, we need the weak RSA problem to be hard for random N, that is, on average. One of the great challenges in cryptography is to find a (seemingly) hard problem X and cryptosystem S so that

(15.31) X worst-case hard $\Longrightarrow$ S secure.

A major breakthrough was Ajtai's (1996) worst-case to average-case reduction for a certain lattice problem; this is the basis for the development in Section 13.13.

15.8. The cost of arithmetic

We usually state the cost of our algorithms. Addition is easy: two n-bit integers, or two elements of $\mathbb{Z}_2^n$, or two polynomials of degree less than

n, can be added in $O(n)$ basic operations. The notion of basic operation depends on the particular environment. For integers it adds two input bits plus a carry bit in the first case, is bit addition (in $\mathbb{Z}_2$) in the second case, and addition in the ground ring for polynomials.

To multiply two n-bit integers, the *classical method* uses $O(n^2)$ bit operations. There are faster methods: Karatsuba's algorithm (from Карацуба & Офман (1962)) takes $O(n^{1.59})$ bit operations, and Schönhage & Strassen (1971) found a fast multiplication algorithm that only uses $O(n \log n \log\log n) \subset O^{\sim}(n)$ bit operations. Similar bounds hold for polynomials over appropriate domains. But for most purposes in cryptography, the classical algorithms are quite sufficient. The power of the more sophisticated algorithms has been harnessed for cryptographic purposes only fairly recently.

In the following, we usually mention those faster algorithms only in the notes, and then $\mathsf{M}(n)$ will refer to the number of basic operations the chosen multiplication algorithm uses for inputs of length n.

We next consider division with remainder (15.8) for polynomials. When $\deg a = n \geq m = \deg b$, then $\deg q = n - m$ and the division can be executed with $O(n^2)$ operations in R. Counting more precisely, one needs at most

$$(15.32) \qquad\qquad (2m + 1)(n - m + 1) \in O(n^2)$$

additions and multiplications in R, plus one division for inverting the leading coefficient of b. If b is monic, at most $2m(n + 1)$ additions and multiplications suffice. In many applications, we have $n < 2m$, and then the cost is at most $2m^2 + O(m)$ ring operations plus an inversion, which is essentially the same as that for multiplying two polynomials of degree at most m.

Similarly, division with remainder of an integer a by b can be performed with $O(m(n - m))$ bit operations if a and b have bit lengths n and m, respectively. Due to the carries, the details are somewhat more complicated than in the polynomial case.

For the cost of the Extended Euclidean Algorithm 15.19 on two polynomials $a, b \in F[x]$ of degree at most n, we have the division with remainder $r_{i-1} = q_i r_i + r_{i+1}$ in step 4, with $\deg r_{i+1} < \deg r_i$. Thus the remainder degree decreases by at least 1 in each step, and there are $\ell \leq n$ many steps. Each step takes $O(n^2)$ operations in F by (15.32), for a total of $O(n^3)$ operations. A tighter analysis using (15.32) yields a cost of $O(n^2)$ operations.

For two n-bit integers and b, it may happen that r_{i+1} is only a little smaller than r_i. The argument above would only yield a bound of roughly

2^n on the number of executions of Step 4 in Algorithm 15.19. However, for $1 \leq i \leq \ell$, we have

$$r_{i-1} = q_i r_i + r_{i+1} \geq r_i + r_{i+1} > 2r_{i+1}.$$

Thus in two steps, the remainder length decreases by at least 1, and $\ell \leq 2n$. Each step takes $O(n^2)$ bit operations, and the whole algorithm $O(n^3)$ operations. Using (15.32), one finds again that the EEA can be executed with $O(n^2)$ operations.

THEOREM 15.33. *The Extended Euclidean Algorithm 15.19 for positive integers a and b with bit lengths n and m, respectively, can be performed with $O(nm)$ bit operations. For polynomials of degree at most n over a field, it takes $O(n^2)$ field operations.*

15.9. Uniqueness of rational approximations

The material of this section is only relevant to the Wiener attack on RSA (Section 3.6) and quantum factorization (Section 14.4). From (15.13), we find

$$\frac{35}{126} = \frac{35}{3 \cdot 35 + 21} = \frac{1}{3 + \frac{21}{35}} = \frac{1}{3 + \frac{1}{1 + \frac{14}{21}}} = \frac{1}{3 + \frac{1}{1 + \frac{1}{1 + \frac{1}{2}}}}.$$

This expression is called the *continued fraction expansion* of $35/126$. For any positive integers a and b with $b < a$, the Extended Euclidean Algorithm for (a, b) yields an expression of this form for b/a. If we leave out some trailing fractions, say the $1/2$ above, we obtain

$$\frac{1}{3 + \frac{1}{1+1}} = \frac{2}{7} = \frac{-s_4}{t_4},$$

with s_4 and t_4 from Example 15.20 (i). This is a reasonable approximation of $35/126$, with

$$\left| \frac{35}{126} - \frac{2}{7} \right| = \left| \frac{-1}{18 \cdot 7} \right| = \frac{1}{18 \cdot 7} < \frac{1}{2 \cdot 7^2} = \frac{1}{2t_4^2}.$$

The corresponding inequality holds in general. We do not need this for our purposes, but rather the converse. Namely, a famous theorem in number theory says that if we have an approximation s/t to a number x with $|x - s/t| \leq 1/(2t^2)$, then s and t actually appear in the Extended Euclidean Algorithm for numerator and denominator of x.

This theorem is central for the proof of correctness of the Wiener attack. For completeness, we now show a result that can be used in its stead and whose (easy) proof follows the approach of Section 15.4.

We fix the notation from Algorithm 15.19 for integers:

$$(15.34) \qquad R = \mathbb{Z}, a, b, r_i, s_i, t_i, q_i, \ell, \text{ and assume } 0 \le b < a.$$

We first note that $r_{i+1} < r_i$ for $i \ge 0$ and hence $0 = r_{\ell+1} < r_\ell < \cdots < r_0 = a$. Thus for every integer u with $0 \le u < a$ there exists a unique $i \le \ell + 1$ with $r_i \le u < r_{i-1}$. Furthermore, we claim that the t_i for $i \ge 1$ alternate in sign; see Example 15.20 (i). For an inductive proof of this, we start with $t_1 = 1$ and $t_2 = t_0 - q_1 t_1 = -q_1 t_1 < 0$. Inductively, $-q_i t_i$ has the same sign as t_{i-1}, and hence also $t_{i+1} = t_{i-1} - q_i t_i$, as claimed. From this fact, it follows that $-t_i r_{i+1}$ has the same sign as $t_{i+1} r_i$, and from Lemma 15.21 (vii) we find

$$(15.35) \qquad a = |t_{i+1} r_i - t_i r_{i+1}| \ge |t_{i+1} r_i| = |t_{i+1}| r_i.$$

LEMMA 15.36. *Assume the notation (15.34), and let r, s, and t be integers with $r = sa + tb \ge 0$ and $4r|t| \le a$. Furthermore, define $i \le \ell + 1$ by $r_i \le 2r < r_{i-1}$. Then there exists an integer u with $1 \le |u| < r_{i-1}/2r_i$, $r = u \cdot r_i$, $s = u \cdot s_i$, and $t = u \cdot t_i$. If $\gcd(s, t) = 1$, then $u \in \pm 1$.*

PROOF.   From (15.35), we have

$$(15.37) \qquad |t_i| \le \frac{a}{r_{i-1}} < \frac{a}{2r}.$$

Eliminating the multiples of b in

$$r = sa + tb,$$
$$r_i = s_i a + t_i b,$$

where the latter is Lemma 15.21 (iv), we obtain

$$(15.38) \qquad r_i t - r t_i = (s t_i - s_i t)a,$$

which means that a divides $r_i t - r t_i$. Inequality (15.37) yields

$$|r t_i| < r \cdot \frac{a}{2r} = \frac{a}{2},$$
$$|r_i t| \le 2r|t| \le \frac{a}{2},$$
$$|r_i t - r t_i| < \frac{a}{2} + \frac{a}{2} = a.$$

It follows that $r_i t - r t_i = s t_i - s_i t = 0$ and $s t_i = s_i t$. Lemma 15.21 (v) implies that $\gcd(t_i, s_i) = 1$, and hence t_i divides t. Thus $u = t/t_i \in \mathbb{Z}$, and it follows that $s/s_i = r/r_i = u$. Furthermore

$$|u| = \frac{|r|}{r_i} < \frac{r_{i-1}}{2r_i}.$$

For the last claim, we note that u divides $\gcd(s, t)$. $\square$

THEOREM 15.39 (Uniqueness of good approximations).   *Let $a > b$ and s, t be positive integers with $\gcd(s, t) = 1$ and*

$$\left| \frac{b}{a} - \frac{s}{t} \right| \le \frac{1}{4t^2}.$$

Then $(s, -t)$ or $(-s, t)$ appears as some (s_i, t_i) in the Extended Euclidean Algorithm for (a, b).

PROOF.   Let $t^* = -t$ and $r = |sa + t^*b| = |sa - tb|$. If $r = 0$, $c = \gcd(a, b)$, and ε is the sign of $s_{\ell+1}$, then $\gcd(a/c, b/c) = 1$, and from Lemma 15.21 (iv) we find $s = b/c = \varepsilon s_{\ell+1}$ and $t = a/c = -\varepsilon t_{\ell+1}$, which implies the claim. We may now assume $r > 0$. Then

$$4r|t^*| = 4\frac{|sa - tb|}{at} \cdot at \cdot t = 4at^2 \left| \frac{b}{a} - \frac{s}{t} \right| \le 4at^2 \cdot \frac{1}{4t^2} = a.$$

Lemma 15.36 says that $(s, t^*) \in \pm(s_i, t_i)$ for some i, from which the claim follows. $\square$

The theorem is actually true with the weaker quality assumption $1/(2t^2)$ rather than $1/(4t^2)$. For our applications, this does not matter.

15.10. Polynomial interpolation

The basis for Shamir's secret sharing algorithm in Section 2.9 is *polynomial interpolation*. This problem is a special case of and useful introduction to the Chinese remaindering in the next section. We have a field F, pairwise distinct arguments $u_1, \ldots, u_r \in F$, and values $a_1, \ldots, a_r \in F$. A polynomial $f \in F[x]$ with

$$f(u_i) = a_i \quad \text{for all } i \le r$$

is called an *interpolating polynomial*. A fundamental fact is that there is a *unique* interpolating polynomial of degree less than r. The *Lagrange interpolants*

$$\ell_j = \prod_{\substack{1 \le k \le r \\ k \ne j}} \frac{x - u_k}{u_j - u_k} \in F[x]$$

for $1 \leq j \leq r$ have degree less than r and the property that

$$\ell_j(u_i) = \begin{cases} 1 & \text{if } j = i, \\ 0 & \text{otherwise.} \end{cases}$$

Then

$$f = \sum_{1 \leq i \leq r} a_i \ell_i$$

is the unique interpolation polynomial of degree less then r, since $f(u_i) = \sum_{1 \leq j \leq r} a_j \ell_j(u_i) = a_i$ for all i.

In fact, any interpolating polynomial $g \in F[x]$, with $g(u_i) = a_i$ for all i, is of the form $g = f + h \prod_{1 \leq i \leq r}(x - u_i)$ for some polynomial $h \in F[x]$. Thus $g = f$ corresponds to $h = 0$, and $h = 1$ gives the unique monic interpolating polynomial of degree exactly r.

Quite generally, *Viète's rule* expresses the coefficients of a polynomial $f = x^n + f_{n-1}x^{n-1} + \cdots f_0$ in terms of its roots $u_1, \ldots, u_n$. In particular,

$$(15.40) \qquad\qquad f_{n-1} = -(u_1 + \cdots u_n).$$

15.11. The Chinese Remainder Algorithm

A basic tool to deal with the rings $\mathbb{Z}_N$ is the important theorem and algorithm that lends this section its name. Among other things, it explains the choice of exponents in RSA and reduces the calculation of Euler's totient function φ to prime power moduli.

We recall the notions of ring homomorphism and ring isomorphism from Section 15.2. From an algebraic point of view, there is no difference between isomorphic rings R and S; all ring-theoretic properties of R are shared by S. But for cryptography, isomorphic objects may be substantially different if their objects are represented differently. An important example is the exp-log isomorphism of finite groups in Figure 4.1.

For positive integers n and m, we can consider the mapping

$$(15.41) \qquad\qquad f: \begin{array}{ccc} \mathbb{Z}_n & \longrightarrow & \mathbb{Z}_m, \\ a & \longmapsto & a \text{ rem } m \end{array}.$$

When is this a ring homomorphism?

EXAMPLE 15.42.   (i) For $n = 6$ and $m = 3$, we have for f as defined by (15.41)

i	0	1	2	3	4	5
$f(i)$	0	1	2	0	1	2

This is a ring homomorphism; for example

$$f(2+2) = f(4) = 1 = 2+2 = f(2) + f(2).$$

(ii) Neither $f\colon \mathbb{Z}_2 \longrightarrow \mathbb{Z}_3$ nor $f^*\colon \mathbb{Z}_3 \longrightarrow \mathbb{Z}_2$, as in (15.41), are ring homomorphisms. For example,

$$f(1+1) = f(0) = 0 \neq 2 = 1+1 = f(1) + f(1),$$

$$f^*(1+2) = f^*(0) = 0 \neq 1 = 1+0 = f^*(1) + f^*(2).$$

In fact, these maps are not even well-defined. We have $1 = 3$ in $\mathbb{Z}_2$, but $f(1) = 1 \neq 0 = 3 = f(3)$ in $\mathbb{Z}_3$. $\diamond$

For any $a, b \in \mathbb{Z}_n$, we need

$$((a+b) \operatorname{rem} n) \operatorname{rem} m = (a \operatorname{rem} m + b \operatorname{rem} m) \operatorname{rem} m.$$

Both sides are integers between 0 and $m-1$, and with $r = (a+b) \operatorname{rem} n$, $s = a \operatorname{rem} m$, $t = b \operatorname{rem} m$, it is sufficient to have

$$r = s+t \text{ in } \mathbb{Z}_m.$$

Writing

$$a+b = q_1 n + r, \quad a = q_2 m + s, \quad b = q_3 m + t,$$

we need that m divides

$$
\begin{aligned}
r - (s+t) &= a+b - q_1 n - (a - q_2 m + b - q_3 m) \\
&= -q_1 n + (q_2 + q_3)m.
\end{aligned}
$$

The second summand is divisible by m, so $q_1 n$ should also be. If m divides n, that is indeed the case. A similar argument works for multiplication, and we have shown the following.

LEMMA 15.43. *If m divides n, then f in (15.41) is a ring homomorphism.*

If m does not divide n, then in fact f is not a ring homomorphism; see Exercise 15.9. For two rings S and T, their product

$$S \times T = \{(a,b)\colon a \in S, b \in T\}$$

becomes a ring when we perform operations componentwise. Thus

$$(a,b) \cdot (a',b') = (a \cdot a', b \cdot b'),$$

and similarly for addition. If we have, furthermore, two ring homomor-
phisms $f\colon R \longrightarrow S$ and $g\colon R \longrightarrow T$, then we get a ring homomorphism

$$f \times g\colon \begin{array}{ccc} R & \longrightarrow & S \times T, \\ a & \longmapsto & (f(a), g(a)). \end{array}$$

This also works with more than two factors.

Putting all this together, we now have a ring homomorphism

$$\mathbb{Z}_N \longrightarrow \mathbb{Z}_{q_1} \times \cdots \times \mathbb{Z}_{q_r},$$

whenever $q_1, \dots, q_r$ divide N. The following important theorem says that
under a reasonable condition, this is an isomorphism.

CHINESE REMAINDER THEOREM 15.44. *Let $N = q_1 \cdots q_r$ with pairwise
coprime integers $q_1, \dots, q_r$. Then the ring homomorphism*

$$\begin{array}{ccc} \mathbb{Z}_N & \longrightarrow & \mathbb{Z}_{q_1} \times \cdots \times \mathbb{Z}_{q_r}, \\ x \bmod N & \longmapsto & (x \bmod q_1, \dots, x \bmod q_r) \end{array}$$

*is an isomorphism. In other words: given integers $a_1, \dots, a_r$, there exists
an integer $x \in \mathbb{Z}$ that solves the congruences*

$$x = a_i \text{ in } \mathbb{Z}_{q_i} \text{ for } 1 \le i \le r$$

*simultaneously, and two such solutions x and x' differ by a multiple of N,
so that $x = x'$ in $\mathbb{Z}_N$.*

PROOF.   We leave it as an exercise to verify that the map is well-defined
and a ring morphism and to infer that it is an isomorphism using the
second statement.

We prove the second statement. The Lagrange interpolation for poly-
nomials uses polynomials $\ell_i(x) = \prod_{j \ne i}(x - u_j)/\prod_{j \ne i}(u_i - u_j)$ to find
a polynomial f that satisfies equations $f(u_i) = a_i$ by just summing:
$f = \sum_i a_i \ell_i$. This works, since $\ell_i(u_j)$ is either 1 or 0 depending on
whether $i = j$ or not. We imitate this and define $s_i = \prod_{j \ne i} q_j$. Of
course, $s_i = 0$ in $\mathbb{Z}_{q_j}$ if $j \ne i$. And since q_i is coprime to any other q_j, we
have $\gcd(s_i, q_i) = 1$. Thus by Corollary 15.26, s_i is invertible in $\mathbb{Z}_{q_i}$, say
$y_i s_i = 1$ in $\mathbb{Z}_{q_i}$. We take any $z_i \in \mathbb{Z}_N$ with $z_i = y_i$ in $\mathbb{Z}_{q_i}$. Since $q_i \le N$,
we have $\{0, \dots, q_i - 1\} \subseteq \{0, \dots, N - 1\}$ and may pick $z_i = y_i$. Now we let
$\ell_i = y_i s_i$. Then $\ell_i = 0$ in $\mathbb{Z}_{q_j}$ if $j \ne i$ and $\ell_i = 1$ in $\mathbb{Z}_{q_i}$. Thus $x = \sum_i a_i \ell_i$
is a simultaneous solution to the congruences in the assertion.

Now suppose x and x' both solve all congruences. Then $x' - x = 0$
in $\mathbb{Z}_{q_i}$ for any i. Thus every q_i divides $x' - x$. Since the q_i are pairwise
coprime, their product N divides $x' - x$.   □

This proof also provides an efficient method to calculate a simultaneous
solution.

ALGORITHM 15.45. Chinese Remainder Algorithm.

Input: Moduli $q_1, \ldots, q_r$ and residues $a_1, \ldots, a_r$ with all $a_i \in \mathbb{Z}$.
Output: A simultaneous solution $x \in \mathbb{Z}$ to the congruences $x = a_i$ in $\mathbb{Z}_{q_i}$
       or the error message "the q_i are not pairwise coprime".

1. $\ell_i \leftarrow \prod_{j \neq i} q_j$ for all i.
2. Try to compute $y_i = \ell_i^{-1}$ in $\mathbb{Z}_{q_i}$ for all i, using Algorithm 15.24. If it outputs "ℓ_i is not invertible" for some i, then return "the q_i are not pairwise coprime".
3. $m \leftarrow q_1 \cdots q_r$.
4. $x \leftarrow \sum_{i=0}^{r} a_i y_i \ell_i$ in $\mathbb{Z}_m$.
5. Return x.

In step 4, y_i is the integer computed as $y_i \in \mathbb{Z}_{q_i}$ in step 2.

THEOREM 15.46. *Algorithm 15.45 works correctly as specified.*

The Chinese Remainder Algorithm 15.45 also works for polynomials in $F[x]$, by replacing $\mathbb{Z}_m$ by $F[x]/(m)$. In fact, interpolation is the special case with moduli $q_i = x - u_i$. These are pairwise coprime: $c \cdot (x - u_i) - c \cdot (x - u_j) = 1$ for $i \neq j$, where $c = (u_j - u_i)^{-1}$. Here we need the pairwise distinctness of $u_1, \ldots, u_r$. Then we have the isomorphism

$$\chi \colon F[x]/\left((x - u_1) \cdots (x - u_r)\right) \longrightarrow F[x]/(x - u_1) \times \cdots \times F[x]/(x - u_r).$$

For $u \in F$, the map $F[x] \longrightarrow F[x]/(x - u)$ takes $f \in F[x]$ to $f(u)$. Surjectivity of χ says that for any $a_1, \ldots, a_r \in F$ there is a solution $f \in F[x]$ with $f(u_i) = a_i$ for all i, and injectivity says that f is uniquely determined modulo $(x - u_1) \cdots (x - u_r)$. In particular, there is exactly one f with $\deg f < r$. We have seen in Section 15.10 that this solution can be computed by Lagrange interpolation. In the special case where all a_i are zero, we conclude that $(x - u_1) \cdots (x - u_r)$ divides f.

COROLLARY 15.47. *For a field or integral domain F, a nonzero polynomial in $F[x]$ of degree n has at most n roots. In particular, $\sqrt{1} = \pm 1$ in F.*

The last statement holds, in particular, in $\mathbb{F}_q$ for a prime power q. It fails in a general ring $\mathbb{Z}_m$, but see Corollary 15.67 for an important special case.

The Chinese Remainder Algorithm is an invitation to *distributed computing*. If an expensive calculation modulo the large number N has to be performed, it may pay to do it modulo the small q_i's and construct the result modulo N via the Chinese Remainder Algorithm. We see some examples of such *modular algorithms*, for discrete logarithms in Section 4.6 and in Schoof's algorithm for the size of elliptic curves in Section 5.7.

Suppose that we want to compute some integer $a \in \mathbb{Z}$ for which we have an a priori bound $|a| \leq B$, and also some modulus N and $b \in \{0, \ldots, N-1\}$ with $a = b$ in $\mathbb{Z}_N$. If $N > 2B$, then $a = b$ in $\mathbb{Z}$. This follows since $|b + kN| > B$ for any $k \in \mathbb{Z} \setminus \{0\}$. If we take b in the symmetric remainder system R_N as in (15.12), then even $N > B$ is sufficient. We might obtain b by some Chinese Remainder computation.

15.12. Efficient exponentiation

The power x^e of an integer x has at least e bits (unless x is -1, 0, or 1), and if x is a nonconstant polynomial, it has degree at least e. For large e, these quantities are too big to be relevant for us. But we are interested in the problem in finite domains. Computing powers with huge (exponentially large in the security parameter n) exponents is a basic task in many cryptosystems, such as RSA and Diffie-Hellman .

The simplest idea for an efficient algorithm is to compute the powers x^{2^i} and multiply them together along the binary representation

$$(e)_2 = (e_{n-1}, \ldots, e_0)$$

of the positive integer e, with

$$e = \sum_{0 \leq i < n} e_i 2^i, \quad e_{n-1}, \ldots, e_0 \in \{0, 1\}, \text{ and } n = \lfloor \log_2 e \rfloor + 1,$$

so that n is the binary length (or number of bits) of e. This representation is unique, and indeed

$$x^e = x^{\sum_{0 \leq i < n} e_i 2^i} = \prod_{0 \leq i < n} (x^{2^i})^{e_i}.$$

As described, the algorithm requires the storage of all the relevant powers x^{2^i}. However, it is sufficient to square, at each step, the previous value x^j to get x^{2j}, and then multiply it by x if and only if the corresponding bit e_i is 1. An example is found in Figure 2.14. We describe the algorithm in the general context of a group; see Definition 15.51 below.

ALGORITHM 15.48.   Repeated squaring.

Input: A group G, a base $x \in G$, and an exponent $e \in \mathbb{Z}$ with $1 \leq e < \#G$.
Output: $x^e \in G$.

1. Let $\sum_{0 \leq i < n} e_i 2^i$ be the binary representation of e with $e_0, \ldots, e_{n-1} \in \{0, 1\}$, $e_{n-1} = 1$, and n its bit length.
2. $y \leftarrow x$.
3. For i from $n - 2$ downto 0 do steps 4-5
4. $y \leftarrow y^2$.
5.   If $e_i = 1$, then $y \leftarrow y \cdot x$.
6. Return y.

The number of multiplications is determined by the number $h = \#\{i : e_i = 1\}$ of ones in the binary representation of e, also called the *Hamming weight* of e. Clearly $h \leq n$.

THEOREM 15.49. *Algorithm 15.48 works correctly as specified. It computes x^e with $n - 1$ squarings and $h - 1$ further multiplications in G. The total number of operations in G is $n + h - 2 \leq 2n - 2$, where n is the bit length of the group order $\#G$.*

PROOF. For correctness, we show the loop invariant $y = x^{s_i}$ with $s_i = \sum_{i \leq j < n} e_j 2^{j-i}$, corresponding to an initial segment of the binary representation, as illustrated in Figure 2.14. In step 2 we have $s_{n-1} = 1 = e_{n-1} 2^{n-1-(n-1)}$ and $y = x^{s_{n-1}}$. Now we assume that the invariant holds for a given $i > 0$ and denote by y and y' the values of y before and after both steps 4 and 5, respectively. Then $y = x^{s_i}$ and $y^2 = x^{2s_i}$ with $2s_i = \sum_{i \leq j < n} e_j 2^{j-(i-1)}$. If $e_{i-1} = 0$, then

$$2s_i = e_{i-1} + \sum_{i \leq j < n} e_j 2^{j-(i-1)} = \sum_{i-1 \leq j < n} e_j 2^{j-(i-1)} = s_{i-1},$$

and $y' = x^{2s_i}$. If $e_{i-1} = 1$, then $y' = x^{2s_i+1}$ and

$$2s_i + 1 = \sum_{i \leq j < n} e_j 2^{j-(i-1)} + e_{i-1} = \sum_{i-1 \leq j < n} e_j 2^{j-(i-1)} = s_{i-1},$$

and the invariant is proven. When $i = 0$ is reached, then we return $x^{s_0} = x^{\sum_{0 \leq i < n} e_i 2^i} = x^e$.

For each e_i we have one squaring in step 4, which yields a total of $n - 1$ squarings. The multiplications in step 5 are only computed for $e_i = 1$, which sums up to $h - 1$ multiplications, since $e_{n-1} = 1$ causes no further operation. $\square$

The next section shows that for an arbitrary exponentiation in G, we may assume the exponent to be less than the group order $\#G$. If this is an n-bit number, then random exponents have Hamming weight $n/2$ on average, and the exponentiation is expected to use about $3n/2$ group operations on average. For the important case $G = \mathbb{Z}_N^\times$, squaring and multiplying modulo an n-bit number N can be done with $O(n^2)$ bit operations, so that an n-bit exponentation takes time $O(n^3)$.

COROLLARY 15.50. *Let N and e have bit length at most n, and $x \in \mathbb{Z}_N$. Then x^e in $\mathbb{Z}_N$ can be computed with $O(n^3)$ bit operations.* □

Algorithm 15.48 obviously works in any *semigroup*, where one has a binary operation "multiplication" which is associative. An example is multiplication in $\mathbb{N}$ or in $\mathbb{Z}_N$ for any $N \in \mathbb{N}$. Not every element has an inverse, so this is not a group. But this does not bother us, since inversion is not used in the algorithm.

15.13. Fermat, Euler, and Lagrange

Taking powers modulo N has some surprising properties. In the integers, high powers usually are very large numbers. No proper power of an integer $x \geq 2$ is 1. But in $\mathbb{Z}_N$, the powers x^1, x^2, x^3, ... of a given integer x cannot be all different, since there are only N possible outcomes. Eventually, the sequence has to repeat, and we find $x^i = x^j$ in $\mathbb{Z}_N$ for some $i \neq j$. Letting i be the larger of the two values, we have $i - j \geq 1$ and $x^{i-j} = 1$ in $\mathbb{Z}_N$. Let us be more specific.

We suppose that our modulus N is prime, and take $x \in \mathbb{Z}_N$. Then already Fermat knew that calculating x^{N-1} in $\mathbb{Z}_N$ is very easy: if $x = 0$ the result is, of course, 0; otherwise the result is 1, *regardless* of x. Thus if ALICE made the unlucky choice $s_A = 2578 = N - 1$ in the Diffie-Hellman Example 2.18, she would send $x = 2^{N-1} = 1$ to BOB, and an eavesdropper can deduce s_A from this, since 2 is a generator of $\mathbb{Z}_N^\times$.

We now study this issue in its proper setting, namely for groups, and derive a number of facts which are useful in several places.

DEFINITION 15.51. *A group is a nonempty set G with a two-input operation $\cdot\colon G \times G \to G$ and an element $1 \in G$ so that the following holds for all $a, b, c \in G$:*

(i) *associativity:*
$$(a \cdot b) \cdot c = a \cdot (b \cdot c),$$

(ii) *identity:*
$$a \cdot 1 = 1 \cdot a = a,$$

(iii) *inverse:*
$$\text{there exists some } d \in G \text{ with } a \cdot d = d \cdot a = 1.$$

The inverse d of a in (iii) is written as a^{-1}. The group can be denoted as $(G; \cdot, 1, ^{-1})$, but usually just the set name G is sufficient.

It is usual, for convenience of notation, to omit the symbol $\cdot$ from products. Thus $a \cdot b$ becomes the simpler ab. It does not matter which symbol we use to denote the operation. In Chapter 5, we use the alternative notation $+$ for $\cdot$, 0 for 1, and $-a$ for a^{-1}. The first notation is called a multiplicative group and the new one an additive group, denoted as $(G; +, 0, -)$. In permutation groups (Section 15.2), the operation is composition denoted as $\circ$.

For historical reasons, both the multiplicative and the additive notation play a role in cryptography. The group-based cryptographic systems of Chapter 4 were invented with the groups $G = \mathbb{Z}_p^\times$, using the multiplicative language. Already here, the additive exponent group $\mathbb{Z}_{p-1}$, isomorphic to $\mathbb{Z}_p^\times$ as in Figure 4.1, also occurs. But for an elliptic curve group $G = E$ (Chapter 5), the additive notation is historical standard stemming from a general concept (the Jacobian) in algebraic geometry. Thus multiplicative and additive groups are abstractly the same objects, described in different languages.

In fact, the addition in a ring R (Definition 15.2) gives R the structure of an additive group. This group is commutative (Definition 15.2 (iv)). In general, a group with this property is called an *Abelian group*.

Familiar examples are the additive groups of $\mathbb{Z}$, $\mathbb{Q}$, $\mathbb{R}$, and $\mathbb{C}$, already mentioned, the multiplicative groups $\mathbb{Q}^\times$, $\mathbb{R}^\times$, and $\mathbb{C}^\times$ of units in these three fields, and for any $N \in \mathbb{N}$, the additive group $\mathbb{Z}_N = \{0, 1, 2, \ldots, N - 1\}$ with addition modulo N, and the multiplicative group $\mathbb{Z}_N^\times = \{1 \le a < N : \gcd(a, N) = 1\}$ with multiplication modulo N.

A nonempty subset H of a group G is a *subgroup* of G if it is closed under multiplication and inversion, that is, if $ab \in H$ and $a^{-1} \in H$ for all $a, b \in H$. A subset S of a group G generates the set $\langle S \rangle \subseteq G$ consisting of all finite products of elements in S and their inverses. If $S = \{g_1, \ldots, g_s\}$ is finite, then we also write $\langle g_1, \ldots, g_s \rangle$ instead of $\langle \{g_1, \ldots, g_s\} \rangle$. $\langle S \rangle$ is the smallest (with respect to inclusion) subgroup of G containing S. For $G = \mathbb{Z}_{11}^\times = \{1, 2, 3, 4, 5, 6, 7, 8, 9, 10\}$ with multiplication modulo 11, we have $3^5 = 1$ and $\langle 3 \rangle = \{3, 9, 5, 4, 1\}$, and $\langle 2 \rangle = G$. If there is an element $g \in G$ that generates the entire group $G = \langle g \rangle$, then the group is called *cyclic*, and g a *generator* of G. Each of the four elements $2, 6, 7$, and 8 generates $\mathbb{Z}_{11}^\times$. The additive group of $\mathbb{R}$ does not have a finite generating set.

For a finite set A, $\#A$ is its *cardinality*, that is, the number of elements in it. The cardinality of a finite group is its *order*.

We now build up a sequence of some classical facts that are fundamental for computations in groups, and are used over and over again in this book.

DEFINITION 15.52. *Let G be a group, $x \in G$, and*

$$\langle x \rangle = \{ \ldots, x^{-3}, x^{-2}, x^{-1}, 1, x, x^2, x^3, \ldots \} = \{ x^j \colon j \in \mathbb{Z} \} \subseteq G$$

be the subgroup generated by x. Then either $\langle x \rangle$ is infinite (and then so is G), or it is finite. In the latter case, $\operatorname{ord} x = \#\langle x \rangle$ is the order of x.

LAGRANGE'S THEOREM 15.53. *Let G be a finite group and $H \subseteq G$ a subgroup.*

(i) *The order $\#H$ of H divides $\#G$.*

(ii) *If $H \neq G$, then $\#H \leq \#G/2$.*

(iii) *For any element x in G, we have $x^{\#G} = 1$.*

Lagrange's Theorem 15.53 says that $g^d = 1$, where 1 denotes the neutral element of the group G, and $d = \#G$. If $e = e'$ in $\mathbb{Z}_d$, then there is some integer u so that $e = e' + ud$, and $g^e = g^{e'+ud} = g^{e'} \cdot (g^d)^u = g^{e'}$. Hence we can restrict the exponent e to the case $0 \leq e < d$. This is also a reason why we should choose a group G of large order; the number of possible secret keys is restricted by d.

PROOF. (i) We consider the *cosets*

$$xH = \{ xy \colon y \in H \} \subseteq G$$

of H in G, for $x \in G$. Each coset has exactly $\#H$ many elements, since $xy = xz$ implies that $y = x^{-1}(xy) = x^{-1}(xz) = z$. Thus the mapping

$$\begin{aligned} H &\longrightarrow xH, \\ y &\longmapsto xy, \end{aligned}$$

is injective, hence bijective. Two cosets are either equal or disjoint:

$$x_1 H = x_2 H \text{ or } x_1 H \cap x_2 H = \varnothing.$$

Namely, if they have a common element, say $x_1 y = x_2 z$ with $y, z \in H$, then $x_2 = x_1(yz^{-1}) \in x_1 H$, since $yz^{-1} \in H$, and for any $y' \in H$ we have $x_2 y' = x_1(yz^{-1}y') \in x_1 H$. Hence $x_2 H \subseteq x_1 H$, and by symmetry, they are equal. Thus the cosets form a partition of G, and if there are k distinct ones, we have $k \cdot \#H = \#G$.

(ii) $\#H$ is a proper divisor of $\#G$, hence at most $\#G/2$.

(iii) We consider the subgroup $H = \langle x \rangle$ generated by x. According to the proof of (i), $xH \subseteq H$ has $\#H$ many elements, and thus $xH = H$. We take the product over the elements of each set:

$$x^{\#H} \prod_{y \in H} y = \prod_{y \in H} (xy) = \prod_{y \in H} y.$$

We can cancel the product on both sides, and find that $x^{\#H} = 1$. Since $\#G$ is an integer multiple of $\#H$ by (i), the claim follows. $\square$

COROLLARY 15.54. *Let G be a group, and $x \in G$ with $\langle x \rangle$ finite. Then*

$$\{i \in \mathbb{Z} : x^i = 1\} = \langle \operatorname{ord} x \rangle = \{j \cdot \operatorname{ord} x : j \in \mathbb{Z}\}$$

is the additive subgroup of all integer multiples of $\operatorname{ord} x$ in $\mathbb{Z}$. Thus

$$\operatorname{ord} x = \min\{i \geq 1 : x^i = 1\}, \quad x^{\operatorname{ord} x} = 1.$$

If G is finite, then $\operatorname{ord} x$ divides $\#G$. If $x^i = 1$, then $\operatorname{ord} x$ divides i.

PROOF.   Let k be the smallest positive integer with $x^k = 1$; such a k exists because $\langle x \rangle$ is finite. We can write any $i \in \mathbb{Z}$ as $i = qk + r$ with $0 \leq r = i \operatorname{rem} k < k$. Then $x^i = (x^k)^q \cdot x^r = x^r$. Thus $\langle x \rangle \subseteq \{1, x, \ldots, x^{k-1}\}$ and $\operatorname{ord} x \leq k$. On the other hand, $1, x, \ldots, x^{k-1}$ are pairwise distinct, since $x^i = x^j$ with $i > j$ would lead to an exponent $i - j < k$ with $x^{i-j} = 1$. Thus $\operatorname{ord} x \geq k$, and in fact $\operatorname{ord} x = k$.
For $i \in \mathbb{Z}$, we have

$$x^i = 1 \iff x^{i \operatorname{rem} k} = 1 \iff i \operatorname{rem} k = 0 \iff k \text{ divides } i.$$

This implies all claims except the last one, which follows from Lagrange's Theorem 15.53. $\square$

In order to apply Lagrange's Theorem to the group of units modulo an integer, we define the *Euler totient function* $\varphi(N)$ as the number $\varphi(N) = \#\mathbb{Z}_N^\times$ of units in $\mathbb{Z}_N$. Now we have the following consequence of Lagrange's Theorem 15.53 (iii).

EULER'S THEOREM 15.55. *For any $x \in \mathbb{Z}_N^\times$, we have $x^{\varphi(N)} = 1$ in $\mathbb{Z}_N^\times$.* $\square$

$\mathbb{Z}_N$ is a ring, and in the important special case when N is prime, $\mathbb{Z}_N$ is even a field, by Corollary 15.27, and hence

$$\varphi(N) = \#\mathbb{Z}_N^\times = \#(\mathbb{Z}_N \setminus \{0\}) = N - 1.$$

This implies the following.

FERMAT'S LITTLE THEOREM 15.56. *If N is a prime number, then for any nonzero $x \in \mathbb{Z}_N$ we have $x^{N-1} = 1$ in $\mathbb{Z}_N^\times$.*   □

The three fundamental theorems proven above imply the following.

COROLLARY 15.57.   *(i) Let G be a finite group, $d = \#G$, and $x \in G$. Then $\operatorname{ord} x$ divides d, and for any integer multiple m of d we have $x^m = 1$. For integers e and e' with $e = e'$ in $\mathbb{Z}_d$ we have $x^e = x^{e'}$.*

(ii) Let $x, y \in G$ with $xy = yx$ have coprime orders a and b, respectively. Then $\operatorname{ord}(xy) = ab$.

(iii) Let N be an integer and $x \in \mathbb{Z}_N^\times$. Then $\operatorname{ord} x$ divides $\varphi(N)$, and for any integer multiple m of $\varphi(N)$ we have $x^m = 1$.

(iv) Let N be a prime, and $x \in \mathbb{Z}_N^\times$. Then $\operatorname{ord} x$ divides $N - 1$, and for any integer multiple m of $N - 1$, we have $x^m = 1$. For any $x \in \mathbb{Z}_N$, we have $x^{m+1} = x$, provided that the exponent is positive. In particular, $x^N = x$ for all $x \in \mathbb{Z}_N$.

(v) If N is prime and $x \in \mathbb{Z}_N$, then $x^N = x$.

PROOF.   For (ii), we observe that $\operatorname{ord}(xy) \mid ab$. Let p be a prime divisor of ab with $\operatorname{ord}(xy)$ dividing ab/p. Then p divides one of the factors, say a. Then $1 = (xy)^{ab/p} = x^{ab/p} \cdot y^{b \cdot a/p} = x^{ab/p}$, and $a = \operatorname{ord}(x)$ divides $a/p \cdot b$ by Corollary 15.57. Since $\gcd(a, b) = 1$, it follows that a divides a/p, a contradiction.

In (iv), we only have to verify the second claim. For $x \in \mathbb{Z}_N^\times$, it follows from $x^m = 1$, and it clearly holds for the only remaining element $x = 0$ of $\mathbb{Z}_N$.   □

A *finite field* $\mathbb{F}_q$ is a field with a finite number q of elements. Such a field exists if and only if q is a power of a prime. When q is prime, the notions $\mathbb{F}_q$ and $\mathbb{Z}_q$ coincide, and we use them interchangeably in this text. But beware: $\mathbb{F}_q$ and $\mathbb{Z}_q$ are quite different when q is not prime. Corollary 15.57 (v) is a reformulation of Fermat's Little Theorem 15.56, and the analog also holds in general:

(15.58) $x^q = x$ for all $x \in \mathbb{F}_q$.

COROLLARY 15.59. *Let x be an element of a group, and $d \in \mathbb{N}_{\geq 1}$ such that $x^d = 1$. Then the order of x is d if and only if $x^{d/p} \neq 1$ for all prime divisors p of d.*

PROOF. Assuming the last condition, by Corollary 15.54, $\operatorname{ord} x$ is a positive integer which divides d and does not divide d/p for any prime divisor p of d. Thus $\operatorname{ord} x = d$.

Let k be the order of x. If $k = d$, then $x^{d/p} \neq 1$ for any p by definition. If $k < d$, then there exists some prime divisor p of d so that k divides d/p, say $ka = d/p$ for some integer a. Then $x^{d/p} = (x^k)^a = 1$. □

This corollary facilitates the calculation of element orders in small groups. We start with $d = \#G$, so that $x^d = 1$ by Lagrange's Theorem 15.53 (iii) and factor d. If we find $x^{d/p} = 1$ for some prime divisor p of d, we replace d by d/p and continue. This reduces order finding to factorization, provided the group order $\#G$ is known. See Section 3.14 for a reduction in the other direction, without known group order.

It is useful to know the order of a power of an element whose order is given. For example, in Schnorr's signature scheme (Section 8.4) we have to construct an element of small order from one of large order.

THEOREM 15.60. *Let $x \in G$ have order d, and let a and b be positive integers.*

(i) *The order of x^a is $d/\gcd(a, d)$.*

(ii) *If $\gcd(a, d) = 1$, then x^a has order d.*

(iii) *If $x^a = x^b = 1$, then $x^{\gcd(a,b)} = 1$.*

(iv) *If a divides d, then x^a has order d/a, and*

$$\langle x^a \rangle = \{y^a : y \in \langle x \rangle\} = \{z \in \langle x \rangle : z^{d/a} = 1\}$$

is the unique subgroup of $\langle x \rangle$ of order d/a.

PROOF. (i) We let $c = \gcd(a, d)$, and have $(x^a)^{d/c} = (x^d)^{a/c} = 1^{a/c} = 1$. Now we take integers s and t with $sa + td = c$, as provided by Algorithm 15.19, and a prime divisor p of d/c, and let $y = (x^a)^{(d/c)/p}$. If $y = 1$, then

$$1 = y^s = (x^{sa/c})^{d/p} = (x^{(g-td)/c})^{d/p} = x^{d/p},$$

contradicting Corollary 15.59. Thus x^a has order d/c, again by the corollary. Claim (ii) is a special case of (i). In (iii), d divides a and b by Corollary 15.54, hence also $\gcd(a, b)$. The first claim in (iv) follows from (i). Let $H = \langle x^a \rangle$ and $K = \{z \in \langle x \rangle : z^{d/a} = 1\}$. The first equality is clear. If $y = x^i \in \langle x \rangle$, then $(y^a)^{d/a} = y^d = (x^d)^i = 1$ and $y \in K$. If $z = x^j \in K$, then $x^{jd/a} = z^{d/a} = 1$. By Corollary 15.54, d divides jd/a, so that $j/a = (jd/a)/d$ is an integer, say b. Then $z = x^j = x^{ab} = (x^b)^a \in H$. It follows that $H = K$. Furthermore, H is clearly a subgroup of $\langle x \rangle$ and consists of $1 = x^0, x^a, x^{2a}, \ldots, x^{(d/a-1)a}$. These elements are pairwise distinct, and $\#H = d/a$. The uniqueness follows from the description of K. $\qquad\square$

COROLLARY 15.61. *Let $G = \langle g \rangle$ be cyclic of finite order d.*

(i) *Let a be a positive integer with $\gcd(a, d) \neq 1$. Then $G^a = \langle g^a \rangle$ is a proper subgroup of G, contains exactly those elements of G whose order divides $d/\gcd(a, d)$, and $\#G^a = d/\gcd(a, d) \leq \#G/2$.*

(ii) *Let 2^j with $j \geq 0$ be the maximal power of 2 dividing d. Then exactly $d/2^j$ elements of G have odd order. If d is even, these are at most half the elements of G.*

PROOF.   (i) Let $c = \gcd(a, d) = sa + tb$ with integers c, s, and t. Theorem 15.60 (i) says that g^a has order d/c and thus $G^a = \langle x^a \rangle$ is a subgroup of G of order $d/\gcd(a, d) \leq d/2 = \#G/2$. From Theorem 15.60 (iv) we find $G^a = G^c = \{z \in G \colon z^{d/c} = 1\}$. Furthermore, we have $g^a = g^{c \cdot a/c} \in G^c$ and $g^c = g^{sa+tb} = (g^a)^s \cdot 1 \in G^a$, so that $G^a = G^c$. (ii) If $x = g^a$ has odd order b, then 2^j divides ab in $\mathbb{Z}_d$, so that 2^j divides a. Thus $x \in \langle g^{2^j} \rangle$ and $\#G^{2^j} = d/2^j$ by (i). $\qquad\square$

We usually identify $\mathbb{Z}_N$ with $\{0, 1, \ldots, N-1\}$. For an integer a, $a \in \mathbb{Z}_N$ means the image of a under this identification.

COROLLARY 15.62. *Let G be a group of order d and $a \in \mathbb{Z}$ coprime to d. Then the exponentiation map $\pi_a \colon x \mapsto x^a$ is a permutation of G.*

PROOF.   We take integers s and t with $sa + td = 1$. Then

$$x = x^1 = x^{sa+td} = x^{as} \cdot (x^d)^t = (x^a)^s,$$

using Lagrange's Theorem 15.53 (iii). Therefore the map $\pi_s \colon y \mapsto y^s$ is an inverse of π_a, and hence π_a is a bijection. $\qquad\square$

The Chinese Remainder Theorem 15.44 deals with rings, but also transfers to their unit groups.

COROLLARY 15.63. *Suppose that the integer N factors as $N = q_1 \cdots q_r$ with pairwise coprime $q_1, \ldots, q_r$. Then the group homomorphism*

$$\mathbb{Z}_N^\times \quad \to \quad \mathbb{Z}_{q_1}^\times \times \cdots \times \mathbb{Z}_{q_r}^\times$$
$$x \bmod N \quad \mapsto \quad (x \bmod q_1, \ldots, x \bmod q_r)$$

is an isomorphism.

Knowing this, it is fairly easy to calculate $\varphi(N)$, provided that one has the prime factorization of N. But still nobody knows a polynomial-time algorithm to find the factorization of a given integer N.

LEMMA 15.64. *Suppose that $N = p_1^{e_1} \cdots p_r^{e_r}$ with pairwise different primes $p_1, \ldots, p_r$, and all $e_i \geq 1$. Then*

$$\varphi(N) = p_1^{e_1-1}(p_1 - 1) \cdots p_r^{e_r-1}(p_r - 1) = N \cdot \left(1 - \frac{1}{p_1}\right) \cdots \left(1 - \frac{1}{p_r}\right).$$

PROOF.   Suppose p is prime and $e \geq 1$. Then precisely the multiples of p are not invertible modulo p^e (Corollary 15.26). There are p^{e-1} such multiples among $0, \ldots, p^e - 1$, and hence $\varphi(p^e) = p^e - p^{e-1} = p^{e-1}(p-1)$.

By Corollary 15.63, $\varphi(N) = \varphi(p_1^{e_1}) \cdots \varphi(p_r^{e_r})$, which proves the stated formula.   □

As an example, $\varphi(12) = \varphi(2^2 \cdot 3) = 2 \cdot (2 - 1) \cdot (3 - 1) = 4$, and the four units are given in Example 15.23 (iii). We can easily read off what happens for a product N of two primes, as it occurs in RSA.

COROLLARY 15.65. *Let p and q be different primes. Then $\varphi(p \cdot q) = (p - 1) \cdot (q - 1)$.*   □

An important example of a cyclic group is the multiplicative group $\mathbb{F}_q^\times$ of a finite field with q elements. In other words, $\mathbb{F}_q^\times$ contains an element of order $q - 1$.

THEOREM 15.66. *Let q be a prime power.*

(i) *The multiplicative group $\mathbb{F}_q^\times$ is cyclic.*

(ii) *If furthermore q is odd, then the multiplicative group $\mathbb{Z}_q^\times$ is cyclic and at most half its elements have odd order.*

PROOF.   (i) We take the prime factorization $q - 1 = \prod_{1 \leq i \leq m} \ell_i^{e_i}$, where the ℓ_i are pairwise distinct primes and all $e_i \geq 1$. For $i \leq m$, let $r_i = \ell_i^{e_i}$

and $G_i = G^{(q-1)/\ell_i} \subseteq \mathbb{F}_q^\times$. Then $\#G_i = (q-1)/\ell_i < q-1$ by Theorem 15.60 (iv) and there is some $x_i \in \mathbb{F}_q^\times$ with $x_i^{(q-1)/\ell_i} \neq 1$ (Corollary 15.47). Let $y_i = x_i^{(q-1)/r_i}$. Then $y_i^{r_i} = 1$ and $\operatorname{ord} y_i \mid r_i$. By the property of x_i, we have $\operatorname{ord} y_i \nmid r_i/\ell_i$. Thus $\operatorname{ord} y_i = r_i$. For $y = \prod_{1 \le i \le m} y_i$, we find $\operatorname{ord} y = \prod_{1 \le i \le m} \operatorname{ord} y_i = q-1$ by Corollary 15.57 (ii).

(ii) We write $q = p^e$ with an odd prime p and $e \ge 1$, so that $\varphi(q) = \#\mathbb{Z}_q^\times = (p-1)p^{e-1}$ (Lemma 15.64). From (i) we know that $\mathbb{Z}_p^\times = \mathbb{F}_p^\times$ is cyclic, take some $g \in \mathbb{Z}_q^\times$ so that $(g \bmod p) \in \mathbb{Z}_p^\times$ generates $\mathbb{Z}_p^\times$, and $h = g^{p^{e-1}} \in \mathbb{Z}_q^\times$. Then $h = g$ in $\mathbb{Z}_p$ (Corollary 15.57 (v)), $h^{p-1} = 1$, and for a divisor $r < p-1$ of $p-1$, we have $h^r \neq 1$, since otherwise $g^r = (g^{p^{e-1}})^r = h^r = 1$ in $\mathbb{Z}_p$. For $i \in \{e-2, e-1\}$, we have $(1+p)^{p^i} = 1 + p^{i+1} + \binom{p^i}{2}p^2 + \cdots = 1 + p^{i+1}$ in $\mathbb{Z}_q$ and hence $1+p$ has order p^{e-1}. By Corollary 15.57 (ii), $h \cdot (1+p)$ has order $(p-1)p^{e-1}$ in $\mathbb{Z}_q^\times$ and is therefore a generator of this group. Since $p-1$ is even, the second claim follows from Corollary 15.61 (ii). $\qquad\square$

COROLLARY 15.67. *Let q be a power of an odd prime. Then $\sqrt{1} = \pm 1$ in $\mathbb{Z}_q$.*

PROOF.   Clearly $\pm 1 \subseteq \mathbb{Z}_q^\times$ is a subgroup of order 2. Since $\mathbb{F}_q^\times$ is cyclic (Theorem 15.66) of order $q-1$ and all $z \in \sqrt{1}$ satisfy $z^2 = 1$, the claim follows from Theorem 15.60 (iv) with $a = (q-1)/2$. $\qquad\square$

For $q = 2^e$ with $e \ge 3$, we have $\mathbb{Z}_q^\times \cong \mathbb{Z}_2 \times \mathbb{Z}_{2^{e-2}}$ and $\sqrt{1} = \pm 1 \cup (2^{e-1} \pm 1)$, but we do not need this.

15.14. Squares and the Jacobi symbol

For integers or real numbers, it is quite easy to say which ones are squares, and to find a square root if one exists. It turns out that the same is true when we compute modulo a prime number (or a power of a prime), but not modulo a large number which has at least two distinct prime factors. The material of this section is fundamental for the HKS cryptosystem (Section 9.9) and the Blum-Blum-Shub generator (Section 11.8).

We start with the easiest case, namely where p is a prime. Figure 15.2 shows the squaring function in two cases and for the composite number 21. The following describes the situation for prime powers.

THEOREM 15.68. *Let p be an odd prime, $e \ge 1$ and $\square_{p^e} = \{b^2 : b \in \mathbb{Z}_{p^e}^\times\}$ in $\mathbb{Z}_{p^e}^\times$. Then $\#\mathbb{Z}_{p^e}^\times = \varphi(p^e) = p^{e-1}(p-1)$, and*

(i) $\#\square_{p^e} = \varphi(p^e)/2$.

(ii) For any $a \in \mathbb{Z}_{p^e}^{\times}$, $a \in \square_{p^e} \Longleftrightarrow a^{\varphi(p^e)/2} = 1$.

(iii) Any $a \in \square_{p^e}$ has exactly two square roots b_1 and b_2, and $b_1 + b_2 = 0$.

(iv) There is a probabilistic polynomial-time algorithm which, on input p^e and $a \in \mathbb{Z}_{p^e}^{\times}$, determines whether $a \in \square_{p^e}$, and if so, computes a square root $b \in \mathbb{Z}_{p^e}^{\times}$ with $a = b^2$.

In particular, for an odd prime p and $a \in \mathbb{Z}_p^{\times}$ we have

(15.69) a is a square $\Longleftrightarrow a^{(p-1)/2} = 1 \Longleftrightarrow a$ is a root of $x^{(p-1)/2} - 1$,

and there are exactly $(p-1)/2$ squares. We have the same number of nonsquares, and these are the roots of $x^{(p-1)/2} + 1$.

There is a concise way of associating to each number a the value of an indicator yes/no telling whether a is a square or not. Taking ± 1 for yes/no, we have the *Legendre symbol* for $a \in \mathbb{Z}$ and a prime p:

$$\left(\frac{a}{p}\right) = \begin{cases} 0 & \text{if } p \mid a, \\ 1 & \text{if } p \nmid a \text{ and } a \text{ is a square in } \mathbb{Z}_p^{\times}, \\ -1 & \text{otherwise.} \end{cases}$$

Fermat's Little Theorem 15.56 says that $a^{p-1} = 1$ in $\mathbb{Z}_p$ for all $a \neq 0$. By (15.69), $x^{(p-1)/2} - 1$ has the $(p-1)/2$ nonzero squares as its roots, and since its degree is $(p-1)/2$, there are no others. It follows that $\left(\frac{a}{p}\right) = a^{(p-1)/2}$ in $\mathbb{Z}_p$. If $p \equiv 1 \bmod 4$, then the exponent is even and $-1 \in \mathbb{Z}_p^{\times}$ is a square. Otherwise, $p \equiv 3 \bmod 4$ (excepting $p = 2$), the exponent is odd and -1 is a nonsquare.

Now $(p-1)/2$ elements of $\mathbb{Z}_p^{\times}$ are squares, and $(p-1)/2$ are nonsquares, so that half of the elements of $\mathbb{Z}_p^{\times}$ have Legendre symbol 1, and half have -1. We have

$$\left(\frac{-1}{p}\right) = 1 \Longleftrightarrow p = 1 \bmod 4.$$

When $N = p \cdot q$ is the product of two distinct odd primes, then the situation is much more interesting. On the one hand, we can again consider the set

$$\square_N = \{b^2 : b \in \mathbb{Z}_N^{\times}\}$$

of squares modulo N. The Chinese Remainder Theorem 15.44 decomposes $\mathbb{Z}_N^{\times}$ into two constituents:

$$\mathbb{Z}_N^{\times} \cong \mathbb{Z}_p^{\times} \times \mathbb{Z}_q^{\times}.$$

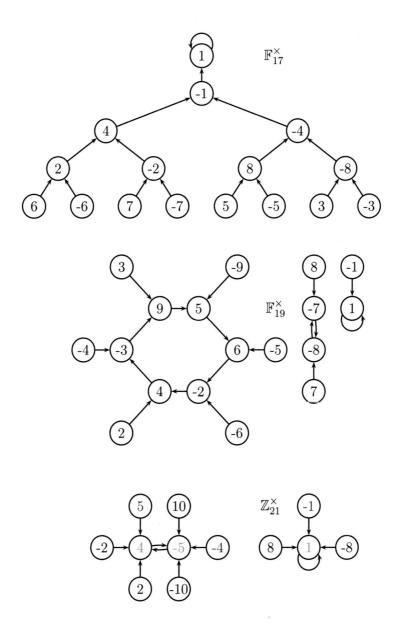

Figure 15.2: Squaring modulo 17, 19, and 21. The elements of $\square_{21}$ are shown in green, each with four square roots, and those of $\boxtimes_{21}$ in red.

Now some $a \in \mathbb{Z}_N^\times$ is a square if and only if it is a square both in $\mathbb{Z}_p^\times$ and in $\mathbb{Z}_q^\times$. These conditions are independent, and therefore only a quarter of the $\varphi(N) = (p-1)(q-1)$ elements of $\mathbb{Z}_N^\times$ are squares.

On the other hand, we can generalize the Legendre symbol to our situation, where it is now called the *Jacobi symbol* of $a \in \mathbb{Z}$:

$$\left(\frac{a}{N}\right) = \left(\frac{a}{p}\right)\left(\frac{a}{q}\right).$$

It is easy to compute $\left(\frac{a}{N}\right)$ by a method similar to the Euclidean Algorithm 15.19, without factoring N. This takes $O(n^2)$ bit operations if a and N are n-bit numbers.

While modulo prime numbers, being a square and having Legendre symbol 1 are synonymous, it now turns out that all squares have Jacobi symbol 1, but not the other way around. We denote as $\boxtimes_N$ the set of the nonsquares in $\mathbb{Z}_N^\times$ with Jacobi symbol 1. The reader should understand these two notions clearly; if we substitute p for N in the definition of $\boxtimes_N$, we obtain the empty set. So for a prime p, we define $\boxtimes_p = \mathbb{Z}_p^\times \smallsetminus \square_p$ as the set of antisquares.

The *quadratic residuosity problem* in $\mathbb{Z}_N$ is to decide on input $a \in \square_N \cup \boxtimes_N$ whether $a \in \square_N$. Of course, given the factors p and q, this becomes easy, since we can compute $\left(\frac{a}{p}\right)$ and $\left(\frac{a}{q}\right)$. But no polynomial-time algorithm is known if these factors are not provided, and in Section 11.8 we show that in fact this is a hard problem if factoring is hard.

Under the isomorphism

$$\chi\colon \mathbb{Z}_N^\times \longrightarrow \mathbb{Z}_p^\times \times \mathbb{Z}_q^\times$$

of Corollary 15.63, we have

(15.70)
$$\begin{aligned}
\chi(\square_N) &= \square_p \times \square_q, \\
\chi(\boxtimes_N) &= \boxtimes_p \times \boxtimes_q.
\end{aligned}$$

We consider the squaring map $\sigma_p\colon \mathbb{Z}_p^\times \longrightarrow \square_p \subseteq \mathbb{Z}_p^\times$ with $\sigma_p(a) = a^2$; see Figure 15.2 for illustrations. If p is 3 modulo 4, then -1 is not a square modulo p, and exactly one of the two square roots a and $-a$ of a^2 is a square.

We now assume that p and q are both 3 modulo 4. Then $N = pq$ is called a *Blum integer*, after Manuel Blum. If $\chi(a) = (u, v)$, then $\chi(a^2)$ has the four square roots

$$(u, v), (-u, v), (u, -v), (-u, -v).$$

Exactly one of them is a square, and χ^{-1} of this square is called the *principal (square) root* of a^2. If, say, $u \in \square_p$ and $v \in \boxtimes_p$, then $(u, -v)$ is the square among the four. This situation is illustrated in Figure 15.3.

EXAMPLE 15.71. We let $p = 3$ and $q = 7$, so that $N = 21$ and $\mathbb{Z}_{21}^{\times} = \{-10, -8, -5, -4, -2, -1, 1, 2, 4, 5, 8, 10\}$ in the symmetric system. Then $\square_3 = \{1\}$, $\boxtimes_3 = \{-1\}$, $\square_7 = \{-3, 1, 2, \}$, and $\boxtimes_7 = \{-2, -1, 3\}$. Figure 15.3 now becomes Figure 15.4. Not surprisingly, 1 is the principal root of 1, and 4 that of $-5 = 16$. But also -5 (and not 2) is the principal root of 4. In other words, $-5 = 16$ is the principal root of $25 = 4$ in $\mathbb{Z}_{21}$. ◊

	$\left(\frac{a}{q}\right) = 1$	$\left(\frac{a}{q}\right) = -1$
$\left(\frac{a}{p}\right) = 1$	$\left(\frac{a}{N}\right) = 1$ $\square_N$ $\square_p \times \square_q$ $(u, -v)$	$\left(\frac{a}{N}\right) = -1$ $\square_p \times \boxtimes_q$ (u, v)
$\left(\frac{a}{p}\right) = -1$	$\left(\frac{a}{N}\right) = -1$ $\boxtimes_p \times \square_q$ $(-u, -v)$	$\left(\frac{a}{N}\right) = 1$ $\boxtimes_N$ $\boxtimes_p \times \boxtimes_q$ $(-u, v)$

Figure 15.3: The values u and v are explained in the text.

	$-3, 1, 2$	$-2, -1, 3$
1	$4 \leftrightarrow (1, -3)$ $1 \leftrightarrow (1, 1)$ $-5 \leftrightarrow (1, 2)$	$-2 \leftrightarrow (1, -2)$ $-8 \leftrightarrow (1, -1)$ $10 \leftrightarrow (1, 3)$
-1	$-10 \leftrightarrow (-1, -3)$ $8 \leftrightarrow (-1, 1)$ $2 \leftrightarrow (-1, 2)$	$5 \leftrightarrow (-1, -2)$ $-1 \leftrightarrow (-1, -1)$ $-4 \leftrightarrow (-1, 3)$

Figure 15.4: Figure 15.3 for $p = 3, q = 7$, and $N = 21$.

In the language of groups in Section 15.13, $\square_N$ is a subgroup of $\mathbb{Z}_N^{\times}$ with $\phi(N)/4$ elements. The squaring map $\sigma_N \colon \mathbb{Z}_N^{\times} \longrightarrow \square_N$ is a homomorphism which always maps four elements $(\pm u, \pm v)$ to one, namely to

(u^2, v^2). Then $\square_N$ has four cosets $\square_N$, $\boxtimes_N$, and, say, C_0 and C_1, and exactly one of these four square roots lies in each coset. In particular, σ induces a bijection on $\square_N$. Multiplication by -1 gives a bijection between $\square_N$ and $\boxtimes_N$ (and between C_0 and C_1). The residue class group $\mathbb{Z}_N^\times / \square_N$ is isomorphic to $\pm 1 \times \pm 1 \cong \mathbb{Z}_2 \times \mathbb{Z}_2$, with $\square_N \leftrightarrow (1,1)$ and $\boxtimes_N \leftrightarrow (-1,-1)$. Here ± 1 is the "multiplicative version" of $\mathbb{Z}_2$. The corresponding mapping $\mathbb{Z}_N^\times \longrightarrow \pm 1 \times \pm 1$ is given by $a \longmapsto \left(\left(\frac{a}{p}\right), \left(\frac{a}{q}\right) \right)$.

15.15. Linear algebra

We recall some basic concepts from linear algebra. Let $M \in F^{n \times n}$ be an $n \times n$ matrix over a field F. Then M is *nonsingular* if its *determinant* $\det M$ is nonzero. In this case, for any $y \in F^n$ one can compute the unique $x \in F^n$ with $Mx = y$ by Gaussian elimination, using $O(n^3)$ operations in F.

If M is a $k \times n$ matrix with entries M_{ij} for $1 \le i \le k$ and $1 \le j \le n$, then its *transpose* M^T is the $n \times k$ matrix with entries M_{ji}. For $k = n$ and an invertible M, we have $(M^{-1})^T = (M^T)^{-1}$.

If F is a field and $S \subseteq F^n$, then the *span* $\operatorname{span}_F(S)$ of S over F is the set of all finite linear combinations of elements of S with coefficients from F. The following is only used in Chapter 13. For $x, y \in \mathbb{R}^n$, we consider the *inner product* $x \star y = \sum_i x_i y_i$ and the *norm* $\|x\| = (x \star y)^{1/2}$. Then x and y are *orthogonal* if and only if $x \star y = 0$.

THEOREM 15.72. *Let $x, y \in \mathbb{R}^n$. Then the following hold.*

(i) *(Cauchy-Schwarz inequality)* $|x \star y| \le \|x\| \|y\|$.

(ii) $\sum_{1 \le i \le n} |x_i| \le n^{1/2} \|x\|$.

PROOF. (i) If x and y are linearly dependent, we can write $y = c \cdot x$ for some real number $c \ne 0$. Then

$$x \star y = c \cdot (x \star x) = c \cdot \|x\| \cdot \|x\| = \|x\| \cdot \|y\|.$$

If x, y are linearly independent, we consider the quadratic function

$$f(c) = \|c \cdot x + y\|^2 = \|x\|^2 c^2 + 2(x \star y)c + \|y\|^2$$

of $c \in \mathbb{R}$. The roots of f are

$$c_\pm = \frac{1}{\|x\|^2} \left(-(x \star y) \pm \sqrt{(x \star y)^2 - \|x\|^2 \|y\|^2} \right).$$

Since $c \cdot x + y \ne 0$ and hence $f(c) > 0$ for every real c, these roots are not real. This implies that $(x \star y)^2 - \|x\|^2 \|y\|^2 < 0$ and the claim.

(ii) For $1 \leq i \leq n$, let

$$y_i = \begin{cases} 1 & \text{if } x_i > 0, \\ 0 & \text{if } x_i = 0, \\ -1 & \text{if } x_i < 0 \end{cases}$$

be the sign of x_i, and $y = (y_1, \ldots, y_n)$. Then by (i)

$$\sum_{1 \leq i \leq n} |x_i| = |x \star y| \leq \|x\| \cdot \|y\| \leq n^{1/2} \cdot \|x\|. \qquad \square$$

We briefly review the Gram-Schmidt orthogonalization procedure from linear algebra. Given an arbitrary basis $(b_1, \ldots, b_\ell)$ of an ℓ-dimensional subspace of $\mathbb{R}^n$, it computes an orthogonal basis $(b_1^*, \ldots, b_\ell^*)$ of the same subspace by essentially performing Gaussian elimination on the *Gram matrix* $(b_i \star b_j)_{1 \leq i,j \leq \ell} \in \mathbb{R}^{\ell \times \ell}$. The b_i^* are defined inductively as follows.

(15.73) $b_i^* = b_i - \displaystyle\sum_{1 \leq j < i} \mu_{ij} b_j^*$, where $\mu_{ij} = \dfrac{b_i \star b_j^*}{b_j^* \star b_j^*} = \dfrac{b_i \star b_j^*}{\|b_j^*\|^2}$ for $1 \leq j < i$.

In particular, $b_1^* = b_1$. Then $(b_1^*, \ldots, b_\ell^*)$ is the *Gram-Schmidt orthogonal basis* of $(b_1, \ldots, b_\ell)$, and the b_i^* together with the μ_{ij} form the *Gram-Schmidt orthogonalization* (or GSO for short) of $b_1, \ldots, b_\ell$. The GSO has rational coefficients if $b_1, \ldots, b_\ell$ have, and then the cost for computing the GSO is $O(n^3)$ arithmetic operations in $\mathbb{Q}$, since $\ell \leq n$.

We can rewrite (15.73) as

(15.74) $$b_i = \sum_{1 \leq j \leq i} \mu_{ij} b_j^* \text{ with } \mu_{ii} = 1.$$

Since the b_j^* are linearly independent, the $\mu_{ij} \in \mathbb{R}$ are uniquely determined by (15.74). We consider the b_i and b_i^* to be row vectors in $\mathbb{R}^n$, and define two matrices $B, B^* \in \mathbb{R}^{\ell \times n}$ and a matrix M in $\mathbb{R}^{\ell \times \ell}$:

(15.75) $$B = \begin{pmatrix} b_1 \\ \vdots \\ b_\ell \end{pmatrix}, \quad B^* = \begin{pmatrix} b_1^* \\ \vdots \\ b_\ell^* \end{pmatrix}, \quad M = (\mu_{ij})_{1 \leq i,j \leq \ell},$$

where $\mu_{ii} = 1$ for $i \leq \ell$, and $\mu_{ij} = 0$ for $1 \leq i < j \leq \ell$. Then M is lower triangular with ones on the diagonal, and (15.73) reads:

(15.76) $$B = \begin{pmatrix} b_1 \\ \vdots \\ b_\ell \end{pmatrix} = \begin{pmatrix} 1 & & 0 \\ \vdots & \ddots & \\ \mu_{n1} & \cdots & 1 \end{pmatrix} \begin{pmatrix} b_1^* \\ \vdots \\ b_\ell^* \end{pmatrix} = M \cdot B^*.$$

EXAMPLE 15.77. We let $\ell = n = 3$, $b_1 = (1,1,0)$, $b_2 = (1,0,1)$, $b_3 = (0,1,1)$, and calculate $b_1^* = b_1 = (1,1,0)$,

$$\mu_{21} = \frac{b_2 \star b_1^*}{b_1^* \star b_1^*} = \frac{1}{2}, \quad b_2^* = b_2 - \mu_{21}b_1^* = \left(\frac{1}{2}, -\frac{1}{2}, 1\right),$$

$$\mu_{31} = \frac{b_3 \star b_1^*}{b_1^* \star b_1^*} = \frac{1}{2}, \quad \mu_{32} = \frac{b_3 \star b_2^*}{b_2^* \star b_2^*} = \frac{1}{3},$$

$$b_3^* = b_3 - \mu_{31}b_1^* - \mu_{32}b_2^* = \left(-\frac{2}{3}, \frac{2}{3}, \frac{2}{3}\right),$$

$$B = \begin{pmatrix} 1 & 1 & 0 \\ 1 & 0 & 1 \\ 0 & 1 & 1 \end{pmatrix} = \begin{pmatrix} 1 & 0 & 0 \\ \frac{1}{2} & 1 & 0 \\ \frac{1}{2} & \frac{1}{3} & 1 \end{pmatrix} \cdot \begin{pmatrix} 1 & 1 & 0 \\ \frac{1}{2} & -\frac{1}{2} & 1 \\ -\frac{2}{3} & \frac{2}{3} & \frac{2}{3} \end{pmatrix} = M \cdot B^*.$$

We have $\|b_1\|^2 = \|b_2\|^2 = \|b_3\|^2 = 2$, $\|b_1^*\|^2 = 2$, $\|b_2^*\|^2 = 3/2$, $\|b_3^*\|^2 = 4/3$ and $\det B^* = -2$.   ◇

The following theorem collects the properties of the Gram-Schmidt orthogonalization that we will need. The proof is left as Exercise 13.2.

THEOREM 15.78. Let $b_1, \ldots, b_\ell \in \mathbb{R}^n$ be linearly independent, and $b_i^*, \ldots, b_\ell^*$ their Gram-Schmidt orthogonalization. For $0 \le k \le \ell$ let $U_k = \sum_{1 \le i \le k} \mathbb{R}b_i \subseteq \mathbb{R}^n$ be the $\mathbb{R}$-subspace spanned by $b_1, \ldots, b_k$.

(i) $\sum_{1 \le i \le k} \mathbb{R}b_i^* = U_k$.

(ii) b_k^* is the projection of b_k onto the orthogonal complement

$$U_{k-1}^\perp = \{b \in \mathbb{R}^n : b \star u = 0 \text{ for all } u \in U_{k-1}\}$$

of U_{k-1}, and hence in particular $\|b_k^*\| \le \|b_k\|$.

(iii) $b_1^*, \ldots, b_\ell^*$ are pairwise orthogonal, that is, $b_i^* \star b_j^* = 0$ if $i \ne j$.

(iv) $\det \begin{pmatrix} b_1 \\ \vdots \\ b_\ell \end{pmatrix} = \det \begin{pmatrix} b_1^* \\ \vdots \\ b_\ell^* \end{pmatrix}$.

EXAMPLE 13.5 CONTINUED. We have $b_1^* = b_1 = (12, 2)$,

$$\mu_{21} = \frac{b_2 \star b_1^*}{b_1^* \star b_1^*} = \frac{41}{37}, \quad b_2^* = b_2 - \mu_{21}b_1^* = \left(-\frac{11}{37}, \frac{66}{37}\right).$$

This is illustrated in Figure 15.5: the vector b_2^* (green) is the projection of b_2 (blue) onto the orthogonal complement of b_1 (red).

An immediate consequence of Theorem 15.78 is the following famous inequality.

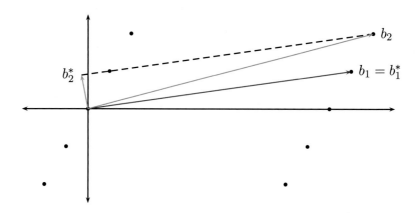

Figure 15.5: The Gram-Schmidt orthogonal basis of $(12, 2)$ and $(13, 4)$.

THEOREM 15.79 (Hadamard's inequality). *Let $B \in \mathbb{R}^{n \times n}$ have row vectors $b_1, \ldots, b_n \in \mathbb{R}^{1 \times n}$, and all entries at most R in absolute value for some $R \in \mathbb{R}$. Then*

$$|\det B| \le \|b_1\| \cdots \|b_n\| \le n^{n/2} R^n.$$

PROOF.   We may assume that B is nonsingular and the b_i are linearly independent. If $(b_1^*, \ldots, b_n^*)$ is their Gram-Schmidt orthogonal basis, then Theorem 15.78 implies that

$$\left| \det \begin{pmatrix} b_1 \\ \vdots \\ b_n \end{pmatrix} \right| = \left| \det \begin{pmatrix} b_1^* \\ \vdots \\ b_n^* \end{pmatrix} \right| = \|b_1^*\| \cdots \|b_n^*\| \le \|b_1\| \cdots \|b_n\|.$$

The second inequality follows from noting that $\|b_i\| \le n^{1/2} R$ for all i.   □

15.16. Finite probability spaces

A *finite probability space* is a nonempty finite set A with a *probability distribution* (or probability measure) prob: $A \to [0 .. 1] \subseteq \mathbb{R}$ such that $\sum_{a \in A} \text{prob}(a) = 1$. For example, if we think of rolling a fair die, then $A = \{1, 2, 3, 4, 5, 6\}$ and $\text{prob}(a) = 1/6$ for all $a \in A$ gives a finite probability space describing the possible outcomes of the experiment. When $\text{prob}(a) = 1/\#A$ for all a, as in the example, then prob is the *uniform probability distribution* on A.

An *event* is a subset $B \subseteq A$, and the probability of B is $\text{prob}(B) = \sum_{a \in B} \text{prob}(a)$. In the above example, the probability of the event "odd roll", that is, $B = \{1, 3, 5\}$, is $1/2$. We have $\text{prob}(\varnothing) = 0$, $\text{prob}(A \setminus B) = 1 - \text{prob}(B)$, and $\text{prob}(B \cup C) = \text{prob}(B) + \text{prob}(C) - \text{prob}(B \cap C)$ for all $B, C \subseteq A$. In particular, $\mathrm{prob}(B \cup C) = \mathrm{prob}(B) + \mathrm{prob}(C)$ if B and C are disjoint.

The *conditional probability* $\mathrm{prob}(B \colon C) = \mathrm{prob}(B \cap C)/\mathrm{prob}(C)$ for two events B and C with $\mathrm{prob}(C) \neq 0$ is the probability of B under the condition that also C happens. This makes $(C, \mathrm{prob}(\cdot \colon C))$ into a finite probability space. The events B and C are *independent* if $\mathrm{prob}(B \cap C) = \mathrm{prob}(B) \cdot \mathrm{prob}(C)$. In that case, we have $\mathrm{prob}(B \colon C) = \mathrm{prob}(B)$ if $\mathrm{prob}(C) \neq 0$. In the above example, the two events $B = \{a \in A \colon a \text{ is odd}\} = \{1, 3, 5\}$ and $C = \{a \in A \colon a \leq 2\} = \{1, 2\}$ are independent, while B and $D = \{a \in A \colon a \leq 3\} = \{1, 2, 3\}$ are not. Intuitively, if two events are independent, then the occurrence of one of them has no impact on the probability of the other one to happen. Events $B_1, \ldots, B_n$ are independent if for any $I \subseteq \{1, \ldots, n\}$, $\mathrm{prob}(\bigcap_{i \in I} B_i) = \prod_{i \in I} \mathrm{prob}(B_i)$.

A *random variable* X on a finite probability space (A, prob) is a function $X \colon A \to \mathbb{R}$. The *expected value* (or expectation, mean value, average) of X is

$$\mathcal{E}(X) = \sum_{a \in A} X(a) \cdot \mathrm{prob}(a) = \sum_{x \in X(A)} x \cdot \mathrm{prob}(X^{-1}(x)),$$

where $X^{-1}(x) = \{a \in A \colon X(a) = x\}$. If $X(a) = a$ in our running example, then the expected value of X is

$$\mathcal{E}(X) = \sum_{1 \leq a \leq 6} a \cdot \frac{1}{6} = \frac{21}{6} = 3.5.$$

If $X(a) = a^2$ for $a \in A$, then

$$\mathcal{E}(X) = \frac{1}{6}(1 + 4 + 9 + 16 + 25 + 36) = \frac{91}{6}.$$

If $Y \colon A \to \mathbb{R}$ is another random variable and $u, v \in \mathbb{R}$, then $\mathcal{E}(uX + vY) = u\mathcal{E}(X) + v\mathcal{E}(Y)$, so that the expected value is linear. The *variance* of a random variable X is $\mathrm{var}(X) = \mathcal{E}((X - \mathcal{E}(X))^2) = \mathcal{E}(X^2) - (\mathcal{E}(X))^2$, so that $\mathrm{var}(uX) = u^2 \mathrm{var}(X)$, and the *standard deviation* of X is $\sigma(X) = \mathrm{var}(X)^{1/2}$. It expresses by how much the actual value of X differs from its mean value on average. If X takes values in $\{0, 1\}$, then $X^2 = X$ and $\mathrm{var}(X) = \mathcal{E}(X)(1 - \mathcal{E}(X))$. For *pairwise independent* random variables

$X_1, \ldots, X_n$ we have $\mathcal{E}(X_i X_j) = \mathcal{E}(X_i)\mathcal{E}(X_j)$ for $i \neq j$ and thus

$$
\begin{aligned}
\mathrm{var}\Big(\sum_{1 \leq i \leq n} X_i \Big) &= \mathcal{E}\Big(\big(\sum_{1 \leq i \leq n} X_i \big)^2 \Big) - \Big(\mathcal{E}\big(\sum_{1 \leq i \leq n} X_i \big) \Big)^2 \\
&= \mathcal{E}\Big(\sum_{1 \leq i,j \leq n} X_i X_j \Big) - \sum_{1 \leq i,j \leq n} \mathcal{E}(X_i)\,\mathcal{E}(X_j) \\
&= \sum_{1 \leq i,j \leq n} \mathcal{E}(X_i X_j) - \sum_{1 \leq i,j \leq n} \mathcal{E}(X_i)\,\mathcal{E}(X_j) \\
&= \sum_{1 \leq i \leq n} \mathcal{E}(X_i^2) - \sum_{1 \leq i \leq n} \mathcal{E}(X_i)^2 = \sum_{1 \leq i \leq n} \mathrm{var}(X_i).
\end{aligned}
$$

(15.80)

If all X_i have the same variance v, then the variance $\mathrm{var}(\frac{1}{n} \sum_i X_i) = v/n$ of the average decreases linearly in n.

Chebyshev's inequality bounds the probability that X assumes a value which is at least a certain multiple of $\sigma(X)$ away from its average:

(15.81)
$$
\mathrm{prob}\{|X - \mathcal{E}(X)| \geq k \cdot \sigma(X)\} \leq \frac{1}{k^2}
$$

for any real $k \geq 1$.

If $X_1, \ldots, X_n$ are independent random variables with values in $[0\mathrel{..}1]$, $X = X_1 + \cdots + X_n$, and $\varepsilon > 0$, then *Hoeffding's bound* says that

(15.82)
$$
\mathrm{prob}\{|X - \mathcal{E}(X)| \geq \varepsilon\} \leq 2e^{-2\varepsilon^2/n}.
$$

This is often a much smaller bound than (15.81), but the latter has the charm of not assuming independence.

The function $\mathrm{prob}_X \colon \mathbb{R} \to [0,1]$ defined by $\mathrm{prob}_X(x) = \mathrm{prob}(X^{-1}(x))$ is called the *probability distribution of the random variable* X, and $(X(A), \mathrm{prob}_X)$ is again a finite probability space. Two random variables X and Y are *independent* if $\mathrm{prob}(X^{-1}(x) \cap Y^{-1}(y)) = \mathrm{prob}(X^{-1}(x)) \cdot \mathrm{prob}(Y^{-1}(y)))$ for all $x, y \in \mathbb{R}$, so that the events $x \xleftarrow{\;\boxtimes\;} X$ and $y \xleftarrow{\;\boxtimes\;} Y$ are independent. Similarly, n random variables $X_1, \ldots, X_n$ are independent if for any $x_1, \ldots, x_n \in \mathbb{R}$ and $I \subseteq \{1, \ldots, n\}$, the probability that $x_i \xleftarrow{\;\boxtimes\;} X_i$ for all $i \in I$ equals $\prod_{i \in I} \mathrm{prob}(x_i \xleftarrow{\;\boxtimes\;} X_i)$. A simple example is $X(a) = \mathrm{prob}(a)$.

An important special case is given by a probabilistic algorithm $\mathcal{A}$ that takes an input a from A and returns a real value $\mathcal{A}(a)$, typically either 0 or 1. Even for a fixed $a \in A$, $\mathcal{A}(a)$ is a random variable depending on the internal random choices in $\mathcal{A}$. For a probability distribution X on A,

(15.83)
$$
\mathcal{E}_{\mathcal{A}}(X) = \sum_{a \in A} \mathcal{A}(a) \cdot X(a)
$$

is the *expectation of $\mathcal{A}$ on inputs according to X*. The underlying probability space consists of (A, X) and the internal randomization of $\mathcal{A}$. In particular, when $\mathcal{A}$ always returns 0 or 1, then

(15.84) $$\mathcal{E}_{\mathcal{A}}(X) = \text{prob}\{1 \leftarrow \mathcal{A}(a) \colon a \xleftarrow{\boxtimes} X\}.$$

Let X_0 and X_1 be two probability distributions on a finite set A. Their *statistical distance* $\Delta(X_0, X_1)$ is

(15.85)
$$\Delta(X_0, X_1) = \max\{|\text{prob}\{B \xleftarrow{\boxtimes} X_0\} - \text{prob}\{B \xleftarrow{\boxtimes} X_1\}| \colon B \subseteq A\}$$
$$= \frac{1}{2}\sum_{a \in A} |\text{prob}\{a \xleftarrow{\boxtimes} X_0\} - \text{prob}\{a \xleftarrow{\boxtimes} X_1\}|.$$

Then $0 \leq \Delta(X_0, X_1) \leq 1$.

DEFINITION 15.86. *We take two distributions X_0 and X_1 on the same finite set A and an algorithm $\mathcal{A}$ which on input an element from A returns 0 or 1. Then the* advantage $\text{adv}_{\mathcal{A}}(X_0, X_1)$ *of $\mathcal{A}$ on X_1 over X_0 (or its* distinguishing power *between X_0 and X_1) is*

(15.87) $$\text{adv}_{\mathcal{A}}(X_0, X_1) = |\mathcal{E}_{\mathcal{A}}(X_0) - \mathcal{E}_{\mathcal{A}}(X_1)|.$$

If $\text{adv}_{\mathcal{A}}(X_0, X_1) \geq \epsilon > 0$, then $\mathcal{A}$ is an ϵ-*distinguisher between X_0 and X_1*, and X_0 and X_1 are ϵ-*distinguishable*.

When X_0 is the uniform distribution on A, we call $\text{adv}_{\mathcal{A}}(X_0, X_1)$—linguistically somewhat gauche—the advantage of $\mathcal{A}$ on X_1. By flipping the output bit of $\mathcal{A}$ if necessary, we may assume that $\text{adv}_{\mathcal{A}}(X_0, X_1) = \mathcal{E}_{\mathcal{A}}(X_1) - \mathcal{E}_{\mathcal{A}}(X_0)$.

Let X be a random variable with values in $[0..1]$ on a finite set A with $\mathcal{E}(X) \geq \varepsilon$, and $B = \{a \in A \colon X(a) \geq \varepsilon/2\}$. Then

(15.88) $$\#B > \varepsilon/2 \cdot \#A$$

by an *averaging argument*:

$$\varepsilon \leq \mathcal{E}(X) = \frac{1}{\#A}\sum_{a \in A} X(a) = \frac{1}{\#A}\left(\sum_{a \in B} X(a) + \sum_{a \in A \setminus B} X(a)\right)$$
$$< \frac{1}{\#A} \cdot (\#B \cdot 1 + (\#A - \#B) \cdot \varepsilon/2) \leq \frac{\#B}{\#A} + \frac{\varepsilon}{2}.$$

Given two finite probability spaces (A_1, p_1) and (A_2, p_2), we may form their *direct product* $(A_1 \times A_2, p_1 \cdot p_2)$ by letting $(p_1 \cdot p_2)(a_1, a_2) = p_1(a_1) \cdot$

$p_2(a_2)$. If X_i is a random variable on A_i and π_i denotes the projection from $A_1 \times A_2$ to A_i for $i = 1, 2$, then $X_1 \circ \pi_1$ and $X_2 \circ \pi_2$ are independent random variables. We have the nth power (A^n, p^n) of (A, p) with $p^n(a_1, \ldots, a_n) = p(a_1) \cdots p(a_n)$, and if X is a random variable on A, then the n random variables $X_i = X \circ \pi_i$ on A^n defined by

$$(15.89) \qquad\qquad X_i(a_1, \ldots, a_n) = X(a_i)$$

are independent and have the same probability distribution.

For example, we consider the third power (A^3, p^3) of the experiment "rolling a die". Thus, for example, $p^3(3, 4, 3) = 6^{-3}$. Let X be the random variable counting the result of a roll of one die. The average sum of three independent rolls is

$$\mathcal{E}(Y) = \mathcal{E}(X_1) + \mathcal{E}(X_2) + \mathcal{E}(X_3) = 3\mathcal{E}(X) = 10.5,$$

where $Y = X_1 + X_2 + X_3$ and X_i is the result of die number i for $i = 1, 2, 3$. More generally, if $X_1, \ldots, X_n$ are as in (15.89), then $\mathcal{E}(X_1 + \cdots + X_n) = n\mathcal{E}(X)$.

Let (A, p) be a finite probability space, $B \subseteq A$ an event, $0 < q = p(B) < 1$ and $r = 1 - q$, and X the random variable indicating whether B has occurred or not, so that $X(a) = 1$ if $a \in B$ and $X(a) = 0$ otherwise. Such an X is called a *Bernoulli random variable*. We are interested in whether the outcomes of our random experiment are in B or not. Suppose that we repeat the experiment (say, rolling dice) potentially infinitely often and want to know the expected number of trials until $X = 1$ happens (say, a 6 occurs) for the first time. Each new trial is independent of the previous ones, and intuitively we expect to need about $1/q$ trials (that is, six rolls). We will now see that this is in fact the case.

In order to avoid infinite probability spaces, we model this situation by considering the probability space (A^n, p^n) for some $n \in \mathbb{N}$ and random variables X_i for $1 \leq i \leq n$ as in (15.89). Later, we will let n tend to infinity. Then $\mathrm{prob}\{X_i = 1\} = q$ and $\mathrm{prob}\{X_i = 0\} = r$ for all i, and $X_1, \ldots, X_n$ are independent. Now let $Y^{(n)}$ be the random variable counting the first occurrence of $X_i = 1$. Thus $Y^{(n)} = i$ if and only if $X_1 = X_2 = \cdots = X_{i-1} = 0$ and $X_i = 1$ for $1 \leq i \leq n$, and $Y^{(n)} = n+1$ if $X_i = 1$ never happens, that is, if $X_1 = X_2 = \cdots = X_n = 0$. Then

$$\mathrm{prob}\{i \xleftarrow{\boxtimes} Y^{(n)}\} = \mathrm{prob}\{0 \xleftarrow{\boxtimes} X_1\} \cdot \mathrm{prob}\{0 \xleftarrow{\boxtimes} X_2\} \cdots$$
$$\mathrm{prob}\{X_{i-1} = 0\} \cdot \mathrm{prob}\{X_i = 1\} = r^{i-1}q$$

for $1 \leq i \leq n$. This is a *geometric distribution*. Furthermore,

$$\mathrm{prob}\{n+1 \xleftarrow{\boxtimes} Y^{(n)}\} = \mathrm{prob}\{0 \xleftarrow{\boxtimes} X_1\} \cdot \mathrm{prob}\{0 \xleftarrow{\boxtimes} X_2\}$$
$$\cdots \mathrm{prob}\{0 \xleftarrow{\boxtimes} X_n\} = r^n,$$

and hence

$$\mathcal{E}(Y^{(n)}) = \sum_{1 \le i \le n+1} i \cdot \text{prob}\{i \xleftarrow{\square} Y^{(n)}\} = (n+1)r^n + \sum_{1 \le i \le n} ir^{i-1}q$$

. Now

$$q \sum_{1 \le i \le n} ir^{i-1} = (1-r) \sum_{1 \le i \le n} ir^{i-1} = -nr^n + \sum_{0 \le i < n} r^i = -nr^n + \frac{1-r^n}{1-r},$$

by the formula for the geometric sum, and we obtain

$$\mathcal{E}(Y^{(n)}) = (n+1)r^n - nr^n + \frac{1-r^n}{1-r} = -\frac{r^{n+1}}{1-r} + \frac{1}{q}.$$

Finally, $\lim_{n \to \infty} \mathcal{E}(Y^{(n)}) = 1/q$, as expected, since $|r| < 1$ implies that $\lim_{n \to \infty} r^{n+1} = 0$. As an example, the waiting time for a 6 to be rolled with a fair die is close to $1/(1/6) = 6$. More precisely, the value of $\mathcal{E}(Y^{(n)})$ with $q = 1/6$ is the expected number of rolls until a 6 shows up, if that happens with no more than n rolls, and counting $n+1$ if no 6 shows up at all. This value gets close to 6 when n is large; the difference $r^{n+1}/(1-r)$ is about 0.13 for $n = 20$ and about $0.6 \cdot 10^{-7}$ for $n = 100$.

The probability that we need at least $k \le n$ trials until B happens for the first time is

$$\text{prob}\{Y^{(n)} \ge k\} = \text{prob}\{0 \xleftarrow{\square} X_1\} \cdots \text{prob}\{0 \xleftarrow{\square} X_{k-1}\} = r^{k-1}$$

for $k \ge 1$, independent of n. It is exponentially decreasing with k. For example, the probability that we need at least 10 rolls until a 6 occurs is $(5/6)^9 \approx 19.38\,\%$. Summing up, we have the following result.

THEOREM 15.90. *If an event B occurs in an experiment with probability q and we repeat the experiment until B occurs (independent Bernoulli trials), then the expected number of executions until B happens is $1/q$.* □

For $k < tq$, the probability s that B occurs at most k times in t independent experiments satisfies

(15.91) $$s \le e^{-2(tq-k)^2/t}.$$

Below and in some other places, we use the simple fact from calculus that $(1-\varepsilon)^{1/\varepsilon}$ tends monotonically from below to e^{-1} as ε approaches 0, and more precisely

(15.92) $$2^{-2\varepsilon x} \le (1-\varepsilon)^x \le e^{-\varepsilon x} \text{ for } 0 \le \varepsilon \le 0.5 \text{ and } x > 0.$$

Similarly,

(15.93) $2.25^{\varepsilon x} \leq (1+\varepsilon)^x \leq e^{\varepsilon x}$ for $0 \leq \varepsilon \leq 0.5$ and $x > 0$.

We describe the useful method of *success amplification* which turns a small advantage of an algorithm into a huge one, by performing independent runs repeatedly. So we have a probabilistic algorithm $\mathcal{A}$ for deciding whether its input x is in some given set A, and some $\epsilon > 0$. We consider two cases:

$$\text{one-sided error: } \mathrm{prob}\{\mathcal{A}(x) = \text{``true''}\} \begin{cases} \geq \epsilon & \text{if } x \in A \\ = 0 & \text{otherwise,} \end{cases}$$

$$\text{two-sided error: } \mathrm{prob}\{\mathcal{A}(x) = \text{``true''}\} \begin{cases} \geq \frac{1}{2} + \epsilon & \text{if } x \in A, \\ \leq \frac{1}{2} - \epsilon & \text{otherwise.} \end{cases}$$

The probability space consists of the internal random choices of $\mathcal{A}$. An example with one-sided error is the set A of composite integers and the "primality" test 3.7 with the convention that $\mathcal{A}(x) = \text{``true''}$ if it returns "composite". By Theorem 3.17, we can take $\epsilon = 1/2$. As an example with two-sided error, we can take a distinguisher such as the DDH distinguisher in Reduction 9.24.

Now we run $\mathcal{A}$ several times independently, say t times, and call this $\mathcal{A}^t$. In the one-sided case, we return "true" if all runs yield "true", and "false" otherwise. In the two-sided case, we return the majority of the individual results. To circumvent the necessity of a tie-breaking rule, here we assume $t = 2m + 1$ to be odd, and also $t \geq \varepsilon^{-1}$.

THEOREM 15.94.   *(i) In the case of one-sided errors, we have*

$$\mathrm{prob}\{\mathcal{A}^t(x) = \text{``true''}\} \begin{cases} \geq 1 - e^{-\epsilon t} & \text{if } x \in A, \\ = 0 & \text{otherwise.} \end{cases}$$

(ii) In the case of the two-sided errors, we have

$$\mathrm{prob}\{\mathcal{A}^t(x) = \text{``true''}\} \begin{cases} > 1 - (1 - 4\epsilon^2)^m/2 & \text{if } x \in A, \\ < (1 - 4\epsilon^2)^m/2 & \text{otherwise.} \end{cases}$$

PROOF.   (i) If $x \in A$, then

$$\delta = \mathrm{prob}\{\mathcal{A}^t \text{ returns ``false''}\} \leq (1 - \epsilon)^t = (1 - \epsilon)^{1/\epsilon \cdot \epsilon t}.$$

It follows from (15.92) that $\delta \leq e^{-\epsilon t} \leq 1/2$ and

(15.95) $\mathrm{prob}\{\mathcal{A}^t \text{ returns ``true''}\} \geq 1 - e^{-\epsilon t}.$

(ii) We assume that the correctness probability of $\mathcal{A}$ on any input x is exactly $1/2 + \epsilon$ with $\epsilon > 0$ both when $x \in A$ and when $x \notin A$. The probability of obtaining exactly i correct answers in t trials is

$$\binom{t}{i}\left(\frac{1}{2}+\epsilon\right)^i\left(\frac{1}{2}-\epsilon\right)^{t-i}.$$

$\mathcal{A}^t$ answers incorrectly if at most m correct answers were given by $\mathcal{A}$. We set $p = 1/2 + \epsilon$ and $q = 1/2 - \epsilon = 1 - p$. Thus $q/p < 1$, $t - m = m + 1$, $\sum_{0 \le i \le t} \binom{t}{i} = (1+1)^t = 2^t$, and for the sum over the first half of the indices we have $\sum_{0 \le i \le m} \binom{t}{i} = 2^t/2 = 2^{2m}$. The probability that $\mathcal{A}^t$ answers incorrectly is at most

$$\sum_{0 \le i \le m} \binom{t}{i} p^i q^{t-i} = p^m q^{t-m} \sum_{0 \le i \le m} \binom{t}{i} (q/p)^{m-i}$$

$$\le (pq)^m q \sum_{0 \le i \le m} \binom{t}{i} = \left(\frac{1}{4} - \epsilon^2\right)^m q \cdot 2^{2m}$$

$$= (1 - 4\epsilon^2)^m q < (1 - 4\epsilon^2)^m/2. \qquad \square$$

COROLLARY 15.96. *In order to improve the correctness probability for one- or two-sided errors from ϵ or $1/2 + \epsilon$, respectively, to $1 - \delta$ by t repetitions, it is sufficient to take*

(i) $t \ge \epsilon^{-1} \ln(\delta^{-1})$ *for one-sided errors,*

(ii) $t \ge \epsilon^{-2} \ln((2\delta)^{-1})/2$ *for two-sided errors.*

PROOF. (i) is clear from Theorem 15.94 (i). For (ii), it is sufficient to choose $m = (t-1)/2$ so that

$$(1 - 4\epsilon^2)^m/2 \le \delta,$$

$$m \ge \frac{\ln 2\delta}{\ln(1 - 4\epsilon^2)}.$$

The Taylor expansion of the natural logarithm gives $\ln(1 - x) = -x + \frac{x^2}{2} - \frac{x^3}{3} + - \ldots \le -x/2$ for $0 \le x < 1$. (Note that the two logarithms have negative values.) Hence

$$\frac{\ln 2\delta}{\ln(1 - 4\epsilon^2)} \le \frac{2}{4\epsilon^2} \ln((2\delta)^{-1}).$$

The claim (ii) follows. $\qquad \square$

In particular, a polynomial number of repetitions suffices to bump up the correctness probability from nonnegligible to exponentially close to 1. This technique of *amplification* occurs several times in this text, for example in the primality test (Theorem 3.18).

For an asymptotic statement, we now take families $(A_n)_{n \in \mathbb{N}}$, $(\mathcal{A}_n)_{n \in \mathbb{N}}$, and $(\epsilon_n)_{n \in \mathbb{N}}$ of the items above.

THEOREM 15.97. *If ϵ_n is nonnegligible, then there exists $t_n \in \text{poly}(n)$ so that for inputs x of length n, the following holds both for one-sided and two-sided errors:*

$$\text{prob}\{\mathcal{A}_n^{t_n}(x) = \text{``true''}\} \text{ is } \begin{cases} \textit{overwhelming} & \textit{if } x \in A, \\ \textit{negligible} & \textit{otherwise.} \end{cases}$$

Notes 15.1. One sometimes considers rings without multiplicative identity, but they do not occur in this text.

15.2. The *sign* of a cycle $\sigma = (a_1 a_2 \ldots a_i)$ is $\text{sign}(\sigma) = (-1)^{i-1}$. One can extend this multiplicatively to $\text{sign}(\sigma)$ for any permutation σ using the cycle representation and show that $\text{sign}(\sigma \circ \rho) = \text{sign}(\sigma) \cdot \text{sign}(\rho)$. The permutations $\sigma \in \text{Sym}_n$ with $\text{sign}(\sigma) = 1$ form the *alternating group* with $n!/2$ elements.

15.4. Euclid stated his algorithm in his famous book *Elements* around 300 BC. Its fascinating history is explained in Knuth (1998), §4.5, and von zur Gathen & Gerhard (2013), Notes to Chapter 3. One way of dealing with the subtle difficulties caused by the non-uniqueness of the gcd is described in von zur Gathen & Gerhard (2013), Chapter 3. There are many variants of the Euclidean Algorithm. The relationship between the gcd in $\mathbb{Z}[x]$ and $\mathbb{Q}[x]$ is explained by Gauß' Lemma, and the most efficient approach is to calculate modulo small random prime numbers. For details, see von zur Gathen & Gerhard (2013), Sections 6.7 and 11.1.

15.6. Sometimes ε is called overwhelming if $1 - \varepsilon$ is negligible, a weaker condition than the one after Definition 15.30.

For deterministic algorithms, general notions of cost and complexity are defined in Strassen (1972a). In the literature, one often reads about the "complexity of an algorithm". This can lead to a confusion with the different notion of the complexity of a problem. For complexity theory, see Goldreich (2008) and Arora & Barak (2009). Some complexity classes that we do not discuss here appear in Section 13.12 and the Notes to Section 11.7. The technical details of P vs. P_{circ} and P/poly are discussed in Goldreich (2001), Section 1.3.3.

15.8. The notation $\text{M}(n)$ is standard for the cost of any multiplication routine. The Schönhage & Strassen algorithm for integer multiplication was improved to $n \log n \, 2^{O(\log^* n)}$ by Fürer (2009) and to $O(n \log n 16^{\log^* n})$ by Harvey *et al.* (2014a), and similarly for polynomials over finite fields by Harvey *et al.* (2014b). In software, one has to implement several of them. The classical one will be the fastest for small input sizes n due to the reduced overhead. Then one has to determine the breakpoint where faster methods take over. The asymptotically fastest methods will win for large n, often beyond the range of cryptographic interest. The value $\log^* n$ is the smallest integer k so that the k-fold iterated logarithm $\log \log \cdots \log(n)$ is less than 1. For practical values of n, $\log^* n \leq 5$. The first implementation of fast arithmetic on cryptographic hardware is in Grabbe, Bednara, Shokrollahi, Teich & von zur Gathen (2003), and the use of Karatsuba's algorithm is quite standard today. For software, see von zur Gathen & Shokrollahi (2005), von zur Gathen, Shokrollahi & Shokrollahi (2007), and the literature cited there. The reader interested in the details of integer division with remainder is referred to the comprehensive discussion in Knuth (1998), Section 4.3.1.

More detailed information, including a discussion of "long integers" and explicit constants in some of the O-estimates, can be found in von zur Gathen & Gerhard (2013), and also in Cormen *et al.* (1990). Using fast arithmetic, the EEA can be performed with $O(\mathsf{M}(n) \log n)$ basic operations for inputs of length n, both for polynomials and integers. For details, see von zur Gathen & Gerhard (2013), Chapter 11, and Section 3.3 for the (easier) $O(n^2)$ bound. A precise version says that the classical algorithm for polynomials $a, b \in F[x]$ with $\deg a = n \geq \deg b = m$ can be performed with at most $m + 1$ inversions and $6nm + O(n)$ additions and multiplications in F.

15.9. A special case of Lemma 15.36 goes back to Thue (1902); see also Kaltofen & Rolletschek (1989). Theorem 15.39 with the assumption $|b/a - s/t| \leq 1/(2t^2)$ is Proposition 1.4 in Hensley (2006). The theory of continued fractions can actually deal with real numbers x rather than just rationals, by executing an analog of the Extended Euclidean Algorithm with $q_i = \lfloor r_{i-1}/r_i \rfloor$. This leads, in general, to infinitely many continued fraction approximations. *Dirichlet's approximation theorem*, improved by Hurwitz (1891), says that every irrational real number x has infinitely many approximations s/t with $|x - s/t| \leq 1/\sqrt{5}t^2$, and that $\sqrt{5}$ is the largest number with this property.

15.10. Interpolation at r points can be performed with $O(r^2)$ operations in F; with fast arithmetic, it takes $O(\mathsf{M}(r) \log r)$ operations; see von zur Gathen & Gerhard (2013), Corollary 10.12.

15.11. Algorithm 15.45 uses $O(n^3)$ bit operations when $q_1 \cdots q_r$ has n bits. According to Theorem 10.25 of von zur Gathen & Gerhard (2013), a fast variant runs in $O(\mathsf{M}(n) \log n)$ bit operations.

15.12. There is a substantial literature on efficient algorithms for exponentiation. When we abstract away the group G and just look at what happens to exponents, then we get the notion of *addition chains*, where we can add two previously computed integers in one step. Knuth 1998, Section 4.6.3, gives a highly readable introduction to this topic. For an n-bit exponent e, Theorem 15.49 gives an upper bound of $2n - 2$ operations. Brauer's 1939 algorithm reduces this to $n+n/\log n+O(n \log\log n/(\log n)^2)$. Exercise 15.15 shows the explicit estimate $n + 3n/\log n$ for $n \geq 6464$. Since the size of the exponents computed can at most double in one step, there is also a lower bound of $\log_2 e \geq n$. Thus the leading term in Brauer's cost estimate is optimal. If the base x is fixed for many exponentiations, then Brickell *et al.* (1992) have shown how the precomputation and storage of a few powers of x helps to reduce the time. Large finite fields $\mathbb{F}_{2^n}$ of characteristic 2 are of interest for some cryptosystems; Diffie-Hellman and ElGamal can be implemented here. In this setting, even better results are available: the time can be reduced to about $n/\log_2 n$, below the lower bound of n in the general model! This is achieved with clever data structures to represent the elements of $\mathbb{F}_{2^n}$; one useful tool here are *Gauß periods*. The algorithms are described in Gao *et al.* (2000) and von zur Gathen *et al.* (2007).

The famous Swiss mathematician Leonhard Euler (1761) used repeated squaring to calculate 7^{160} in $\mathbb{Z}_{641}$ by computing $7^2, 7^4, 7^8, 7^{16}, 7^{32}, 7^{64}, 7^{128}, 7^{160} = 7^{128} \cdot 7^{32}$, reducing modulo 641 after each step. (He also listed, unnecessarily, 7^3.) He had discovered in 1732/33 that 641 divides the fifth Fermat number $F_5 = 2^{2^5} + 1$.

The not so famous German mathematician Carl Friedrich Hindenburg (1794) explains addition chains very clearly:

> Hier entstehen, nachdem man die erste Eins heruntergesetzt hat, die dekadischen Zahlen nach einander, durch Verdoppeln der vorhergehenden und Zusetzen von 0 oder 1, nachdem eine dieser Ziffern in der oberen Stelle steht, unter die man die so gefundene dekadische Zahl nach ihrer Ordnung setzt.[1]

He uses it for converting a binary representation into decimal, and also gives the reverse conversion, by repeated division with remainder by 2. Here are his three examples:

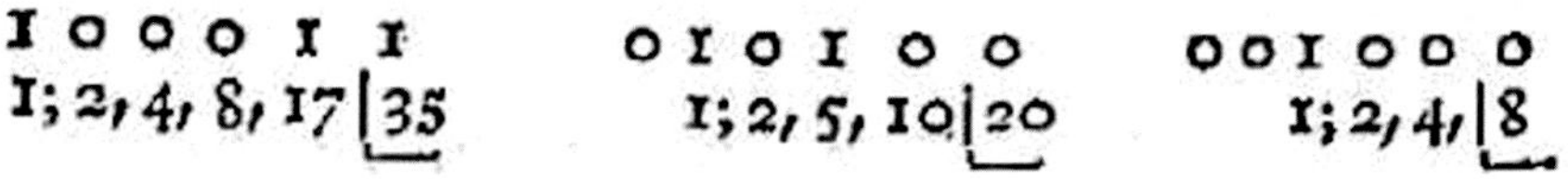

Figure 15.6: Hindenburg's addition chains for 35, 20, and 8.

His context is cryptographic, namely the representation of a grille cipher (see Section F.4) in bases 2, 16, and 64.

15.13. For general groups, the multiplicative notation is standard, and the additive notation is only used for some Abelian groups. Most groups occurring in this text are commutative.

We have proven the three fundamental theorems in the sequence: Lagrange $\Rightarrow$ Euler $\Rightarrow$ Fermat. (In fact, we had to work a bit only for Lagrange's Theorem.) This is the reverse of the chronological order: Fermat stated his Theorem without proof in 1640, and Leibniz and Euler later proved it; see von zur Gathen & Gerhard (2013), Notes to Section 4.3 and 4.4. Euler (1760/61) defines his φ function and shows his Theorem 15.55 and Lagrange's Theorem 15.53 for $G = \mathbb{Z}_N^\times$. Lagrange (1770/71) states his theorem for the symmetric group Sym_5 on 5 elements. The first complete proof is in Gauß (1801), article 49, for $G = \mathbb{Z}_p^\times$. Gauß also shows this for some other groups of his interest, but did not have the general notion of groups. See Roth (2001) for a discussion.

15.14. For a proof of Theorem 15.68 (iv) in the case $e = 1$, see von zur Gathen & Gerhard (2013), Section 14.3; for general e, one can then apply Newton iteration (Algorithm 9.22 and Exercise 9.34 in the cited book). In a finite field $\mathbb{F}_q$ of characteristic 2, squaring is a linear map and every element $a \in \mathbb{F}_q$ is a square, namely $a = (a^{q/2})^2$ by Fermat's Little Theorem 15.56.

For more about the Jacobi symbol and computing it efficiently, in time $O(\mathsf{M}(n) \log n)$, see Schönhage et al. (1994), §7.2.3, Bach & Shallit (1996), Section 5.9, and Brent & Zimmermann (2010). For composite N, the elements of $\boxtimes_N$ are also called *pseudosquares* in the literature, but this is a different use of "pseudo" from pseudoprimes—which are usually primes—and "pseudorandom" elements—which behave like random elements; the elements of $\boxtimes_N$ are never squares. I learned the term *antisquares* from Michael Nüsken.

15.16. An introduction to probability theory is in Chapter 8 of Graham et al. (1994), and Feller (1971) is a classical reference. Inequality (15.81) is from Hoeffding (1963).

---

[1] Here arise, after putting the first 1 under the leading digit, the numbers in decimal representation [in the bottom line] one after another, by doubling the previous one and adding 0 or 1, according to which of these digits is in the top row at the place into which the newly found decimal number is written.

When A is infinite, we may take as the statistical difference of X_0 and X_1 the value $\Delta(X_0, X_1) = \sup\{|X_0(B) - X_1(B)|\colon B \subseteq A\}$ over the "measurable" sets B on which $X_0(B)$ and $X_1(B)$ are defined. For the statistical distance and (15.85), see Levin *et al.* (2009), Section 4.1.

Yao (1982) and Goldwasser & Micali (1984) define (in)distinguishability of two distributions.

Exercises.

EXERCISE 15.1 (Power of 3). *Compute* $3^{1\,000\,003}$ mod 101 *by hand. Note: Only a small calculation is needed!*

EXERCISE 15.2 (AES rings). *We want to show that the rings R and S of AES are not fields.*

(i) *Show that* $(z + 1)^8 = 0$ *in R.*

(ii) *Name a zero divisor in R.*

(iii) *Show that $z + 1$ does not have an inverse in R.*

(iv) *Show that* $(y + 1)^4 = 0$ *in S.*

(v) *Name a zero divisor in S.*

(vi) *Show that $y + 1$ does not have an inverse in S.*

EXERCISE 15.3 ($\mathbb{Z}_N$ a ring). *Let $N \in \mathbb{Z}$ be nonzero.*

(i) *Show that the operations on $\mathbb{Z}_N$ defined in Section 15.3 are independent of the chosen representatives: if $x, x', y, y', r, r' \in \mathbb{Z}$ are such that in $\mathbb{Z}_N$ we have $x = x'$, $y = y'$, $r = x + y$, and $r = x' + y'$, then $r = r'$ in $\mathbb{Z}_N$. Similarly for multiplication.*

(ii) *Use the fact that $\mathbb{Z}$ is a ring to conclude that $\mathbb{Z}_N$ is a ring.*

EXERCISE 15.4 (Null ring).

(i) *A pathological example of a ring is $R = \{0\}$, consisting of the single element $0 = 1$. It is called the null ring. Show that $\mathbb{Z}_1$ is the null ring.*

(ii) *Let R be a ring. Show that the following are equivalent:*

    ○ $0 = 1$,

    ○ R is the null ring,

    ○ 0 is a unit,

    ○ 1 is a zero divisor.

EXERCISE 15.5 (Easy powers).

(i) *Let G be a group with d elements, and $e > d$ an integer. Determine an integer e' so that $0 \le e' < d$ and $x^e = x^{e'}$ for all $x \in G$.*

(ii) *Calculate by hand the last two decimal digits of $23^{1\,000\,005}$. Show your computations.*

EXERCISE 15.6 (Testing an unknown group order for primality). *Let G be a finite group of (unknown) order d, and $x \in G$ with $x \neq 1$.*

(i) *Show that if d is prime, then x is a generator of G.*

(ii) *We are given an integer D with $d \leq D$ and a prime p with $D/2 < p \leq D$. Show that $\#G = p$ if and only if $x^p = 1$.*

EXERCISE 15.7. *Prove the claims of Lemma 15.18.*

EXERCISE 15.8 (Polynomial units).

(i) *Let R be an integral domain. Show that $(R[x])^{\times} = R^{\times}$.*

(ii) *Give an example of a ring R where $(R[x])^{\times} \neq R^{\times}$.*

EXERCISE 15.9 (Residue rings).

(i) *Show that f in (15.41) respects multiplication, and hence is a ring homomorphism, if m divides n.*

(ii) *Show that if m does not divide n, then f in (15.41) is not well-defined.*

EXERCISE 15.10 (Easy gcd). *Let a and e be positive integers. Compute $g = \gcd(a^e - 1, a^e - a^{e-1})$ and $s, t \in \mathbb{Z}$ so that $g = s(a^e - 1) + t(a^e - a^{e-1})$.*

EXERCISE 15.11 (Factor refinement). *In Section 4.8, we mention factor refinement. This procedure takes a partial factorization of an integer, computes gcds and exact division of integers (where the divisor divides the dividend), and yields a factorization into powers of pairwise coprime integers, which are not necessarily prime. This is called a coprime basis. In contrast to the full factorization of integers, this process can be performed efficiently; see Bach & Shallit (1996), Section 4.8. As an example, take the prime $p = 368\,947\,264\,001$, then $p - 1$ and its factor $336\,140$. Perform iteratively gcds and exact integer divisions on previously available numbers to arrive at a coprime basis of $p - 1$. Along your computation, keep track of the current partial factorization of $p - 1$. It can be done in 11 steps; can you do better?*

    The example is small and special, and you "see" factors like 1000 immediately; but you are not allowed to use such information—just like a computer executing a program.

EXERCISE 15.12 (Coprime basis). *Exercise 15.11 presents an example of factor refinement. You are to study this in greater generality.*

    When we have a factorization $a = b \cdot c$ with integers at least 2, we can apply the Chinese Remainder Theorem 15.44 only if b and c are coprime. The following approach makes this possible. Suppose that $b = (b_1, \dots, b_m)$ is a sequence of integers all at least 2, whose product equals a.

    For $1 \leq i < j \leq m$, a refinement step computes $c = \gcd(b_i, b_j)$ and replaces b_i and b_j by c^2, b_i/c, b_j/c in some order, removing 1 if it occurs.

(i) *Devise a procedure which from an initial sequence b as above produces a sequence whose entries are pairwise coprime and have product a.*

(ii) *Associate positive integers exponents, initially all 1, to the entries of each intermediate sequence so that a is the corresponding power product.*

(iii) *Derive a cost estimate for your procedure in terms of the bit length of a.*

(iv) *Generalize your procedure to the case of initial factorizations of several integers a, obtaining a single sequence of pairwise coprime integers so that each a is a power product of them. You might proceed recursively along the list of a's. Estimate the cost of your procedure. (Bach & Shallit (1996), §4.8, deal with this problem under the designation of gcd-free basis.)*

EXERCISE 15.13 (Continued fractions). *The continued fraction of a positive real number x can be computed in the following way. If x is not an integer, write $x = q_1 + 1/x_1$ where $q_1 = \lfloor x \rfloor$ and $x_1 > 1$. If x_1 is not an integer, write $x_1 = q_2 + 1/x_2$ where $q_2 = \lfloor x_1 \rfloor$ and $x_2 > 1$ and so forth, as long as the remaining number x_i is not an integer. For example we arrive at the equality*

$$\frac{15}{11} = 1 + \frac{1}{2 + \frac{1}{1+\frac{1}{3}}}.$$

The continued fraction approximations are obtained by truncating the continued fraction. (Choose one of the plus signs and replace the next numerator by zero.) For example 1 is the zeroth, $1 + \frac{1}{2} = \frac{3}{2}$ the first and $1 + \frac{1}{2+\frac{1}{1}} = \frac{4}{3}$ the second continued fraction approximation of $\frac{15}{11}$. The third and all further approximations are equal to $\frac{15}{11}$.

(i) *Execute the Extended Euclidean Algorithm for $a = 127$ and $b = 105$.*

(ii) *Compute the continued fraction of $127/105$ and the corresponding continued fraction approximations.*

(iii) *Describe similarities of the results obtained in parts (i) and (ii).*

(iv) *Compute the continued fraction of $x = \sqrt{3}$.*

EXERCISE 15.14 (Computing in finite fields).
	Consider the quotient ring $F = \mathbb{F}_3[x]/(x^3 - x + 1)$ and write a for $x \bmod x^3 - x + 1 \in F$.

(i) *Prove that F is a field. How many elements does it have?*

(ii) *How are sums, products, and inverses computed in F? Compute $a^2 \cdot a^2$, $(a-1)^3$ and a^{-1} in F.*

An element $b \in F$ is called normal (in F over $\mathbb{F}_3$) if (b, b^3, b^{3^2}) is an $\mathbb{F}_3$-basis of F. (It is then called a normal basis.) If b is a normal element, then every element $c \in F$ can be uniquely written as $c = c_0 b^{3^0} + c_1 b^{3^1} + c_2 b^{3^2}$ with $c_0, c_1, c_2 \in \mathbb{F}_3$.

(iii) *Is a normal?*

(iv) *Verify that $b = a^2$ is normal. How can you multiply elements of F which are written in terms of the normal basis given by b? Compute c^3 for an arbitrary c given by its representation in the normal basis.*

EXERCISE 15.15 (Addition chains).   *Exponentiation algorithms such as the repeated squaring Algorithm 15.48 correspond to addition chains, where starting from $a_0 = 1$ one builds up integers by addition only:*

$$a_k \longleftarrow a_i + a_j, \text{ for some } i \text{ and } j \text{ with } 0 \le i, j < k.$$

Such a chain computes any element in it. The addition chain corresponding to Figure 2.14 is:

instruction	value
$a_0 \longleftarrow 1$	1
$a_1 \longleftarrow a_0 + a_0$	2
$a_2 \longleftarrow a_1 + a_0$	3
$a_3 \longleftarrow a_2 + a_2$	6
$a_4 \longleftarrow a_3 + a_3$	12
$a_5 \longleftarrow a_4 + a_0$	13

The total number of additions is called the length *of the chain. The example has length 5. In the following, let e be a positive integer, and $n = \lfloor \log_2 e \rfloor + 1$ its binary length.*

(i) *Describe the addition chain computing e that corresponds to repeated squaring, and prove that it has length at most $2(n-1)$.*

(ii) *Prove that any addition chain computing e has length at least $n - 1$. [Hint: prove by induction the lower bound 2^k on the value of a_k.]*

(iii) *(Brauer's windowing method). Instead of the binary representation of e, we use the 2^s-ary representation for some positive window size s. We compute $1, 2, 3, 4, 5, \ldots, 2^s - 1$, and if $e = \sum_{0 \le i \le m} e_i 2^{si}$ with $0 \le e_i < 2^s$ for all i and $e_m \neq 0$, we alternate the addition of the next digit e_i and s doublings. Determine m and prove that this method computes e with at most $n + 2^s + (n+1)/s$ steps.*

(iv) *Set $s = \lfloor \log n - 2 \log\log n \rfloor$ and check that $s \ge 1$ for $n \ge 80$. Show that the resulting addition chain for e has length at most $n + 3n/\log n$ for $n \ge 6464$.*

(v) *Determine the thresholds for n beyond which the upper bounds $n + 5n/\log n$, $n + 2n/\log n$, and $n + n/\log n + 3n\log\log n/\log^2 n$ hold.*

This windowing method is due to Brauer (1939), see also Knuth (1998), and has been rediscovered many times.

EXERCISE 15.16 (Multiplication).

(i) *Implement in your favorite programming environment an algorithm to multiply two polynomials in $\mathbb{Z}_2[x]$ of degree less than n.*

(ii) *Determine roughly the largest n for which your code takes less than a minute (say, of CPU time) to execute.*

(iii) *For n starting at $n = 10$ and doubling until you reach (roughly) the value n from (ii) run 100 experiments each with random inputs of degree n. Record the average CPU time for each n.*

(iv) *Check if the time is roughly cn^2 for some constant c, and determine a good value for c. Run more experiments filling the gaps of (iii). Is your estimate good for these new experiments?*

(v) *Do (i)-(iv) for division with remainder in $\mathbb{Z}_2[x]$.*

(vi) *Do (i)-(iv) for the Extended Euclidean Algorithm in $\mathbb{Z}_2[x]$.*

EXERCISE 15.17 (Order of a product). *Let G be a (multiplicative) commutative group, x and y elements of order a and b, respectively. What is the order c of xy?*

(i) *Take $G = \mathbb{Z}_{1321}^{\times}$, $x = 17$ and $y = 53$. Check that $a = 24$ and $b = 33$. Compute the orders of xy, y^2, xy^2, y^3, and xy^3.*

Now to the general case.

(ii) *Show that c divides $\mathrm{lcm}(a, b)$.*

(iii) *Show that if $\gcd(a, b) = 1$, then $c = \mathrm{lcm}(a, b) = a \cdot b$.*

(iv) *Show that if $a = ms$ and $b = mt$ with coprime s and t, then st divides c. Does m divide c?*

EXERCISE 15.18 (Order in $\mathbb{Z}_N^{\times}$).

(i) *Let M and N be coprime, and $x \in \mathbb{Z}_{MN}^{\times}$. Let e be the order of x in $\mathbb{Z}_M^{\times}$ and f the order of x in $\mathbb{Z}_N^{\times}$. Show that the least common multiple $\mathrm{lcm}(e, f)$ of e and f is the order of x in $\mathbb{Z}_{MN}$.*

(ii) *Compute the (multiplicative) order of 3 in $\mathbb{Z}_{100}^{\times}$. Show your calculations.*

(iii) *Generalize (i) to allow for more pairwise coprime factors $N_1, \ldots, N_r$ and compute the order of 3 in $\mathbb{Z}_{100100}^{\times}$.*

EXERCISE 15.19 (-1 a square). *Let p be an odd prime and g a generator of $\mathbb{Z}_p^{\times}$. You are to show that -1 is a square in $\mathbb{Z}_p$ if and only if $p \equiv 1 \bmod 4$.*

(i) *Show that $-1 = g^{(p-1)/2}$.*

(ii) *If $p \equiv 1 \bmod 4$, then $(p - 1)/4$ is an integer. Calculate some $x \in \mathbb{Z}_p^{\times}$ with $x^2 = -1$.*

(iii) *Suppose that $p \equiv 3 \bmod 4$ and $x = g^a \in \mathbb{Z}_p^{\times}$ is arbitrary, with an integer a. Consider $\mathrm{dlog}_g(-1)$ and $\mathrm{dlog}_g(x^2)$ in $\mathbb{Z}_{p-1}$ and show that $-1 \neq x^2$.*

EXERCISE 15.20 (Square roots in cyclic groups). *Let G be a cyclic group of order d. If $a = b^2$ in G, then a is a square and b a square root of a. Show the following.*

(i) *If d is odd, then every element of G has exactly one square root. Find an integer e so that a^e is a square root of a, for all $a \in G$.*

(ii) *For which primes p does $\mathbb{Z}_p^{\times}$ have odd order?*

(iii) *If d is even, then each element of G has either two or zero square roots in G.*

Now let $d - 2$ be divisible by 4, so that d is 2 modulo 4.

(iv) *If $a \in G$ is a square, then $a^{(d+2)/4}$ is a square root of a.*

(v) *For the nonsquares a, which values does $a^{(d+2)/4}$ take?*

(vi) *Specify a group circuit (Section 4.9) of polynomial size that computes square roots in G.*

(vii) *Now consider $G = \mathbb{Z}_p^{\times}$ with an n-bit prime p which is 3 modulo 4, so that 4 divides $p-3$. Determine the number of bit operations of your square root circuit from (vi) over G. [Recall that the group $\mathbb{Z}_p^{\times}$ is cyclic, since p is prime.]*

(viii) *Now consider $G = \mathbb{Z}_p^{\times}$ with a prime p which is 1 modulo 4. Can you conceive of a similar square root algorithm? That is, some integer e so that a^e is a square root of any square a. If yes, describe it, if not, explain why. Is there a polynomial time square root algorithm for G?*

EXERCISE 15.21 (Square roots modulo composites). Let $p, q \in \mathbb{N}$ be two different odd prime numbers and $N = p \cdot q$.

(i) Show that $a \in \mathbb{Z}_p^\times$ is a square if and only if $a^{(p-1)/2} = 1$.

(ii) Prove that $a \in \mathbb{Z}_N^\times$ is a square if and only if a is a square in $\mathbb{Z}_p^\times$ and $\mathbb{Z}_q^\times$.

(iii) Given square roots of $a \in \mathbb{Z}_N^\times$ in $\mathbb{Z}_p^\times$ and $\mathbb{Z}_q^\times$, show how to compute a square root of a in $\mathbb{Z}_N^\times$.

(iv) How many squares are there in $\mathbb{Z}_p^\times$? In $\mathbb{Z}_N^\times$? How many square roots does a square have?

(v) Reduce factoring N to finding square roots in $\mathbb{Z}_N^\times$.

(vi) Suppose someone presents you a polynomial-time algorithm for computing a square root of any square $a \in \mathbb{Z}_N^\times$, just given N and a. Which conclusions do you draw?

EXERCISE 15.22 (Computing the Jacobi symbol). You may assume the following properties of the Jacobi symbol, where a, a', b, and b' are integers, and b and b' are odd.

(i) $\left(\frac{a}{b}\right) = \left(\frac{a \text{ rem } b}{b}\right)$, and $\left(\frac{a}{b}\right) = 0$ if and only if $\gcd(a, b) \neq 1$.

(ii) $\left(\frac{1}{b}\right) = +1$, $\left(\frac{a'a}{b}\right) = \left(\frac{a'}{b}\right) \cdot \left(\frac{a}{b}\right)$, $\left(\frac{a}{b'b}\right) = \left(\frac{a}{b'}\right) \cdot \left(\frac{a}{b}\right)$.

(iii) $\left(\frac{-1}{b}\right) = (-1)^{\frac{b-1}{2}}$. This equals $+1$ for $b \equiv 1 \bmod 4$ and -1 for $b \equiv -1 \bmod 4$.

(iv) $\left(\frac{2}{b}\right) = (-1)^{\frac{b^2-1}{8}}$. This equals $+1$ for $b \equiv \pm 1 \bmod 8$ and -1 for $b \equiv \pm 3 \bmod 8$.

(v) The law of quadratic reciprocity states that, if a is also an odd natural number, then

$$\left(\frac{a}{b}\right) = (-1)^{\frac{a-1}{2}\frac{b-1}{2}}\left(\frac{b}{a}\right).$$

Thus the two Jacobi symbols differ in sign if and only if both a and b are 3 modulo 4.

(i) Develop an algorithm, similar to the Euclidean algorithm, for computing the Jacobi symbol using polynomial time. [Hint: It can be done in $O(n^2)$.]

(ii) Which numbers a have $\left(\frac{a}{N}\right) = 1$, when $N = pq$ is the product of two distinct odd primes? Compare with the two properties "a is a square modulo p" and "a is a square modulo q".

EXERCISE 15.23 (Chebyshev). For an event A, let I_A denote its indicator function, with $I_A = 1$ if A occurs and $I_A = 0$ otherwise. Let X be a random variable with $e = \mathcal{E}(X)$ and $s = \sigma(X)$, and $k \geq 1$. Explain and verify the steps in the following proof of Chebyshev's inequality (15.81):

$$\text{prob}\{|X - e| \geq ks\} = \mathcal{E}(I_{|X-e|\geq ks}) = \mathcal{E}(I_{((X-e)/ks)^2 \geq 1})$$

$$\leq \mathcal{E}\left(\left(\frac{X-e}{ks}\right)^2\right) = \frac{1}{k^2}\frac{\mathcal{E}((X-e)^2)}{s^2} = \frac{1}{k^2}.$$

Notation and sources of quotations, images, and ornaments

Notation

$\mathbb{N}$, $\mathbb{N}_{>n}$	set of nonnegative integers, set of integers greater than $n \in \mathbb{N}$
$\mathbb{Z}$	ring of integers
$\mathbb{Q}$, $\mathbb{Q}_{>0}$	field of rational numbers, set of positive rational numbers
$\mathbb{R}$, $\mathbb{R}_{>r}$	field of real numbers, set of real numbers greater than $r \in \mathbb{R}$
$\mathbb{C}$	field of complex numbers
$\mathbb{B}$	set $\{0,1\}$ of Boolean values
$\mathbb{F}_q$	finite field with q elements
$\mathbb{Z}_N$	integers modulo N, usually $\{0,\ldots,N-1\}$
R_N	symmetric representatives $\{-\lfloor N/2 \rfloor, \ldots, \lfloor (N-1)/2 \rfloor\}$ of $\mathbb{Z}_N$
$a = b$ in $\mathbb{Z}_N$, a equals b modulo N, $a \equiv b \bmod N$	N divides $a - b$
$a = (a \operatorname{quo} N) \cdot N + (a \operatorname{rem} N)$, $0 \le a \operatorname{rem} N < N$	division with remainder
$a = (a \operatorname{squo} N) \cdot N + (a \operatorname{srem} N)$, $a \operatorname{srem} N \in R_N$	division with symmetric remainder
$\varnothing$	empty set
$A \cup B$	union of the sets A and B
$A \cap B$	intersection of the sets A and B
$A \setminus B$	set-theoretic difference of A and B
$A \times B$	product (set of pairs) of the sets A and B
A^n	vectors of length $n \in \mathbb{N}$ over the set A
$\#A$	cardinality (number of elements) of the set A
$\langle A \rangle$	subgroup, ideal, or subspace generated by the elements of A
$A \cong B$	A and B are isomorphic groups or rings
$R^{\times}$	group of units of the ring R
$R[x]$	ring of polynomials in the variable x over the ring R
$R^{n \times m}$	ring of $n \times m$ matrices over the ring R for $n, m \in \mathbb{N}$
$R/(m)$	residue class ring of the ring R modulo (the ideal generated by) $m \in R$
$F(x)$	field of rational functions in the variable x over the field F
$\exp(a)$	exponential function, e^a for $a \in \mathbb{R}$
$\exp_b(a)$	real exponential b^a for positive $a, b \in \mathbb{R}$
$\ln a$	natural (base e) logarithm of $a \in \mathbb{R}_{>0}$
$\log a$	binary (base 2) logarithm of $a \in \mathbb{R}_{>0}$
$\operatorname{dlog}_g x$, $\operatorname{dexp}_g x$	discrete logarithm, exponential of x in base g
$|a|$	absolute value of $a \in \mathbb{C}$

© Springer-Verlag Berlin Heidelberg 2015
J. von zur Gathen, *CryptoSchool*, DOI 10.1007/978-3-662-48425-8

$\lfloor a \rfloor$	greatest integer less than or equal to $a \in \mathbb{R}$
$\lceil a \rceil$	smallest integer greater than or equal to $a \in \mathbb{R}$
$\lceil a \rfloor$	integer nearest to $a \in \mathbb{R}$, $\lfloor a + 1/2 \rfloor$
$\pm a$	set $\{a, -a\}$
$\sqrt{a}$	positive square root in $\mathbb{R}$, set of square roots in other rings
$a \star b$	inner product of vectors a and b
$\|a\|$	Euclidean norm of a vector or a polynomial a
$\mathcal{B}_n(z, r)$	n-dimensional ball with center $z \in \mathbb{R}^n$ and radius $r > 0$
$\mathcal{S}_n$	n-dimensional real unit sphere
$a \mid b$	a divides b, $\exists c\ b = ac$
f'	formal derivative of the polynomial or rational function f
$\partial f / \partial x$	formal derivative of the multivariate polynomial f with respect to x
$\binom{n}{k}$	binomial coefficient for $n, k \in \mathbb{N}$
$O, O^{\sim}, \Omega, o, \omega$	asymptotic notation, Section 15.6
P, NP, BPP	complexity classes, Section 15.5
XOR, $\oplus$	exclusive or = addition modulo 2
$\vee, \wedge$	Boolean OR, AND
$a \xleftarrow{\boxdot} X$	random choice of a according to distribution X
$a \xleftarrow{\boxdot} A$	uniform random choice of a from finite set A
$a \longleftarrow b$	assignment of the value of b to the variable a
$\mathcal{E}(X), \mathrm{var}(X)$	expected value (average), variance of random variable X
$\Delta(X, Y)$	statistical distance between distributions X and Y
$\square_N, \boxtimes_N, \square_N^+$	sets of squares, antisquares, absolute value of squares modulo N
$\langle x|y \rangle, \langle x|, |y\rangle$	Dirac bra-ket
$\square$	end of proof
$\Diamond$	end of example

Sources of quotations

The quotes preserve their original spelling.

Chapter 1: Jonathan Swift, *Lemuel Gulliver, Travels into Several Remote Nations of the World*, Part III: A voyage to Laputa, Balribarbi, Glubbdubdrib, Luggnag, and Japan, Benjamin Motte, London, 1726. Quote in Chapter V The grand academy of Lagado.

Lord Byron, *English Bards, and Scotch Reviewers. A Satire*, 1809. Quote from the third edition, James Cawthorn, 1810, page 4, lines 51–52.

Edgar Allan Poe, *A few words on secret writing*, Graham's Lady's and Gentleman's Magazine, Philadelphia, 1840, page 38.

Johannes Balthasar Friderici, *Cryptographia oder Geheime schrifftmünd- und würckliche Correspondenß welche lehrmäßig vorstellet eine hochschätzbare Kunst verborgene Schrifften zu machen und auffzulösen*, Georg Rebenlein, Hamburg, 1685, Vorrede.

William P. Bundy, *Some of my wartime experiences*, Cryptologia, Volume XI (2), Taylor & Francis, April 1987, 65–77. Quote on page 67, reprinted with permission by Taylor & Francis Ltd, http://www.tandfonline.com.

David Kahn, *Cryptology goes public*, in *Foreign Affairs*, Volume 50 (**1**), Council on Foreign Relations, 1979, URL https://www.foreignaffairs.com/articles/1979-09-01/cryptology-goes-public, page 159.

Chapter 2: William Jones, *New Introduction to the Mathematics: Containing the Principles of Arithmetic & Geometry*. Printed by J. Matthews for Jeff. Wale at the Angel in St. Paul's Church-Yard, 1706, page 6.

Godfrey Harold Hardy, *A mathematician's apology*, Cambridge University Press, 1940, page 80.

Edgar Allan Poe, *The Gold-Bug*, In *The Dollar Newspaper*, 1843. For publication details, see Notes to Section A.4.

Richard Wilmer Rowan, *The Spy Menace—An Exposure of International Espionage*, Thornton Butterworth Limited, London, 1934, page 172.

Fletcher Pratt, *Secret and Urgent*, The Bobbs-Merrill Company, 1939, page 15.

Steven Levy, *Crypto Rebels*, Wired, Issue 1.02, May/June 1993, Condé, page 55.

David Hilbert, *Mathematische Probleme*, Göttinger Nachrichten, 1900, pages 253–297.

Chapter A: Caesar, *Commentarii de bello gallico* (The Gallic War), Book V, Chapter 48. Edition by Harvard University Press, 1970, page 296.

Sir Arthur Conan Doyle, *The Adventures of Sherlock Holmes. XVII. — The Adventure of the "Gloria Scott."*, The Strand Magazine, John Newnes (editor), 1893, page 400.

Matteo Argenti, *Trattato che insegna di formare cifre di varie sorti che s'appartiene a secretarii et a gran signori*, cited in Aloys Meister (1906), *Die Geheimschrift im Dienste der Päpstlichen Kurie*, page 152.

Carl Benjamin Boyer, *A History of Mathematics*, John Wiley & Sons, New York, 1968, page viii. Copyright ©1968 by John Wiley & Sons, Inc.

Rudyard Kipling, *The Man who would be King*, in *The Phantom 'Rickshaw and other Eerie Tales*, A. H. Wheeler & Co., 1888. Edition by Oxford University Press, Oxford, 1987, page 264.

Chapter 3: Kurt Gödel, Letter to John von Neumann, 20.III.1956. Reproduced with kind permission of the Institute for Advanced Study, Princeton NJ, USA. The letter is in the collection *John Von Neumann and Klara Dan Von Neumann papers, 1912-2000*, Library of Congress, Washington DC, USA.

William Stanley Jevons, *The principles of science*, Macmillan and Co., London and New York, 1877, page 122.

Joseph Diaz Gergonne, *Questions proposées. Problème d'analise indéterminée*, Annales de Mathématiques pures et appliquées, tome onzième, Nismes, de l'imprimerie de P. Durand-Belle, 1820 et 1821, page 316.

Derrick Henry Lehmer, *On the Converse of Fermat's Theorem*, The American Mathematical Monthly, Volume 43, Number 6 (Jun.–Jul.), 1936, page 347. Reprinted with the kind permission of the Mathematical Association of America.

Chapter B: Ho Chi Minh, to the cadre and students of the 1950 Viet Bac Combat Sector army cryptographic class. Quoted in *Essential Matters. A History of the Cryptographic Branch of the People's Army of Viet-Nam*, 1945 - 1975, Center for Cryptologic History, NSA, 1994, page xiii.

George Ballard Mathews, *Theory of Numbers*, Deighton, Bell and Co., 1892, Part 1, sect. 29.

Sir Alfred Ewing, in Jones, *Alfred Ewing and Room 40*, 1979, page 75.

Johann Samuel Halle, *Magie, oder, die Zauberkäfte der Natur, so auf den Nutzen und die Belustigung angewandt wurden*, 1786, page 602.

Samuel Lindner, *Elementa artis decifratoriae*, Litteris Hartvngianis, Regiomonti [Kaliningrad], 1770, page 2.

Thomas Edward Lawrence (of Arabia), *Seven Pillars of Wisdom*, printed by Roy Manning Pike and Herbert John Hodgson, London, 1926, Chapter III, page 38. Unpublished edition 1922.

Benjamin Franklin, *Reply to the Governor*, written for the Pennsylvania Assembly, 11 November 1755, printed in *Votes and Proceedings of the House of Representatives, 1755–1756*, Philadelphia, 1756, pages 19–21.

Chapter 4: Գարեգին Նժդեհ, (Garegin Nzhdeh or Garegin Ter-Harutunyan) is an Armenian national hero. The quotation is attributed to him, without a specific source. Thanks go to Vahe Hakobyan for help with the Armenian.

John Napier **of Merchiston (Ioannis Neperus)**, *Rabdologiæ sev nvmerationis per Virgulas libri dvo*, Andreas Hart, Edinburgh, 1617, Epist. Dedicat.

William Jones, *Synopsis palmariorum matheseos: or, New Introduction to the Mathematics: Containing the Principles of Arithmetic & Geometry.* Printed by J. Matthews for Jeff. Wale at the Angel in St. Paul's Church-Yard, 1706, page 179.

Thomas Jefferson, Letter dated Monticello, 18 June 1799. Published in *Scripta Mathematica*, **1**(1), 1933, 88–90.

Augustus De Morgan, *Study and Difficulties of Mathematics*, Chicago, 1902, Chapter 12.

Chapter 5: Isaac Newton, *A Treatise of the Method of Fluxions and Infinite Series With its Application to the Geometry of Curve Lines*, T. Woodman and J. Millan, 1737, Quote in anonymous preface, page xiii–xiv. Newton (1642–1727) wrote this book in 1671, but it was only published posthumously in 1736.

David Eugene Smith, *The Teaching of Elementary Mathematics*, The Macmillan Company, New York, 1900, page 219.

Bernard Le Bouyer de Fontenelle, *Entretiens sur la pluralité des mondes*, Michael Brunet, Paris, 1686. Quote in *Cinquième soir*, pages 178–179 of the 7th edition, 1724.

Chapter C: Jean de la Fontaine, *L' Ours et l'Amateur des Jardins*, Libre VIII in the *Fables*, Fable 10, Claude Barbin, Paris, 1678.

Friedrich August Leo (publisher), *Mysterienbuch alter und neuer Zeit oder Anleitung geheime Schriften lesen zu können, geschwind und kurz schreiben zu lernen, ingleichen Chiffern aufzulösen. Nebst einem Anhange die Blumenchiffern der Morgenländerinnen zu verstehen und nachzunahmen*, 1797, page 64.

Paul Louis Eugène Valèrio, *De la cryptographie. Essai sur les méthodes de déchiffrement*, Journal des Sciences Militaires, Librairie militaire de L. Baudoin, Paris, 1893, page 46.

Willem Jacob 's Gravesande, *Introduction à la Philosophie*, Jean and Herm. Verbeek, Leiden, 1748, section 1056, page 408.

Filson Young, *With the battle cruisers*, Cassell and Company, Ltd, 1922, page 173.

Roger Bacon (c. 1213–c. 1294), *Epistola de Secretis Operibus Artis et Naturæ, et de Nullitate Magiæ*. The origin and dates of this work are contentious. It is printed in Fr. Rogeri Bacon, *Opera quædam hactenus inedita*, Vol. I, Longman, Green, Longman, and Roberts, London, 1859, Appendix, page 544.

Chapter 6: The Who, Single *The Relay* (3:52), Pete Townshend, Track Records/ MCA, 1972. Today interpreted as being about the internet.

Voltaire (François-Marie Arouet), *Dictionnaire philosophique*, Gabriel Grasset, Genève, 1764. Quote in article *Poste*.

Lothar Philipp, *Die Geheimschrift*, Albert Otto Paul, Leipzig, 1931, page 35.

Carl Arnold Kortum, *Anfangsgründe der Entzifferungskunst deutscher Zifferschriften*. Hellwingische Universitätsbuchhandlung, Duisburg, 1782, 144 pages. Quote in §7. This book was essentially reprinted anonymously as *Die Kunst, geheime Schriften zu entziffern, für Feldherrn, Gesandtschafts-Secretaire, Beamten bey geheimen Kabinetten, Archivare, Polizeybeamten, Postofficianten, Diplomatiker u. a. m.*, Joachim's

Literarisches Magazin, Leipzig, 1808.

Chapter 7: Arthur Charles Clarke, *Profiles of the Future*, Harper & Row, Publishers, New York and Evanston, 1958, page xi.

Alan Matheson Turing, Report *Intelligent Machinery*, National Physics Laboratory, 1948, page 9. Courtesy of NPL.

Confucius, *The Analects of Confucius*, Book II *Wei Chang*, written ca. 500 BC. Translated by James Legge, http://www.gutenberg.org/files/4094/4094-0.txt. Quote in Chapter 17. It is often cited as "Real knowledge is to know the extent of one's ignorance."

Rick Cook, *The Wizardry Compiled*, Baen Books, 1989. Quote in Chapter Six. Excerpt reprinted by permission of Baen Books. Copyright ©1989 by Rick Cook.

Jules Verne, *Voyage au centre de la terre*, J. Hetzel et Cie, Paris, 1867, page 10.

David Shulman, *An Annotated Bibliography of Cryptography*, Garland Publishing, Inc., New York & London, 1976, part II, page 145. The quote refers to Richard Deacon's 1969 book *A History of the British Secret Service*.

Chapter D: Franz Stix, *Berichte und Studien zur Geschichte Karls V.*, Nachrichten von der Gesellschaft der Wissenschaften zu Göttingen, Philologisch-Historische Klasse, Neue Folge, Volume 1, Number 6, Vandenhoeck u. Ruprecht, 1936, pages 207–226. Quote on page 207.

François Viète, *Deschifrement d'vne lettre escrite par le commandevr Moreo av Roy d'Espaigne son maistre du 28 Octobre 1589*. See Figure D.5.

Alexander d'Agapeyeff, *Codes and Ciphers*, Oxford University Press, 1939, page 7.

Chapter 8: Anonymous.

Jules Verne, *Mathias Sandorf*, J. Hetzel et Cie, Paris, 1885, Chapitre II, *Le Comte Sandorf*, page 28.

Tom Clancy and Steve Pieczenik, *Tom Clancy's Op-Center*, Berkley Books, New York, 1995, page 177.

Henry David Thoreau, *A Week on the Concord and Merrimac Rivers*, James Munroe and Company, Boston and Cambridge, 1849, page 381.

Chapter 9: Creedence Clearwater Revival, *Penthouse Pauper* (3:39), in Album: *Bayou Country*, Fantasy Records. John Fogerty (Producer), 1969.

Charles Babbage, *Passages from the Life of a Philosopher, Various applications*, Longman, Green, Longman, Roberts & Green, 1864, page 137.

George Edwin Collins, *Computer Algebra of Polynomials and Rational Functions*, The American Mathematical Monthly, Vol. 80, Number 7, 1973, page 726.

John Michael Crichton, *The Lost World*, Knopf/Random House, 1995. Quote from edition by Ballantine Books, New York NY, 1996, page 211.

Auguste Kerckhoffs, *La Cryptographie Militaire*, Journal des Sciences Militaires, Janvier 1883.

Philip Thicknesse, *A treatise on the art of decyphering, and of writing in cypher. With an harmonic alphabet*, W. Brown, London, 1772, page 12.

Chapter E: Napoléon Bonaparte, Letter of thanks concerning his membership in the Institut national, *Gazette Nationale ou Le Moniteur Universel*, Vol. 99, 1797 (an VI).

Helen Fouché Gaines, *Elementary Cryptanalysis: a study of ciphers and their solution*, American Photographic Publishing Co., Boston, 1939. Quote in Chapter XV *Miscellaneous Phases of Vigenère Decryptment, page 144.*

Daniel Defoe, *An essay upon literature: or, An enquiry into the antiquity and original of letters; proving that the two tables, written by the finger of God in Mount Sinai, was the first writing in the world, and that all other alphabets derive from the Hebrew ; with a short view of the methods made use of by the antients to supply the want of letters before, and improve the use of them, after they were known,* Printed for Tho. Bowles, John Clark and John Bowles, London, 1726, page 298.

Jacques Casanova de Seingalt, *Mémoires de J. Casanova de Seingalt écrits par lui-même,* édition originale, tome cinquième, Heideloff et Campé, Paris, 1832. Chapitre V, pages 210 and 228.

C. W. P., *Steganographia recens detecta, oder ganz neu entdecktes Kunststück, so geheim zu schreiben, daß es kein Dechiffreur auflösen kan, welches nicht nur leicht zu erlernen, sondern auch ohne Beschwerlichkeit zu practiciren, entdeckt, gezeigt, und in verschiedenen Exempeln deutlich bewiesen,* Bartholomäi, Ulm, 1764. Quote in second edition, 1767, Fünfter Abschnitt, page 31. The author's pseudonym has not been resolved.

Chapter 10: Patrick O'Brian (Richard Patrick Russ), *Master and Commander,* J. B. Lippincott Co., 1969. Quote from 1989 edition by Harper Collins, page 249.

Peter Høeg, *Frøken Smillas fornemmelse for sne* (Miss Smilla's Feeling for Snow), Rosinante & Co., 1992. Quote in Chapter 8.

Niccolò Machiavelli, *Discorsi di Niccolò Machiavelli sopra la prima deca di Tito Livio.* Quote from Volume Two, Chapter XV, Nicolò Bettoni, Milano, 1824, page 143.

Marin Mersenne, *Seconde Partie de l'Harmonie Universelle, Livre de l'Utilité de l'Harmonie,* Pierre Ballard, Paris, 1636. Quote on page 16.

Chapter F: Francis Bacon, *Instauratio magna,* 1640, *On learning,* Book VI, pages 264–265.

Maurice Paléologue, *Journal de l'affaire Dreyfus, 1895–1899, l'affaire Dreyfus et le Quai d'Orsay,* Plon, Paris, 1955, page 13. [Lieutenant-colonel Alfred Dreyfus was Jewish and victim of an antisemitic witch hunt instigated by the French military top brass for alleged high treason. After four years in a penitentiary on the Île du Diable in French Guyana, his innocence was established after seven more years by, among others documents, the cryptanalysis of messages suppressed until then.]

Πλούταρχος (Plutarch), *Parallel lives,* Chapter XXVIII: Alcibiades.

Ξενοφῶν (Xenophon), *Hellenika,* Book V, Chapter 2. Written after 362 BC.

Reichskriegsminister, *Schlüsselanleitung für das Wehrmacht-Handschlüsselverfahren. Vom 5.2.1936,* Band 15, Teil 1 von H. Dv, Reichsdruckerei, 1940, page 8.

Chapter 11: James Russell Lowell, *The Biglow Papers,* The Atlantic Monthly, 1862, *Mason and Slidell: A Yankee idyll,* page 268.

Pierre Simon Laplace (M. le Marquis De Laplace), *Essai philosophique sur les probabilités,* Sixième Edition, Bachelier, Paris, 1840, Chapter *Notice historique sur le Calcul des Probabilités,* pages 272–273.

Robert R. Coveyou, title of his paper printed in Studies in Applied Mathematics 3, SIAM, Philadelphia PA, 1969, 70–111, and presented at the 1967 SIAM *Symposium on applied probability and Monte Carlo methods.*

Celia Margaret Fremlin, *The Jealous One,* 1964. Quote from edition by Academy Chicago Publishers, Chicago, 1985, page 99.

Albert Einstein, letter to Max Born dated 4 December 1926. In Albert Einstein, Hedwig and Max Born, Briefwechsel 1916–1955, Rowohlt, Taschenbuchverlag, Reinbek, 1972, pages 97–98. English translation: *The Born-Einstein Letters* (translated by Irene Born), Walker and Company, New York, 1971. The full quote is: "*Die Quantenmechanik ist sehr achtung-gebietend. Aber eine innere Stimme sagt mir, daß das doch nicht der wahre Jakob ist. Die Theorie liefert viel, aber dem Geheimnis des Alten bringt sie uns kaum näher. Jedenfalls bin ich überzeugt, daß der nicht würfelt."* This quote is often stated as *God does not play dice.*

John Edensor Littlewood, *A Mathematician's Miscellany*, Methuen & Co. Ltd., London, 1953, page 23.

Philip Kerr, *The Grid*, Warner Books, Inc., New York, 1995, page 122.

Elvis Presley, *Jailhouse Rock* (2:23), RCA Records, 1957.

Robert Ludlum, *The Rhinemann Exchange*, Dial Press, New York, 1974. Quote from Dell Publishing Company, 1st edition, New York NY, 1975, page 323.

Chapter G: Ibn ad-Durayhim, Arabic original and English translation in Mrayati *et al.* (2004), pages 98–101.

Blaise de Vigenère, *Traicté des chiffres, ou secrets manieres d'escrire*, Abel l'Angelier, Paris, 1586, page 13.

Frances A. Yates, *Giordano Bruno and the Hermetic Tradition*, University of Chicago Press, 1964, Chapter VIII, page 145.

Aulus Gellius, *Noctes Atticae*, Book 17, Chapter 9. Written around 143 AD.

Chapter 12: Novalis (Georg Philipp Friedrich Freiherr von Hardenberg) in *Freiberger naturwissenschaftliche Studien*, Section 12 *Arythmetika Universalis*, 1798/99. Modern reprint *Novalis Schriften*, Band 3: Das philosophische Werk II, hrsg. von Richard Samuel, Kohlhammer, Stuttgart, 1960.

Simon Singh, *Fermat's Enigma, The Epic Quest to Solve the World's Greatest Mathematical Problem*, Anchor Books, Doubleday, New York, 1997, page x.

Charles Dickens, *Great Expectations*, in *All Year Round*, London, 1861, Chapter 40. Published as a book by Chapman and Hall later in 1861.

Alfred North Whitehead, *A Treatise on Universal Algebra*, Cambridge University Press, 1898, Chapter I.7, page 12.

Joseph Conrad (Józef Teodor Konrad Korzeniowski), *Heart of Darkness*, Blackwood's Magazine, published by William Blackwood, Edinburgh, 1899, Chapter I, page 14.

John Locke, *An Essay concerning Humane Understanding*, Edw. Mory, London, 1690, Book IV *Of Knowledge and Opinion*, Chapter 2 *Of the Degrees of our Knowledge*, §3.

Bertrand Russell, *Mathematics and the Metaphysicians*, in James R. Newman, *The World of Mathematics*, Simon and Schuster, New York, 1956, Volume 3, Part X, p. 1576–1590. Quote on page 1578.

Robert Frost, *The Road not taken*. In *Mountain Interval*, Henry Holt and Company, New York, 1916, page 9. From the book *The Poetry of Robert Frost* edited by Edward Connery Lathem. ©1916 Henry Holt and Company, ©1944 by Robert Frost. Arranged by permission of Henry Holt and Company, LLC. All rights reserved.

Chapter H: Christopher Layer, This is Layer's motto, from Pulteney (1722). See Section H.4.

Philip Kerr, *Dark Matter, The Private Life of Sir Isaac Newton*, Crown Publishers, New York, 2002, page 28.

William Shakespeare, *Henry V*, Act II. Scene II: *Southampton. A Council-chamber.*, 1599.

James Westfall Thompson and Saul K. Padover, *Secret Diplomacy, A Record of Espionage and Double-Dealing: 1500–1815*, Jarrolds, Ltd., London, 1937. New edition *Secret Diplomacy, Espionage and Cryptography*, Frederick Ungar Publishing Co., New York, 1965, quote on page 261.

Erich Hüttenhain, *Die Geheimschriften des Fürstbistums Münster unter Christoph Bernhard von Galen. 1650–1678*, Schriften der Historischen Kommission Westfalens Bd. 9, Verlag Aschendorf, Münster, 1974, page 41.

Samuel Zimmerman, *Neues Titularbuch*, Ingolstadt, 1579. The "pearls" refer to Matthew 7:6.

Chapter 13: Hermann Minkowski, *Diophantische Approximationen. Eine Einführung in die Zahlentheorie*, B. G. Teubner, Leipzig, 1907, Vorwort.

Felix Klein, *Über angewandte Mathematik und Physik in ihrer Bedeutung für den Unterricht an den höheren Schulen. Nebst Erläuterung der bezüglichen Göttinger Universitätseinrichtungen*, mit Eduard Riecke, B. G. Teubner, Leipzig und Berlin, 1900. Aufsatz II, page 228.

Sun Tzu, (544–496 BC), *The Art of War*. Modern edition: Barnes & Noble Books, New York, 1994, page 135.

Colonel Marcel Givierge, *Questions de chiffre*, Revue Militaire Française, 94e année, nouvelle série, volume 13, number 37, 1923, page 78.

Chapter I: Arthur S. Link, *Wilson*, Volume 5: *Campaigns for Progressivism and Peace*, 1916–1917, Princeton University Press, 1965, page 346.

Colonel Edward Mandell House, *The Intimate Papers of Colonel House. Volume II. From Neutrality to War 1915–1917. Arranged as a narrative by Charles Seymour*, Ernest Benn Limited, London, 1926, page 456.

Ellis Mark Zacharias, *Secret Missions. The Story of an intelligence officer*. G. B. Putnam's Sons, New York, 1946. Quote on page 87 of First Bluejacket Books printing 2003.

James Madison, No source seems to exist for this oft-cited quote, and it may be fictitious. In a similar spirit, Madison said "The means of defense agst. foreign danger, have been always the instruments of tyranny at home" at the Federal Convention on 29 June 1787, page 465 of its Records.

Alistair MacLean, *Seawitch*, Collins, London, 1977, page 223.

Chapter 14: James Gleick, *Genius. The Life and Science of Richard Feynman*, Pantheon Books, New York, 1992. Quote in Chapter *MIT*, page 90 of Vintage Books Edition, 1993.

John Burdon Sanderson Haldane, *Possible Worlds and Other Essays*, Chatto & Windus, Chapter *Possible Worlds*, 1927, page 286.

Yuri Ivanovitch Manin, *A Course in Mathematical Logic*, Graduate Texts in Mathematics, Volume 53, Springer, New York, 1977. Quote in Chapter II, *Truth and deducibility*, page 83.

Katherine Coffaro, *Sunward Journey*, Harlequin American Romance, 1984, pages 114–115. Pseudonym of Katherine (Catherine?) Kovacs.

Dionysius Lardner, Review of *Babbage's Calculating Engine*, Edinburgh Review Vol. **59**, No. 120, July 1834, pages 263–327.

Chapter J: Ladislas Farago, *The Game of the Foxes*, David McKay Co., New York, 1971, page 426.

Louis Kruh and Cipher A. Deavours, *The Typex cryptograph*, Cryptologia, Volume VII (2), Taylor & Francis, 1983, 145–165. Quote on page 163, reprinted with permission by Taylor & Francis Ltd, http://www.tandfonline.com.

Ewen Edward Samuel Montagu, *Beyond Top Secret Ultra*, Coward, McCann & Geoghegan, New York, 1977, Chapter 5: The Organization of N.I.D. Section 17M, page 46.

Gustave Bertrand, *Enigma, ou la plus grande énigme de la guerre 1939–1945*, Librairie Plon, Paris, 1973, Chapter II *Une histoire merveilleuse, L'utopie est vaincue!*, page 61.

Bertrand Vivian Bowden (Baron Bowden), *Faster than Thought*, in *A symposium on digital computing machines*, Pitman Publishing, London, 1953, Glossary, page 414.

Chapter 15: Adam Risen, *Rechenbuch auff Linien und Ziphern in allerley Hand*, Franck. Bey. Chr. Egen., Erben 1574, Vorrede.

Robert Recorde, *Whetstone of Witte*, Jhon [sic!] Brugstone, London, 1557, *The Preface*, leaf b.iii^r.

أَبُو جَعْفَر مُحَمَّد بِن مُوسَى الْخَوَارِزِمِي (ʾAbū ǧaʿfar muḥammad ben mūsā al-Ḥwārizmī or Al-Khwarizmi, c. 780-c. 850), *Algorithmi de numero Indorum*, often called *Arithmetic*, c. 830. 13th century Latin manuscript from the library of the Hispanic Society of America, New York. It is probably a copy of a 12th century Latin translation of al-Ḥwārizmī's book on arithmetic whose original is lost. It was written after his *Algebra*. The manuscript was edited by Folkerts (1997). Quote from end of Chapter 7, Plate 8 (f. 20v) and p. 70. Crossley & Henry (1990) translate the Latin text of another surviving manuscript, at Cambridge.

Georges Perec, *La Vie mode d'emploi*, éditions Hachette, Paris, 1978. Quote in Chapitre XV *Chambres de bonne 5, Smautf.*

The last words of this text are from a partially encrypted document of Duke Rudolph IV of Austria (1339–1365) and read *das puchel hat ein ent = das Buch hat ein Ende* (Kürschner 1872, Tafel II, Figure 2, page 796). The kind permission of the library of Klosterneuburg (Codex 1226), Austria, is gratefully acknowledged.

Sources of images

Figures were produced for this text or are © Joachim von zur Gathen 2015, unless indicated otherwise in the text, the notes, or the following. Many thanks go to all persons and institutions who gave permission to reproduce the figures.

Figure 2.2: John J. G. Savard's page
http://www.quadibloc.com/crypto/co040401.htm. Last visited 23 March 2015.

Figures A.5 and A.6: photographs taken in June 2015 at the Tutankhamun Exhibition in the Kleine Olympiahalle, München, Germany. The originals are not allowed to travel outside of Egypt, and the exhibition and our figures show reproductions with smoothed edges and without the original cracks in the gilt wooden boards. Printed with kind permission of Semmel Concerts, Bayreuth, Germany.

Figure A.9: Myer (1879), Plate XV, Fig. 8, opposite page 117.

Figures A.11 and A.12: *Codex Aureus*, Bayerische Staatsbibliothek München, Germany, Clm 14000, leaf 126^r.

Figure A.13: Bayerische Staatsbibliothek München, Germany, Clm 1086, leaf 71^v.

Figure A.14: Alberti (1568), page 213.

Figures A.15 and A.16: ©Jan Braun/Heinz Nixdorf MuseumsForum, Paderborn, Germany.

Figure A.17: Signal Corps Converter M-209-B, serial number 76112.

Figure A.22: this drawing by G. C. Widney appears on page 16 of Poe (1902).

Figures B.2 and B.3 are from Comiers d'Ambrun (1690). The title page in Figure B.2 measures 13.5 cm × 7.5 cm and has some red underlining by hand. Figure B.3 is on page 16. The fold-out table in Figure B.4 from Comiers d'Ambrun (1691) measures 43 cm × 28.5 cm.

Figures B.5 through B.7 are from Anonymous (1770), title page and pages 388 and 389, respectively.

Figure C.7: Porta's table is from Porta (1602), page 130.

Figure D.1 top and preceding letter from the King of Navarra: *Recueil de chroniques relatives à l'histoire de France et d'Angleterre jusqu'en 1434, le tout écrit de la main de Pierre Cochon*, Français 5391, Bibliothèque Nationale de France, Paris, leaves 54^r and 53^v.

Figure D.2: Codex Vindobonensis 2398, from Tranchedino (1970), leaf 37^v.

Figure D.3: Cipher 38 on page 329 of Devos (1950).

Figure D.4: Archivo General de Simancas, Spain, Estado, legajo 2846,41.

Figure D.5: Bibliothèque nationale de France, Paris.

Image in Remark 9.5: an overlay of a NASA picture and an openclipart.

Figure E.1: Bibliothèque nationale de France, Paris, Fonds Bastille de la Bibliothèque de l'Arsenal, Ms 12455, leaf 25. It was published in de Sade (1980), volume 2, page 105, with the transcription on pages 105 and 106.

Figure E.2: Friderici (1685), opposite page 194.

Figure E.4: Bacon (1640), Sixth Book, page 268.

Figure F.1: page 26 of Porta (1602).

Figure F.2: from the first (title) page of the 192-leaf manuscript Latin 1868 from the 9th century at the Bibliothèque Nationale de France, Paris.

Figure F.3: Kantonsbibliothek Thurgau, Frauenfeld, Switzerland, Ms y 38a, leaf 5^v.

Figure F.4: Bayerische Staatsbibliothek München, Germany, Clm 2598, leaf 29^v (modern page numbering).

Figure F.5: Wilkins (1641), Chap. VI, *Secret writing with the common letters, by changing of their places*, page 68.

Figure F.6: Verne (1867), page 8.

Figure F.7: Verne (1885), Chapitre IV, *Le billet chiffré*, page 59.

Figure 11.1: Bergmann (2014).

Figure 11.2: USB stick with the noisy diode generator, kindly provided by Frank Bergmann.

Figure G.1: Mrayati *et al.* (2003), Figure 3.2 on page 113.

Figure G.2: Al-Kadi (1992), Figure 2b, page 109.

Figure G.3: Musée Condé, Château de Chantilly, France.

Figures G.4 through G.6: title page, leaf A^r and leaf o ii^r in Trithemius (1518).

Figure G.7: leaf K ii^r (= 254^r) in Trithemius (1561).

Figure G.8: Liber primus, Cap. XXIV, pages 204-205, in Trithemius (1721).

Figure G.9: Liber secundus, Cap. XII, page 261, in Trithemius (1721).

Figure G.10: Mersenne (1636), first chapter *Liure de la Nature des Sons*, Proposition XXII, page 41.

Figures G.11 and G.12: from Davys (1737), pages 24 and 26.

Figure H.1: Archivo General de Indias, Sevilla, Spain, signatura: Justicia, 1009, N.3, R.1, leaf 457^r.

Figure H.2: Burnet (1828), page 60.

Figure H.3: The National Library of Scotland, shelfmark Adv.MS.81.1.12, leaf 95^r.

Figure H.4: The National Archives, London, United Kingdom.

Figures H.5 and H.6: C.52 and H.36 (page 18) of Pulteney (1722).

Figure H.7: Archivo Nacional de Cuba, Havana, Cuba.

Figure I.1: Militärgeschichtliches Archiv, Freiburg, Germany, RM 5/3514.

Figures I.2 and I.3: Auswärtiges Amt, Politisches Archiv, Berlin, Abteilung IA, Mexico 16 secr., R 16916.

Figure I.4: US National Archives and Records Administration (NARA), College Park MD, Decimal File 862.20212/82A (1910–1929), General Records of the Department of State, Record Group 59.

Figures I.6 and I.5: The National Archives of the UK. ©Crown Copyright, used with permission of Director GCHQ.

Figures J.1 and J.3: ©Jan Braun/Heinz Nixdorf MuseumsForum, Paderborn, Germany.

Figures J.4 and J.5: pictures taken in July 2001.

Figure 15.6: Hindenburg (1794), page 88.

Sources of ornaments

The historical chapters are adorned with a *headpiece* at the top and an ornamental *initial* on their first page, and a *tailpiece* (or *colophon*) at the end.

Title: Frontispiece from Schott (1665).

Chapter A: Headpiece and initial from Porta (1602), Book 1, first page, leaf A^r, and tailpiece from page 138.

Chapter B: Headpiece, initial, and tailpiece from Schott (1665), leaf Yy^r (page 351), page 14, and last page.

Chapter C: Headpiece, initial, and tailpiece from Vignère (1587), leaves 2^r, 2^r, and 50^r.

Chapter D: Headpiece, initial, and tailpiece from Selenus (1624), pages 1, 134, and 8. Selenus is Duke August II. the Younger (1579–1666) of Braunschweig-Lüneburg. His nom de plume is a play on σελήνη = selene = luna = lüne = moon.

Chapter E: Headpiece and initial from Heidel's 1721 edition of Trithemius' *Steganographia* in the Foreword, and tailpiece from page 296.

Chapter F: Headpiece, initial, and tailpiece from Palatino (1548).

Chapter G: Headpiece, initial, and tailpiece from Trithemius (1561), leaves 112^r, 112^r, and last page.

Chapter H: Headpiece, initial, and tailpiece from Bacon (1640), pages 41 (leaf ff^r), 1 (leaf aa^r), and B2^v.

Chapter I: Headpiece, initial, and tailpiece from Schooling (1896), pages 119, 93, and 451.

Chapter J: Headpiece, initial, and tailpiece from Friderici (1685), in *Vorrede* (introduction), page 112, and *Vorrede*.

Bibliography

The numbers in brackets at the end of a reference are the pages on which it is cited. Names of authors and titles are usually given in the same form as on the article or book.

L. M. ADLEMAN (1983). On Breaking Generalized Knapsack Public Key Cryptosystems. In *Proceedings of the Fifteenth Annual ACM Symposium on Theory of Computing,* Boston MA, 402–412. ACM Press. ISBN 0-89791-099-0. URL http://portal.acm.org/citation.cfm?id=800061.808771. [648]

LEONARD M. ADLEMAN (1994). Algorithmic Number Theory—The Complexity Contribution. In *Proceedings of the 35th Annual IEEE Symposium on Foundations of Computer Science,* Santa Fe NM, SHAFI GOLDWASSER, editor, 88–113. IEEE Computer Society Press, Santa Fe NM. [193]

DIVESH AGGARWAL & UELI MAURER (2009). Breaking RSA Generically Is Equivalent to Factoring. In Joux (2009), 36–53. URL http://dx.doi.org/10.1007/978-3-642-01001-9. [144]

MANINDRA AGRAWAL, NEERAJ KAYAL & NITIN SAXENA (2004). PRIMES is in P. *Annals of Mathematics* **160**(2), 781–793. URL http://annals.math.princeton.edu/issues/2004/Sept2004/Agrawal.pdf. [143]

DORIT AHARONOV & ODED REGEV (2005). Lattice problems in NP ∩ coNP. *Journal of the ACM* **52**(5), 749–765. URL http://dx.doi.org/10.1145/1089023.1089025. Preliminary version in *Proceedings of the 45th Annual IEEE Symposium on Foundations of Computer Science,* Rome, Italy (2004). [650]

WILLIAM AIELLO & JOHAN HASTAD (1987). Perfect Zero-Knowledge Languages can be Recognized in Two Rounds. In *Proceedings of the 28th Annual IEEE Symposium on Foundations of Computer Science, Los Angeles CA,* 439–448. IEEE Computer Society Press. URL http://dx.doi.org/10.1109/SFCS.1987.47. [549]

M. AJTAI (1996). Generating Hard Instances of Lattice Problems. *Electronic Colloquium on Computational Complexity* TR96-007. URL http://eccc.hpi-web.de/eccc-reports/1996/TR96-007/Paper.pdf. 29 pages. Final version in *Quaderni di Matematica* 13 (2004). [761]

MIKLÓS AJTAI (2008). Representing Hard Lattices with $O(n \log n)$ Bits. *Chicago Journal of Theoretical Computer Science* **2008**(2). ISSN 1073-0486. URL http://dx.doi.org/10.4086/cjtcs.2008.002. [650]

MIKLOS AJTAI & CYNTHIA DWORK (1996). A Public-Key Cryptosystem with Worst-Case/Average-Case Equivalence. *Electronic Colloquium on Computational Complexity* 3. URL http://citeseer.nj.nec.com/ajtai96publickey.html. 50 pages. [650]

© Springer-Verlag Berlin Heidelberg 2015
J. von zur Gathen, *CryptoSchool,* DOI 10.1007/978-3-662-48425-8

MIKLÓS AJTAI, RAVI KUMAR & D. SIVAKUMAR (2001). A Sieve Algorithm for the Shortest Lattice Vector Problem. In *Proceedings of the Thirty-third Annual ACM Symposium on Theory of Computing*, Hersonissos, Crete, Greece, 601–610. ACM Press, 1515 Broadway, New York, New York 10036. ISBN 1-58113-349-9. [649]

LEON BATTISTA ALBERTI (1568). *Opvscoli Morali Di Leon Batista Alberti Gentil'hvomo Firentino: Ne' quali si contengono molti ammaestramenti, necessarij al viuer de l'Huomo, cosi posto in dignità, come priuato. Tradotti, & parte corretti da M. Cosimo Bartoli*. In Venetia: apresso Francesco Franceschi, Sanese, Venezia, [8], 426, [4], [1] pages. [104, 526, 815]

LEON BATTISTA ALBERTI (1994). *Dello Scrivere in Cifra*. Galimberti Tipografi Editori, 26 Via S. Pio V, 10125 Torino, ITALY, 84 pages. [104]

WERNER ALEXI,BENNY CHOR, ODED GOLDREICH & CLAUS P. SCHNORR (1988). RSA and Rabin functions: Certain Parts are As Hard As the Whole. *SIAM Journal on Computing* **17**(2), 194–209. ISSN 0097-5397. URL http://dx.doi.org/10.1137/0217013. [489–490]

WILLIAM ROBERT ALFORD, A. GRANVILLE & C. POMERANCE (1993). There are infinitely many Carmichael numbers. Mathematical preprint series, University of Georgia. [143]

ROSS J. ANDERSON (2001). *Security Engineering: A Guide to Building Dependable Distributed Systems*. John Wiley & Sons, New York. ISBN 0-471-38922-6. 640 pages. [129]

ANONYMOUS (1770). *Neueröffneter Schauplatz geheimer philosophischer Wissenschaften*. Johann Leopold Montag, Regensburg, Germany. [815]

K. APPEL & W. HAKEN (1977). Every planar map is four colorable. Part I: Discharging. *Illinois Journal of Mathematics* **21**(3), 429–490. [549]

JACOB APPELBAUM, AARON GIBSON, CHRISTIAN GROTHOFF, ANDY MÜLLER-MAGUHN, LAURA POITRAS, MICHAEL SONTHEIMER & CHRISTIAN STÖCKER (2014). Snowden-Dokumente: Was die NSA knacken kann — und was nicht. Webpage. URL http://www.spiegel.de/netzwelt/netzpolitik/snowden-dokument-so-unterminiert-die-nsa-die-sicherheit-des-internets-a-1010588.html. Last accessed 02 July 2015. [55]

KLAUS ARNOLD (1991). Johannes Trithemius—Leben und Werk. In *Johannes Trithemius. Humanismus und Magie im vorreformatorischen Deutschland*, RICHARD AUERNHEIMER & FRANK BARON, editors, 1–16. Profil Verlag, München. ISBN 3-89019-301-3. [525]

SANJEEV ARORA & BOAZ BARAK (2009). *Computational Complexity: A Modern Approach*. Cambridge University Press, NY, USA. ISBN 978-0521424264. URL http://dl.acm.org/citation.cfm?id=1540612. [796]

EMIL ARTIN (1964). *The Gamma Function*. Holt, Rinehart and Winston, New York. Translated from German by Michael Butler. [657]

A. O. L. ATKIN (1992). The number of points on an elliptic curve modulo a prime (II). Available at: https://listserv.nodak.edu/cgi-bin/wa.exe?A2=ind9207&L=NMBRTHRY&F=&S=&P=2628. [233]

L. BABAI (1986). On Lovász' lattice reduction and the nearest lattice point problem. *Combinatorica* **6**(1), 1–13. URL http://dx.doi.org/10.1007/BF02579403. [649]

LÁSZLÓ BABAI & ENDRE SZEMERÉDI (1984). On the complexity of matrix group problems I. In *Proceedings of the 25th Annual IEEE Symposium on Foundations of Computer Science,* Singer Island FL, 229–240. IEEE Computer Society Press. ISBN 0-8186-0591-X. ISSN 0272-5428. [201]

ERIC BACH, GARY MILLER & JEFFREY SHALLIT (1986). Sums of divisors, perfect numbers and factoring. *SIAM Journal on Computing* **15**(4), 1143–1154. [143]

ERIC BACH & JEFFREY SHALLIT (1996). *Algorithmic Number Theory, Vol.1: Efficient Algorithms.* MIT Press, Cambridge MA. [143, 798–801]

FRANCIS BACON (1605). *The Tvvoo Bookes of Francis Bacon. Of the proficience and aduancement of Learning, diuine and humane.* Henrie Tomes, London. 118 pp. [414–416]

FRANCIS BACON (1640). *Instaur[atio] Mag[na] P[ars] I. of the advancement and proficience of learning of the partitions of sciences, ix Bookes Written in Latin by the Most Eminent Illustrious Famous LORD FRANCIS BACON, Baron of Verulam, Vicont St Alban, Counsilour of Estate and Lord Chancellor of England.* Robert Young and Edward Forrest, Oxford. [416, 816–817]

W. BANASZCZYK (1993). New bounds in some transference theorems in the geometry of numbers. *Mathematische Annalen* **296**, 625–635. [648]

ADRIANO BARENCO, CHARLES H. BENNETT, RICHARD CLEVE, DAVID P. DIVIN-CENZO, NORMAN MARGOLUS, PETER SHOR, TYCHO SLEATOR, JOHN A. SMOLIN & HARALD WEINFURTER (1995). Elementary gates for quantum computation. *Physical Review A* **52**(5), 3457–3467. [713]

NIKO BARIĆ & BIRGIT PFITZMANN (1997). Collision-Free Accumulators and Fail-Stop Signature Schemes Without Trees. In Fumy (1997), 480–494. URL http://www.springerlink.com/content/fv4yqvggpaktgwmm/. [401]

WAYNE G. BARKER (1961). *Cryptanalysis of The Single Columnar Transposition Cipher.* Charles E. Tuttle Company, Rutland VT. Reprint 1992 by Aegean Park Press. ISBN 0-89412-192-8 (soft cover), ix, 146 pages. [443]

FRIEDRICH L. BAUER (1995). *Entzifferte Geheimnisse. Methoden und Maximen der Kryptologie.* Springer-Verlag. English translation: *Decrypted Secrets. Methods and Maxims of Cryptology,* 1996. [105]

MIHIR BELLARE (2014). New Proofs for NMAC and HMAC: Security Without Collision-Resistance. *Cryptology ePrint Archive* **2006/043**. URL https://eprint.iacr.org/2006/043. Last revised 9 April 2014. Conference version in *Advances in Cryptology: Proceedings of CRYPTO 2006,* Santa Barbara, volume 4117 of *Lecture Notes in Computer Science,* Springer Verlag. ISSN 0302-9743 (Print), 1611-3349 (Online). [429]

MIHIR BELLARE, RAN CANETTI & HUGO KRAWCZYK (1996). Keying Hash Functions for Message Authentication. In *Advances in Cryptology: Proceedings of CRYPTO 1996,* Santa Barbara, CA, NEAL KOBLITZ, editor, number 1109 in Lecture Notes in Computer Science, 1–15. Springer-Verlag. ISBN 3-540-61512-1. ISSN 0302-9743. URL http://link.springer.de/link/service/series/0558/bibs/1109/11090001.htm. [429]

STEVEN M. BELLOVIN (2011). Frank Miller: Inventor of the One-Time Pad. *Cryptologia* **35**(3), 203–222. URL http://dx.doi.org/10.1080/01611194.2011.583711. [401]

R. L. BENDER & C. POMERANCE (1998). Rigorous discrete logarithm computations in finite fields via smooth polynomials. *AMS/IP Studies in Advanced Mathematics* **7**, 221–232. [201]

PAUL A. BENIOFF (1980). The Computer as a Physical System: A Microscopic Quantum Mechanical Hamiltonian Model of Computers as Represented by Turing Machines. *Journal of Statistical Physics* **22**(5), 563–591. [713]

PAUL A. BENIOFF (1982a). Quantum Mechanical Hamiltonian Models of Discrete Processes That Erase Their Own Histories: Application to Turing Machines. *International Journal of Theoretical Physics* **21**(3/4), 177–201. [713]

PAUL A. BENIOFF (1982b). Quantum Mechanical Models of Turing Machines That Dissipate No Energy. *Physical Review Letters* **48**(23), 1581–1585. [713]

C. H. BENNETT (1973). Logical Reversibility of Computation. *IBM Journal of Research and Development* **17**, 525–532. [713]

ROBERT LOUIS BENSON & MICHAEL WARNER (editors) (1996). *VENONA.* SOVIET ESPIONAGE AND THE AMERICAN RESPONSE 1939-1957. National Security Agency and Central Intelligence Agency, Washington, D.C. ISBN 0894122657, 450 pages. [377]

FRANK BERGMANN (2014). Physikalische Zufallssignalgeneratoren für kryptografische Applikationen. URL http://www.ibbergmann.de/. [489, 816]

E. R. BERLEKAMP (1970). Factoring Polynomials Over Large Finite Fields. *Mathematics of Computation* **24**(11), 713–735. [146]

ELWYN R. BERLEKAMP, ROBERT J. MCELIECE &HENK C. A. VAN TILBORG (1978). On the Inherent Intractability of Certain Coding Problems. *IEEE Transactions on Information Theory* **IT-24**(3), 384–386. [405]

THOMAS M. BERNDT (2010). *CryptCOM — Insuring secure communication on arbitrary GSM phones by applying strong cryptography.* Diploma thesis, Department of Computer Security, b-it – Bonn-Aachen International Center for Information Technology, University of Bonn, Bonn, Germany. [221]

DANIEL J. BERNSTEIN, TANJA LANGE & CHRISTIANE PETERS (2008). Attacking and defending the McEliece cryptosystem. *Cryptology ePrint Archive* **2008/318**, 16. URL http://eprint.iacr.org/2008/318. [405]

ETHAN BERNSTEIN & UMESH VAZIRANI (1993). Quantum Complexity Theory. *Proceedings of the Twenty-fifth Annual ACM Symposium on Theory of Computing*, San Diego CA 11–20. [713]

COUNT BERNSTORFF (1920). *My three years in America.* Charles Scribner's sons, New York. [677–678]

GUIDO BERTONI, JOAN DAEMEN, MICHAËL PEETERS & GILLES VAN ASSCHE (2013). The Keccak sponge function family. online: http://keccak.noekeon.org/. Last visited 2 June 2013. [314]

GUSTAVE BERTRAND (1973). *Enigma—ou la plus grande énigme de la guerre 1939-1945.* Librairie Plon, Paris. 295, [5], 32 pages. [736]

T. BETH, D. JUNGNICKEL & H. LENZ (1999). *Design Theory*, volume 2. Cambridge University Press, Cambridge, UK, 2nd edition. ISBN 0-521-44432-2. [490]

ELI BIHAM (2006). How to Make a Difference: Early History of Differential Crypt-analysis. In *Fast Software Encryption 2006, 13th International Workshop, FSE 2006,* Graz, Austria. URL http://fse2006.iaik.tugraz.at/program.html. [295]

ELI BIHAM & ADI SHAMIR (1991a). Differential Cryptanalysis of DES-like Cryptosys-tems. In *Advances in Cryptology: Proceedings of CRYPTO 1990,* Santa Barbara, CA, ALFRED J. MENEZES & SCOTT A. VANSTONE, editors, volume 537 of *Lecture Notes in Computer Science,* 2–21. Springer-Verlag, Berlin. ISBN 978-3-540-54508-8 (Print) 978-3-540-38424-3 (Online). ISSN 0302-9743. URL http://dx.doi.org/10.1007/3-540-38424-3_1. [295]

ELI BIHAM & ADI SHAMIR (1991b). Differential Cryptanalysis of DES-like Cryp-tosystems. *Journal of Cryptology* **4**(1), 3–72. URL http://dx.doi.org/10.1007/BF00630563. [268, 295]

ELI BIHAM & ADI SHAMIR (1992). Differential Cryptanalysis of the Full 16-round DES. In *Advances in Cryptology: Proceedings of CRYPTO 1992,* Santa Barbara, CA, ERNEST F. BRICKELL, editor, number 740 in Lecture Notes in Computer Science, 487–496. Springer-Verlag. ISSN 0302-9743. [295]

ELI BIHAM & ADI SHAMIR (1993). *Differential Cryptanalysis of the Data Encryption Standard.* Springer-Verlag, New York. ISBN 978-1-4613-9316-0 (Print) 978-1-4613-9314-6 (Online). URL http://dx.doi.org/10.1007/978-1-4613-9314-6. [295]

TIMO BINGMANN (2008). Speedtest and Comparison of Open-Source Cryptography Libraries and Compiler Flags. URL http://panthema.net/2008/0714-cryptography-speedtest-comparison/. Last visited 18 June 2014. [55]

ALEX BIRYUKOV, ADI SHAMIR & DAVID WAGNER (2001). Real Time Cryptanalysis of A5/1 on a PC. In *Fast Software Encryption 2000, 7th International Workshop, FSE 2000,* New York, NY, USA, BRUCE SCHNEIER, editor, volume 1978 of *Lecture Notes in Computer Science,* 37–44. Springer-Verlag. ISBN 3-540-41728-1. ISSN 0302-9743 (Print) 1611-3349 (Online). URL http://dx.doi.org/10.1007/3-540-44706-7_1. [429]

IAN BLAKE, GADIEL SEROUSSI & NIGEL SMART (1999). *Elliptic Curves in Crypto-graphy.* Number 265 in London Mathematical Society Lecture Note Series. Cambridge University Press. ISBN 0-521-65374-6. [233]

JOHANNES BLÖMER & ALEXANDER MAY (2003). New Partial Key Exposure Attacks on RSA. In *Advances in Cryptology - CRYPTO 2003,* DAN BONEH, editor, volume 2729 of *Lecture Notes in Computer Science,* 27–43. Springer-Verlag, Berlin, Heidelberg. ISBN 978-3-540-40674-7. URL http://dx.doi.org/10.1007/b11181710.1007/b11817. [649]

CÉLINE BLONDEAU & KAISA NYBERG (2013). New Links between Differential and Linear Cryptanalysis. In *Advances in Cryptology: Proceedings of EUROCRYPT 2013,* Athens, Greece, THOMAS JOHANSSON & PHONG Q. NGUYEN, editors, volume 7881 of *Lecture Notes in Computer Science,* 388–404. Springer-Verlag. ISBN 978-3-642-38347-2 (Print) 978-3-642-38348-9 (Online). ISSN 0302-9743. URL http://dx.doi.org/10.1007/978-3-642-38348-9_24. [295]

L. BLUM, M. BLUM & M. SHUB (1986). A simple unpredictable pseudo-random number generator. *SIAM Journal on Computing* **15**(2), 364–383. URL http://dx.doi.org/10.1137/0215025. [447, 479–480]

MANUEL BLUM & SHAFI GOLDWASSER (1985). An Efficient Probabilistic Public-Key Encryption Scheme Which Hides All Partial Information. In *Advances in Cryptology: Proceedings of CRYPTO 1984,* Santa Barbara, CA, G. R. BLAKLEY & D. CHAUM, editors, volume 196 of *Lecture Notes in Computer Science*, 289–299. Springer-Verlag, Berlin. ISBN 0-387-15658-5. ISSN 0302-9743. URL http://dx.doi.org/10.1007/3-540-39568-7. [402]

MANUEL BLUM & SILVIO MICALI (1984). How to generate cryptographically strong sequences of pseudo-random bits. *SIAM Journal on Computing* **13**(4), 850–864. ISSN 0097-5397. URL http://dx.doi.org/10.1137/0213053. [490]

ANDREY BOGDANOV, DMITRY KHOVRATOVICH & CHRISTIAN RECHBERGER (2011). Biclique Cryptanalysis of the Full AES. In *Advances in Cryptology: Proceedings of ASIACRYPT 2011,* Seoul, South Korea, DONG HOON LEE & XIAOYUN WANG, editors, volume 7073 of *Lecture Notes in Computer Science*, 344–371. Springer-Verlag. ISBN 978-3-642-25385-0. ISSN 0302-9743. URL http://dx.doi.org/10.1007/978-3-642-25385-0_19. Preprint at http://eprint.iacr.org/2011/449. [33]

THOMAS BOGHARDT (2012). *The Zimmermann Telegram: Diplomacy, Intelligence and the American Entry into World War I.* Naval Institute Press, United States of America. ISBN 978-1-61251-148-1 (Print) 978-1-61251-147-4 (Online). [677]

CHARLES CARROLL BOMBAUGH (1961). *Oddities and curiosities of words and literature. With an introduction by Martin Gardner.* New York, Dover Publications Inc. XV, 375 pages. First edition 1860. [416]

D. BONEH & G. DURFEE (2000). Cryptanalysis of RSA with private key d less than $N^{0.292}$. *IEEE Transactions on Information Theory* **46**(4), 1339–1349. URL http://dx.doi.org/10.1109/18.850673. [144]

DAN BONEH, GLENN DURFEE & YAIR FRANKEL (1998). An Attack on RSA Given a Small Fraction of the Private Key Bits. In *Advances in Cryptology - ASIACRYPT'98,* K. OHTA & D. PEI, editors, volume 1514 of *Lecture Notes in Computer Science*, 25–34. Springer-Verlag, Berlin, Heidelberg. ISBN 978-3-540-65109-3. ISSN 0302-9743 (Print) 1611-3349 (Online). URL http://dx.doi.org/10.1007/3-540-49649-1_3. [649]

DAN BONEH & RICHARD J. LIPTON (1996). Algorithms for Black-Box Fields and their Application to Cryptography. In *Advances in Cryptology: Proceedings of CRYPTO 1996,* Santa Barbara, CA, NEAL KOBLITZ, editor, number 1109 in Lecture Notes in Computer Science, 283–297. Springer-Verlag. ISSN 0302-9743. [201]

DAN BONEH & RAMARATHNAM VENKATESAN (1998). Breaking RSA may not be equivalent to factoring. In *Advances in Cryptology: Proceedings of EUROCRYPT 1998,* Helsinki, Finland, K. NYBERG, editor, volume 1403 of *Lecture Notes in Computer Science*, 59–71. Springer-Verlag, Berlin, Heidelberg. ISBN 3-540-64518-7. ISSN 0302-9743. URL http://dx.doi.org/10.1007/BFb0054117. [144]

RAVI B. BOPPANA & MICHAEL SIPSER (1990). The Complexity of Finite Functions. In *Handbook of Theoretical Computer Science*, J. VAN LEEUWEN, editor, volume A, 757–804. Elsevier Science Publishers B.V., Amsterdam, and The MIT Press, Cambdrige MA. ISBN 978-0262720144 (both volumes: 978-0262720205). [490]

FRANK L. BORCHARDT (1972). Trithemius and the Mask of Janus. In *Traditions and Transitions. Studies in Honor of Harold Jantz*, 37–49. Müchen. ISBN 3-7689-0098-3. [526]

JOPPE W. BOS, MARCELO E. KAIHARA, THORSTEN KLEINJUNG, ARJEN K. LENSTRA & PETER L. MONTGOMERY (2009). PlayStation 3 computing breaks 2^{60} barrier. 112-bit prime ECDLP solved. WWW. URL http://lacal.epfl.ch/page81774.html. Laboratory for Cryptologic Algorithms. [222–223]

G. E. P. BOX & MERVIN E. MULLER (1958). A note on the generation of random normal deviates. *Annals of Mathematical Statistics* **29**(2), 610–611. URL http://dx.doi.org/10.1214/aoms/1177706645. [650]

JOAN BOYAR (1989). Inferring Sequences Produced by Pseudo-Random Number Generators. *Journal of the ACM* **36**(1), 129–141. URL http://dx.doi.org/10.1145/58562.59305. [489]

ZVIKA BRAKERSKI, ADELINE LANGLOIS, CHRIS PEIKERT, ODED REGEV & DAMIEN STEHLÉ (2013). Classical Hardness of Learning with Errors. In *Proceedings of the Fourty-fifth Annual ACM Symposium on Theory of Computing*, Palo Alto, CA, USA, DAN BONEH, TIM ROUGHGARDEN & JOAN FEIGENBAUM, editors, 575–584. ACM Press, Cambridge, MA, USA. ISBN 978-1-4503-2029-0. URL http://dx.doi.org/10.1145/2488608.2488680. [640, 651]

NOEL L. BRANN (1999). Trithemius, Johann. A chapter in the controversy over occult studies in early modern Europe. In *Encyclopedia of the Renaissance*, PAUL F. GRENDLER, editor, volume 6. Charles Scribner's Sons, New York. ISBN 0-684-80514-6. URL http://www.gale.com/servlet/ItemDetailServlet?region=9&imprint=144&titleCode=S81&type=4&id=167883. [525]

GILLES BRASSARD (1994). Cryptology Column — Quantum Computing: The End of Classical Cryptography? *SIGACT News* **25**(4), 15–21. [714]

GILLES BRASSARD & CLAUDE CRÉPEAU (1996). Cryptology Column — 25 Years of Quantum Cryptography. *SIGACT News* **27**(3), 13–24. [714]

GILLES BRASSARD, CLAUDE CRÉPEAU, RICHARD JOZSA & DENIS LANGLOIS (1993). A Quantum Bit Commitment Scheme Provably Unbreakable by both Parties. *Proceedings of the 34th Annual IEEE Symposium on Foundations of Computer Science*, Palo Alto CA 362–371. [714]

A. BRAUER (1939). On addition chains. *Bulletin of the American Mathematical Society* **45**, 736–739. [797, 802]

RICHARD P. BRENT (2000). Recent Progress and Prospects for Integer Factorisation Algorithms. In *6th Annual International Computing and Combinatorics Conference (COCOON 2000)*, Sydney, Australia, July 2000, volume 1858 of *Lecture Notes in Computer Science*, 3–22. Springer-Verlag, Berlin, Heidelberg. ISBN 3-540-67787-9. ISSN 0302-9743. URL http://link.springer.de/link/service/series/0558/tocs/t1858.htm. [145]

RICHARD P. BRENT & PAUL ZIMMERMANN (2010). An $O(M(n)\log n)$ Algorithm for the Jacobi Symbol. In *Algorithmic Number Theory, Ninth International Symposium, ANTS-IX*, Nancy, France, GUILLAUME HANROT, FRANÇOIS MORAIN & EMMANUEL THOMÉ, editors, volume 6197 of *Lecture Notes in Computer Science*, 83–95. Springer-Verlag, Berlin, Heidelberg. ISBN 978-3-642-14517-9. ISSN 0302-9743. URL http://dx.doi.org/10.1007/978-3-642-14518-6. [798]

ERNEST F. BRICKELL, DANIEL M. GORDON, KEVIN S. MCCURLEY & DAVID B. WILSON (1992). Fast Exponentiation with Precomputation. In *Advances in Cryptology: Proceedings of EUROCRYPT 1992*, Balatonfüred, Hungary, R. A. RUEPPEL, editor, number 658 in Lecture Notes in Computer Science, 200–207. Springer-Verlag, Berlin. ISSN 0302-9743. [797]

PETER F. BROWN, VINCENT J. DELLA PIETRA, ROBERT L. MERCER, STEPHEN A. DELLA PIETRA & JENNIFER C. LAI (1992). An Estimate of an Upper Bound for the Entropy of English. *Computational Linguistics* **18**(1), 31–40. URL http://acl.ldc.upenn.edu/J/J92/J92-1002.pdf. [101]

BSI (2014). Kryptographische Verfahren: Empfehlungen und Schlüssellängen. Technische Richtlinie BSI TR-02102-1, Bundesamt für Sicherheit in der Informationstechnik, Bonn, Germany. URL https://www.bsi.bund.de/SharedDocs/Downloads/DE/BSI/Publikationen/TechnischeRichtlinien/TR02102/BSI-TR-02102_pdf.html. [224, 233]

FRIEDERICH JOHANN BUCK (1772). *Mathematischer Beweiß: daß die Algebra zur Entdeckung einiger verborgener Schriften bequem angewendet werden könne.* Königsberg. [82]

ALEKSEI BURLAKOV, JOHANNES VOM DORP, JOACHIM VON ZUR GATHEN, SARAH HILLMANN, MICHAEL LINK, DANIEL LOEBENBERGER, JAN LÜHR, SIMON SCHNEIDER & SVEN ZEMANEK (2015). Comparative analysis of pseudorandom generators. In *Proceedings of the 22nd Crypto-Day, 09–10 July 2015, Munich.* Gesellschaft für Informatik. URL http://fg-krypto.gi.de/fileadmin/fg-krypto/Handout-22.pdf. [451, 489]

MARGARET BURNET (1828). *Letters from Lady Margaret Burnet to John, Duke of Lauderdale.* Bannatyne Club, Edinburgh, 6, vi, 107, 3 pages. [572, 816]

W. T. BUTTLER, R. J. HUGHES, P. G. KWIAT, S. K. LAMOREAUX, G. G. LUTHER, G. L. MORGAN, J. E. NORDHOLT, C. G. PETERSON & C. M. SIMMONS (1998). Practical Free-Space Quantum Key Distribution over 1 km. *Physical Review Letters* **81**(15), 3283–3286. [714]

CHRISTIAN CACHIN (2004). An Information-Theoretic Model for Steganography. *Information and Computation* **192**(1), 41–56. URL http://dx.doi.org/10.1016/j.ic.2004.02.003. [414]

A. R. CALDERBANK, E. M. RAINS, P. SHOR & N. J. A. SLOANE (1997). Quantum Error Correction and Orthogonal Geometry. *Physical Review Letters* **78**(3), 405–408. [713]

A. R. CALDERBANK & PETER W. SHOR (1996). Good quantum error-correcting codes exist. *Physical Review A* **54**(2), 1098–1105. [713]

JAN L. CAMENISCH, JEAN-MARC PIVETEAU & MARKUS A. STADLER (1995). Blind signatures based on the discrete logarithm problem. In *Advances in Cryptology: Proceedings of EUROCRYPT 1994,* Perugia, Italy, ALFREDO DE SANTIS, editor, number 950, 428–432. Springer-Verlag, Berlin. URL http://www.springerlink.com/openurl.asp?genre=article&issn=0302-9743&volume=950&spage=428. [364]

RAN CANETTI, ODED GOLDREICH & SHAI HALEVI (2004). The Random Oracle Methodology, Revisited. *Journal of the ACM (JACM)* **51**(4), 557–594. URL http://portal.acm.org/citation.cfm?coll=GUIDE&dl=GUIDE&id=1008734. Article home page: http://www.wisdom.weizmann.ac.il/~oded/p_rom.html. [402]

E. R. CANFIELD, PAUL ERDŐS & CARL POMERANCE (1983). On a problem of Oppenheim concerning 'Factorisatio Numerorum'. *Journal of Number Theory* **17**, 1–28. [153]

CHRISTOPHE DE CANNIÈRE & CHRISTIAN RECHBERGER (2006). Finding SHA-1 Characteristics: General Results and Applications. In Lai & Chen (2006), 1–20. URL http://dx.doi.org/10.1007/11935230_1. [314]

GIROLAMO CARDANO (1550). *Hieronymi Cardani Mediolanensis, medici, de subtilitate libri XXI.* Nürnberg, Paris. [442–443]

CLAUDE CARLET (2010). Boolean Functions for Cryptography and Error Correcting Codes. *In Crama & Hammer (2010).* URL http://www-roc.inria.fr/secret/ Claude.Carlet/chap-fcts-Bool.pdf. [296]

D. CHAUM, E. VAN HEIJST & B. PFITZMANN (1992). Cryptograhically strong undeniable signatures, unconditionally secure for the signer. In Feigenbaum (1992), 470–484. URL http://dx.doi.org/10.1007/3-540-46766-1. [316]

P. L. CHEBYSHEV (1852). Mémoire sur les nombres premiers. *Journal de Mathématiques Pures et Appliquées, I série* **17**, 366–390. *Mémoires présentées à l'Académie Impériale des sciences de St.-Pétersbourg par divers savants* **6** (1854), 17–33. *Œuvres* I, eds. A. MARKOFF and N. SONIN, 1899, reprint by Chelsea Publishing Co., New York, 49–70.

BENNY CHOR & RONALD L. RIVEST (1988). A knapsack–type public key cryptosystem based on arithmetic in finite fields. *IEEE Transactions on Information Theory* **IT-34**(5), 901–909. *Advances in Cryptology: Proceedings of CRYPTO 1984*, Santa Barbara CA, Lecture Notes in Computer Science **196**, Springer-Verlag, New York, 1985, 54–65. [648]

C. CID, S. MURPHY & M. ROBSHAW (2005). Small Scale Variants of the AES. In *12th International Workshop, FSE 2005, Paris, France, February 21-23, 2005, Revised Selected Papers*, HENRI GILBERT & HELENA HANDSCHUH, editors, volume 3557 of *Lecture Notes in Computer Science*, 145–162. Springer-Verlag. ISBN 978-3-540-26541-2 (Print) 978-3-540-31669-5 (Online). ISSN 0302-9743. URL http://dx.doi.org/ 10.1007/11502760_10. Available at http://www.isg.rhul.ac.uk/~sean/smallAES-fse05.pdf. [295]

STELVIO CIMATO (2011). *Visual Cryptography and Secret Image Sharing (Digital Imaging and Computer Vision)*. CRC Press. ISBN 143983721X. 501 pages. [56]

RICHARD CLEVE, ARTUR K. EKERT, CHIARRA MACCHIAVELLO & MICHELE MOSCA (1998). Quantum algorithms revisited. *Proceedings of the Royal Society, London* **A454**, 339–354. [715]

WILLIAM COBBETT (1811). *Cobbett's Parliamentary History of England. From the Norman Conquest, in 1066, to the year 1803*, volume 8. T. C. Hansard. [559, 572–573]

ALAN COBHAM (1965). The intrinsic computational difficulty of functions. In *Proc. from the second International Congress of Logic, Methodology, and Philosophy of Science*. North Holland. URL http://en.wikipedia.org/wiki/Cobham's_thesis. [400]

PIERRE COCHON (1870). *Chronique Normande*. A. Le Brument, Rouen. Edited by Ch. de Robillard de Beaurepaire. [342]

HENRI COHEN & GERHARD FREY (2006). *Handbook of Elliptic and Hyperelliptic Curve Cryptography*. Discrete Mathematics and its Applications. Chapman & Hall/CRC. ISBN 1-58488-518-1. With the help of Roberto M. Avanzi, Christophe Doche, Tanja Lange, Kim Nguyen, and Frederik Vercauteren. [232–233]

CLAUDE COMIERS (1690). *L'Art d'Écrire et de Parler Occultement et sans Soupçon.* Michel Guerout, Paris, 72. [815]

CLAUDE COMIERS (1691). *Traité de la parole, langues, et écritures, contenant la stéganographie impénétrable, ou L'art d'écrire et de parler occultement.* Jean Léonard, Bruxelles, 16, 276. [815]

STEPHEN A. COOK (1971). The Complexity of Theorem–Proving Procedures. In *Proceedings of the Third Annual ACM Symposium on Theory of Computing,* Shaker Heights OH, 151–158. ACM Press. [370]

DON COPPERSMITH (1992). DES designed to withstand differential cryptanalysis. Appeared in an IBM Internal newsgroup (CRYPTAN FORUM) and reproduced in newsgroup sci.crypt with the approval of the author. URL http://groups.google. com/group/sci.crypt/msg/ba8d55788e174bf9. [268]

DON COPPERSMITH (1993). Modifications of the Number Field Sieve. *Journal of Cryptology* **6**(3), 169–180. URL http://dx.doi.org/10.1007/BF00198464. [140]

D. COPPERSMITH (1994a). The Data Encryption Standard (DES) and its strength against attacks. *IBM Journal of Research and Development* **38**(3), 243–250. URL http://dx.doi.org/10.1147/rd.383.0243. [295]

DON COPPERSMITH (1994b). Solving homogeneous linear equations over GF(2) via block Wiedemann algorithm. *Mathematics of Computation* **62**(205), 333–350. [715]

DON COPPERSMITH (1997). Small Solutions to Polynomial Equations, and Low Exponent RSA Vulnerabilities. *Journal of Cryptology* **10**(4), 233–260. Final version of two articles from Eurocrypt '96. [616]

THOMAS H. CORMEN, CHARLES E. LEISERSON & RONALD L. RIVEST (1990). *Introduction to Algorithms.* MIT Press, Cambridge MA. ISBN 0-262-03141-8, xx+1028. [105, 797]

JEAN-SÉBASTIEN CORON & ALEXANDER MAY (2007). Deterministic Polynomial-Time Equivalence of Computing the RSA Secret Key and Factoring. *Journal of Cryptology* **20**(1), 39–50. ISSN 0933-2790 (Print) 1432-1378 (Online). URL http://dx.doi.org/ 10.1007/s00145-006-0433-6. [144]

MATTHIJS J. COSTER, ANTOINE JOUX, BRIAN A. LAMACCHIA, ANDREW M. ODLYZKO, CLAUS-PETER SCHNORR & JACQUES STERN (1992). Improved low-density subset sum algorithms. *computational complexity* **2**, 111–128. [648]

R. R. COVEYOU (1969). Random Number Generation Is Too Important to Be Left to Chance. In *A collection of papers presented by invitation at the Symposia on Applied Probability and Monte Carlo Methods and Modern Aspects of Dynamics sponsored by the Air Force Office of Scientific Research at the 1967 National Meeting of SIAM in Washington, D. C. , June 11-15, 1967,* volume 3 of *Studies in Applied Mathematics,* 70–111. Society for Industrial and Applied Mathematics, Philadelphia. [489]

MICHAEL J. COWAN (2004). Rasterschlüssel 44—The Epitome of Hand Field Ciphers. *Cryptologia* **XXVIII**(2), 115–148. [443]

YVES CRAMA & PETER L. HAMMER (editors) (2010). *Boolean Models and Methods in Mathematics, Computer Science, and Engineering.* Cambridge University Press. ISBN 978-0-521-84752-0.

RONALD CRAMER (editor) (2005). *Advances in Cryptology: Proceedings of EURO-CRYPT 2005,* Aarhus, Denmark, volume 3494 of *Lecture Notes in Computer Science.* Springer-Verlag, Berlin, Heidelberg. ISBN 978-3-540-25910-7. ISSN 0302-9743. URL http://dx.doi.org/10.1007/b136415.

RONALD CRAMER & VICTOR SHOUP (2004). Design and Analysis of Practical Public-Key Encryption Schemes Secure against Adaptive Chosen Ciphertext Attack. *SIAM Journal on Computing* **33**(1), 167–226. ISSN 0097-5397. URL http://dx.doi.org/10.1137/S0097539702403773. [402]

RICHARD CRANDALL & CARL POMERANCE (2005). *Prime numbers – A computational perspective.* Springer-Verlag, 2nd edition. ISBN 0-387-25282-7. [140, 144]

JOHN N. CROSSLEY & ALAN S. HENRY (1990). Thus Spake al-Khw-arizm-ı: A Translation of the Text of Cambridge University Library Ms. Ii.vi.5. *Historia Mathematica* **17**, 103–131. [815]

JOAN DAEMEN, LARS KNUDSEN & VINCENT RIJMEN (1997). The Block Cipher SQUARE. In *Fast Software Encryption 1997, 4th International Workshop, FSE 1997,* Haifa, Israel, ELI BIHAM, editor, volume 1267 of *Lecture Notes in Computer Science,* 149–165. Springer-Verlag, Heidelberg. ISBN 3-540-63247-6. ISSN 0302-9743. URL http://link.springer.de/link/service/series/0558/bibs/1267/12670149.htm. [32]

JOAN DAEMEN & VINCENT RIJMEN (1999). The Rijndael Block Cipher. AES Proposal: Rijndael. [296]

JOAN DAEMEN & VINCENT RIJMEN (2002a). AES and the Wide Trail Design Strategy. In *Advances in Cryptology: Proceedings of EUROCRYPT 2002,* Amsterdam, The Netherlands, LARS KNUDSEN, editor, volume 2332 of *Lecture Notes in Computer Science,* 108–109. Springer-Verlag, Berlin, Heidelberg. ISBN 3-540-43553-0. ISSN 0302-9743. URL http://dx.doi.org/10.1007/3-540-46035-7_7. [290]

JOAN DAEMEN & VINCENT RIJMEN (2002b). *The Design of Rijndael. AES - The Advanced Encryption Standard.* Springer, Berlin, Heidelberg, New York. ISBN 3-540-42580-2. [33]

IVAN BJERRE DAMGÅRD (1990). A design principle for hash functions. In *Advances in Cryptology: Proceedings of CRYPTO 1989,* Santa Barbara, CA, number 435 in Lecture Notes in Computer Science, 416–427. Springer-Verlag. ISSN 0302-9743. [310]

JOHN COLEMAN DARNELL (2004). *The Enigmatic Netherworld Books of the Solar Osirian Unity. Cryptographic Compositions in the Tombs of Tutankhamun, Ramesses VI, and Ramesses IX.* Academic Press, Vandenhoeck & Ruprecht, Göttingen. XIII, 683 pages. [103]

MARK DAVIES (2008-2012). The Corpus of Contemporary American English: 450 million words, 1990-present. URL http://corpus.byu.edu/coca/. [104]

JOHN DAVYS (1737). *An essay on the art of decyphering.* L. Gilliver and J. Clarke, London. [524, 816]

C. A. DEAVOURS (1981). The Black Chamber: A Column Shutting off the Spigot in 1981. *Cryptologia* **5**(1), 43–45. ISSN 01611194.

DANIEL DEFOE (1726). An Essay Upon Literature: Or, An Enquiry into the Antiquity and Original of Letters. In *Writings on Travel Discovery and History,* P. N. FURBANK, editor, volume 4, 352 pages. Pickering & Chatto, London. ISBN 1-85196-718-4. [519]

D. DEUTSCH (1985). Quantum theory, the Church-Turing principle and the universal quantum computer. *Proceedings of the Royal Society of London Series A* **400**, 97–117. [713]

DAVID DEUTSCH & RICHARD JOZSA (1992). Rapid solution of problem by quantum computation. *Proceedings of the Royal Society, London* **A439**(1907), 553–558. [713]

DEUTSCHE NATIONALVERSAMMLUNG (editor) (1920). *Stenographische Berichte über die öffentlichen Verhandlungen des 15. Untersuchungsausschusses der Verfassunggebenden Nationalversammlung nebst Beilagen*, volume II. Verlag der Norddeutschen Buchdruckerei und Verlagsanstalt, Berlin. [677–678]

J.-P. DEVOS (1946). La cryptographie espagnole durant la seconde moitié du XVIe siècle et le XVIIe siècle. In *Miscellanea historica in honorem Alberti de Meyer*, volume 2, 1025–1035. [342]

J. P. DEVOS (1950). *Les chiffres de Philippe II (1555-1598) et du despacho universal durant le XVIIe siècle*. Palais des Académies, Bruxelles. [339, 342–343, 816]

LUC DEVROYE (1986). *Non-Uniform Random Variate Generation*. Springer-Verlag, New York. ISBN 0-387-96305-7. 843 pages. [650]

WHITFIELD DIFFIE & MARTIN E. HELLMAN (1976). New directions in cryptography. *IEEE Transactions on Information Theory* **IT-22**(6), 644–654. URL http://dx.doi.org/10.1109/TIT.1976.1055638. [14, 38, 55, 401, 588]

WHITFIELD DIFFIE & SUSAN LANDAU (1998). *Privacy on the Line*. MIT Press. [429]

W. DIFFIE, P. VAN OORSCHOT & M. WIENER (1992). Authentication and Authenticated Key Exchanges. In *Designs, Codes and Cryptography*, volume 2, 107–125. Kluwer Academic Publishers. [429]

JOHN F. DILLON (1974). *Elementary Hadamard Difference Sets*. Ph.D. thesis, University of Maryland. Originally in the NSA Technical Journal, Special Issue 1972, pages 191-215. [296]

G. LEJEUNE DIRICHLET (1837). Beweis des Satzes, dass jede unbegrenzte arithmetische Progression, deren erstes Glied und Differenz ganze Zahlen ohne gemeinschaftlichen Factor sind, unendlich viele Primzahlen enthält. *Abhandlungen der Königlich Preussischen Akademie der Wissenschaften* 45–81. *Werke*, Erster Band, ed. L. KRONECKER, 1889, 315–342. Reprint by Chelsea Publishing Co., 1969. [354]

G. LEJEUNE DIRICHLET (1849). Über die Bestimmung der mittleren Werthe in der Zahlentheorie. *Abhandlungen der Königlich Preussischen Akademie der Wissenschaften* 69–83. *Werke*, Zweiter Band, ed. L. KRONECKER, 1897, 51–66. Reprint by Chelsea Publishing Co., 1969. [120]

JOHN D. DIXON (1981). Asymptotically Fast Factorization of Integers. *Mathematics of Computation* **36**(153), 255–260. [136]

HANS DOBBERTIN (1998). Cryptanalysis of MD4. *Journal of Cryptology* **11**(4), 253–271. Conference version in *Fast Software Encryption 1996, 3rd International Workshop, FSE 1996*, Cambridge, UK, volume 1039 of *Lecture Notes in Computer Science*, Springer Verlag, Berlin, Heidelberg, editor Dieter Gollmann, ISSN 0302-9743. [313]

DANNY DOLEV, CYNTHIA DWORK & MONI NAOR (1991). Non-Malleable Cryptography. In *Proceedings of the Twenty-third Annual ACM Symposium on Theory of Computing*, New Orleans LA, 542–552. ACM Press. URL http://delivery.acm.org/10.1145/110000/103474/p542-dolev.pdf. [402]

DR. DORAN (1877). *London in the Jacobite Times.* Richard Bentley & Son, New Burlington Street. In two volumes. [573]

ECC BRAINPOOL (2010). Elliptic Curve Cryptography (ECC) Brainpool Standard — Curves and Curve Generation. Available at http://tools.ietf.org/html/rfc5639. Last visited 24 October 2014. [226]

JACK EDMONDS (1965). Paths, Trees, and Flowers. *Canadian Journal of Mathematics* **17**, 449–467. [400]

ELECTRONIC FRONTIER FOUNDATION (editor) (1998). *Cracking DES: Secrets of Encryption Research, Wiretap Politics & Chip Design.* O'Reilly Media. ISBN 1-56592-520-5, 266. URL http://cryptome.org/jya/cracking-des/cracking-des.htm. [296]

ARTUR K. EKERT & RICHARD JOZSA (1996). Quantum computation and Shor's factoring algorithm. *Reviews of Modern Physics* **68**(3), 733–753. [715]

TAHER ELGAMAL (1985a). A Public Key Cryptosystem and a Signature Scheme Based on Discrete Logarithms. *IEEE Transactions on Information Theory* **IT-31**(4), 469–472. ISSN 0018-9448. URL http://dx.doi.org/10.1109/TIT.1985.1057074. [174]

T. ELGAMAL (1985b). A Subexponential-Time Algorithm for Computing Discrete Logarithms over $GF(p^2)$. *IEEE Transactions on Information Theory* **IT-31**(4). [349]

NOAM D. ELKIES (1998). Elliptic and modular curves over finite fields and related computational issues. In *Computational Perspectives on number theory, Proceedings of a Conference in Honor of A. O. L. Atkin*, D. A. BUELL & J. T. TEITELBAUM, editors, volume 7 of *AMS/IP Studies in Advanced Mathematics*, 21–76. University of Illinois at Chicago, The American Mathematical Society and International Press, Illinois, Chicago. ISBN 0-8218-0880-X. ISSN 1089-3288. [233]

K. L. ELLIS (1958). British communications and diplomacy in the eighteenth century. *Bulletin of the Institute of Historical Research* 159–167. [573]

A. ENGE & P. GAUDRY (2002). A general framework for subexponential discrete logarithm algorithms. *Acta Arithmetica* **102**, 83–103. [201]

THOMAS ERNST (1996). *Schwarzweiße Magie. Der Schlüssel zum dritten Buch der Steganographia des Trithemius.* Rodopi B.V., Amsterdam. 2, 206 pages. [526]

LEONHARD EULER (1732/33). Observationes de theoremate quodam Fermatiano aliisque ad numeros primos spectantibus. *Commentarii academiae scientiarum imperialis Petropolitanae* **6**, 103–107. Eneström 26. *Opera Omnia*, series 1, volume 2, B. G. Teubner, Leipzig, 1915, 1–5. [797]

LEONHARD EULER (1760/61). Theoremata arithmetica nova methodo demonstrata. *Novi commentarii academiae scientiarum imperialis Petropolitanae* **8**, 74–104. Summarium ibidem 15–18. Eneström 271. *Opera Omnia*, series 1, volume 2, B. G. Teubner, Leipzig, 1915, 531–555. [798]

LEONHARD EULER (1761). Theoremata circa residua ex divisione potestatum relicta. *Novi commentarii academiae scientiarum imperialis Petropolitanae* **7**, 49–82. Eneström 262. *Opera Omnia*, series 1, volume 2, B. G. Teubner, Leipzig, 1915, 493–518. [797]

A. W. EWING (1939). *The Man of Room 40, The Life of Sir Alfred Ewing.* Hutching & Co. Ltd., 295 pages. [677–678]

JOHANN ALBERT FABRICIUS (1858). *Bibliotheca Latina mediae et infimae aetatis.* Thomas Baracchi, Florenz. [526]

J. FEIGENBAUM (editor) (1992). *Advances in Cryptology: Proceedings of CRYPTO 1991,* Santa Barbara, CA, volume 576 of *Lecture Notes in Computer Science.* Springer-Verlag, Berlin, Heidelberg. ISBN 978-3-540-55188-1. ISSN 0302-9743 (Print) 1611-3349 (Online). URL http://dx.doi.org/10.1007/3-540-46766-1.

W. FELLER (1971). *An Introduction to Probability Theory and its Applications.* John Wiley & Sons, 2nd edition. [798]

RICHARD P. FEYNMAN (1982). Simulating Physics with Computers. *International Journal of Theoretical Physics* **21**(6/7), 467–488. ISSN 0020-7748 (Print) 1572-9575 (Online). URL http://dx.doi.org/10.1007/BF02650179. [681, 713]

R. FISCHLIN & C. P. SCHNORR (2000). Stronger Security Proofs for RSA and Rabin Bits. *Journal of Cryptology* **13**(2), 221–244. URL http://dx.doi.org/10.1007/s001459910008. Communicated by Oded Goldreich. [490]

PENELOPE FITZGERALD (1977). *The Knox Brothers.* Macmillan London Limited, 294 pages. [678]

MENSO FOLKERTS (1997). *Die älteste lateinische Schrift über das indische Rechnen nach al-Ḫw-arizm-ı.* Number 113 in Abhandlungen der Bayerischen Akademie der Wissenschaften, Philosophisch-historische Klasse, neue Folge. Verlag der Bayerischen Akademie der Wissenschaften, München. C. H. Beck'sche Verlagsbuchhandlung, München. [815]

LANCE FORTNOW (1987). The Complexity of Perfect Zero-Knowledge (extended abstract). In *Proceedings of the Nineteenth Annual ACM Symposium on Theory of Computing,* New York, 204–209. ACM Press. ISBN 0-89791-221-7. URL http://dx.doi.org/10.1145/28395.28418. [549]

OLE IMMANUEL FRANKSEN (1984). *Mr. Babbage's Secret. The Tale of a Cipher— and APL.* Strandberg, Birkerød, Denmark. [159, 258]

OLE IMMANUEL FRANKSEN (1993). Babbage and cryptography. Or, the mystery of Admiral Beaufort's cipher. *Mathematics and Computers in Simulation* **35**, 327–367. [258]

EDWARD FREDKIN & TOMMASO TOFFOLI (1982). Conservative Logic. *International Journal of Theoretical Physics* **21**(3/4), 219–253. [713]

PETER FREEMAN (2006). The Zimmermann Telegram Revisited: A Reconciliation of the Primary Sources. *Cryptologia* **30**(2), 98–150. [677]

GERHARD FREY & HANS-GEORG RÜCK (1994). A remark concerning m-divisibility and the discrete logarithm problem in the divisor class group of curves. *Mathematics of Computation* **62**(206), 865–874. [233]

JOHANNES BALTHASAR FRIDERICI (1685). *Cryptographia oder Geheime schrifft= münd= und würckliche Correspondenß/ welche lehrmäßig vorstellet eine hoch=schätzbare Kunst verborgene Schrifften zu machen und auffzulösen.* Georg Rebenlein, 8, 280 pages. First edition 1684. [414, 816–817]

WILLIAM F. FRIEDMAN (1921). *The Index of Coincidence and Its Applications in Cryptography.* Riverbank Publication No. 22. Imprimerie-Librairie Militaire Universelle, L. Fournier, Paris. 87 pages. [259]

WILLIAM F. FRIEDMAN (1937). Edgar Allan Poe, cryptographer. *Signal Corps Bulletin* **97**, 41–53. Addendum in *Signal Corps Bulletin* **98** (1937), 54–72. Also in Cryptography and cryptanalysis articles, ed. William F. Friedman, c. 1938, reprinted 1976 by Aegean Park Press, Laguna Hills CA, pages 145–156 and 167–189.

WILLIAM F. FRIEDMAN & CHARLES J. MENDELSOHN (1938). *The Zimmermann Telegram of 16 January 1917, and its Cryptographic Background.* War Department, Office of the Chief Signal Officer, US Government Printing Office, Washington DC, [8], 54, [3] pages. Reprint 1976 and 1994 by Aegean Park Press, Laguna Hills CA. [677–678]

ALAN M. FRIEZE (1986). On the Lagarias-Odlyzko Algorithm for the Subset Sum Problem. *SIAM Journal on Computing* **15**(2), 536–539. URL http://www.math.cmu.edu/~af1p/Texfiles/LAGARIAS-ODLYZKO.pdf. [648]

ALAN M. FRIEZE, JOHAN HÅSTAD, RAVI KANNAN, JEFFREY C. LAGARIAS & ADI SHAMIR (1988). Reconstructing truncated integer variables satisfying linear congruences. *SIAM Journal on Computing* **17**(2), 262–280. URL http://epubs.siam.org/doi/pdf/10.1137/0217016. [648]

HJALMAR FRISK (1973). *Griechisches Etymologisches Wörterbuch.* Carl Winter Universitätsverlag, Heidelberg. [525]

FRUS (1931). *Papers Relating to the Foreign Relations of the United States—1917, Supplement 1: The World War.* U.S. Goverment Printing Office, Washington. [678]

W. FUMY (editor) (1997). *Advances in Cryptology: Proceedings of EUROCRYPT 1997,* Konstanz, Germany, volume 1233 of *Lecture Notes in Computer Science.* Springer-Verlag, Berlin. ISSN 0302-9743. URL http://link.springer.de/link/service/series/0558/tocs/t1233.htm.

MARTIN FÜRER (2009). Faster Integer Multiplication. *SIAM Journal on Computing* **39**(3), 979–1005. URL http://dx.doi.org/10.1137/070711761. [796]

DAVID GADDY (2004). Letter to the Editor: Vigenère Decryption. *Cryptologia* **28**(4), 380–381. [167]

GARY W. GALLAGHER (editor) (1989). *Fighting for the Confederacy: The Personal Recollections of General Edward Porter Alexander.* University of North Carolina Press. [104, 167]

SHUHONG GAO, JOACHIM VON ZUR GATHEN, DANIEL PANARIO & VICTOR SHOUP (2000). Algorithms for Exponentiation in Finite Fields. *Journal of Symbolic Computation* **29**(6), 879–889. URL http://dx.doi.org/10.1006/jsco.1999.0309. [797]

MICHAEL R. GAREY & DAVID S. JOHNSON (1979). *Computers and intractability: A Guide to the Theory of NP-Completeness.* W. H. Freeman and Co., San Francisco CA. ISBN 0-7167-1045-5. [549, 738]

JOACHIM VON ZUR GATHEN (2003). Claude Comiers: the first arithmetical cryptography. *Cryptologia* **XXVII**(4), 339–349. URL http://dx.doi.org/10.1080/0161-110391891946. [167]

JOACHIM VON ZUR GATHEN (2004a). Arithmetic Circuits for Discrete Logarithms. In *Proceedings of LATIN 2004,* Buenos Aires, Argentina, MARTIN FARACH-COLTON, editor, volume 2976 of *Lecture Notes in Computer Science,* 557–566. Springer-Verlag, Berlin, Heidelberg. ISBN 978-3-540-21258-4. ISSN 0302-9743 (Print) 1611-3349 (Online). URL http://dx.doi.org/10.1007/978-3-540-24698-5_58. [201]

JOACHIM VON ZUR GATHEN (2004b). Friederich Johann Buck: arithmetic puzzles in cryptography. *Cryptologia* **XXVIII**(4), 309–324. URL http://dx.doi.org/10.1080/0161-110491892953. [82]

JOACHIM VON ZUR GATHEN (2007). Zimmermann Telegram: The Original Draft. *Cryptologia* **31**(1), 2–37. URL http://dx.doi.org/10.1080/01611190600921165. [677–678]

JOACHIM VON ZUR GATHEN & JÜRGEN GERHARD (2013). *Modern Computer Algebra.* Cambridge University Press, Cambridge, UK, Third edition. ISBN 9781107039032. URL http://cosec.bit.uni-bonn.de/science/mca/. Other editions: 1999, 2003, Chinese edition, Japanese translation. [105, 144–146, 201, 233, 400, 404, 586, 649, 741, 796–798]

JOACHIM VON ZUR GATHEN & DANIEL LOEBENBERGER (2015). Computing the entropy of English. Preprint. [88, 104–105, 336]

JOACHIM VON ZUR GATHEN & JAMSHID SHOKROLLAHI (2005). Efficient FPGA-based Karatsuba multipliers for polynomials over $\mathbb{F}_2$. In *Selected Areas in Cryptography (SAC 2005)*, BART PRENEEL & STAFFORD TAVARES, editors, number 3897 in Lecture Notes in Computer Science, 359–369. Springer-Verlag, Kingston, ON, Canada. ISBN 3-540-33108-5. URL http://dx.doi.org/10.1007/11693383. [797]

JOACHIM VON ZUR GATHEN, AMIN SHOKROLLAHI & JAMSHID SHOKROLLAHI (2007). Efficient Multiplication Using Type 2 Optimal Normal Bases. In *International Workshop on the Arithmetic of Finite Fields, WAIFI 2007*, CLAUDE CARLET & BERK SUNAR, editors, volume 4547 of *Lecture Notes in Computer Science*, 55–68. Springer-Verlag, Berlin/ Heidelberg. ISSN 0302-9743 (Print) 1611-3349 (Online). URL http://dx.doi.org/10.1007/978-3-540-73074-3_6. [797]

JOACHIM VON ZUR GATHEN & IGOR E. SHPARLINSKI (2004). Predicting Subset Sum Pseudorandom Generators. In *Selected Areas in Cryptography*, HELENA HANDSCHUH & M. ANWAR HASAN, editors, volume 3357 of *Lecture Notes in Computer Science*, 241–251. Springer-Verlag, Berlin, Heidelberg. ISBN 3-540-24327-5. ISSN 0302-9743. URL http://dx.doi.org/10.1007/978-3-540-30564-4_17. [493]

JOACHIM VON ZUR GATHEN & IGOR E. SHPARLINSKI (2008). Approximate polynomial gcd: small degree and small height perturbations. In *Proceedings of LATIN 2008*, Búzios, Rio de Janeiro, Brazil, EDUARDO SANY LABER, CLAUDSON BORNSTEIN, LOANA TITO NOGUEIRA & LUERBIO FARIA, editors, volume 4957 of *Lecture Notes in Computer Science*, 276–283. Springer-Verlag, Berlin, Heidelberg. ISSN 0302-9743. URL http://dx.doi.org/10.1007/978-3-540-78773-0_24. [493]

JOACHIM VON ZUR GATHEN & IGOR SHPARLINSKI (2013). Generating safe primes. *Journal of Mathematical Cryptology* **7**(4), 333–365. ISSN 1862-2984 (Online) 1862-2976 (Print). URL http://dx.doi.org/10.1515/jmc-2013-5011. [123, 144]

CARL FRIEDRICH GAUSS (1801). *Disquisitiones Arithmeticae.* Gerh. Fleischer Iun., Leipzig. English translation by ARTHUR A. CLARKE, Springer-Verlag, New York, 1986. [63, 159, 201, 648, 798]

CARL FRIEDRICH GAUSS (1809). *Theoria motus corporum coelestium in sectionibus conicis solem ambientum.* Perthes und Besser, Hamburg. *Werke* VII, Königliche Gesellschaft der Wissenschaften, Göttingen, 1906, 1–288. Reprinted by Georg Olms Verlag, Hildesheim New York, 1973. [650]

DANIEL GENKIN, ADI SHAMIR & ERAN TROMER (2013). RSA Key Extraction via Low-Bandwidth Acoustic Cryptanalysis. URL http://tau.ac.il/~tromer/acoustic/. Last visited 19 December 2013. [129]

Rosario Gennaro, Shai Halevi & Tal Rabin (1999). Secure Hash-and-Sign Signatures Without the Random Oracle. In *Advances in Cryptology: Proceedings of EUROCRYPT 1999*, Prague, Czech Republic, J. Stern, editor, volume 1592 of *Lecture Notes in Computer Science*, 123–139. Springer-Verlag, Berlin, Heidelberg. ISBN 3-540-65889-0. ISSN 0302-9743. URL http://www.springerlink.com/content/bryhef8g51vwbl10/?p=8d42fdca76d6472ca31db56d3b833c55&pi=1. [357]

Craig Gentry, Chris Peikert & Vinod Vaikuntanathan (2008). Trapdoors for Hard Lattices and New Cryptographic Constructions (Extended Abstract). In *Proceedings of the Fourtieth Annual ACM Symposium on Theory of Computing*, Victoria, BC, Canada, 197–206. ACM Press. URL http://dx.doi.org/10.1145/1374376.1374407. [650–651]

Oded Goldreich (2001). *Foundations of Cryptography*, volume I: Basic Tools. Cambridge University Press, Cambridge. ISBN 0-521-79172-3. [401, 549, 796]

Oded Goldreich (2004). *Foundations of Cryptography*, volume II: Basic Applications. Cambridge University Press, Cambridge. ISBN 0-521-83084-2. [56, 401–402]

Oded Goldreich (2008). *Computational Complexity*. Cambridge University Press. ISBN 978-0-521-88473-0. URL http://www.wisdom.weizmann.ac.il/~oded/cc-book.html. [796]

Oded Goldreich & Shafi Goldwasser (1998). On the possibility of basing cryptography on the assumption that $P \neq NP$. *Cryptology ePrint Archive* **1998/005**, 5 pages. URL http://eprint.iacr.org/1998/005. [400]

Oded Goldreich & Shafi Goldwasser (2000). On the Limits of Nonapproximability of Lattice Problems. *Journal of Computer and System Sciences* **60**(3), 540–563. URL http://groups.csail.mit.edu/cis/pubs/shafi/2000-jcss.pdf. Conference version in STOC '98 Proceedings of the Fourteenth Annual ACM Symposium on Theory of Computing, pages 1–9. [650]

Oded Goldreich, Shafi Goldwasser & Shai Halevi (1997). Public-Key Cryptosystems from Lattice Reduction Problems. In *Advances in Cryptology: Proceedings of CRYPTO 1997*, Santa Barbara, CA, B. S. Kaliski, Jr., editor, volume 1294 of *Lecture Notes in Computer Science*, 112–131. Springer-Verlag, Berlin, Heidelberg. ISBN 3-540-63384-7. ISSN 0302-9743. URL http://dx.doi.org/10.1007/BFb0052231. See also http://citeseerx.ist.psu.edu/viewdoc/summary?doi=10.1.1.18.2377. [650]

Oded Goldreich, Silvio Micali & Avi Wigderson (1991). Proofs that Yield Nothing But Their Validity or All Languages in NP Have Zero-Knowledge Proof Systems. *Journal of the ACM* **38**(1), 691–729. Extended abstract in *Proceedings of the 27th Annual IEEE Symposium on Foundations of Computer Science*, Toronto, Ontario, Canada, 174–187 (1988). [549]

Oded Goldreich, Amit Sahai & Salil Vadhan (1999). Can Statistical Zero Knowledge Be Made Non-interactive? or On the Relationship of $\mathcal{SZK}$ and $\mathcal{NISZK}$. In *Advances in Cryptology: Proceedings of CRYPTO 1999*, Santa Barbara, CA, M. Wiener, editor, volume 1666 of *Lecture Notes in Computer Science*, 467–484. Springer-Verlag, Berlin, Heidelberg. ISBN 978-3-540-48405-9 (Online) 978-3-540-66347-8 (Print). ISSN 0302-9743. URL http://dx.doi.org/10.1007/3-540-48405-1_30. [105, 489]

Shafi Goldwasser & Mihir Bellare (2008). Lecture Notes on Cryptography. Summer course "Cryptography and Computer Security" at MIT, 1996–2002, 2004, 2005 and 2008. URL http://cseweb.ucsd.edu/~mihir/papers/gb.pdf. [402]

S. GOLDWASSER & S. MICALI (1984). Probabilistic Encryption. *Journal of Computer and System Sciences* **28**(2), 270–299. ISSN 00220000. URL http://groups. csail.mit.edu/cis/pubs/shafi/1984-jcss.pdf. Preliminary version in STOC '82 Proceedings of the Fourteenth Annual ACM Symposium on Theory of Computing, pages 365-377. [402, 490, 799]

SHAFI GOLDWASSER, SILVIO MICALI & CHARLES RACKOFF (1989). The knowledge complexity of interactive proof systems. *SIAM Journal on Computing* **18**(1), 291– 304. Preliminary version in Proceedings of the Seventeenth Annual ACM Symposium on Theory of Computing, Providence RI (1985), 291-304. Early versions date back to 1983. [536, 549]

D. GOLLMANN (editor) (1996). *Fast Software Encryption 1996, 3rd International Workshop, FSE 1996,* Cambridge, UK, volume 1039 of *Lecture Notes in Computer Science.* Springer-Verlag, Berlin, Heidelberg. ISBN 3-540-60865-6. ISSN 0302-9743.

DANIEL M. GORDON (1993). Discrete logarithms in $GF(p)$ using the number field sieve. *SIAM Journal on Discrete Mathematics* **6**(1), 124–138. URL http://siamdl. aip.org/dbt/dbt.jsp?KEY=SJDMEC&Volume=6&Issue=1. [193]

DANIEL GOTTESMAN (1996). Class of quantum error-correcting codes saturating the quantum Hamming bound. *Physical Review A* **54**(3), 1862–1868. [713]

C. GRABBE, M. BEDNARA, J. SHOKROLLAHI, J. TEICH & J. VON ZUR GATHEN (2003). FPGA Designs of parallel high performance $GF(2^{233})$ Multipliers. In *Proc. of the IEEE International Symposium on Circuits and Systems (ISCAS-03),* volume II, 268–271. Bangkok, Thailand. [797]

R. L. GRAHAM, D. E. KNUTH & O. PATASHNIK (1994). *Concrete Mathematics.* Addison-Wesley, Reading MA, 2nd edition. First edition 1989. [657, 798]

J. P. GRAM (1883). Ueber die Entwickelung reeller Functionen in Reihen mittelst der Methode der kleinsten Quadrate. *Journal für die reine und angewandte Mathematik* **94**, 41–73. [648]

ANDREW GRANVILLE (2008). Smooth numbers: computational number theory and beyond. In *Algorithmic volume Theory: Lattices, volume Fields, Curves and Cryptography,* JOSEPH P. BUHLER & PETER STEVENHAGEN, editors, volume 44 of *Mathematical Sciences Research Institute Publications,* 69–82. Cambridge University Press, New York. ISBN 978-0-521-80854-5. URL http://www.math.leidenuniv.nl/~psh/ ANTproc/09andrew.pdf. [55]

G. J. 'S GRAVESANDE (1748). *Introduction à la philosophie, contenant la metaphysique, et la logique.* Jean and Herm. Verbeek, Leiden. URL http://hdl.handle.net/2027/ nnc1.1000327930. [10], 472, [blank] pages. [95]

MATTHEW GREEN (2015). Hopefully the last post I'll ever write on Dual EC DRBG. A Few Thoughts on Cryptographic Engineering blog. URL http://blog.cryptographyengineering.com/2015/01/hopefully-last-post- ill-ever-write-on.html. [489]

NIGEL DE GREY (1917–1945). Manuscript and typescript material concerning the Zimmermann Telegram. US National Archives. [678]

BRUNO GRIESSER (1952). *Carta Caritatis* und *Institutiones capituli generalis* in einer Salemer Handschrift. *Cistercienser Chronik* **59**, 22–26. ISSN 0379-8291. [443]

Lov K. Grover (1996). A fast quantum mechanical algorithm for database search. *Proceedings of the Twenty-eighth Annual ACM Symposium on Theory of Computing*, Philadelphia PA 212–219. [713]

Zvi Gutterman, Benny Pinkas & Tzachy Reinman (2006). Analysis of the Linux Random Number Generator. In *Proceedings of the 2006 IEEE Symposium on Security and Privacy*, 371–385. IEEE Computer Society. ISBN 0-7695-2574-1. URL http://dx.doi.org/10.1109/SP.2006.5. Also available at http://eprint.iacr.org/2006/086. [489]

J. Hadamard (1893). Résolution d'une question relative aux déterminants. *Bulletin des Sciences Mathématiques* **17**, 240–246. [648]

Shai Halevi (editor) (2009). *Advances in Cryptology: Proceedings of CRYPTO 2009*, Santa Barbara, CA, volume 5677 of *Lecture Notes in Computer Science*. Springer-Verlag. ISBN 3-642-03355-5. URL http://dx.doi.org/10.1007/978-3-642-03356-8. 0302-9743.

R. W. Hamming (1980). We Would Know What They Thought When They Did It. In *A History of Computing in the Twentieth Century*, N. Metropolis, J. Howlett & Gian-Carlo Rota, editors. Academic Press, New York. ISBN 0-12-491650-3. [54]

Godfrey Harold Hardy (1940). *A mathematician's apology*. Cambridge University Press, Cambridge, UK. [111]

G. H. Hardy & J. E. Littlewood (1923). Some problems of "Partitio numerorum"; III: On the expression of a number as a sum of primes. *Acta Mathematica* **44**(1), 1–70. ISSN 0001-5962. URL http://dx.doi.org/10.1007/BF02403921. [144]

G. H. Hardy &E. M. Wright (1985). *An introduction to the theory of numbers*. Clarendon Press, Oxford, 5th edition. First edition 1938. [143]

David Harvey, Joris van der Hoeven & Grégoire Lecerf (2014a). Even faster integer multiplication. *e-print arXiv:1407.3360* 28 pages. URL http://arxiv.org/abs/1407.3360. Last accessed 23 July 2014. [796]

David Harvey, Joris van der Hoeven & Grégoire Lecerf (2014b). Faster polynomial multiplication over finite fields. *e-print arXiv:1407.3361* 23 pages. URL http://arxiv.org/abs/1407.3361. Last accessed 23 July 2014. [796]

Helmut Hasse (1933). Beweis des Analogons der Riemannschen Vermutung für die Artinschen und F. K. Schmidtschen Kongruenzzetafunktionen in gewissen elliptischen Fällen. Vorläufige Mitteilung. *Nachrichten von der Gesellschaft der Wissenschaften zu Göttingen, Mathematisch-Physikalische Klasse* **42**, 253–262. [233]

Johan Håstad (1988). Solving simultaneous modular equations of low degree. *SIAM Journal on Computing* **17**(2), 336–341. URL http://dx.doi.org/10.1137/0217019. [144, 150]

Johan Håstad & Mats Näslund (1998). The Security of Individual RSA Bits. In *Proceedings of the 39th Annual IEEE Symposium on Foundations of Computer Science*, Palo Alto CA, 510–519. IEEE Computer Society Press, Los Alamitos CA. [145]

Ishay Haviv & Oded Regev (2012). Tensor-based Hardness of the Shortest Vector Problem to within Almost Polynomial Factors. *Theory of Computing* **8**(23), 513–531. URL http://dx.doi.org/10.4086/toc.2012.v008a023. Preliminary version in *Proceedings of the Thirty-ninth Annual ACM Symposium on Theory of Computing*, San Diego, California, USA (2007). [649]

D. R. HEATH-BROWN (1986). Artin's conjecture for primitive roots. *Quarterly Journal of Mathematics* **37**, 27–38. [122, 144]

VVOLFGANGVS ERNESTVS HEIDEL (1721). *Johannis Trithemii primo Spanheimensis deinde divi Jacobi Peapolitani abbatis steganographia.* Apud Joh. Fridericvm Rvdigervm, Norimbergæ, 364 pp. [817]

JESSICA HELFAND (2002). *Reinventing the Wheel.* Princeton Architectural Press, New York. 160 pages. [526]

TOR HELLESETH (editor) (1994). *Advances in Cryptology: Proceedings of EURO-CRYPT 1993,* Lofthus, Norway, volume 765 of *Lecture Notes in Computer Science.* Springer-Verlag, Heidelberg. ISBN 3-540-57600-2. ISSN 0302-9743. URL http://dx.doi.org/10.1007/3-540-48285-7.

MARTIN E. HELLMAN (1992). Responses to NIST's proposal. *Communications of the ACM* **35**(7), 47–49. ISSN 0001-0782. URL http://dx.doi.org/10.1145/129902.129905. [359]

BURTON J. HENDRICK (1922). *The Life and Letters of Walter H. Page.* Doubleday, Page & Company, Garden City NY. 3 volumes. [677–678]

NADIA HENINGER, ZAKIR DURUMERIC, ERIC WUSTROW & J. ALEX HALDERMAN (2012). Mining Your Ps and Qs: Detection of Widespread Weak Keys in Network Devices. *To appear in Proceedings of the 21st USENIX Security Symposium* 21 pages. URL https://factorable.net/weakkeys12.extended.pdf. [144]

DOUG HENSLEY (2006). *Continued Fractions.* World Scientific Publishing Co. Pte. Ltd., Singapore. ISBN 981-256-477-1, 260 pages. [652, 797]

HOWARD M. HEYS (2001). A Tutorial on Linear and Differential Cryptanalysis. Technical Report CORR 2001-17, Centre for Applied Cryptographic Research, Department of Combinatorics and Optimization, University of Waterloo. URL http://www.engr.mun.ca/~howard/Research/Papers/index.html. [295]

DAVID HILBERT (1897). *Die Theorie der algebraischen Zahlkörper,* volume 4 of *Jahresbericht der Deutschen Mathematiker-Vereinigung.* Georg Reimer, Berlin, 175-546. URL https://www.dmv.mathematik.de/m-publikationen/m-jahresbericht-der-dmv.html. [297]

NICHOLAS HILEY (1987). The Strategic Origins of Room 40. *Intelligence and National Security* **2**(2), 245–273. ISSN 0268-4527. [677]

LESTER S. HILL (1931). Concerning certain linear transformation apparatus of cryptography. *The American Mathematical Monthly* **38**, 135–154. [105, 164]

CARL FRIEDRICH HINDENBURG (1794). Fragen eines Ungenannten über die Art durch Gitter geheim zu schreiben, und vorläufige Beantwortung derselben vom Herausgeber des Archivs. *ARAM* **1**, 347–351. [798, 817]

F. H. HINSLEY et al. (1984). *British Intelligence in the Second World War.* Her Majesty's Stationary Office, London. ISBN 978-0116309334, 978-0116309341, 0-11-630935-0, 978-0116309525, 978-0116309549. [443, 736]

WASSILY HOEFFDING (1963). Probability inequalities for sums of bounded random variables. *Journal of the American Statistical Association* **58**(301), 13–30. ISSN 01621459. URL http://www.jstor.org/stable/2282952. [798]

DENNIS HOFHEINZ, EIKE KILTZ & VICTOR SHOUP (2013). Practical Chosen Ciphertext Secure Encryption from Factoring. *Journal of Cryptology* **26**(1), 102–118. ISSN 0933-2790 (Print) 1432-1378 (Online). URL http://dx.doi.org/10.1007/s00145-011-9115-0. Extended abstract in *Advances in Cryptology: Proceedings of EURO-CRYPT 2009*, Cologne, Germany, volume 5479 of *Lecture Notes in Computer Science*, Springer Verlag, Berlin, Heidelberg, 313-332. [392, 402, 480]

OTTO HOLSTEIN (1917). A New Cipher. *Scientific American Supplement* **83**, 235. [166]

NICHOLAS J. HOPPER, JOHN LANGFORD & LUIS VON AHN (2002). Provably Secure Steganography. In *Advances in Cryptology - CRYPTO 2002*, M. YUNG, editor, volume 2442 of *Lecture Notes in Computer Science*, 77–92. Springer-Verlag, Berlin, Heidelberg. ISBN 978-3-540-44050-5 (Print) 978-3-540-45708-4 (Online). ISSN 0302-9743 (Print) 1611-3349 (Online). URL http://dx.doi.org/10.1007/3-540-45708-9_6. [414]

G. HORNAUER, W. STEPHAN & R. WERNSDORF (1994). Markov ciphers and alternating groups. In Helleseth (1994), 453–460. URL http://dx.doi.org/10.1007/3-540-48285-7_41. [295]

JEREMY HORWITZ & RAMARATHNAM VENKATESAN (2002). Random Cayley Digraphs and the Discrete Logarithm. In *Algorithmic Number Theory, Fifth International Symposium, ANTS-V*, Sydney, Australia, CLAUS FIEKER & DAVID R. KOHEL, editors, volume 2369 of *Lecture Notes in Computer Science*, 416–430. Springer-Verlag. ISBN 3-540-43863-7. ISSN 0302-9743. URL http://dx.doi.org/0.1007/3-540-45455-1_33. [200]

COLONEL HOUSE (1926). *The Intimate Papers of Colonel House. Volume II. From Neutrality to War 1915-1917. Arranged as a narrative by Charles Seymour.* Ernest Benn Limited. [677]

T. B. HOWELL (1816). *A complete Collection of State Trials and Proceedings for high Treason and other Crimes and Misdemeanors from the earliest Period to the Year 1783, with Notes and other Illustrations*, volume 16. Printed by T. C. Hansard for Longman, Hurst, Rees, Orme, Black, Parbury, and others., London. [573]

NICHOLAS HOWGRAVE-GRAHAM (1997). Finding Small Roots of Univariate Modular Equations Revisited. In *Cryptography and Coding*, M. DARNELL, editor, volume 1355 of *Lecture Notes in Computer Science*, 131–142. Springer-Verlag, Berlin, Heidelberg. ISBN 3-540-63927-6. ISSN 0302-9743. URL http://dx.doi.org/10.1007/BFb0024458. [649]

DAVID A. HUFFMAN (1952). A Method for the Construction of Minimum-Redundancy Codes. *Proceedings of the I.R.E.* **40**(9), 1098–1101. [105]

A. HURWITZ (1891). Ueber die angenäherte Darstellung der Irrationalzahlen durch rationale Brüche. *Mathematische Annalen* **39**(2), 279–284. ISSN 1432-1807 (Online) 0025-5831 (Print). URL http://dx.doi.org/10.1007/BF01206656. [797]

RUSSELL IMPAGLIAZZO & DAVID ZUCKERMANN (1989). How to Recycle Random Bits. In *Proceedings of the 30th Annual IEEE Symposium on Foundations of Computer Science*, Research Triangle Park NC, 248–253. IEEE Computer Society Press. ISBN 0-8186-1982-1 (Print). URL http://dx.doi.org/10.1109/SFCS.1989.63486. [647]

SEBASTIAAN INDESTEEGE, NATHAN KELLER, ORR DUNKELMAN, ELI BIHAM & BART PRENEEL (2008). A Practical Attack on KeeLoq. In *Advances in Cryptology:*

Proceedings of EUROCRYPT 2008, Istanbul, Turkey, NIGEL SMART, editor, volume 4965 of *Lecture Notes in Computer Science*, 1–18. Springer-Verlag, Istanbul, Turkey. ISBN 978-3-540-78966-6 (Print) 978-3-540-78967-3 (Online). URL http://dx.doi.org/10.1007/978-3-540-78967-3_1. Slides available at author's homepage: http://www.cosic.esat.kuleuven.be/keeloq/. [103]

C. G. JACOBI (1839). *Canon Arithmeticus sive tabulae quibus exhibentur pro singulis numeris primis vel primorum potestatibus infra 1000 numeri ad datos indices et indices ad datos numeros pertinentes*. Typi Academici, Berlin. Corrected edition, Numerus- und Indextafeln für die ungeraden Primzahlen unter 1000, ed. Heinrich Brandt, Akademie-Verlag, Berlin, 1956. 16 + 347 pages. [201]

ADMIRAL SIR WILLIAM JAMES (1956). *The Eyes of the Navy: A Biographical Study of Admiral Sir William Hall*. Methuen & Co. Ltd., London. xxv, 212 pages. [678–679]

HAMZA JELJELI (2014). Resolution of Linear Algebra for the Discrete Logarithm Problem Using GPU and Multi-core Architectures. In *Euro-Par 2014 Parallel Processing*, FERNANDO SILVA, INÊS DUTRA & VÍTOR SANTOS COSTA, editors, volume 8632 of *Lecture Notes in Computer Science*, 764–775. Springer-Verlag. ISBN 978-3-319-09873-9. URL http://dx.doi.org/10.1007/978-3-319-09873-9_64. Also available at http://arxiv.org/abs/1402.3661. [201]

R. V. JONES (1979). Alfred Ewing and 'Room 40'. *Notes and Records of the Royal Society of London* **34**, 65–90. [677–678]

ANTOINE JOUX (editor) (2009). *Advances in Cryptology: Proceedings of EUROCRYPT 2009*, Cologne, Germany, volume 5479 of *Lecture Notes in Computer Science*. Springer-Verlag, Berlin, Heidelberg. ISBN 978-3-642-01000-2. ISSN 0302-9743 (Print) 1611-3349 (Online). URL http://dx.doi.org/10.1007/978-3-642-01001-9.

DAVID JOYNER (2013). Review of *Cryptographic Boolean Functions and Applications* by Thomas Cusick and Pantelimon Stănică. *Cryptologia* **37**(2), 189–192. ISSN 0161-1194 (print). URL http://dx.doi.org/10.1080/01611194.2013.767683. [296]

PASCAL JUNOD (2001). On the Complexity of Matsui's Attack. In *8th Annual International Workshop, Selected Areas in Cryptography 2001*, Toronto, Ontario, Canada, S. VAUDENAY & A. M. YOUSSEF, editors, volume 2259 of *Lecture Notes in Computer Science*, 199–211. Springer-Verlag, Berlin, Heidelberg. ISBN 3-540-43066-0. ISSN 0302-9743. URL http://dx.doi.org/10.1007/3-540-45537-X_16. [296]

IBRAHIM A. AL-KADI (1992). Origins of cryptology: the Arab contributions. *Cryptologia* **16**(2), 97–126. [524, 816]

DAVID KAHN (1967). *The Codebreakers*. The Macmillan Company, New York. xvi, 1164 pages. [61, 104, 259, 342, 400–401, 572–573, 677–678]

DAVID KAHN (1974). Enigma unwrapped; The Ultra Secret. *The New York Times Book Review*, 5. ISSN 0028-7806.

DAVID KAHN (1991). *Seizing the Enigma. The Race to Break the German U-Boat Codes, 1939–1943*. Houghton Mifflin Company, Boston. [103, 677]

DAVID KAHN (1999). Edward Bell and his Zimmermann Telegram Memoranda. *Intelligence and National Security* **14**(3), 143–159. ISSN 0268-4527. [677–678]

DAVID KAHN (2004). *The Reader of Gentlemen's Mail - Herbert O. Yardley and the Birth of American Codebreaking*. Yale University Press, New Haven & London.

GIL KALAI (2011). How Quantum Computers Fail: Quantum Codes, Correlations in Physical Systems, and Noise Accumulation. *eprint arXiv:1106.0485v1* (1106.0485v1), 17 pages. URL http://arxiv.org/abs/1106.0485v1. [714]

ERICH KALTOFEN & HEINRICH ROLLETSCHEK (1989). Computing greatest common divisors and factorizations in quadratic number fields. *Mathematics of Computation* **53**(188), 697–720. [797]

RAVI KANNAN (1987). Algorithmic geometry of numbers. *Annual Review of Computer Science* **2**, 231–267. [649]

АНАТО́ЛИЙ КАРАЦУ́БА & Ю́. ОФМАН (1962). Умножение многозначных чисел на автоматах. Доклады Академии Наук СССР **145**, 293–294. A. KARATSUBA and YU. OFMAN, Multiplication of multidigit numbers on automata, Soviet Physics–Doklady **7** (1963), 595–596. [762]

F. W. KASISKI (1863). *Die Geheimschriften und die Dechiffrir-Kunst.* E. S. Mittler und Sohn, Berlin. viii + 95 pp. + 6 tables. [158, 164, 241]

MARKUS KASPER, TIMO KASPER, AMIR MORADI & CHRISTOF PAAR (2009). Breaking KEELOQ in a Flash: On Extracting Keys at Lightning Speed. In *Progress in Cryptology: Proceedings of AFRICACRYPT 2009*, Casablanca, Morocco, B. PRENEEL, editor, volume 5580 of *Lecture Notes in Computer Science*, 403–420. Springer-Verlag. ISBN 978-3-642-02383-5. ISSN 0302-9743 (Print) 1611-3349 (Online). URL http://dx.doi.org/10.1007/978-3-642-02384-2_25. Also available at KeeLoq homepage: http://www.emsec.rub.de/keeloq. [103]

THOMAS KELLY (1998). The myth of the skytale. *Cryptologia* **XXII**(3), 244–260. [442]

AUGUSTE KERCKHOFFS (1883). La Cryptographie Militaire. *Journal des Sciences Militaires* **9**, 161–191. URL http://www.cl.cam.ac.uk/users/fapp2/kerckhoffs/la_cryptographie_militaire_ii.htm. [62]

PHILIP KERR (2002). *Dark Matter, The Private Life of Sir Isaac Newton.* Crown Publishers, New York. ISBN 0-609-60981-5. [573]

SUBHASH KHOT (2005). Hardness of Approximating the Shortest Vector Problem in Lattices. *Journal of the ACM* **52**(5), 789–808. URL http://dx.doi.org/10.1145/1089023.1089027. Preliminary version in *Proceedings of the 45th Annual IEEE Symposium on Foundations of Computer Science*, Rome, Italy (2004). [649]

SUBHASH KHOT (2010). *Inapproximability Results for Computational Problems on Lattices*, chapter 14, 453–473. In Nguyen & Vallée (2010). URL http://dx.doi.org/10.1007/978-3-642-02295-1_14. [649]

AGGELOS KIAYIAS, YONA RAEKOW & ALEXANDER RUSSELL (2005). Efficient Steganography with Provable Security Guarantees. In *Information Hiding*, MAURO BARNI, JORDI HERRERA-JOANCOMARTÍ, STEFAN KATZENBEISSER & FERNANDO PÉREZ-GONZÁLEZ, editors, volume 3727 of *LNCS*, 118–130. Springer-Verlag, Berlin/Heidelberg. ISBN 978-3-540-29039-1. ISSN 0302-9743 (Print) 1611-3349 (Online). URL http://dx.doi.org/10.1007/11558859_10. [414]

WOLFGANG KILLMANN & WERNER SCHINDLER (2008). A Design for a Physical RNG with Robust Entropy Estimators. In *CHES 2008*, ELISABETH OSWALD & PANKAJ ROHATGI, editors, volume 5154 of *LNCS*, 146–163. ISBN 978-3-540-85052-6. URL http://dx.doi.org/10.1007/978-3-540-85053-3_10. [449, 489]

WOLFGANG KILLMANN & WERNER SCHINDLER (2011). A proposal for: Functionality classes for random number generators. Anwendungshinweise und Interpretationen zum Schema AIS 20/AIS 31, Bundesamt für Sicherheit in der Informationstechnik, Bonn, Germany. URL https://www.bsi.bund.de/SharedDocs/Downloads/DE/BSI/ Zertifizierung/Interpretationen/AIS_31_Functionality_classes_for_random_ number_generators_e.pdf?__blob=publicationFile. Version 2.0. [489]

JEONG HAN KIM, RAVI MONTENEGRO, YUVAL PERES & PRASAD TETALI (2010). A Birthday Paradox for Markov chains with an optimal bound for collision in the Pollard Rho algorithm for discrete logarithm. *Annals of Applied Probability* **20**(2), 495–521. URL http://dx.doi.org/10.1214/09-AAP625. [184, 200]

ATHANASIUS KIRCHER (1663). *Polygraphia nova et universalis ex combinatoria arte detecta.* Varesius, 148 pages.

MELVILLE KLEIN (2006). Securing Record Communications: The TSEC/KW-26. URL https://www.nsa.gov/about/_files/cryptologic_heritage/publications/ misc/tsec_kw26.pdf. National Security Agency/Central Security Service. Center for Cryptologic History. United States. [401]

THORSTEN KLEINJUNG, KAZUMARO AOKI, JENS FRANKE, ARJEN K. LENSTRA, EMMANUEL THOMÉ, JOPPE W. BOS, PIERRICK GAUDRY, ALEXANDER KRUPPA, PETER L. MONTGOMERY, DAG ARNE OSVIK, HERMAN TE RIELE, ANDREY TIMOFEEV & PAUL ZIMMERMANN (2010). Factorization of a 768-bit RSA modulus. URL http://eprint.iacr.org/2010/006.pdf. [140]

VLASTIMIL KLÍMA (2005). Finding MD5 Collisions — a Toy For a Notebook. *Crytology ePrint Archive* **2005/075**, 7. URL http://eprint.iacr.org/2005/075. [316]

DONALD E. KNUTH (1973). *The Art of Computer Programming, vol.3: Sorting and Searching.* Addison-Wesley, Reading MA. [145]

DONALD E. KNUTH (1997). *The Art of Computer Programming, vol. 1, Fundamental Algorithms.* Addison-Wesley, Reading MA, 3rd edition. First edition 1969. [105, 152]

DONALD E. KNUTH (1998). *The Art of Computer Programming, vol. 2, Seminumerical Algorithms.* Addison-Wesley, Reading MA, 3rd edition. ISBN 0-201-89684-2. First edition 1969. [452, 650, 796–797, 802]

JOHANNES KÖBLER, UWE SCHÖNING & JACOBO TORÁN (1993). *The Graph Isomorphism Problem: Its Structural Complexity.* Birkhäuser Verlag, Boston. ISBN 0-8176-3680-3. 160 pages. [549]

NEAL KOBLITZ (1987a). *A Course in Number Theory and Cryptography.* Number 114 in Graduate Texts in Mathematics. Springer-Verlag, New York. [233]

NEAL KOBLITZ (1987b). Elliptic Curve Cryptosystems. *Mathematics of Computation* **48**(177), 203–209. [233]

SERGEI V. KONYAGIN, FLORIAN LUCA, BERNARD MANS, LUKE MATHIESON & IGOR E. SHPARLINSKI (2013). Functional Graphs of Polynomials over Finite Fields. *e-print arXiv:/1307.2718* 31 pages. URL http://arxiv.org/abs/1307.2718. [145]

WŁADYSŁAW KOZACZUK (1967). *Bitwa o tajemnice (Battle on Secrets). Służby wywiadowcze Polski i Rzeszy Niemieckiej 1922-1939. Wydanie drugie rozszerszone.* Książka i Wiedza, Warszawa. 466 pages. [736]

HUGO KRAWCZYK, KENNETH G. PATERSON & HOETECK WEE (2013). On the Security of the TLS Protocol: A Systematic Analysis. *Cryptology ePrint Archive* **2013**/339, 49 pages. URL http://eprint.iacr.org/2013/339. Last visited 2 march 2015. [429]

FRANZ KÜRSCHNER (1872). Herzog Rudolph's IV. Schriftdenkmale. *Mitteilungen der k.k. Zentralkommission (Wien)* **17**, 71–80. [815]

J. C. LAGARIAS, H. W. LENSTRA, JR. & C. P. SCHNORR (1990). Korkin-Zolotarev bases and successive minima of a lattice and its reciprocal lattice. *Combinatorica* **10**(4), 333–348. ISSN 0209-9683 (Print) 1439-6912 (Online). URL http://dx.doi.org/10.1007/BF02128669. [648–650]

J. C. LAGARIAS & A. M. ODLYZKO (1985). Solving Low-Density Subset Sum Problems. *Journal of the ACM* **32**(1), 229–246. Conference version in *Proceedings of the 24th Annual IEEE Symposium on Foundations of Computer Science, FOCS 1983*, Tucson AZ, 1-10. [648]

JOSEPH LOUIS DE LAGRANGE (1770/71). Réflexions sur la résolution algébrique des équations. *Œuvres complètes* **3**, 205–421. URL http://gallica.bnf.fr/ark:/12148/bpt6k229222d/f206. Nouveaux Mémoires de l'Académie royale des Sciences et Belles-Lettres de Berlin. [798]

JOSEPH-LOUIS LAGRANGE (1773). Nouvelle méthode pour résoudre les équations littérales par le moyen des séries. *Extraits des recueils de l'académie royale des sciences et belles-lettres de Berlin* URL http://gallica.bnf.fr/ark:/12148/bpt6k229222d. Reprinted in Œuvres de Lagrange, publiées par les Soins De M. J.-A. Serret, sous les auspices de son excellence le ministre de l'instruction publique. Tome Troisième 1867–1892. [648]

XUEJIA LAI & KEFEI CHEN (editors) (2006). *Advances in Cryptology: Proceedings of ASIACRYPT 2006*, Shanghai, China, volume 4284 of *Lecture Notes in Computer Science*. Springer-Verlag, Berlin, Heidelberg. ISBN 978-3-540-49475-1. ISSN 0302-9743 (Print) 1611-3349 (Online). URL http://dx.doi.org/10.1007/11935230.

ANDRÉ LANGE & É.-A. SOUDART (1935). *Traité de cryptographie*. Librairie Félix Alcan, Paris. Xv + 366 + vi pp. [104]

ROBERT LANSING (1935). *War Memoirs*. Bobbs-Merrill, Indianapolis and New York, 383 pages. [678]

CHRISTOPHER LAYER (1723). *Criminal-Proces des vor der sogenannten Königs-Banck in dem grossen Saal zu Westmünster angeklagten und wegen überzeugten Hohen-Verraths am 27ten Novembr. 1722. styl. vet. daselbst zum Tode verurtheilten Christopher Layer Esqr.* Wiering, Hamburg. 304 pages. [573]

ALBERT LECOY DE LA MARCHE (1875). *Le roi René. Sa vie, son administration, ses travaux artistiques et littéraires*, volume 1. Slatkine Reprints, 1969, 559 pages. [443]

PAUL LEHMANN (1961). Merkwürdigkeiten des Abtes Johannes Trithemius,. *Sitzungsberichte der Bayerischen Akademie der Wissenschaften, Philosophisch-historische Klasse* **2**. [525]

D. H. LEHMER (1951). Mathematical methods in large-scale computing units. In *Proceedings of a Second Symposium on Large-Scale Digital Calculating Machinery, 13–16 September 1949*, volume 26 of *Annals of the Computation Laboratory of Harvard University*, 141–146. Harvard University Press, Cambridge, Massachusetts. URL https://archive.org/details/proceedings_of_a_second_symposium_on_large-scale_. [453, 489, 648]

GOTTFRIED WILHELM LEIBNIZ (1679). De progressione dyadica. Manuscript letter dated 15 March 1679. [415]

GOTTFRIED WILHELM LEIBNIZ (1703). Continuatio analyseos quadraturarum rationalium. *Acta eruditorum* 19–26. [415]

GILBERT LÉLY (1961). *The Marquis de Sade, a biography.* Grove Press. ISBN ASIN: B000NKT11Q.

ARJEN K. LENSTRA (2006). Key Lengths. In *Handbook of Information Security, Volume 2*, HOSSEIN BIDGOLI, editor, 617–635. John Wiley & Sons, Inc. ISBN 978-0-471-64833-8. Also available at ftp://cm.bell-labs.com/who/akl/key_lengths.pdf. [233, 400]

ARJEN K. LENSTRA, JAMES P. HUGHES, MAXIME AUGIER, JOPPE W. BOS, THORSTEN KLEINJUNG & CHRISTOPHE WACHTER (2012). Ron was wrong, Whit is right. *Cryptology ePrint Archive: Report 2012/064* URL http://eprint.iacr.org/2012/064. [144]

A. K. LENSTRA & H. W. LENSTRA, JR. (editors) (1993). *The development of the number field sieve.* Springer Verlag, Berlin. *Lecture Notes in Mathematics*, volume 1554. [140]

A. K. LENSTRA, H. W. LENSTRA, JR. & L. LOVÁSZ (1982). Factoring Polynomials with Rational Coefficients. *Mathematische Annalen* **261**, 515–534. [578, 649]

ARJEN LENSTRA, XIAOYUN WANG & BENNE DE WEGER (2005). Colliding X.509 Certificates. *Cryptology ePrint Archive* **2005/067**, 5. URL http://eprint.iacr.org/2005/067. [316]

H. W. LENSTRA, JR. (1987). Factoring integers with elliptic curves. *Annals of Mathematics* **126**, 649–673. [233, 238]

H. W. LENSTRA, JR. & CARL POMERANCE (1992). A rigorous time bound for factoring integers. *Journal of the Australian Mathematical Society* **5**(3), 483–516. URL http://dx.doi.org/10.1090/S0894-0347-1992-1137100-0. [140]

H. W. LENSTRA, JR. & CARL POMERANCE (2011). Primality testing with Gaussian periods. Author's homepage: https://www.math.dartmouth.edu/~carlp/aks041411.pdf. 47 pages. [143]

J. H. LEOPOLD (1900). De scytala laconica. *Mnemosyne* **2**, 365–391. [442]

DAVID A. LEVIN, YUVAL PERES & ELIZABETH L. WILMER (2009). *Markov Chains and Mixing Times.* American Mathematical Society, Providence, RI. ISBN 978-0-8218-4739-8, xviii+371 pages. [799]

L. A. LEVIN (1973). Universal sequential search problems. *Problems of Information Transmission* **9**, 265–266. Translated from *Problemy Peredachi Informatsii* **9**(3) (1973), 115–116. [400]

RUDOLF LIDL & HARALD NIEDERREITER (1983). *Finite Fields.* Number 20 in Encyclopedia of Mathematics and its Applications. Addison-Wesley, Reading MA. [33]

STEN G. LINDBERG (1979). Mobiles in Books. Volvelles, inserts, pyramids, divinations and children's games. *The Private Library (Third Series)* **2**(2), 49–82. [526]

I. B. LINDENFELS (1819). *Den hemmelige Skrivekonst eller: Chiffrer= og Dechiffrer=Konsten.* Brummer, Kjøbenhavn, 340 pages. [164]

J. E. LITTLEWOOD (1953). *A Mathematician's Miscellany.* Methuen & Co. Ltd., London, 136. [455]

YI-KAI LIU, VADIM LYUBASHEVSKY & DANIELE MICCIANCIO (2006). On Bounded Distance Decoding for General Lattices. In *9th International Workshop on Approximation Algorithms for Combinatorial Optimization Problems, APPROX 2006 and 10th International Workshop on Randomization and Computation, RANDOM 2006, Barcelona, Spain, August 28-30 2006.*, JOSEP DÍAZ, KLAUS JANSEN, JOSÉ D. P. ROLIM & URI ZWICK, editors, Lecture Notes in Computer Science, 450–461. Springer-Verlag. ISBN 978-3-540-38044-3 (Print) 978-3-540-38045-0 (Online). ISSN 0302-9743. URL http://dx.doi.org/10.1007/11830924_41. [651]

HOI-KWONG LO & H. F. CHAU (1997). Is Quantum Bit Commitment Really Possible? *Physical Review Letters* **78**(17), 3410–3413. [714]

HOI-KWONG LO & H. F. CHAU (1998). Why quantum bit commitment and ideal quantum coin tossing are impossible. *Physica D* **120**, 177–187. [714]

DANIEL LOEBENBERGER & MICHAEL NÜSKEN (2014). Notions for RSA integers. *International Journal of Applied Cryptography* **3**(2), 116–138. ISSN 1753-0571 (online), 1753-0563 (print). URL http://dx.doi.org/10.1504/IJACT.2014.062723. [55]

PIERRE DE SEGUSSON LONGLÉE (1912). *Dépêches diplomatiques de M. de Longlée, résident de France en Espagne (1582–1590). Ouvrage publié pour la Sociéte d'Histoire diplomatique par Albert Mousse, archiviste-paléographe. Avec un facsimilé.* Librairie Plon, Paris, 59, 432. [342]

M. LUBIN (1901). Méthode de correspondance chiffrée. *Revue Scientifique, Paris* **16**(Sem. 2, Sér. 4), 809–812. [166]

MARTIN LUTHER (1529). *Von heimliche[n] vnd gestolen brieffen Sampt einem Psalm ausgelegt widder Hertzog Georgen zu Sachsen.* Hans Lufft, Wittenberg. [556]

VADIM LYUBASHEVSKY & DANIELE MICCIANCIO (2009). On Bounded Distance Decoding, Unique Shortest Vectors, and the Minimum Distance Problem. In Halevi (2009), 577–594. URL http://dx.doi.org/10.1007/978-3-642-03356-8_34. 0302-9743. [651]

F. J. MACWILLIAMS & N. J. A. SLOANE (1977). *The Theory of Error-Correcting Codes.* Number 16 in Mathematical Library. North-Holland, Amsterdam. [33, 405]

YURI I. MANIN (1980). *Vychislimoe i nevychislimoe [Computable and Noncomputable].* Sovetskoye radio, Moscow. URL http://publ.lib.ru/ARCHIVES/M/MANIN_Yuriy_Ivanovich/_Manin_Yu.I..html. 131 pages. [681]

MITSURU MATSUI (1994a). The First Experimental Cryptanalysis of the Data Encryption Standard. In *Advances in Cryptology: Proceedings of CRYPTO 1994,* Santa Barbara, CA, YVO G. DESMEDT, editor, volume 839 of *Lecture Notes in Computer Science,* 1–11. Springer-Verlag, Berlin, Heidelberg. ISBN 0-387-58333-5, 3-540-58333-5. ISSN 0302-9743. URL http://dx.doi.org/10.1007/3-540-48658-5_1. [295–296]

MITSURU MATSUI (1994b). Linear Cryptanalysis Method for DES Cipher. In Helleseth (1994), 386–397. URL http://dx.doi.org/10.1007/3-540-48285-7_33. [279, 296]

UELI M. MAURER & STEFAN WOLF (1998). On the Hardness of the Diffie-Hellman Decision Problem. URL http://www.crypto.ethz.ch/~maurer/publications.html. Department of Computer Science, Swiss Federal Institute of Technology (ETH), Zürich.

UELI M. MAURER & STEFAN WOLF (1999). The relationship between breaking the Diffie-Hellman protocol and computing discrete logarithms. *SIAM Journal on Computing* **28**(5), 1689–1721. [55, 201]

ALEXANDER MAY (2004). Computing the RSA Secret Key is Deterministic Polynomial Time Equivalent to Factoring. In *Advances in Cryptology: Proceedings of CRYPTO 2004*, Santa Barbara, CA, MATT. FRANKLIN, editor, volume 3152 of *Lecture Notes in Computer Science*, 213–219. Springer-Verlag, Berlin, Heidelberg. ISBN 3-540-22668-0. ISSN 0302-9743. URL http://dx.doi.org/10.1007/b99099. [144]

ALEXANDER MAY (2010). *Using LLL-Reduction for Solving RSA and Factorization Problems*, chapter 10, 315–348. In Nguyen & Vallée (2010). URL http://dx.doi.org/10.1007/978-3-642-02295-1_10. [144, 623, 649]

J. PETER MAY (2003). Munshi's Proof of the Nullstellensatz. *The American Mathematical Monthly* **110**(2), 133–140. [616]

DOMINIC MAYERS (1997). Unconditionally Secure Quantum Bit Commitment is Impossible. *Physical Review Letters* **78**(17), 3414–3417. [714]

ERNST W. MAYR (1997). Some complexity results for polynomial ideals. *Journal of Complexity* **13**, 303–325. [400]

ALASDAIR MCANDREW (2011). *Introduction to Cryptography with Open-Source Software*. Discrete Mathematics and Its Applications. CRC Press. ISBN 9781439825709. URL http://www.crcpress.com/product/isbn/9781439825709. [742]

GEORGE E. MCCRACKEN (1948). Athanasius Kirchner's Universal Polygraphy. *Isis* **39**, 215–228. [526]

R. J. MCELIECE (1978). A Public-Key Cryptosystem Based On Algebraic Coding Theory. *Deep Space Network Progress Report* **44**, 114–116. URL http://ipnpr.jpl.nasa.gov/progress_report2/42-44/44N.PDF. [405]

CHARLES J. MENDELSOHN (1939). Cardan on cryptography. *Scripta Mathematica* **6**, 157–168. [259]

A. J. MENEZES, T. OKAMOTO & S. A. VANSTONE (1993). Reducing elliptic curve logarithms to a finite field. *IEEE Transactions on Information Theory* **39**(4), 1639–1646. [233]

RALPH C. MERKLE (1988). A digital signature based on a conventional encryption function. In *Advances in Cryptology: Proceedings of CRYPTO 1987*, Santa Barbara, CA, C. POMERANCE, editor, volume 293 of *Lecture Notes in Computer Science*, 369–378. Springer-Verlag, Berlin. ISSN 0302-9743. URL http://link.springer.de/link/service/series/0558/tocs/t0293.htm. [316]

RALPH C. MERKLE (1990). A Fast Software One-Way Hash Function. *Journal of Cryptology* **3**, 43–58. [310]

RALPH C. MERKLE & MARTIN E. HELLMAN (1978). Hiding information and signatures in trapdoor knapsacks. *IEEE Transactions on Information Theory* **IT-24**(5), 525–530. URL http://ieeexplore.ieee.org/xpls/abs_all.jsp?arnumber=1055927. [588]

MARIN MERSENNE (1636). *Harmonie universelle contenant la théorie et la pratique de la musique*. Sebastien Cramoisy, Paris. Reprinted by Centre National de la Recherche Scientifique, Paris, 1975. [816]

SIHEM MESNAGER (2015). *Binary bent functions: fundamentals and results*. Springer-Verlag. To appear. [296]

CARL H. MEYER & STEPHEN M. MATYAS (1982). *Cryptography: A new Dimension in Computer Data Security*. John Wiley & Sons. [88, 104]

ULRIKE MEYER & SUSANNE WETZEL (2004). A man-in-the-middle attack on UMTS. In *Proceedings of the 3rd ACM workshop on Wireless security*, MARKUS JAKOBSSON & ADRIAN PERRIG, editors, WiSe '04, 90–97. ACM, New York, NY, USA. ISBN 1-58113-925-X. URL http://dx.doi.org/10.1145/1023646.1023662. [429]

DANIELE MICCIANCIO (2012). Inapproximability of the Shortest Vector Problem: Toward a Deterministic Reduction. *Theory of Computing* **8**(22), 487–512. URL http://dx.doi.org/10.4086/toc.2012.v008a022. [649]

DANIELE MICCIANCIO & SHAFI GOLDWASSER (2002). *Complexity of Lattice Problems: A Cryptographic Perspective*, volume 671. Kluwer Academic Publishers. ISBN 0-7923-7688-9. URL http://www.cs.ucsd.edu/users/daniele/papers/book.html. [649–651]

DANIELE MICCIANCIO & ODED REGEV (2007). Worst-case to Average-case Reductions based on Gaussian Measures. *SIAM Journal on Computing* **37**(1), 267–302. URL http://www.cims.nyu.edu/~regev/papers/average.pdfàĂŨ. Preliminary version in FOCS 2004. [650–651]

A. RAY MILLER (2001). The Cryptographic Mathematics of Enigma. In *NSA Historical Publications*, 6. Public and Media Affairs Office. [734]

GARY L. MILLER (1976). Riemann's Hypothesis and Tests for Primality. *Journal of Computer and System Sciences* **13**, 300–317. [143]

GEORGE A. MILLER & ELIZABETH A. FRIEDMAN (1957). The Reconstruction of Mutilated English Texts. *Information and Control* **1**, 38–55.

J. C. P. MILLER (1975). On Factorisation, with a Suggested New Approach. *Mathematics of Computation* **29**(129), 155–172. [145]

STEPHEN D. MILLER & RAMARATHNAM VENKATESAN (2006). Spectral Analysis of Pollard Rho Collisions. In *Algorithmic Number Theory, Seventh International Symposium, ANTS-IV,* Berlin, Germany, FLORIAN HESS, SEBASTIAN PAULI & MICHAEL POHST, editors, volume 4076 of *Lecture Notes in Computer Science*. Berlin, Heidelberg. ISBN 978-3-540-36075-9. ISSN 0302-9743 (Print) 1611-3349 (Online). URL http://dx.doi.org/10.1007/11792086_40. Preprint at http://arxiv.org/abs/math/0603727v2. [200]

STEPHEN D. MILLER & RAMARATHNAM VENKATESAN (2009). Non-degeneracy of Pollard Rho Collisions. *International Mathematics Research Notices* **2009**(1), 1–10. ISSN 10737928. URL http://dx.doi.org/10.1093/imrn/rnn114. [184, 200]

VICTOR S. MILLER (1986). Use of Elliptic Curves in Cryptography. In *Advances in Cryptology: Proceedings of CRYPTO 1985,* Santa Barbara, CA, HUGH C. WILLIAMS, editor, volume 218 of *Lecture Notes in Computer Science*, 417–426. Springer-Verlag, Berlin. ISBN 978-3-540-16463-0 (Print) 978-3-540-39799-1 (Online). ISSN 0302-9743. URL http://link.springer.com/book/10.1007/3-540-39799-X. [233]

WALTER MILLIS (1935). *Road to War. America: 1914–1917*. Houghton Mifflin Company, Boston and New York, xiv, 466 pages. [669]

H. MINKOWSKI (1910). *Geometrie der Zahlen*. B. G. Teubner, Leipzig. [648]

C. J. MITCHELL, F. PIPER & P. WILD (1992). Digital Signatures. In *Contemporary Cryptology, The Science of Information Integrity*, GUSTAVUS J. SIMMONS, editor, chapter 6, 325–378. IEEE Press, New York. [359]

LOUIS MONIER (1980). Evaluation and comparison of two efficient probabilistic primality testing algorithms. *Theoretical Computer Science* **12**, 97–108. [143]

FRANCISCO MONTERDE GARCÍA ICAZBALCETA (1925). La carta cifrada de Don Hernán Cortés. *Anales del Museo Nacional de Arqueología, Historia y Etnografía, Mexico* **4**, 436–443. [558]

H. MORANVILLÉ (editor) (1891). *Chronographia Regum Francorum*, volume 2. Librairie Renouard, H. Laurens, Paris. [342]

MICHAEL A. MORRISON & JOHN BRILLHART (1975). A Method of Factoring and the Factorization of F_7. *Mathematics of Computation* **29**(129), 183–205. [131, 136, 144]

MICHELE MOSCA & CHRISTOF ZALKA (2003). Exact quantum Fourier transforms and discrete logarithm algorithms. *International Journal of Quantum Information* **2**, 91–100. [715]

M. MRAYATI, Y. MEER ALAM & M. H. AL-TAYYAN (2003). Al-Kindi's Treatise on Cryptanalysis. In *Series on Arabic Origins of Cryptology*, volume 1. King Faisal Center for Research and Islamic Studies and King Abdulaziz City for Science and Technology, Riyadh. ISBN 9960-890-08-02. [816]

M. MRAYATI, Y. MEER ALAM & M. H. AL-TAYYAN (2004). Ibn ad-Durayhim's Treatise on Cryptanalysis. In *Series on Arabic Origins of Cryptology*, volume 3. King Faisal Center for Research and Islamic Studies and King Abdulaziz City for Science and Technology, Riyadh. ISBN 9960-890-20-0. [813]

M. MRAYATI, YAHYA MEER ALAM & M. HASSAN AL-TAYYAN (1987). *Origins of Arab Cryptography and Cryptanalysis. Volume One. Analysis and Editing of Three Arabic Manuscripts: Al-Kindi, Ibn-Adlan, Ibn-Al-Durahim*. Arab Academy of Damascus publications. [162, 501, 524]

D. E. MULLER (1956). Complexity in Electronic Switching Circuits. *IRE Transactions on Electronic Computers* **5**, 15–19. [490]

MOHAMMAD A. MUSA, EDWARD F. SCHAEFER & STEPHEN WEDIG (2003). A simplified AES algorithm and its linear and differential cryptanalyses. *Cryptologia* **17**(2), 148–177. [295]

ALEXEI MYASNIKOV, VLADIMIR SHPILRAIN & ALEXANDER USHAKOV (2011). *Noncommutative Cryptography and Complexity of Group-theoretic Problems*, volume 177 of *Mathematical Surveys and Monographs*. American Mathematical Society. ISBN 0-8218-5360-0, 385 pp. [200]

BVT. BRIG. GEN. ALBERT J. MYER (1879). *Manual of Signals for the Use of Signal Officers in the Field and for Military and Naval Students, Military Schools, etc.* Government Printing Office, Washington. First edition 1864. [75, 103, 815]

MONI NAOR & ADI SHAMIR (1995). Visual cryptography. In *Advances in Cryptology: Proceedings of EUROCRYPT 1994*, Perugia, Italy, ALFREDO DE SANTIS, editor, number 950 in Lecture Notes in Computer Science, 1–12. Springer-Verlag. ISBN 3-540-60176-7. ISSN 0302-9743. [53, 56]

MONI NAOR & MOTI YUNG (1990). Public-key Cryptosystems Provably Secure against Chosen Ciphertext Attacks. In *Proceedings of the Twenty-second Annual ACM Symposium on Theory of Computing*, Baltimore MD, 424–437. ACM Press. ISBN 0-89791-361-2. URL http://dx.doi.org/10.1145/100216.100273. [402]

MATS NÄSLUND (1998). *Bit Extraction, Hard-Core Predicates, and the Bit Security of RSA*. Ph.D. thesis, Department of Numerical Analysis and Computing Science, Kungl Tekniska Högskolan (Royal Institute of Technology), Stockholm. [145]

MARTIN NASSUA (1992). *"Gemeinsame Kriegsführung. Gemeinsamer Friedensschluß." Das Zimmermann-Telegramm vom 13. Januar 1917 und der Eintritt der USA in den 1. Weltkrieg*, volume 520 of *Europäische Hochschulschriften, Reihe III, Geschichte und ihre Hilfswissenschaften*. Peter Lang, Frankfurt am Main, [8], 163 pages. [677]

V. I. NECHAEV (1994). Complexity of a determinate algorithm for the discrete logarithm. *Mathematical Notes* **55**(2), 165–172. ISSN 0001-4346. Translation of Нечаев (1994).

В. И. НЕЧАЕВ (1994). К вопросу о сложности детерминированного алгоритма для дискретного логарифма. Российская Академия Наук. Математические Заметки **55**(2), 91–101, 189. ISSN 0025-567X. English translation see Nechaev (1994). [201]

JOHN VON NEUMANN (1951). Various techniques used in connection with random digits. Monte Carlo methods. *National Bureau of Standards, Applied Mathematics Series* **12**, 36–38.

MINH VAN NGUYEN (2009). *Exploring Cryptography Using the Sage Computer Algebra System*. Bachelor thesis, School of Engineering and Science, Victoria University, Melbourne, Australia. URL http://www.sagemath.org/files/thesis/nguyen-thesis-2009.pdf. [742]

PHONG Q. NGUYEN (2010). *Hermite's Constant and Lattice Algorithms*, chapter 2, 19–69. In Nguyen & Vallée (2010). URL http://dx.doi.org/10.1007/978-3-642-02295-1_2. [648]

P. Q. NGUYEN & D. STEHLÉ (2009). An LLL algorithm with quadratic complexity. *SIAM Journal on Computing* **39**(3), 874–903. URL http://dx.doi.org/10.1137/070705702. [648]

PHONG Q. NGUYEN & JACQUES STERN (2000). Lattice Reduction in Cryptology: An Update. In *Algorithmic Number Theory, Fourth International Symposium, ANTS-IV*, Leiden, The Netherlands, WIEB BOSMA, editor, number 1838 in Lecture Notes in Computer Science, 85–112. Springer-Verlag. ISSN 0302-9743. URL http://www.di.ens.fr/~stern/publications.html. [649]

PHONG Q. NGUYEN & BRIGITTE VALLÉE (editors) (2010). *The LLL Algorithm. Survey and Applications*. Springer-Verlag. ISBN 978-3-642-02294-4. URL http://dx.doi.org/10.1007/978-3-642-02295-1. [648]

MICHAEL A. NIELSEN & ISAAC L. CHUANG (2004). *Quantum Computation and Quantum Information*. Cambridge University Press, Cambridge, UK. ISBN 0-521-63503-9. [714]

NOAM NISAN & AVI WIGDERSON (1994). Hardness vs Randomness. *Journal of Computer and System Sciences* **49**, 149–167. Conference version in *Proceedings of the 29th Annual IEEE Symposium on Foundations of Computer Science*, White Plains, NY, 2–11. IEEE Computer Society Press. [447]

NIST (1993). *Federal Information Processing Standards Publication 180-1 — Secure Hash Standard.* National Institute of Standards and Technology. URL `www.itl.nist.gov/fipspubs/fip180-1.htm`. Federal Information Processing Standards Publication 180-1. [316]

NIST (1999a). *Federal Information Processing Standards Publication 46 - Data Encryption Standard (DES).* National Institute of Standards and Technology. URL `http://csrc.nist.gov/publications/fips/archive/fips46-3/fips46-3.pdf`. Federal Information Processing Standards Publication 46.

NIST (1999b). Recommended Elliptic Curves for Federal Government Use. Technical report, National Institute of Standards and Technology. Last download 7 January 2010. [225]

NIST (2000). *Digital Signature Standard (DSS).* U.S. Department of Commerce / National Institute of Standards and Technology. URL `http://csrc.nist.gov/publications/fips/fips186-2/fips186-2.pdf`. Federal Information Processing Standards Publication 186-2. [359]

NIST (2010). *A Statistical Test Suite for Random and Pseudorandom Number Generators for Cryptographic Applications.* U.S. Department of Commerce / National Institute of Standards and Technology. URL `http://csrc.nist.gov/groups/ST/toolkit/rng/documents/SP800-22rev1a.pdf`. Special Publication. [452]

NIST (2012a). *Federal Information Processing Standards Publication 180-4 — Secure Hash Standard.* National Institute of Standards and Technology. URL `http://csrc.nist.gov/publications/fips/fips180-4/fips-180-4.pdf`. Federal Information Processing Standards Publication 180-4. [316]

NIST (2012b). NIST Special Publication 800-57: Recommendation for Key Management – Part 1: General (Revision 3). URL `http://csrc.nist.gov/publications/nistpubs/800-57/sp800-57_part1_rev3_general.pdf`. [224]

NIST (2012c). NIST Special Publication 800-90A: Recommendation for Random Number Generation Using Deterministic Random Bit Generators. URL `http://csrc.nist.gov/publications/nistpubs/800-90A/SP800-90A.pdf`. [491]

NIST (2014). *Federal Information Processing Standards Publication – SHA-3 Standard: Permutation-Based Hash and Extendable-Output Functions.* National Institute of Standards and Technology. URL `http://csrc.nist.gov/publications/drafts/fips-202/fips_202_draft.pdf`. Federal Information Processing StandardsPublication 202. [316]

KAISA NYBERG (1994). Differentially uniform mappings for cryptography. In Helleseth (1994), 55–64. URL `http://link.springer.de/link/service/series/0558/bibs/0765/07650055.htm`. [33, 296]

KAISA NYBERG & LARS RAMKILDE KNUDSEN (1993). Provable Security Against Differential Cryptanalysis. In *Advances in Cryptology: Proceedings of CRYPTO 1992,* Santa Barbara, CA, ERNEST F. BRICKELL, editor, volume 740 of *Lecture Notes in Computer Science,* 566–574. Springer-Verlag, Berlin. ISBN 978-3-540-57340-1 (Print), 978-3-540-48071-6 (Online). ISSN 0302-9743. URL `http://dx.doi.org/10.1007/3-540-48071-4_41`. [296]

KAISA NYBERG & RAINER A. RUEPPEL (1996). Message Recovery for Signature Schemes Based on the Discrete Logarithm Problem. *Designs, Codes and Cryptography* **7**(1–2), 61–81. ISSN 0925-1022 (Print) 1573-7586 (Online). URL `http://dx.doi.org/10.1007/BF00125076`. [362]

A. M. ODLYZKO (1990). The Rise and Fall of Knapsack Cryptosystems. In *Cryptology and Computational Number Theory*, CARL POMERANCE, editor, number 42 in Proceedings of Symposia in Applied Mathematics, 75–88. American Mathematical Society. [649]

TATSUAKI OKAMOTO (1993). Provably secure and practical identification schemes and corresponding signature schemes. In *Advances in cryptology: Proceedings of CRYPTO 1992*, Santa Barbara CA, ERNEST F. BRICKELL, editor, number 740 in Lecture Notes in Computer Science, 31–53. Springer-Verlag, Berlin. ISBN 3-540-57340-2. ISSN 0302-9743. [422]

DAG ARNE OSVIK, JOPPE W. BOS, DEIAN STEFAN & DAVID CANRIGHT (2010). Fast Software AES Encryption. In *Fast Software Encryption 2010, 17th International Workshop, FSE 2010*, Seoul, Korea, SEOKHIE HONG & TETSU IWATA, editors, volume 6147 of *Lecture Notes in Computer Science*, 75–93. Springer-Verlag. ISBN 978-3-642-13857-7 (Print) 978-3-642-13858-4 (Online). ISSN 0302-9743. URL http://dx.doi.org/10.1007/978-3-642-13858-4_5. [55]

GIOVANNI BATTISTA PALATINO (1548). *Libro di M. Giovambattista. Palatino cittadino romano, Nel qual s'insegna à Seriuere ogni forte lettera, antica, et moderna, di qualunque natione, con le sue Regole, et misure, et essempi, et con un breve et util discorso de le cifre, Riveduto nuovamente, & correto dal proprio Autore, Con la giunta di qvindici Tavole bellissime.* Antonio Blado Solano, Roma. 63 leaves. Editions 1540, 1545, 1548, 1550, 1553. [817]

CHRIS PEIKERT (2009). Public-Key Cryptosystems from the Worst-Case Shortest Vector Problem: Extended Abstract. In *Proceedings of the Fourty-first Annual ACM Symposium on Theory of Computing*, Bethesda, Maryland, USA, MICHAEL MITZENMACHER, editor, 333–342. ACM Press, Bethesda, Maryland, USA. ISBN 978-1-60558-506-2. URL http://doi.acm.org/10.1145/1536414.1536461. Submitted by invitation to J ACM. [650]

CHRIS PEIKERT (2014). A Decade of Lattice Cryptography. 54 pages. [649–651]

P.-M. PERRET (1890). Les règles de Cicco Simonetta pour le déchiffrement des écritures secrètes. *Bibliothèque de l'École des Chartes* **51**, 515–525. Revue d'Érudition. [342, 524]

RAPHAEL CHUNG-WEI PHAN (2002). Mini Advanced Encryption Standard (Mini-AES): A Testbed for Cryptanalysis Students. *Cryptologia* **26**(4), 283–306. [295]

EDGAR ALLAN POE (1902). *The Gold Bug.* Rand, McNally & Company, Chicago, New York, London. Edited by Theda Gildemeister, Illustrated by G. C. Widney. [104, 815]

STEPHEN C. POHLIG & MARTIN E. HELLMAN (1978). An Improved Algorithm for Computing Logarithms over $GF(p)$ and Its Cryptographic Significance. *IEEE Transactions on Information Theory* **IT-24**(1), 106–110. URL http://dx.doi.org/10.1109/TIT.1978.1055817. [184, 201]

JOHN M. POLLARD (1974). Theorems on factorization and primality testing. *Proceedings of the Cambridge Philosophical Society* **76**, 521–528. [145]

JOHN M. POLLARD (1975). A Monte Carlo method for factorization. *BIT* **15**, 331–334. [132, 135, 179]

J. M. POLLARD (1978). Monte Carlo Methods for Index Computation (mod p). *Mathematics of Computation* **32**(143), 918–924. [179, 184, 201]

C. POMERANCE (1982). Analysis and comparison of some integer factoring algorithms. In *Computational Methods in Number Theory,* Part 1, H. W. LENSTRA, JR. & R. TIJDEMAN, editors, number 154 in Mathematical Centre Tracts, 89–139. Mathematisch Centrum, Amsterdam. [140]

CARL POMERANCE (1985). The quadratic sieve factoring algorithm. In *Advances in Cryptology: Proceedings of EUROCRYPT 1984,* Paris, France, THOMAS BETH, NORBERT COT & INGEMAR INGEMARSSON, editors, volume 209 of *Lecture Notes in Computer Science,* 169–182. Springer-Verlag, Berlin. ISBN 3-540-16076-0, 0-387-16076-0. ISSN 0302-9743. URL http://dx.doi.org/10.1007/3-540-39757-4_17. [145]

CARL POMERANCE (1990). Factoring. In *Cryptology and Computational Number Theory,* CARL POMERANCE, editor, number 42 in Proceedings of Symposia in Applied Mathematics, 27–47. American Mathematical Society. [112]

CARL POMERANCE (1996). A tale of two sieves. *Notices of the American Mathematical Society* **43**(12), 1473–1485. [145]

GIOVANNI BATTISTA DELLA PORTA (1602). *De Furtivis Literarvm Notis, Vvlgo De Ziferis Libri Quinque.* Ioannis Baptista Subtilis, Neapoli. First edition 1563, 1591 (pirated), 1593. [104, 249, 434, 526, 816–817]

BART PRENEEL, VINCENT RIJMEN & ANTOON BOSSELAERS (1998). Recent Developments in the Design of Conventional Cryptographic Algorithms. In *State of the Art in Applied Cryptography,* B. PRENEEL & V. RIJMEN, editors, volume 1528 of *Lecture Notes in Computer Science,* 105–130. Springer-Verlag, Berlin, Heidelberg. ISBN 3-540-65474-7. ISSN 0302-9743. URL http://dx.doi.org/10.1007/3-540-49248-8_4. [296]

PULTENEY, WILLIAM (1722). *A report from the committee Appointed by Order of the House of Commons to examine Christopher Layer, and Others.* House of Commons, London. [564, 573, 813, 816]

JEAN-JACQUES QUISQUATER, MYRIAM QUISQUATER, MURIEL QUISQUATER, MICHAËL QUISQUATER, LOUIS GUILLOU, MARIE ANNICK GUILLOU, GAÏD GUILLOU, ANNA GUILLOU, GWENDOLÉ GUILLOU, SOAZIG GUILLOU & TOM BERSON (1989). How to Explain Zero-Knowlegde Protocols to Your Children. In *Advances in Cryptology: Proceedings of CRYPTO 1989,* Santa Barbara, CA, number 435 in Lecture Notes in Computer Science, 628–631. Springer-Verlag. ISSN 0302-9743. URL http://dx.doi.org/10.1007/0-387-34805-0_60. [550]

MICHAEL O. RABIN (1976). Probabilistic algorithms. In *Algorithms and Complexity,* J. F. TRAUB, editor, 21–39. Academic Press, New York. [143]

MICHAEL O. RABIN (1979). Digitalized signatures and public key-functions as intractable as factorization. Technical report, Massachusetts Institute of Technology, Laboratory for Computer Science, Cambridge, Massachusetts 02139. URL http://publications.csail.mit.edu/lcs/pubs/pdf/MIT-LCS-TR-212.pdf. [362, 404, 490]

MICHAEL O. RABIN (1980). Probabilistic Algorithms for Testing Primality. *Journal of Number Theory* **12**(1), 128–138. [143]

CHARLES RACKOFF (2012). Fundamentals of Cryptography. Fall course "Computer Science Courses (CSC2426)" at the University of Toronto. URL http://www.cs.toronto.edu/~rackoff/2426f12/. [56, 402]

CHARLES RACKOFF & DANIEL R. SIMON (1992). Non-Interactive Zero-Knowledge Proof of Knowledge and Chosen Ciphertext Attack. In Feigenbaum (1992), 433–444. URL http://dx.doi.org/10.1007/3-540-46766-1_35. [402]

JAMES REEDS (1977). "Cracking" a random number generator. *Cryptologia* **1**(1), 20–26. [489]

JIM REEDS (1998). Solved: The Ciphers in Book III of Trithemius's *Steganographia*. URL http://dx.doi.org/10.1080/0161-119891886948. 28 pages. [526]

ODED REGEV (2004). New Lattice-Based Cryptographic Constructions. *Journal of the ACM* **51**(6), 899–942. URL http://portal.acm.org/citation.cfm?id=1039490. [651]

ODED REGEV (2009). On Lattices, Learning with Errors, Random Linear Codes, and Cryptography. *Journal of the ACM* **56**(6), 34.1–34.40. ISSN 0004-5411. URL http://doi.acm.org/10.1145/1568318.1568324. Other version on the author's homepage: http://www.cs.tau.ac.il/~odedr/papers/qcrypto.pdf. Conference version in *Proceedings of the Thirty-Seventh Annual ACM Symposium of Theory of Computing*, 2005, Baltimore MD, pages 84–93. [627, 638, 642, 647, 650, 658]

ODED REGEV (2010). *On the Complexity of Lattice Problems with Polynomial Approximation Factors*, chapter 15, 475–496. In Nguyen & Vallée (2010). URL http://dx.doi.org/10.1007/978-3-642-02295-1_15. [649]

MARIAN REJEWSKI (1980). An application of the theory of permutations in breaking the Enigma cipher. *Zastosowania matematyki* **16**(4), 543–559. [736]

MARIAN REJEWSKI (1981). How Polish Mathematicians deciphered the Enigma. *Annals of the History of Computing* **3**(3), 213–234. URL http://ieeexplore.ieee.org/stamp/stamp.jsp?tp=&arnumber=4640685. [736]

MARIAN REJEWSKI (1982a). Mathematical solution of the Enigma cipher. *Cryptologia* **6**(1), 1–18. [736]

MARIAN REJEWSKI (1982b). Remarks on Appendix 1 to British Intelligence in the Second World War by F. H. Hinsley. *Cryptologia* **6**(1), 75–83. [736]

GEORG FRIEDRICH BERNHARD RIEMANN (1859). Ueber die Anzahl der Primzahlen unter einer gegebenen Grösse. *Monatsberichte der Berliner Akademie* 145–153. *Gesammelte Mathematische Werke*, ed. HEINRICH WEBER, Teubner Verlag, Leipzig, 1892, 177-185. [143]

V. RIJMEN, J. DAEMEN, B. PRENEEL, A. BOSSELAERS & E. DE WIN (1996). The cipher SHARK. In Gollmann (1996), 99–111. [32]

VINCENT RIJMEN & ELISABETH OSWALD (2005). Update on SHA-1. Cryptology ePrint Archive, Report 2005/010. URL http://eprint.iacr.org/2005/010.

RONALD L. RIVEST (1991). The MD4 Message Digest Algorithm. In *Advances in Cryptology: Proceedings of CRYPTO 1990,* Santa Barbara, CA, number 537 in Lecture Notes in Computer Science, 303–311. Springer-Verlag. ISSN 0302-9743.

RONALD L. RIVEST (1992). Responses to NIST's proposal. *Communications of the ACM* **35**(7), 41–47. ISSN 0001-0782. URL http://dx.doi.org/10.1145/129902.129905. [359]

RONALD RIVEST, ADI SHAMIR & LEN ADLEMAN (1977). A Method for Obtaining Digital Signatures and Public-Key Cryptosystems. Technical Report MIT/LCS/TM-82, April 1977, Massachusetts Institute of Technology, Laboratory for Computer Science, Cambridge, Massachusetts. Final version in *Communications of the ACM* **21**(2), 120–126, 1978. [38, 144, 359]

ADOLPHE ROCHAS (1860). *Biographie du Dauphiné contenant l'histoire des hommes nés dans cette province qui se sont fait remarquer dans les Lettres, les Sciences, les Arts, etc. avec le catalogue de leurs ouvrages et la Description de leurs Portraits.* Reprint 1971 by Slatkine Reprints, Genéve. 2 volumes, Paris, Charavay.

PHILLIP ROGAWAY & THOMAS SHRIMPTON (2004). Cryptographic Hash-Function Basics: Definitions, Implications, and Separations for Preimage Resistance, Second-Preimage Resistance, and Collision Resistance. In *Fast Software Encryption 2004, 11th International Workshop, FSE 2004,* Delhi, India, BIMAL ROY & WILLI MEIER, editors, volume 3017 of *Lecture Notes in Computer Science,* 371–388. Springer-Verlag. ISBN 978-3-540-22171-5. ISSN 0302-9743 (Print) 1611-3349 (Online). URL http://dx.doi.org/10.1007/978-3-540-25937-4_24. Full version: http://www.cs.ucdavis.edu/~rogaway/papers/relates.html. [315]

J. BARKLEY ROSSER & LOWELL SCHOENFELD (1962). Approximate formulas for some functions of prime numbers. *Illinois Journal of Mathematics* **6**, 64–94. [143]

RICHARD L. ROTH (2001). A History of Lagrange's Theorem on Groups. *Mathematics Magazine* **74**(2), 99–108. URL http://dx.doi.org/10.2307/2690624. [798]

O. S. ROTHAUS (1976). On "Bent" Functions. *Journal of Combinatorial Theory, Series A* **20**, 300–305. URL http://dx.doi.org/10.1016/0097-3165(76)90024-8. [293]

MICHAEL ROTHSTEIN (1976). *Aspects of symbolic integration and simplification of exponential and primitive functions.* Ph.D. thesis, University of Wisconsin-Madison. [296]

JOANNE K. ROWLING (1998). *Harry Potter and the Philosopher's Stone.* Bloomsbury Children's Books. ISBN 0747542988, 223 pages. URL http://www.bloomsbury.com/uk/harry-potter-and-the-philosophers-stone-9780747532699/. [88]

É. ROY (1926). Un emploi de scytales en 1431. *Mémoires de l'Académie des Sciences, Arts et Belles-Lettres de Dijon. Recueil mensuel.* 227–240. [443]

RAINER A. RUEPPEL & JAMES L. MASSEY (1985). The Knapsack as a Non-Linear Function (abstract). *IEEE International Symposium on Information Theory* 46. [493]

MARQUIS DE SADE (1980). *Lettres et Mélanges littéraires écrits à Vincennes et à la Bastille avec des lettres de Madame de Sade, de Marie-Dorothée de Rousset et de diverses personnes,* volume 1. Éditions Borderie. Edited by Georges Daumas et Gilbert Lely. [816]

TAKAKAZU SATOH & KIYOMICHI ARAKI (1997). Fermat Quotients and the Polynomial Time Discrete Log Algorithm for Anomalous Elliptic Curves. Unpublished. [233]

DOROTHY L. SAYERS (1932). *Have his carcase.* Victor Gollancz Ltd, London. ISBN 0-575-00685-4. Reissued 1971. 448 pp.

KLAUS SCHMEH (2004). Als deutscher Code-Knacker im Zweiten Weltkrieg. *TELEPOLIS* **23.09.2004**. URL http://www.heise.de/bin/tp/issue/r4/dl-artikel2.cgi?artikelnr=18371&mode=html&zeilenlaenge=72. [736]

ERHARD SCHMIDT (1907). Zur Theorie der linearen und nichtlinearen Integralgleichungen, I. Teil: Entwicklung willkürlicher Funktionen nach Systemen vorgeschriebener. *Mathematische Annalen* **63**, 433–476. Reprint of Erhard Schmidt's Dissertation, Göttingen, 1905. [648]

C. P. SCHNORR (1982). Refined Analysis and Improvements on Some Factoring Algorithms. *Journal of Algorithms* **3**, 101–127. [145]

C. P. SCHNORR (1987). A hierarchy of polynomial time lattice basis reduction algorithms. *Theoretical Computer Science* **53**, 201–224. [649]

C. P. SCHNORR (1989). Efficient Identification and Signatures for Smart Cards. In *Advances in Cryptology - EUROCRYPT '89*, JEAN-JACQUES QUISQUATER & JOOS VANDEWALLE, editors, number 434 in Lecture Notes in Computer Science, 688–689. Springer-Verlag, Berlin, Heidelberg. ISBN 3-540-53433-4. URL http://link. springer.de/link/service/series/0558/bibs/0434/04340688.htm. [359]

C. P. SCHNORR (1991a). Efficient Signature Generation by Smart Cards. *Journal of Cryptology* **4**(3), 161–174. ISSN 0933-2790. Conference version in *Advances in Cryptology: Proceedings of CRYPTO 1989*, Santa Barbara CA, *Lecture Notes in Computer Science*, Springer Verlag, volume 435, 239–252. [354, 361]

CLAUS-P. SCHNORR (1991b). Method for identifying subscribers and for generating and verifying electronic signatures in a data exchange system. US Patent Number 4,995,082. URL http://www.delphion.com. [359]

C. P. SCHNORR (1997). On the coverage of the DSA by US patent 4,995,082. Note, Universität Frankfurt/Main. [359]

CLAUS PETER SCHNORR (2001). Small Generic Hardcore Subsets for the Discrete Logarithm: Short Secret DL-Keys. *Information Processing Letters* **79**(2), 93–98. URL http://dx.doi.org/10.1016/S0020-0190(00)00173-3. [201]

ARNOLD SCHÖNHAGE, ANDREAS F. W. GROTEFELD & EKKEHART VETTER (1994). *Fast Algorithms – A Multitape Turing Machine Implementation.* BI Wissenschaftsverlag, Mannheim. [798]

A. SCHÖNHAGE & V. STRASSEN (1971). Schnelle Multiplikation großer Zahlen. *Computing* **7**, 281–292. [762]

RENÉ SCHOOF (1985). Elliptic Curves over Finite Fields and the Computation of Square Roots mod p. *Mathematics of Computation* **44**(170), 483–494. [227–228]

RENÉ SCHOOF (1995). Counting points on elliptic curves over finite fields. *Journal de Théorie des Nombres de Bordeaux* **7**, 219–254. [234]

JOHN HOLT SCHOOLING (1896). Secrets in Cipher. *The Pall Mall Magazine* **8**, 119–129, 245–256, 452–462, 608–618. [817]

GASPAR SCHOTT (1665). *Schola steganographica, in classes octo distributa.* Heirs of Johannes Andreas and Wolfgang Endter, Nürnberg, [34], 346, [4 index], [2 errata], [1 blank] pages. [526, 817]

THORSTEN W. SCHRÖDER (2012). *SmartCom — Secure SMS Encryption with High Usability.* Diploma thesis, Department of Computer Security, b-it – Bonn-Aachen International Center for Information Technology, University of Bonn, Bonn, Germany. [221]

HUGH SEBAG-MONTEFIORE (2001). *Enigma. The battle for the code.* Phoenix, London. 507 pages.

SECOUSSE (1755). *Recueil de pièces servant de preuves aux mémoires Sur les Troubles exités en France par Charles II. Dit le mauvais. Roi de Navarre et comte d'Evreux.* Durand, 677 pages. [342]

GUSTAVUS SELENUS (1624). *Gustavi Seleni cryptomenytices et cryptographiæ Libri IX.* Exscriptum typis & impensis Johannis & Henrici fratrum, Lunæburgi, 493 pp. [817]

IGOR A. SEMAEV (1998). An algorithm for evaluation of discrete logarithms in some nonprime finite fields. *Mathematics of Computation* **67**(223), 1679–1689. [233]

ADI SHAMIR (1979). How to Share a Secret. *Communications of the ACM* **22**(11), 612–613. [51, 148]

ADI SHAMIR (1984). A polynomial-time algorithm for breaking the basic Merkle-Hellman cryptosystem. *IEEE Transactions on Information Theory* **IT-30**(5), 699–704. URL http://citeseerx.ist.psu.edu/viewdoc/summary?doi=10.1.1.123.5840. Extended abstract in *Proceedings of the 23rd Annual IEEE Symposium on Foundations of Computer Science*, 1982, Chicago IL, 145–152. [588, 648]

ADI SHAMIR (1992). $IP = PSPACE$. *Journal of the ACM* **39**(4), 869–877. [533]

DANIEL SHANKS (1971). Class number, a theory of factorization, and genera. *Proceedings of Symposia in Pure Mathematics* **20**, 415–440. URL http://www.ams.org/mathscinet-getitem?mr=0316385. [145]

C. E. SHANNON, (1949). Communication Theory of Secrecy Systems. *Bell System Technical Journal* **28**, 656–715. [62, 290]

PETER W. SHOR (1994). Algorithms for Quantum Computation: Discrete Logarithms and Factoring. In *Proceedings of the 35th Annual IEEE Symposium on Foundations of Computer Science,* Santa Fe NM, SHAFI GOLDWASSER, editor, 124–134. IEEE Computer Society Press, Los Alamitos CA. [713]

PETER W. SHOR (1996). Fault-Tolerant Quantum Computation. *Proceedings of the 37th Annual IEEE Symposium on Foundations of Computer Science*, Burlington VT 56–65. [713]

PETER W. SHOR (1999). Polynomial-Time Algorithms for Prime Factorization and Discrete Logarithms on a Quantum Computer. *SIAM Review* **41**(2), 303–332. [713]

VICTOR SHOUP (1997). Lower Bounds for Discrete Logarithms and Related Problems. In Fumy (1997), 256–266. URL http://www.shoup.net/papers/. [201, 205]

IGOR E. SHPARLINSKI (2002). Playing "Hide-and-Seek" in Finite Fields: The Hidden Number Problem and Its Applications. *Proceedings of the 7th Spanish Meeting on Cryptology and Information Security* **1**, 49–72. URL http://www.comp.mq.edu.au/~igor/Hide-Seek.ps. [649]

DAN SHUMOW & NIELS FERGUSON (2007). On the Possibility of a Back Door in the NIST SP800-90 Dual EC PRNG. URL http://rump2007.cr.yp.to/15-shumow.pdf. [489]

SIGNAL SECURITY AGENCY (1945). German Cryptographic Systems During the First World War. Prepared under the Direction of the Chief Signal Officer. Now in the National Archives, Washington, NARA RG 459 Historical Cryptographic Collection, Box 1059; Folder German Cryptographic Systems During WWI. [678]

JOSEPH H. SILVERMAN (1986). *The Arithmetic of Elliptic Curves*. Number 106 in Graduate Texts in Mathematics. Springer-Verlag, New York. ISBN 0-387-96203-4, 3-540-96203-4. [233]

DANIEL R. SIMON (1994). On the Power of Quantum Computation. In *Proceedings of the 35th Annual IEEE Symposium on Foundations of Computer Science*, Santa Fe NM, 116–123. IEEE Computer Society Press, Santa Fe NM. ISBN 0-8186-6580-7. URL http://doi.ieeecomputersociety.org/10.1109/SFCS.1994.365701. [713]

HUGH SKILLEN (editor) (1995). *The Enigma Symposium*. H. Skillen. ISBN 0951519077. [443]

N. P. SMART (1999). Elliptic Curve Cryptosystems over Small Fields of Odd Characteristic. *Journal of Cryptology* **12**, 141–151. [233]

R. SOLOVAY & V. STRASSEN (1977). A fast Monte-Carlo test for primality. *SIAM Journal on Computing* **6**(1), 84–85. Erratum in **7** (1978), p. 118. [143]

OTTO SPAMER (1894). *Spamers Illustrierte Weltgeschichte. Illustrierte Geschichte der neueren Zeit. Erster Teil*, volume 5. Verlag und Druck von Otto Spamer, Leipzig. URL https://archive.org/details/spamersillustrie05kaem. [524]

PHILIP HENRY STANHOPE, LORD MAHON (1839). *History of England from the peace of Utrecht to the peace of Aix-La-Chapelle. In three Volumes*. [573]

A. M. STEANE (1996). Error Correcting Codes in Quantum Theory. *Physical Review Letters* **77**(5), 793–797. [713]

DAMIEN STEHLÉ (2004). Breaking Littlewood's Cipher. *Cryptologia* **XXVIII**(4), 341–357. URL http://dx.doi.org/10.1080/0161-110491892971. [455]

RON STEINFELD, JOSEF PIEPRZYK & HUAXIONG WANG (2006). On the Provable Security of an Efficient RSA-Based Pseudorandom Generator. In Lai & Chen (2006), 194–209. URL http://dx.doi.org/10.1007/11935230_13. [489]

JACQUES STERN (1987). Secret linear congruential generators are not cryptographically secure. *Proceedings of the 28th Annual IEEE Symposium on Foundations of Computer Science, Los Angeles CA* 421–426. [489]

MARC STEVENS, ALEXANDER SOTIROV, JACOB APPELBAUM, ARJEN LENSTRA, DAVID MOLNAR, DAG ARNE OSVIK & BENNE WEGER (2009). Short Chosen-Prefix Collisions for MD5 and the Creation of a Rogue CA Certificate. In Halevi (2009), 55–69. URL http://dx.doi.org/10.1007/978-3-642-03356-8_4. 0302-9743. [313]

DOUGLAS R. STINSON (1995). *Cryptography, Theory and Practice*. CRC Press, Boca Raton FL. ISBN 0849385210. [549]

ARNE STORJOHANN (1996). Faster Algorithms for Integer Lattice Basis Reduction. Technical Report 249, Eidgenössische Technische Hochschule Zürich. URL ftp://ftp.inf.ethz.ch/pub/publications/tech-reports/2xx/249.ps.gz. 24 pages. [648]

V. STRASSEN (1972a). Berechnung und Programm. I. *Acta Informatica* **1**, 320–335. [796]

V. STRASSEN (1972b). Evaluation of rational Functions. In *Complexity of Computer Computations*, RAYMOND E. MILLER & JAMES W. THATCHER, editors. Plenum Press, New York. [201]

VOLKER STRASSEN (1976). Einige Resultate über Berechnungskomplexität. *Jahresberichte der DMV* **78**, 1–8. [131, 145]

FRENCH STROTHER (1918). German Codes and Ciphers. *The World's Work* **36**, 143–153. [677]

GAIUS SUETON (1615). *De xii Cæsaribvs libri viii.* Petrus & Iacobus Chouët, Colonia Allobrogum (Grenoble). Editor Isaac Casaubon, 1192 pages. [85, 104]

GEORG SWARZENSKI (1969). *Die Regensburger Buchmalerei des X. und XI. Jahrhunderts.* Anton Hiersemann, Stuttgart, 228 pages. [103]

RICHARD TAYLOR & ANDREW WILES (1995). Ring-theoretic properties of certain Hecke algebras. *Annals of Mathematics* **141**, 553–572. [216]

A. THUE (1902). Et par andtydninger til en talteoretisk methode. *Videnskabers Selskab Forhandlinger Christiana* **7**. [797]

TICOM (1948). German Cryptanalytic Device for Solution of M-209 Traffic. Technical Report DF 114, Document 2785, TICOM. [734]

TOMMASO TOFFOLI (1980). Reversible computing. *Proceedings of the 7th International Colloquium on Automata, Languages and Programming ICALP* 632–644. [713]

TOMMASO TOFFOLI (1981). Bicontinuous Extensions of Invertible Combinatorial Functions. *Mathematical Systems Theory* **14**, 13–23. URL http://. [713]

FRANCESCO TRANCHEDINO (1970). *Diplomatische Geheimschriften.* Akademische Druck- und Verlagsanstalt, Graz-Austria, 43, 338 pages. [324, 338, 342–343, 816]

IOANNES TRITHEMIUS (1518). *Polygraphiæ libri sex.* Adam Petri, Basel, xi, 244, 14 leaves. [69, 525, 816]

IOANNES TRITHEMIUS (1561). *Polygraphie, et Vniuerselle escriture Cabalistique de M. I. Tritheme Abbé. Traduicte par Gabriel de Collange.* Jacques Kerver, Paris, 18, 300 leaves. [82, 816–817]

IOANNES TRITHEMIUS (1571). *Polygraphiae libri sex, Ioannis Trithemii, abbatis Peapolitani, quondam Spanheimensis, ad Maximilianum I. Caesarem. Accessit clauis Polygraphiae liber vnus, eodem authore... Additae sunt etiam aliqvot locorum explicationes, eorum præsertim in quibus admirandi operis Steganographiæ principia latent, quae quidem ingeniosis occasionem præbent, longè maiora & subtiliora inueniende, Per virum erudisissimum D. Adolphum à Glauburg, Patricium Francofortensem.* Theodor Baum, Cologne, 554, 54 pages. [514]

IOANNES TRITHEMIUS (1616). *Clauis Steganographiae.* Ederianus, Ingolstadt. [526]

IOANNES TRITHEMIUS (1721). *Primo Spanheimensis deinde divi Jacobi Peapolitani abbatis Steganographia quae hvcvsqve a nemine intellecta, ...nvnc tandem vindicata reserata et illvstrata... avtore VVolfgango Ernesto Heidel.* Joh. Fridericus Rudigerus, Nürnberg. [8], 394, [3]. [816]

BARBARA W. TUCHMAN (1958). *The Zimmermann Telegram.* Macmillan Publishing Company, New York. xii, 244 pages. Reprinted 1966. [677–679]

A. M. TURING (1937). On computable numbers, with an application to the Entscheidungsproblem. *Proceedings of the London Mathematical Society, Second Series* **42**, 230–265, and **43**, 544–546. [726]

LESLIE G. VALIANT (2001). Quantum Computers that can be Simulated Classically in Polynomial Time. In *Proceedings of the Thirty-third Annual ACM Symposium on Theory of Computing,* Hersonissos, Crete, Greece, 114–122. ACM Press, 1515 Broadway, New York, New York 10036. ISBN 1-58113-349-9. [713]

SERGE VAUDENAY (1998). Cryptanalysis of the Chor-Rivest Cryptosystem. In *Advances in Cryptology: Proceedings of CRYPTO 1998,* Santa Barbara, CA, H. KRAWCZYK, editor, volume 1462 of *Lecture Notes in Computer Science,* 243–256. Springer-Verlag, Berlin, Heidelberg. ISBN 3-540-64892-5. ISSN 0302-9743. URL http://dx.doi.org/10.1007/BFb0055732. [648]

VLATKO VEDRAL, ADRIANO BARENCO & ARTUR EKERT (1996). Quantum networks for elemantary arithmetic operations. *Physical Review A* **54**(1), 147–153. [715]

GILBERT SANDFORD VERNAM (1919). USPatent 1310719 A: Secret Signaling System. Online. URL http://www.google.com/patents/US1310719. [377]

G. S. VERNAM (1926). Cipher Printing Telegraph Systems. *Journal of the American Institute of Electrical Engineers* **45**, 109–115. [377, 401]

JULES VERNE (1867). *Voyage au centre de la terre.* J. Hetzel, editeur, Bibliothèque d'éducation et de récréation. [816]

JULES VERNE (1881). La Jangada. *Magasin d'éducation et de récréation* **33**.

JULES VERNE (1885). *Mathias Sandorf.* J. Hetzel, Paris. [440, 816]

FRANÇOIS VIÈTE (1646). *Opera mathematica ed. Franciscus Schooten.* Elzevier, Leyden. Reprint 1970 ed. by Joseph E. Hofmann, Georg Olms Verlag, Hildesheim. [342]

BLAISE DE VIGENÈRE (1587). *Traicté des chiffres, ou secrets manieres d'escrire.* Abel L'Angelier, Paris. Second edition, first 1586. [51, 257, 513, 817]

PAUL VOLK (1950). Der Familienname des Abtes Johannes Trithemius. *Archiv für mittelrheinische Kirchengeschichte* **2**, 309–311. [525]

CHRYSOGONUS WADDELL (1999). *Narrative and legislative texts from early Cîteaux. Latin text in dual edition with English translation and notes,* volume 9 of *Cîteaux / Studia et documenta.* Abbaye de Cîteaux. ISBN 90-805439-1-8. 524 pages. [443]

JOHN WALLIS (1697). Letter to Menken, 1/11 Jan. 1696/7. In *Opera Mathematica,* 659–660.

XIAOYUN WANG, DENGGUO FENG, XUEJIA LAI & HONGBO YU (2004). Collisions for Hash Functions MD4, MD5, HAVAL–128 and RIPEMD. *Cryptology ePrint Archive* **2004**/199, 4. URL http://eprint.iacr.org/2004/199. [313]

XIAOYUN WANG, XUEJIA LAI, DENGGUO FENG, HUI CHEN & XIUYUAN YU (2005a). Cryptanalysis of the Hash Functions MD4 and RIPEMD. In Cramer (2005), 1–18. URL http://dx.doi.org/10.1007/11426639_1. [316]

XIAOYUN WANG, YIQUN LISA YIN & HONGBO YU (2005b). Finding Collisions in the Full SHA-1. In *Advances in Cryptology - CRYPTO 2005,* VICTOR SHOUP, editor, volume 3621 of *Lecture Notes in Computer Science,* 17–36. Springer-Verlag, Berlin, Heidelberg. ISBN 3-540-28114-2. ISSN 0302-9743. URL http://dx.doi.org/10.1007/11535218_2. [314]

XIAOYUN WANG & HONGBO YU (2005). How to Break MD5 and Other Hash Functions. In Cramer (2005), 19–35. URL http://dx.doi.org/10.1007/11426639_2. [316]

LAWRENCE C. WASHINGTON (2008). *Elliptic Curves — Number Theory and Cryptography.* Discrete Mathematics and its Applications. CRC Press, Boca Raton, FL, USA. ISBN 978-1-4200-7146-7. First Edition 2003. [231–233]

WILLIAM C. WATERHOUSE (1969). Abelian varieties over finite fields. *Annales scientifiques de l'École Normale Supérieure, 4ᵉ série* **2**, 521–560. [233]

JOHN WATROUS (2009). Quantum Computational Complexity. *Encyclopedia of Complexity and Systems Science* 7174–7201. URL http://dx.doi.org/10.1007/978-0-387-30440-3_428. Also available at http://arxiv.org/abs/0804.3401. [715]

ERICH WENGER & PAUL WOLFGER (2014). Solving the Discrete Logarithm of a 113-bit Koblitz Curve with an FPGA Cluster. In *21st International Conference, Selected Areas in Cryptography 2014,* Montreal, QC, Canada, ANTOINE JOUX & AMR YOUSSEF, editors, number 8781 in Lecture Notes in Computer Science, 363–379. Springer-Verlag, Heidelberg, Dordrecht, London, New York. ISBN 978-3-319-13050-7, e-ISBN 978-3-319-13051-4. ISSN 0302-9743, e-ISSN 1611-3349. URL http://dx.doi.org/10.1007/978-3-319-13051-4_22. Also available at https://eprint.iacr.org/2014/368. [222]

R. WERNSDORF (2002). The round functions of Rijndael generate the alternating group. In *9th International Workshop, FSE 2002 Leuven, Belgium, February 4–6,* JOAN DAEMEN & VINCENT RIJMEN, editors, volume 2365 of *Lecture Notes in Computer Science,* 143–148. Springer-Verlag, Berlin Heidelberg. ISBN 978-3-540-44009-3 (Print) 978-3-540-45661-2 (Online). URL http://dx.doi.org/10.1007/3-540-45661-9_11. [295]

MICHAEL WERTHEIMER (2015). Encryption and the NSA Role in International Standards. *Notices of the American Mathematical Society* **62**(02), 165–167. URL http://www.ams.org/notices/201502/rnoti-p165.pdf. [489]

STEPHANIE WEST (1988). Archilochus' Message-Stick. *Classical Quarterly* **38**, 42–48. [442]

MICHAEL J. WIENER (1990). Cryptanalysis of short RSA secret exponents. *IEEE Transactions on Information Theory* **36**(3), 553–558. ISSN 0018-9448. [127]

ANDREW WILES (1995). Modular elliptic curves and Fermat's Last Theorem. *Annals of Mathematics* **142**, 443–551. [216]

JOHN WILKINS (1641). *Mercury or, the Secret and Swift Messenger. Shewing how a man may with privacy and speed communicate his thoughts to a friend at a distance.* Printed by I. Norton for John Maynard and Timothy Wilkins, London. 172 pages. [436, 443, 816]

D. B. WILSON (1979). Littlewood's cipher. *Cryptologia* **3**, 120–121 and 172–176. [455]

W. K. WIMSATT, JR. (1943). What Poe knew about cryptography. *Publications of the Modern Language Association of America* **58**, 754–779. [95]

F. W. WINTERBOTHAM (1974). *The Ultra Secret. The first account of the most astounding cryptanalysis coup of World War II—how the British broke the German code and read most of the signals between Hitler and his generals throughout the war.* Harper & Row, Publishers, New York. Xiii, 199 pages. Review in Kahn (1974). [736]

NANYANG XU, JING ZHU, DAWEI LU, XIANYI ZHOU, XINHUA PENG & JIANGFENG DU (2011). Quantum Factorization of 143 on a Dipolar-Coupling NMR system. URL http://arxiv.org/abs/1111.3726. Preprint, 5 pages. [707]

ANDREW C. YAO (1982). Theory and Applications of Trapdoor Functions. In *Proceedings of the 23rd Annual IEEE Symposium on Foundations of Computer Science*, Chicago IL, 80–91. IEEE Computer Society Press. URL http://dx.doi.org/10.1109/SFCS.1982.95. [402, 447, 490, 799]

ANDREW C.-C. YAO (1993). Quantum circuit complexity. *Proceedings of the 34th Annual IEEE Symposium on Foundations of Computer Science*, Palo Alto CA 352–361. [713]

FRANCES AMELIA YATES (1972). *The Rosicrucian Enlightenment*. Routledge & Kegan Paul, London, xvi, 269 pages with a map plus 31 pages of plates. [525]

FILSON YOUNG (1922). *With the Battle Cruisers*. Cassell and Company, Ltd., London, New York, Toronto, Melbourne. [18], 296 pages. [677]

D. ZAGIER (1997). Newman's Short Proof of the Prime Number Theorem. *The American Mathematical Monthly* **104**(8), 705–708. URL http://www.jstor.org/stable/2975232. [143]

KONSTANTIN ZIEGLER & JOHANNES ZOLLMANN (2013). Fast and uniform generation of safe RSA moduli — Extended Abstract. In *WEWoRC 2013 — Book of Abstracts*, 15–19. Karlsruher Institut für Technologie (KIT), Karlsruhe. URL http://digbib.ubka.uni-karlsruhe.de/volltexte/1000036384. [144]

JOHANNES ZOLLMANN (2013). *Factoring-Based Cryptography*. Master's thesis, Rheinische Friedrich-Wilhelms-Universität Bonn. [490]

Any inaccuracies in this index may be explained by the fact
that it has been prepared with the help of a computer.

DONALD ERVIN KNUTH (1969)

If 'publish or perish' were really true,
Leonard Euler would still be alive.

ERIC BACH

Si nous avons réussi à intéresser le lecteur à nos recherches,
nous serons hautement récompensé de l'effort
que nous nous sommes imposé.[2]

MAURICE KRAÏTCHIK (1926)

I'm sorry to say that the subject I most disliked was
mathematics. I have thought about it. I think the reason
was that mathematics leaves no room for argument.
If you made a mistake, that was all there was to it.

MALCOLM X (1965)

I have been happy in decyphering what the President of Congress
sends by this Opportunity. The use of the same Cypher
by all the British Commanders is now pretty fairly concluded.
The Enemy play a grand Stake, May the Glory
redound to the Allied Force under your Excellencys Command!

JAMES LOWELL (1781)

Don't know much about geography,
don't know much about trigonometry,
don't know much about algebra,
don't know what a slide rule is for.

But I do know one and one is two,
and if this one could be with you,
what a wonderful world this would be.

SAM COOKE (1960)

---

[2]If we have succeeded in interesting the reader in our research, we will be highly
compensated for the effort required.

Index

A page number is underlined (for example: <u>21</u>) when it represents the definition or the main source of information about the index entry. A page range indicates that most, but maybe not all, pages within the range contain the index term. Page 386$^+$ means pages 386–387.

© Springer-Verlag Berlin Heidelberg 2015

J. von zur Gathen, *CryptoSchool*, DOI 10.1007/978-3-662-48425-8

ᛏᛣᛒ·ᛪᚤᛣᛃ≠ᛝ·ᛃᛯ᛫
≠ᚾᛢ·≠ᛦᛍ··

This book comes to an end.